WORLD HEALTH ORGANIZATION

INTERNATIONAL AGENCY FOR RESEARCH ON CANCER

IARC MONOGRAPHS
ON THE
EVALUATION OF CARCINOGENIC RISKS TO HUMANS

Some Naturally Occurring Substances: Food Items and Constituents, Heterocyclic Aromatic Amines and Mycotoxins

VOLUME 56

This publication represents the views and expert opinions
of an IARC Working Group on the
Evaluation of Carcinogenic Risks to Humans,
which met in Lyon,

9–16 June 1992

1993

IARC MONOGRAPHS

In 1969, the International Agency for Research on Cancer (IARC) initiated a programme on the evaluation of the carcinogenic risk of chemicals to humans involving the production of critically evaluated monographs on individual chemicals. In 1980 and 1986, the programme was expanded to include the evaluation of the carcinogenic risk associated with exposures to complex mixtures and other agents.

The objective of the programme is to elaborate and publish in the form of monographs critical reviews of data on carcinogenicity for agents to which humans are known to be exposed, and on specific exposure situations; to evaluate these data in terms of human risk with the help of international working groups of experts in chemical carcinogenesis and related fields; and to indicate where additional research efforts are needed.

This project is supported by PHS Grant No. 5-UO1 CA33193-11 awarded by the US National Cancer Institute, Department of Health and Human Services. Additional support has been provided since 1986 by the Commission of the European Communities.

©International Agency for Research on Cancer 1993

ISBN 92 832 1256-8

ISSN 0250-9555

Publications of the World Health Organization enjoy copyright protection in accordance with the provisions of Protocol 2 of the Universal Copyright Convention. All rights reserved. Application for rights of reproduction or translation, in part or *in toto*, should be made to the International Agency for Research on Cancer.

Distributed for the International Agency for Research on Cancer
by the Secretariat of the World Health Organization

PRINTED IN THE UK

CONTENTS

NOTE TO THE READER 5

LIST OF PARTICIPANTS 7

PREAMBLE

    Background 13
    Objective and Scope 13
    Selection of Topics for Monographs 14
    Data for Monographs 15
    The Working Group 15
    Working Procedures 15
    Exposure Data 16
    Studies of Cancer in Humans 17
    Studies of Cancer in Experimental Animals 21
    Other Relevant Data 23
    Summary of Data Reported 24
    Evaluation 26
    References 30

GENERAL REMARKS 33

THE MONOGRAPHS

    Food items and constituents

        Salted fish 41
        Pickled vegetables 83
        Caffeic acid 115
        d-Limonene 135

    Heterocyclic aromatic amines

        IQ (2-Amino-3-methylimidazo[4,5-*f*]quinoline) 165
        MeIQ (2-Amino-3,4-dimethylimidazo[4,5-*f*]quinoline) 197
        MeIQx (2-Amino-3,8-dimethylimidazo[4,5-*f*]quinoxaline) 211
        PhIP (2-Amino-1-methyl-6-phenylimidazo[4,5-*b*]pyridine) 229

CONTENTS

Mycotoxins

    Aflatoxins: B_1, B_2, G_1, G_2, M_1 245
    Toxins derived from *Fusarium graminearum*, *F. culmorum* and *F. crookwellense*:
        zearalenone, deoxynivalenol, nivalenol and fusarenone X 397
    Toxins derived from *Fusarium moniliforme*: fumonisins B_1 and B_2 and
        fusarin C 445
    Toxins derived from *Fusarium sporotrichioides*: T-2 toxin 467
    Ochratoxin A 489

SUMMARY OF FINAL EVALUATIONS 523

APPENDIX 1. SUMMARY TABLES OF GENETIC AND RELATED EFFECTS . 527

APPENDIX 2. ACTIVITY PROFILES FOR GENETIC AND RELATED
    EFFECTS 543

CUMULATIVE INDEX TO THE *MONOGRAPHS* SERIES 573

NOTE TO THE READER

The term 'carcinogenic risk' in the *IARC Monographs* series is taken to mean the probability that exposure to an agent will lead to cancer in humans.

Inclusion of an agent in the *Monographs* does not imply that it is a carcinogen, only that the published data have been examined. Equally, the fact that an agent has not yet been evaluated in a monograph does not mean that it is not carcinogenic.

The evaluations of carcinogenic risk are made by international working groups of independent scientists and are qualitative in nature. No recommendation is given for regulation or legislation.

Anyone who is aware of published data that may alter the evaluation of the carcinogenic risk of an agent to humans is encouraged to make this information available to the Unit of Carcinogen Identification and Evaluation, International Agency for Research on Cancer, 150 cours Albert Thomas, 69372 Lyon Cedex 08, France, in order that the agent may be considered for re-evaluation by a future Working Group.

Although every effort is made to prepare the monographs as accurately as possible, mistakes may occur. Readers are requested to communicate any errors to the Unit of Carcinogen Identification and Evaluation, so that corrections can be reported in future volumes.

IARC WORKING GROUP ON THE EVALUATION OF CARCINOGENIC RISKS TO HUMANS: SOME NATURALLY OCCURRING SUBSTANCES: FOOD ITEMS AND CONSTITUENTS, HETEROCYCLIC AROMATIC AMINES AND MYCOTOXINS

Lyon, 9–16 June 1992

LIST OF PARTICIPANTS

Members

P. Bannasch, Department of Cytopathology, German Cancer Research Center, Im Neuenheimer Feld 280, 6900 Heidelberg 1, Germany

F. Berrino, Tumour Registry of Lombardia, National Institute for the Study and Treatment of Tumours, via Venezian 1, 20133 Milan, Italy

G.A. Boorman, Tumor Pathology Section, National Toxicology Program, National Institute of Environmental Health Sciences, PO Box 12233, Research Triangle Park NC 27709, USA

I.N. Chernozemsky, Bisser 2, PO Box 35, Sofia 1421, Bulgaria (*Vice-Chairman*)

G. Dirheimer, Institute of Molecular and Cellular Biology, National Center for Scientific Research, 15, rue René Descartes, 67084 Strasbourg Cedex, France

J.D. Groopman, Johns Hopkins University, 615 North Wolfe St, Baltimore, MD 21205-2179, USA

K. Hemminki, Center for Nutrition and Toxicology, CNT Novum, 14152 Huddinge, Sweden

W.M.F. Jongen, Agrotechnological Research Institute, Haagsteeg 6, Postbus 17, 6700 AA Wageningen, Netherlands

M. Knize, Lawrence Livermore National Laboratory, University of California, PO Box 5507, Livermore, CA 94550, USA

S.H. Lu, Cancer Institute, Chinese Academy of Medical Sciences, Panjiayuan, Chaoyang District, PO Box 2258, Beijing 100021, China

W.F.O. Marasas, Environmental Health and Toxicology, Research Institute for Nutritional Diseases of the Medical Research Council, PO Box 19070, Tygerberg 7505, South Africa

A.B. Miller, Department of Preventive Medicine and Biostatistics, Faculty of Medicine, McMurrich Building, University of Toronto, Toronto, Ontario, Canada M5S 1A8 (*Chairman*)

J.D. Miller, Agriculture Canada, Plant Research Centre, Ottawa, Ontario, Canada K1A OC6

M. Nagao, National Cancer Center Research Institute, Tsukiji 5-chome, Chuo-ku, Tokyo 104, Japan

G.E. Neal, Toxicology Unit, Medical Research Council Laboratories, Woodmansterne Road, Carshalton, Surrey SM5 4EF, United Kingdom

H. Norppa, Department of Industrial Hygiene and Toxicology, Institute of Occupational Health, Topeliuksenkatu 41 a A, 00250 Helsinki, Finland

S. Olin, International Life Sciences Institutes, Risk Science Institute, 1126 Sixteenth Street NW, Washington DC 20036, USA

J.H. Olsen, Danish Cancer Registry, Rosenvaengets Hovedvej 35, Box 839, 2100 Copenhagen Ø, Denmark

R. Schulte-Hermann, Institute for Tumour Biology–Cancer Research, University of Vienna, Borschkegasse 8a, 1090 Vienna, Austria

R. Stahlmann, Institute for Toxicology and Embryopharmacology (WE18), Free University of Berlin, Garystrasse 5, 1000 Berlin 33, Germany

M. Tatematsu, Laboratory of Ultrastructure Research, Aichi Cancer Center Research Institute, Kanokoden, Chikusa-ku, Nagoya 464, Japan

S. Venitt, Institute of Cancer Research, Royal Cancer Hospital, The Haddow Laboratories, 15 Cotswold Road, Belmont, Sutton, Surrey SM2 5NG, United Kingdom

A. Visconti, Institute of Toxins and Mycotoxins of Vegetable Parasites, viale Einaudi 51, 70125 Bari, Italy

A. Wu-Williams, Reproductive and Cancer Hazard Assessment Section, Office of Environmental Health Hazard Assessment, California Environmental Protection Agency, 2151 Berkeley Way, Annex 11, Berkeley, CA 94704, USA

D.G. Zaridze, Institute of Carcinogenesis, Cancer Research Centre, Russian Academy of Medical Sciences, Kashirskoye Shosse 24, 115478 Moscow, Russian Federation

Representative and observers

Representative of the National Cancer Institute

S.M. Sieber, Division of Cancer Etiology, National Cancer Institute, Building 31, Room 11A03, Bethesda MD 20892, USA

US Flavor and Extract Manufacturers' Association

R.L. Smith, Department of Pharmacology and Toxicology, St Mary's Hospital Medical School, Norfolk Place, London W2 1PG, United Kingdom

Food and Drug Administration

R. Scheuplein, Office of Toxicological Science, Food and Drug Administration, Federal Office Building 8, Room 2025, 202 C Street SW, Washington DC 20204, USA

National Coffee Association

A. Sivak, Environmental Health Sciences, PO Box 1038, Kendall Square, Cambridge, MA 02142, USA

IARC Secretariat

B. Armstrong, Deputy Director
P. Boffetta, Unit of Analytical Epidemiology
F.X. Bosch, Unit of Field and Intervention Studies
J.R.P. Cabral, Unit of Mechanisms of Carcinogenesis
E. Cardis, Director's Office
M. Castegnaro, Unit of Environmental Carcinogens and Host Factors
M. Friesen, Unit of Environmental Carcinogens and Host Factors
M.-J. Ghess, Unit of Carcinogen Identification and Evaluation
E. Heseltine, Lajarthe, 24290 St Léon-sur-Vézère, France
V. Krutovskikh, Unit of Multistage Carcinogenesis
J. Little, Unit of Analytical Epidemiology
D. McGregor, Unit of Carcinogen Identification and Evaluation
D. Mietton, Unit of Carcinogen Identification and Evaluation
H. Møller, Unit of Carcinogen Identification and Evaluation
I. O'Neill, Unit of Environmental Carcinogens and Host Factors
C. Partensky, Unit of Carcinogen Identification and Evaluation
I. Peterschmitt, Unit of Carcinogen Identification and Evaluation, Geneva, Switzerland
R. Pleština, International Programme on Chemical Safety, World Health Organization, 1211 Geneva 27, Switzerland
D. Shuker, Unit of Environmental Carcinogens and Host Factors
L. Tomatis, Director
H. Vainio, Chief, Unit of Carcinogen Identification and Evaluation
J. Wilbourn, Unit of Carcinogen Identification and Evaluation
C. Wild, Unit of Mechanisms of Carcinogenesis
H. Yamasaki, Unit of Multistage Carcinogenesis

Secretarial assistance

M. Lézère
J. Mitchell
S. Reynaud

PREAMBLE

IARC MONOGRAPHS PROGRAMME ON THE EVALUATION OF CARCINOGENIC RISKS TO HUMANS[1]

PREAMBLE

1. BACKGROUND

In 1969, the International Agency for Research on Cancer (IARC) initiated a programme to evaluate the carcinogenic risk of chemicals to humans and to produce monographs on individual chemicals. The *Monographs* programme has since been expanded to include consideration of exposures to complex mixtures of chemicals (which occur, for example, in some occupations and as a result of human habits) and of exposures to other agents, such as radiation and viruses. With Supplement 6 (IARC, 1987a), the title of the series was modified from *IARC Monographs on the Evaluation of the Carcinogenic Risk of Chemicals to Humans* to *IARC Monographs on the Evaluation of Carcinogenic Risks to Humans*, in order to reflect the widened scope of the programme.

The criteria established in 1971 to evaluate carcinogenic risk to humans were adopted by the working groups whose deliberations resulted in the first 16 volumes of the *IARC Monographs* series. Those criteria were subsequently updated by further ad-hoc working groups (IARC, 1977, 1978, 1979, 1982, 1983, 1987b, 1988, 1991a; Vainio *et al.*, 1992).

2. OBJECTIVE AND SCOPE

The objective of the programme is to prepare, with the help of international working groups of experts, and to publish in the form of monographs, critical reviews and evaluations of evidence on the carcinogenicity of a wide range of human exposures. The *Monographs* may also indicate where additional research efforts are needed.

The *Monographs* represent the first step in carcinogenic risk assessment, which involves examination of all relevant information in order to assess the strength of the available evidence that certain exposures could alter the incidence of cancer in humans. The second step is quantitative risk estimation. Detailed, quantitative evaluations of epidemiological data may be made in the *Monographs*, but without extrapolation beyond the range of the data

---

[1]This project is supported by PHS Grant No. 5-UO1 CA33193-11 awarded by the US National Cancer Institute, Department of Health and Human Services. Since 1986, the programme has also been supported by the Commission of the European Communities.

available. Quantitative extrapolation from experimental data to the human situation is not undertaken.

The term 'carcinogen' is used in these monographs to denote an exposure that is capable of increasing the incidence of malignant neoplasms; the induction of benign neoplasms may in some circumstances (see p. 22) contribute to the judgement that the exposure is carcinogenic. The terms 'neoplasm' and 'tumour' are used interchangeably.

Some epidemiological and experimental studies indicate that different agents may act at different stages in the carcinogenic process, and several different mechanisms may be involved. The aim of the *Monographs* has been, from their inception, to evaluate evidence of carcinogenicity at any stage in the carcinogenesis process, independently of the underlying mechanisms. Information on mechanisms may, however, be used in making the overall evaluation (IARC, 1991a; Vainio *et al.*, 1992; see also pp. 28-29).

The *Monographs* may assist national and international authorities in making risk assessments and in formulating decisions concerning any necessary preventive measures. The evaluations of IARC working groups are scientific, qualitative judgements about the evidence for or against carcinogenicity provided by the available data. These evaluations represent only one part of the body of information on which regulatory measures may be based. Other components of regulatory decisions may vary from one situation to another and from country to country, responding to different socioeconomic and national priorities. **Therefore, no recommendation is given with regard to regulation or legislation, which are the responsibility of individual governments and/or other international organizations.**

The *IARC Monographs* are recognized as an authoritative source of information on the carcinogenicity of a wide range of human exposures. A users' survey, made in 1988, indicated that the *Monographs* are consulted by various agencies in 57 countries. Each volume is generally printed in 4000 copies for distribution to governments, regulatory bodies and interested scientists. The *Monographs* are also available *via* the Distribution and Sales Service of the World Health Organization.

3. SELECTION OF TOPICS FOR MONOGRAPHS

Topics are selected on the basis of two main criteria: (a) there is evidence of human exposure, and (b) there is some evidence or suspicion of carcinogenicity. The term 'agent' is used to include individual chemical compounds, groups of related chemical compounds, physical agents (such as radiation) and biological factors (such as viruses). Exposures to mixtures of agents may occur in occupational exposures and as a result of personal and cultural habits (like smoking and dietary practices). Chemical analogues and compounds with biological or physical characteristics similar to those of suspected carcinogens may also be considered, even in the absence of data on a possible carcinogenic effect in humans or experimental animals.

The scientific literature is surveyed for published data relevant to an assessment of carcinogenicity. The IARC surveys of chemicals being tested for carcinogenicity (IARC, 1973-1992) and directories of on-going research in cancer epidemiology (IARC, 1976-1992) often indicate those exposures that may be scheduled for future meetings. Ad-hoc working groups convened by IARC in 1984, 1989 and 1991 gave recommendations as to which agents should be evaluated in the *IARC Monographs* series (IARC, 1984, 1989, 1991b).

As significant new data on subjects on which monographs have already been prepared become available, re-evaluations are made at subsequent meetings, and revised monographs are published.

4. DATA FOR MONOGRAPHS

The *Monographs* do not necessarily cite all the literature concerning the subject of an evaluation. Only those data considered by the Working Group to be relevant to making the evaluation are included.

With regard to biological and epidemiological data, only reports that have been published or accepted for publication in the openly available scientific literature are reviewed by the working groups. In certain instances, government agency reports that have undergone peer review and are widely available are considered. Exceptions may be made on an ad-hoc basis to include unpublished reports that are in their final form and publicly available, if their inclusion is considered pertinent to making a final evaluation (see pp. 26 *et seq.*). In the sections on chemical and physical properties, on analysis, on production and use and on occurrence, unpublished sources of information may be used.

5. THE WORKING GROUP

Reviews and evaluations are formulated by a working group of experts. The tasks of the group are: (i) to ascertain that all appropriate data have been collected; (ii) to select the data relevant for the evaluation on the basis of scientific merit; (iii) to prepare accurate summaries of the data to enable the reader to follow the reasoning of the Working Group; (iv) to evaluate the results of epidemiological and experimental studies on cancer; (v) to evaluate data relevant to the understanding of mechanism of action; and (vi) to make an overall evaluation of the carcinogenicity of the exposure to humans.

Working Group participants who contributed to the considerations and evaluations within a particular volume are listed, with their addresses, at the beginning of each publication. Each participant who is a member of a working group serves as an individual scientist and not as a representative of any organization, government or industry. In addition, nominees of national and international agencies and industrial associations may be invited as observers.

6. WORKING PROCEDURES

Approximately one year in advance of a meeting of a working group, the topics of the monographs are announced and participants are selected by IARC staff in consultation with other experts. Subsequently, relevant biological and epidemiological data are collected by IARC from recognized sources of information on carcinogenesis, including data storage and retrieval systems such as BIOSIS, Chemical Abstracts, CANCERLIT, MEDLINE and TOXLINE—including EMIC and ETIC for data on genetic and related effects and reproductive and developmental effects, respectively.

For chemicals and some complex mixtures, the major collection of data and the preparation of first drafts of the sections on chemical and physical properties, on analysis, on production and use and on occurrence are carried out under a separate contract funded by the US National Cancer Institute. Representatives from industrial associations may assist in the preparation of sections on production and use. Information on production and trade is obtained from governmental and trade publications and, in some cases, by direct contact with industries. Separate production data on some agents may not be available because their publication could disclose confidential information. Information on uses may be obtained from published sources but is often complemented by direct contact with manufacturers. Efforts are made to supplement this information with data from other national and international sources.

Six months before the meeting, the material obtained is sent to meeting participants, or is used by IARC staff, to prepare sections for the first drafts of monographs. The first drafts are compiled by IARC staff and sent, prior to the meeting, to all participants of the Working Group for review.

The Working Group meets in Lyon for seven to eight days to discuss and finalize the texts of the monographs and to formulate the evaluations. After the meeting, the master copy of each monograph is verified by consulting the original literature, edited and prepared for publication. The aim is to publish monographs within nine months of the Working Group meeting.

The available studies are summarized by the Working Group, with particular regard to the qualitative aspects discussed below. In general, numerical findings are indicated as they appear in the original report; units are converted when necessary for easier comparison. The Working Group may conduct additional analyses of the published data and use them in their assessment of the evidence; the results of such supplementary analyses are given in square brackets. When an important aspect of a study, directly impinging on its interpretation, should be brought to the attention of the reader, a comment is given in square brackets.

7. EXPOSURE DATA

Sections that indicate the extent of past and present human exposure, the sources of exposure, the people most likely to be exposed and the factors that contribute to the exposure are included at the beginning of each monograph.

Most monographs on individual chemicals, groups of chemicals or complex mixtures include sections on chemical and physical data, on analysis, on production and use and on occurrence. In monographs on, for example, physical agents, biological factors, occupational exposures and cultural habits, other sections may be included, such as: historical perspectives, description of an industry or habit, chemistry of the complex mixture or taxonomy.

For chemical exposures, the Chemical Abstracts Services Registry Number, the latest Chemical Abstracts Primary Name and the IUPAC Systematic Name are recorded; other synonyms are given, but the list is not necessarily comprehensive. For biological agents, taxonomy and structure are described, and the degree of variability is given, when applicable.

Information on chemical and physical properties and, in particular, data relevant to identification, occurrence and biological activity are included. For biological agents, mode of replication, life cycle, target cells, persistence and latency, host response and description of nonmalignant disease caused by them are given. A description of technical products of chemicals includes trades names, relevant specifications and available information on composition and impurities. Some of the trade names given may be those of mixtures in which the agent being evaluated is only one of the ingredients.

The purpose of the section on analysis is to give the reader an overview of current methods, with emphasis on those widely used for regulatory purposes. Methods for monitoring human exposure are also given, when available. No critical evaluation or recommendation of any of the methods is meant or implied. The IARC publishes a series of volumes, *Environmental Carcinogens: Methods of Analysis and Exposure Measurement* (IARC, 1978–92), that describe validated methods for analysing a wide variety of chemicals and mixtures. For biological agents, methods of detection and exposure assessment are described, including their sensitivity, specificity and reproducibility.

The dates of first synthesis and of first commercial production of a chemical or mixture are provided; for agents which do not occur naturally, this information may allow a reasonable estimate to be made of the date before which no human exposure to the agent could have occurred. The dates of first reported occurrence of an exposure are also provided. In addition, methods of synthesis used in past and present commercial production and different methods of production which may give rise to different impurities are described.

Data on production, international trade and uses are obtained for representative regions, which usually include Europe, Japan and the USA. It should not, however, be inferred that those areas or nations are necessarily the sole or major sources or users of the agent. Some identified uses may not be current or major applications, and the coverage is not necessarily comprehensive. In the case of drugs, mention of their therapeutic uses does not necessarily represent current practice nor does it imply judgement as to their therapeutic efficacy.

Information on the occurrence of an agent or mixture in the environment is obtained from data derived from the monitoring and surveillance of levels in occupational environments, air, water, soil, foods and animal and human tissues. When available, data on the generation, persistence and bioaccumulation of the agent are also included. In the case of mixtures, industries, occupations or processes, information is given about all agents present. For processes, industries and occupations, a historical description is also given, noting variations in chemical composition, physical properties and levels of occupational exposure with time and place. For biological agents, the epidemiology of infection is described.

Statements concerning regulations and guidelines (e.g., pesticide registrations, maximal levels permitted in foods, occupational exposure limits) are included for some countries as indications of potential exposures, but they may not reflect the most recent situation, since such limits are continuously reviewed and modified. The absence of information on regulatory status for a country should not be taken to imply that that country does not have regulations with regard to the exposure. For biological agents, legislation and control, including vaccines and therapy, are described.

8. STUDIES OF CANCER IN HUMANS

(a) Types of studies considered

Three types of epidemiological studies of cancer contribute to the assessment of carcinogenicity in humans—cohort studies, case–control studies and correlation (or ecological) studies. Rarely, results from randomized trials may be available. Case reports of cancer in humans may also be reviewed.

Cohort and case–control studies relate individual exposures under study to the occurrence of cancer in individuals and provide an estimate of relative risk (ratio of incidence in those exposed to incidence in those not exposed) as the main measure of association.

In correlation studies, the units of investigation are usually whole populations (e.g., in particular geographical areas or at particular times), and cancer frequency is related to a summary measure of the exposure of the population to the agent, mixture or exposure circumstance under study. Because individual exposure is not documented, however, a causal relationship is less easy to infer from correlation studies than from cohort and case–control studies. Case reports generally arise from a suspicion, based on clinical experience, that the concurrence of two events—that is, a particular exposure and occurrence of a cancer—has happened rather more frequently than would be expected by chance. Case reports usually lack complete ascertainment of cases in any population, definition or enumeration of the population at risk and estimation of the expected number of cases in the absence of exposure. The uncertainties surrounding interpretation of case reports and correlation studies make them inadequate, except in rare instances, to form the sole basis for inferring a causal relationship. When taken together with case–control and cohort studies, however, relevant case reports or correlation studies may add materially to the judgement that a causal relationship is present.

Epidemiological studies of benign neoplasms, presumed preneoplastic lesions and other end-points thought to be relevant to cancer are also reviewed by working groups. They may, in some instances, strengthen inferences drawn from studies of cancer itself.

(b) Quality of studies considered

The *Monographs* are not intended to summarize all published studies. Those that are judged to be inadequate or irrelevant to the evaluation are generally omitted. They may be mentioned briefly, particularly when the information is considered to be a useful supplement to that in other reports or when they provide the only data available. Their inclusion does not imply acceptance of the adequacy of the study design or of the analysis and interpretation of the results, and limitations are clearly outlined in square brackets at the end of the study description.

It is necessary to take into account the possible roles of bias, confounding and chance in the interpretation of epidemiological studies. By 'bias' is meant the operation of factors in study design or execution that lead erroneously to a stronger or weaker association than in fact exists between disease and an agent, mixture or exposure circumstance. By 'confounding' is meant a situation in which the relationship with disease is made to appear stronger or to appear weaker than it truly is as a result of an association between the apparent causal factor and another factor that is associated with either an increase or decrease in the incidence of the disease. In evaluating the extent to which these factors have been minimized in an individual study, working groups consider a number of aspects of design and analysis as described in the report of the study. Most of these considerations apply equally to case–control, cohort and correlation studies. Lack of clarity of any of these aspects in the reporting of a study can decrease its credibility and the weight given to it in the final evaluation of the exposure.

Firstly, the study population, disease (or diseases) and exposure should have been well defined by the authors. Cases of disease in the study population should have been identified in a way that was independent of the exposure of interest, and exposure should have been assessed in a way that was not related to disease status.

Secondly, the authors should have taken account in the study design and analysis of other variables that can influence the risk of disease and may have been related to the exposure of interest. Potential confounding by such variables should have been dealt with either in the design of the study, such as by matching, or in the analysis, by statistical adjustment. In cohort studies, comparisons with local rates of disease may be more appropriate than those with national rates. Internal comparisons of disease frequency among individuals at different levels of exposure should also have been made in the study.

Thirdly, the authors should have reported the basic data on which the conclusions are founded, even if sophisticated statistical analyses were employed. At the very least, they should have given the numbers of exposed and unexposed cases and controls in a case–control study and the numbers of cases observed and expected in a cohort study. Further tabulations by time since exposure began and other temporal factors are also important. In a cohort study, data on all cancer sites and all causes of death should have been given, to reveal the possibility of reporting bias. In a case–control study, the effects of investigated factors other than the exposure of interest should have been reported.

Finally, the statistical methods used to obtain estimates of relative risk, absolute rates of cancer, confidence intervals and significance tests, and to adjust for confounding should have been clearly stated by the authors. The methods used should preferably have been the generally accepted techniques that have been refined since the mid-1970s. These methods have been reviewed for case–control studies (Breslow & Day, 1980) and for cohort studies (Breslow & Day, 1987).

(c) Inferences about mechanism of action

Detailed analyses of both relative and absolute risks in relation to temporal variables, such as age at first exposure, time since first exposure, duration of exposure, cumulative exposure and time since exposure ceased, are reviewed and summarized when available. The analysis of temporal relationships can be useful in formulating models of carcinogenesis. In particular, such analyses may suggest whether a carcinogen acts early or late in the process of carcinogenesis, although at best they allow only indirect inferences about the mechanism of action. Special attention is given to measurements of biological markers of carcinogen exposure or action, such as DNA or protein adducts, as well as markers of early steps in the carcinogenic process, such as proto-oncogene mutation, when these are incorporated into epidemiological studies focused on cancer incidence or mortality. Such measurements may allow inferences to be made about putative mechanisms of action (IARC, 1991a; Vainio *et al.*, 1992).

(d) Criteria for causality

After the quality of individual epidemiological studies of cancer has been summarized and assessed, a judgement is made concerning the strength of evidence that the agent, mixture or exposure circumstance in question is carcinogenic for humans. In making their judgement, the Working Group considers several criteria for causality. A strong association (i.e., a large relative risk) is more likely to indicate causality than a weak association, although it is recognized that relative risks of small magnitude do not imply lack of causality and may be important if the disease is common. Associations that are replicated in several studies of the same design or using different epidemiological approaches or under different circumstances of exposure are more likely to represent a causal relationship than isolated observations from single studies. If there are inconsistent results among investigations, possible reasons are sought (such as differences in amount of exposure), and results of studies judged to be of high quality are given more weight than those from studies judged to be methodologically less sound. When suspicion of carcinogenicity arises largely from a single study, these data are not combined with those from later studies in any subsequent reassessment of the strength of the evidence.

If the risk of the disease in question increases with the amount of exposure, this is considered to be a strong indication of causality, although absence of a graded response is not necessarily evidence against a causal relationship. Demonstration of a decline in risk after cessation of or reduction in exposure in individuals or in whole populations also supports a causal interpretation of the findings.

Although a carcinogen may act upon more than one target, the specificity of an association (i.e., an increased occurrence of cancer at one anatomical site or of one morphological type) adds plausibility to a causal relationship, particularly when excess cancer occurrence is limited to one morphological type within the same organ.

Although rarely available, results from randomized trials showing different rates among exposed and unexposed individuals provide particularly strong evidence for causality.

When several epidemiological studies show little or no indication of an association between an exposure and cancer, the judgement may be made that, in the aggregate, they show evidence of lack of carcinogenicity. Such a judgement requires first of all that the studies giving rise to it meet, to a sufficient degree, the standards of design and analysis described above. Specifically, the possibility that bias, confounding or misclassification of exposure or outcome could explain the observed results should be considered and excluded with reasonable certainty. In addition, all studies that are judged to be methodologically sound should be consistent with a relative risk of unity for any observed level of exposure and, when considered together, should provide a pooled estimate of relative risk which is at or near unity and has a narrow confidence interval, due to sufficient population size. Moreover, no individual study nor the pooled results of all the studies should show any consistent tendency for relative risk of cancer to increase with increasing level of exposure. It is important to note that evidence of lack of carcinogenicity obtained in this way from several epidemiological studies can apply only to the type(s) of cancer studied and to dose levels and intervals between first exposure and observation of disease that are the same as or less than those observed in all the studies. Experience with human cancer indicates that, in some cases, the period from first exposure to the development of clinical cancer is seldom less than 20 years; latent periods substantially shorter than 30 years cannot provide evidence for lack of carcinogenicity.

9. STUDIES OF CANCER IN EXPERIMENTAL ANIMALS

For several agents (e.g., aflatoxins, 4-aminobiphenyl, bis(chloromethyl)ether, diethylstilboestrol, melphalan, 8-methoxypsoralen (methoxsalen) plus ultra-violet radiation, mustard gas and vinyl chloride), evidence of carcinogenicity in experimental animals preceded evidence obtained from epidemiological studies or case reports. Information compiled from the first 41 volumes of the *IARC Monographs* (Wilbourn *et al.*, 1986) shows that, of the 44 agents and mixtures for which there is *sufficient* or *limited* evidence of carcinogenicity to humans (see p. 26), all 37 that have been tested adequately produce cancer in at least one animal species. Although this association cannot establish that all agents and mixtures that cause cancer in experimental animals also cause cancer in humans, nevertheless, **in the absence of adequate data on humans, it is biologically plausible and prudent to regard agents and mixtures for which there is sufficient evidence (see p. 27) of carcinogenicity in experimental animals as if they presented a carcinogenic risk to humans.** The possibility that a given agent may cause cancer through a species-specific mechanism which does not operate in humans, see p. 28, should also be taken into consideration.

The nature and extent of impurities or contaminants present in the chemical or mixture being evaluated are given when available. Animal strain, sex, numbers per group, age at start of treatment and survival are reported.

Other types of studies summarized include: experiments in which the agent or mixture was administered in conjunction with known carcinogens or factors that modify carcinogenic effects; studies in which the end-point was not cancer but a defined precancerous lesion; and experiments on the carcinogenicity of known metabolites and derivatives.

For experimental studies of mixtures, consideration is given to the possibility of changes in the physicochemical properties of the test substance during collection, storage, extraction, concentration and delivery. Chemical and toxicological interactions of the components of mixtures may result in nonlinear dose–response relationships.

An assessment is made as to the relevance to human exposure of samples tested in experimental animals, which may involve consideration of: (i) physical and chemical characteristics, (ii) constituent substances that indicate the presence of a class of substances, (iii) the results of tests for genetic and related effects, including genetic activity profiles, DNA adduct profiles, proto-oncogene mutation and expression and suppressor gene inactivation. The relevance of results obtained with viral strains analogous to that being evaluated in the monograph must also be considered.

(a) Qualitative aspects

An assessment of carcinogenicity involves several considerations of qualitative importance, including (i) the experimental conditions under which the test was performed, including route and schedule of exposure, species, strain, sex, age, duration of follow-up; (ii) the consistency of the results, for example, across species and target organ(s); (iii) the spectrum of neoplastic response, from preneoplastic lesions and benign tumours to malignant neoplasms; and (iv) the possible role of modifying factors.

As mentioned earlier (p. 15), the *Monographs* are not intended to summarize all published studies. Those studies in experimental animals that are inadequate (e.g., too short a duration, too few animals, poor survival; see below) or are judged irrelevant to the evaluation are generally omitted. Guidelines for conducting adequate long-term carcinogenicity experiments have been outlined (e.g., Montesano *et al.*, 1986).

Considerations of importance to the Working Group in the interpretation and evaluation of a particular study include: (i) how clearly the agent was defined and, in the case of mixtures, how adequately the sample characterization was reported; (ii) whether the dose was adequately monitored, particularly in inhalation experiments; (iii) whether the doses and duration of treatment were appropriate and whether the survival of treated animals was similar to that of controls; (iv) whether there were adequate numbers of animals per group; (v) whether animals of both sexes were used; (vi) whether animals were allocated randomly to groups; (vii) whether the duration of observation was adequate; and (viii) whether the data were adequately reported. If available, recent data on the incidence of specific tumours in historical controls, as well as in concurrent controls, should be taken into account in the evaluation of tumour response.

When benign tumours occur together with and originate from the same cell type in an organ or tissue as malignant tumours in a particular study and appear to represent a stage in the progression to malignancy, it may be valid to combine them in assessing tumour incidence (Huff *et al.*, 1989). The occurrence of lesions presumed to be preneoplastic may in certain instances aid in assessing the biological plausibility of any neoplastic response observed. If an agent or mixture induces only benign neoplasms that appear to be end-points that do not readily undergo transition to malignancy, it should nevertheless be suspected of being a carcinogen and it requires further investigation.

(b) Quantitative aspects

The probability that tumours will occur may depend on the species, sex, strain and age of the animal, the dose of the carcinogen and the route and length of exposure. Evidence of an increased incidence of neoplasms with increased level of exposure strengthens the inference of a causal association between the exposure and the development of neoplasms.

The form of the dose–response relationship can vary widely, depending on the particular agent under study and the target organ. Since many chemicals require metabolic activation before being converted into their reactive intermediates, both metabolic and pharmacokinetic aspects are important in determining the dose–response pattern. Saturation of steps such as absorption, activation, inactivation and elimination may produce nonlinearity in the dose–response relationship, as could saturation of processes such as DNA repair (Hoel *et al.*, 1983; Gart *et al.*, 1986).

(c) Statistical analysis of long-term experiments in animals

Factors considered by the Working Group include the adequacy of the information given for each treatment group: (i) the number of animals studied and the number examined histologically, (ii) the number of animals with a given tumour type and (iii) length of survival. The statistical methods used should be clearly stated and should be the generally accepted techniques refined for this purpose (Peto *et al.*, 1980; Gart *et al.*, 1986). When there is no difference in survival between control and treatment groups, the Working Group usually compares the proportions of animals developing each tumour type in each of the groups. Otherwise, consideration is given as to whether or not appropriate adjustments have been made for differences in survival. These adjustments can include: comparisons of the proportions of tumour-bearing animals among the effective number of animals (alive at the time the first tumour is discovered), in the case where most differences in survival occur before tumours appear; life-table methods, when tumours are visible or when they may be considered 'fatal' because mortality rapidly follows tumour development; and the Mantel-Haenszel test or logistic regression, when occult tumours do not affect the animals' risk of dying but are 'incidental' findings at autopsy.

In practice, classifying tumours as fatal or incidental may be difficult. Several survival-adjusted methods have been developed that do not require this distinction (Gart *et al.*, 1986), although they have not been fully evaluated.

10. OTHER RELEVANT DATA

(a) Absorption, distribution, metabolism and excretion

Concise information is given on absorption, distribution (including placental transfer) and excretion in both humans and experimental animals. Kinetic factors that may affect the dose–response relationship, such as saturation of uptake, protein binding, metabolic activation, detoxification and DNA repair processes, are mentioned. Studies that indicate the metabolic fate of the agent in humans and in experimental animals are summarized briefly, and comparisons of data from humans and animals are made when possible. Comparative information on the relationship between exposure and the dose that reaches the target site may be of particular importance for extrapolation between species.

(b) Toxic effects

Data are given on acute and chronic toxic effects (other than cancer), such as organ toxicity, increased cell proliferation, immunotoxicity and endocrine effects. The presence and toxicological significance of cellular receptors is described.

(c) Reproductive and developmental effects

Effects on reproduction, teratogenicity, fetotoxicity and embryotoxicity are also summarized briefly.

(d) Genetic and related effects

Tests of genetic and related effects are described in view of the relevance of gene mutation and chromosomal damage to carcinogenesis (Vainio *et al.*, 1992).

The adequacy of the reporting of sample characterization is considered and, where necessary, commented upon; with regard to complex mixtures, such comments are similar to those described for animal carcinogenicity tests on p. 21. The available data are interpreted critically by phylogenetic group according to the end-points detected, which may include DNA damage, gene mutation, sister chromatid exchange, micronucleus formation, chromosomal aberrations, aneuploidy and cell transformation. The concentrations employed are given, and mention is made of whether use of an exogenous metabolic system *in vitro* affected the test result. These data are given as listings of test systems, data and references; bar graphs (activity profiles) and corresponding summary tables with detailed information on the preparation of the profiles (Waters *et al.*, 1987) are given in appendices.

Positive results in tests using prokaryotes, lower eukaryotes, plants, insects and cultured mammalian cells suggest that genetic and related effects could occur in mammals. Results from such tests may also give information about the types of genetic effect produced and about the involvement of metabolic activation. Some end-points described are clearly genetic in nature (e.g., gene mutations and chromosomal aberrations), while others are to a greater or lesser degree associated with genetic effects (e.g., unscheduled DNA synthesis). In-vitro tests for tumour-promoting activity and for cell transformation may be sensitive to changes that are not necessarily the result of genetic alterations but that may have specific relevance to the process of carcinogenesis. A critical appraisal of these tests has been published (Montesano *et al.*, 1986).

Genetic or other activity manifest in experimental mammals and humans is regarded as being of greater relevance than that in other organisms. The demonstration that an agent or mixture can induce gene and chromosomal mutations in whole mammals indicates that it may have carcinogenic activity, although this activity may not be detectably expressed in any or all species. Relative potency in tests for mutagenicity and related effects is not a reliable indicator of carcinogenic potency. Negative results in tests for mutagenicity in selected tissues from animals treated *in vivo* provide less weight, partly because they do not exclude the possibility of an effect in tissues other than those examined. Moreover, negative results in short-term tests with genetic end-points cannot be considered to provide evidence to rule out carcinogenicity of agents or mixtures that act through other mechanisms (e.g., receptor-mediated effects, cellular toxicity with regenerative proliferation, peroxisome proliferation) (Vainio *et al.*, 1992). Factors that may lead to misleading results in short-term tests have been discussed in detail elsewhere (Montesano *et al.*, 1986).

When available, data relevant to mechanisms of carcinogenesis that do not involve structural changes at the level of the gene are also described.

The adequacy of epidemiological studies of reproductive outcome and genetic and related effects in humans is evaluated by the same criteria as are applied to epidemiological studies of cancer.

(e) Structure–activity considerations

This section describes structure–activity relationships that may be relevant to an evaluation of the carcinogenicity of an agent.

11. SUMMARY OF DATA REPORTED

In this section, the relevant epidemiological and experimental data are summarized. Only reports, other than in abstract form, that meet the criteria outlined on p. 15 are considered for evaluating carcinogenicity. Inadequate studies are generally not summarized: such studies are usually identified by a square-bracketed comment in the preceding text.

(a) Exposures

Human exposure is summarized on the basis of elements such as production, use, occurrence in the environment and determinations in human tissues and body fluids. Quantitative data are given when available.

(b) *Carcinogenicity in humans*

Results of epidemiological studies that are considered to be pertinent to an assessment of human carcinogenicity are summarized. When relevant, case reports and correlation studies are also summarized.

(c) *Carcinogenicity in experimental animals*

Data relevant to an evaluation of carcinogenicity in animals are summarized. For each animal species and route of administration, it is stated whether an increased incidence of neoplasms or preneoplastic lesions was observed, and the tumour sites are indicated. If the agent or mixture produced tumours after prenatal exposure or in single-dose experiments, this is also indicated. Negative findings are also summarized. Dose–response and other quantitative data may be given when available.

(d) *Other data relevant to an evaluation of carcinogenicity and its mechanisms*

Data on biological effects in humans that are of particular relevance are summarized. These may include toxicological, kinetic and metabolic considerations and evidence of DNA binding, persistence of DNA lesions or genetic damage in exposed humans. Toxicological information, such as that on cytotoxicity and regeneration, receptor binding and hormonal and immunological effects, and data on kinetics and metabolism in experimental animals are given when considered relevant to the possible mechanism of the carcinogenic action of the agent. The results of tests for genetic and related effects are summarized for whole mammals, cultured mammalian cells and nonmammalian systems.

When available, comparisons of such data for humans and for animals, and particularly animals that have developed cancer, are described.

Structure–activity relationships are mentioned when relevant.

For the agent, mixture or exposure circumstance being evaluated, the available data on end-points or other phenomena relevant to mechanisms of carcinogenesis from studies in humans, experimental animals and tissue and cell test systems are summarized within one or more of the following descriptive dimensions:

(i) Evidence of genotoxicity (i.e., structural changes at the level of the gene): for example, structure–activity considerations, adduct formation, mutagenicity (effect on specific genes), chromosomal mutation/aneuploidy

(ii) Evidence of effects on the expression of relevant genes (i.e., functional changes at the intracellular level): for example, alterations to the structure or quantity of the product of a proto-oncogene or tumour suppressor gene, alterations to metabolic activation/-inactivation/DNA repair

(iii) Evidence of relevant effects on cell behaviour (i.e., morphological or behavioural changes at the cellular or tissue level): for example, induction of mitogenesis, compensatory cell proliferation, preneoplasia and hyperplasia, survival of premalignant or malignant cells (immortalization, immunosuppression), effects on metastatic potential

(iv) Evidence from dose and time relationships of carcinogenic effects and interactions between agents: for example, early/late stage, as inferred from epidemiological studies; initiation/promotion/progression/malignant conversion, as defined in animal carcinogenicity experiments; toxicokinetics

These dimensions are not mutually exclusive, and an agent may fall within more than one of them. Thus, for example, the action of an agent on the expression of relevant genes could be summarized under both the first and second dimension, even if it were known with reasonable certainty that those effects resulted from genotoxicity.

12. EVALUATION

Evaluations of the strength of the evidence for carcinogenicity arising from human and experimental animal data are made, using standard terms.

It is recognized that the criteria for these evaluations, described below, cannot encompass all of the factors that may be relevant to an evaluation of carcinogenicity. In considering all of the relevant data, the Working Group may assign the agent, mixture or exposure circumstance to a higher or lower category than a strict interpretation of these criteria would indicate.

(a) *Degrees of evidence for carcinogenicity in humans and in experimental animals and supporting evidence*

These categories refer only to the strength of the evidence that an exposure is carcinogenic and not to the extent of its carcinogenic activity (potency) nor to the mechanisms involved. A classification may change as new information becomes available.

An evaluation of degree of evidence, whether for a single agent or a mixture, is limited to the materials tested, as defined physically, chemically or biologically. When the agents evaluated are considered by the Working Group to be sufficiently closely related, they may be grouped together for the purpose of a single evaluation of degree of evidence.

(i) *Carcinogenicity in humans*

The applicability of an evaluation of the carcinogenicity of a mixture, process, occupation or industry on the basis of evidence from epidemiological studies depends on the variability over time and place of the mixtures, processes, occupations and industries. The Working Group seeks to identify the specific exposure, process or activity which is considered most likely to be responsible for any excess risk. The evaluation is focused as narrowly as the available data on exposure and other aspects permit.

The evidence relevant to carcinogenicity from studies in humans is classified into one of the following categories:

Sufficient evidence of carcinogenicity: The Working Group considers that a causal relationship has been established between exposure to the agent, mixture or exposure circumstance and human cancer. That is, a positive relationship has been observed between the exposure and cancer in studies in which chance, bias and confounding could be ruled out with reasonable confidence.

Limited evidence of carcinogenicity: A positive association has been observed between exposure to the agent, mixture or exposure circumstance and cancer for which a causal interpretation is considered by the Working Group to be credible, but chance, bias or confounding could not be ruled out with reasonable confidence.

Inadequate evidence of carcinogenicity: The available studies are of insufficient quality, consistency or statistical power to permit a conclusion regarding the presence or absence of a causal association, or no data on cancer in humans are available.

Evidence suggesting lack of carcinogenicity: There are several adequate studies covering the full range of levels of exposure that human beings are known to encounter, which are mutually consistent in not showing a positive association between exposure to the agent, mixture or exposure circumstance and any studied cancer at any observed level of exposure. A conclusion of 'evidence suggesting lack of carcinogenicity' is inevitably limited to the cancer sites, conditions and levels of exposure and length of observation covered by the available studies. In addition, the possibility of a very small risk at the levels of exposure studied can never be excluded.

In some instances, the above categories may be used to classify the degree of evidence related to carcinogenicity in specific organs or tissues.

(ii) *Carcinogenicity in experimental animals*

The evidence relevant to carcinogenicity in experimental animals is classified into one of the following categories:

Sufficient evidence of carcinogenicity: The Working Group considers that a causal relationship has been established between the agent or mixture and an increased incidence of malignant neoplasms or of an appropriate combination of benign and malignant neoplasms in (a) two or more species of animals or (b) in two or more independent studies in one species carried out at different times or in different laboratories or under different protocols.

Exceptionally, a single study in one species might be considered to provide sufficient evidence of carcinogenicity when malignant neoplasms occur to an unusual degree with regard to incidence, site, type of tumour or age at onset.

Limited evidence of carcinogenicity: The data suggest a carcinogenic effect but are limited for making a definitive evaluation because, e.g., (a) the evidence of carcinogenicity is restricted to a single experiment; or (b) there are unresolved questions regarding the adequacy of the design, conduct or interpretation of the study; or (c) the agent or mixture increases the incidence only of benign neoplasms or lesions of uncertain neoplastic potential, or of certain neoplasms which may occur spontaneously in high incidences in certain strains.

Inadequate evidence of carcinogenicity: The studies cannot be interpreted as showing either the presence or absence of a carcinogenic effect because of major qualitative or quantitative limitations, or no data on cancer in experimental animals are available.

Evidence suggesting lack of carcinogenicity: Adequate studies involving at least two species are available which show that, within the limits of the tests used, the agent or mixture is not carcinogenic. A conclusion of evidence suggesting lack of carcinogenicity is inevitably limited to the species, tumour sites and levels of exposure studied.

(b) *Other data relevant to the evaluation of carcinogenicity*

Other evidence judged to be relevant to an evaluation of carcinogenicity and of sufficient importance to affect the overall evaluation is then described. This may include data on preneoplastic lesions, tumour pathology, genetic and related effects, structure–activity relationships, metabolism and pharmacokinetics, and physicochemical parameters.

Data relevant to mechanisms of the carcinogenic action are also evaluated. The strength of the evidence that any carcinogenic effect observed is due to a particular mechanism is assessed, using terms such as weak, moderate or strong. Then, the Working Group assesses if that particular mechanism is likely to be operative in humans. The strongest indications that a particular mechanism operates in humans come from data on humans or biological specimens obtained from exposed humans. The data may be considered to be especially relevant if they show that the agent in question has caused changes in exposed humans that are on the causal pathway to carcinogenesis. Such data may, however, never become available, because it is at least conceivable that certain compounds may be kept from human use solely on the basis of evidence of their toxicity and/or carcinogenicity in experimental systems.

For complex exposures, including occupational and industrial exposures, chemical composition and the potential contribution of carcinogens known to be present are considered by the Working Group in its overall evaluation of human carcinogenicity. The Working Group also determines the extent to which the materials tested in experimental systems are related to those to which humans are exposed.

(c) Overall evaluation

Finally, the body of evidence is considered as a whole, in order to reach an overall evaluation of the carcinogenicity to humans of an agent, mixture or circumstance of exposure.

An evaluation may be made for a group of chemical compounds that have been evaluated by the Working Group. In addition, when supporting data indicate that other, related compounds for which there is no direct evidence of capacity to induce cancer in humans or in animals may also be carcinogenic, a statement describing the rationale for this conclusion is added to the evaluation narrative; an additional evaluation may be made for this broader group of compounds if the strength of the evidence warrants it.

The agent, mixture or exposure circumstance is described according to the wording of one of the following categories, and the designated group is given. The categorization of an agent, mixture or exposure circumstance is a matter of scientific judgement, reflecting the strength of the evidence derived from studies in humans and in experimental animals and from other relevant data.

Group 1—The agent (mixture) is carcinogenic to humans.
The exposure circumstance entails exposures that are carcinogenic to humans.

This category is used when there is *sufficient evidence* of carcinogenicity in humans. Exceptionally, an agent (mixture) may be placed in this category when evidence in humans is less than sufficient but there is *sufficient evidence* of carcinogenicity in experimental animals and strong evidence in exposed humans that the agent (mixture) acts through a relevant mechanism of carcinogenicity.

Group 2

This category includes agents, mixtures and exposure circumstances for which, at one extreme, the degree of evidence of carcinogenicity in humans is almost sufficient, as well as those for which, at the other extreme, there are no human data but for which there is evidence of carcinogenicity in experimental animals. Agents, mixtures and exposure circumstances are assigned to either group 2A (probably carcinogenic to humans) or group 2B (possibly carcinogenic to humans) on the basis of epidemiological and experimental evidence of carcinogenicity and other relevant data.

Group 2A—The agent (mixture) is probably carcinogenic to humans.
The exposure circumstance entails exposures that are probably carcinogenic to humans.

This category is used when there is *limited evidence* of carcinogenicity in humans and *sufficient evidence* of carcinogenicity in experimental animals. In some cases, an agent (mixture) may be classified in this category when there is *inadequate evidence* of carcinogenicity in humans and *sufficient evidence* of carcinogenicity in experimental animals and strong evidence that the carcinogenesis is mediated by a mechanism that also operates in humans. Exceptionally, an agent, mixture or exposure circumstance may be classified in this category solely on the basis of *limited evidence* of carcinogenicity in humans.

Group 2B—The agent (mixture) is possibly carcinogenic to humans.
The exposure circumstance entails exposures that are possibly carcinogenic to humans.

This category is used for agents, mixtures and exposure circumstances for which there is *limited evidence* of carcinogenicity in humans and less than *sufficient evidence* of carcinogenicity in experimental animals. It may also be used when there is *inadequate evidence* of carcinogenicity in humans but there is *sufficient evidence* of carcinogenicity in experimental animals. In some instances, an agent, mixture or exposure circumstance for which there is *inadequate evidence* of carcinogenicity in humans but *limited evidence* of carcinogenicity in experimental animals together with supporting evidence from other relevant data may be placed in this group.

Group 3—The agent (mixture or exposure circumstance) is not classifiable as to its carcinogenicity to humans.

This category is used most commonly for agents, mixtures and exposure circumstances for which the evidence of carcinogenicity is inadequate in humans and inadequate or limited in experimental animals.

Exceptionally, agents (mixtures) for which the evidence of carcinogenicity is inadequate in humans but sufficient in experimental animals may be placed in this category when there is strong evidence that the mechanism of carcinogenicity in experimental animals does not operate in humans.

Agents, mixtures and exposure circumstances that do not fall into any other group are also placed in this category.

Group 4—The agent (mixture) is probably not carcinogenic to humans.

This category is used for agents or mixtures for which there is *evidence suggesting lack of carcinogenicity* in humans and in experimental animals. In some instances, agents or mixtures for which there is *inadequate evidence* of carcinogenicity in humans but *evidence suggesting lack of carcinogenicity* in experimental animals, consistently and strongly supported by a broad range of other relevant data, may be classified in this group.

References

Breslow, N.E. & Day, N.E. (1980) *Statistical Methods in Cancer Research*, Vol. 1, *The Analysis of Case-control Studies* (IARC Scientific Publications No. 32), Lyon, IARC

Breslow, N.E. & Day, N.E. (1987) *Statistical Methods in Cancer Research*, Vol. 2, *The Design and Analysis of Cohort Studies* (IARC Scientific Publications No. 82), Lyon, IARC

Gart, J.J., Krewski, D., Lee, P.N., Tarone, R.E. & Wahrendorf, J. (1986) *Statistical Methods in Cancer Research*, Vol. 3, *The Design and Analysis of Long-term Animal Experiments* (IARC Scientific Publications No. 79), Lyon, IARC

Hoel, D.G., Kaplan, N.L. & Anderson, M.W. (1983) Implication of nonlinear kinetics on risk estimation in carcinogenesis. *Science*, *219*, 1032–1037

Huff, J.E., Eustis, S.L. & Haseman, J.K. (1989) Occurrence and relevance of chemically induced benign neoplasms in long-term carcinogenicity studies. *Cancer Metastasis Rev.*, *8*, 1–21

IARC (1973–1992) *Information Bulletin on the Survey of Chemicals Being Tested for Carcinogenicity/-Directory of Agents Being Tested for Carcinogenicity*, Numbers 1–15, Lyon

> Number 1 (1973) 52 pages
> Number 2 (1973) 77 pages
> Number 3 (1974) 67 pages
> Number 4 (1974) 97 pages
> Number 5 (1975) 88 pages
> Number 6 (1976) 360 pages
> Number 7 (1978) 460 pages
> Number 8 (1979) 604 pages
> Number 9 (1981) 294 pages
> Number 10 (1983) 326 pages
> Number 11 (1984) 370 pages
> Number 12 (1986) 385 pages
> Number 13 (1988) 404 pages
> Number 14 (1990) 369 pages
> Number 15 (1992) 317 pages

IARC (1976–1992)

> *Directory of On-going Research in Cancer Epidemiology 1976*. Edited by C.S. Muir & G. Wagner, Lyon
>
> *Directory of On-going Research in Cancer Epidemiology 1977* (IARC Scientific Publications No. 17). Edited by C.S. Muir & G. Wagner, Lyon
>
> *Directory of On-going Research in Cancer Epidemiology 1978* (IARC Scientific Publications No. 26). Edited by C.S. Muir & G. Wagner, Lyon
>
> *Directory of On-going Research in Cancer Epidemiology 1979* (IARC Scientific Publications No. 28). Edited by C.S. Muir & G. Wagner, Lyon
>
> *Directory of On-going Research in Cancer Epidemiology 1980* (IARC Scientific Publications No. 35). Edited by C.S. Muir & G. Wagner, Lyon
>
> *Directory of On-going Research in Cancer Epidemiology 1981* (IARC Scientific Publications No. 38). Edited by C.S. Muir & G. Wagner, Lyon
>
> *Directory of On-going Research in Cancer Epidemiology 1982* (IARC Scientific Publications No. 46). Edited by C.S. Muir & G. Wagner, Lyon

Directory of On-going Research in Cancer Epidemiology 1983 (IARC Scientific Publications No. 50). Edited by C.S. Muir & G. Wagner, Lyon

Directory of On-going Research in Cancer Epidemiology 1984 (IARC Scientific Publications No. 62). Edited by C.S. Muir & G. Wagner, Lyon

Directory of On-going Research in Cancer Epidemiology 1985 (IARC Scientific Publications No. 69). Edited by C.S. Muir & G. Wagner, Lyon

Directory of On-going Research in Cancer Epidemiology 1986 (IARC Scientific Publications No. 80). Edited by C.S. Muir & G. Wagner, Lyon

Directory of On-going Research in Cancer Epidemiology 1987 (IARC Scientific Publications No. 86). Edited by D.M. Parkin & J. Wahrendorf, Lyon

Directory of On-going Research in Cancer Epidemiology 1988 (IARC Scientific Publications No. 93). Edited by M. Coleman & J. Wahrendorf, Lyon

Directory of On-going Research in Cancer Epidemiology 1989/90 (IARC Scientific Publications No. 101). Edited by M. Coleman & J. Wahrendorf, Lyon

Directory of On-going Research in Cancer Epidemiology 1991 (IARC Scientific Publications No. 110). Edited by M. Coleman & J. Wahrendorf, Lyon

Directory of On-going Research in Cancer Epidemiology 1992 (IARC Scientific Publications No. 117). Edited by M. Coleman, J. Wahrendorf & E. Demaret, Lyon

IARC (1977) *IARC Monographs Programme on the Evaluation of the Carcinogenic Risk of Chemicals to Humans. Preamble* (IARC intern. tech. Rep. No. 77/002), Lyon

IARC (1978) *Chemicals with* Sufficient Evidence *of Carcinogenicity in Experimental Animals*—IARC Monographs *Volumes 1–17* (IARC intern. tech. Rep. No. 78/003), Lyon

IARC (1978–1993) *Environmental Carcinogens. Methods of Analysis and Exposure Measurement*:

Vol. 1. *Analysis of Volatile Nitrosamines in Food* (IARC Scientific Publications No. 18). Edited by R. Preussmann, M. Castegnaro, E.A. Walker & A.E. Wasserman (1978)

Vol. 2. *Methods for the Measurement of Vinyl Chloride in Poly(vinyl chloride), Air, Water and Foodstuffs* (IARC Scientific Publications No. 22). Edited by D.C.M. Squirrell & W. Thain (1978)

Vol. 3. *Analysis of Polycyclic Aromatic Hydrocarbons in Environmental Samples* (IARC Scientific Publications No. 29). Edited by M. Castegnaro, P. Bogovski, H. Kunte & E.A. Walker (1979)

Vol. 4. *Some Aromatic Amines and Azo Dyes in the General and Industrial Environment* (IARC Scientific Publications No. 40). Edited by L. Fishbein, M. Castegnaro, I.K. O'Neill & H. Bartsch (1981)

Vol. 5. *Some Mycotoxins* (IARC Scientific Publications No. 44). Edited by L. Stoloff, M. Castegnaro, P. Scott, I.K. O'Neill & H. Bartsch (1983)

Vol. 6. *N-Nitroso Compounds* (IARC Scientific Publications No. 45). Edited by R. Preussmann, I.K. O'Neill, G. Eisenbrand, B. Spiegelhalder & H. Bartsch (1983)

Vol. 7. *Some Volatile Halogenated Hydrocarbons* (IARC Scientific Publications No. 68). Edited by L. Fishbein & I.K. O'Neill (1985)

Vol. 8. *Some Metals: As, Be, Cd, Cr, Ni, Pb, Se, Zn* (IARC Scientific Publications No. 71). Edited by I.K. O'Neill, P. Schuller & L. Fishbein (1986)

Vol. 9. *Passive Smoking* (IARC Scientific Publications No. 81). Edited by I.K. O'Neill, K.D. Brunnemann, B. Dodet & D. Hoffmann (1987)

Vol. 10. *Benzene and Alkylated Benzenes* (IARC Scientific Publications No. 85). Edited by L. Fishbein & I.K. O'Neill (1988)

Vol. 11. *Polychlorinated Dioxins and Dibenzofurans* (IARC Scientific Publications No. 108). Edited by C. Rappe, H.R. Buser, B. Dodet & I.K. O'Neill (1991)

Vol. 12. *Indoor Air* (IARC Scientific Publications No. 109). Edited by B. Seifert, H. van de Wiel, B. Dodet & I.K. O'Neill (1993)

IARC (1979) *Criteria to Select Chemicals for* IARC Monographs (IARC intern. tech. Rep. No. 79/003), Lyon

IARC (1982) *IARC Monographs on the Evaluation of the Carcinogenic Risk of Chemicals to Humans,* Supplement 4, *Chemicals, Industrial Processes and Industries Associated with Cancer in Humans (IARC Monographs, Volumes 1 to 29),* Lyon

IARC (1983) *Approaches to Classifying Chemical Carcinogens According to Mechanism of Action* (IARC intern. tech. Rep. No. 83/001), Lyon

IARC (1984) *Chemicals and Exposures to Complex Mixtures Recommended for Evaluation in* IARC Monographs *and Chemicals and Complex Mixtures Recommended for Long-term Carcinogenicity Testing* (IARC intern. tech. Rep. No. 84/002), Lyon

IARC (1987a) *IARC Monographs on the Evaluation of Carcinogenic Risks to Humans,* Supplement 6, *Genetic and Related Effects: An Updating of Selected* IARC Monographs *from Volumes 1 to 42,* Lyon

IARC (1987b) *IARC Monographs on the Evaluation of Carcinogenic Risks to Humans,* Supplement 7, *Overall Evaluations of Carcinogenicity: An Updating of* IARC Monographs *Volumes 1 to 42,* Lyon

IARC (1988) *Report of an IARC Working Group to Review the Approaches and Processes Used to Evaluate the Carcinogenicity of Mixtures and Groups of Chemicals* (IARC intern. tech. Rep. No. 88/002), Lyon

IARC (1989) *Chemicals, Groups of Chemicals, Mixtures and Exposure Circumstances to be Evaluated in Future IARC Monographs, Report of an ad hoc Working Group* (IARC intern. tech. Rep. No. 89/004), Lyon

IARC (1991a) *A Consensus Report of an* IARC Monographs *Working Group on the Use of Mechanims of Carcinogenesis in Risk Identification* (IARC intern. tech. Rep. No. 91/002), Lyon

IARC (1991b) *Report of an Ad-hoc* IARC Monographs *Advisory Group on Viruses and Other Biological Agents Such as Parasites* (IARC intern. tech. Rep. No. 91/001), Lyon

Montesano, R., Bartsch, H., Vainio, H., Wilbourn, J. & Yamasaki, H., eds (1986) *Long-term and Short-term Assays for Carcinogenesis—A Critical Appraisal* (IARC Scientific Publications No. 83), Lyon, IARC

Peto, R., Pike, M.C., Day, N.E., Gray, R.G., Lee, P.N., Parish, S., Peto, J., Richards, S. & Wahrendorf, J. (1980) Guidelines for simple, sensitive significance tests for carcinogenic effects in long-term animal experiments. In: *IARC Monographs on the Evaluation of the Carcinogenic Risk of Chemicals to Humans,* Supplement 2, *Long-term and Short-term Screening Assays for Carcinogens: A Critical Appraisal,* Lyon, pp. 311–426

Vainio, H., Magee, P., McGregor, D. & McMichael, A., eds (1992) *Mechanisms of Carcinogenesis in Risk Identification* (IARC Scientific Publications No. 116), Lyon, IARC

Waters, M.D., Stack, H.F., Brady, A.L., Lohman, P.H.M., Haroun, L. & Vainio, H. (1987) Appendix 1. Activity profiles for genetic and related tests. In: *IARC Monographs on the Evaluation of Carcinogenic Risks to Humans,* Suppl. 6, *Genetic and Related Effects: An Updating of Selected IARC Monographs from Volumes 1 to 42,* Lyon, IARC, pp. 687–696

Wilbourn, J., Haroun, L., Heseltine, E., Kaldor, J., Partensky, C. & Vainio, H. (1986) Response of experimental animals to human carcinogens: an analysis based upon the IARC Monographs Programme. *Carcinogenesis,* 7, 1853–1863

GENERAL REMARKS

This fifty-sixth volume of *IARC Monographs* comprises a series of individual monographs on some naturally occurring substances, including two food items (salted fish and pickled vegetables), two naturally occurring plant substances (caffeic acid and d-limonene), some substances that occur in cooked meat and fish (IQ, MeIQ, MeIQx and PhIP) and some mycotoxins (aflatoxins, *Fusarium* toxins and ochratoxin A). IQ (IARC, 1986a), MeIQ (IARC, 1986b), MeIQx (IARC, 1986c), aflatoxins (IARC, 1972, 1976, 1987), T-2 toxin (T_2-trichothecene) (IARC, 1983a), zearalenone (IARC, 1983b) and ochratoxin A (IARC, 1983c) have been the subjects of previous IARC monographs. New data on carcinogenicity and other relevant aspects of these agents are summarized and evaluated in the present monographs.

Prior to the meeting of the Working Group, draft monographs were prepared on sesamol (a minor component of sesame oil), black pepper and piperine, chilli and capsaicin, estragole, and opium pyrolysis products. Because limited data and time were available to the present Working Group, these agents will be evaluated by future groups considering other spices and certain drugs of abuse.

In the monograph on salted fish, the Working Group attempted to separate the effects of Chinese-style salted fish, prepared in a manner which involves putrefaction, from those of salted fish in which putrefaction is either absent or minimal. This separation was facilitated by the fact that Chinese-style salted fish is generally consumed only by populations in southern China or by southern Chinese populations living in other areas.

Many of the substances implicated in the mutagenicity of pickled vegetables, such as the hydroxy flavones quercetin (IARC, 1983d) and kaempferol (IARC, 1983e), are present in vegetables that have not been pickled. Studies that included comparisons of pickled and unpickled vegetables would therefore be critical to a proper evaluation of pickled vegetables; such studies have not been done. As the traditional methods of preparing these foods vary from one region to another, epidemiological studies are difficult to carry out.

No epidemiological study was available for evaluation by the Working Group on caffeic acid or d-limonene (to which human exposures in fruit, vegetables and beverages are at levels of milligrams per day); however, caffeic acid was considered by a previous Working Group as a constituent of coffee (IARC, 1991). The present Working Group decided not to evaluate the large number of studies on consumption of citrus fruits and fruit juices (which contain d-limonene), although they noted that reviews by other groups have suggested that citrus fruits are protective against gastric and other cancers (US National Research Council, 1982, 1989).

Cooked muscle meats appear to be a major source of bacterial mutagenic activity and heterocyclic amines in the human diet. Known heterocyclic amines, and particularly MeIQx, are responsible for about 85% of the bacterial mutagenic activity in cooked beef; PhIP and

MeIQx constitute the majority of heterocyclic amines in cooked meats in general. Cooking method, temperature and time are determinants of bacterial mutagenic activity; frying and barbecueing at high temperatures and for long times produce the greatest activity. The relationship between heterocyclic amine formation and cooking conditions is less clear and requires further study.

The Working Group considered a number of mycotoxins. The presence of one mycotoxin in food and feed should automatically alert investigators for co-contamination by others: Numerous reports are available of multiple contaminations because (i) a single fungus can produce several mycotoxins and (ii) food or feed can be contaminated by several mycotoxin-producing fungi simultaneously. Finally, some mycotoxins have additive or synergistic effects.

Much human exposure to aflatoxins occurs in populations in which there is a high prevalence of hepatitis B virus infection. Hepatitis viruses are not evaluated in this volume of monographs but will be discussed by another IARC working group.

The mycotoxins considered in this volume are presented in the order of the three groups of *Fusarium* species involved, that is, *Fusarium graminearum, F. culmorum* and *F. crookwellense* (which produce zearalenone, deoxynivalenol, nivalenol and fusarenone X); *F. moniliforme* (which produces fumonisins B_1 and B_2 and fusarin C); and *F. sporotrichioides* (which produces T-2 toxin). The trichothecenes (deoxynivalenol, nivalenol, T-2 toxin and fusarenone X) are thus covered in relation to their sources.

The most widely distributed toxigenic species is *F. graminearum*, which causes disease in wheat and maize all over the world, except in dryland wheat and subtropical maize production. This fungus produces deoxynivalenol, zearalenone and nivalenol, depending on the strain. The closely related species, *F. culmorum* and *F. crookwellense*, produce the same toxins and occur in cooler and slightly warmer areas, respectively. Another important toxigenic *Fusarium* species is *F. moniliforme* and related species, such as *F. proliferatum*; these are ubiquitous in maize kernels wherever the plant is grown. *F. sporotrichioides*, which produces T-2 toxin, occurs rarely in wheat and maize.

Two naturally occurring compounds, ochratoxin A and d-limonene, the carcinogenicity of which is considered in this volume, have been linked with chronic nephropathies; however, the diseases differ in many respects. Ochratoxin A is the major cause of a chronic nephropathy in pigs; the renal lesions include degeneration of the proximal tubules, interstitial fibrosis and hyalinization of the glomeruli. This pathological picture is strikingly similar to that of human Balkan endemic nephropathy, which has been closely correlated with a high incidence of urinary tract tumours. A basically different chronic nephropathy has been observed in male but not in female rats after administration of d-limonene. In this case, a characteristic accumulation of hyaline droplets containing both $\alpha_{2\mu}$-globulin and d-limonene has been observed in the proximal tubules. In addition, cellular necrosis and increased cell proliferation have been found in this segment of the renal tubular system.

In the monographs in the present volume, the circumstances of human exposure were often derived from different studies from those in which effects were measured. Evaluation of the effects of *Fusarium* toxins in humans, for example, can be attempted only on the basis of data on the contamination of foods. In order to delineate specifically the effects on humans of many of the substances considered, future studies should incorporate relevant biomarkers of exposure, mechanism or early effect. Such techniques are available.

References

IARC (1972) *IARC Monographs on the Evaluation of Carcinogenic Risk of Chemicals to Man*, Vol. 1, Lyon, pp. 145–156

IARC (1976) *IARC Monographs on the Evaluation of Carcinogenic Risk of Chemicals to Man*, Vol. 10, *Some Naturally Occurring Substances*, Lyon, pp. 51–72

IARC (1983a) *IARC Monographs on the Evaluation of the Carcinogenic Risk of Chemicals to Humans*, Vol. 31, *Some Food Additives, Feed Additives and Naturally Occurring Substances*, Lyon, pp. 265–278

IARC (1983b) *IARC Monographs on the Evaluation of the Carcinogenic Risk of Chemicals to Humans*, Vol. 31, *Some Food Additives, Feed Additives and Naturally Occurring Substances*, Lyon, pp. 279–291

IARC (1983c) *IARC Monographs on the Evaluation of the Carcinogenic Risk of Chemicals to Humans*, Vol. 31, *Some Food Additives, Feed Additives and Naturally Occurring Substances*, Lyon, pp. 191–206

IARC (1983d) *IARC Monographs on the Evaluation of the Carcinogenic Risk of Chemicals to Humans*, Vol. 31, *Some Food Additives, Feed Additives and Naturally Occurring Substances*, Lyon, pp. 213–229

IARC (1983e) *IARC Monographs on the Evaluation of the Carcinogenic Risk of Chemicals to Humans*, Vol. 31, *Some Food Additives, Feed Additives and Naturally Occurring Substances*, Lyon, pp. 171–178

IARC (1986a) *IARC Monographs on the Evaluation of the Carcinogenic Risk of Chemicals to Humans*, Vol. 40, *Some Naturally Occurring and Synthetic Food Components, Furocoumarins and Ultraviolet Radiation*, Lyon, pp. 261–273

IARC (1986b) *IARC Monographs on the Evaluation of the Carcinogenic Risk of Chemicals to Humans*, Vol. 40, *Some Naturally Occurring and Synthetic Food Components, Furocoumarins and Ultraviolet Radiation*, Lyon, pp. 275–281

IARC (1986c) *IARC Monographs on the Evaluation of the Carcinogenic Risk of Chemicals to Humans*, Vol. 40, *Some Naturally Occurring and Synthetic Food Components, Furocoumarins and Ultraviolet Radiation*, Lyon, pp. 283–288

IARC (1987) *IARC Monographs on the Evaluation of Carcinogenic Risks to Humans*, Suppl. 7, *Overall Evaluations of Carcinogenicity: An Updating of* IARC Monographs *Volumes 1 to 42*, Lyon, pp. 83–87

IARC (1991) *IARC Monographs on the Evaluation of Carcinogenic Risks to Humans*, Vol. 51, *Coffee, Tea, Mate, Methylxanthines and Methylglyoxal*, Lyon, pp. 41–206

US National Research Council (1982) *Diet, Nutrition, and Cancer*, Washington DC, National Academy Press

US National Research Council (1989) *Diet and Health. Implications for Reducing Chronic Disease Risk*, Washington DC, National Academy Press

THE MONOGRAPHS

FOOD ITEMS AND CONSTITUENTS

SALTED FISH

1. Exposure Data

1.1 Production

Salted fish (salt fish) are treated by brining, dry-salting, pickle curing or a combination of these treatments, increasing the amount of salt in the fish substantially beyond that ordinarily found in the fresh product. Brining is the process of placing fish in a solution of salt (sodium chloride) in water for a period of sufficient length for the fish tissue to absorb the required amount of salt. Dry-salting is the process of mixing fish with dry salt and allowing the resultant brine (from dissolution of the salt in the water present in the fish) to drain away. Pickling or pickle curing is the process whereby fish is mixed with salt and is stored under the brine (pickle) which is formed when the salt dissolves in the water extracted from the fish tissue.

Processes for the preparation of salted fish have been reviewed by the FAO/WHO Codex Alimentarius Commission (1983). The following general description is taken from that review, unless otherwise noted. Salt acts upon fish as upon other foods by withdrawing water from the tissue. Fish flesh contains 75–80% water (in the case of very fatty fish, 60–65%), and this water can be replaced partly by salt. In the preparation of salted fish, water diffusing from the fish becomes saturated with the surrounding salt and is termed 'pickle'. Dry-salting results in a rapid loss of the weight of the fish, while with 'wet' salting, after an initial weight loss, there is a gradual weight gain. Salt uptake and water loss are influenced by the fat content of the fish, the thickness of the flesh, freshness, temperature, the chemical purity of the salt and other factors. Fat acts as a barrier to both the entry of salt and withdrawal of water; thus, water loss is slower from more fatty fish. The salting process may be terminated when the fish have achieved the required salinity and acquired the desired taste, consistency and odour.

Salting may be divided into salt preservation, as such, and ripening. Ripening, which is desirable for some fatty fish products, is a process that causes changes in the chemical and physical characteristics of fish flesh, generally by some enzymatic process. The rate of ripening depends on the fish, the salt composition employed, the temperature and the amount of salt absorbed by the fish tissues. These variables give rise to many different and uniquely characteristic products. Spoilage of fish is brought about chiefly by autolysis and microbial decomposition. Most enzymes and micro-organisms are inactivated by high salt concentrations, and the reduced moisture content of salted fish also results in an unfavourable environment for the multiplication of micro-organisms. If poor quality raw fish is used and/or salting takes place at elevated temperature, however, decomposition may proceed faster than the penetration of salt into the tissues, and spoilage of the fish will occur.

While salting reduces the rate of autolysis, it does not completely stop enzymatic action, which increases with increasing temperature. Salting enhances fat oxidation, and fat hydrolysis and the development of rancidity may contribute to the spoilage of fish. Certain halophilic micro-organisms can multiply under the conditions of dry-salting and can also spoil the product. Salted fish should therefore be cured and stored under cool conditions, and some fatty fish should be kept in the absence of air.

The origin, and thus the composition, of the salt used in the salting process in different countries varies. Mine salt or rock salt is usually almost pure sodium chloride, but solar salt of marine origin (sea salt) contains several impurities, including calcium sulfate, magnesium sulfate, magnesium chloride and nitrates or nitrites (Armstrong & Eng, 1983). Too much calcium (> 0.35%) may reduce the rate of salt penetration to the extent that spoilage occurs. Magnesium salt levels above 0.15% may lead to an unpleasant flavour and/or spoilage.

The commercial processing of salted fish, as practised in Europe and North America, has been described by the FAO/WHO Codex Alimentarius Commission (1983), Anon. (1986) and Bjarnason (1987). Figure 1 is a flow chart of the commercial processes. Ripening is not a part of the process. A number of steps, including heading and splitting, are largely done by machines, and the fish are moved through the processes on conveyor belts (FAO/WHO Codex Alimentarius Commission, 1983).

Traditional processes used in the preparation of salted fish in various regions have been described. Almost 30% of the fish caught in Southeast Asia is preserved by curing (salting, drying or smoking) (Ah-Weng *et al.*, 1985).

1.1.1 *South China*

In South China, fish are generally not gutted prior to salting, and only when bigger fish such as red snapper are salted are the guts drawn out through the throat, without making an incision in the belly of the fish. Salting is done in wooden vats, and the fish are arranged in alternate layers with coarse rock salt. After a few days, the fish are immersed in brine and weights (often large stones placed on top of grass mats) are placed on the surface to prevent the fish from floating. The length of salting ranges from one to five days, after which the fish are taken out to dry in direct sunlight, usually spread on woven grass mats. They are turned every few hours and left out for one to seven days, depending on the size of the fish and the weather. During drying, insect infestation can be a serious problem, especially in damp weather. Unfortunately, in South China, the relative humidity is over 80% for about eight months of the year, and it often reaches 96–99% during the summer months.

Sometimes, fish is allowed to soften by decomposition before salting, to produce *muihum* salted fish. Fish that is not previously decomposed is known as *sud yoke* (literally, tough meat) salted fish (Tannenbaum *et al.*, 1985).

1.1.2 *Malaysia*

In Malaysia, fish is usually gutted and cleaned by fishermen in the port. The degree of cleaning of the fish varies considerably and is generally unregulated; refrigeration is virtually non-existent between the point of catching the fish and selling it on the market. Salted fish is prepared only when the markets have enough fresh fish, when there is a failure of transport

Figure 1. Flow chart for the commercial processing of salted fish

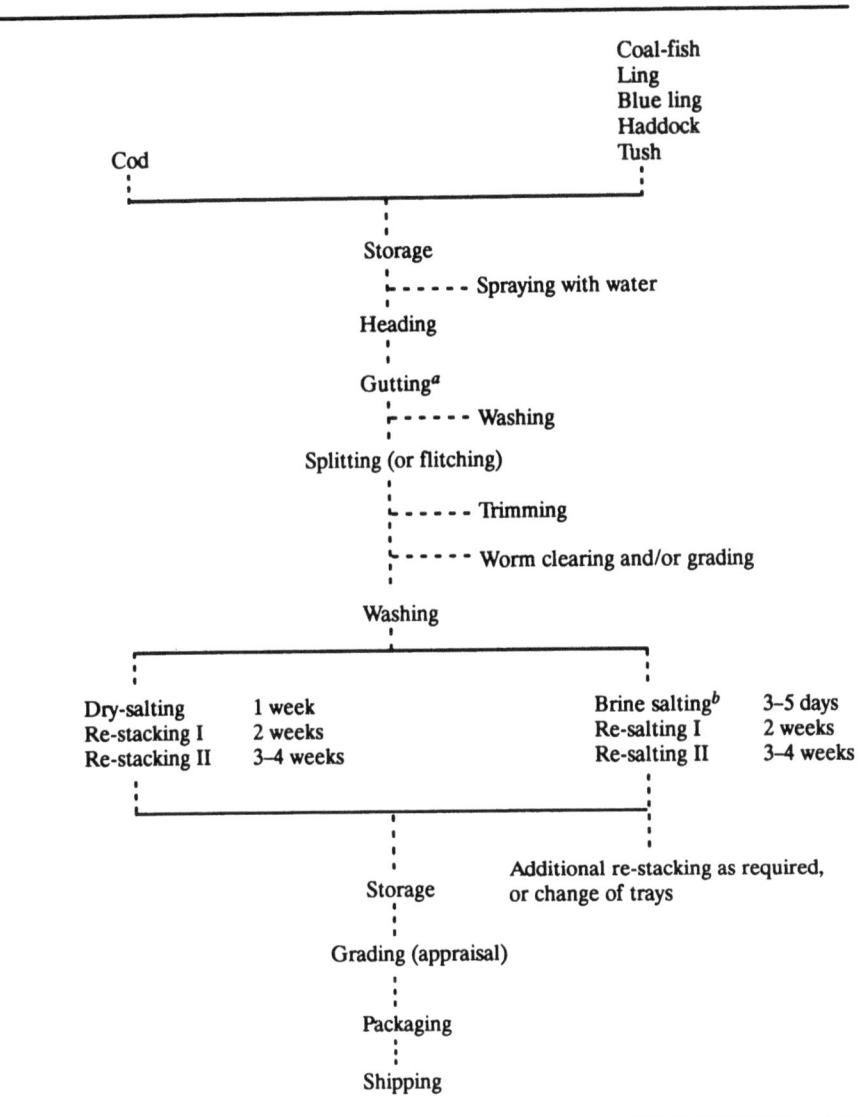

From Bjarnason (1987)
[a]During the summer fishing season in Iceland, gutting is done aboard the fishing vessels.
[b]Brine salting in Iceland is dry-salting and keeping the fish in the resultant brine (pickling).

between the port and the market or some other delay. Thus, fresh fish undergoes varying degrees of deterioration before it is salted down for pickling. Most fish for salting are sprinkled and stuffed with crude sea salt and placed in a tank. When the tank is full, the fish are covered and compressed with bricks or stones. After two to three days of pickling, the fish are either smoked or left as is and put to dry in the sun. The more expensive varieties of salted fish are sometimes marinated in jars of cooking oil (usually palm or coconut oil) before dispatch to market. Depending on the weather, the process before the fish go into storage or to market takes about a week (Armstrong & Eng, 1983).

Malaysian Chinese have a distinct preference for what is known locally as 'Kuantan salted fish'—salted red snapper which takes its name from the port that first gained the reputation for supplying the best quality, although today an equivalent product may come from other ports in Malaysia. Salted red snapper is processed and sold in three ways, chiefly for Chinese consumers. The bulk of the fish is salted, smoked and dried in the usual way to produce 'Kuantan hard fish' or *ikan merah jerok*. Smaller proportions are preserved in salt chips for three days, rather than two, before drying, to produce an exceptionally salty fish (*ikan merah jerok asin*). Another small proportion is first preserved for two to three days in tanks of vinegar and salt and then dried in the sun. An alternative, perhaps older, procedure is to allow the fish to soften by decomposition then stuff it with salt, compress for three days and dry in the sun. The latter two processes yield the expensive and famous *moi heong* (literally, 'red plum') or 'Kuantan soft-type' salted fish (*ikan merah jerok asam*). The procedures vary widely among fish handlers. In salting inferior, less expensive grades of fish, only enough salt to prevent gross deterioration may be used while the fish is drying (Armstrong & Eng, 1983).

Although smoking of salted fish is common in Malaysia, duration of smoking and kinds of smokehouses and fuels used have not been studied (Armstrong & Eng, 1983).

1.1.3 *Northeastern Thailand*

Salted fish consumed in northeastern Thailand can be placed in one of three categories, according to the type of fish (fresh-water or sea-water) and curing process used. *Pla-ra* is a fresh-water fish that has been left to ferment in salted water for at least one month. This is the traditional salted-fish dish of northeastern Thailand and is prepared at home and kept wet. Other fresh-water fish undergo shorter fermentation (usually one week) and may be kept wet or dried. The third type comprises sea fish fermented with salt for several weeks and allowed to dry: This typical Chinese-style fish is purchased on the market rather than prepared at home (Sriamporn *et al.*, 1992).

1.1.4 *Tunisia*

In Tunisia, salted anchovies and sardines are prepared by alternating layers of salt and fish in glass or earthernware containers, which are then sealed and stored for up to three years. The fish is used mainly in salads (Hubert, 1984).

1.1.5 *Egypt*

In Egypt, sand-salted fish is prepared by covering the gills and body of fish with salt, wrapping it in canvas and burying it in sand for 15–30 days. Tin-salted fish is made by covering the gills and body with salt and then leaving the fish for several hours in the sun until the body has swollen; the salt is then renewed, and the fish is arranged in alternate layers with salt in a special tin or barrel. Treated fish is consumed after at least 10 days' storage (Elmossalami & Sedik, 1972).

1.1.6 *Sudan*

Alestes baremose is the fish species most commonly used for producing wet-salted fish in the Sudan. Salting is done by fishermen in small, temporary sheds or huts made of reed mats. Whole fish are washed and placed on mats made of palm leaves, sometimes in perforated wooden or steel barrels, with layers of fine-grain salt (solar sodium chloride). The fish are then covered with a thick layer of salt and wrapped with the same mats; weights are placed on top to press down the fish. The fluid formed is allowed to run off. Depending on climatic conditions, and especially the ambient temperature, the fish are left for three to five days to ferment. They are then transferred to tins in which heavy layers of salt are alternated with layers of fish. After about three days, the fluid is decanted and additional layers of fish and salt are applied. The process is continued for 7–10 days, after which time the tins are sealed. At this stage, the fish has lost about one-third of its initial weight. The average amount of salt needed to ferment 1 kg of fresh fish is about 0.1 kg. The final product is called *fassiekh*. The technique is believed to have been introduced from Egypt during the Turkish rule in the nineteenth century (Yousif, 1989).

1.2 Worldwide production, trade and consumption

1.2.1 *Production and import/export*

About 15% of the world fish catch is preserved by curing, i.e., salting, drying or smoking, or a combination of those treatments (Ah-Weng *et al.*, 1985). World production of salted fish in 1980–89 is presented in Table 1.

Global production (by region) of dried, salted and smoked fish in 1989 was 2 621 757 tonnes in Asia, 769 300 tonnes in the USSR, 514 316 tonnes in Europe, 342 892 tonnes in Africa, 106 063 tonnes in North America, 72 921 tonnes in South America and 10 175 tonnes in Oceania (FAO, 1989).

World production of salted (and dried) cod and related products is on average 300–400 thousand tonnes per annum. Europe produces about two-thirds of the total, the largest producers in order of importance being Norway, Iceland, Spain, Portugal, the Faroe Islands and France. The most important producing countries outside Europe are Canada and the USA. Curing of salted fish is a traditional industry in Iceland, and exports of salted dried fish in 1980–82 were 2539, 840 and 1460 tonnes, respectively. Major importers of Icelandic salted dried fish in those years were Brazil, France, Martinique, Panama, Portugal and Zaire (Bjarnason, 1987).

Almost 40% of the total annual fish catch in Indonesia is converted into dried salted products. In 1985, the fish catch in Indonesia was about two million tonnes (Esser *et al.*, 1990). Approximately 34% of fish from inland waters and 48% of marine fish were cured, 71% of the cured fish being processed by salting and drying (Buckle *et al.*, 1988). The major

Table 1. World production of salted fish, 1980–89

Year	Production (thousand tonnes)	
	Dried, salted or smoked	Dried, salted or in brine
1980	3997	3103
1981	3892	2951
1982	4031	3079
1983	4032	3091
1984	4025	3075
1985	4194	3232
1986	4203	3223
1987	4312	3313
1988	4287	3408
1989	4437	3432

From FAO (1989)

producing islands are Sumatra, South Kalimantan and South Sulawesi. In 1987, 4200 tonnes of dried salted fish from Jakarta were exported to Saudi Arabia, and small amounts are regularly exported to Australia (Wibowo et al., 1990).

Production of wet-salted fish in the Sudan in 1986 was estimated to have been 1160 tonnes. Nearly 80% of the product was exported to Egypt (Yousif, 1989).

The growth of the Asian populations of Europe, the USA and Australia is expected to stimulate importation of dried salted fish from the traditional producing regions (Wibowo et al., 1990).

1.2.2 *Consumption*

Chinese-style ('Cantonese marine') salted fish is a favourite dish along the South China coast and in Southeast Asian countries. In Selangor, Malaysia, over 90% of Chinese households and about 40% of Malay and Indian households reported regular consumption of salted fish (Armstrong & Eng, 1983). Although the amount consumed at any one time is small (not more than 10 g), the dish may appear at every meal; some people actually prefer the spoiled parts (Fong & Chan, 1973a).

Some populations are exposed to Chinese-style salted fish from weaning. Seventeen mothers in Guangzhou (Canton City), China, reported that they fed salted fish mixed with rice at least five times a week to their children both during and after weaning. The median ratio of salted fish to rice was 1:9 (Yu et al., 1989a). In a study of nasopharyngeal carcinoma in Hong Kong Chinese, 250 Chinese controls were interviewed regarding their dietary habits three years previously and at the age of 10, and the mothers of 155 controls were questioned about the diets of the study subjects at the age of 10, between the ages of one and two and during weaning (Yu et al., 1986; Yu & Henderson, 1987). The frequency of consumption of Chinese-style salted fish in this population is shown in Table 2.

Table 2. Frequency of salted fish consumption among Chinese in Hong Kong

Frequency	No. of subjects
Three years previously	
Rarely	164
Monthly[a]	66
Weekly[b]	19
Daily	1
At age 10	
Rarely	108
Monthly	101
Weekly	39
Between ages 1–2	
Never	83
Sometimes	34
Often[c]	8
During weaning	
Never	96
Ever	31

From Yu et al. (1986); Yu & Henderson (1987)
[a]Once a month to less than once a week
[b]Once a week to less than daily
[c]Considered by mothers to be a typical meal

Poirier et al. (1987) reported that the frequency of consumption of salted fish varied from once or twice a week to more than three times a week in Tunisia and South China.

Consumption of salted fish more than 10 times a month was reported by 52 and 70% of the adult farm populations and 36 and 57% of the adult non-farm populations of Hiroshima and Miyagi, Japan, respectively (Haenszel et al., 1976). The Japanese National Nutrition Surveys for 1975 indicated that the daily per-caput dietary intake of salted fish was 6.2 g (Omura et al., 1987). In Indonesia, fish provides approximately 70% of the per-caput dietary protein intake (Esser et al., 1990).

1.3 Regulations and guidelines

The FAO/WHO Codex Alimentarius Commission (1983) adopted the Recommended International Code of Practice for Salted Fish in December 1979. The Code covers the technological and essential hygienic requirements for the preparation of high-quality salted fish products, but the drying of salted fish is not covered. Pirimiphos methyl has been cleared by FAO/WHO for use on salted–dried fish to protect against blowfly infestation during processing and against further insect infestation during storage (Esser et al., 1990).

1.4 Compounds present in salted fish

1.4.1 Nitrosamines and related contaminants

Ho (1972) speculated some time ago that the presence of nitrosamines in salted fish might be significant. Fish are rich sources of secondary and tertiary amines, and nitrate and possibly nitrite occur in the crude sea salt used to pickle them. Furthermore, pickling and drying are done in the open, so that the fish are susceptible to bacterial contamination which might contribute to nitrosation (Fong & Chan, 1976a,b). Subsequent analyses revealed the presence of nitrosamines in salted fish; levels of N-nitrosodimethylamine (NDMA) (see IARC, 1978a, 1987) are summarized in Table 3.

Kawabata *et al*. (1984) conducted a survey of the occurrence of total N-nitroso compounds, total N-nitrosamides, volatile N-nitrosamines (NDMA) and nitrite in the Japanese diet. No appreciable amount, or only trace quantities, of nitrosamines were detected in uncooked fish products, but the content of volatile nitrosamines (especially in dried squid samples) increased upon broiling on a city gas range. Table 4 presents the occurrence of the four groups of compounds in salt-dried fish.

Samples of uncooked, steamed and fried salted yellow croaker, purchased at three southern Chinese markets on six occasions were analysed for volatile nitrosamines. The mean levels found were: NDMA, 100–600 ng/kg (uncooked fish and steamed fish), 200–1400 ng/kg (fried fish); N-nitrosodiethylamine (see IARC, 1978b, 1987), not detected to 50 ng/kg (uncooked fish), not detected to 100 ng/kg (steamed fish), not detected to 10 ng/kg (fried fish); N-nitrosodi-n-propylamine (see IARC, 1978c, 1987), mean in one batch, 50 ng/kg (steamed fish), 30 ng/kg (fried fish); N-nitrosodi-n-butylamine (see IARC, 1978d, 1987), same batch, 50 ng/kg (fried fish); and N-nitrosomorpholine (see IARC, 1978e, 1987), 200 ng/kg (uncooked fish). NDMA and N-nitrosodiethylamine were found in both uncooked salted fish heads (60–100 ng/kg) and soups prepared from them (10–60 ng/l); NDMA was found in all batches, and N-nitrosodiethylamine in only one batch (Huang *et al*., 1981).

Samples of seven species of salt-dried fish purchased in the Tokyo area were analysed for nitrosamines before and after broiling. Levels of NDMA in uncooked samples ranged from not detected to 5.0 µg/kg, while levels in cooked samples ranged from trace to 26.1 µg/kg. N-Nitrosopyrrolidine (see IARC, 1978f, 1987) was not detected in either uncooked or cooked samples. The NDMA content of the samples apparently increased after broiling on a gas range; however, considerable variation in the increase was observed, depending on the species of fish. It has been reported that covering fish with aluminium foil or broiling it in an electric range decreases NDMA formation during cooking (Kawabata *et al*., 1980).

Food extracts from high-risk areas for nasopharyngeal carcinoma (NPC) were analysed for nitrosamines; and, since N-nitroso compounds can be formed from ingested foods and nitrate or nitrite, aqueous food extracts were also examined after acid-catalysed nitrosation *in vitro*. Samples of hard salted and dried grouper from China contained 388 µg/kg NDMA, 81 µg/kg N-nitrosopiperidine (see IARC, 1978g, 1987) and 30 µg/kg N-nitrosopyrrolidine. After nitrosation *in vitro*, the samples contained 1191 µg/kg NDMA, a trace amount of N-nitrosopiperidine and 98 µg/kg N-nitrosopyrrolidine. Volatile nitrosamines were not detected before nitrosation in soft salted and dried Japanese mackerel from China; after nitrosation, 377 µg/kg NDMA and 20 µg/kg N-nitrosopyrrolidine were found. Salted anchovies from Tunisia contained NDMA at 299 µg/kg before and 43 µg/kg after nitrosation (Poirier *et al.*, 1989).

Table 3. *N*-Nitrosodimethylamine (NDMA) levels in uncooked salted fish

Fish product[a] (place of purchase)	No. of samples	NDMA level (µg/kg)	Analytical method	Reference
Salted fish (Hong Kong)		ND–300	GC or GC-MS	Fong & Chan (1973a)
White herring	7	40–300[b]		
Yellow croaker	5	10–200		
Anchovies	2	20–100		
Croaker	2	20–30		
Pomfret	1	ND		
Salted fish (Hong Kong)		< 1–35	GC-MS	Huang *et al.* (1978a)
Anchovy	2	2–35		
Croaker	3	ND–8		
Red snapper	4	< 1		
White herring	4	< 1–8		
Yellow croaker	6	< 1–18		
Salted fish (Hong Kong)		ND–2.8	GC-TEA	Tannenbaum *et al.* (1985)
Yellow croaker	1	0.64		
Mackerel	1	ND		
Croaker	1	2.0		
Red snapper	1	ND		
Black pomfret	1	ND		
White croaker	1	ND		
Unknown import from Japan	1	2.8		
Salted-dried fish (China)		< 0.1–133	HPLC-TEA	Poirier *et al.* (1987)
Grouper	1	133		
Croaker	4	< 0.1–2		
Japanese mackerel	1	1.1		
Spotted mackerel	1	1.0		
Silver pomfret	1	0.2		
Dolphin fish	1	9.2		
Soldier croaker	1	0.3		
Ribbon fish	1	6.4		
Sprats, small fry	1	1.5		
Ribbon, small fry	1	14		
Croaker, small fry	1	1.0		
Salted fish[c] (China)	27	ND–24.4	GC-TEA and/or GC-MS	Song & Hu (1988)
Hard-salted and dried grouper (China)	1	388	GC-TEA	Poirier *et al.* (1989)
Soft-salted and dried Japanese mackerel (China)	1	ND	GC-TEA	Poirier *et al.* (1989)

Table 3 (contd)

Fish product[a] (place of purchase)	No. of samples	NDMA level (µg/kg)	Analytical method	Reference
Salted fish (China)		13–323	GC-TEA	Zou et al. (1992)
Sihui County	5	45–323		
Guiping County	4	15–36		
Henshan County	3	14–188		
Changsha City	2	91–112		
Shanghai City	6	13–59		
Salt-dried fish (Japan)		< 0.5–5	GC	Kawabata et al. (1980)
Horse-mackerel	3	1.0–4.9		
Shishyamo	2	0.5–4		
Pacific saury	1	trace		
Flounder	1	< 0.5		
Chub mackerel	1	0.5		
Round herring	1	5		
Salted/dried fish (Canada)		0.4–4.2	GC-MS or GC-TEA	Sen et al. (1985)
Cod	3	0.4–0.9		
Hake	1	4.2		
Caplin	1	0.9		
Salted fish (Canada)		0.2–1.1	GC-MS or GC-TEA	Sen et al. (1985)
Mackerel	2	0.2 each		
Herring	5	0.4–1.1		
Turbot	2	0.6–0.8		
Cod	1	0.3		
Salted anchovies (Tunisia)	1	299	GC-TEA	Poirier et al. (1989)

ND, not detected; GC, gas chromatography; HPLC, high-performance liquid chromatography; MS, mass spectrometry; TEA, thermal energy analysis

[a]Author's terminology used
[b]Level reached 1 ppm (mg/kg) in one sample of spoiled herring
[c]Squid, octopus, cuttlefish, hairtail fish and fish sauce

Fong and Chan (1973b) reported an increase in the level of NDMA in fish broth after it was inoculated with *Staphylococcus aureus* isolated from salted fish obtained from a Chinese market. They suggested that the amount of NDMA present in salted fish was dependent on storage conditions, degree of contamination by nitrate-reducing bacteria and the levels of precursors present.

Fong and Chan (1976a,b) investigated the effects on the presence of NDMA in salted fish of preservation with crude sea salt containing 40 ppm (mg/kg) nitrate or with sodium chloride. The NDMA levels in two species of marine fish were 25–40 µg/kg after treatment with crude salt and 6–7 µg/kg after preservation with sodium chloride. The levels in two species of freshwater fish were 19–20 µg/kg after treatment with crude salt and not detected to 5 µg/kg after treatment with sodium chloride.

Table 4. Levels of total *N*-nitroso compounds (TNC), total *N*-nitrosamides (TNAd), volatile *N*-nitrosamines (VNA) and nitrite in the Japanese diet

Dietary item	TNC (μg NO/kg)	TNAd (μg NO/kg)	VNA (NDMA, μg/kg)	Nitrite (mg/kg)
Salt-dried fish				
Sardine, *maruboshi iwashi* (uncooked)	0.86	0.63	0.6	< 0.05
Sardine, *maruboshi iwashi* (city gas-broiled)	4.53	2.04	6.1	< 0.05
Pacific saury, *samma hiraki* (uncooked)	1.87	1.58	0.7	< 0.05
Pacific saury, *samma hiraki* (city gas-broiled)	6.74	5.42	3.3	0.28
Mackerel, *aji hiraki* (uncooked)	1.45	1.14	0.9	< 0.05
Mackerel, *aji hiraki* (city gas-broiled)	6.61	3.62	7.4	0.40
Air-dried squid, *hoshi surume* (uncooked)	26.6	22.6	9.9	< 0.05
Air-dried squid, *hoshi surume* (city gas-broiled)	37.7	22.5	37.5	0.77
Other products				
Salted-fermented squid, *ika shiokara*	15.8	–	3.5	< 0.05
Fish sausage 1	2.5	–	2.1	0.07
Fish sausage 2	3.4	–	2.6	< 0.05

From Kawabata *et al.* (1984). Detection limits: TNC and TNAd, 0.5 μg/kg NO equivalent; VNA, 0.1 μg/kg; nitrite, 0.05 mg/kg; –, not analysed

1.4.2 *Micro-organisms and toxins*

Fong and Walsh (1971) obtained viable cultures on salt-agar of nitrate-reducing halobacteria and salt-tolerant *S. aureus* from all samples of Cantonese salt-dried fish that they examined. Onishi *et al.* (1980) isolated a variety of halophilic bacteria from salted fish, including salmon, salmon roe, codfish, cod roe and guts of cuttlefish.

Bacteria isolated from Egyptian sand-salted fish included micrococci, gram-positive bacilli, *Proteus vulgaris*, *P. mirabilis* and *Aeromonas liquefaciens*. All of the same microorganisms except *A. liquefaciens* were isolated from tin-salted fish; *Serratia marcescens*, *P. rettgeri*, *P. morganii*, *Enterobacter aerogenes* and *Corynebacterium freundii* were isolated from tin-salted fish only. Total bacterial counts were lower in sand-salted fish than in tin-salted fish (Elmossalami & Sedik, 1972).

Species of fungi isolated from Indonesian dry salted fish included *Paecilomyces variotii*, *Eurotium amstelodami*, *Aspergillus candidus* and *A. sydowii* (Wheeler & Hocking, 1988).

Aflatoxin B_1, produced by the fungi *A. flavus* and *A. parasiticus*, was reported in cured fish (see also the monograph on aflatoxins). Okonkwo and Nwokolo (1978) found a mean aflatoxin B_1 concentration of 650 μg/kg in Nigerian dried fish. Stockfish, a dried imported fish from Scandinavia, contained no aflatoxin. Shank *et al.* (1972) identified aflatoxins in 5% of 139 samples of dried fish/shrimp purchased in markets in Thailand but in none of 35 samples purchased in Hong Kong. In the contaminated samples, the mean aflatoxin (B_1, B_2, G_1 and G_2) concentration was 166 μg/kg.

Total bacterial count and coliforms were estimated in fresh Egyptian *bolti* fish after brining, after drying and during storage at room temperature for three months. The total bacterial count decreased after drying, owing to the high salt content and the lack of free water in fish tissues. Coliforms were not present after brining, drying or throughout storage (Zaki *et al.*, 1976).

1.4.3 *Other*

Polynuclear aromatic hydrocarbons were identified by high-performance liquid chromatography in uncooked, salted and sun-dried fish commonly consumed in South India. Chrysene (see IARC, 1983a, 1987) was found at 3.6–18.6 µg/g and benzo[*a*]pyrene (see IARC, 1983b, 1987) at 5.7–60.8 µg/g; one sample contained 1,2:5,6-dibenz[*a*]anthracene [dibenz[*a,h*]anthracene] (see IARC, 1983c, 1987) at 12.34 µg/g (Sivaswamy *et al.*, 1990).

The mean concentration of sodium ion in five samples of salted fish purchased in Hong Kong was 46.5 mg/g of fish (Yu *et al.*, 1989a).

1.5 Analysis

Selected methods for the analysis of *N*-nitroso compounds in various matrices have been reviewed (Walker *et al.*, 1978, 1980; Bartsch *et al.*, 1982; Preussmann *et al.*, 1983; O'Neill *et al.*, 1984).

2. Studies of Cancer in Humans

2.1 Chinese-style salted fish

A number of studies considered the effects of salted fish, as traditionally prepared in southern China, in different Chinese population throughout the world. As this preparation is quite different from preparations of salted fish elsewhere in the world, epidemiological studies in those populations were considered separately. It cannot be excluded, however, that salted fish products consumed by other populations are similar to Chinese preparations.

2.1.1 *Nasopharyngeal carcinoma*

(a) *Correlation (ecological) studies*

Nasopharyngeal carcinoma (NPC) is a rare cancer in most parts of the world; the annual age-standardized incidence rate for either sex is generally less than 1 per 100 000 population (Hirayama, 1978; Muir *et al.*, 1987; Ning *et al.*, 1990). A few populations, however, have very high rates of NPC; these include Chinese living in China and in other parts of world (e.g., Hong Kong, Malaysia, Singapore, USA). Nevertheless, there is substantial variation among the rates of NPC within China: The incidence rates generally increase from North (2–3/100 000) to South China (25–40/100 000), so that rates in the southernmost province of Guangdong may be 10–20 times higher than those in the northernmost provinces (Chinese National Cancer Control Office/Nanjing Institute of Geography, 1979; Yu *et al.*, 1981; Muir *et al.*, 1987). The three regions in which rates of NPC are particularly high are Guangdong Province, Guangxi Autonomous Region and Fujian Province (Fig. 2).

Fig. 2. Map of China, showing the provinces and other areas that have been studied with respect to consumption of salted fish

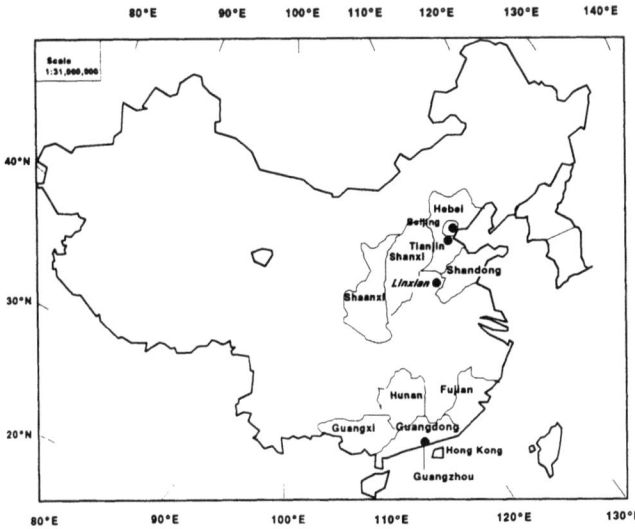

From Li *et al.* (1985)

This north–south gradient in NPC rates in China corresponds to a traditionally low intake of salted fish in northern China and a high intake of this food in southern China (Yu & Henderson, 1993). There are also distinct differences in rates of NPC by ethnic and dialect groups of southern Chinese. Among the Cantonese in Hong Kong, the highest rates of NPC are observed among the Tankas (boat people), representing 1–2% of the Cantonese population, who are fishermen and consume large quantities of salted fish in their daily diet. The rates of NPC among the Tankas are about twice those of the land-dwelling Cantonese (Ho, 1967, 1971a, 1978; Li *et al.*, 1985). In turn, the rate of NPC among the land-dwelling Cantonese is twice that of the Hakka and Chiu Chau (Teochew) dialect groups who reside in north-eastern Guangdong (Ho, 1967, 1971b; Yu *et al.*, 1981; Ho, 1975; Li *et al.*, 1985). It is noteworthy that the people of Fujian Province, who are culturally similar to the Chiu Chau people in Guangdong Province, have similar rates of NPC. Hakkas, who originated from northern China many years ago, also have rates of NPC that are similar to those of their Chiu Chau neighbours and not to those of their low-risk ancestors in the north (Ho, 1971a, 1975; Li *et al.*, 1985). Thus, the distinct pattern of NPC incidence among different ethnic or dialect groups in southern China coincides with the pattern of their consumption of salted fish (Ho, 1978; Yu *et al.*, 1981). Cantonese who migrate to other parts of southeast Asia continue to exhibit a risk for NPC that is twice that of Hakkas, Chiu Chaus and Fujianese who have migrated (Armstrong *et al.*, 1979; Armstrong & Eng, 1983).

The high rates of NPC in provinces that neighbour Guangdong Province are probably due to the adoption of some of the dietary habits of people in Guangdong. Guangxi Autonomous Region, which borders Guangdong Province on the west, has the second highest rates of mortality from NPC in China. These high rates are observed mainly among the Han people of the Guangxi Region (Yu et al., 1988), who are ethnically similar to the Cantonese in Guangdong; the Zhuang people in western Guangxi have about one-fifth the rates of Cantonese. In areas of Hunan Province that border Guangxi and Guangdong to the north, the rates of NPC are, not surprisingly, also high (Yu & Henderson, 1993).

Among the high-risk southern Chinese, people in the lower social strata have higher rates of NPC than those in the higher social strata (Armstrong et al., 1978; Geser et al., 1978; Yu et al., 1981). Exposure to salted fish may explain this observed inverse association, as salted fish is traditionally one of the cheapest foods in southern China.

The incidence of NPC in southern Chinese peaks at the ages of 45–54 and declines clearly thereafter (Ho, 1971b). This observation suggests that exposure to the etiological agent occurs early in life and that either exposure declines or the target tissue becomes less susceptible (or both) at older ages. Salted fish mixed with soft rice is commonly, and sometimes the only food, fed to infants in the weaning and post-weaning period (Topley, 1973). Exposure to this food item very early in life is thus consistent with the age distribution of NPC (Ho, 1979; Yu & Henderson, 1993).

(b) Case–control studies (see Table 5)

Since Ho published his hypothesis in the early 1970s, eight case–control studies have been conducted to investigate the association between consumption of salted fish and the occurrence of NPC among Chinese living in different parts of the world and displaying distinct risks for NPC. The evidence is derived from studies conducted among the very high-risk Cantonese Chinese in Hong Kong (Geser et al., 1978; Yu et al., 1986) and Guangzhou (Yu et al., 1989b), among southern Chinese in Guangxi Autonomous Region, who have intermediate rates of this disease (Yu et al., 1988), and among Chinese in Tianjin, who have relatively low rates (Ning et al., 1990). In addition, the association between consumption of salted fish and NPC has been observed among southern Chinese living outside of China or Hong Kong but who have maintained this custom (Henderson et al., 1976; Henderson & Louie, 1978; Armstrong et al., 1983).

A case–control study of NPC was conducted among Chinese, whites and other ethnic groups in California, USA (Henderson et al., 1976). Cases were identified in 1971–74 from the population-based tumour registries of Los Angeles County (n = 150) and the San Francisco Bay area (n = 53 Chinese); a further 27 cases originally diagnosed in 1960–70 were obtained from the California Tumor Registry. Of the 230 cases, 156 were interviewed; 74 were Chinese, 65 of whom had been born in China. Controls were identified from hospitals and clinics in the same geographical areas as the cases; 267 were interviewed, 110 were Chinese, of whom 91 had been born in China [response rates not given]. Questions on life style, medical and family histories and use of salted fish (for Chinese subjects only) were included in the interview. More cases than controls currently used salted fish (odds ratio [OR], 2.1; 95% confidence interval [CI], 0.6–6.3), and a significant, positive association was found between intake frequency and risk for NPC: OR, 1.6 for at least monthly use, 2.1 for weekly use and 3.1 for more than weekly use as compared to no use (p for trend = 0.02) (Henderson & Louie, 1978). [The Working Group noted that it is not clear whether consumption of salted fish was adjusted for other risk factors for NPC.]

Geser et al. (1978) conducted a hospital-based case–control study of NPC in Hong Kong which included 150 cases selected 'by rotation' from among about 350 cases and an equal number of controls (mainly with other cancers), matched to cases by sex, age (plus or minus five years) and hospital ward. The NPC patients were identified at one of two major hospitals in Hong Kong while undergoing radiotherapy treatment for this disease between 1973–74. Interviews were conducted with each case and control and also with senior women in the households of 108 cases and 103 controls, in order to determine socioeconomic conditions, religious practices and dietary habits in the households. Current intake of salted fish was similar for cases and controls [details not presented]; however, significantly more senior women in the households of NPC cases (75%) reported having fed their babies salted fish after weaning than in those of controls (53%) (OR, 2.6; $p < 0.01$). The effect of intake of salted fish remained after adjusting for traditional life style practices. [The Working Group noted that it is not clear how the 150 NPC patients were selected from the 350 admissions or whether the information about weaning referred directly to the study subjects or was an indication of general household practice.]

Armstrong et al. (1983) conducted a case–control study of NPC (100 cases, 100 controls) among Chinese residing in 27 census districts in Selangor, Malaysia. Controls were individually matched to cases on sex, age and neighbourhood of residence. All cases were histologically confirmed, had been resident in the study area for at least five years and had been diagnosed between 1973 and 1980 in the only hospital that offered radiotherapy treatment for NPC in Malaysia. The interviews covered information on use of alcohol, tobacco and nasal ointments and dietary habits. Current intake of salted fish as well as intake during childhood and adolescence were assessed. Salted fish consumption during childhood was a significant risk factor (OR for any consumption *versus* no consumption, 3.0; $p = 0.04$); the OR was 17.4 (95% CI, 2.7–111.1) for daily intake compared to no intake. Salted fish intake during adolescence was also a risk factor, but its effect was weaker than that observed for childhood intake. The OR for daily consumption of salted fish compared to less than weekly consumption during adolescence was 3.5 (95% CI, 1.2–10.7). Current intake of salted fish was not a risk factor. The effect of salted fish intake remained when the effects of other risk factors (i.e., occupational exposure to smoke and/or dust) were accounted for in the analysis.

Yu et al. (1986) conducted a population-based case–control study of NPC in Hong Kong. Eligible cases were Chinese residents of Hong Kong with histologically confirmed NPC diagnosed when they were less than 35 years of age, identified from four hospitals in Hong Kong covering over 90% of all new cases in the area. A total of 266 cases were identified, and interviews were successfully completed with 250 (245 confirmed histologically; 16 refused). Controls were friends of cases, and each was matched to the index case by sex and date of birth (within five years); in two instances, the first control refused and a second eligible control was interviewed. In addition, 182 case and 155 control mothers participated in the study. The index subjects and their mothers were interviewed by one interviewer (one of the authors of the study), who could thus not be blinded to the status of the respondent.

Consumption of salted fish three years before interview/diagnosis and at the age of 10, as reported by the index subjects, and at the age of 10, between the ages of one and two and during weaning, as reported by the mothers of index subjects, was significantly associated with the occurrence of NPC, and a strong dose–response relationship was observed, based on frequency of consumption. For the four time periods assessed, the risk for NPC was up to 38 times higher for people with the highest frequency of intake than for those who had never or rarely ate the food. The association was stronger for consumption in childhood than in adulthood, and the strongest association was with intake before the age of two. There was strong concordance in the pattern of intake of salted fish at age 10 as reported by the index subject and by mothers of index subjects. The consumption patterns of male and female controls were similar, and the ORs for corresponding levels of consumption were alike. The ORs for salted fish intake remained highly significant after account was taken of domestic and occupational exposure to smoke, dust or fumes. A number of preserved foods were significantly associated with the occurrence of NPC, but none remained significant after salted fish intake was adjusted for.

Yu *et al.* (1988) conducted a case–control study in Yulin Prefecture in Guangxi Autonomous Region, China, among Han Chinese under the age of 45. This study area was chosen because it borders Guangdong Province and represents an intermediate-risk population within southern China. Eligible cases were residents of Yulin Prefecture with histologically confirmed incident cases of NPC (231 cases), diagnosed between 1984 and 1986 in one of eight hospitals located in Yulin and neighbouring prefectures. Controls of the Han race were chosen from the general population of Yulin Prefecture and matched to cases by sex, age (within five years) and neighbourhood; about 60% of the controls were the first that were eligible. Interviews were conducted with 128 case and 174 control mothers of the 231 cases and 231 controls identified, giving a total of 109 matched case–control mother pairs. All interviews were conducted by trained interviewers, who used a structured questionnaire to enquire about dietary habits, use of nasal oils, incense and mosquito coils and other life style practices. Mothers were asked about the index subjects' consumption of 10 food items during weaning, of 19 food times between the ages of one and two and of 42 food items at around the age of 10. The mother's own pattern of consumption of salted fish during pregnancy with and lactation of the index subject was also assessed. Mothers were asked to describe consumption patterns at the different time points using three frequency categories: rarely, at least once a month to less than once a week, once a week to less than daily. Subjects' intake of salted fish during weaning as well as the mothers' intake during pregnancy and nursing was significantly related to the occurrence of NPC. When compared to mothers who rarely ate salted fish during pregnancy and nursing, mothers who ate this food weekly during pregnancy had an OR of 3.1 (95% CI, 1.1–8.8) (which was reduced after adjustment for other dietary factors to 1.7 [$p = 0.017$]), and those who ate it weekly during lactation had an OR of 4.9 (95% CI, 1.5–15.9). Subjects' consumption of salted fish during weaning was associated with about a three-fold increase in risk (OR, 2.6; 95% CI, 1.2–5.6). Consumption of salted fish between the ages of one and two years and consumption around the age of 10 were also positively associated with risks for NPC, but these results did not reach statistical significance. The findings clearly suggest that early exposure (before the age of two years) is more important than later exposure (around the age of 10 years). Because of the extremely close correlations between mothers' intake during pregnancy and nursing and subjects' intake during weaning, however, it was not possible to assess the relative importance of exposure during the three time periods.

Yu et al. (1989b) conducted a case–control study of NPC in Guangzhou (Canton City), China, where most of the population are Cantonese and from where most of the Chinese in Hong Kong originated. The rates of NPC in Guangzhou are believed to be comparable to those in Hong Kong. In the primary treatment facility for NPC in Guangzhou, 329 histologically confirmed incident cases diagnosed under the age of 50 were identified between 1983 and 1985, of whom 306 (209 men, 97 women) were alive and were interviewed. An equal number of age-, sex- and neighbourhood-matched controls were identified and interviewed. The mothers of 110 cases (71 men, 38 women) and 139 controls (90 men, 49 women) who were under the age of 45 were interviewed in person about dietary habits by four trained interviewers using structured questionnaires; subjects were also interviewed. Intake of salted fish by mothers during pregnancy and nursing and by subjects during weaning, at the ages of one to two, at the age of around 10 and three years before interview were significantly associated with the risk for NPC. At the highest frequency of intake at each time period (i.e., daily intake during nursing and pregnancy for mothers; daily intake for subjects at the age of around 10 and three years previously; weekly intake for subjects at the ages of one to two; and during weaning), there was a significant, two-fold increased risk. The effect of salted fish was independent of other significant dietary risk factors—mouldy bean curd and *chan pai mui* (salted, dried plums).

Ning et al. (1990) conducted a case–control study of NPC in Tianjin City, in northern China, a low-risk region for NPC, with rates of about 2–3 per 100 000. Cases were identified through the population-based tumour registry of Tianjin. Subjects were considered to be eligible if they were aged 64 or less at the time of diagnosis and had histologically confirmed NPC diagnosed on or after 1 January 1981. Three neighbourhood controls were matched individually to the patients for age (within five years), sex and race (Han). Of the 163 eligible cases identified, 100 patients (68 men, 32 women) and 300 controls were interviewed about diet, mostly by one of the authors of the study. Exposure to salted fish was significantly associated with an increased risk for NPC (OR, 2.2; 95% CI, 1.3–3.7). Four characteristics of exposure to salted fish contributed independently to the increased risk: decreasing age at first exposure, increasing duration of consumption, increasing frequency of consumption at the age of 10 and consumption of salted fish after steaming (OR, 4.2; 95% CI, 2.2–8.3) rather than after frying, grilling or boiling (OR, 1.6; 95% CI, 0.8–3.2). The effect of salted fish intake was independent of the increased risk associated with consumption of salted shrimp paste. [The Working Group noted that the study population was older than those in the studies of Yu et al. and that no information was available on exposure to salted fish at weaning or in early childhood.]

A further case–control study of NPC was conducted in a hospital in north-eastern Thailand, a region in which the risk for this neoplasm is intermediate (Sriamporn et al., 1992). Data on current salted fish consumption, cigarette smoking, alcohol drinking and occupational exposure to smoke or dust were collected for 120 NPC cases diagnosed during 1987–90 (67.5% males) and for the same number of hospital controls matched by sex and age. As consumption of *pla-ra* (freshwater fish left to ferment in salted water) was reported by all but four study subjects, it could not be evaluated. After adjustment for alcohol and cigarette consumption, occupation, level of education, residence, age and sex, consumption of Chinese-style sea-salted fish less than once a week gave an OR of 1.5 (95% CI, 0.6–3.5), and consumption at least once a week gave an OR of 2.5 (95% CI, 1.2–5.2), when compared to no consumption.

[The Working Group noted that the case–control studies on NPC and consumption of salted fish did not examine the concomitant role of Epstein-Barr virus (EBV) infection in the etiology of this cancer. Although NPC patients, regardless of ethnic or geographic origin, show a characteristic EBV antibody pattern and the presence of viral fingerprints in tumour cells, the nature of the association between EBV and NPC is unclear (de-Thé et al., 1982). Southern Chinese at high risk for NPC do not differ from low-risk populations in China or elsewhere in the prevalence of EBV infection or in the age at primary infection with EBV (Zeng, 1985). The dissimilarities in the geographical and racial distributions of EBV infection and NPC thus suggest that the observed effect of salted fish cannot be explained by EBV infection.]

2.1.2 *Stomach cancer* (see Table 5)

The role of salted fish in the etiology of stomach cancer was investigated in a study, described in detail in the monograph on pickled vegetables (p. 97), conducted in 1984–86 in a high-risk area in China (Linqu, a rural county in Shandong Province) (You et al., 1988). People who had the highest level of intake of salted fish (> 1 kg/year; about 2.8 g/day) had a nonsignificant increase in risk compared with people who consumed 0.5 kg or less per year. The authors noted that the effect of salted fish was enhanced in logistic regression analyses when the effects of other risk factors were accounted for [details not given].

2.1.3 *Oesophageal cancer* (see Table 5)

The role of salted fish in the etiology of oesophageal cancer was investigated in a study, described in detail in the monograph on pickled vegetables (p. 99), conducted among Chinese in Hong Kong in 1989–90 (Cheng et al., 1992). Current intake of salted fish was associated with an increased risk for oesophageal cancer, reaching an OR of 4.73 (95% CI, 2.11–10.60) for daily or more frequent intake. The association observed in univariate analyses was, however, greatly weakened [details not given] when the effects of other dietary variables (e.g., pickled vegetables) were accounted for. The authors proposed that the dilution of the effect of current intake of salted fish might have been related to a stronger effect of exposure in early childhood than currently and to the generally smaller amounts of salted fish consumed compared to pickled vegetables.

Table 5. Summary of case–control studies of cancer and consumption of Chinese-style salted fish

Geographical area (reference)	No. of cases and no. and type of controls	Intake	Odds ratio	95% CI or p value	Comments
Nasopharyngeal cancer					
USA, California (Henderson et al., 1976; Henderson & Louis, 1978)	Chinese subjects: 74 cases and 110 hospital/clinic controls	Any current use At least 1/month } vs never 1/week > 1/week	2.1 1.6 2.1 3.1 p for trend = 0.02	0.6–6.3	Chinese subjects represented 47% of cases and 41% of controls interviewed.
Hong Kong (Geser et al., 1978)	Chinese subjects: 150 cases and 150 hospital controls	Current use Salted fish after weaning	No association 2.6	p < 0.01	Analysis of use around weaning based on interview of senior women in 108 case and 103 control households
Malaysia (Armstrong et al., 1983)	Chinese subjects: 100 cases and 100 population controls	Childhood use: Yes vs no < 1/day 1/day Adolescent use: < 1/week 1/day Current use	3.0 2.8 17.4 p for trend < 0.001 1.0 3.5 No association	p = 0.04 1.2–10.7	Mean age of cases and controls: 45 years Effect of salted fish remained when occupational exposures were accounted for.
Hong Kong (Yu et al., 1986)	Chinese subjects: 250 cases and 250 friend controls; 182 case and 155 control mothers also interviewed	Three years before interview 1/month–< 1/week ≥ 1/week but < 1/month } vs rarely 1/day At 10 years of age 1/month but < 1/week } vs rarely ≥ 1/week Between age 1 and 2 years (reported by mother) Sometimes } vs never Often During weaning (reported by mother) Ever vs never	2.3 3.2 7.5 15.0 37.7 6.1 20.2 7.5	1.5–3.5 1.7–6.1 0.9–65.3 6.0–37.2 14.1–100.4 3.0–12.5 6.8–60.2 3.9–14.8	Subjects aged < 35 years; interviewer aware of status of patient. Effect of salted fish remained when effects of potential confounders (including other foods, domestic/occupational exposures to smoke, dust or fumes) were accounted for. Intake of salted fish was important during all time periods, particularly during childhood.

Table 5 (contd)

Geographical area (reference)	No. of cases and no. and type of controls	Intake	Odds ratio	95% CI or p value	Comments
Guangxi Region, China (Yu et al., 1988)	231 cases and 231 population controls; 128 case and 174 control mothers also interviewed	*Mothers' intake* During pregnancy: 1/month } vs rarely 1/week During nursing: 1/month } vs rarely 1/week *Subjects' intake* During weaning: yes vs no Between ages 1 and 2: 1/month } vs rarely 1/week Around age of 10: 1/month } vs rarely 1/week	1.9 3.1 1.3 4.9 2.6 1.2 2.2 1.5 1.5	1.0–3.6 1.1–8.8 0.6–2.6 1.5–15.9 1.2–5.6 0.5–2.7 0.7–7.6 0.9–2.7 0.5–4.3	Subjects < 45 years of age. Risks adjusted for subject's sex and age (< 35, ≥ 35 years). 1/month = at least 1/month–< 1/week; 1/week = at least 1/week to < 1/day. Intake of many salted and preserved foods during weaning and between ages 1 and 2 was assessed: intake of salted fish during weaning, strongest risk factor; intake during pregnancy remained significant after adjustment for other foods in regression analyses. Other foods with a significant effect on risk included salted ducks' eggs, salted mustard greens and *chung choi* during weaning; dried fish, fermented black bean paste and fermented soya bean paste between ages 1 and 2
Guangzhou, China (Yu et al., 1989b)	306 cases and 306 population controls; 110 case and 139 control mothers also interviewed	*Subjects' intake* (reported by mothers) During weaning: yes vs no Between ages 1 and 2: 1/month } vs rarely 1/week Around age 10: 1/month } vs rarely 1/week 1/day	2.1 1.6 2.0 1.3 1.1 2.4	1.2–3.6 0.8–3.4 1.1–3.6 0.6–2.8 0.5–2.3 1.0–6.0	Subjects < 50 years of age. 1/month = at least 1/month to < 1/week. 1/week = at least 1/week to < 1/day. Only intake of salted fish during weaning was significant when intakes during other time periods were examined simultaneously. Results for salted fish remained significant when intake of other foods was accounted for.

Table 5 (contd)

Geographical area (reference)	No. of cases and no. and type of controls	Intake	Odds ratio	95% CI or p value	Comments
Guangzhou, China (Yu et al., 1989b) (contd)		*Subjects' intake (reported by subjects)* Around age 10:			Other foods with an independent effect on risk included mouldy bean curd and *chan pai mui*. Intake prevalence among cases was similar to that in Hong Kong (Yu et al., 1986), but intake among controls was twice as common in Guangzhou as in Hong Kong.
		1/month ⎫	1.1	0.7–1.9	
		1/week ⎬ vs rarely	1.4	0.9–2.2	
		1/day ⎭	2.1	1.2–3.6	
		Three years previously:			
		1/month ⎫	0.9	0.6–1.4	
		1/week ⎬ vs rarely	1.4	0.9–2.1	
		1/day ⎭	1.8	0.9–3.6	
		Mothers' intake During pregnancy:			
		1/month ⎫	0.9	0.4–1.8	
		1/week ⎬ vs rarely	1.7	0.9–3.2	
		1/day ⎭	2.2	1.1–4.6	
		During nursing:			
		1/month ⎫	0.8	0.4–1.7	
		1/week ⎬ vs rarely	1.1	0.6–2.2	
		1/day ⎭	2.3	1.1–4.6	
Tianjin, China (Ning et al., 1990)	Han Chinese subjects: 100 cases and 300 neighbourhood controls	*Any intake* Ever vs never	2.2	1.3–3.7	Subjects were < 65 years of age; mean age: cases, 44.9 years; controls, 45.2 years. 39% (63/163) cases could not be interviewed. Most interviews were conducted by one of the authors of the study. The effect of salted fish remained when other factors were adjusted for. Intake of salted shrimp paste was another risk factor.
		Age at first exposure (years)			
		≥ 21	1.5	0.7–3.3	
		11–20	1.9	0.9–4.0	
		1–10	2.6	1.5–4.6	
		Duration of consumption (years)			
		1–10	1.6	0.9–3.1	
		11–20	2.8	1.4–5.4	
		≥ 21	2.8	1.4–5.6	

Table 5 (contd)

Geographical area (reference)	No. of cases and no. and type of controls	Intake	Odds ratio	95% CI or p value)	Comments
Tianjin, China (Ning et al., 1990) (contd)		*Frequency of consumption at age of 10:*			
		1/year	1.6	0.8–3.2	
		1/month	3.5	1.6–7.4	
		1/week or 1/day	6.7	2.2–20.7	
		Cooking method for consumption at age 10:			
		Steaming	4.2	2.2–8.3	
		Other (frying, grilling or broiling)	1.6	0.8–3.2	
North-eastern Thailand (Sriamporn et al., 1992)	120 cases and 120 hospital controls	No consumption	1.0		Logistic estimates adjusted for occupation and area of residence, alcohol intake, type of cigarettes smoked and level of education
		< 1/week	1.5	0.6–3.5	
		≥ 1/week	2.5	1.2–5.2	
Stomach cancer					
Shandong, China (You et al., 1988)	564 cases and 1131 population controls	≤ 1 kg/year } vs ≤ 0.5 kg/year	1.0	0.8–1.4	Questionnaire included 85 food items; intake habits about 1965 and 1980; adjusted for sex, age and family income
		> 1 kg/year	1.4	0.8–1.5	
Oesophageal cancer					
Hong Kong (Cheng et al., 1992)	Chinese subjects: 400 cases and 1598 controls (800 hospital, 798 general practice)	< 1/year	1.0		Adjusted for age, level of education and birthplace. Univariate analysis much weakened by adjustment for other variables.
		< 1/month	0.99	0.56–1.77	
		1–3/month	1.17	0.72–1.92	
		1–3/week	2.03	1.23–3.34	
		4–6/week	3.15	1.58–6.29	
		1/day or more	4.73	2.11–10.60	

CI, confidence interval

2.2 Other salted fish

2.2.1 *Nasopharyngeal carcinoma*

Case–control studies (see Table 6)

Jeannel *et al.* (1990) carried out a case–control study on all incident cases of NPC that had been histologically confirmed and treated at the Institute of Cancer in Tunis, Tunisia, in 1986–87. Eighty cases and 160 age- and sex-matched neighbourhood controls were interviewed about socioeconomic conditions, diet during the year preceding diagnosis and, with the help of their families, diet during the first month of life, weaning, childhood and adolescence. Tunisian families live under the same roof, and in most households there is someone of an older generation who can answer questions about childhood diet and weaning. Living conditions, such as living in a *gourbi* (makeshift one-room dwelling) without separate beds for parents and children and without windows, and cooking in the main room during childhood, were significantly associated with risk for NPC. All of these variables were pooled in a single score for multivariate matched analysis of dietary variables. Subjects who had been weaned directly from mother's milk on to an adult diet were at higher risk, regardless of living conditions. Intake of salted anchovies during childhood was associated with a small increase in risk (crude OR, 2.6; $p = 0.08$), but this association was no longer significant when living conditions were accounted for in the analysis (adjusted OR, 1.5; $p = 0.5$).

In a study of NPC among Alaskan native populations (Eskimos, Indians and Aleuts), Lanier *et al.* (1980) collected information on the dietary habits of 13 of 31 cancer patients diagnosed during 1966–76 and 13 controls individually matched on age, sex, ethnic group and residence. More cases than controls reported having eaten salted fish frequently in childhood (four *versus* one discordant pairs).

2.2.2 *Stomach cancer*

Most of the available studies on salted fish and stomach cancer were conducted in populations at high risk for cancer at this site.

(a) Correlation studies

Two of three correlational studies conducted in Japan suggested that consumption of salted fish is positively associated with risk for stomach cancer. (For a detailed description of the studies, see the monograph on pickled vegetables, pp. 90–91.) The two studies with positive results used the rates of mortality from stomach cancer and dietary information from 46 prefectures. In the study of Hirayama (1971), standardized death rates for stomach cancer in 1955 were correlated with marketing data obtained in 1959 and nutritional survey data obtained in 1958 by the Ministry of Agriculture. A significant positive association was found between salted fish intake and stomach cancer mortality (salted salmon, $p < 0.01$; salted trout, $p < 0.05$; other salted fish, $p > 0.10$). In a later study (Hara *et al.*, 1985), cancer mortality data for 1969–79 were related to consumption data obtained from a national survey in 1974 [specific type of dietary data obtained not described]. Of the many categories of foods compiled in the survey, 26 food items were selected for this correlation study. The standardized mortality ratio (SMR) for stomach cancer was significantly associated with intake of salted or dried fish ($p < 0.05$). In a study in which mortality rates for 1969-74 were related to dietary data from 1040 census tracts in Japan sampled in 1974, 1975 and 1976 (Nagai *et al.*, 1982), consumption of dried or salted fish was not associated with risk for stomach cancer ($r = 0.052$ for men; $r = -0.026$ for women). In this study, the nutritional data were based on foods consumed over a three-day period by approximately 70 people in about 20 households in 300 areas of Japan.

Kolonel *et al.* (1980, 1981, 1983) conducted a series of correlation studies among Japanese and other ethnic groups in Hawaii. The results of these studies support a positive association between intake of salted or dried fish and risk for stomach cancer. A decreasing gradient in the rates of stomach cancer was seen for Japanese in Japan, first-generation Japanese in Hawaii, second-generation Japanese in Hawaii and Caucasians in Hawaii. A corresponding gradient was seen for intake of salted or dried fish.

(b)   Cohort studies (see Table 7)

In a cohort study conducted on Hawaiian men of Japanese ancestry (Chyou *et al.*, 1990), described in detail in the monograph on pickled vegetables (p. 93), incident cases of stomach cancer identified in 1966-83 were compared with a random sample of the cohort who had not developed cancer at the end of the follow-up period. The two groups did not differ significantly in their intake of dried fish: the mean daily intake was 0.4 g for cases and 1.3 g for the subcohort. [The Working Group noted that it was not clear whether salted fish was included in the category of dried fish. The authors describe the study as a case-cohort study; however, the controls were not selected from the total cohort at its inception.]

Kneller *et al.* (1991) investigated the effect of salted fish in a cohort of 17 633 white US men aged 35 and older, mainly of Scandinavian and German descent, who responded to a mailed questionnaire in 1966. After 20 years of follow-up, 23% of the cohort (4027 respondents) were lost to follow-up, and there were 1033 cancer deaths, including 75 from stomach cancer. The questionnaire covered demographic factors, use of tobacco and alcohol, dietary pattern (including 35 food items) and other factors. After control for age and smoking status, salted fish emerged as a risk factor for stomach cancer. Subjects who ate salted fish at least once a month showed about a two-fold increase in risk over those with less frequent intake. This increase was most apparent among residents of the north-central states (relative risk, 2.1; 95% CI, 1.08-4.13) and among immigrants and first-generation Americans (relative risk, 2.2; 95% CI, 0.97-4.77). The results for salted fish were adjusted for year of birth and current smoking habits but not for other significant dietary factors.

(c)   Case-control studies (see Table 6)

Intake of dried and salted fish emerged as a risk factor for stomach cancer in a study conducted among Japanese in Hawaii (Haenszel *et al.*, 1972), described in detail in the monograph on pickled vegetables (pp. 93-94). About a two-fold increase in risk for stomach cancer was associated with eating dried or salted fish at least four times per month, when compared with nonconsumers; however, the risks associated with intake once or two to three times per month were not significantly different from those of nonconsumers. The risk associated with consumption of dried and salted fish combined was no higher than the risks associated with eating dried and salted fish individually. Analyses were not conducted for dried and salted fish simultaneously, so that it is difficult to determine the independent effect of each of these food items. The association between dried/salted fish and stomach cancer was observed among both migrants (Issei) and second-generation (Nisei) Japanese Americans, but the effects were slightly more pronounced among the Issei. The effects of other foods were not accounted for in the analysis.

In a study conducted in Hiroshima and Miyagi prefectures in Japan (see Figure 3) (Haenszel et al., 1976), described in detail in the monograph on pickled vegetables (p. 94), intake of salted/dried fish was associated with a small, nonsignificant increased risk for stomach cancer in both study areas. The ORs were 1.06 and 1.20, respectively, for consumption four to nine and 10 or more times per month compared to less than four times per month. When the results were examined separately for the two prefectures, the increase was somewhat higher in Miyagi (where traditional customs persisted) than in Hiroshima. The association between intake of salted fish and stomach cancer risk was similar for diffuse and intestinal types of stomach cancers. The effects of other foods were not accounted for in the analysis.

Figure 3. Map of Japan showing the islands, prefectures and cities that have been studied with respect to consumption of salted fish

From Hara et al. (1985)

In a study conducted in Nagoya, Japan (Tajima & Tominaga, 1985), described in detail in the monograph on pickled vegetables (p. 94), a significant, two-fold increase in risk for stomach cancer was associated with intake of dried or salted fish at least once a week compared with less frequent intake, after adjustment for sex and age (1.99; $p < 0.05$). [The Working Group noted that although multivariate analyses were conducted, it is not clear whether the results for salted fish were adjusted for other dietary factors.]

In a study described in detail in the monograph on pickled vegetables (pp. 94, 97), conducted in Saga Prefecture, a low-risk rural area in Kyushu, Japan (Kono *et al.*, 1988), intake of salted fish did not differ between gastric cancer patients and the two control groups (hospital and population controls).

Buiatti *et al.* (1989) conducted a population-based case–control study of stomach cancer in Italy, comprising seven centres in two high- and two low-risk areas displaying a three-fold range of rates for stomach cancer mortality. Eligible subjects were aged 75 or less, residents of the study areas and with histologically confirmed gastric cancers diagnosed in 1985–87 in one of the hospitals in the study areas. Personal interviews were conducted with 1016 of 1229 eligible patients. Of the 1159 general population controls of similar age, sex and study area, 266 were second controls. The interviews were extensive, covering the intake of 146 food items and beverages consumed during a 12-month period approximately two years before the interview. Over 90% of patient interviews were conducted in hospital, whereas over 60% of interviews with controls took place at home. Risk increased with increasing tertile of intake of salted and dried anchovies, cod and herring (ORs, 1.0, 1.1, and 1.4, respectively, for tertiles; p for trend = 0.001), even after adjustment for matching variables and the main potential confounders (excluding other dietary variables). The ORs for stomach cancer and intake of salted/dried fish were not reported separately for the high- and low-risk areas, but intake of these products was not consistent in the two areas.

In a subsequent analysis by histological type of stomach cancer, a significantly increased risk was found to be associated with the highest compared to the lowest tertile of intake of salted/dried fish for both intestinal (OR, 1.5; 95% CI, 1.1–1.9) and diffuse (1.5, 1.1–2.1) types of stomach cancer (Buiatti *et al.*, 1991). In a further analysis, 68 cases of cancer of the cardia were separated from the remaining sites (Palli *et al.*, 1992). No difference was seen in the risk associated with consumption of salted/dried fish from that of gastric cancer at other sites.

(d) Precursor lesions

Nomura *et al.* (1982) conducted a study in Japan, described in detail in the monograph on pickled vegetables (pp. 97–98), to identify risk factors for intestinal metaplasia of the stomach, an established precursor stage of stomach cancer. The dietary item most strongly related to metaplasia score was dried fish ($p = 0.002$, multiple linear regression analysis), consisting mainly of dried sardine, cod, mackerel or herring, with salt added. This finding was made only among males.

2.2.3 *Other cancers* (see Table 6)

The association between salted fish and oesophageal cancer was investigated in a number of correlation studies conducted in Japan, described in detail in the monograph on pickled vegetables (pp. 98–99). In the study of Nagai *et al.* (1982), dried or salted fish was one

Table 6. Summary of results of case–control studies on consumption of salted fish other than Cantonese-style

Geographical area (reference)	No. of cases and no. and type of controls	Intake level	Odds ratio	95% CI or p value	Comments
Nasopharyngeal carcinoma					
Tunisia (Jeannel et al., 1990)	80 cases and 160 neighbourhood controls	*During childhood* Salted anchovies: yes vs no	1.5	$p = 0.5$	Adjusted for living conditions; crude RR was 2.6 ($p = 0.08$)
USA, Alaska (Lanier et al., 1980)	13 cases and 13 controls	*During childhood* Salted fish	4 vs 1		Discordant pairs
Stomach cancer					
USA, Hawaii (Haenszel et al., 1972)	Japanese subjects: 220 cases and 440 hospital controls	*Dried fish* Any use vs no use	1.8		Questionnaire included a number of Japanese and western foods. RR adjusted for sex and place of birth. Any use significant for Issei not Nisei. Results elevated but not significant for Issei and Nisei separately
		1/month	1.5	$p < 0.05$	
		2–3/month } vs no use	1.6	$p < 0.05$	
		≥ 4/month	2.8	$p > 0.05$	
				$p < 0.05$	
		Salted fish Any use vs no use	1.6	$p < 0.05$	
		1/month	1.4	$p < 0.05$	
		2–3/month } vs no use	2.0	$p > 0.05$	
		≥ 4/month	2.0	$p < 0.05$	
		Dried and salted fish combined Use of both vs no use	2.0	$p < 0.05$	Results significant for Issei
		Total, 2/month	1.5	$p < 0.05$	
		Total, 3–5/month } vs no use	2.5	$p > 0.05$	
		Total, 6/month	2.6	$p < 0.05$	
				$p < 0.05$	
Japan, Hiroshima and Miyagi (Haenszel et al., 1976)	783 cases and 1566 hospital controls	*Salted/dried fish* ≥ 4/month	1.12	$p > 0.05$	Questionnaire included a number of Japanese and western foods. Increase greater in Miyagi than in Hiroshima
		4–9/month } vs < 4/month	1.06		
		≥ 11/month	1.20	$p > 0.05$	

Table 6 (contd)

Geographical area (reference)	No. of cases and no. and type of controls	Intake level	Odds ratio	95% CI or p value	Comments
Japan, Aichi Cancer Center, Nagoya (Tajima & Tominaga, 1985)	93 cases and 186 hospital controls	*Dried/salted fish* $\geq$ 1/week vs < 1/week	1.99	$p < 0.05$	Questionnaire included a list of specific foods and eating frequencies. Results adjusted for sex and age. Increase greater for dried or salted fish in 40-45 age group than in 56-70 age group.
Japan, northern Kyushu (Kono et al., 1988)	139 cases, 2574 hospital controls, 278 general population controls	*Salted fish* 1-3/month } vs none $\geq$ 3/month	0.9 1.0	$p > 0.05$ $p > 0.05$	Questionnaire pertained to current habits only (year preceding interview); 25 food items. Some differences in the administration of questionnaire, even though all interviewers were trained together.
Italy (Buiatti et al., 1989)	1016 cases and 1159 population controls	*Salted and dried fish* Tertiles of weekly intake: Intermediate } vs low High	1.1 1.4 p for trend = 0.001		Questionnaire included frequency of intake of 146 food items. OR adjusted for age, sex, study area, social class, residence, migration, family history and Quetelet's index
(Buiatti et al., 1991)		*By histological type* Highest vs lowest tertile: Intestinal Diffuse Unclassified	1.5 1.5 1.3	1.1-1.9 1.1-2.1 0.8-2.0	Salted/dried fish included anchovies, cod, herring. Tertile cutoffs not presented
Brain tumour					
Canada, southern Ontario (Burch et al., 1987)	215 cases and 215 hospital controls	*Salted fish* $\geq$ 1 vs < 1/month	1.4	0.4-4.4	Only 58 cases were self-respondents. Only 124 case-control pairs were included in the analysis for salted fish, owing to missing information.

Table 6 (contd)

Geographical area (reference)	No. of cases and no. and type of controls	Intake level	Odds ratio	95% CI or p value	Comments
Liver cancer					
Thailand, north-east (Parkin et al., 1991)	103 cases and 103 hospital controls	*Salted fish* ≥ monthly vs < monthly	0.5	0.2–0.9	Dietary information pertains to one year prior to interview; 54 dietary items included. Results no longer significant in multivariate analysis.

CI, confidence interval

Table 7. Summary of cohort studies on stomach cancer and consumption of salted or dried fish

Geographical area (reference)	No. of subjects, no. of cases and length of follow-up	Intake level	Relative risk	95% CI or p value	Comments
USA, Hawaii (Chyou et al., 1990)	8006 men of Japanese descent 111 stomach cancers (361 cancer-free controls) 18 years of follow-up	*Dried fish:* **Mean intake** Cases: 0.4 g/day Controls: 0.5 g/day		$p = 0.44$	24-h dietary intake data were recoded to allow comparison by specific foods or food groups. p values calculated by comparing two mean values, adjusting for age
USA (Kneller et al., 1991)	17 633 men of Scandinavian and German descent 75 stomach cancers 20 years of follow-up	*Salted fish* < 1/month ≥ 1/month } vs never Used previously but not currently	1.0 1.9 1.3	0.6–1.8 1.0–3.6 0.6–2.6 p for trend, NS	Mailed questionnaire, 35 food items included. Results adjusted for year of birth and current cigarette smoking. Salted fish primarily a risk factor for immigrants and second-generation Americans and for residents of north-central states

CI, confidence interval

of three of the 20 foods or food groups investigated which was significantly and positively correlated with the occurrence of oesophageal cancer in both men and women. The standardized partial regression coefficient was 0.09 ($p < 0.01$) for men and 0.069 ($p < 0.05$) for women. Hara et al. (1985) correlated cancer mortality rates in 46 Japanese prefectures and data on food consumption from a national survey. Of the 26 food items investigated, salted or dried fish was significantly associated with SMRs for oesophageal cancer ($p < 0.01$).

Burch et al. (1987) conducted an exploratory study of brain tumours in adults in southern Ontario, Canada, between 1979 and 1982. A total of 328 histologically confirmed brain tumours diagnosed during 1977 and 1981 were identified through the medical record departments of 33 hospitals in the study area; 247 patients were selected for interview, but a high proportion of the interviews were conducted with relatives (only 58 were self-respondents). One hospital control without cancer was matched to each case on the basis of sex, area of residence, marital status, year of birth (within five years), date of diagnosis (within one year) for live cases and date of death (within one year) for deceased cases, resulting in 215 appropriately matched case–control pairs. Intake of a variety of processed meat and fish products containing nitrite was investigated. A small, nonsignificantly increased risk for brain tumour (OR, 1.4; 95% CI, 0.4–4.4) was found to be associated with regular consumption (at least once a month) of salted fish compared to less frequent consumption.

Parkin et al. (1991) conducted a case–control study of cholangiocarcinoma of the liver in north-eastern Thailand, described in detail in the monograph on aflatoxins (p. 262). In a univariate analysis, there was a significant deficit of cases who were frequent (monthly or more) consumers of salted fish; this result was no longer significant in a multivariate analysis.

In the study of Tajima and Tominaga (1985), described above and in detail in the monograph on pickled vegetables (p. 94), nonsignificant increases in risk for cancers of the colon and rectum were associated with consumption of dried or salted fish.

In the study of Lanier et al. (1980), described above, no difference in intake of salted fish in childhood was found between 17 patients with tumours at sites other than the nasopharynx (salivary glands and other cancers of the head and neck, excluding thyroid) and their matched controls.

3. Studies of Cancer in Experimental Animals

3.1 Oral administration

3.1.1 *Rat*

Ten female and 10 male inbred WA albino rats, one month of age and weighing 150 g, were each given 30 g per day of steamed Chinese-style salted fish for six months and then 20 ml of an extract of Cantonese salted fish heads as drinking-water (500 g of salted fish heads boiled in 1 l of water to give a final concentration of 500 ml of extract, which was diluted 1:5 with fresh drinking-water to give a sodium chloride concentration of 0.9 g/l) on five days per week. A group of three males and three females served as controls. The animals were sacrificed after one to two years or when moribund [number of animals alive at the time of diagnosis of the first tumour and number sacrificed at each time period not given]. The heart, lung, liver, spleen, kidney, trachea, oesophagus, stomach and six sections of nasal cavity were examined. In the treated group, 4/10 females developed carcinomas in the nasal or paranasal regions: two developed adenocarcinomas of the nasal cavity after 12 and 24 months, respectively, one an undifferentiated carcinoma of the paranasal sinus after 15 months and another a squamous-cell carcinoma of the bucco-alveolar sulcus after 24 months. No nasal or paranasal tumour was seen in treated males or in controls. The two rats with nasal adenocarcinomas also developed a mammary adenocarcinoma and a nodular liver, respectively (Huang *et al.*, 1978b). [The Working Group noted the small number of animals.]

Groups of 37 female and 37 male inbred Wistar-Kyoto rats, aged 21 days, were fed a powdered diet to which one part of Chinese-style salted fish (from Hong Kong; 48% soft-type fish and 52% hard-salted fish of nine common species) was added to five parts of diet (low dose) or to three parts of diet (high dose) for 18 months. A group of 37 males and 36 females given only rat chow served as controls. The salted fish was air-mailed to the study site (Los Angeles, USA) and then kept frozen; every two weeks, fish was thawed, steamed for 45 min, dried in a laminar flow hood, then ground to form a powder which was mixed with the rodent diet. After 18 months, all rats were given standard rat chow pellets, and at three years of age all remaining rats were killed and examined. Lung, kidney, liver, stomach and nasal cavity were examined. Median lengths of survival were 121, 123 and 127 weeks for low-dose, high-dose and control females and 131, 130 and 129 weeks for males. Three of 74 rats given the high dose of salted fish had epithelial tumours of the nasal cavity: a squamous-cell carcinoma during week 97 in a female, an undifferentiated carcinoma during week 44 in a male and a spindle-cell carcinoma during week 99 in a female. One low-dose male had a spindle-cell tumour of the nasal cavity [not otherwise specified] during week 118 of the experiment. Nasal cavity tumours were not found in low-dose females or in controls of either sex. No significant difference in the occurrence of other tumours was observed between treated and control animals (Yu *et al.*, 1989a). [The Working Group noted the limited number of tissues examined.]

3.1.2 *Hamster*

Six female and eight male Syrian golden hamsters, aged one month and weighing 100 g, were each given 20 g per animal of steamed Chinese-style salted fish for six months and then 20 ml of an extract of Cantonese salted fish heads as drinking-water (prepared as described in section 3.1.1) on five days per week. Three male and three females served as controls. The animals were sacrificed after one to two years or when moribund [numbers sacrificed at each time period not given]. The incidences of tumours of the heart, lung, liver, spleen, kidney, trachea, oesophagus, stomach and nasal cavity were not increased (Huang *et al.*, 1978b). [The Working Group noted the small number of animals.]

3.2 Administration with known carcinogens

Mouse

A group of 19 female mice (weighing 20–22 g) was given N-nitrososarcosine ethyl ester at 20 ml/kg bw in the diet five times over 200 days, and a group of 22 mice was fed the nitrosamine plus Chinese-style dry-salted fish [concentration of salted fish unspecified]. Carcinomas of the forestomach developed in 6/19 mice (31.6%) given the nitrosamine alone and in 14/22 (63.6%) also given the salted fish ($p < 0.05\%$). All mice had papillomas and carcinomas of the forestomach (Lin *et al.*, 1986).

4. Other Relevant Data

4.1 Absorption, distribution, metabolism and excretion

No data were available to the Working Group.

4.2 Toxic effects

No data were available to the Working Group.

4.3 Reproductive and developmental toxicity

No data were available to the Working Group.

4.4 Genetic and related effects

4.4.1 *Humans*

The urine of four nonsmoking Japanese became mutagenic to *Salmonella typhimurium* on days 2–3 and 6–7 following meals of fried, salted salmon. The urine was tested over seven consecutive days. Urinary mutagenicity increased up to 12–14 h after initial intake, reached a plateau at about 20 h and remained at that level up to 24 h. The excretion of mutagenic substances thus appeared to be complete within about 20 h (Ohyama *et al.*, 1987).

4.4.2 *Experimental systems* (see also Table 8 and Appendices 1 and 2)

In one study, rats fed dried salted fish from southern China produced urine that was mutagenic to bacteria. A variety of extracts prepared from several different kinds of salted fish from various countries were mutagenic to bacteria. In one study of salmon, cooking rather than salting produced bacterial mutagens.

Table 8. Genetic and related effects of salted fish

Test system	Result		Dose (LED/HID; standardized to weight of original material)	Reference
	Without exogenous metabolic system	With exogenous metabolic system		
Hawaiian dried salted fish				
ERC, *Escherichia coli rec* strains, differential toxicity	–		100 μl fish extract	Ichinotsubo & Mower (1982)
SA7, *Salmonella typhimurium* TA1537, reverse mutation	–	–	1000 μl fish extract	Ichinotsubo & Mower (1982)
SA0, *Salmonella typhimurium* TA100, reverse mutation	+	–	400 μl fish extract	Ichinotsubo & Mower (1982)
SA9, *Salmonella typhimurium* TA98, reverse mutation	–	–	1000 μl fish extract	Ichinotsubo & Mower (1982)
EC2, *Escherichia coli* WP2, reverse mutation	–	–	1000 μl fish extract	Ichinotsubo & Mower (1982)
Southern Chinese salted fish				
SA0, *Salmonella typhimurium* TA100, reverse mutation	+	+	1.25 mg (aqueous)	Ho *et al.* (1978)
SA9, *Salmonella typhimurium* TA98, reverse mutation	+	+	1.25 mg (aqueous)	Ho *et al.* (1978)
BFA, Urine from rats, microbial mutagenicity	+	+	Not given	Ho *et al.* (1978)
Japanese salted salmon (fried)				
SA0, *Salmonella typhimurium* TA100, reverse mutation	–	+	1 g[a]	Ohyama *et al.* (1987)
SA5, *Salmonella typhimurium* TA1535, reverse mutation	–	–	1 g[a]	Ohyama *et al.* (1987)
SA8, *Salmonella typhimurium* TA1538, reverse mutation	(+)	+	1 g[a]	Ohyama *et al.* (1987)
SA9, *Salmonella typhimurium* TA98, reverse mutation	(+)	+	50 mg[a]	Ohyama *et al.* (1987)
BFH, Urine from humans, microbial mutagenicity	+	+	Not given	Ohyama *et al.* (1987)
Chinese salted fish (*pak wik*) from Canada				
SA5, *Salmonella typhimurium* TA1535, reverse mutation (suspension assay)	–	0	500 mg/ml (methanol extract)	Stich *et al.* (1982)

+, positive; (+), weakly positive; –, negative; 0, not tested
[a]Dichloromethane–XAD-2 ion-exchange resin eluate of aqueous extract

5. Summary of Data Reported and Evaluation

5.1 Exposure data

Salted fish is prepared by treating fish with dry salt or an aqueous salt solution and is often subsequently dried in the sun. It is produced and consumed primarily in Southeast Asia and northern Europe. Chinese-style salted fish is usually softened by partial decomposition before or during salting. High levels of N-nitrosodimethylamine have been reported in some samples of Chinese-style salted fish.

5.2 Human carcinogenicity data

The pattern of nasopharyngeal carcinoma incidence in China reflects the pattern of consumption of salted fish. Eight case–control studies consistently demonstrate that consumption of Chinese-style salted fish is strongly related to risk for nasopharyngeal carcinoma. The effect remained in studies that controlled for other risk factors. A significant dose–response relationship is seen between frequency of intake and risk for nasopharyngeal carcinoma, and the association is especially strong for intake of salted fish during childhood. Two further case–control studies, on oesophageal cancer and stomach cancer, found nonsignificant associations with consumption of Chinese-style salted fish.

The association between cancer and the consumption of other types of salted fish was examined in several studies. A study in Tunisia and one in Alaska suggested a relationship between intake of salted fish and nasopharyngeal carcinoma. Ecological studies in Japan showed a correlation between consumption of dried or salted fish and cancers of the stomach and oesophagus. One cohort study in the USA and three case–control studies in Hawaii, Japan and Italy showed positive associations between intake of dried or salted fish and risk for stomach cancer; a cohort study from Hawaii and two case–control studies from Japan found no association. In none of these studies was the effect of salt consumption evaluated independently.

Studies on cancers at other sites were not considered informative for the evaluation.

5.3 Animal carcinogenicity data

Chinese-style salted fish was tested in two studies in rats by administration in the diet or in the diet and drinking-water. A small number of carcinomas was oberved in the nasal, paranasal and oral cavities in each of the studies in rats, mostly in females.

5.4 Other relevant data

Extracts of Chinese-style salted fish are mutagenic to bacteria.

5.5 Evaluation[1]

There is *sufficient evidence* in humans for the carcinogenicity of Chinese-style salted fish.
There is *inadequate evidence* in humans for the carcinogenicity of other salted fish.
There is *limited evidence* in experimental animals for the carcinogenicity of Chinese-style salted fish.

Overall evaluation

Chinese-style salted fish *is carcinogenic to humans (Group 1)*.
Other salted fish *is not classifiable as to its carcinogenicity to humans (Group 3)*.

6. References

Ah-Weng, P., Hanson, S.W. & McGuire, K.J. (1985) Water activity data in relation to quality loss for Southeast Asian cured fish. In: Reilly, A., ed., *Spoilage of Tropical Fish and Product Development* (FAO Fisheries Report No. 317 (Suppl.)), Rome, FAO, pp. 306–314

Anon. (1986) *Regulations for Salted Fish and Klipfish* (Fisheries and Aquatic Sciences No. 5282), Ottawa, Canada Institute for Scientific and Technical Information, National Research Council

Armstrong, R.W. & Eng, A.C.S. (1983) Salted fish and nasopharyngeal carcinoma in Malaysia. *Soc. Sci. Med.*, 17, 1559–1567

Armstrong, R.W., Kutty, M.K. & Armstrong, M.J. (1978) Self-specific improvements associated with nasopharyngeal carcinoma in Selangor, Malaysia. *Soc. Sci. Med.*, 12D, 149–156

Armstrong, R.W., Kutty, M.K., Dharmalingam, S.K. & Ponnudurai, J.R. (1979) Incidence of nasopharyngeal carcinoma in Malaysia, 1968–1977. *Br. J. Cancer*, 40, 557–567

Armstrong, R.W., Armstrong, M.J., Yu, M.C. & Henderson, B.E. (1983) Salted fish and inhalants as risk factors for nasopharyngeal carcinoma in Malaysian Chinese. *Cancer Res.*, 43, 2967–2970

Bartsch, H., Castegnaro, M., O'Neill, I.K. & Okada, M., eds (1982) N-*Nitroso Compounds: Occurrence and Biological Effects* (IARC Scientific Publications No. 41), Lyon, IARC

Bjarnason, J. (1987) *A Manual of Fish Processing. I. Curing of Salted Fish* (Fisheries and Aquatic Sciences No. 5323), Ottawa, Canada Institute for Scientific and Technical Information, National Research Council

Buckle, K.A., Souness, R.A., Putro, S. & Wuttijumnong, P. (1988) Studies on the stability of dried salted fish. In: Teng, T.T. & Quah, C.H., eds, *Food Preservation by Moisture Control*, London, Elsevier, pp. 103–115

Buiatti, E., Palli, D., Decarli, A., Amadori, D., Avellini, C., Bianchi, S., Biserni, R., Cipriani, F., Cocco, P., Giacosa, A., Marubini, E., Puntoni, R., Vindigni, C., Fraumeni, J.F., Jr & Blot, W.J. (1989) A case–control study of gastric cancer and diet in Italy. *Int. J. Cancer*, 44, 611–616

Buiatti, E., Palli, D., Bianchi, S, Decarli, A., Amadori, D., Avellini, C., Cipriani, F., Cocco, P., Giacosa, A., Lorenzini, L., Marubini, E., Puntoni, R., Saragoni, A., Fraumeni, J.F., Jr & Blot, W.J. (1991) A case–control study of gastric cancer and diet in Italy. III. Risk patterns by histologic type. *Int. J. Cancer*, 48, 369–374

---

[1]For definition of the italicized terms, see Preamble, pp. 26–29.

Burch, J.D., Craib, K.J.P., Choi, B.C.K., Miller, A.B., Risch, H.A. & Howe, G.R. (1987) An exploratory case–control study of brain tumors in adults. *J. natl Cancer Inst.*, **78**, 601–609

Cheng, K.K., Day, N.E., Duffy, S.W., Lam, T.H., Fox, M. & Wong, J. (1992) Pickled vegetables in the aetiology of oesophageal cancer in Hong Kong Chinese. *Lancet*, **339**, 1314–1318

Chinese National Cancer Control Office/Nanjing Institute of Geography (1979) *Atlas of Cancer Mortality in the People's Republic of China*, Shanghai, China Map Press

Chyou, P.-H., Nomura, A.M.Y., Hankin, J.H. & Stemmermann, G.N. (1990) A case–cohort study of diet and stomach cancer. *Cancer Res.*, **50**, 7501–7504

Elmossalami, E. & Sedik, M.F. (1972) Studies on sand- and tin-salted fish 'Mugil cephalus'. *Zbl. vet. Med. B.*, **19**, 521–531

Esser, J.R., Hanson, S.W., Wiryante, J. & Tausin, S. (1990) Prevention of insect infestation and losses of salted–dried fish in Indonesia by treatment with an insecticide approved for use on fish. In: Souness, R.A., ed., *FAO Fisheries Report No. FAO-FIIU-401* (Suppl.), Rome, FAO, pp. 168–179

FAO (1989) *FAO Yearbook—Fishery Statistics Commodities—Vol. 69* (FAO Fisheries Series No. 37/FAO Statistics Series No. 101), Rome, pp. 146–149

FAO/WHO Codex Alimentarius Commission (1983) *Codex Alimentarius Volume B. Recommended International Code of Practice for Salted Fish* (CAC/RCP 26-1979), Rome

Fong, Y.Y. & Chan, W.C. (1973a) Dimethylnitrosamine in Chinese marine salt fish. *Food Cosmet. Toxicol.*, **11**, 841–845

Fong, Y.Y. & Chan, W.C. (1973b) Bacterial production of dimethylnitrosamine in salted fish. *Nature*, **243**, 421–422

Fong, Y.Y. & Chan, W.C. (1976a) Reduction of nitrosodimethylamine content in Cantonese salt fish. In: Walker, E.A., Bogovski, P. & Griciute, L., eds, *Environmental N-Nitroso Compounds. Analysis and Formation* (IARC Scientific Publications No. 14), Lyon, IARC, pp. 465–471

Fong, Y.Y. & Chan, W.C. (1976b) Methods for limiting the content of dimethylnitrosamine in Chinese marine salt fish. *Food Cosmet. Toxicol.*, **14**, 95–98

Fong, Y.Y. & Walsh, E.O'F. (1971) Carcinogenic nitrosamines in Cantonese salt–dried fish (Letter to the Editor). *Lancet*, **ii**, 1032

Geser, A., Charnay, N., Day, N.E., de Thé, G. & Ho, H.C. (1978) Environmental factors in the etiology of nasopharyngeal carcinoma: report on a case–control study in Hong Kong. In: de Thé, G. & Ito, Y., eds, *Nasopharyngeal Carcinoma: Etiology and Control* (IARC Scientific Publications No. 20), Lyon, IARC, pp. 213–229

Haenszel, W., Kurihara, M., Segi, M. & Lee, R.K.C. (1972) Stomach cancer among Japanese in Hawaii. *J. natl Cancer Inst.*, **49**, 969–988

Haenszel, W., Kurihara, M., Locke, F.B., Shimuzu, K. & Segi, M. (1976) Stomach cancer in Japan. *J. natl Cancer Inst.*, **56**, 265–274

Hara, N., Sakata, K., Nagai, M., Fujita, Y., Hashimoto, T. & Yanagawa, H. (1985) Statistical analyses on the pattern of food consumption and digestive-tract cancers in Japan. *Nutr. Cancer*, **6**, 220–228

Henderson, B.E. & Louie, E. (1978) Discussion of risk factors for nasopharyngeal carcinoma. In: de Thé, G. & Ito, Y., eds, *Nasopharyngeal Carcinoma: Etiology and Control* (IARC Scientific Publications No. 20), Lyon, IARC, pp. 251–260

Henderson, B.E., Louie, E., Jing, J.S.-H., Buell, P. & Gardner, M.B. (1976) Risk factors associated with nasopharyngeal carcinoma. *New Engl. J. Med.*, **295**, 1101–1106

Hirayama, T. (1971) Epidemiology of stomach cancer. *Gann Monogr. Cancer Res.*, **11**, 3–19

Hirayama, T. (1978) Descriptive and analytical epidemiology of nasopharyngeal cancer. In: de Thé, G. & Ito, Y., eds, *Nasopharyngeal Cancer: Etiology and Control* (IARC Scientific Publications No. 20), Lyon, IARC, pp. 167-189

Ho, H.C. (1967) Nasopharyngeal carcinoma in Hong Kong. In: Muir, C.S. & Shanmugaratnam, K., eds, *Cancer of the Nasopharynx* (UICC Monograph Series 1), Copenhagen, Munksgaard, pp. 58-63

Ho, J.H.C. (1971a) Genetic and environmental factors in nasopharyngeal carcinoma. In: Nakahara, W., Nishioka, N., Hirayama, T. & Ito, Y., eds, *Recent Advances in Tumor Virology and Immunology*, Baltimore, MD, University Park Press, pp. 275-295

Ho, H.C. (1971b) Incidence of nasopharyngeal cancer in Hong Kong. *UICC Bull. Cancer*, **9**, 305

Ho, J.H.C. (1972) Nasopharyngeal carcinoma (NPC). *Adv. Cancer Res.*, **15**, 57-92

Ho, H.C. (1975) Epidemiology of nasopharyngeal carcinoma. *J. R. Coll. Surg. Edinburgh*, **20**, 223-235

Ho, J.H.C. (1978) An epidemiologic and clinical study of nasopharyngeal carcinoma. *Int. J. Radiat. Oncol. Biol. Phys.*, **4**, 183-198

Ho, J.H.C. (1979) Some epidemiologic observations on cancer in Hong Kong. *Natl Cancer Inst. Monogr.*, **53**, 35-47

Ho, J.H.C., Huang, D.P. & Fong, Y.Y. (1978) Salted fish and nasopharyngeal carcinoma in southern Chinese [Letter to the Editor]. *Lancet*, **ii**, 626

Huang, D.P., Ho, J.H.C. & Gough, T.A. (1978a) Analysis for volatile nitrosamines in salt-preserved foodstuffs traditionally consumed by southern Chinese. In: de Thé, G. & Ito, Y., eds, *Nasopharyngeal Carcinoma: Etiology and Control* (IARC Scientific Publications No. 20), Lyon, IARC, pp. 309-314

Huang, D.P., Ho, J.H.C., Saw, D. & Teoh, T.B. (1978b) Carcinoma of the nasal and paranasal regions in rats fed Cantonese salted marine fish. In: de Thé, G. & Ito, Y., *Nasopharyngeal Carcinoma: Etiology and Control* (IARC Scientific Publications No. 20), Lyon, IARC, pp. 315-328

Huang, D.P., Ho, J.H.C., Webb, K.S., Wood, B.J. & Gough, T.A. (1981) Volatile nitrosamines in salt-preserved fish before and after cooking. *Food Cosmet. Toxicol.*, **19**, 167-171

Hubert, A. (1984) *Le Pain et l'Olive. Aspects de l'Alimentation en Tunisie* [Bread and the olive. Aspects of feeding in Tunisia], Lyon, Centre National de la Recherche Scientifique, Centre Régional de Publication, p. 36

IARC (1978a) *IARC Monographs on the Evaluation of the Carcinogenic Risk of Chemicals to Humans*, Vol. 17, *Some N-Nitroso Compounds*, Lyon, pp. 125-175

IARC (1978b) *IARC Monographs on the Evaluation of the Carcinogenic Risk of Chemicals to Humans*, Vol. 17, *Some N-Nitroso Compounds*, Lyon, pp. 83-124

IARC (1978c) *IARC Monographs on the Evaluation of the Carcinogenic Risk of Chemicals to Humans*, Vol. 17, *Some N-Nitroso Compounds*, Lyon, pp. 177-189

IARC (1978d) *IARC Monographs on the Evaluation of the Carcinogenic Risk of Chemicals to Humans*, Vol. 17, *Some N-Nitroso Compounds*, Lyon, pp. 51-75

IARC (1978e) *IARC Monographs on the Evaluation of the Carcinogenic Risk of Chemicals to Humans*, Vol. 17, *Some N-Nitroso Compounds*, Lyon, pp. 263-280

IARC (1978f) *IARC Monographs on the Evaluation of the Carcinogenic Risk of Chemicals to Humans*, Vol. 17, *Some N-Nitroso Compounds*, Lyon, pp. 313-326

IARC (1978g) *IARC Monographs on the Evaluation of the Carcinogenic Risk of Chemicals to Humans*, Vol. 17, *Some N-Nitroso Compounds*, Lyon, pp. 287-301

IARC (1983a) *IARC Monographs on the Evaluation of the Carcinogenic Risk of Chemicals to Humans*, Vol. 32, *Polynuclear Aromatic Compounds, Part 1, Chemical, Environmental and Experimental Data*, Lyon, pp. 247–261

IARC (1983b) *IARC Monographs on the Evaluation of the Carcinogenic Risk of Chemicals to Humans*, Vol. 32, *Polynuclear Aromatic Compounds, Part 1, Chemical, Environmental and Experimental Data*, Lyon, pp. 211–224

IARC (1983c) *IARC Monographs on the Evaluation of the Carcinogenic Risk of Chemicals to Humans*, Vol. 32, *Polynuclear Aromatic Compounds, Part 1, Chemical, Environmental and Experimental Data*, Lyon, pp. 299–308

IARC (1987) *IARC Monographs on the Evaluation of Carcinogenic Risks to Humans*, Suppl. 7, *Overall Evaluations of Carcinogenicity: An Updating of IARC Monographs Volumes 1 to 42*, Lyon, pp. 58, 60–61, 67–68

Ichinotsubo, D.Y. & Mower, H.F. (1982) Mutagens in dried/salted Hawaiian fish. *J. agric. Food Chem.*, **30**, 937–939

Jeannel, D., Hubert, A., de Vathaire, F., Ellouz, R., Camoun, M., Ben Salem, M., Sancho-Garnier, H. & de Thé, G. (1990) Diet, living conditions and nasopharyngeal carcinoma in Tunisia—a case-control study. *Int. J. Cancer*, **46**, 421–425

Kawabata, T., Uibu, J., Ohshima, H., Matsui, M., Hamano, M. & Tokiwa, H. (1980) Occurrence, formation, and precursors of *N*-nitroso compounds in the Japanese diet. In: Walker, E.A., Griciute, L., Castegnaro, M. & Börzsönyi, M., eds, *N-Nitroso Compounds: Analysis, Formation and Occurrence* (IARC Scientific Publications No. 31), Lyon, IARC, pp. 481–492

Kawabata, T., Matsui, M., Ishibashi, T. & Hamano, M. (1984) Analysis and occurrence of total *N*-nitroso compounds in the Japanese diet. In: O'Neill, I.K., von Borstel, R.C., Miller, C.T., Long, J. & Bartsch, H., eds, *N-Nitroso Compounds: Occurrence, Biological Effects and Relevance to Human Cancer* (IARC Scientific Publications No. 57), Lyon, IARC, pp. 25–31

Kneller, R.W., McLaughlin, J.K., Bjelke, E., Schuman, L.M., Blot, W.J., Wacholder, S., Gridley, G., CoChien, H.T. & Fraumeni, J.F., Jr (1991) A cohort study of stomach cancer in a high-risk American population. *Cancer*, **68**, 672–678

Kolonel, L.N., Ward Hinds, M. & Hankin, J.H. (1980) Cancer patterns among migrant and native-born Japanese in Hawaii in relation to smoking, drinking, and dietary habits. In: Gelboin, H.V., MacMahon, B., Matsushima, T., Sugimura, T., Takayama, S. & Takebe, H., eds, *Genetic and Environmental Factors in Experimental and Human Cancer*, Tokyo, Japan Scientific Societies Press, pp. 327–340

Kolonel, L.N., Nomura, A.M.Y., Hirohata, T., Hankin, J.H. & Hinds, M.W. (1981) Association of diet and place of birth with stomach cancer incidence in Hawaii Japanese and Caucasians. *Am. J. clin. Nutr.*, **34**, 2478–2485

Kolonel, L.N., Nomura, A.M.Y., Hinds, M.W., Hirohata, T., Hankin, J.H. & Lee, J. (1983) Role of diet in cancer incidence in Hawaii. *Cancer Res.*, **43** (Suppl.), 2397s–2402s

Kono, S., Ikeda, M., Tokudome, S. & Kuratsune, M. (1988) A case–control study of gastric cancer and diet in northern Kyushu, Japan. *Jpn. J. Cancer Res. (Gann)*, **79**, 1067–1074

Lanier, A., Bender, T., Talbot, M., Wilmeth, S., Tschopp, C., Henle, W., Henle, G., Ritter, D. & Terasaki, P. (1980) Nasopharyngeal carcinoma in Alaskan Eskimos, Indians, and Aleuts: a review of cases and study of Epstein-Barr virus, HLA, and environmental risk factors. *Cancer*, **46**, 2100–2106

Li, C.-C., Yu, M.C. & Henderson, B.E. (1985) Some epidemiologic observations of nasopharyngeal carcinoma in Guangdong, People's Republic of China. *Natl Cancer Inst. Monogr.*, **69**, 49–52

Lin, P., Zhang, J., Ding, Z. & Cai, H. (1986) Carcinogenic and promoting effects of fish juice, preserved rice and salted dry fish on the forestomach epithelium of mice and esophageal epithelium of rats (Chin.). *Zhonghua Zhongliu Zazhi*, **8**, 332–335

Muir, C., Waterhouse, J., Mack, T., Powell, J. & Whelan, S., eds (1987) *Cancer Incidence in Five Continents, Vol. V* (IARC Scientific Publications No. 88), Lyon, IARC

Nagai, M., Hashimoto, T., Yanagawa H., Yokoyama, H. & Minowa, M. (1982) Relationship of diet to the incidence of esophageal and stomach cancer in Japan. *Nutr. Cancer*, **3**, 257–268

Ning, J.-P., Yu, M.C., Wang, Q.-S. & Henderson, B.E. (1990) Consumption of salted fish and other risk factors for nasopharyngeal carcinoma (NPC) in Tianjin, a low-risk region for NPC in the People's Republic of China. *J. natl Cancer Inst.*, **82**, 291–296

Nomura, A., Yamakawa, H., Ishidate, T., Kamiyama, S., Masuda, H., Stemmermann, G.N., Heilbrun, L.K. & Hankin, J.H. (1982) Intestinal metaplasia in Japan: association with diet. *J. natl Cancer Inst.*, **68**, 401–405

Ohyama, S., Inamasu, T., Ishizawa, M., Ishinishi, N. & Matsuura, K. (1987) Mutagenicity of human urine after the consumption of fried salted salmon. *Food. chem. Toxicol.*, **25**, 147–153

Okonkwo, P.O. & Nwokolo, C. (1978) Aflatoxin B_1: simple procedures to reduce levels in tropical foods. *Nutr. Rep. int.*, **17**, 387–395

Omura, T., Hisamatsu, S., Takizawa, Y., Minowa, M., Yanagawa, H. & Shigematsu, I. (1987) Geographical distribution of cerebrovascular disease mortality and food intakes in Japan. *Soc. Sci. Med.*, **24**, 401–407

O'Neill, I.K., von Borstel, R.C., Miller, C.T., Long, J. & Bartsch, H., eds (1984) *N-Nitroso Compounds: Occurrence, Biological Effects and Relevance to Human Cancer* (IARC Scientific Publications No. 57), Lyon, IARC

Onishi, H., Fuchi, H., Konomi, K., Hidaka, O. & Kamekura, M. (1980) Isolation and distribution of a variety of halophilic bacteria and their classification by salt-response. *Agric. Biol. Chem.*, **44**, 1253–1258

Palli, D., Bianchi, S., Decarli, A., Cipriani, F., Avellini, C., Cocco, P., Falcini, F., Puntoni, R., Russo, A., Vindigni, C., Fraumeni, J.F., Jr, Blot, W.J. & Buiatti, E. (1992) A case–control study of cancers of the gastric cardia in Italy. *Br. J. Cancer*, **65**, 263–266

Parkin, D.M., Srivatanakul, P., Khlat, M., Chenvidhya, D., Chotiwan, P., Insiripong, S., L'Abbé, K.A. & Wild, C.P. (1991) Liver cancer in Thailand. I. A case–control study of cholangiocarcinoma. *Int. J. Cancer*, **48**, 323–328

Poirier, S., Ohshima, H., de Thé, G., Hubert, A., Bourgade, M.C. & Bartsch, H. (1987) Volatile nitrosamine levels in common foods from Tunisia, south China and Greenland, high-risk areas for nasopharyngeal carcinoma (NPC). *Int. J. Cancer*, **39**, 293–296

Poirier, S., Bouvier, G., Malaveille, C., Ohshima, H., Shao, Y.M., Hubert, A., Zeng, Y., de Thé, G. & Bartsch, H. (1989) Volatile nitrosamine levels and genotoxicity of food samples from high-risk areas for nasopharyngeal carcinoma before and after nitrosation. *Int. J. Cancer*, **44**, 1088–1094

Preussmann, R., O'Neill, I.K., Eisenbrand, G., Spiegelhalder, B. & Bartsch, H., eds (1983) *Environmental Carcinogens. Selected Methods of Analysis*, Vol. 6, *N-Nitroso Compounds* (IARC Scientific Publications No. 45), Lyon, IARC

Sen, N.P., Tessier, L., Seaman, S.W. & Baddoo, P.A. (1985) Volatile and nonvolatile nitrosamines in fish and the effect of deliberate nitrosation under simulated gastric conditions. *J. agric. Food Chem.*, **33**, 264–268

Shank, R.C., Wogan, G.N., Gibson, J.B. & Nondasuta, A. (1972) Dietary aflatoxins and human liver cancer. II. Aflatoxins in market foods and foodstuffs of Thailand and Hong Kong. *Food Cosmet. Toxicol.*, **10**, 61–69

Sivaswamy, S.N., Balachandran, B. & Sivaramakrishnan, V.M. (1990) Polynuclear aromatic hydrocarbons in South Indian diet. *Curr. Sci.*, **59**, 480–481

Song, P.J. & Hu, J.F. (1988) *N*-Nitrosamines in Chinese foods. *Food chem. Toxicol.*, **26**, 205–208

Sriamporn, S., Vatanasapt, V., Pisani, P., Yongchaiyudha, S. & Rungpitarangsri, V. (1992) Environmental risk factors for nasopharyngeal carcinoma: a case–control study in northeastern Thailand. *Cancer Epidemiol. Biomarkers Prev.*, **1**, 345–348

Stich, H.F., Chan, P.K.L. & Rosin, M.P. (1982) Inhibitory effects of phenolics, teas and saliva on the formation of mutagenic nitrosation products of salted fish. *Int. J. Cancer*, **30**, 719–724

Tajima, K. & Tominaga, S. (1985) Dietary habits and gastro-intestinal cancers: a comparative case–control study of stomach and large intestinal cancers in Nagoya, Japan. *Jpn. J. Cancer Res. (Gann)*, **76**, 705–716

Tannenbaum, S.R., Bishop, W., Yu, M.C. & Henderson, B.E. (1985) Attempts to isolate *N*-nitroso compounds from Chinese-style salted fish. *Natl Cancer Inst. Monogr.*, **69**, 209–211

de Thé, G., Ho, J.H.C. & Muir, C.S. (1982) Nasopharyngeal carcinoma. In: Evans, A.S., ed., *Viral Infections of Humans. Epidemiology and Control*, New York, Plenum Medical Book Co., pp. 621–652

Topley, M. (1973) Cultural and social factors related to Chinese infant feeding and weaning. In: Field, C.E. & Barber, F.M., eds, *Growing Up in Hong Kong*, Hong Kong, Hong Kong University Press, pp. 56–65

Walker, E.A., Castegnaro, M., Griciute, L. & Lyle, R.E., eds (1978) *Environmental Aspects of N-Nitroso Compounds* (IARC Scientific Publications No. 19), Lyon, IARC

Walker, E.A., Castegnaro, M., Griciute, L. & Börzsönyi, M., eds (1980) *N-Nitroso Compounds: Analysis, Formation and Occurrence* (IARC Scientific Publications No. 31), Lyon, IARC

Wheeler, K.A. & Hocking, A.D. (1988) Water relations of *Paecilomyces variotii*, *Eurotium amstelodami*, *Aspergillus candidus*, and *Aspergillus sydowii*, xerophilic fungi isolated from Indonesian dried fish. *Int. J. Food Microbiol.*, **7**, 73–78

Wibowo, S., Poernomo, A. & Putro, S. (1990) Dried salted fish marketing and distribution in Indonesia. In: Souness, R.A., ed., *FAO Fisheries Report* (No. FAO-FIIU-401 Suppl.), Rome, FAO, pp. 214–218

You, W.-C., Blot, W.J., Chang, Y.-S., Ershow, A.G., Yang, Z.-T., An, Q., Henderson, B.E., Xu, G.-W., Fraumeni, J.F., Jr & Wang, T.-G. (1988) Diet and high risk of stomach cancer in Shandong, China. *Cancer Res.*, **48**, 3518–3523

Yousif, O.M. (1989) Wet-salted freshwater fish (*fassiekh*) production in Sudan. In: Souness, R.A., ed., *FAO Fisheries Report* (No. FAO-FIIU-400 Suppl.), Rome, FAO, pp. 176–180

Yu, M.C. & Henderson, B.E. (1987) Intake of Cantonese-style salted fish as a cause of nasopharyngeal carcinoma. In: Bartsch, H., O'Neill, I.K. & Schulte-Hermann, R., eds, *The Relevance of N-Nitroso Compounds to Human Cancer: Exposures and Mechanisms* (IARC Scientific Publications No. 84), Lyon, IARC, pp. 547–549

Yu, M.C. & Henderson, B.E. (1993) Nasopharynx. In: Schottenfeld, D. & Fraumeni, J.F., Jr, eds, *Cancer Epidemiology and Prevention*, Philadelphia, W.B. Saunders (in press)

Yu, M.C., Ho, J.H.C., Ross, R.K. & Henderson, B.E. (1981) Nasopharyngeal carcinoma in Chinese-salted fish or inhaled smoke? *Prev. Med.*, **10**, 15–24

Yu, M.C., Ho, J.H.C., Lai, S.-H. & Henderson, B.E. (1986) Cantonese-style salted fish as a cause of nasopharyngeal carcinoma: report of a case–control study in Hong Kong. *Cancer Res.*, **46**, 956–961

Yu, M.C., Mo, C.-C., Chong, W.-X., Yeh, F.-S. & Henderson, B.E. (1988) Preserved foods and nasopharyngeal carcinoma: a case–control study in Guangxi, China. *Cancer Res.*, **48**, 1954–1959

Yu, M.C., Nichols, P.W., Zou, X.-N., Estes, J. & Henderson, B.E. (1989a) Induction of malignant nasal cavity tumours in Wistar rats fed Chinese salted fish. *Br. J. Cancer*, **60**, 198–201

Yu, M.C., Huang, T.-B. & Henderson, B.E. (1989b) Diet and nasopharyngeal carcinoma: a case–control study in Guangzhou, China. *Int. J. Cancer*, **43**, 1077–1082

Zaki, M.S., Hassan, Y.M. & Rahma, E.H.A. (1976) Technological studies on the dehydration of the Nile *bolti* fish (*Tilapia nilotica*). *Nahrung*, **20**, 467–474

Zeng, Y. (1985) Seroepidemiological studies on nasopharyngeal carcinoma in China. *Adv. Cancer Res.*, **44**, 121–138

Zou, X.-N., Li, J.-Y., Lu, S.-X., Wang, X.-R., Guo, L.-P. & Lin, Q.-N. (1992) Analysis of volatile N-nitrosamines in salted fish samples from high- and low-risk areas for NPC in China (Abstract). In: O'Neill, I.K. & Bartsch, H., eds, *Nitroso Compounds, Biological Mechanisms, Exposures and Cancer Etiology* (IARC Technical Report No. 11), Lyon, IARC

PICKLED VEGETABLES

1. Exposure Data

1.1 Production and consumption

1.1.1 *Introduction*

Pickling, broadly defined, is the use of brine, vinegar or a spicy solution to preserve and give a unique flavour to a food adaptable to the process. Numerous vegetables and fruits can be pickled not only to preserve them but also to modify their flavour. The categories of pickled products are many, the most common being those of cucumbers and other vegetables; fruits; nuts; relishes of all kinds; cured meats, fish and poultry; and such special products as pickled mushrooms and pickled cherries (Peterson, 1977).

The pickling processes of particular interest in this monograph are the traditional methods used in some parts of China and Japan where elevated risks for oesophageal and gastric cancer have been observed. For example, an unusual variety of pickled vegetables is made in Linxian, China, by fermenting turnips, sweet potato leaves and other vegetables in water without salt or vinegar (Li *et al.*, 1989). Several special processes, with and without salt, are used in preparing certain types of Japanese and Korean vegetable products (Shin, 1978; Itabashi, 1983; Uda *et al.*, 1984; Itabashi, 1985; Itabashi & Takamura, 1985).

1.1.2 *Production/preparation*

Traditionally, pickled vegetables are popular in some areas of China where there are high incidence rates of oesophageal cancer. Among the vegetables commonly prepared in this way are Chinese cabbage, turnip, soya bean, sweet potato, sesame (Yang, 1980), potherb mustard (Zhang *et al.*, 1983) and others. They are prepared each autumn by chopping, washing and dipping briefly in boiling water the roots, stems and/or leaves, as appropriate, and cooling and packing the vegetables tightly in a large earthenware (ceramic) jar. The vegetables are covered with water, a heavy stone is placed on them and they are allowed to ferment for several weeks or months (Yang, 1980).

The leaves of *takana* (*Brassica juncea* L.), a popular cruciferous vegetable in Japan, are mainly processed by salting. The salted products are divided into two types. One is called *shinzuke-takana*, processed to contain 3 or 4% (w/w) salt in the final product. It has a pungent flavour owing to the presence of isothiocyanates, which are formed enzymatically from the corresponding glucosinolates during the salting process. In recent years, the product has been stored under refrigeration or frozen to retain the pungent flavour and its green appearance. The other product, *furuzuke-takana*, contains about 10% (w/w) salt. It is stored for over six months after the salting process, and during this time the salted materials undergo changes in volatile constituents and pigments; the final product has a characteristic flavour and amber colour. Other popular pickled products in Japan are *nozawana-zuke* and *hiroshimana-zuke*, produced by salting the fresh leaves of *nozawana* (*Brassica campestris* L. var. *rapa*) and *hiroshimana* (*B. campestris* L. var. *pekinensis*) to achieve a salt concentration of 3–4% in the final products, similar to *shinzuke-takana*. The products are fermented at 3–12 °C for five days (Uda *et al.*, 1988).

Sunki, a pickle produced in Kiso district, Nagano Prefecture, from the leaves of green vegetables, is prepared without salt owing to lactic acid fermentation by lactobacilli contained in the 'pickling seeds'. The pickling seeds generally used are dried *sunki* pickles produced in the previous year which contain several species of lactobacillus. The pickles are subsequently dried and preserved throughout the year. Some farmers make *sunki* from cooked leaves of *otaki* turnip, not only by adding dried *sunki* as the source of lactobacillus but also by adding wild fruits or berries (Itabashi, 1983, 1985; Itabashi & Takamura, 1985).

Kimchi, spiced, lactic acid-fermented vegetables, are one of the commonest traditional side dishes in the daily meals of Koreans. They are prepared by salting Chinese cabbages and radishes, washing the salted vegetables in fresh water, adding spices and seasonings and then leaving the spiced vegetables to undergo a process of natural lactic acid fermentation. The amount of salt used in the preparation of *kimchi* corresponds to about 10% of the weight of the fresh vegetables, and salting time usually ranges between 8 and 15 h before washing. The seasoning mixture used includes cayenne pepper, garlic, ginger and pickled seafood. Many kinds of microorganisms are involved in *kimchi* fermentation, the principal ones being *Lactobacillus plantarum* and *L. brevis*. *Kimchi* is usually prepared in the home; however, it is also now produced and distributed on a commercial basis in Japan and the USA (Shin, 1978).

In the USA and northern Europe, the manufacture of cucumber pickles consists of a cure in a 10% salt solution, during which fermentation by halophilic (salt-tolerant) bacteria takes place. The curing process takes from 28 to 42 days, and the salt prevents growth of spoilage organisms. When the curing process is completed, the product is placed in a more concentrated salt solution and stored until final processing, which includes immersion in water to remove the salt, the addition of vinegar and a final bath in water that contains calcium chloride, a firming agent and turmeric, a colour enhancer. To make sweet pickles, a spiced sweet vinegar is added to the final soak. For dill pickles, the dill plant or its seeds are used as flavouring. The shelf-life of cucumber pickles is dependent upon the presence of preservatives and, when used, on pasteurization. Vinegar is the most commonly used preservative. Pickled vegetables are packed, for the most part, in sealed glass jars, usually under vacuum (Peterson, 1977).

1.1.3 *Consumption*

In China, pickled vegetables and juice are eaten either as such or cooked in a gruel. During the summer, some of the juice is also consumed as a drink. In some families, pickled vegetables are eaten daily for as many as 9–12 months in a year and are an important part of the diet (Yang, 1980).

In a survey in Japan in 1962–63, consumption of pickled vegetables more than 60 times a month was reported by 60 and 81% of the adult farm populations and 43 and 70% of the adult non-farm populations of Hiroshima and Miyagi prefectures, respectively (Haenszel et al., 1976). In a later survey (Yan, 1989), the dietary habits of Chinese residing in Japan were compared with those of native Japanese. Consumption of pickled vegetables three or more times a week was reported by 35% of 346 Chinese men and by 41% of 288 Chinese women interviewed and by 75% of 8071 Japanese men and 78% of 9932 Japanese women. The average daily per-caput ingestion of various salt-fermented vegetables in Japan was estimated to be about 37 g (Kawabata et al., 1980).

Per-caput consumption of *kimchi* by Koreans is 200–300 g daily (Shin, 1978).

1.2 Chemical composition

1.2.1 *General aspects*

Volatile constituents of the two Japanese pickled products, *nozawana-zuke* and *hiroshimana-zuke*, were studied by gas chromatography with and without mass spectrometry after fractionation into basic, acidic, phenolic and neutral fractions. A total of 57 constituents were identified in the latter three fractions (the basic had almost no odour), consisting of three carbonyls, eight esters, two sulfides, four alcohols, seven phenols, seven nitriles, eight isothiocyanates, 17 hydrocarbons and one acid. Little difference was observed between the two pickled products in the kinds of volatile constituents present; most were found in the neutral fractions where the degradation products of glucosinolates, namely isothiocyanates and nitriles, were the major components, together with some methyl esters of lower-molecular-weight fatty acids (C_{10}–C_{14}) (Uda et al., 1988).

Uda et al. (1984) investigated changes in the relative amounts of volatile compounds in *takana-zuke* five days, three months and six months after salting and after storage at 2–3 °C for three to six months. (Relative amounts were the percentages represented by the peak area of each component out of the total peak area of all the components detected in the acidic, phenolic and neutral fractions, which contained 11 acids, nine phenols, eight esters, four alcohols, two carbonyls, 12 hydrocarbons, two sulfides, four nitriles and eight isothiocyanates.) The relative amounts of isothiocyanates decreased from 83% at five days to 16% following six months' storage. The relative amounts of alcohols, acids, phenols, nitriles and hydrocarbons were increased after three or six months of cold storage.

The main organic acids produced by *kimchi* fermentation are the nonvolatile compounds, lactic acid and succinic acid. They are produced in greater amounts when *kimchi* is fermented for a long period at low temperatures (6–7 °C) than when it is fermented for a short period at high temperatures (22–23 °C). Oxalic, malic, tartaric, fumaric, malonic, maleic and glycolic acids are produced in smaller quantities at low temperatures. Volatile compounds contained in aged *kimchi* are formic acid and acetic acid. Carbon dioxide generated in the process of fermentation gives *kimchi* a tart taste. More acetic acid and carbon dioxide are contained in *kimchi* with a low salt content (~1%) than in that with a high salt content (~3%), and also in *kimchi* fermented at low temperatures (4–5 °C) than in that fermented at high temperatures (20–25 °C) (Shin, 1978).

1.2.2 Compounds present in pickled vegetables

(a) Nitrosamines

Samples of Chinese pickled vegetables fermented in brine were found to contain *N*-nitrosodimethylamine (NDMA) at < 0.1–15 μg/kg (Poirier *et al.*, 1987; Song & Hu, 1988) and *N*-nitrosopyrrolidine at < 0.5–96 μg/kg (Poirier *et al.*, 1987), < 0.1–25.5 ppb (μg/kg) (Song & Hu, 1988) and 62 μg/kg (Poirier *et al.*, 1989). *N*-Nitrosopiperidine was found in one study at < 0.5–14 μg/kg (Poirier *et al.*, 1987) but not in two others (Song & Hu, 1988; Poirier *et al.*, 1989); *N*-nitrosodiethylamine was found in one study at < 0.1–1.1 ppb (μg/kg) (Song & Hu, 1988).

Two of 49 samples of Japanese salt-fermented vegetables contained NDMA, at levels of 1.4–1.5 μg/kg; six contained *N*-nitrosopyrrolidine, at levels of 1.2–32 μg/kg. The authors calculated that the average daily intake of volatile nitrosamines *per caput* from salt-fermented vegetables was 0.002 μg NDMA and 0.04 μg *N*-nitrosopyrrolidine (Kawabata *et al.*, 1980). Kawabata *et al.* (1984) surveyed the occurrence of total *N*-nitroso compounds, total *N*-nitrosamides, volatile *N*-nitrosamines, nitrates and nitrites in the Japanese diet (Table 1). Relatively high levels of total *N*-nitroso compounds were detected in salt-fermented vegetables, the highest being found in *hakusai-zuke* (Chinese cabbage) and in *takuan* (radish roots); the levels of total *N*-nitrosamides were similar. Only trace quantities of volatile *N*-nitrosamines were detected in these vegetable products.

Table 1. Occurrence of total *N*-nitroso compounds (TNC), total *N*-nitrosamides (TNAd), volatile *N*-nitrosamines (VNA), nitrite and nitrate in salt-fermented vegetables in the Japanese diet

Product	TNC (μg NO/kg)	TNAd (μg NO/kg)	VNA (μg/kg)		Nitrite (mg/kg)	Nitrate (mg/kg)
			NDMA	NPYR		
Radish root (*takuan*) (1)	85.0	92.6	–	–	–	–
Radish root (*takuan*) (2)	252.6	266.2	ND	ND	3.43	126.2
Radish root (*takuan ume-zuke*)	110.0	102.2	Trace	0.5	1.73	264.3
Radish root (*bettara-zuke*)	118.1	128.6	Trace	ND	1.03	15.35
Oriental melon dipped in *sake* lees (*nara-zuke*)	5.52	–	Trace	ND	1.96	29.35
Chinese cabbage (*hakusai-zuke*)	2466.8	2325.4	ND	ND	31.0	258.5
Turnip leaves (*nozawana*)	100.0	107.4	ND	Trace	12.9	283.7
Broad-leaved mustard (*takana*)	57.1	49.9	ND	ND	6.47	221.4
Pot herb mustard (*kyona*)	98.0	101.8	ND	ND	14.3	165.4
Mixed fermented vegetables seasoned with soya sauce (*fukujin-zuke*)	7.74	–	ND	ND	2.59	3.70

From Kawabata *et al.* (1984). Limits of detection: TNC and TNAd, 0.5 μg NO/kg; VNA, 0.1 μg/kg; nitrate and nitrite, 0.05 mg/kg. ND, not detected; Trace, 0.1–0.5 μg/kg; –, not analysed

Samples of foods consumed frequently in Kashmir, India, a high-risk area for oesophageal cancer, were analysed for the presence of volatile N-nitrosamines. In eight samples of mixed pickled vegetables, NDMA levels ranged from not detected to 6.1 µg/kg; N-nitrosopiperidine and N-nitrosopyrrolidine were not detected (Siddiqi et al., 1991).

In Tunisia, turnips fermented in brine were found to contain NDMA at 3 µg/kg in one study (Poirier et al., 1987), but none was found in a later study (Poirier et al., 1989) in which N-nitrosopyrrolidine was found at 31 µg/kg.

(b) Roussin red methyl ester

In a study of the etiological factors of oesophageal cancer in Linxian County, North China, a pure compound was isolated from several kinds of pickled vegetables and identified as Roussin red methyl ester (RRME; bis[µ-(methanethiolato)]tetranitrosodiiron; see Table 2). RRME was first synthesized in 1858 but had not previously been reported in Nature. It has been shown to react readily with secondary amines to form nitrosamines both *in vitro* and *in vivo* (Wang et al., 1980; Zhang et al., 1983). Pickled vegetables from Linxian County contained 0.1–4.5 mg/kg RRME and pickles from Beijing contained less than 0.005 mg/kg (Zhang et al., 1983). The compound was detected in 55% of 69 samples of pickles from Linxian (Li et al., 1986).

(c) Flavonoids

Flavonol aglycones were not detected in pickled vegetables prepared by industrial methods in Italy, but flavonol glycosides were found at 31.9 mg/kg in pickled peppers, 1.2 mg/kg in cauliflower and 7.8 mg/kg in carrots. The vegetables had been pickled in a boiling water solution containing 7% acetic acid and 3% sodium chloride (Fieschi et al., 1989).

Quercetin (see IARC, 1983a) and rhamnetin (see Table 2) were identified as the principal mutagenic substances in samples of Japanese pickle. The most mutagenic samples of carrots and radishes, from Akita Prefecture, Japan, contained quercetin at 6.60 mg/g of crude extract and rhamnetin at 1.96 mg/g (Takenaka et al., 1989). Mutagenic substances in Japanese pickled vegetables purchased in Tokyo were isolated and identified as the flavonoids kaempferol (see IARC, 1983b) and isorhamnetin (see Table 2) (Takahashi et al., 1979). The quantities of quercetin and kaempferol found in *takana* increased with increasing duration of pickling. The flavonols contained in the fresh vegetables as glycosides were found to be freed by hydrolysis during pickling (Mizuta & Kanamori, 1983).

(d) Phorbides

The photosensitizers pheophorbide a and pyropheophorbide a, decomposition products of chlorophyll, were found in samples of salted green vegetables; pheophorbide a was also detected in fresh vegetables, but pyropheophorbide a was not. In salted vegetables stored for more than three months, almost all the chlorophyll had been converted to pyropheophorbide a. The concentrations of pheophorbide a and pyropheophorbide a determined in Japanese vegetables are shown in Table 3 (Takeda et al., 1985).

Table 2. Nomenclature and formulae of compounds found in pickled vegetables

Roussin red methyl ester
Chem. Abstr. Serv. Reg. No.: 16071-96-8
Deleted CAS Reg. Nos.: 15696-35-2, 79408-10-9, 110658-92-9
Chem. Abstr. Name: Bis[μ-(methanethiolato)]tetranitrosodiiron
Synonyms: Bis(methanethiolato)tetranitrosodiiron; Roussin's red methyl ester

$C_2H_6Fe_2N_4O_4S_2$ Mol. wt: 325.92

Rhamnetin
Chem. Abstr. Serv. Reg. No.: 90-19-7
Chem. Abstr. Name: 2-(3,4-Dihydroxyphenyl)-3,5-dihydroxy-7-methoxy-4H-1-benzopyran-4-one
Synonyms: C.I. 75690; 7-methoxyquercetin; 7-methylquercetin; 7-O-methylquercetin; quercetin 7-methyl ether; β-rhamnocitrin; 3,3',4',5-tetrahydroxy-7-methoxyflavone; 3,5,3'4'-tetrahydroxy-7-methoxyflavone

$C_{16}H_{12}O_7$ Mol. wt: 316.27

Isorhamnetin
Chem. Abstr. Serv. Reg. No.: 480-19-3
Chem. Abstr. Name: 3,5,7-Trihydroxy-2-(4-hydroxy-3-methoxyphenyl)-4H-1-benzopyran-4-one
Synonyms: C.I. 75680; isorhamnetol; 3'-methoxyquercetin; 3'-methylquercetin; 3'-O-methylquercetin; quercetin 3'-methyl ether; 3,4',5,7-tetrahydroxy-3'-methoxyflavone

$C_{16}H_{12}O_7$ Mol. wt: 316.27

Table 3. Mean concentrations (μg/g) of pheophorbide a and pyropheophorbide a found in fresh and salted vegetables from a Tokyo market

Sample	No. of samples	Storage (days)	Plant part	Pheophorbide a	Pyropheophorbide a
Takana[a] (fresh)	3	0	Leaf	22.5	ND
			Stem	5.9	ND
			Whole	18.4	ND
Takana[a]-zuke	3	20	Leaf	51.4	182.0
			Stem	5.3	2.8
			Whole	26.2	82.7
Takana[a]-zuke	4	90	Leaf	ND	534.8
			Stem	5.5	28.3
			Whole	2.0	235.3
Nozawana[b]-zuke	5	7	Leaf	129.3	235.1
			Stem	2.6	1.6
			Whole	53.3	95.0
Hiroshimana[c]-zuke	5	7	Leaf	140.4	165.5
			Stem	4.1	3.3
			Whole	95.0	111.4

From Takeda et al. (1985); ND, not detected
[a]*Brassica juncea* var. *integrifolia*
[b]*Brassica* spp.
[c]*B. pekinensis*

(e) Isothiocyanates

The steam-volatile isothiocyanates that occur in raw and salted Japanese cruciferous vegetables (*hiroshimana-zuke* (leaves), *nozawana-zuke* (leaves), *hinonakabu-zuke* (roots), *takana-zuke* (leaves), *takanafuru-zuke* (leaves), *zasai-zuke* (stalks)) were investigated by gas chromatography–mass spectrometry. The vegetables, which had undergone autolysis during the process of salting at their unadjusted pH, were subjected to steam distillation. A relatively large percentage of total isothiocyanates was observed in the steam-volatile fractions obtained from the salted vegetables, except from *zasai-zuke*. The isothiocyanates identified included: pentyl, 2-propenyl, 3-butenyl, 4-pentenyl, 2-phenethyl, 4-methylthiobutyl and 5-methylthiopentyl isothiocyanate (Maeda *et al.*, 1979).

(f) Fungi

Chinese pickled vegetables are contaminated by fungi. Among 24 samples analysed in one study, 20 contained *Geotrichum candidum*, some samples contained *Mucor* spp. and yeasts, and a few samples contained *Aspergillus flavus*, *A. niger*, *A. fumigatus*, *A. nidulans* and *Fusarium* spp. (see monographs in this volume) (Yang, 1980).

1.3 Analysis

Selected methods for the analysis of *N*-nitroso compounds in various matrices have been reviewed (Walker *et al.*, 1978, 1980; Bartsch *et al.*, 1982; Preussmann *et al.*, 1983; O'Neill *et al.*, 1984).

A method for the quantitative analysis of RRME in pickled vegetables by gas chromatography–mass spectrometry has been described. The detection limit was 5 μg/kg (Wang *et al.*, 1980; Zhang *et al.*, 1983).

Reverse-phase high-performance liquid chromatography with ultraviolet detection at 415 nm has been used to determine pheophorbide a and pyropheophorbide a in pickled vegetables (Takeda *et al.*, 1985).

2. Studies of Cancer in Humans

2.1 Stomach cancer

Studies on pickled vegetables and stomach cancer have been conducted in high- and low-risk areas in Japan, among high-risk Japanese in Hawaii and in high-risk areas in China. The rates of stomach cancer in Japan are among the highest in the world: the average age-adjusted mortality rates in 1972 were 40.5 per 100 000 in men and 25.5 per 100 000 in women (Hirayama, 1984). High-risk areas in Japan, located in the northern Honshu prefectures, generally show mortality rates which are 25–30% above the Japanese average; the low-risk areas, located in southern Kyushu, show mortality rates about 40% below the Japanese average (see Fig. 1) (Haenszel *et al.*, 1976). Japanese-Americans in Hawaii have age-adjusted incidence rates for stomach cancer (men, 34.0/100 000; women, 15.1/100 000 in 1973–77 (Waterhouse *et al.*, 1982; Hirayama, 1984)) that are lower than the Japanese average. Mortality rates among whites in the USA in 1970 were 9.0/100 000 for men and 4.4/100 000 for women (Sandler & Holland, 1987).

Studies in which the risk for cancer was examined by intake of all pickles (or pickled vegetables) combined may suffer from problems of misclassification of the exposure; therefore, the true effect of the exposure may be diluted. In only a few studies was the type of pickled vegetables included in the investigation stated; in most of the other studies the terms 'pickles', 'pickled vegetables', 'other pickles' and 'salted vegetables' were used, without specification of the food items included or the precise pickling or storing methods.

2.1.1 *Correlation (ecological) studies*

The intake of various foods, as determined in a food consumption survey conducted by the Japanese Ministry of Agriculture in 1958–59, was correlated with stomach cancer mortality rates in 1955 for 46 prefectures in Japan. Positive but nonsignificant associations were observed between mortality from this cancer and intake of pickled radish or other pickles (Hirayama, 1971).

Nagai *et al.* (1982) conducted a nationwide correlation study which covered 1040 census tracts in Japan. Standardized mortality ratios for cancer of the stomach and oesophagus in 1969–74 in the cities, towns and villages covered in the nutrition survey were related to data on household nutrition gathered between 1974 and 1976. In a simple correlation analysis and a multiple regression analysis, consumption of pickled vegetables was positively related to death from stomach cancer in men (standardized partial regression coefficient, $r = 0.152$; $p = 0.01$) but not in women ($r = 0.073$; $p > 0.05$).

Fig. 1. Age-standardized death rates from cancer of the stomach, Japan, 1955

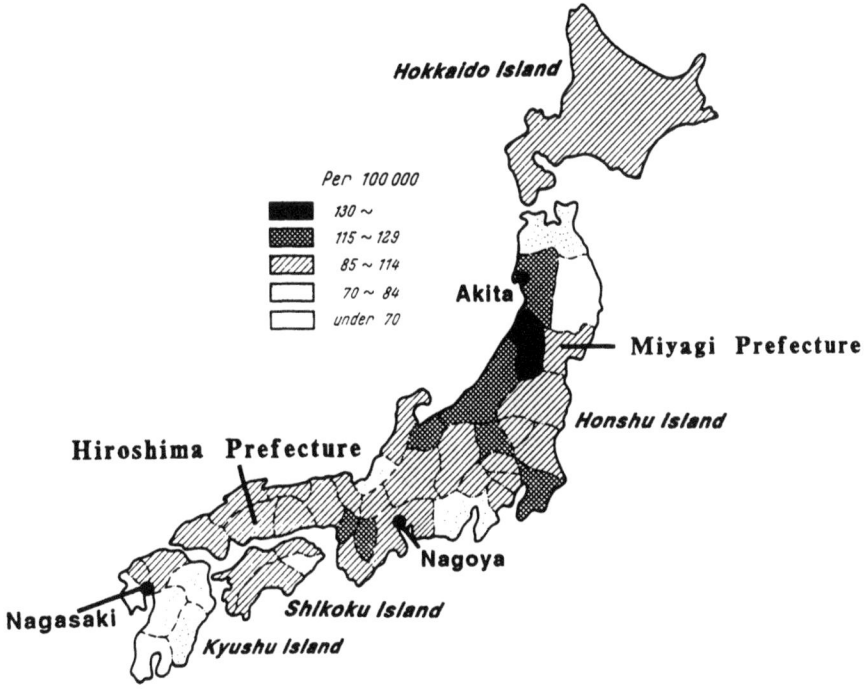

From Hirayama (1967)

Kolonel et al. (1980, 1981, 1983) related incidence rates for stomach cancer among Caucasians in Hawaii, in Japanese migrants from Japan to Hawaii and in Japanese born in Hawaii, all aged 45 years or more, to smoking, drinking and dietary habits. Both the rates of stomach cancer and intake of pickled vegetables were lowest among Caucasians in Hawaii, intermediate among Japanese born in Hawaii and highest among Japanese migrants.

2.1.2 *Cohort studies* (see Table 4)

Hirayama (1971, 1979) initiated a prospective study in 1965 in 29 health centre districts in six prefectures in Japan in which 122 261 men and 142 857 women aged 40 years and over were interviewed about their diet, smoking, drinking and occupational histories. A record-linkage system with death registrations was established for annual follow-up. After 10 years of follow-up, there were 2757 cases of gastric cancer (1823 men, 934 women). Men who ate pickles at every meal had a significantly increased risk for stomach cancer when compared with those who rarely or occasionally ate this food (age-adjusted relative risk, 1.16; $p < 0.01$), but women had no increase in risk associated with intake of pickles at every meal (age-adjusted relative risk, 0.92; $p > 0.05$). [The Working Group noted that the types of vegetables included under 'pickles' were not specified and that it is not clear whether the association with pickle consumption at every meal remained significant when other risk

Table 4. Summary of cohort studies on stomach cancer and consumption of pickled vegetables

Geographical area (reference)	No. of subjects, no. of cases and length of follow-up	Intake level	Relative risk	95% CI or p value	Comments
Japan (Hirayama, 1979)	All subjects: 122 261 M 142 857 F Stomach cancer deaths: 1823 M, 934 F 10 years of follow-up	*Very salted pickles (tsukemono)* Males Rarely or occasionally Every meal Females Rarely or occasionally Every meal Pickles	1.00 1.16 1.00 0.92 No association	$p < 0.01$ $p < 0.05$	No information on how dietary information was collected Types of vegetables included under pickles not specified
Hiroshima, Nagasaki, Japan (Ikeda *et al.*, 1983)	All subjects: 7553 M and F 79 stomach cancer deaths 11 years of follow-up	*Salted pickles*	No significant association; results positive in Hiroshima, negative in Nagasaki.		Part of the Adult Health study; not clear if 'salted pickles' includes pickled vegetables. Information obtained on 11 variables, 5 on diet
Hawaii, USA (Chyou *et al.*, 1990)	All subjects: 8006 Japanese men 111 stomach cancers (361 cancer-free controls) 18 years of follow-up	Pickles	*Mean intake* Cases: 15.7 g/day Controls: 15.6 g/day	p value = 0.97	24-h dietary intake data recoded to allow comparison by specific foods or food groups; p value calculated by comparing mean values, adjusting for age

CI, confidence interval; M, males; F, females

factors (smoking daily, hot tea frequently, fish daily) and protective factors (green-yellow vegetables daily, milk daily) were adjusted for.]

A study by Ikeda et al. (1983) involved 11 203 subjects already enrolled in the Adult Health Study of the Radiation Effects Research Foundation in Hiroshima and Nagasaki between 1968 and 1970. Complete personal histories and information on current dietary and other habits were available for 7553 of these subjects (1781 men and 3341 women in Hiroshima; 965 men and 1466 women in Nagasaki). After 11 years of follow-up, there were 244 deaths from cancer, 79 of which were from stomach cancer. Intake of five foods was assessed in a multivariate analysis, which included radiation exposure; salted pickles were not associated with death from stomach cancer. [The Working Group noted that it is not clear whether 'salted pickles' included only pickled vegetables.]

Chyou et al. (1990) examined the association between consumption of pickles and risk for stomach cancer using 24-h dietary recall data from a case–cohort study of 8006 Japanese men in Hawaii which was initiated between 1965 and 1968. In order to investigate specific foods and food groups (e.g., dried fish and pickles), data from the questionnaire were recoded. After almost complete follow-up of the cohort for 18 years, 111 incident cases of stomach cancer were identified; controls comprised 361 cancer-free men selected from the remainder of the cohort. Cases and controls did not differ in their intake of pickles; the mean intake of pickles was 15.7 and 15.6 g per day for cases and controls, respectively. The authors noted that the consumption of these food items was low and 24-h dietary recall is a crude method. [The Working Group noted that the controls were not selected from the total cohort at its inception.]

2.1.3 *Case–control studies* (see Table 5)

In 1963, the dietary pattern three years earlier was compared between 1524 stomach cancer patients and 3792 control patients, matched by sex, age and occupation in six selected prefectures in Japan (Hirayama, 1967, 1971). The author stated that stomach cancer patients followed 'the conventional diet pattern more frequently than the controls', but results were not presented by intake of specific foods. In order to standardize various host and environmental conditions, a second set of controls was selected to match the study group with respect not only to sex, age and occupation but also to place of residence and personal and family medical histories. Thus, a second analysis was conducted of 652 stomach cancer patients and an equal number of matched controls. Cases and controls did not differ in their intake of pickles; however, consumption of very salty pickles (*tsukemono*) was significantly more frequent among stomach cancer patients than controls in all socioeconomic classes, and the effect was independent of intake of other foods. Crude odds ratios (ORs) [calculated by the Working Group] were 1.0 for consumption of *tsukemono* never or rarely, 1.1 for consumption occasionally, 2.6 for daily and 2.1 at every meal. [The Working Group noted that the design of the study was not clearly explained. No adjustment was made for other potential confounders. Types of vegetables included under the category of pickles were not specified].

Haenszel et al. (1972) interviewed 220 Japanese patients with stomach cancer (96% histologically confirmed) and 440 hospital controls in Honolulu, Hawaii, over a six-year period starting in 1963. Two controls were interviewed for each case, namely the next oldest and next youngest Japanese of the same sex in the same hospital service but excluding those with diagnoses of gastric ulcers, other diseases of the stomach or other cancers of the digestive system. Controls and cases were not matched on place of birth but were compared in their frequency of intake of five types of pickled vegetables: Japanese radish, *hakusai* (cabbage), plum, cucumber and aubergine. Intake of pickled Japanese radish, *hakusai*, plum and cucumber was each associated with a small, nonsignificant increase in risk for stomach cancer. Those who reported the highest level of intake (≥ 21 times per month) of pickled Japanese radish and *hakusai* showed about two-fold increased risks for stomach cancer when compared to non-users. When intake of all five types of pickled vegetables was combined, there was a significant, two-fold increase in risk for the highest frequency of intake (two or more types of vegetables consumed ≥ 21 times per month) when compared to non-use. The increase was observed among both migrants and second-generation Japanese-Americans, but the effect was stronger among the migrants. The observation remained unchanged when the analyses accounted for intake of western vegetables.

Using the same study design and a comparable questionnaire, Haenszel *et al.* (1976) conducted a companion study in Japan in 1962–65 which included 783 stomach cancer patients (367 from Hiroshima, 416 from Miyagi prefectures) and 1566 hospital controls. Hiroshima prefecture was selected because a high proportion of the original migrants to Hawaii came from this and neighbouring prefectures, whereas Miyagi prefecture was selected because of its very high rates of stomach cancer and the persistence in the area of traditional Japanese customs. Two controls were interviewed for each patient, chosen by the same selection criteria as in the Hawaiian study (Haenszel *et al.*, 1972). In Hiroshima, the risk for stomach cancer was significantly lower among people who consumed pickled vegetables frequently, whereas in Miyagi it was higher, but not significantly so. The authors suggested that lack of background variation in food habits may explain their failure to detect case–control differences in consumption patterns.

Tajima and Tominaga (1985) conducted a case–control study in Nagoya, a high-risk area for stomach cancer in Japan. Ninety-three stomach cancer patients (59 males, 34 females) and 186 controls (111 males, 75 females) were identified at Aichi Cancer Center Hospital in 1981–83. Hospital controls free of cancer were matched individually to each case on sex, age (within five years) and time of interview; however, 18% of the controls had chronic gastritis, 20% had gastric and duodenal ulcers and a further 6% had other conditions of the gastrointestinal tract. Cases and controls were interviewed by one of the authors of the study using a standardized questionnaire which covered 34 food items and other habits in the one or two years before the patient went to hospital. Frequent intake of pickled *hakusai* was associated with an increased risk for stomach cancer: the sex- and age-adjusted ORs were 1.40 for intake 1–3 times per week and 3.04 ($p < 0.01$) for intake ≥ 4 times per week as compared to less than once a week. Intake of other pickles was not associated with risk for stomach cancer, the corresponding adjusted ORs being 1.01 and 0.87. [The Working Group noted that the method of selection of controls may have introduced bias, which would have affected the results in an unpredictable direction.]

Kono *et al.* (1988) conducted a study in a low-risk area for stomach cancer in rural northern Kyushu, Japan. Between 1979 and 1982, 139 newly diagnosed cases of gastric

Table 5. Summary of case–control studies of stomach cancer and consumption of pickled vegetables

Geographical area (reference)	No. of cases and no. and type of controls	Intake	Odds ratio	95% CI or p value	Comments
USA, Hawaii (Haenszel et al., 1972)	220 Japanese cases 440 hospital controls	*All pickled vegetables combined* ≥ 3 pickled vegetables vs < 2 *No. of pickled vegetables used ≥ 21 times/month* ≥ 1 ≥ 1 only } vs non-use ≥ 2	1.09 1.47 1.02 2.7	$p > 0.05$ $p < 0.05$ $p > 0.05$ $p < 0.05$	Questionnaire included a number of Japanese and western foods. Odds ratio adjusted for sex and place of birth. Intake of Japanese radish, *hakusai*, plum, cucumber and aubergine was studied.
Japan, Hiroshima and Miyagi (Haenszel et al., 1976)	783 cases 1566 hospital controls	*Pickled vegetables used* > 60 times ≥ 1/month 1–2/month } vs non-use ≥ 3/month	0.79 0.87 0.76	$p < 0.05$ $p > 0.05$ $p < 0.05$	Questionnaire included a number of Japanese and western foods. Results were positive in Miyagi and negative in Hiroshima.
Japan, Nagoya (Tajima & Tominaga, 1985)	93 cases 186 hospital controls	*Pickled hakusai* < 1/week 1–3/week ≥ 4/week ≥ 1/week *Other pickles* < 1/week 1–3/week ≥ 4/week	1.0 1.40 3.04 2.03 1.0 1.01 0.87	 $p < 0.01$ $p < 0.01$ Not significant	Questionnaire included a list of specific foods and eating frequencies. Results on *hakusai* pickles somewhat stronger in 56–70 age group than in 40–55 age group.
Japan, northern Kyushu (Kono et al., 1988)	139 cases 2574 hospital controls 278 general population controls	*Pickled green vegetables* 1/day ≥ 2/day } vs ≤ 1–3/week *Pickled radish* 1/day ≥ 2/day } vs ≤ 1–3/week	1.1 1.1 0.9 1.1	Not significant Not significant	Questions pertained to current habits only (year preceding interview); 25 food items. Some differences in the administration of questionnaire, even though all interviewers were trained together.

Table 5 (contd)

Geographical area (reference)	Subjects (cases, controls) control type	Intake	Odds ratio	95% CI or p value	Comments
China, Shandong Province (You et al., 1988)	564 cases 1131 population controls	*Salted vegetables* daily vs < daily	1.1	0.7–1.8	Questionnaire included 85 food items; intake habits during 1965 and 1980 studied Adjusted for sex, age and family income
China, Heilongjiang Province (Hu et al., 1988)	241 cases 241 hospital controls	*Salted vegetables*	No association		Questionnaire included 25 food items; intake habits during 1965 and 1980 studied

CI, confidence interval

cancer (88% histologically confirmed) were identified among 4729 subjects who had visited a referral centre in the area for the diagnosis of gastrointestinal diseases. Cases were compared with two sets of controls: 2574 hospital controls free of gastrointestinal disease and 278 general population controls who were similar to the cases by sex, year of birth and residence. Two different groups of people, using a standard questionnaire, interviewed patients before diagnostic procedures at the referral centre and the general population about dietary habits in the year preceding the interview or before a change in dietary habits. Intake of pickles was not associated with risk for gastric cancer when either hospital or population controls were used. The OR associated with intake of pickled green vegetables was 1.1 for once a day and 1.1 for two or more times a day, as compared to one to three times a week or less; the ORs for pickled radish were 0.9 and 1.1, respectively.

You et al. (1988) conducted a large, population-based case–control study of stomach cancer in Linqu, a rural county in Shandong Province, China (see Fig. 2). The annual age-adjusted stomach cancer mortality rates (China standard) in this area were 55 and 19 per 100 000 for men and women, respectively, in 1980–82. Over a 2.5-year period during 1984–86, 685 incident cases of stomach cancer were identified among long-term ($\geq$ 10 years) residents at county and commune hospitals in Linqu and neighbouring Yidu County. Interviews were completed with 564 stomach cancer patients and with 1131 of 1132 controls who were randomly selected from age and sex strata of the Linqu population. A structured questionnaire was used to gather information on demographic variables, medical history, occupation, smoking and other items, including frequency of consumption and portion size of 85 food items consumed during 1980 and 1965. No strong association was found between intake of salted vegetables and risk for stomach cancer. When daily intake of salted vegetables was compared with less than daily intake, the OR was 1.1 (95% confidence interval, 0.7–1.8).

A case–control study of stomach cancer, comprising patients from two hospitals in Heilongjiang Province in north-east China (see Fig. 2), was conducted by Hu et al. (1988). A total of 241 (170 men, 71 women) patients newly diagnosed in 1985–86 with histologically confirmed stomach cancer and 241 control patients with non-neoplastic diseases were interviewed. Interviewers, trained for the study, asked about prior disease history, economic status, occupation, tobacco and alcohol intake and average frequency and quantity of intake of about 25 food items at about the time of interview and in 1966. The risk for stomach cancer was not associated with intake of salted vegetables in either time period. [The Working Group noted that the study was presented in insufficient detail.]

2.1.4 *Precursor lesions*

Nomura et al. (1982) examined the association between dietary factors and intestinal metaplasia, a precursor lesion that has been strongly associated with stomach cancer. The study was conducted in Akita Prefecture, a high-risk area in Japan; 387 subjects from a rural community responded to a dietary questionnaire (including 33 specific food items) and had gastric biopsy specimens taken from five sites in the stomach. Each specimen was assessed for the presence of intestinal metaplasia and was given a grade from 0 to 3; grade 0 designated no intestinal metaplasia, whereas a score of 3 denoted intestinal metaplasia of the entire specimen. The questionnaire asked about occupation, smoking, alcohol consumption and recent diet. Eating pickled plums (*umeboshi*) was not related to metaplasia score in men [details not given] and was negatively related to metaplasia score in women (standardized regression coefficient, $r = -0.142$; $p = 0.045$).

Fig. 2. Areas of China in which intake of pickled vegetables has been studied in relation to cancer

From Li *et al.* (1980a)

2.2 Oesophageal cancer

2.2.1 *Correlation studies*

The relationship between intake of pickled vegetables and the occurrence of oesophageal cancer has been examined in a few studies conducted in Japan and China. In the correlation study conducted in Japan described on p. 90, Nagai *et al.* (1982) reported no association between eating pickled vegetables and risk for oesophageal cancer. In China, the association between pickled vegetables and oesophageal cancer was evaluated in 30 communes in high-risk provinces (Henan, Hebei and Shanxi and the northwestern region of Sichuan Province) and in eight communes in a low-risk province (Guangdong) (see Fig. 2). When the commune was used as a unit, a positive correlation was seen between mortality from oesophageal cancer and frequency of intake of pickled vegetables in the high-risk provinces but not in the low-risk province (Li *et al.*, 1980a; Yang, 1980).

2.2.2 Case–control studies (see Table 6)

Li et al. (1989) conducted a large population-based case–control study of cancers of the oesophagus and gastric cardia in Linxian County, Henan Province, China (Fig. 2), to investigate the role of pickled vegetables. The study area is a rural county in north-central China with one of the world's highest rates of mortality from these tumours, the age-adjusted rates (World) being 310/100 000 in northern communes and 180/100 000 in southern communes. These rates exceed the national levels by nearly 10 times (Chinese National Cancer Control Office/Nanjing Institute of Geography, 1979). Interviews were completed with 1244 patients with cancer of oesophagus or gastric cardia and with 1314 population-based controls; the response rate was 98% for cases and 100% for controls. Eligible cases included all diagnoses of oesophageal cancer among residents of this area, aged 35–64 years, identified from all hospitals in the County over a 21-month period in 1984–85. Controls were randomly selected from the general population of Linxian and were similar to cases on age and sex but were free of cancer. All interviews were conducted by trained interviewers using a structured questionnaire which asked about occupation, smoking, diet and food preparation and storage methods in the late 1950s and late 1970s. No association was seen between intake of pickled vegetables during adult life and risk for oesophageal or gastric cardia cancer in either males or females. The results were similar when the analyses were conducted separately for the two tumour sites and separately for the lower-risk communes in the south and the higher-risk communes in the north. Exposure was more prevalent in lower-risk than in higher-risk communes. [The Working Group noted that the results for pickled vegetables were not adjusted for consumption of other foods that showed significant associations with risk.]

Cheng et al. (1992) conducted a hospital-based case–control study of oesophageal cancer in Hong Kong. Cases were consecutive admissions of patients with histologically confirmed diagnosis of oesophageal cancer to surgical departments of four general hospitals in Hong Kong during a 22-month period between 1989 and 1990. Of the 461 patients, 400 were successfully interviewed. For each case, four controls were selected—two from the same surgical departments (excluding those with tobacco- or alcohol-related malignancies) and two from the general practice clinic in which the case was initially seen. Both types of controls were matched to cases by age (within five years) and sex. A total of 1598 controls were interviewed from the 1682 individuals selected. Both cases and controls were interviewed in the hospital/clinic by trained interviewers using a structured questionnaire, which asked about smoking and drinking habits, tea and coffee consumption, personal and family history and dietary intake of 22 food items at age 20–30; dietary intake of these items was also recorded prior to onset of illness for cases and in the current diet for controls. Intake of pickled vegetables was associated with a significantly increased risk for oesophageal cancer, and a clear dose–response relationship was seen. The effect of pickled vegetables remained statistically significant when other significant risk factors (preference for hot drinks, tobacco and alcohol intake) and factors associated with a reduction in risk (such as citrus fruits, any green leafy vegetables and high level of education) were accounted for in the analysis. The adjusted ORs for pickled vegetable intake were 1.66 for < 1/month, 1.51 for

Table 6. Summary of case–control studies of oesophageal cancer and consumption of pickled vegetables

Geographical area (reference)	No. of cases and no. and type of controls	Intake	Odds ratio	95% CI	Comments
China, Linxian County (Li et al., 1989)	1244 cases (782 oesophagus, 397 gastric cardia, 54 mixed, 11 unknown) 1314 population controls	*Pickled vegetables* Men ≤ 1/day vs never > 1/day vs never Women ≤ 1/day vs never > 1/day vs never	0.6 0.9 0.9 1.1	0.4–0.9 0.6–1.3 0.5–1.5 0.7–1.7	Questionnaire asked about intake of 72 food items during the late 1950s and 1970s Results are for 1970s intake in high-risk northern communes. Results were similar for lower-risk southern communes, and for intake in the 1950s. Results adjusted for age and (for men) smoking
Hong Kong (Cheng et al., 1992)	400 cases 1598 hospital controls	*Pickled vegetables* < 1/year < 1/month 1–3/month 1–3/week 4–6/week Daily or more	1.00 1.66 1.51 2.09 6.27 13.12	0.06–4.43 0.67–3.39 0.92–4.47 2.03–19.39 2.57–66.93	Questionnaire asked about 22 food items. Odds ratios adjusted for preference for hot drinks, tobacco, alcohol intake and citrus fruits, any green leafy vegetables and high level of education
China, Shanxi Province (Wang et al., 1992)	326 cases, 396 population controls	*Pickled vegetable juice* Yangcheng County sometimes, often vs never, rarely Linqen County sometimes, often vs never, rarely	3.6 11.6	1.1–18.4 6.3–21.6	Questionnaire asked about 84 food items. Pickled vegetable intake not associated with risk

CI, confidence interval

1–3 times/month, 2.09 for 1–3 times/week, 6.27 for 4–6 times/week and 13.12 for daily or more frequent consumption compared to once per year.

Wang *et al.* (1992) conducted a case–control study of oesophageal cancer in two counties in Shanxi, northern China (see Fig. 2). Cases and controls were derived from a high-risk (Yangcheng County; age-adjusted mortality rate in men, 143.1) and an intermediate-risk (Linfen County; age-adjusted mortality rate in men, 33.1) county. Eligible patients were identified in the major tumour hospital in each study area over a 13-month period between 1988 and 1989. Population controls were selected by frequency matching to cases on gender, age and residence within the two counties. A total of 326 oesophageal cancer patients and 396 controls were interviewed (210 cases and 203 controls in Yangcheng, 116 cases and 193 controls in Linfen), representing about 15–20% of oesophageal cancer patients in the two areas. Twenty-eight interviewers in Yangcheng and 15 in Linfen were trained to administer the structured questionnaire, which assessed relevant factors only in the more recent time period, after 1977. The factors included family and medical history, occupation, smoking and alcohol intake, and dietary habits (84 food items, methods of preparation and eating). Results were presented separately for the two counties. No significant increase in risk was associated with intake of pickled vegetables [details not presented], but a significantly increased risk was associated with 'some or often' consumption of pickled vegetable juice compared to 'never or rare' intake. The OR (adjusted for age, gender, farm/non-farm occupation) for pickle juice consumption was 3.6 (95% confidence interval, 1.1–18.4) in Yangcheng and 11.6 (6.3–21.6) in Linfen. [The Working Group noted that the results for consumption of pickled vegetable juice were not adjusted for other dietary factors (millet gruel, millet soup with noodles, boiled vegetables, mouldy foods, soya beans) or for non-dietary factors (family history of oesophageal cancer).]

2.3 Nasopharyngeal cancer (see Table 7)

Yu *et al.* (1986) conducted a case–control study, described in detail in the monograph on salted fish (see pp. 55–56), of nasopharyngeal carcinoma (NPC) in Hong Kong. Interviewers enquired about intake of several salt-preserved foods, including salted vegetables (*mui choi*, *harm choi*), roots (*choi po*, *harm choi*, *chung choi*) and olives (*larm gok*). Consumption of salted mustard greens (*mui choi*) was significantly associated with risk for NPC, but the result was no longer significant when intake of salted fish was accounted for [details not given].

Yu *et al* (1988) conducted a case–control study, described in detail in the monograph on salted fish (see pp. 56–57), of NPC in Yulin Prefecture, Guangxi Autonomous Region, China, where salted vegetables are frequently consumed early in life. Intake of several salted vegetables during weaning, between the ages of one and two and around the age of 10 was assessed. Consumption of salted mustard greens during weaning, between the ages of one and two and around the age of 10 was each associated with an increased risk for NPC. The strongest association was found with exposure during weaning. Similarly, intake of *chung choi* during weaning and at the age of 10 was each associated with a significantly increased risk for NPC after adjustment for intake of salted fish; exposure during weaning gave the higher risk.

In the study conducted by Jeannel *et al.* (1990) in Tunisia, described in detail in the monograph on salted fish (see p. 63), intake of pickled vegetables during childhood, current

Table 7. Summary of case–control studies of nasopharyngeal carcinoma (NPC) and consumption of salted/pickled vegetables

Geographical area (reference)	No. of cases and no. and type of controls	Intake	Odds ratio	95% CI or p value	Comments
Hong Kong (Yu et al., 1986)	250 cases 250 (friends) controls	Salted mustard greens (mui choi)	Not significant (detailed data not presented)		Questionnaire asked about salt-preserved foods, including salted vegetables (mui choi, harm choi), roots (choi po, harm choi, chung choi) and olives (larm gok). Muichoi significantly associated with risk for NPC; not significant after adjustment for salted fish intake
China, Guangxi Province (Yu et al., 1988)	231 cases 231 population controls 128 case and 174 control mothers also interviewed	*During weaning* Salted mustard greens Yes vs No	5.4	1.2–23.8	Questionnaire asked about salt-preserved foods, including salted mustard greens, salted cabbage, *chung choi*, salted radish and salted olives during weaning, between ages one and two and around 10 years of age
		Chung choi Yes vs No	2.0	1.3–3.2	
					Only intake of salted mustard greens and *chung choi* during weaning remained significant in multivariate analysis.
		Between 1–2 years of age Salted mustard greens Monthly vs rarely	1.6	0.5–4.7	
		At age of 10 years Salted mustard greens Monthly } vs rarely Weekly	0.9 1.3	0.4–2.0 0.3–6.7	
		Chung choi Monthly } vs rarely Weekly Daily	1.4 1.8 3.2	0.7–2.9 0.9–3.6 0.6–17.6	

Table 7 (contd)

Geographical area (reference)	No. of cases and no. and type of controls	Intake	Odds ratio	95% CI or p value	Comments
Tunisia (Jeannel et al., 1990)	80 cases 160 neighborhood controls	*During childhood* Pickled vegetables 1/month–2/week } vs < 1/week > 2/week Fungus on pickles Yes vs No *Year before diagnosis* Pickled olives 1/month–2/week } vs < 1/week > 2/week Fungus on pickles Yes vs No	6.7 2.2 3.8 9.7 8.7 3.3	p = 0.04 p, NR p = 0.03 p = 0.02 p, NR p = 0.03	Dietary habits during the first month of life, weaning, childhood, adolescence and the year prior to diagnosis of NPC were assessed. Homemade pickled vegetables containing a layer of fungus Results presented are significant and adjusted only for living conditions. Results not significant when adjusted for other dietary factors

CI, confidence interval; NR, not reported

intake of pickled olives and recent and childhood intake of mouldy pickles was each associated with a significantly increased risk for NPC. These results remained significant when living conditions (presumably to control for socioeconomic status) were accounted for; however, the results on pickled foods were no longer significant when intake of other foods (stewing mixture, snacks with *harissa*, orange-flower water and castor plant poultices) were accounted for in the analysis.

3. Studies of Cancer in Experimental Animals

3.1 Oral administration

3.1.1 *Pickled vegetables*

(a) Mouse

In a review, papillomas were reported to have developed after 143 days of treatment in the forestomachs of a group of mice [number, sex, age and strain unspecified] each given a concentrated fluid of pickles (about 50 ml weekly) by oral gavage (Li *et al.*, 1980a,b). [The Working Group noted the incomplete reporting of the results.]

(b) Rat

In a review, it was reported that 39 Wistar rats [sex and age unspecified] were fed extracted or concentrated fluid from pickled vegetables [doses unspecified] for 330–730 days. One developed an adenocarcinoma of the glandular stomach, four had fibrosarcomas of the liver and another had an angioendothelioma of the thoracic wall. No tumour was noted in control rats [number unspecified] (Li *et al.*, 1980a,b). [The Working Group noted the incomplete reporting of the results.]

3.1.2 *Roussin red methyl ester*

Mouse

Gastric intubation of mice [number, sex, strain and age unspecified] with Roussin red methyl ester (RRME) [purity unspecified] at 4 mg twice weekly alone or in combination with sarcosine ethyl ester (50 mg) was reported to have induced epithelial hyperplasia of the oesophagus and forestomach; papillomas of the forestomach developed within 194–269 days [incidences unspecified] (Li *et al.*, 1980a,b). [The Working Group noted the incomplete reporting of the results.]

In a further review, RRME administered to mice at 8 mg/week by intubation was reported to have induced epithelial hyperplasia of the upper digestive tract and papillomas of the forestomach after 135–615 days; no malignant neoplasm was found (Li & Cheng, 1984). [The Working Group noted the incomplete reporting of the results]

3.2 Administration with known carcinogens

3.2.1 *Oral administration*

(a) Mouse

Groups of 30 or 40 male mice [strain and age unspecified], weighing 20–25 g, were given RRME with or without pretreatment with *N*-nitrosomethyl-*N*-benzylamine (NMBzA).

Thirty control animals each received 2 mg RRME (purity, ~95%) dissolved in peanut oil six times a week; a second group of 30 mice received 1 mg/kg NMBzA in peanut oil by gavage three times a week; a third group, of 40 mice, received RRME following the NMBzA treatment. Papillomas of the forestomach were found in 4/19 mice in the group receiving NMBzA and in 15/29 mice in the group receiving both NMBzA and RRME, after 131 days; no such tumour was found in controls (0/11). The authors suggested that RRME promotes tumorigenesis initiated by NMBzA in the mouse forestomach (Lu et al., 1985). [The Working Group noted that the effective numbers of animals, especially in the group receiving RRME alone, were insufficient.]

Groups of female mice [strain unspecified], weighing 20–22 g, were given RRME with or without pretreatment with N-nitrososarcosine ethyl ester (nitrite plus sarcosine ethyl ester). Mice in group 1 were each given 1.5 mg RRME [purity unspecified] in peanut oil by gavage three times a week until the end of the experiment; mice in group 2 were gavaged with 10 ml/kg bw of 3% nitrite in distilled water followed by 10 ml/kg bw of 20% sarcosine ethyl ester in distilled water; mice in group 3 received RRME after the nitrite and sarcosine ethyl ester treatments. No forestomach carcinoma was found in group 1 (0/6) after 150–200 days, but these tumours were found in 16/39 mice in group 3 after 90–200 days (41%; $p < 0.01$) and in 4/42 mice in group 2 after 90–300 days (9.5%) (Lin et al., 1986). [The Working Group noted the incomplete reporting of the study.]

(b) Rat

Groups of female Wistar rats weighing 100–120 g received RRME with or without prior exposure to NMBzA or nitrite plus sarcosine ethyl ester. In 15 female Wistar rats given 20 mg/kg bw RRME in peanut oil by gavage three times a week until the end of the experiment, no tumour developed after 400–820 days. In a group of 19 rats intubated daily with 1 mg/kg bw NMBzA in peanut oil for six days, one papilloma of the oesophagus was observed. When 29 rats were intubated with RRME after NMBzA, five papillomas and one carcinoma of the oesophagus were observed after 400–820 days. In a further group of 19 rats intubated with RRME following administration by gavage of 3% nitrite in distilled water and 10 ml/kg bw of 20% sarcosine ethyl ester in distilled water, seven times every other day, three papillomas and 12 oesophageal carcinomas were found after 350–520 days ($p < 0.01$). In 19 rats intubated with nitrite and the ethyl ester only, 10 oesophageal papillomas but no carcinoma were found (Lin et al., 1986). [The Working Group noted that the initial number of animals in each group was not recorded.]

In a study reported as a short communication, a group of Wistar rats [initial number, sex and age unspecified] received 25 mg N-nitrosodiethylamine once a week by gavage for two successive weeks and, starting from week 4, 6 mg RRME twice a week for 114–669 days. A control group of 25 rats was treated with the nitrosamine alone for two weeks. Four forestomach papillomas (4/10) and four forestomach dysplasias developed in the treated group, and one papilloma (1/11) and two dysplasias (2/11) developed in the control group. Such lesions tended to develop earlier in the group treated with N-nitrosodiethylamine plus RRME than in those given the nitrosamine alone (Liu & Li, 1989). [The Working Group noted the small number of effective animals.]

3.2.2 *Skin application*

Mouse

In a study reported as a short communication, groups of female BALB/c mice, aged seven to nine weeks, received applications of 7,12-dimethylbenz[*a*]anthracene at 100 nmol (2.6 µg) dissolved in 0.2 ml acetone on hairless skin. Three doses of RRME (150 nmol [49 µg], 300 nmol [98 µ] and 450 nmol [147 µg]) were subsequently given twice a week for 37 weeks. Skin papillomas and carcinomas appeared only in mice that received the two higher doses of RRME. No skin tumour was found in the 20 mice treated with 150 nmol RRME; 20 skin papillomas and three carcinomas occurred in 25 mice given 300 nmol (12%), and 24 skin papillomas and five carcinomas developed in 20 mice given 450 nmol (25%). The latent period for appearance of the first skin papillomas was 17 weeks in mice that received 450 nmol RRME and 21 weeks in those given 300 nmol. No skin tumour developed in 20 mice treated with RRME in acetone alone (Liu & Li, 1989).

4. Other Relevant Data

4.1 Absorption, distribution, metabolism and excretion

No data were available to the Working Group.

4.2 Toxic effects

4.2.1 *Humans*

Ingestion of 30 oz (about 850 g) of Japanese pickle (*fukujinzuke*; assorted vegetables pickled in soya sauce) or vinegared gherkins by three volunteers over a three-day period caused marked changes in gastric surface epithelium and gastric pits. The abnormalities consisted of loss of cellular mucus, nuclear enlargement, prominent nuclear chromatin and an increased number of mitotic figures (MacDonald *et al.*, 1967).

4.2.2 *Experimental systems*

Rats were administered combinations of proline, RRME and nitrite orally. Ingestion of freshly synthesized RRME resulted in a small increase in the urinary excretion of *N*-nitrosoproline. The effect was, however, much smaller than that of an equivalent amount of nitrite. When an old, partially decomposed sample of RRME was administered, more proline was nitrosated than with an equivalent amount of nitrite, but the nitrosating derivatives of RRME were not identified (Croisy *et al.*, 1984).

4.3 Reproductive and developmental toxicity

No data were available to the Working Group.

4.4 Genetic and related effects

4.4.1 *Humans*

No data were available to the Working Group.

4.4.2 *Experimental systems* (see also Table 8)

Methanol and methanol–chloroform extracts of pickled vegetables from Linxian County, Henan Province, China, induced gene mutation in bacteria. In a single study, a dichloromethane extract of pickled vegetables from the same region induced cell transformation in Syrian hamster embryo cells *in vitro*. The extract slightly increased the induction of sister chromatid exchange in Syrian and Chinese hamster cells and of gene mutation in Chinese hamster lung cells *in vitro*.

Organic solvent extracts of Japanese pickles and purified fractions of the extracts induced gene mutation in bacteria. Kaempferol, quercetin and isorhamnetin were identified as major mutagenic components of these fractions (Takahashi *et al.*, 1979).

5. Summary of Data Reported and Evaluation

5.1 Exposure data

Traditional processes for pickling vegetables in some regions of China, Japan and Korea involve fermentation of local vegetables, with or without salting. Such preparations are often eaten daily or several times a week. Among the many compounds found at low levels in pickled vegetables are *N*-nitrosamines and Roussin red methyl ester, which reacts with secondary amines to form *N*-nitrosamines.

5.2 Human carcinogenicity data

A cohort study from Japan suggests that intake of pickled vegetables is positively associated with risk for stomach cancer, but further cohort studies from Japan and Hawaii do not support an association. The methods used to determine dietary intake differed in these studies, and the types of pickled vegetables included may also have differed.

Seven case–control studies of stomach cancer have been conducted that included data on consumption of pickled vegetables. Three conducted in Japan gave negative results and another gave positive results. One study of Japanese in Hawaii showed an association, but two conducted in China did not.

A large case–control study of oesophageal cancer in Hong Kong showed a significant dose–response relationship between consumption of pickled vegetables and oesophageal cancer, after potential confounding factors were taken into account. A study in a high- and an intermediate-risk area in China showed an association with consumption of pickled vegetable juice, although there was no association with consumption of pickled vegetables; a population-based study in a high-risk area of northern China also gave negative results for pickled vegetables.

Intake of salted/pickled vegetables (leafy vegetables, roots and olives) has been investigated in two case–control studies of nasopharyngeal carcinoma from China and in one from Tunisia. One of these studies, from Guangxi, China, showed a significant association with eating salted/pickled vegetables.

Table 8. Genetic and related effects of pickled vegetables

Test system	Result without exogenous metabolic system	Result with exogenous metabolic system	Dose (LED/HID; standardized to weight of original material)[a]	Reference
Chinese pickles				
SA0, *Salmonella typhimurium* TA100, reverse mutation	–	+	12 mg (MeOH extract)	Takahashi et al. (1979)
SA0, *Salmonella typhimurium* TA100, reverse mutation	–	+	230 mg (CHCl$_3$:MeOH extract)	Takahashi et al. (1979)
SA0, *Salmonella typhimurium* TA100, reverse mutation	–	+	2.8 g pickle (ether extract)	Lu et al. (1981)
SA9, *Salmonella typhimurium* TA98, reverse mutation	–	+	23 mg (MeOH extract)	Takahashi et al. (1979)
SA9, *Salmonella typhimurium* TA98, reverse mutation	–	+	230 mg (CHCl$_3$:MeOH extract)	Takahashi et al. (1979)
SA9, *Salmonella typhimurium* TA98, reverse mutation	–	(+)	0.28 g pickle (ether extract)	Lu et al. (1981)
G9H, Gene mutation, Chinese hamster lung V79 cells, *hprt* locus	–	0	50 mg (DCM extract)	Cheng et al. (1980)
SIC, Sister chromatid exchange, Chinese hamster lung V79 cells *in vitro*	(+)	0	50 mg (DCM extract)	Cheng et al. (1980)
SIS, Sister chromatid exchange, Syrian hamster embryo cells *in vitro*	(+)	0	100 mg (DCM extract)	Cheng et al. (1980)
TFS, Cell transformation, Syrian hamster embryo cells, focus assay	+	0	150 mg (DCM extract)	Cheng et al. (1980)
TCM, Cell transformation, C3H 10T½ mouse cells	–	0		
Japanese pickles				
SA0, *Salmonella typhimurium* TA100, reverse mutation	–	–	580 mg (MeOH extract)	Takahashi et al. (1979)
SA0, *Salmonella typhimurium* TA100, reverse mutation	–	–	2000 mg (CHCl$_3$:MeOH extract)	Takahashi et al. (1979)
SA0, *Salmonella typhimurium* TA100, reverse mutation	–	+	~100 g[b]	Takenaka et al. (1989)
SA9, *Salmonella typhimurium* TA98, reverse mutation	–	+	58 mg (MeOH extract)	Takahashi et al. (1979)
SA9, *Salmonella typhimurium* TA98, reverse mutation	+	+	1000 mg (CHCl$_3$:MeOH extract)	Takahashi et al. (1979)
SA9, *Salmonella typhimurium* TA98, reverse mutation	0	+	~100 g[b]	Takahashi et al. (1979)
SA9, *Salmonella typhimurium* TA98, reverse mutation	–	+	~100 g[b]	Takenaka et al. (1989)

+, positive; (+), weakly positive; –, negative; 0, not tested

[a]MeOH, methanol; CHCl$_3$, chloroform; ether, diethyl ether; DCM, dichloromethane
[b]High-performance liquid chromatography fraction of CHCl$_3$:MeOH extract

Two correlation studies carried out in Japan and one carried out in Hawaii suggest a relationship between consumption of pickled vegetables and stomach cancer, but the results are not completely consistent. The results of correlation studies on oesophageal cancer were also inconsistent.

No data were available on pickled vegetables made elsewhere in the world.

5.3 Animal carcinogenicity data

No adequate study on the carcinogenicity of pickled vegetables to experimental animals was available to the Working Group.

5.4 Other relevant data

In a single study, extracts of pickled vegetables from northern China induced morphological transformation of Syrian hamster embryo cells in culture. Extracts of pickled vegetables from northern China and Japan are mutagenic to bacteria.

5.5 Evaluation[1]

There is *limited evidence* in humans for the carcinogenicity of pickled vegetables as prepared traditionally in Asia.

There is *inadequate evidence* in experimental animals for the carcinogenicity of pickled vegetables.

Overall evaluation

Pickled vegetables (traditional Asian) are *possibly carcinogenic to humans (Group 2B).*

6. References

Bartsch, H., Castegnaro, M., O'Neill, I.K. & Okada, M., eds (1982) *N-Nitroso Compounds: Occurrence and Biological Effects* (IARC Scientific Publications No. 41), Lyon, IARC

Cheng, S.-J., Sala, M., Li, M.H., Wang, M.-Y., Pot-Deprun, J. & Chouroulinkov, I. (1980) Mutagenic, transforming and promoting effect of pickled vegetables from Linxian County, China. *Carcinogenesis*, 1, 685–692

Cheng, K.K., Day, N.E., Duffy, S.W., Lam, T.H., Fok, M. & Wong, J. (1992) Pickled vegetables in the aetiology of oesophageal cancer in Hong Kong Chinese. *Lancet*, 339, 1314–1318

Chinese National Cancer Control Office/Nanjing Institute of Geography (1979) *Atlas of Cancer Mortality in the People's Republic of China*, Shanghai, China Map Press

---

[1]For definition of the italicized terms, see Preamble, pp. 26–29.

Chyou, P.-H., Nomura, A.M.Y., Hankin, J.H. & Stemmermann, G.N. (1990) A case–cohort study of diet and stomach cancer. *Cancer Res.*, **50**, 7501–7504

Croisy, A., Ohshima, H. & Bartsch, H. (1984) Nitrosating properties of bis-methylthio-diiron-tetranitrosyl (Roussin's red methyl ester), a nitroso compound isolated from pickled vegetables consumed in northern China. In: O'Neill, I.K., von Borstel, R.C., Miller, C.T., Long, J. & Bartsch, H., eds, *N-Nitroso Compounds: Occurrence, Biological Effects and Relevance to Human Cancer* (IARC Scientific Publications No. 57), Lyon, IARC, pp. 327–335

Fieschi, M., Codignola, A. & Luppi Mosca, A.M. (1989) Mutagenic flavonol aglycones in infusions and in fresh and pickled vegetables. *J. Food Sci.*, **54**, 1492–1495

Haenszel, W., Kurihara, M., Segi, M. & Lee, R.K.C. (1972) Stomach cancer among Japanese in Hawaii. *J. natl Cancer Inst.*, **49**, 969–988

Haenszel, W., Kurihara, M., Locke, F.B., Shimuzu, K. & Segi, M. (1976) Stomach cancer in Japan. *J. natl Cancer Inst.*, **56**, 265–274

Hirayama, T. (1967) The epidemiology of cancer of the stomach in Japan with special reference to the role of diet. In: Harris, R.J.C., ed., *Pathogenesis and Epidemiology of Cancer of the Stomach* (UICC Monograph Series Vol. 10), New York, Springer Verlag, pp. 37–49

Hirayama, T. (1971) Epidemiology of stomach cancer. *Gann Monogr. Cancer Res.*, **11**, 3–19

Hirayama, T. (1979) The epidemiology of gastric cancer in Japan. In: Pfeiffer, C.J., ed., *Gastric Cancer Etiology and Pathogenesis*, New York, Gerhard Witzstrock, pp. 60–82

Hirayama, T. (1984) Epidemiology of stomach cancer in Japan with special reference to the strategy for the primary prevention. *Jpn. J. clin. Oncol.*, **14**, 159–168

Hu, J., Zhang, S., Jia, E., Wang, Q., Liu, S., Liu, Y., Wu, Y. & Cheng, Y. (1988) Diet and cancer of the stomach: a case–control study in China. *Int. J. Cancer*, **41**, 331–335

IARC (1983a) *IARC Monographs on the Evaluation of the Carcinogenic Risk of Chemicals to Humans*, Vol. 31, *Some Food Additives, Feed Additives and Naturally Occurring Substances*, Lyon, pp. 213–229

IARC (1983b) *IARC Monographs on the Evaluation of the Carcinogenic Risk of Chemicals to Humans*, Vol. 31, *Some Food Additives, Feed Additives and Naturally Occurring Substances*, Lyon, pp. 171–178

Ikeda, M., Yoshimoto, K., Yoshimura, T., Kono, S., Kato, H. & Kuratsune, M. (1983) A cohort study on the possible association between broiled fish intake and cancer. *Gann*, **74**, 640–648

Itabashi, M. (1983) Studies on the Japanese pickles *sunki*. 3. Dietetic components of *mizuna* pickled by *sunki*-pickling method (Jpn.). *Nippon Shokuhin Kogyo Gakkaishi*, **30**, 719–721

Itabashi, M. (1985) Studies on the Japanese pickles *sunki*. 5. *Sunki* pickles using pickle starter as the source of lactic acid bacteria (Jpn.). *Nippon Shokuhin Kogyo Gakkaishi*, **32**, 124–126

Itabashi, M. & Takamura, N. (1985) Studies on the Japanese pickles *sunki*. 9. *Sunki*-pickles prepared by adding fruits of '*zumi*' and '*yamabudo*' (Jpn.). *Nippon Shokuhin Kogyo Gakkaishi*, **32**, 859–863

Jeannel, D., Hubert, A., de Vathaire, F., Ellouz, R., Camoun, M., Ben Salem, M., Sancho-Garnier, H. & de Thé, G. (1990) Diet, living conditions and nasopharyngeal carcinoma in Tunisia—a case–control study. *Int. J. Cancer*, **46**, 421–425

Kawabata, T., Uibu, J., Ohshima, H., Matsui, M., Hamano, M. & Tokiwa, H. (1980) Occurrence, formation and precursors of *N*-nitroso compounds in the Japanese diet. In: Walker, E.A., Griciute, L., Castegnaro, M. & Börszönyi, M., eds, *N-Nitroso Compounds: Analysis, Formation and Occurrence* (IARC Scientific Publications No. 31), Lyon, IARC, pp. 481–492

Kawabata, T., Matsui, M., Ishibashi, T. & Hamano, M. (1984) Analysis and occurrence of total N-nitroso compounds in the Japanese diet. In: O'Neill, I.K., von Borstel, R.C., Miller, C.T., Long, J. & Bartsch, H., eds, N-*Nitroso Compounds: Occurrence, Biological Effects and Relevance to Human Cancer* (IARC Scientific Publications No. 57), Lyon, IARC, pp. 25-31

Kolonel, L.N., Ward Hinds, M. & Hankin, J.H. (1980) Cancer patterns among migrant and native-born Japanese in Hawaii in relation to smoking, drinking, and dietary habits. In: Gelboin, H.V., MacMahon, B., Matsushima, T., Sugimura, T., Takayama, S. & Takebe, H., eds, *Genetic and Environmental Factors in Experimental and Human Cancer*, Tokyo, Japan Scientific Societies Press, pp. 327-340

Kolonel, L.N., Nomura, A.M.Y., Hirohata, T., Hankin, J.H. & Ward Hinds, M. (1981) Association of diet and place of birth with stomach cancer incidence in Hawaii Japanese and Caucasians. *Am. J. clin. Nutr.*, 34, 2478-2485

Kolonel, L.N., Nomura, A.M.Y., Ward Hinds, M., Hirohata, T., Hankin, J.H. & Lee, J. (1983) Role of diet in cancer incidence in Hawaii. *Cancer Res.*, 43 (Suppl.), 2397s-2402s

Kono, S., Ikeda, M., Tokudome, S. & Kuratsune, M. (1988) A case-control study of gastric cancer and diet in northern Kyushu, Japan. *Jpn. J. Cancer Res. (Gann)*, 79, 1067-1074

Li, M.-H. & Cheng, S.-J. (1984) Carcinogenesis of esophageal cancer in Linxian, China. *Chin. med. J.*, 97, 311-316

Li, M.-H., Li, P. & Li, P.-J. (1980a) Recent progress in research on esophageal cancer in China. *Adv. Cancer Res.*, 33, 173-249

Li, M.-H., Lu, S.-H., Ji, C., Wang, Y., Wang, M., Cheng, S. & Tian, G. (1980b) Experimental studies on the carcinogenicity of fungus-contaminated food from Linxian County. In: Gelboin, H.V., MacMahon, B., Matsushima, T., Sugimura, T., Takayama, S. & Takebe, H., eds, *Genetic and Environmental Factors in Experimental and Human Cancer*, Tokyo, Japan Scientific Societies Press, pp. 139-148

Li, M.-H., Ji, C. & Cheng, S.-J. (1986) Occurrence of nitroso compounds in fungi-contaminated foods: a review. *Nutr. Cancer*, 8, 63-68

Li, J.-Y., Ershow, A.G., Chen, Z.-J., Wacholder, S., Li, G.-Y., Guo, W., Li, B. & Blot, W.J. (1989) A case-control study of cancer of the esophagus and gastric cardia in Linxian. *Int. J. Cancer*, 43, 755-761

Lin, P., Lu, S., Zhang, J. & Ding, Z. (1986) Carcinogenic and promoting effects of Roussin red methyl ester (RRME) on the forestomach epithelium of mice and esophageal epithelium of rats, and its inhibition with retinamide and vitamin C (Chin.). *Chin. J. Oncol.*, 8, 405-408

Liu, J.-G. & Li, M.-H. (1989) Roussin red methyl ester, a tumor promoter isolated from pickled vegetables. *Carcinogenesis*, 10, 617-620

Lu, S.-H., Camus, A.-M., Tomatis, L. & Bartsch, H. (1981) Mutagenicity of extracts of pickled vegetables collected in Linshien county, a high-incidence area for esophageal cancer in northern China. *J. natl Cancer Inst.*, 66, 33-36

Lu, S.-X., Lin, P.-Z., Li, F.M., Wang, Y.-L. & Wang, M.-Y. (1985) Promoting effect of Roussin red methyl ester (RRME) in the pickled vegetable in Linxian county on the forestomach epithelial of mice (Chin.). *Chin. J. Oncol.*, 7, 241-243

MacDonald, W.C., Anderson, F.H. & Hashimoto, S. (1967) Histological effect of certain pickles on the human gastric mucosa: a preliminary report. *Can. med. Assoc. J.*, 96, 1521-1525

Maeda, Y., Ozawa, Y. & Uda, Y. (1979) Steam volatile isothiocyanates of raw and salted cruciferous vegetables (Jpn.). *Nippon Nôgeikagaku Kaishi*, 53, 261-268

Mizuta, M. & Kanamori, H. (1983) Time-course of mutagenicity due to flavonols in a pickled vegetable. *Mutat. Res.*, **122**, 287–291

Nagai, M., Hashimoto, T., Yanagawa, H., Yokoyama, H. & Minowa, M. (1982) Relationship of diet to the incidence of esophageal and stomach cancer in Japan. *Nutr. Cancer*, **3**, 257–268

O'Neill, I.K., von Borstel, R.C., Miller, C.T., Long, J. & Bartsch, H., eds (1984) *N-Nitroso Compounds: Occurrence, Biological Effects and Relevance to Human Cancer* (IARC Scientific Publications No. 57), Lyon, IARC

Nomura, A., Yamakawa, H., Ishidate, T., Kamiyama, S., Masuda, H., Stemmermann, G.N., Heilbrun, L.K. & Hankin, J.H. (1982) Intestinal metaplasia in Japan: association with diet. *J. natl Cancer Inst.*, **68**, 401–405

Peterson, M.S. (1977) Pickles. In: Desrosier, N.W., ed., *Elements of Food Technology*, Westport, CT, Avi Publications, pp. 690–691

Poirier, S., Ohshima, H., de Thé, G., Hubert, A., Bourgade, M.C. & Bartsch, H. (1987) Volatile nitrosamine levels in common foods from Tunisia, South China, and Greenland, high-risk areas for nasopharyngeal carcinoma (NPC). *Int. J. Cancer*, **39**, 293–296

Poirier, S., Bouvier, G., Malaveille, C., Ohshima, H., Shao, Y.M., Hubert, A., Zeng, Y., de Thé, G. & Bartsch, H. (1989) Volatile nitrosamine levels and genotoxicity of food samples from high-risk areas for nasopharyngeal carcinoma before and after nitrosation. *Int. J. Cancer*, **44**, 1088–1094

Preussmann, R., O'Neill, I.K., Eisenbrand, G., Spiegelhalder, B. & Bartsch, H., eds (1983) *Environmental Carcinogens, Selected Methods of Analysis*, Vol. 6, N-Nitroso Compounds (IARC Scientific Publications No. 45), Lyon, IARC

Sandler, R.S. & Holland, K.L. (1987) Trends in gastric cancer sex ratio in the United States. *Cancer*, **59**, 1032–1035

Shin, D.H. (1978) Preservation of vegetables in the Republic of Korea: the processing of *kimchi*. In: *International Forum on Appropriate Industrial Technology, New Dehli/Anand, India, 20–30 November 1978* (Report No. ID/WG. 282/89; US NTIS PB-297205), Vienna, UNIDO

Siddiqi, M.A., Tricker, A.R., Kumar, R., Fazili, Z. & Preussmann, R. (1991) Dietary sources of *N*-nitrosamines in a high-risk area for oesophageal cancer—Kashmir, India. In: O'Neill, I.K., Chen, J. & Bartsch, H., eds, *Relevance to Human Cancer of* N-Nitroso Compounds, Tobacco Smoke and Mycotoxins (IARC Scientific Publications No. 105), Lyon, IARC, pp. 210–213

Song, P.J. & Hu, J.F. (1988) *N*-Nitrosamines in Chinese foods. *Food chem. Toxicol.*, **26**, 205–208

Tajima, K. & Tominaga, S. (1985) Dietary habits and gastro-intestinal cancers: a comparative case-control study of stomach and large intestinal cancers in Nagoya, Japan. *Jpn. J. Cancer Res. (Gann)*, **76**, 705–716

Takahashi, Y., Nagao, M., Fujino, T., Yamaizumi, Z. & Sugimura, T. (1979) Mutagens in Japanese pickle identified as flavonoids. *Mutat. Res.*, **68**, 117–123

Takeda, Y., Uchiyama, S. & Saito, Y. (1985) High performance liquid chromatography of pheophorbide a and pyropheophorbide a in salted vegetables and chlorella. *J. Food Hyg. Soc. Jpn.*, **26**, 56–60

Takenaka, S., Sera, N., Tokiwa, H., Hirohata, I. & Hirohata, T. (1989) Identification of mutagens in Japanese pickles. *Mutat. Res.*, **223**, 35–40

Uda, Y., Ikawa, H., Ishibashi, O. & Maeda, Y. (1984) Studies on the flavor of salted cruciferous vegetables. I. Volatile constituents of processed *takana* (*takana-zuke*) and their changes during cold storage. *Nippon Shokuhin Kogyo Gakkaishi*, **31**, 371–378

Uda, Y., Suzuki, K. & Maeda, Y. (1988) Volatile constituents of pickled cruciferous vegetables, *Brassica campestris* var. *rapa* and *B. campestris* var. *pekinensis* (*nozawana-zuke* and *hiroshimana-zuke*). *Nippon Shokuhin Kogyo Gakkaishi*, **35**, 352–359

Walker, E.A., Castegnaro, M., Griciute, L. & Lyle, R.E., eds (1978) *Environmental Aspects of N-Nitroso Compounds* (IARC Scientific Publications No. 19), Lyon, IARC

Walker, E.A., Castegnaro, M., Griciute, L. & Börzsönyi, M., eds (1980) *N-Nitroso Compounds: Analysis, Formation and Occurrence* (IARC Scientific Publications No. 31), Lyon, IARC

Wang, G.-H., Zhang, W.-X. & Chai, W.-G. (1980) The identification of Roussin red methyl ester—a product isolated from pickled vegetables. *Adv. Mass Spectrom.*, **8B**, 1369–1374

Wang, Y.-P., Han, X.-Y., Su, W., Wang, Y.-L., Zhu, Y.-W., Sasaba, T., Nakachi, K., Hoshiyama, Y. & Tagashira, Y. (1992) Esophageal cancer in Shanxi Province, People's Republic of China: a case–control study in high and moderate risk areas. *Cancer Causes Control*, **3**, 107–113

Waterhouse, J., Muir, C., Shanmugaratnam, K. & Powell, J., eds (1982) *Cancer Incidence in Five Continents, Vol. IV* (IARC Scientific Publications No. 42), Lyon, IARC, pp. 634–635

Yan, S. (1989) A socio-medical study of adult diseases related to the life style of Chinese in Japan (Jpn). *Nippon Eiseigaku Zasshi*, **44**, 877–886

Yang, C.S. (1980) Research on esophageal cancer in China: a review. *Cancer Res.*, **40**, 2633–2644

You, W.-C., Blot, W.J., Chang, Y.-S., Ershow, A.G., Yang, Z.-T., An, Q., Henderson, B.E., Xu, G.-W., Fraumeni, J.F., Jr & Wang, T.-G. (1988) Diet and high risk of stomach cancer in Shandong, China. *Cancer Res.*, **48**, 3518–3523

Yu, M.C., Ho, J.H.C., Lai, S.-H. & Henderson, B.E. (1986) Cantonese-style salted fish as a cause of nasopharyngeal carcinoma: report of a case–control study in Hong Kong. *Cancer Res.*, **46**, 956–961

Yu, M.C., Mo, C.-C., Chong, W.-X., Yeh, F.-S. & Henderson, B.E. (1988) Preserved foods and nasopharyngeal carcinoma: a case–control study in Guangxi, China. *Cancer Res.*, **48**, 1954–1959

Zhang, W.-X., Xu, M.-S., Wang, G.-H. & Wang, M.-Y. (1983) Quantitative analysis of Roussin red methyl ester in pickled vegetables. *Cancer Res.*, **43**, 339–341

CAFFEIC ACID

1. Exposure Data

1.1 Chemical and physical data

1.1.1 Synonyms, structural and molecular data

Chem. Abstr. Serv. Reg. No.: 331-39-5
Chem. Abstr. Name: 3-(3,4-Dihydroxyphenyl)-2-propenoic acid
Synonyms: Caffeic acid; 5(4)-(2-carboxyethenyl)-1,2-dihydroxybenzene; 4-(2′-carboxyvinyl)-1,2-dihydroxybenzene; 3,4-dihydroxybenzeneacrylic acid; 3,4-dihydroxycinnamic acid; 3-(3,4-dihydroxyphenyl)propenoic acid; 3-(3,4-dihydroxyphenyl)-2-propenoic acid

$C_9H_8O_4$ Mol. wt: 180.15

1.1.2 Chemical and physical properties

(a) *Description*: Yellow prisms or plates from water (Lide, 1991)
(b) *Melting-point*: Decomposes at 225 °C (Lide, 1991)
(c) *Solubility*: Sparingly soluble in cold water; very soluble in hot water and cold ethanol (Budavari, 1989)
(d) *Stability*: Caffeic acid exists in *cis* and *trans* forms, *trans* being the predominant naturally occurring form (Janssen Chimica, 1991). Solutions of caffeic acid and its derivatives (e.g., chlorogenic and isochlorogenic acids) are unstable in sunlight and ultraviolet light. The *trans* form of caffeic acid is partially converted to the *cis* form, which in turn is partially converted to the lactone, aesculetin (Grodzinska-Zachwieja *et al.*, 1973; Hartley & Jones, 1975; Borges & Pinto, 1989).
(e) *Reactivity*: Caffeic acid inhibited the formation of *N*-nitrosodimethylamine *in vitro* from simulated gastric juice and nitrite and *in vivo* in rats given aminopyrine and nitrite (Kuenzig *et al.*, 1984). With lower concentrations of caffeic acid than of reactants, caffeic acid can, however, catalyse nitrosamine formation (Dikun *et al.*, 1991).

1.1.3 Trade names, technical products and impurities

Caffeic acid is not known to be a significant commercial product. It is available as the *trans* isomer in research quantities at purities ranging from 97% to > 99%; the *cis* isomer of caffeic acid is not available commercially (Fluka Chemie AG, 1990; Riedel-de Haen, 1990; Janssen Chimica, 1991; Lancaster Synthesis, 1991; TCI America, 1991; Aldrich Chemical Co., 1992).

1.1.4 *Analysis*

Since ultraviolet light partially converts the *trans* isomers of several cinnamic acids, including caffeic acid, to their respective *cis* forms (Kahnt, 1967), it is important in any study of cinnamic acid derivatives in plant materials to have an accurate, rapid method for the determination of both the *cis* and *trans* isomers. Hartley and Jones (1975) separated *cis* from *trans* isomers of caffeic acid as their trimethylsilyl derivatives by gas chromatography with flame ionization detection. Conkerton and Chapital (1983) reported the separation of *cis* from *trans* isomers of caffeic acid by reverse-phase high-performance liquid chromatography (HPLC), using ultraviolet and electrochemical detection in series. Borges and Pinto (1989) reported the separation of *cis*- from *trans*-caffeic acids and aesculetin by isocratic HPLC with ultraviolet detection at 350 nm. Hydroxycinnamic acids, including caffeic acid, were separated from maize tissue samples on a reverse-phase HPLC column with ultraviolet detection at 254 nm (Hagerman & Nicholson, 1982).

An improved capillary gas chromatography procedure with flame ionization detection was developed for the analysis of nonvolatile organic acids and fatty acids in flue-cured tobacco hydrolysates. Two major nonvolatile organic acids, citric and malic acids, and two minor acids, quinic and caffeic acids, were readily quantified as their trimethylsilyl derivatives in an aqueous extract of tobacco (Court & Hendel, 1986).

Caffeic acid has been determined in grape juice, apple juice and pear juice using reverse-phase HPLC with diode array detection (Spanos & Wrolstad, 1990a,b; Spanos *et al.*, 1990).

Ficarra *et al.* (1990) reported the use of HPLC with diffuse infrared reflectance spectroscopy for qualitative and quantitative analysis of naturally occurring phenolic compounds, including caffeic acid, in the medicinal plant *Crataegus oxyacantha* L.

1.2 Production and use

1.2.1 *Production*

No information was available to the Working Group about the worldwide production, trade or consumption of caffeic acid.

1.2.2 *Use*

Several studies have been done to develop drugs based on caffeic acid for use in the treatment of asthma and allergies (Koshihara *et al.*, 1984; Murota & Koshihara, 1985). The antibacterial effect of caffeic acid and of its oxidation products was studied in four bacteria, *Staphylococcus aureus*, *Streptococcus faecalis*, *Salmonella typhimurium* and *Escherichia coli*. Bactericidal activity appeared only after oxidation of caffeic acid and under alkaline conditions (Cuq & Jaussan, 1991).

1.3 Occurrence

1.3.1 *Plants*

Caffeic acid occurs naturally in a wide range of plants, free and in various combined forms (Conkerton & Chapital, 1983). Caffeic acid and chlorogenic acids are constituents of numerous species, including Umbelliferae, Cruciferae, Cucurbitaceae, Polygonaceae, Compositae, Labiatae, Solanaceae, Leguminosae, Saxifragaceae, Caprifoliaceae, Theaceae and Valerianaceae (Herrmann, 1956; Litvinenko *et al.*, 1975).

Caffeic acid has been identified in plants used for medicinal purposes, including *Davallia mariesii* Moore (Davalliaceae), a fern used in Korean folk medicine for the treatment of the common cold, neuralgia and stomach cancer and in China as a traditional medicine for treatment of lumbago, rheumatism, toothache and tinnitus (Cui *et al.*, 1990); the roots of *Carissa spinarum* L. (Apocynaceae), a thorny, evergreen shrub used medicinally in India as a purgative and for the treatment of rheumatism (Raina *et al.*, 1971); the flowers of *Ixora javanica*, also used in Indian medicine, as an antitumour agent, gastric sedative, intestinal antiseptic and astringent (Nair & Panikkar, 1990); *Centaurium umbellatum* Gil. (Gentianaceae), a medicinal plant used in numerous countries combined with other plants (Hatjimanoli & Debelmas, 1977); and *Artemisia rubripes* Nakai, a Chinese plant used medically (Koshihara *et al.*, 1984).

Caffeic acid has been found in the flowers, leaves and buds of the medicinal plant *Crataegus oxyacantha* L. (a Rosaceae with cardiovascular effects) (Ficarra *et al.*, 1990); in the flowers of *Tussilago farfara* L. (an antispasmodic) with caffeoyl tartric acid (Didry *et al.*, 1980); in the essential oil of flowers of *Cytisus scoparius* L. (Kurihara & Kikuchi, 1980); in the leaves of *Melissa officinalis* L. (a Labiatae which inhibits viral development and tumour cell division) with chlorogenic acid (Chlabicz & Gałasiński, 1986) and in timothy grass (*Phleum pratense* L.) with chlorogenic acid isomers (Mino & Harada, 1974); in the seeds of *Argyreia speciosa* Sweet (elephant creeper, a Convolvulaceae with hypotensive and spasmolytic activity (Agarwal & Rastogi, 1974); in the essential oil of *Foeniculum vulgare* (Umbelliferae) (Trenkle, 1971); in the herb *Veronica chamaedrys* L. (Światek *et al.*, 1971); and in the roots of *Arctium lappa* L. (burdock, a Compositae or Asteraceae used as a diuretic) with chlorogenic acid (Leung, 1980).

1.3.2 *Fruits, vegetables and seasonings*

Caffeic acid is present in a variety of fruits, vegetables and seasonings, predominantly in the form of ester conjugates, including chlorogenic acids (esters of caffeic acid and quinic acid) and related compounds. These conjugates may be hydrolysed upon ingestion, leading to variable uptakes of caffeic acid. Table 1 summarizes the levels of caffeic acid and its conjugates in various fruits and vegetables, as determined by thin-layer chromatography following enzymic, acid and/or alkaline hydrolysis.

Caffeic acid (free and conjugated) has been found at concentrations > 1000 ppm (mg/kg) in thyme, basil, aniseed, caraway, rosemary, tarragon, marjoram, savory, sage, dill and absinthe (Ames *et al.*, 1991). Other agricultural products that have been found to contain caffeic acid and conjugates include sweet potatoes (Hayase & Kato, 1984), sunflower seeds and meal (Pomenta & Burns, 1971; Felice et al., 1976), soya beans (Pratt & Birac, 1979), tobacco (Andersen & Vaughn, 1970), spinach, red peppers, apricots, coconut and rolled oats (Kusnawidjaja et al., 1969; Ames et al., 1991).

Table 1. Occurrence of caffeic acid and its conjugates in various fruit and vegetables

Botanical species	Plant part	Concentration as caffeic acid (mg/kg, fresh weight)	Type of hydrolysis
Vegetables			
Bean, broad bush	Hulls	12–14	Enzymatic
	Unripe fruit	< 0.5–9	Enzymatic
Beetroot, red	Whole vegetable	5–17	Enzymatic
	Outside	5	Enzymatic
	Heart	4	Enzymatic
Beetroot, sugar	Whole vegetable	3–4	Enzymatic
Broccoli	Florets	8	Enzymatic
		10	Enzymatic followed by alkaline/acid
Brussels sprouts	–	34–44	Enzymatic
		35–50	Alkaline/acid
Cabbage, Chinese	Outer leaves	11–42	Enzymatic
		52	Enzymatic followed by alkaline/acid
	Inner leaves	4–11	Enzymatic
		11	Enzymatic followed by alkaline/acid
Red	Outer leaves	6–11	Enzymatic
		16–24	Alkaline/acid
	Head	12–16	Enzymatic
		16–17	Alkaline/acid
Savoy	Outer leaves	9–14	Enzymatic
		14–36	Alkaline/acid
	Head	4–7	Enzymatic and alkaline/acid
White	Outer leaves	< 0.5–31	Enzymatic
		< 0.5–62	Alkaline/acid
	Head	< 0.5–10	Enzymatic
		< 0.5–12	Alkaline/acid
Carrot	Whole vegetable	18–96	Enzymatic
	Rind	27–141	Enzymatic
	Central cylinder	8–73	Enzymatic
Cauliflower	Leaves	9–29	Enzymatic
		58	Alkaline/acid
		90	Enzymatic followed by alkaline/acid
	Florets	1–3	Enzymatic
		4	Alkaline/acid
		6	Enzymatic followed by alkaline/acid

Table 1 (contd)

Botanical species	Plant part	Concentration as caffeic acid (mg/kg, fresh weight)	Type of hydrolysis
Vegetables (contd)			
Celery	Whole vegetable	89–104	Enzymatic
	Outside	87–122	Enzymatic
	Heart	84–109	Enzymatic
	Root	168	Enzymatic
Chives	Green leaves	< 0.5	Enzymatic or enzymatic followed by alkaline/acid
Fennel	Tuber	100	Enzymatic followed by alkaline/acid
Garlic	Dry bulb skin	< 20	Enzymatic or enzymatic followed by alkaline/acid
	Fleshy tissue of bulb	14	Enzymatic
		7	Enzymatic followed by alkaline/acid
Horseradish	Whole vegetable	10–11	Enzymatic
	Peel	14	Enzymatic
	Heart	4	Enzymatic
Kale	Stalk and midrib	9	Enzymatic
		13	Enzymatic followed by alkaline/acid
	Leaf and stalk	77	Enzymatic
		92	Alkaline/acid
	Leaf and blade	125	Alkaline
		51–163	Enzymatic
		305	Enzymatic followed by alkaline/acid
Kohlrabi	Leaves	15–66	Enzymatic
		34–113	Alkaline/acid
	Tuber	< 0.5–2	Enzymatic
		2–5	Alkaline/acid
Lettuce	Leaves	767–1570	Enzymatic
		804–1440	Alkaline/acid
Onion	Green leaves	< 0.5–15	Enzymatic
		< 0.5	Alkaline/acid
		19	Enzymatic followed by alkaline/acid
Parsley	Whole plant	6	Enzymatic
	Leaf	< 0.5	Enzymatic
Pea	Unripe seeds	< 0.5–1	Enzymatic
	Hulls	< 0.5	Enzymatic

Table 1 (contd)

Botanical species	Plant part	Concentration as caffeic acid (mg/kg, fresh weight)	Type of hydrolysis
Vegetables (contd)			
Potato	Peel	163–280	Enzymatic
		167–196	Alkaline/acid
		205	Enzymatic followed by alkaline/acid
	Whole tuber	3–30	Enzymatic
		5–33	Alkaline/acid
		8	Enzymatic followed by alkaline/acid
Radish	Whole vegetable	13–17	Enzymatic
	Peel	47–52	Enzymatic
	Heart	3–9	Enzymatic
	Leaves	376–417	Enzymatic
		394–396	Enzymatic followed by alkaline/acid
Radish, black	Whole vegetable	5–8	Enzymatic
	Peel	6–7	Enzymatic
	Heart	7	Enzymatic
	Leaves	163–247	Enzymatic
		156–220	Enzymatic followed by alkaline/acid
Rhubarb	Leaves	8–13	Enzymatic
		6	Alkaline/acid
		16	Enzymatic followed by alkaline/acid
	Outer stem	< 0.5	Enzymatic
Rutabaga	Whole vegetable	2	Enzymatic
	Peel	3	Enzymatic
	Heart	4	Enzymatic
Scorzonera	Whole vegetable	49–212	Enzymatic
	Outside	106	Enzymatic
	Heart	62	Enzymatic
Fruit			
Blueberry	Fruit	83–588	Enzymatic
Courgette (zucchini)	Whole fruit	10	Enzymatic
Currant, black	Fruit	14–93	Enzymatic
red	Fruit	8–16	Enzymatic
white	Fruit	10–21	Enzymatic
Eggplant (aubergine)	Ripe fruit	360–436	Alkaline/acid
Gooseberry,			
green	Fruit	24–32	Enzymatic
red	Fruit	29–32	Enzymatic
yellow	Fruit	23–27	Enzymatic

Table 1 (contd)

Botanical species	Plant part	Concentration as caffeic acid (mg/kg, fresh weight)	Type of hydrolysis
Fruit (contd)			
Grapefruit	Fruit	11–40	Enzymatic
	Peel	14–51	Enzymatic
Lemon	Fruit	13–27	Enzymatic
	Peel	16–35	Enzymatic
Orange	Fruit	19–50	Enzymatic
	Peel	12–36	Enzymatic
Pepper, sweet	Green fruit	3–7	Enzymatic
	Red fruit	4–10	Enzymatic
Strawberry	Fruit	< 0.5–14	Enzymatic
Sweet melon	Peel	3	Enzymatic
	Fruit	< 0.5	Enzymatic
Tomato	Unripe green fruit	13–79	Enzymatic
		44	Alkaline/acid
	Ripe red fruit	32–97	Enzymatic
	Peel	97	Enzymatic
	Pulp	41	Enzymatic
	Seeds	119	Enzymatic
Watermelon	Peel or fruit	< 0.5	Enzymatic

From Schmidtlein & Herrmann (1975a,b,c); Stöhr & Herrmann (1975a,b,c,d)

Cynara scolymus L. (Compositae or Asteraceae, artichoke) contains up to 2% *O*-diphenolic derivatives such as caffeic acid. 1,3-Dicaffeoylquinic acid and 3-caffeoylquinic acid (chlorogenic acid) are believed to be among the active constituents (Leung, 1980; Hinou *et al.*, 1989),

The content of caffeic acid in the peel of potato tubers changes with the physical state of the tuber: It is highest during profound dormancy, drops during emergence from dormancy and remains low during sprouting of the tubers. During dormancy, the largest amount is contained in the eyes of the tubers (18.7 µg/g crude weight), a smaller amount in the peel (12.5 µg/g crude weight) and the smallest amount in the pulp (2.3 µg/g crude weight). At the end of dormancy, the peel contains the largest amount of caffeic acid (3.7 µg/g crude weight), the eyes a smaller amount (2.8 µg/g crude weight) and the pulp the least (0.7 µg/g crude weight) (Morozova *et al.*, 1975). Increased biosynthesis of caffeic acid and chlorogenic acid was reported in Irish potato tubers and of caffeic acid, chlorogenic acid and isochlorogenic acid in sweet potato roots after infection or stress (Kuc, 1972).

Chlorogenic acid, which is metabolized to caffeic acid, is present in apricots, cherries, plums and peaches at concentrations ranging approximately from 50 to 500 ppm (mg/kg) (Ames *et al.*, 1991). It is also present in coffee beans, potatoes, apples and tobacco leaves in significant quantities: 34–140 mg/kg fresh weight in several varieties of potato, 120–310 mg/l juice produced from apples, 890 mg/kg fresh, mature apples and 5590–6740 mg/kg dry tea shoots (Iwahashi *et al.*, 1990). Neochlorogenic acid, which is also metabolized to caffeic acid, is present at concentrations ranging from approximately 50 to 500 ppm (mg/kg) in apples, pears, peaches, apricots, plums, cherries, Brussels sprouts, kale, cabbage and broccoli (Ames *et al.*, 1991).

Relatively small changes occurred in the caffeic and quinic acid contents of sunflower kernels during storage; however, the chlorogenic acid content decreased during storage at 5 °C, 15 °C and 40 °C for 120 days (Pomenta & Burns, 1971).

The occurrence of caffeic acid conjugates in coffee beans has been reviewed (IARC, 1991). The conjugates of caffeic acid are present in green and roasted beans, the total percentage of chlorogenic acids in commercial roasted coffee samples ranging from 0.2 to 3.8% (Trugo, 1984; Maier, 1987).

1.3.3 *Beverages*

Caffeic acid is present in free and conjugated forms in beverages made from agricultural plants of which it is a constituent. Assuming that one cup of coffee contains 10 g of ground coffee, a level of 15–325 mg chlorogenic acids per cup can be calculated on the basis of the percentage range given above (Viani, 1988). Actual data from the USA give an average of 190 mg total chlorogenic acids per cup of brewed coffee (Clinton, 1985).

Caffeic acid is also found in fruit juices and wine. Apple juice has been reported to contain caffeic acid at 0–10 mg/l (Kusnawidjaja *et al.*, 1969) and chlorogenic acid at 20–60 mg/l, but the levels may be much higher (~200 mg/l), especially in fermented juice (Kusnawidjaja *et al.*, 1969; Brause & Raterman, 1982; Cilliers *et al.*, 1990).

In 50 samples of commercial white wines, the average concentration of caffeic acid was 2.5 mg/l. German Riesling wines contained the highest concentration (4.1 mg/l), followed by Japanese *koshu* wine (3.1 mg/l), Chardonnay wine (1.7 mg/l) and Sémillon wine (0.9 mg/l) (Okamura & Watanabe, 1981). Italian wines and a sherry were also found to contain caffeic acid (Cartoni *et al.*, 1991).

The influence of variety, maturity, processing and storage on the phenolic composition of pear, grape and apple juice has been investigated (Spanos & Wrolstad, 1990a,b; Spanos *et al.*, 1990). The concentration of chlorogenic acid increases by about six fold when apple juices are processed by diffusion extraction at different temperatures over that found by conventional pressing. In grape juices, the addition of sulfur dioxide during processing resulted in higher levels of caffeic acid.

1.3.4 *Other sources*

Caffeic acid has also been identified in wood smoke condensates (Ohshima *et al.*, 1989) and in the bee product, propolis, presumably from resin gathered from caffeic acid-containing plants (Čižmárik & Matel, 1970).

1.4 Regulations and guidelines

No information was available to the Working Group concerning the regulatory status of caffeic acid.

2. Studies of Cancer in Humans

No data were available to the Working Group.

3. Studies of Cancer in Experimental Animals

3.1 Oral administration

3.1.1 *Mouse*

Groups of 30 male and 30 female B6C3F$_1$ (C57Bl/6N × C3H/HeN F$_1$) mice, six weeks of age, were fed a diet containing 0 or 2% [20 g/kg of diet] caffeic acid (purity, ≥ 98%) for 96 weeks (intakes, 2120 mg/kg bw per day for males and 3126 mg/kg bw per day for females). Squamous-cell papillomas and carcinomas occurred in the forestomachs of treated mice (4/30 papillomas and 3/30 carcinomas in males and 0/29 papillomas and 1/29 carcinomas in females), but not in controls. The incidences of epithelial hyperplasia of the forestomach were increased significantly ($p < 0.01$) in both treated males and females. Renal-cell adenomas were found in 8/29 ($p < 0.01$) females and in 0/29 controls; one renal adenocarcinoma was seen in a treated male. Significant increases in the incidence of renal tubular-cell hyperplasia were seen in both treated males and treated females. An increased incidence of alveolar type II-cell tumours (adenomas plus carcinomas) of the lung (8/30; $p < 0.05$) was reported in males but not in females. Spontaneous incidences of alveolar-cell tumours in male B6C3F$_1$ mice have been reported to be 2.2–13.9% (Hagiwara *et al.*, 1991).

3.1.2 *Rat*

Groups of 30 male and 30 female Fischer 344 rats, six weeks of age, were fed a diet containing 0 or 2% [20 g/kg of diet] caffeic acid (purity, ≥ 98%) for 104 weeks (intakes, 678 mg/kg bw per day for males and 814 mg/kg bw per day for females). Squamous-cell papillomas and carcinomas occurred in the forestomach in 23/30 (papillomas) and 17/30 (carcinomas) males and in 24/30 (papillomas) and 15/30 (carcinomas) females. No neoplastic change was noted in the forestomachs of an untreated group of 30 males and 30 females. The frequency of forestomach hyperplasia was also increased significantly in animals of each sex as compared to the control level ($p < 0.01$). Renal tubular-cell adenomas were seen in four treated males but in no untreated male or in females. Significant increases in the incidence of renal tubular-cell hyperplasia were seen in both treated males and treated females (Hagiwara *et al.*, 1991).

3.2 Administration with known carcinogens

3.2.1 *Sequential exposure*

(a) Mouse

Three groups of 30 female CD-1 mice, seven weeks of age, were treated topically with 200 nmol [51 mg] 7,12-dimethylbenz[a]anthracene (DMBA) in 200 μl acetone. After one week, the mice were treated topically with 5 nmol [3 mg] 12-O-tetradecanoylphorbol 13-acetate (TPA) with or without simultaneous caffeic acid [purity unspecified] at 10 or 20 μmol [1.8 or 3.6 mg] twice a week for 19 weeks. The mean number of skin tumours per mouse treated with TPA and the high dose of caffeic acid was significantly lower than that in the group treated with TPA alone: 1.12, 4.43 and 6.18 tumours/mouse in the groups treated with TPA plus 20 μmol caffeic acid, TPA plus 10 μmol caffeic acid and TPA alone, respectively (Huang et al., 1988).

(b) Rat

Two groups of 20 female Sprague-Dawley rats, 50 days of age, received 25 mg/kg bw DMBA in 0.5 ml sesame oil by gavage. One week later, the animals were fed a diet containing 0.5% [5 g/kg of diet] caffeic acid (purity, > 99%) for 51 weeks. The mammary glands, ear ducts, stomach, liver and kidneys were examined. The incidence of papillomas of the forestomach was significantly increased (6/19) in animals treated with DMBA plus caffeic acid as compared to rats treated with DMBA alone (0/19; $p < 0.01$) No other significant increase in tumour incidence was found (Hirose et al., 1988).

Three groups of 15 male Fischer 344 rats, six weeks of age, were administered 150 mg/kg bw N-methyl-N'-nitro-N-nitrosoguanidine (MNNG) in dimethyl sulfoxide by gavage. One week later, animals were fed a diet containing 0.5% [5 g/kg of diet] caffeic acid (purity, > 98%) or caffeic acid plus other phenolic antioxidants (0.2% catechol plus 0.5% butylated hydroxyanisole plus 0.25% 2-*tert*-butyl-4-methylphenol) for 35 weeks or the basal diet. The oesophagus, stomach, intestines, liver and kidneys were examined. Squamous-cell carcinomas of the forestomach occurred in 1/15 animals in the group treated with MNNG alone, in 4/15 in the group receiving MNNG plus caffeic acid (no significant difference from controls given MNNG alone) and in 12/15 in the group receiving MNNG plus caffeic acid together with phenolic antioxidants ($p < 0.01$). No other increase in tumour incidence was found (Hirose et al., 1991).

3.2.2 *Prior or concomitant exposure*

Mouse

Female ICR/Ha mice, nine weeks of age, were fed a diet containing 0.06 mmol/g [10 g/kg of diet] caffeic acid (purity, 99%). From experimental day 8, the mice were also given 1 mg benzo[a]pyrene by gavage twice a week for four weeks. The diet containing caffeic acid was removed three days after the last benzo[a]pyrene treatment. Mice were killed at 211 days of age. In the 17 effective mice, the number of forestomach tumours (≥ 1 mm)/mouse [histology unspecified] was significantly decreased by caffeic acid ($p < 0.05$) (3.1 *versus* 5.0 tumours/mouse among 38 mice treated with benzo[a]pyrene alone) (Wattenberg et al., 1980). [The Working Group noted the limited reporting.]

4. Other Relevant Data

4.1 Absorption, distribution, metabolism and excretion

In rats, chlorogenic acid is hydrolysed in the stomach and intestine to caffeic and quinic acids (Czok *et al.*, 1974). A number of metabolites have been identified (Fig. 1). Glucuronides of *meta*-coumaric acid and *meta*-hydroxyhippuric acid appear to be the main metabolites in humans. After oral administration of caffeic acid to human volunteers, *O*-methylated derivatives (ferulic, dihydroferulic and vanillic acids) were excreted rapidly in the urine, while the *meta*-hydroxyphenyl derivatives appeared later. The dehydroxylation reactions were ascribed to the action of intestinal bacteria (Arnaud, 1988).

4.2 Toxic effects

4.2.1 *Humans*

No data were available to the Working Group.

4.2.2 *Experimental data*

Five male Fischer 344 rats, six weeks of age, were given 20 g/kg of diet caffeic acid (purity, > 98%) in basal diet for four weeks. The forestomachs of all treated animals showed epithelial hyperplasia. No hyperplasia was detected in the five untreated controls (Hirose *et al.*, 1987).

A group of 15 male Syrian golden hamsters, seven weeks of age, was fed a diet containing 1% [10 g/kg diet] caffeic acid (purity, > 98%) for 20 weeks. The dose level of 1% was selected as one-fourth the LD_{50} in rats. The stomachs and urinary bladders were processed for histopathological and autoradiographic examinations. Mild epithelial hyperplasia of the forestomach was noted in 14/15 treated animals (severe in one) and in 7/15 untreated animals ($p < 0.001$). Assessment of ^{3}H-thymidine incorporation revealed an increase in the number of labelled cells in the forestomach and pyloric region of the glandular stomach as compared with untreated rats, but this was not statistically significant (Hirose *et al.*, 1986).

Caffeic acid effectively inhibits lipoxygenase and thereby inhibits the biosynthesis of leukotrienes, which are involved in immunoregulation and in a variety of diseases, including asthma, inflammation and allergic conditions. It blocks platelet aggregation by inhibiting the production of thromboxane A2, which can cause bronchoconstriction (Koshihara *et al.*, 1984; Murota & Koshihara, 1985). Caffeic acid has been widely used as a pharmacological inhibitor in a variety of organ and cell systems, and Sugiura *et al.* (1989) have shown that caffeic acid and several of its derivatives (at IC_{50} of 10^{-8}–10^{-7}M) specifically inhibit the 5-lipoxygenase. Caffeic acid (at 10^{-4}M [0.2 g]) inhibited the proliferation of malignant human haematopoietic cell lines, as measured by ^{3}H-thymidine incorporation (Snyder *et al.*, 1989); at a concentration of 17.5 µM [3.2 mg], it modified the differentiation of HL-60 cells to mature granulocytes (Miller *et al.*, 1990) by affecting leukotriene formation. The cloning efficiency of T-lymphocyte progenitor cells is also inhibited *in vitro* by caffeic acid and restored by leukotriene B4, indicating a regulatory role for arachidonic acid metabolites

Fig. 1. Proposed metabolic pathways of caffeic acid

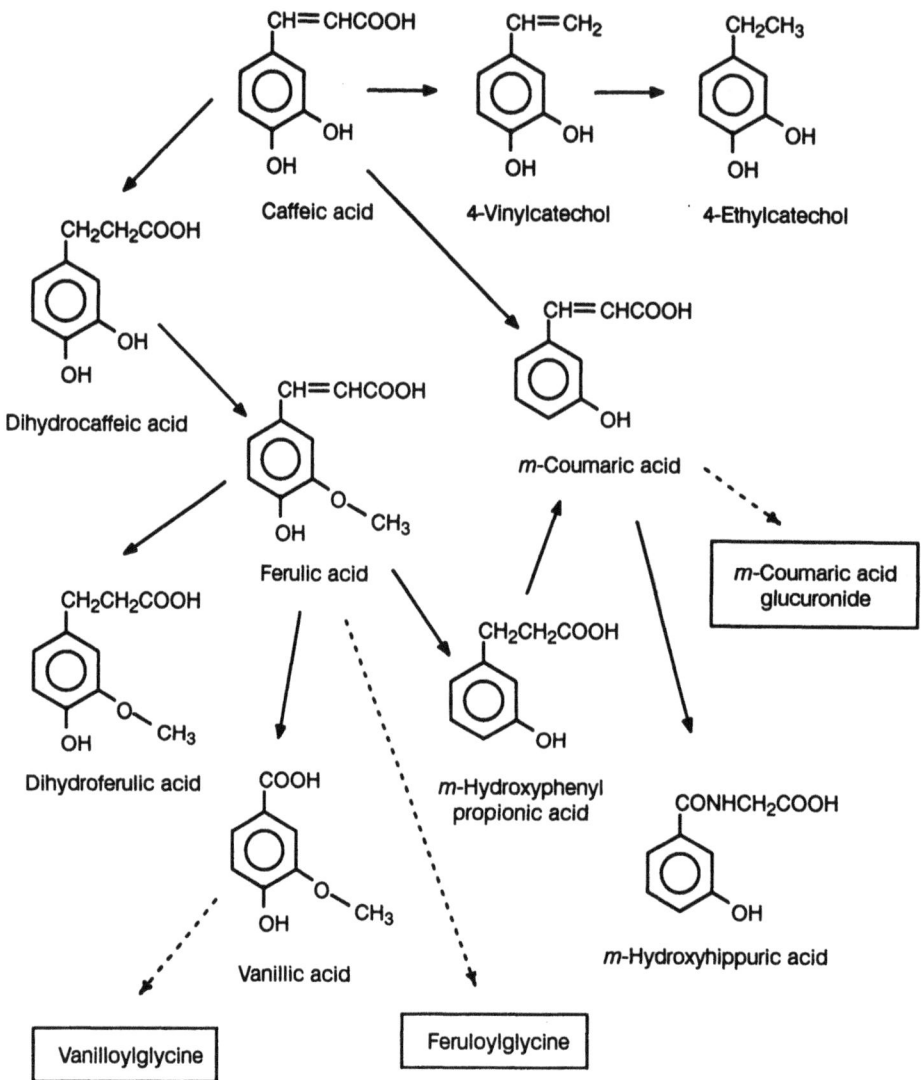

From Arnaud (1988)

(Miller et al., 1989). Several investigators have demonstrated that caffeic acid at 10^{-4}M inhibits lipid peroxidation induced by superoxide anion and formation of hydroxyl radicals *in vitro* (Iwahashi et al., 1990; Toda et al., 1991; Zhou & Zheng, 1991).

4.3 Reproduction and developmental toxicity

No data were available to the Working Group.

4.4 Genetic and related effects

4.4.1 *Humans*

No data were available to the Working Group.

4.4.2 *Experimental systems* (see also Table 2)

When tested alone, caffeic acid was not mutagenic to *Salmonella typhimurium* and did not induce gene conversion in *Saccharomyces cerevisiae* D7. Forward mutations were induced in cultured mouse lymphoma L5178Y cells, and chromosomal aberrations were induced in cultured Chinese hamster ovary cells. The clastogenic effect could not be ascribed to the hydrogen peroxide that was present at low concentrations in freshly prepared caffeic acid solutions (Hanham et al., 1983). Caffeic acid did not induce micronuclei in bone-marrow or intestinal cells of mice treated *in vivo*.

The effects of simultaneous exposure of cells to some transition elements and caffeic acid have also been studied. In the presence of manganese ($MnCl_2$) chelated with glycine, caffeic acid induced mutations in *S. typhimurium*, gene conversion in *S. cerevisiae* D7 and chromosomal aberrations in Chinese hamster ovary cells (Stich et al., 1981a,b).

The phenethyl ester of caffeic acid inhibited adenovirus-induced transformation in rat embryo fibroblasts (Su et al., 1991).

Caffeic acid decreased MNNG-induced mutation in *S. typhimurium* (Francis et al., 1989). Treatment of Chinese hamster ovary K-1 cells with caffeic acid increased the incidence of sister chromatid exchange induced by mitomycin C and ultraviolet radiation. In contrast, the number of x ray-induced sister chromatid exchanges was reduced by post-treatment with caffeic acid during the G_1 phase of the cell cycle (Sasaki et al., 1989).

5. Summary of Data Reported and Evaluation

5.1 Exposure data

Caffeic acid is found in many fruits, vegetables, seasonings and beverages consumed by humans, principally in conjugated forms such as chlorogenic acid.

5.2 Human carcinogenicity data

No data were available to the Working Group.

Table 2. Genetic and related effects of caffeic acid

Test system	Result		Dose[a] (LED/HID)	Reference
	Without exogenous metabolic system	With exogenous metabolic system		
SA0, *Salmonella typhimurium* TA100, reverse mutation	−[b]	−	5000.0000	Stich et al. (1981a)
SA0, *Salmonella typhimurium* TA100, reverse mutation	−	−	5000.0000	Fung et al. (1988)
SA5, *Salmonella typhimurium* TA1535, reverse mutation	−	−	5000.0000	Fung et al. (1988)
SA7, *Salmonella typhimurium* TA1537, reverse mutation	−	−	5000.0000	Fung et al. (1988)
SA8, *Salmonella typhimurium* TA1538, reverse mutation	−	−	5000.0000	Fung et al. (1988)
SA9, *Salmonella typhimurium* TA98, reverse mutation	−	−	150.0000	MacGregor & Jurd (1978)
SA9, *Salmonella typhimurium* TA98, reverse mutation	−[b]	−	5000.0000	Stich et al. (1981a)
SA9, *Salmonella typhimurium* TA98, reverse mutation	−	−	5000.0000	Fung et al. (1988)
SCG, *Saccharomyces cerevisiae* D7, gene conversion	(+)[b]	−	40000.0000	Stich et al. (1981a)
G5T, Gene mutation, mouse lymphoma L5178Y cells, *tk* locus *in vitro*	+	−	307.0000	Fung et al. (1988)
CIC, Chromosomal aberrations, Chinese hamster ovary cells *in vitro*	+[c]	−	200.0000	Stich et al. (1981b)
CIC, Chromosomal aberrations, Chinese hamster ovary cells *in vitro*	+	0	250.0000	Hanham et al. (1983)
MVM, Micronucleus test, B6C3F₁ mouse bone marrow *in vivo*	−		2400.0000, diet	Raj et al. (1983)
MVM, Micronucleus test, C56Bl/6J mouse intestine *in vivo*	−		4800.0000, diet	Wargovich et al. (1985)
MVM, Micronucleus test, C56Bl/6J mouse intestine *in vivo*	−	4	800.0000, diet	Wargovich et al. (1983)

+, positive; (+), weak positive; −, negative; 0, not tested
[a]In-vitro tests, μg/ml; in-vivo tests, mg/kg bw
[b]Positive responses in the presence of Mn^{++}-glycine complex (10^{-4} M Mn^{++})
[c]Positive response enhanced in the presence of Mn^{++}-glycine complex

5.3 Animal carcinogenicity data

Caffeic acid was tested for carcinogenicity by oral administration in the diet in one study in mice and one study in rats. In mice, it produced renal-cell adenomas in females and a high incidence of renal tubular-cell hyperplasia in animals of each sex. An increase in the combined incidence of squamous-cell papillomas and carcinomas of the forestomach was seen in male mice, and a high incidence of hyperplasia of the forestomach was seen in both males and females. In rats, it produced squamous-cell papillomas and carcinomas of the forestomach in animals of each sex and a few renal-cell adenomas in males.

Oral administration of caffeic acid in combination with known carcinogens resulted in enhancing or inhibiting effects depending upon the carcinogen and the time of administration.

5.4 Other relevant data

Humans and experimental animals metabolize caffeic acid to the same metabolites and hydrolyse chlorogenic acid to caffeic acid.

Caffeic acid did not induce micronuclei in mice treated *in vivo*. It produced gene mutation and chromosomal aberrations in cultured rodent cells. It did not induce gene mutation in bacteria.

5.5 Evaluation[1]

No data were available on the carcinogenicity of caffeic acid to humans.

There is *sufficient evidence* in experimental animals for the carcinogenicity of caffeic acid.

Overall evaluation

Caffeic acid is *possibly carcinogenic to humans (Group 2B)*.

6. References

Agarwal, S.K. & Rastogi, R.P. (1974) Ergometrine and other constituents of *Argyreia speciosa* Sweet. *Indian J. Pharm.*, **36**, 118–119

Aldrich Chemical Co. (1992) *Aldrich Catalog/Handbook of Fine Chemicals 1992–1993*, Milwaukee, WI, p. 480

Ames, B.N., Profet, M. & Gold, L.S. (1991) Dietary carcinogens and mutagens from plants. In: Hayatsu, H., ed., *Mutagens in Food: Detection and Prevention*, Boca Raton, FL, CRC Press, pp. 29–50

---

[1]For definition of the italicized terms, see Preamble, pp. 26–29.

Andersen, R.A. & Vaughn, T.H. (1970) Rapid electron capture determination of caffeic acid and quercetin moieties in plants. *J. Chromatogr.*, **52**, 385–392

Arnaud, M.J. (1988) The metabolism of coffee constitutents. In: Clarks, R.J. & Macrae, R., eds, *Coffee*, Vol. 3, *Physiology*, London, Elsevier, pp. 33–55

Borges, M.F.M. & Pinto, M.M.M. (1989) Isocratic high performance liquid chromatography separation of esculetin and *cis/trans* isomers of caffeic acid. *J. liq. Chromatogr.*, **12**, 2345–2354

Brause, A.R. & Raterman, J.M. (1982) Verification of authenticity of apple juice. *J. Assoc. off. anal. Chem.*, **65**, 846–849

Budavari, S., ed. (1989) *The Merck Index*, 11th ed., Rahway, NJ, Merck & Co., p. 248

Cartoni, G.P., Coccioli, F., Pontelli, L. & Quattrucci, E. (1991) Separation and identification of free phenolic acids in wines by high-performance liquid chromatography. *J. Chromatogr.*, **537**, 93–99

Chlabicz, J. & Gałasiński, W. (1986) The components of *Melissa officinalis* L. that influence protein biosynthesis *in-vitro*. *J. Pharm. Pharmacol.*, **38**, 791–794

Cilliers, J.J.L., Singleton, V.L. & Lamuela-Raventos, R.M. (1990) Total polyphenols in apples and ciders; correlation with chlorogenic acid. *J. Food Sci.*, **55**, 1458–1459

Čižmárik, J. & Matel, I. (1970) Examination of the chemical composition of propolis. I. Isolation and identification of the 3,4-dihydroxycinnamic acid (caffeic acid) from propolis. *Experentia*, **26**, 713

Clinton, W.P. (1985) The chemistry of coffee. In: *11e Colloque Scientifique International sur le Café, Lomé, 1985*, Paris, Association Scientifique International du Café, pp. 87–92

Conkerton, E.J. & Chapital, D.C. (1983) High-performance liquid chromatography separation of the *cis-trans* isomers of cinnamic acid derivatives. Ultraviolet and electrochemical detection. *J. Chromatogr.*, **281**, 326–329

Court, W.A. & Hendel, J.G. (1986) Capillary gas chromatography of nonvolatile organic acids, fatty acids and certain carbohydrates in flue-cured tobacco. *Tobacco int.*, **188**, 58–61

Cui, C.-B., Tezuka, Y., Kikuchi, T., Nakano, H., Tamaoki, T. & Park, J.-H. (1990) Constituents of a fern, *Davallia mariesii* Moore. I. Isolation and structures of davallialactone and a new flavanone glucuronide. *Chem. pharm. Bull.*, **38**, 3218–3225

Cuq, J.L. & Jaussan, V. (1991) Caffeic acid oxidation and antibacterial effects (Fr.). *Sci. Alim.*, **11**, 25–36

Czog, G., Walter, W., Knoche, K. & Degener, H. (1974) Reabsorption of chlorogenic acid in the rat (Ger.). *Z. Ernährungswiss.*, **13**, 108–112

Didry, N., Pinkas, M. & Torck, M. (1980) Phenolic components from *Tussilago farfara* (Fr.). *Ann. pharm. Fr.*, **38**, 237–241

Dikun, P.P., Ermilov, V.B. & Shendrikova, I.A. (1991) Some approaches to prevention of endogenous formation of *N*-nitrosamines in humans. In: O'Neill, I.K., Chen, J. & Bartsch, H., eds, *Relevance to Human Cancer of N-Nitroso Compounds, Tobacco Smoke and Mycotoxins* (IARC Scientific Publications No. 105), Lyon, IARC, pp. 552–557

Felice, L.J., King, W.P. & Kissinger, P.T. (1976) A new liquid chromatography approach to plant phenolics. Application to the determination of chlorogenic acid in sunflower meal. *J. agric. Food Chem.*, **24**, 380–382

Ficarra, P., Ficarra, R., Villari, A., De Pasquale, A., Monforte, M.T. & Calabrò, M.L. (1990) High-performance liquid chromatography and diffuse reflectance spectroscopy of flavonoids in *Crataegus oxyacantha* L. III. Analysis of 2-phenyl-chroman derivatives and caffeic acid. *Farmaco*, **45**, 237–245

Fluka Chemie AG (1990) *Fluka Chemika-BioChemika 1990/91*, Buchs, p. 266

Francis, A.R., Shetty, T.K. & Bhattacharya, R.K. (1989) Modification of the mutagenicity of aflatoxin B_1 and N-methyl-N'-nitro-N-nitrosoguanidine by certain phenolic compounds. *Cancer Lett.*, **45**, 177–182

Fung, V.A., Cameron, T.P., Hughes, T.J., Kirby, P.E. & Dunkel, V.C. (1988) Mutagenic activity of some coffee flavor ingredients. *Mutat. Res.*, **204**, 219–228

Grodzinska-Zachwieja, Z., Kahl, W. & Klimczak, M. (1973) Spectrophotometric investigation of changes of caffeic, chlorogenic and isochlorogenic acids under the influence of some physicochemical factors. *Pol. Pharmacol. Pharm.*, **25**, 299–305

Hagerman, A.E. & Nicholson, R.L. (1982) High-performance liquid chromatographic determination of hydroxycinnamic acids in the maize mesocotyl. *J. agric. Food Chem.*, **30**, 1098–1102

Hagiwara, A., Hirose, M., Takahashi, S., Ogawa, K., Shirai, T. & Ito, N. (1991) Forestomach and kidney carcinogenicity of caffeic acid in F344 rats and C57Bl/6N × C3H/HeN F_1 mice. *Cancer Res.*, **51**, 5655–5660

Hanham, A.F., Dunn, B.P. & Stich, H.F. (1983) Clastogenic activity of caffeic acid and its relationship to hydrogen peroxide generated during autooxidation. *Mutat. Res.*, **116**, 333–339

Hartley, R.D. & Jones, E.C. (1975) Effect of ultraviolet light on substituted cinnamic acids and the estimation of their *cis* and *trans* isomers by gas chromatography. *J. Chromatogr.*, **107**, 213–218

Hatjimanoli, M. & Debelmas, A.-M. (1977) Phenolic acids from *Centaurium umbellatum* Gil. (Fr.). *Ann. pharm. Fr.*, **35**, 107–111

Hayase, F. & Kato, H. (1984) Antioxidative components of sweet potatoes. *J. nutr. Sci. Vitaminol.*, **30**, 37–46

Herrmann, K. (1956) On caffeic acid and chlorogenic acid (Ger.). *Pharmazie*, **11**, 433–449

Hinou, J., Harvala, C. & Philianos, S. (1989) Polyphenolic substances from *Cynara scolymus* L. leaves (Fr.). *Ann. pharm. Fr.*, **47**, 95–98

Hirose, M., Inoue, T., Asamoto, M., Tagawa, Y. & Ito, N. (1986) Comparison of the effects of 13 phenolic compounds in induction of proliferative lesions of the forestomach and increase in the labelling indices of the glandular stomach and urinary bladder epithelium of Syrian golden hamsters. *Carcinogenesis*, **7**, 1285–1289

Hirose, M., Masuda, A., Imaida, K., Kagawa, M., Tsuda, H. & Ito, N. (1987) Induction of forestomach lesions in rats by oral administrations of naturally occurring antioxidants for 4 weeks. *Jpn. J. Cancer Res. (Gann)*, **78**, 317–321

Hirose, M., Masuda, A., Fukushima, S. & Ito, N. (1988) Effects of subsequent antioxidant treatment on 7,12-dimethylbenz[a]anthracene-initiated carcinogenesis of the mammary gland, ear duct and forestomach in Sprague-Dawley rat. *Carcinogenesis*, **9**, 101–104

Hirose, M., Mutai, M., Takahashi, S., Yamada, M., Fukushima, S. & Ito, N. (1991) Effects of phenolic antioxidants in low dose combination on forestomach carcinogenesis in rats pretreated with N-methyl-N'-nitrosoguanidine. *Cancer Res.*, **51**, 824–827

Huang, M.-T., Smart, R.C., Wong, C.-Q. & Conney, A.H. (1988) Inhibitory effect of curcumin, chlorogenic acid, caffeic acid, and ferulic acid on tumor promotion in mouse skin by 12-*O*-tetradecanoylphorbol-13-acetate. *Cancer Res.*, **48**, 5941–5946

IARC (1991) *IARC Monographs on the Evaluation of Carcinogenic Risks to Humans*, Vol. 51, *Coffee, Tea, Mate, Methylxanthines and Methylglyoxal*, Lyon, pp. 41–197

Iwahashi, H., Ishii, T., Sugata, R. & Kido, R. (1990) The effects of caffeic acid and its related catechols on hydroxyl radical formation by 3-hydroxyanthranilic acid, ferric chloride, and hydrogen peroxide. *Arch. Biochem. Biophys.*, **276**, 242–247

Janssen Chimica (1991) *Catalog Handbook of Fine Chemicals*, Beerse, p. 449

Kahnt, G. (1967) *trans-cis*-Equilibrium of hydroxycinnamic acids during irradiation of aqueous solutions at different pH. *Phytochemistry*, **6**, 755–758

Koshihara, Y., Neichi, T., Murota, S.-I., Lao, A.-N., Fujimoto, Y. & Tatsuno, T. (1984) Caffeic acid is a selective inhibitor for leukotriene biosynthesis. *Biochim. biophys. Acta*, **792**, 92–97

Kuc, J. (1972) Compounds accumulating in plants after infection. In: Aje, S.J. & Kadis, S., eds, *Microbial Toxins*, New York, Academic Press, pp. 211–247

Kuenzig, W., Chau, J., Norkus, E., Holowaschenko, H., Newmark, H., Mergens, W. & Conney, A.H. (1984) Caffeic and ferulic acid as blockers of nitrosamine formation. *Carcinogenesis*, **5**, 309–313

Kurihara, T. & Kikuchi, M. (1980) Studies on the constituents of flowers. XIII. On the components of the flower of *Cytisus scoparius* Link (Jpn.). *Yakugaku Zasshi*, **100**, 1054–1057

Kusnawidjaja, K., Thomas, H. & Dirscherl, W. (1969) Occurrence of vanillic acid, ferulic acid, and caffeic acid in food plants (Ger.). *Z. Ernährungswiss.*, **9**, 290–300

Lancaster Synthesis (1991) *MTM Research Chemicals/Lancaster Catalogue 1991/92*, Windham, NH, p. 504

Leung, A.Y. (1980) *Encyclopedia of Common Natural Ingredients Used in Food, Drugs and Cosmetics*, New York, John Wiley & Sons, pp. 35–36

Lide, D.R., ed. (1991) *CRC Handbook of Chemistry and Physics*, 72nd ed., Boca Raton, FL, CRC Press, p. 3–181

Litvinenko, V.I., Popova, T.P., Simonjan, A.V., Zoz, I.G. & Sokolov, V.S. (1975) Tannins and derivatives of hydroxycinnamic acid in *Labiatae* (Ger.). *Planta med.*, **27**, 372–380

MacGregor, J.T. & Jurd, L. (1978) Mutagenicity of plant flavonoids: structural requirements for mutagenic activity in *Salmonella typhimurium*. *Mutat. Res.*, **54**, 297–309

Maier, H.G. (1987) The acids of coffee. In: *12e Colloque Scientific International sur le Café, Montreux, 1987*, Paris, Association Scientifique Internationale du Café, pp. 229–237

Miller, A.M., Elfenbein, G.J. & Barth, K.C. (1989) Regulation of T-lymphopoiesis by arachidonic acid metabolites. *Exp. Hematol.*, **17**, 198–202

Miller, A.M., Kobb, S.M. & McTiernan, R. (1990) Regulation of HL-60 differentiation by lipoxygenase pathway metabolites *in vitro*. *Cancer Res.*, **50**, 7257–7260

Mino, Y. & Harada, T. (1974) The occurrence of caffeic acid and its derivatives in the leaves of timothy, *Phleum pratense* L. *J. Jpn. Grassl. Sci.*, **20**, 193–198

Morozova, É.V., Korableva, N.P. & Metlitski, L.V. (1975) Determination of the content of natural growth inhibitors of the potato tuber-caffeic acid and scopoletin. *Appl. Biochem. Microbiol.*, **10**, 397–401

Murota, S. & Koshihara, Y. (1985) New lipoxygenase inhibitors isolated from Chinese plants. Development of new anti-allergic drugs. *Drugs exp. clin. Res.*, **11**, 641–644

Nair, S.C. & Panikkar, K.R. (1990) Antitumour principles from *Ixora javanica*. *Cancer Lett.*, **49**, 121–126

Ohshima, H., Friesen, M., Malaveille, C., Brouet, I., Hautefeuille, A. & Bartsch, H. (1989) Formation of direct-acting genotoxic substances in nitrosated smoked fish and meat products: identification of simple phenolic precursors and phenyldiazonium ions as reactive products. *Food chem. Toxicol.*, **27**, 193–203

Okamura, S. & Watanabe, M. (1981) Determination of phenolic cinnamates in white wine and their effect on wine quality. *Agric. Biol. Chem.*, **45**, 2063–2070

Pomenta, J.V. & Burns, E.E. (1971) Factors affecting chlorogenic, quinic, and caffeic acid levels in sunflower kernels. *J. Food Sci.*, **36**, 490–492

Pratt, D.E. & Birac, P.M. (1979) Source of antioxidant activity of soybeans and soy products. *J. Food Sci.*, **44**, 1720–1722

Raina, M.K., Bhatnagar, J.K. & Atal, C.K. (1971) Isolation of caffeic acid from the roots of *Carissa spinarium* L. *Indian J. Pharm.*, **33**, 76–77

Raj, A.S., Heddle, J.A., Newmark, H.L. & Katz, M. (1983) Caffeic acid as an inhibitor of DMBA-induced chromosomal breakage in mice assessed by bone-marrow micronucleus test. *Mutat. Res.*, **124**, 247–253

Riedel-de Haen (1990) *Laboratory Chemicals 1990*, Seelze, p. 481

Sasaki, Y.F., Imanishi, H., Ohta, T. & Shirasu, Y. (1989) Modifying effects of components of plant essence on the induction of sister-chromatid exchanges in cultured Chinese hamster ovary cells. *Mutat. Res.*, **226**, 103–110

Schmidtlein, H. & Herrmann, K. (1975a) On the phenolic acids in vegetables. I. Hydroxycinnamic acids and hydroxybenzoic acids of Brassica species and leaves of other Cruciferae (Ger.). *Z. Lebensmittel. Untersuch.-Forsch.*, **159**, 139–148

Schmidtlein, H. & Herrmann, K. (1975b) On the phenolic acids of vegetables. II. Hydroxycinnamic acids and hydroxybenzoic acids of fruit and seed vegetables (Ger.). *Z. Lebensmittel. Untersuch.-Forsch.*, **159**, 213–218

Schmidtlein, H. & Herrmann, K. (1975c) On the phenolic acids of vegetables. IV. Hydroxycinnamic acids and hydroxybenzoic acids of vegetables and potatoes (Ger.). *Z. Lebensmittel. Untersuch.-Forsch.*, **159**, 255–263

Snyder, D.S., Castro, R. & Desforges, J.F. (1989) Antiproliferative effects of lipoxygenase inhibitors on malignant human hematopoietic cell lines. *Exp. Hematol.*, **17**, 6–9

Spanos, G.A. & Wrolstad, R.E. (1990a) Influence of variety, maturity, processing, and storage on the phenolic composition of pear juice. *J. agric. Food Chem.*, **38**, 817–824

Spanos, G.A. & Wrolstad, R.E. (1990b) Influence of processing and storage on the phenolic composition of Thompson seedless grape juice. *J. agric. Food Chem.*, **38**, 1565–1571

Spanos, G.A., Wrolstad, R.E. & Heatherbell, D.A. (1990) Influence of processing and storage on the phenolic composition of apple juice. *J. agric. Food Chem.*, **38**, 1572–1579

Stich, H.F., Rosin, M.P., Wu, C.H. & Powrie, W.D. (1981a) A comparative genotoxicity study of chlorogenic acid (3-*O*-caffeoylquinic acid). *Mutat. Res.*, **90**, 201–212

Stich, H.F., Rosin, M.P., Wu, C.H. & Powrie, W.D. (1981b) The action of transition metals on the genotoxicity of simple phenols, phenolic acids and cinnamic acids. *Cancer Lett.*, **14**, 251–260

Stöhr, H. & Herrmann, K. (1975a) On the phenolic acids in vegetables. III. Hydroxycinnamic acids and hydroxybenzoic acids of root vegetables (Ger.). *Z. Lebensmittel. Untersuch.-Forsch.*, **159**, 218–224

Stöhr, H. & Herrmann, K. (1975b) The phenolics of fruits. V. The phenolics of strawberries and their changes during development and ripeness of the fruits (Ger.). *Z. Lebensmittel. Untersuch.-Forsch.*, **159**, 341–348

Stöhr, H. & Herrmann, K. (1975c) The phenolics of fruits. VI. The phenolics of currants, gooseberries and blueberries. Changes in phenolic acids and catechins during development of black currants (Ger.). *Z. Lebensmittel. Untersuch.-Forsch.*, **159**, 31–37

Stöhr, H. & Herrmann, K. (1975d) On the occurrence of derivatives of hydroxycinnamic acids, hydroxybenzoic acids, and hydroxycoumarins in citrus fruits (Ger.). *Z. Lebensmittel. Untersuch.-Forsch.*, **159**, 305–306

Su, Z.-Z., Grunberger, D. & Fisher, P.B. (1991) Suppression of adenovirus type 5E1A-mediated transformation and expression of the transformed phenotype by caffeic acid phenethyl ester (CAPE). *Mol. Carcinog.*, **4**, 231–242

Sugiura, M., Naito, Y., Yamaura, Y., Fukaya, C. & Yokoyama, K. (1989) Inhibitory activities and inhibition specificities of caffeic acid derivatives and related compounds toward 5-lipoxygenase. *Chem. pharm Bull.*, **37**, 1039–1043

Świątek, L., Broda, B. & Frej, E. (1971) Chemical constituents of *Veronica chamaedrys* L. (Pol.). *Acta pol. pharm.*, **28**, 189–194

TCI America (1991) *TCI America Organic Chemicals 91/92 Catalog*, Portland OR, p. 238

Toda, S., Kumura, M. & Ohnishi, M. (1991) Effects of phenolcarboxylic acids on superoxide anion and lipid peroxidation induced by superoxide anion. *Planta med.*, **57**, 8–10

Trenkle, K. (1971) *Foeniculum vulgare*: organic acids, especially phenylcarbonic acids (Ger.). *Planta med.*, **20**, 289–301

Trugo, L.C. (1984) *HPLC in Coffee Analysis*, PhD Thesis, University of Reading

Viani, R. (1988) Physiologically active substances in coffee. In: Clarke, R.J. & Macrae, R., eds, *Coffee*, Vol. 3, *Physiology*, London, Elsevier Applied Science, pp. 1–31

Wargovich, M.J., Goldberg, M.T., Newmark, H.L. & Bruce, W.R. (1983) Nuclear aberrations as a short-term test for genotoxicity to the colon: evaluation of nineteen agents in mice. *J. natl Cancer Inst.*, **71**, 133–137

Wargovich, M.J., Eng, V.W.S. & Newmark, H.L. (1985) Inhibition by plant phenols of benzo[a]pyrene-induced nuclear aberrations in mammalian intestinal cells: a rapid in vivo assessment method. *Food chem. Toxicol.*, **23**, 47–49

Wattenberg, L.W., Coccia, J.B. & Lam, L.K.T. (1980) Inhibitory effects of phenolic compounds on benzo[a]pyrene-induced neoplasia. *Cancer Res.*, **40**, 2820–2823

Zhou, Y.-C. & Zheng, R.-L. (1991) Phenolic compounds and an analog as superoxide anion scavengers and antioxidants. *Biochem. Pharmacol.*, **42**, 1177–1179

d-LIMONENE

1. Exposure Data

1.1 Chemical and physical data

Limonene is, with the possible exception of α-pinene, the most frequently occurring natural monoterpene. It is a major constituent of the oils of citrus fruit peel and is found at lower levels in many fruits and vegetables. It occurs naturally in the *d* (or R)- and *l* (or S) optically active forms and as *dl* mixtures including the optically inactive racemate (dipentene). For example, the *d* form comprises 98–100% of the limonene in most citrus oils (family Rustaceae), whereas that in oil of citronella and oil of lemongrass (family Gramineae) is 96–100% *l*-limonene (Furia & Bellanca, 1975; Clayton & Clayton, 1981; Sax & Lewis, 1987; Mosandl *et al.*, 1990).

1.1.1 *Synonyms, structural and molecular formulae*

Chem. Abstr. Serv. Reg. No.: 5989-27-5
Deleted CAS Reg. Nos.: 7705-13-7; 95327-98-3
Chem. Abstr. Name: (R)-1-Methyl-4-(1-methylethenyl)cyclohexene
Synonyms: Cajaputene; carvene; cinene; (+)-dipentene; *d*-(+)-limonene; D-(+)-limonene; (+)-limonene; (R)-limonene; (R)-(+)-limonene; (+)-*para*-mentha-1,8-diene; (R)-(+)-*para*-mentha-1,8-diene; 1-methyl-4-isopropenyl cyclohexene-1; Refchole

$C_{10}H_{16}$ Mol. wt: 136.24

1.1.2 *Chemical and physical properties*

(a) *Description*: Colourless liquid (Sax & Lewis, 1987; Budavari, 1989) with a pleasant, lemon-like odour (US National Toxicology Program, 1990)
(b) *Melting-point*: −74.3 °C (Lide, 1991)
(c) *Boiling-point*: 175.5–176 °C (Budavari, 1989)

(d) *Density*: 0.8411 g/cm^3 at 20 °C/4 °C (Sax & Lewis, 1987)
(e) *Solubility*: Insoluble in water; soluble in benzene, carbon tetrachloride, diethyl ether, ethanol and petroleum ether (Lide, 1991; STN International, 1992); slightly soluble in glycerine (Flavor & Extract Manufacturers' Association, 1991)
(f) *Refractive index*: n_D^{20}, 1.4730 (Lide, 1991)
(g) *Optical rotation*: $[\alpha]_D^{20}$ + 125.6° (Lide, 1991)
(h) *Spectroscopy data*: Infrared, nuclear magnetic resonance and mass spectral data have been reported (Aldrich Chemical Co., 1992; STN International, 1992).
(i) *Stability*: Oxidizes to film in air (Sax & Lewis, 1987); must be stored away from light and air at −18 °C (Ranganna *et al.*, 1983)

1.1.3 *Trade names, technical products and impurities*

d-Limonene is available commercially in an untreated technical grade (purity, 95%) as a clear liquid, which is variably colourless to yellow cast with a strong citrus odour; as a food grade (purity, 97%), a clear water-white liquid with a mild orange odour; and as a lemon-lime grade (purity, 70%), a clear water-white liquid with a lemon-lime odour (Florida Chemical Co., 1991a,b,c).

1.1.4 *Analysis*

Bertsch *et al.* (1974) described an analytical method in which trace quantities of organic materials, including limonene, in air are adsorbed on a porous polymer and separated by capillary gas chromatography (GC).

d-Limonene has been measured in a range of natural products, such as orange juice, by GC and head-space analysis (Massaldi & King, 1974; Marsili, 1986) and in packaging materials by thermal desorption (Lloyd, 1984). Searle (1989) described a procedure for monitoring airborne limonene vapour by GC (detection limit, 5 μg). GC with flame ionization detection has been used to analyse carrot volatiles collected on porous polymer traps. Samples were ground (blending), sliced or grated, and volatiles were collected on the polymer traps and eluted for analysis (Simon *et al.*, 1980).

Oil recoverable by distillation from orange, tangerine and grapefruit juices is at least 98% *d*-limonene. The *d*-limonene content of such oils has been determined by co-distillation with isopropanol, acidification and titration with potassium bromide–potassium bromate solution (Boland, 1984). The distribution of optical isomers of limonene has been determined in various essential oils using multidimensional GC, by coupling chiral and nonchiral columns (Mosandl *et al.*, 1990).

1.2 Production and use

1.2.1 *Production*

d-Limonene was first recovered as a commercial product during the 1941–42 Florida (USA) citrus season, from the steam evaporater condensate in the production of citrus molasses. By 1946, commercial production in Florida was common (Schulz, 1972).

The principal sources of d-limonene are the oils of orange, grapefruit and lemon (Verghese, 1968). It is the main volatile constituent of citrus peel oil, and the collected volatile portion of oil is usually referred to as d-limonene in the trade (Gerow, 1974). d-Limonene may be obtained by steam distillation of citrus peels and pulp resulting from the production of juice and cold-pressed oils or from deterpenation of citrus oils. It is sometimes redistilled (Furia & Bellanca, 1975).

Citrus peel oil can contain up to 95% d-limonene and stripper oil over that amount. Stripper oil is the oil recovered during concentration of the liquor which separates from the peel during pressing. The press liquor is concentrated to give citrus molasses, and the vapour which separates during concentration is condensed to yield stripper oil. In commercial practice, the essential peel oil is extracted by mechanical rupturing of oil sacs in the sub-epidermal layer (flavedo) of the peel and expression of the oil as an aqueous emulsion, from which it is separated by centrifuging (Ranganna et al., 1983).

d-Limonene also occurs in other oils and essences obtained during the processing of citrus juice, including: juice oil, deoiler oil (oil separated from juice by centrifuging or decantation), essence oil (oil obtained from the recovery unit during concentration of fruit juices) and aroma oil (oil obtained by distillation of the aqueous discharge from the centrifuge used to separate pressed oil) (Ranganna et al., 1983).

Annual worldwide production of d-limonene and orange oil/essence oil (95% d-limonene) has recently been approximately 45 000 tonnes. Citrus plantings under way in southern Florida, Brazil, Venezuela, Mexico, the Caribbean basin and elsewhere are expected to increase that figure to 73 000 tonnes annually within a decade (Florida Chemical Co., 1991a). The production of d-limonene in Florida in 1989 and 1990 was estimated to be 8600 and 6800–7700 tonnes, respectively (Anon., 1989). Limonene production in Florida in 1971 was estimated at approximately 4500 tonnes; an additional 450 tonnes of terpenes were recovered from 'folding' cold pressed oils (Schulz, 1972). In 1990, 450 tonnes of d-limonene were produced in California (USA), and Brazilian production was estimated to have been between 4100 and 8200 tonnes. Brazilian d-limonene is used almost exclusively by resin producers (Topfer, 1990).

In 1990, Brazil was the largest producer of orange oil, while Florida led in production of tangerine and grapefruit oil (Anon., 1990); Mexico supplies $\geq$ 80% of the world's lime oil (Anon., 1988a). In 1979, Italian production of essential oil from citrus fruit was as follows: oranges, 380 tonnes; lemon, 550 tonnes; bergamot, 100 tonnes; and mandarine, 30 tonnes (Anon., 1981).

Table 1 shows US imports of four citrus oils in 1981–90 and the major sources of the oils. In 1985–87, average US exports of orange oil were 1500 tonnes per year (Anon., 1988b). In 1983, Japan imported 200 tonnes of lemon oil (Anon., 1984).

1.2.2 Use

For nearly 50 years, d-limonene and orange oil/essence oil (95% d-limonene) have been used widely as flavour and fragrance additives in perfume, soap, food and beverages. d-Limonene has been used in non-alcoholic beverages, ice cream and ices, sweets, baked goods, gelatins and puddings, and chewing gum. It is also used as a chemical intermediate in the production of l-carvone, in terpene resin manufacture as a wetting and dispersing agent

Table 1. US imports of citrus oils

Year	Type of oil	Amount (thousand tonnes)	Major sources (decreasing order)
1981	Orange	2.1	Brazil
	Lemon	0.3	Argentina, Italy
	Lime	0.5	Mexico, Brazil, Haiti, Peru
	Grapefruit	0.04	Israel, Belize
1985	Orange	4.5	Brazil, Israel, Belize
	Lemon	1.0	Argentina, Italy
	Lime	0.6	Mexico, Peru, Brazil, Haiti
	Grapefruit	0.09	Israel, Brazil, Belize
1990	Orange	6.5	Brazil, Mexico
	Lemon	1.4	Argentina, Italy, Spain
	Lime	0.9	Mexico, Peru
	Grapefruit	0.3	Israel

From Yokoyama *et al.* (1988); US Department of Commerce (1991)

and in the preparation of sulfurized terpene lubricating oil additives. *d*-Limonene is also an important organic monomer in the synthesis of tackifying resins for adhesives. It has been used as a solvent, cleaner and odour in, e.g., the petroleum industry (Schulz, 1972; Furia & Bellanca, 1975; Sax & Lewis, 1987; Florida Chemical Co., 1991a,b).

Because *d*-limonene is a natural product with low toxicity for mammals and high acute toxicity for bark beetles, fruit flies and cat fleas, it has been proposed as an alternative to synthetic insecticides. Karr and Coats (1988) found that *d*-limonene had limited insecticidal properties against German cockroaches, house flies, rice weevils and corn rootworms. It has been used in shampoos and sprays for the control of fleas on dogs and cats; one such product reportedly contained 78.2% *d*-limonene (Hooser *et al.*, 1986; Hooser, 1990).

d-Limonene has been used to dissolve retained cholesterol gallstones postoperatively (Igimi *et al.*, 1991).

1.3 Occurrence

1.3.1 *Foods and botanical species*

Limonene is widely distributed among citrus and other plant species. It has been reported in more than 300 essential oils, at concentrations up to 90–95%, and at lesser, although still appreciable, concentrations in foods (e.g., 800 mg/l in non-alcoholic beverages, 3000 mg/kg in chewing gum) (Flavor and Extract Manufacturers' Association, 1991) (Table 2). Botanical species in which limonene occurs and which are used in pharmaceutical and para-pharmaceutical products, cosmetics, foods and beverages are presented in Table 3.

Table 2. Some food products containing d-limonene

Source	Concentration (ppm [mg/kg or mg/l])
Orange juice	0.4–219
Orange peel oil	740 000–970 000
Lemon oil	484 000
Lemon peel oil	520 000–810 000
Grapefruit juice	15.7–86
Grapefruit peel oil	837 000–973 000
Mandarine peel oil	660 000
Lime peel oil (cold press)	437 000–681 000
Lime peel oil (distilled)	428 000–523 000
Pomelo peel oil	861 900
Bilberry	0.08
Bilberry juice	Trace–0.007
Cranberry juice	0.02–0.12
Black currant	0.30
Currant leaves	5.6
Currant buds	77
Guava pulp	0.0002–30
Muscat grape	0.07–0.11
Papaya pulp	0.1–0.5
Peach	0.26–2600
Raspberry	Trace–0.1
Carrot	0.06–15.2
Celery leaves (fresh)	214
Celery root	0.8–26.8
Celery oil	128 000–150 000
Capsicum annuum (bell pepper)	0.2
Aniseed oil	3100 (67 in seeds)
Cinnamomum zeylanicum bark oil	Trace–5000
Nutmeg oil	20 000–130 000
Cumin oil	5000
Yellow ginger	7000
Pepper	2550–9950
Pepper oil	222 000
Black pepper oil	263 000
Chicken (heated)	0.0006–0.007
Coffee	1.7
Green tea	1.0
Mango	0.004–40
Dill herb	3.3–51
Dill root	48
Dill seed	2522
Kiwifruit	Trace
Illicium anisatum oil	15 000
Mentha pulegium oil	410

Table 2 (contd)

Source	Concentration (ppm [mg/kg or mg/l])
Origanum oil	4000
Bergamot oil	250 000–320 000

From Flavor and Extract Manufacturers' Association (1975); Shaw (1979); Saleh et al. (1985); Maarse & Visscher (1988); Flavor and Extract Manufacturers' Association (1991)

Table 3. Occurrence of limonene in various botanical species

Common name Scientific name Synonyms Family	Plant part used	Volatile oil (%) (limonene)[a]	Country or region of cultivation or growth
Angelica *Angelica archangelica* L. Garden angelica, European angelica Umbelliferae or Apiaceae	Rhizome, root, fruit, stem	0.3–1% (major)	Belgium, Hungary, Germany
Anise *Pimpinella anisum* L. Aniseed, sweet cumin, illicium, Chinese anise Umbelliferae or Apiaceae	Dried fruit	1–4% (constituent)	Widely cultivated
Sweet bay *Laurus nobilis* L. Laurel, Grecian laurel Lauraceae	Dried leaf	0.3–3.1% (minor)	Widely cultivated
West Indian bay *Pimenta racemosa* L. Mill. Myrcia, bay rum tree Myrtaceae	Leaf	3.9% (minor)	Venezuela, Puerto Rico, Caribbean Islands
Bergamot *Citrus bergamia* Rutaceae	Fruit peel	– (minor)	Italy
Bois de Rose oil *Aniba rosaeodora* Ducke Rosewood oil, Cayenne rosewood oil Lauraceae	Wood	– (minor)	Amazon region, wild
Buchu *Agathosma betulia* or *A. crenulata* Bookoo, buku, diosma Rutaceae	Dried leaf	1.0–3.5% (major, d-)	Cape Province, South Africa
Cananga oil *Cananga odorata* Annonaceae	Flower	– (major)	Java, Malaysia, Philippines, Moluccas

Table 3 (contd)

Common name Scientific name Synonyms Family	Plant part used	Volatile oil (%) (limonene)[a]	Country or region of cultivation or growth
Caraway *Carum carvi* L. Caraway fruit, carum Umbelliferae or Apiaceae	Dried fruit	2–8% (40%)	Widely cultivated
Cardamom *Elettaria cardamomum* L. Cardamom seed Zingiberaceae	Dried fruit, seed	3–8% (minor)	India, Sri Lanka, Lao People's Democratic Republic, Guatemala, El Salvador
Carrot *Daucus carota* L. Queen Anne's lace, carrot, wild carrot Umbelliferae or Apiaceae	Dried fruit, root	– (constituent)	Widely cultivated
Cascarilla *Croton eluteria* L. Sweetwood bark, sweetbark Euphorbiaceae	Dried bark	1.5–3% (major, *d*-)	West Indies, Mexico, Colombia, Ecuador
Celery *Apium graveolens* L. Celery seed, celery fruit Umbelliferae or Apiaceae	Seed (dried fruit)	2% (60%, *d*-)	France, India
Citronella *Cymbopogon nardus* L., *C. winterianus* Ceylon or Lenabatu citronella oil, Java or Maha Pengiri citronella oil Gramineae	Dried grass	– (constituent)	Sri Lanka, Java, Taiwan, Hainan Island, Malaysia, Africa, Central America, South America
Clary sage *Salvia sclarea* L. Clary wort, muscatel sage, clear eye, see bright Labiatae or Lamiaceae	Flowering top, leaf	0.1–0.15% (minor)	Widely cultivated
Coriander *Coriandrum sativum* L. Umbelliferae or Apiaceae	Dried fruit, leaf	0.2–2.6% (constituent, *d*-)	Widely cultivated
Cubebs *Piper cubeba* L. Cubeba, tailed pepper Piperaceae	Dried fruit	10–20% (constituent)	Southeast Asia
Cumin *Cuminum cyminum* L. Cummin, cumin seed Umbelliferae or Apiaceae	Dried fruit	2–5% (constituent)	Egypt, Iran, India, Morocco, Turkey, former USSR

Table 3 (contd)

Common name Scientific name Synonyms Family	Plant part used	Volatile oil (%) (limonene)[a]	Country or region of cultivation or growth
Dill, Indian dill *Anethum graveolens* L., *A. sowa* European Indian dill, East Indian dill Umbelliferae or Apiaceae	Dried fruit/ whole herb	1.2–7.7%/ 2.5–4% (major, *d*-)	Widely cultivated, India, Japan
Eucalyptus *Eucalyptus globulus* Blue gum, fever tree, gum tree Myrtaceae	Leaf	0.5–3.5% (major, *d*-)	Widely cultivated
Fennel *Foeniculum vulgare* Florence fennel, finocchio Umbelliferae or Apiaceae	Dried fruit	1.5–8.6% (2–6%) (constituent)	Widely cultivated
Galbanum *Ferula gummosa* Galbanum resin, galbanum gum Umbelliferae or Apiaceae	Exudate from stem	5–26% (major, *d*-)	Middle East, western Asia
Grapefruit *Citrus paradisi* Rutaceae	Fruit peel	– (90%)	USA, West Indies, Brazil, Israel, Portugal, Nigeria
Hops *Humulus lupulus* L. European hops, common hops Moraceae or Cannabaceae	Strobile	0.3–1% (minor)	Widely cultivated
Horehound *Marrubium vulgare* L. Marrubium, hoarhound Labiatae or Lamiaceae	Flowering top, dried leaf	Trace (constituent)	Europe, Asia, North America
Juniper *Juniperus communis* L. Juniper berries Cupressaceae	Dried cone	0.2–3.42% (1–2%) (minor)	Italy, Hungary, France, former Yugoslavia, Austria, former Czechoslovakia, former USSR, Germany, Poland, Spain
Labdanum *Cistus ladaniferus* L. Ambreine, rockrose Cistaceae	Leaf, twig	– (constituent)	Mediterranean region
Lavender, spike lavender *Lavandula angustifolia*, *L. latifolia* Garden lavender, aspic Labiatae or Lamiaceae	Flowering top/ dried flower	0.5–1.5%/ 0.5–1% (constituent)	Mediterranean region

Table 3 (contd)

Common name Scientific name Synonyms Family	Plant part used	Volatile oil (%) (limonene)[a]	Country or region of cultivation or growth
Lemon *Citrus limon* L. Cedro oil Rutaceae	Fruit, peel, leaf, twig	– (70%)	Widely cultivated, especially in USA, Italy, Cyprus, Guinea
Lime *Citrus aurantifolia* Rutaceae	Fruit, peel	– (major, *d-*)	Florida, West Indies, Central America
Mint *Mentha piperita* L., *M. spicata* L. Peppermint, spearmint Labiatae or Lamiaceae	Dried leaf/ whole herb	0.1–1%/ 0.7% (minor/minor)	Widely cultivated, especially in USA, Japan, Taiwan, Brazil
Myrrh *Commiphora molmol* Myrrha, gum myrrh Burseraceae	Exudate from bark	1.5–17% (constituent)	Northeast Africa, southwest Asia, especially in Yemen, Somalia, Ethiopia
Olibanum *Boswellia carteri* Frankincense, olibanum gum Burseraceae	Exudate from bark	3–10% (major)	Red Sea region, northeast Africa
Orange (bitter) *Citrus aurantium* L., *C. vulgaris*, *C. bigaradia* Rutaceae	Fruit, peel, flower, leaf, twig	1–2.5% (major, *d-*)	China, southern Europe, USA
Orange (sweet) *Citrus sinensis* L. Rutaceae	Fruit, peel	1.5–2% ($\geq$ 90%, *d-*)	Widely cultivated, especially in the USA, Mediterranean countries, Brazil
Pepper *Piper nigrum* L. Black pepper, white pepper Piperaceae	Dried fruit	2–4% (constituent)	India, Indonesia, Malaysia, China
Pine needle *Pinus mugo*, *P. sylvestris* L. Dwarf pine, Scotch pine Pinaceae	Leaf, twig	– (major, *d-*)	Europe, USA, western Asia
Rosemary *Rosmarinus officinalis* L. Labiatae or Lamiaceae	Flowering top, dried leaf	0.5% (major)	Widely cultivated, especially in California, United Kingdom, France, Spain, Portugal, Yugoslavia, Morocco, China
Rue *Ruta graveolens* L. Common rue, garden rue Rutaceae	Whole herb	0.1% (minor)	Widely cultivated

Table 3 (contd)

Common name Scientific name Synonyms Family	Plant part used	Volatile oil (%) (limonene)[a]	Country or region of cultivation or growth
Sage *Salvia lavandulaefolia* Spanish sage, Dalmatian sage Labiatae or Lamiaceae	Leaf	– (1–41%)	Spain, France
Savory *Satureja hortensis* L., *S. montana* L. Summer savory, winter savory Labiatae or Lamiaceae	Dried leaf/stem	1%/1.6% (major)	Europe, USA, Mediterranean region
Tagetes *Tagetes erecta, T. patula, T. minuta* African marigold, Aztec marigold, French marigold, Mexican marigold Compositae or Asteraceae	Whole herb	– (constituent)	Widely cultivated
Tamarind *Tamarindus indica* L. Tamarindo Leguminosae or Fabaceae	Dried fruit	– (constituent)	Widely cultivated

From Leung (1980); –, not available

[a]Limonene content in the volatile oil, expressed as percentage (when available), otherwise noted as a constituent, major (constituent) or minor (constituent); stereoisomer not given if unspecified

Frozen reconstituted orange juice samples contained 219 ppm (mg/l) limonene [isomer unspecified]. Oxidation did not occur to a significant extent over four weeks of storage in glass (Marsili, 1986). When citrus juices are packed aseptically into laminated cartons, the d-limonene content of the juice is reduced by about 25% within 14 days' storage owing to absorption by the polyethylene (Mannheim *et al.*, 1987). Limonene has been found in packaging material at 25 ppm (µg/g) (Lloyd, 1984).

Limonene [isomers unspecified] was found to represent 32.4% of total terpenes in *Pinus greggii* and 0.5% in *P. pringlei*, two pine species indigenous to Mexico (Lockhart, 1990).

Daily US per-caput consumption of d-limonene, as a result both of its natural occurrence in food and of its presence as a flavour, was estimated to be 0.27 mg/kg bw per day for a 60-kg individual (Flavor and Extract Manufacturers' Association, 1991). Intake of d-limonene can vary considerably, however, depending on the types of food consumed. Citrus juice products are among the richest sources of d-limonene: intake owing to consumption of these products may approach 1 mg/kg bw per day for adults and 2 mg/kg bw per day for young children (US Department of Agriculture, 1982).

Annual US consumption of d-limonene from a variety of foods was calculated as follows: carrots, 5879 kg; celery leaves, 165 379 kg; heated chicken, 3.5 kg; roasted coffee, 1896 kg; cranberries, 3.0 kg; muscat grapes, 1.2 kg; grapefruit juice, 254 320 kg; lemon oil, 465 750 kg; mango, 736 kg; nutmeg, 17 250 kg; orange juice, 154 560 kg; oregano, 508 kg; peaches,

209 kg); pepper, 234 312 kg; raspberries, 2.5 kg; and green tea, 184 kg. Total annual US consumption of *d*-limonene as a result of its natural occurrence in these foods was 1300 tonnes. A survey in 1982 indicated that annual US consumption of *d*-limonene as a flavouring additive was 68 tonnes (Stofberg & Grundschober, 1987).

1.3.2 *Air*

d-Limonene was detected in the air of 81% of the mobile homes surveyed in Texas (USA) during a survey of air quality. Levels of *d*-limonene analysed by GC–mass spectrometry were 0.01–29 ppb [0.06–162 μg/m^3], with a mean of 2.2 ppb [13 μg/m^3] (Connor *et al.*, 1985). Limonene [isomer unspecified] was detected at 0–5.7 ppb [32 μg/m^3] in air samples taken at various locations around Houston, Texas. It was present in all of the more than 150 samples analysed by GC–mass spectrometry over a 15-month period (Bertsch *et al.*, 1974).

1.3.3 *Biological fluids*

Limonene [isomer unspecified] has been detected in human urine (Zlatkis & Liebich, 1971), as have its metabolites (Kodama *et al.*, 1974).

1.4 Regulations and guidelines

d-Limonene is generally recognized as safe for human consumption as a synthetic flavouring substance by the US Food and Drug Administration (1991).

2. Studies of Cancer in Humans

No data were available to the Working Group.

3. Studies of Cancer in Experimental Animals

3.1 Oral administration

3.1.1 *Mouse*

Groups of 50 male and 50 female B6C3F$_1$ mice, eight to nine weeks of age, received 0, 250 or 500 (males) and 0, 500 or 1000 (females) mg/kg bw *d*-limonene (> 99% pure) in corn oil by gavage on five days a week for 103 weeks. The experiment was terminated after 105 weeks. No significant increase in the incidence of neoplasms was observed. The incidence of neoplasms (adenomas and carcinomas combined) of the anterior pituitary was lower in high-dose females than in controls (2/48 *versus* 12/49) (US National Toxicology Program, 1990).

3.1.2 *Rat*

Groups of 50 male and 50 female Fischer 344/N rats, seven to eight weeks of age, received 0, 75 or 150 (males) and 0, 300 or 600 (females) mg/kg bw *d*-limonene (> 99%

pure) in corn oil by gavage on five days a week for 103 weeks. The experiment was terminated after 105 weeks. In males, treatment-related increases were observed in the incidences of renal tubular hyperplasia (vehicle control, 0/50; low-dose, 4/50; high-dose, 7/50), renal tubular-cell adenoma (vehicle control, 0/50; low-dose, 4/50; high-dose, 8/50; $p < 0.01$, trend test) and renal tubular-cell adenocarcinoma (vehicle control, 0/50; low-dose, 4/50; high-dose, 3/50). The incidence of lesions of the kidney was not increased in female rats (US National Toxicology Program, 1990).

3.2 Intraperitoneal administration

Mouse

In a screening assay based on the accelerated induction of lung tumours in a strain highly susceptible to development of this neoplasm, groups of 15 male and 15 female A/He mice, six to eight weeks old, received intraperitoneal injections of 0.2 g/kg bw or 1 g/kg bw (maximal tolerated dose) *d*-limonene in tricaprylin [purity 85-99%] three times per week for eight weeks (total doses, 4.8 and 24 g/kg bw). Vehicle control groups of 80 males and 80 females received intraperitoneal injections of 0.1 ml tricaprylin on the same schedule. The experiment was terminated 24 weeks after the first injection, and the lungs were removed and surface nodules counted. Survival was comparable between the groups. Lung tumour incidence was not increased; males—control, 22/77 (28%), low-dose, 1/15 (7%); and high-dose, 3/15 (20%); females—control, 15/77 (20%); low-dose, 2/15 (13%); and high-dose, 2/13 (15%) (Stoner *et al.*, 1973).

3.3 Administration with known carcinogens

3.3.1 *Oral administration*

(a) *Mouse*

Groups of 20–30 [exact number unspecified] male and female (both sexes being represented almost equally) stock albino mice [strain unspecified], > 6–8 weeks of age, received a single dose of 50 μg benzo[*a*]pyrene in 0.2 ml polyethylene glycol or polyethylene glycol alone by stomach tube and no further treatment, or they subsequently received 40 weekly intubations of 0.05 ml *d*-limonene [concentration unspecified] contaminated with < 0.1% *para*-cymene. A further group of 20–30 male and female mice served as untreated controls. In the group that received benzo[*a*]pyrene plus *d*-limonene, 5/23 mice that survived > 60 days and were autopsied within 24 h after death had a total of eight forestomach papillomas; in the group that received benzo[*a*]pyrene alone, 2/17 mice had a total of two forestomach papillomas; and in the group that received polyethylene glycol and *d*-limonene, 2/15 mice had a total of three forestomach papillomas. None of the 18 mice that survived more than 60 days in the untreated control group had a forestomach tumour (Field & Roe, 1965). [The Working Group noted the limited reporting.]

Two groups of 15 female A/J mice, nine weeks of age, were administered 0.2 mmol [27.3 mg] *d*-limonene (99% pure) in 0.2 ml cottonseed oil or cottonseed oil alone by oral gavage once a week for eight weeks; 1 h later they received a gavage of 20 mg/kg bw

N-nitrosodiethylamine (NDEA) in 0.2 ml water. Mice were autopsied 26 weeks after the first dose of NDEA, but only forestomachs and lungs were examined for tumours. All of the animals that received cottonseed oil plus NDEA had forestomach papillomas; 11/15 (73%) had more than 30 papillomas/stomach, and 4/15 (27%) also had carcinomas of the forestomach. In addition, the average number of pulmonary adenomas/mouse in this group was 10.4. Of the animals that received d-limonene prior to NDEA, only 10/15 (67%) had stomach papillomas ($p < 0.05$, two-sided U-test of Wilcoxon, Mann and Whitney); no mouse had more than 30 papillomas/stomach ($p < 0.001$), and none was shown to have a carcinoma of the forestomach ($p < 0.05$). The average number of pulmonary adenomas/mouse in this group was 6.5 ($p < 0.05$) (all p values are *versus* vehicle controls) (Wattenberg *et al.*, 1989). [The Working Group noted the limited number of tissues examined.]

Two groups of female A/J mice, nine weeks of age, received oral intubations of 0.5 mg/mouse 4-(methylnitrosamino)-1-(3-pyridyl)-1-butanone (NNK), in 0.1 ml tricaprylin twice a week for eight weeks. One hour prior to each administration of NNK, one group of 15 mice each received 25 mg d-limonene (99% pure) in 0.2 ml cottonseed oil by oral gavage and one group of 20 mice received cottonseed oil alone. Mice were autopsied 28 weeks after the initial administration of NNK, but only stomachs and lungs were examined for tumours. Of the animals that received cottonseed oil plus NNK, 90% (18/20) had forestomach papillomas, with an average of 2.8 papillomas/mouse; one mouse also had a stomach carcinoma. In the group that received d-limonene plus NNK, no forestomach papilloma or carcinoma was observed ($p < 0.001$). Mice that received cottonseed vehicle plus NNK had 50.8 ± 2.6 (standard error) pulmonary adenomas/mouse, while those that received d-limonene plus NNK had only 15.3 ± 2.1 pulmonary adenomas/mouse ($p < 0.001$) (Wattenberg & Coccia, 1991). [The Working Group noted the limited number of tissues examined.]

Two groups of female A/J mice, nine weeks of age, each received an intraperitoneal injection of 2 mg NNK in 0.1 ml saline. One hour prior to the injection of NNK, one group of 14 mice each received 25 mg d-limonene (99% pure) in 0.2 ml cottonseed oil by oral gavage and one group of 15 mice received cottonseed oil alone. Mice were autopsied 28 weeks after the initial injection of NNK, but only stomachs and lungs were examined. Neither group of mice developed stomach papillomas, but mice that received cottonseed oil plus NNK had an average of 11.2 ± 1.1 pulmonary adenomas/mouse, while those that received d-limonene plus NNK had an average of 2.5 ± 0.7 pulmonary adenomas/mouse ($p < 0.001$) (Wattenberg & Coccia, 1991). [The Working Group noted the limited number of tissues examined.]

(b) *Rat*

Groups of 25 female Sprague-Dawley rats, 47 days old, were fed diets containing 0, 1000 or 10 000 mg/kg d-limonene (> 99% pure) one week prior to a single oral administration by gavage of 65 mg/kg bw 7,12-dimethylbenz[*a*]anthracene (DMBA) in sesame oil [volume unspecified]. The diets were continued for a further 27 weeks, during which time the animals were weighed weekly and palpated for tumours. The experiment was terminated 27 weeks after DMBA administration. In animals fed d-limonene, the time to development of the first mammary tumour was reported to have been longer than that in controls (Elegbede *et al.*, 1984). [The Working Group noted that the number of tumour-bearing animals was not given.]

Groups of 30 six-week-old female Sprague-Dawley rats were fed diets containing 5% d-limonene [purity unspecified] for one week before administration by gastric intubation of a single dose of 65 mg/kg bw DMBA. The d-limonene diet was continued for a further week, after which time animals were returned to basal diet. A second group of 30 females received DMBA by gastric intubation, followed one week later by administration of 5% d-limonene in the diet for 25 weeks. A third group of 30 females received DMBA and was maintained on basal diet, thus serving as controls. The experiment was terminated 25 weeks after the DMBA treatment. Tumours larger than 350 mm^3 were surgically resected, and all resected tumours and those found at necropsy were examined histologically. More than 95% of the mammary tumours were carcinomas. In animals that received d-limonene one week prior to and one week after DMBA, tumour latency was significantly increased ($p < 0.005$); no effect on latency was seen in animals fed d-limonene one week after DMBA for 25 weeks. The number of tumours per rat was reduced in both groups fed d-limonene ($p < 0.05$) (Elson et al., 1988). [The Working Group noted the lack of detailed reporting.]

In a study reported as a short communication, groups of female Wistar Furth rats [number per group unspecified], 64–69 days old, were fed diets containing 5% d-limonene [purity unspecified], 5% orange oil [percentage of d-limonene unspecified] or basal diet throughout the study. After two weeks, the animals received a single intravenous injection of 50 mg/kg bw N-methyl-N-nitrosourea (MNU). All animals survived to the end of the experiment at 23 weeks. Feeding of both orange oil and d-limonene decreased tumour incidence ($p < 0.001$): At the end of the experiment, 80% of controls had mammary carcinomas, while the incidence in orange oil-treated animals was 47% and that in d-limonene-treated animals was 45% (Maltzman et al., 1989). [The Working Group noted the lack of detailed reporting.]

In a study reported in a brief communication, groups of female Wistar Furth rats [numbers unspecified], 64–69 days old, were fed diets containing 5% d-limonene [purity unspecified] or basal diet. After two weeks, the animals received a single intravenous injection of 50 mg/kg bw MNU, and the d-limonene diet was continued for a further week, after which time the rats were returned to basal diet. A further group of females [number unspecified] received a single intravenous injection of MNU and one week later were fed diets containing 5% d-limonene until the end of the experiment. All animals survived to the end of the experiment at 23 weeks. Administration of d-limonene two weeks before and one week after MNU treatment did not affect the incidence or number of mammary tumours. Rats given d-limonene from one week after MNU treatment until the end of the experiment had about half the average number of tumours per rat as MNU-treated controls (Maltzman et al., 1989). [The Working Group noted the lack of detailed reporting.]

Two groups, of 31 and 32 female Sprague-Dawley rats, six weeks old, were fed diets containing 0 and 1% (w/w) d-limonene (99.9% pure), respectively, for two weeks and then received 65 mg/kg bw DMBA (> 95% pure) once by gastric intubation; the two groups were continued on their respective diets for 20 weeks. d-Limonene caused a significant reduction in the incidence of mammary tumours: 58 in rats fed d-limonene plus DMBA (1.8 tumours/rat; median latency, 84 days) and 81 in rats fed DMBA alone (2.6 tumours/rat; median latency, 70 days). Two further groups of 52 female Sprague-Dawley rats, six weeks old, were fed diets containing 0 or 0.5% (w/w) d-limonene (99.9% pure) for two weeks and then received 65 mg/kg bw DMBA once by gastric intubation; the diets were continued for a further week, after which time both groups were continued on basal diet until 20 weeks. Administration of d-limonene for two weeks prior to and one week after DMBA did not significantly reduce the incidence of mammary tumours: 156 in the d-limonene plus DMBA-treated rats (3.0 tumours/rat; median latency, 61 days) compared to 129 in rats fed DMBA alone (2.5 tumours/rat; median latency, 58 days) (Russin et al., 1989).

Groups of 31–38 male Fischer 344 and 31–37 male NBR rats, eight weeks of age, were given 0 or 0.05% (500 mg/l) N-nitrosoethylhydroxyethylamine (NEHEA) in the drinking-water for two weeks; they were then given tap-water and treated by oral gavage with 150 mg/kg bw d-limonene (> 99% pure) in 3 ml/kg corn oil daily on five days a week for 30 weeks. The livers and kidneys were examined, and other tumours were noted grossly. In Fischer 344 rats, 9/31 treated with NEHEA and d-limonene had renal adenomas, compared with 1/30 treated with NEHEA alone and none of 31 rats given d-limonene alone and none of 31 untreated controls ($p < 0.05$). The numbers of atypical hyperplasias in the kidney were 15.5 ± 1.5/rat treated with NEHEA and d-limonene, 1.2 ± 0.2/rat treated with NEHEA alone, 0.4 ± 0.1/rat treated with d-limonene alone and none in untreated controls. NBR rats did not develop renal adenomas, and no difference in the number of atypical hyperplasias was seen after subsequent administration of d-limonene (0.2 ± 0.1/rat in both groups), after feeding of d-limonene alone (0.1 ± 0.0/rat) or after no treatment (0.1 ± 0.1/rat). In Fischer rats, but not in NBR rats, the number of liver tumorus was reduced by administration of d-limonene (Dietrich & Swenberg, 1991).

3.3.2 Skin application

The Working Group was aware of a series of studies on orange oil (which contains d-limonene) in which a promoting effect on mouse skin carcinogenesis initiated by DMBA was reported (Roe, 1959; Roe & Peirce, 1960). Because of limitations in the conduct and reporting of these studies, they were not reviewed.

Mouse

Groups of 50 female ICR/Ha Swiss mice, six to eight weeks of age, each received topical applications on shaved back skin three times a week for 440 days of 10 mg limonene [stereo-chemistry and purity unspecified] in 0.1 ml acetone, 10 mg limonene simultaneously with 5 μg benzo[a]pyrene in 0.1 ml acetone, 5 μg benzo[a]pyrene alone or 0.1 ml acetone alone. A group of 100 females served as untreated controls. Only tumours that persisted 30 days or more were counted in the cumulative totals. Animals that developed skin carcinomas were killed approximately two months after the tumours had been classified as malignant or when they were moribund. No skin tumour was seen at 440 days in untreated controls, vehicle controls or mice that received limonene alone. In the benzo[a]pyrene-treated group, 16/50 mice [number of survivors not specified] had a total of 26 papillomas and 12 carcinomas (first papilloma seen at 210 days). Of mice that received limonene plus benzo[a]pyrene, 13/50 [number of survivors unspecified] had a total of 13 papillomas and four carcinomas (first papilloma seen at 295 days). The authors concluded that limonene had slightly inhibited benzo[a]pyrene carcinogenesis (Van Duuren & Goldschmidt, 1976). [The Working Group noted that no statistical analysis was performed and that this conclusion appears to be based on the lowered multiplicity of tumours and decreased number of malignant tumours in those given limonene.]

3.3.3 *Topical application and feeding*

Mouse

Groups of 24 female CD-1 mice, eight weeks of age, each received a single topical application of 0.2 μmol (51.2 μg) DMBA in 0.2 ml acetone or acetone alone on the shaved back. On day 7 after DMBA treatment, one DMBA-treated and one acetone-treated group were fed a diet containing 1% *d*-limonene (> 99% pure). On day 14 after DMBA treatment, three other groups received topical applications of 0.2 ml *d*-limonene (630 mmol, 1:1) in acetone, 10 nmol 12-*O*-tetradecanoylphorbol 13-acetate (TPA) in acetone or acetone alone twice a week. Beginning seven weeks after DMBA treatment, the number of mice bearing papillomas and the number of papillomas were recorded weekly. The experiment was terminated 40 weeks after DMBA treatment. All mice that received DMBA plus TPA rapidly developed skin tumours, whereas mice that received TPA alone or DMBA alone did not. DMBA-treated mice fed 1% *d*-limonene in the diet did not develop skin tumours, nor did untreated mice that received an application of *d*-limonene alone on the skin. In DMBA-treated mice that received skin applications of *d*-limonene, the incidence of skin papillomas/mouse was increased slightly (Elegbede *et al.*, 1986). [The Working Group noted the lack of detailed reporting.]

3.3.4 *Subcutaneous injection*

Mouse

Three groups of 50 male C57Bl/6 Jax mice [age unspecified] each received a subcutaneous injection of 25 μg benzo[*rst*]pentaphene (dibenzo[*a,i*]pyrene; DBP) in 0.1 ml tricaprylin and, 24 h later, a subcutaneous injection (at about the same site) of 0.2 ml 75% (v/v) orange oil (containing about 45–50% *d*-limonene) or 75% (v/v) orange oil (containing about 85–90% *d*-limonene) or no additional treatment. A vehicle-control group of 50 males received an initial subcutaneous injection of 0.1 tricaprylin only. Two further groups of 50 male C57Bl/6 Jax mice [age unspecified] received a single subcutaneous injection of 0.2 ml 75% (v/v) orange oil (containing about 45–50% limonene) or 75% (v/v) orange oil (containing about 85–90%) with no DBP pretreatment. The animals were observed for up to two years. Neither tricaprylin alone nor either of the two orange oils alone caused subcutaneous tumours (survival at two years: tricaprylin, 33/50; 45% limonene, 26/50; 84% limonene, 30/50). Mice that received DBP alone had a higher incidence of subcutaneous tumours than did the group that also received orange oil containing 45–50% limonene, and the group that also received orange oil containing 85–90% limonene had an even lower incidence (Homburger *et al.*, 1971).

Groups of 50 male C57Bl/6 Jax mice [age unspecified] each received a subcutaneous injection of 25 μg DBP in 0.1 ml tricaprylin and, 24, 48, 72 and 96 h later, subcutaneous injections (at about the same site) of either 0.05 ml 75% (v/v) *d*-limonene in tricaprylin, 0.05 ml 75% (v/v) autoxidized *d*-limonene (containing 6% hydroperoxides) in tricaprylin, 0.05 ml hydroperoxides alone [75% solution in tricaprylin unspecified], 0.05 ml orange oil (containing 85–90% limonene; control 1), 0.05 ml tricaprylin alone (control 2) or no further treatment (positive control). Tumour growth was reported to be reduced by d-limonene and by the hydroperoxides of d-limonene (Homburger *et al.*, 1971). [The Working Group noted the inadequate reporting.]

3.3.5 *Intravenous and intraperitoneal injection*

Mouse

Groups of 50 female A/Jax mice, two to three months old, each received a single subcutaneous injection of 500 µg DBP in 0.1 ml peanut oil or 0.1 ml peanut oil alone, followed after 24 h by weekly intravenous injections into the tail vein (later, intraperitoneal injections) of either 1% (v/v) orange oil (containing 45–50% limonene [stereoisomer, volume and diluent unspecified]), 1% (v/v) orange oil (containing 85–90% limonene) [volume and diluent unspecified] or 0.1 ml diluent. At the end of 13 weeks, the incidence of lung tumours was 21% in mice receiving peanut oil and diluent, 74% in mice receiving DBP plus diluent and 43 and 44% in mice receiving DBP plus either of the orange oils. In a second experiment, groups of 50 female A/Jax mice, seven weeks old, received a single subcutaneous injection of 500 µg/mouse DBP in 0.1 ml peanut oil or 0.1 ml peanut oil alone, followed after 24 h by weekly intravenous injections into the tail vein of either 1% (v/v) d-limonene, 1% (v/v) of its hydroperoxide or 0.1 ml diluent. At the end of 16 weeks, the incidences of lung adenomas were 27% in mice receiving peanut oil and diluent, 75% in mice receiving DBP and diluent, 40% in mice receiving DBP plus d-limonene and 50% in mice receiving DBP plus d-limonene hydroperoxide. The two orange oils and d-limonene hydroperoxide given alone without DBP pretreatment had no significant effect on the incidence of lung adenomas, whereas d-limonene alone reduced the incidence of lung adenomas from 27% (diluent controls) to 7% ($p < 0.05$) (Homburger *et al.*, 1971).

4. Other Relevant Data

4.1 Absorption, distribution, metabolism and excretion

4.1.1 *Humans*

d-Limonene is absorbed in the gastrointestinal tract. Two male volunteers administered ^{14}C-d-limonene at 1.6 g orally excreted 55–83% of the dose in their urine within 48 h. The major urinary metabolite isolated was 8-hydroxy-*para*-menth-1-en-9-yl-β-D-glucopyranosiduronic acid (M-VI, Fig. 1) (Kodama *et al.*, 1976).

4.1.2 *Experimental systems*

^{14}C-d-Limonene was absorbed rapidly following administration (800 mg/kg; 4.15 µCi/animal) by stomach tube to male Wistar rats. Radiolabel levels were maximal in the blood after 2 h; large amounts of radiolabel were also observed in the liver (maximal after 1 h) and the kidneys (maximal after 2 h). Negligible concentrations were found in blood and organs after 48 h (Igimi *et al.*, 1974).

Figure 1. Possible metabolic pathways of *d*-limonene

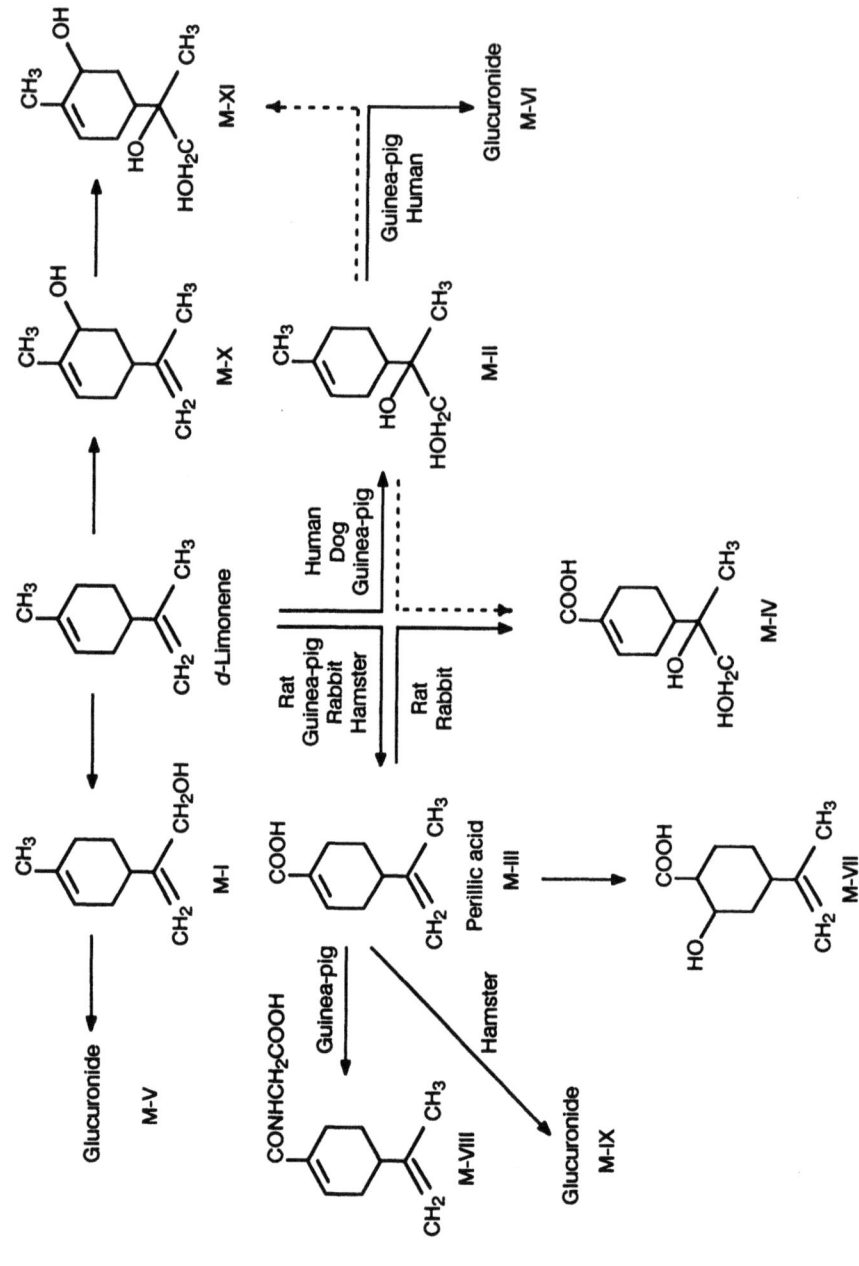

From Kodama *et al.* (1976)
M-I, *p*-Mentha-1,8-dien-10-ol; M-II, *p*-menth-1-ene-8,9-diol; M-IV, perillic acid-8,9-diol; M-V, *p*-mentha-1,8-dien-10-yl-β-D-glucopyranosiduronic acid; M-VI, 8-hydroxy-*p*-menth-1-en-9-yl-β-D-glucopyranosiduronic acid; M-VII, 2-hydroxy-*p*-menth-8-en-7-oic acid; M-VIII, perillylglycine; M-IX, perillyl-β-D-glucopyranosiduronic acid; M-X, *p*-mentha-1,8-dien-6-ol; M-XI, *p*-menth-1-ene-6,8,9-triol

Urinary recovery of ^{14}C-d-limonene was 77–96% within three days in rats, guinea-pigs, hamsters and dogs. Faecal recovery was 2–9% within three days (Kodama et al., 1976). Bile-duct cannulated rats administered d-limonene orally excreted 25% of the dose in the bile within 24 h (Igimi et al., 1974).

Following oral administration of d-limonene to rabbits, the urinary metabolites isolated were *para*-mentha-1,8-dien-10-ol (M-I), *para*-menth-1-ene-8,9-diol (M-II), perillic acid (M-III), perillic acid-8,9-diol (M-IV), *para*-mentha-1,8-dien-10-yl-β-D-glucopyranosiduronic acid (M-V) and 8-hydroxy-*para*-menth-1-en-9-yl-β-D-glucopyranosiduronic acid (M-VI) (Kodama et al., 1974) (see Fig. 1). Following oral administration of d-limonene to dogs and rats, a further five urinary metabolites were isolated: 2-hydroxy-*para*-menth-8-en-7-oic acid (M-VII), perillylglycine (M-VIII), perillyl-β-D-glucopyranosiduronic acid (M-IX), *para*-mentha-1,8-dien-6-ol (M-X) and probably *para*-menth-1-ene-6,8,9-triol (M-XI). The major urinary metabolite was M-IV in rats and rabbits, M-IX in hamsters, M-II in dogs and M-VI in guinea-pigs (Kodama et al., 1976).

d-Limonene was metabolized by rat liver microsomes *in vitro* to the glycols d-limonene 8,9-diol and d-limonene 1,2-diol *via* the 8,9- and 1,2-epoxides (Watabe et al., 1980, 1981) (See Fig. 2).

Figure 2. Oxidation of d-limonene double bonds by rat liver microsomes

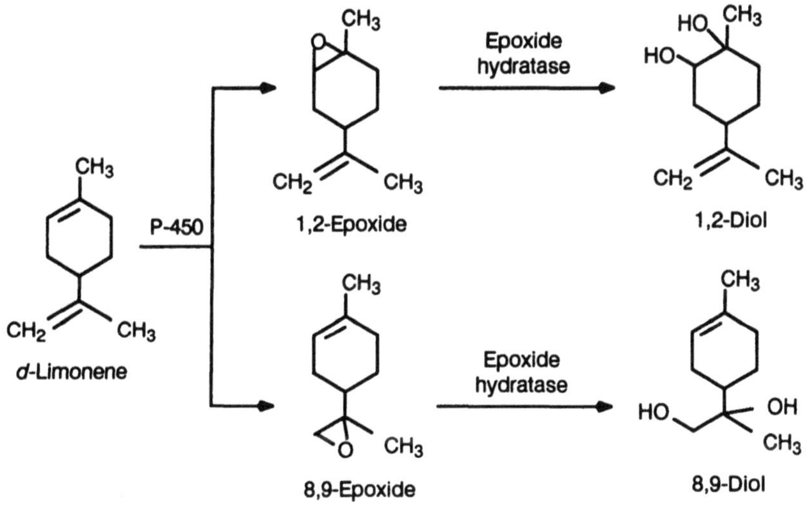

From Watabe et al. (1980)

4.2 Toxic effects

4.2.1 Humans

Five healthy male adult volunteers who received a single oral dose of 20 g d-limonene all developed transient proteinuria, non-bloody diarrhoea and tenesmus. The results of other functional tests of the liver, kidney and pancreas were normal (Igimi et al., 1976).

4.2.2 Experimental systems

LD_{50} values for d-limonene were reported in male and female mice to be 5.6 and 6.6 (oral), 1.3 and 1.3 (intraperitoneal) and > 41.5 and > 41.5 (subcutaneous) g/kg bw, respectively; those in male and female rats were reported to be 4.4 and 5.1 (oral), 3.6 and 4.5 (intraperitoneal), > 20.2 and > 20.2 (subcutaneous) and 0.125 and 0.11 (intravenous) g/kg bw, respectively (Tsuji et al., 1975a). The acute oral LD_{50} in rats and the acute dermal LD_{50} in rabbits were reported to exceed 5 g/kg bw (Opdyke, 1975).

After daily oral administration of d-limonene at 277–2770 mg/kg bw to male and female Sprague-Dawley rats for one month, the highest dose was found to have caused a slight decrease in body weight and food consumption. On histological examination, granular casts were observed in the kidney of males, but no significant change was found in the other organs (Tsuji et al., 1975a).

d-Limonene did not cause renal disease in NCI Black Reiter (NBR) male rats. These rats do not synthesize α_{2u}-globulin, which is normally present in the hyaline droplets that are found in male Fischer 344 rats with d-limonene-induced nephrotoxicity (Dietrich & Swenberg, 1991).

A dose-related increase in relative liver and kidney weights was observed in young adult male Fischer 344 rats administered 75, 150 or 300 mg/kg bw d-limonene daily by gavage on five days per week and killed on study days 6 or 27. Dose-related formation of hyaline droplets was also observed in the kidneys. α_{2u}-Globulin was detected in larger amounts in the renal cortical tissue of animals treated with d-limonene than in controls. Alterations considered to be sequelae of the hyaline droplet response, including granular casts in the outer zone of the medulla and multiple cortical changes collectively classified as chronic nephrosis, were observed in the kidneys of all rats killed on day 27 (Kanerva et al., 1987).

Chronic oral administration of 75 or 150 mg d-limonene to male Fischer 344/N rats on five days per week for two years was associated with dose-related alterations to the kidney, such as increased incidences of mineralization and epithelial hyperplasia and increased severity of spontaneous nephropathy (US National Toxicology Program, 1990).

After single administration of 14, 41, 136 or 409 mg/kg bw d-limonene by gavage to male and female Sprague-Dawley rats, a dose-related accumulation of hyaline droplets was observed in proximal renal tubules only in male rats. After administration of ^{14}C-labelled d-limonene at 409 mg/kg bw, 2.5 times more radiolabel was accumulated in renal tissue of males than in females; the label was found to be reversibly bound to protein (Lehman-McKeeman et al., 1989). In contrast, adult male and female beagle dogs administered d-limonene at 100 or 1000 mg/kg bw (maximal tolerated dose for emesis) per day by gavage twice daily for six months had increased kidney weights but showed no histopathological change, hyaline droplet accumulation or nephropathy (Webb et al., 1990).

In a chronic toxicity study in beagle dogs, oral doses of more than 340 mg/kg bw (females) and 1000 mg/kg bw (males) per day for six months resulted in protein casts in the renal tubules. Daily doses of more than 1000 mg/kg bw (females) and 3024 mg/kg bw (males) resulted in slight weight loss in some animals due to frequent vomiting (Tsuji et al., 1975b).

d-Limonene is a mildly toxic skin irritant when applied at full strength to intact or scratched rabbit skin for 24 h under occlusion (Opdyke, 1975). Continuous perfusion of

0.5% d-limonene (0.5–0.6 ml [420–504 mg]/min) into the gall-bladder of rabbits for 6 h irritated particularly the mucous membranes and the common bile duct (Tsuji et al., 1975c).

4.2.3 Mechanisms of toxicity

Treatment of male rats with d-limonene leads to a characteristic nephrotoxicity, a key feature of which is the accumulation in proximal tubule cells of hyaline droplets containing $\alpha_{2\mu}$-globulin. $\alpha_{2\mu}$-Globulin is the major low-molecular-weight protein excreted in male rat urine; it is present at much lower levels in females.

A number of studies have shown that d-limonene and d-limonene-1,2-oxide bind specifically, but reversibly, to $\alpha_{2\mu}$-globulin; the binding of d-limonene-1,2-oxide resulted in reduced degradation of the protein by lysosomal proteinases *in vitro* (Lehman-McKeeman et al., 1990). Cell death and proliferation were enhanced in renal tubules of Fischer 344 rats treated with d-limonene; no enhancement of cell proliferation occurred in NBR rats, which do not synthesize $\alpha_{2\mu}$-globulin (Dietrich & Swenberg, 1991).

In mature male rats, approximately 50 mg of $\alpha_{2\mu}$-globulin are filtered per day, about 40% being excreted in urine and 60% being reabsorbed and catabolized (Neuhaus et al., 1981; Baetcke et al., 1991). Female rats excrete 100–300 times less $\alpha_{2\mu}$-globulin in urine (Borghoff et al., 1990; Baetcke et al., 1991; Dietrich & Swenberg, 1991). Mice excrete large amounts of a structurally similar protein in a sex-dependent manner; however, this protein does not bind to d-limonene, nor is it reabsorbed in the kidney (Lehman-McKeeman & Caudill, 1992a).

$\alpha_{2\mu}$-Globulin belongs to a superfamily of proteins called lipocalins, which are widely distributed among mammalian species and bind, stabilize and transport hydrophobic ligands such as retinol and steroid hormones (Lehman-McKeeman & Caudill, 1992b). Normal human urine contains very little of this class of proteins, although a sex-dependent protein (urine protein 1) has been identified, which occurs in male urine at concentrations five times higher than in female urine. The concentration in male human urine is four to five times lower than that of $\alpha_{2\mu}$-globulin in male rat urine (Bernard et al., 1989; Baetcke et al., 1991). It is structurally related to rabbit uteroglobulin and not to the rat protein (Jackson et al., 1988). It does not bind to d-limonene. Of the lipocalins studied, only that in male rat kidney binds to limonene, whereas those of mice, hamsters, guinea-pigs, dogs and humans do not (Lehmann-McKeeman & Caudill, 1992b).

4.3 Reproductive and developmental toxicity

4.3.1 Humans

No data were available to the Working Group.

4.3.2 Experimental systems

Mice, rats and rabbits were treated orally during the period of organogenesis with daily doses of d-limonene up to 2363 mg/kg bw (mice), 2869 mg/kg bw (rats) and 1000 mg/kg bw (rabbits). The highest dose was lethal to < 40% of pregnant rabbits, and several rat dams died. The studies consistently showed impaired weight gain in the dams and delayed prenatal development, which was restored to normal during the postnatal period. In mice and rabbits, anomalies of the ribs were observed in the offspring (Tsuiji *et al.*, 1975d; Kodama *et al.*, 1977a, b).

4.4 Genetic and related effects

4.4.1 *Humans*

No data were available to the Working Group.

4.4.2 *Experimental systems* (see also Table 4 and Appendices 1 and 2)

d-Limonene was not mutagenic to *Salmonella typhimurium*. In single studies, it did not induce differential toxicity in *Bacillus subtilis* strains, sister chromatid exchange or chromosomal aberrations in Chinese hamster ovary cells, trifluorothymidine resistance in mouse lymphoma L5178Y cells or transformation in rat tracheal cells *in vitro*. *d*-Limonene-1,2-oxide did not induce unscheduled DNA synthesis in primary cultures of rat hepatocytes.

5. Summary of Data Reported and Evaluation

5.1 Exposure data

d-Limonene is found widely in citrus and many other plant species and is a major constituent of many essential oils. It is used extensively as a component of flavourings and fragrances, as a chemical intermediate and as an insect repellant. Widespread exposures occur through consumption of fruits, vegetables and products containing essential oils. Consumption of *d*-limonene has been estimated to be 0.2–2 mg/kg bw per day.

5.2 Human carcinogenicity data

No data were available to the Working Group.

5.3 Animal carcinogenicity data

d-Limonene has been tested for carcinogenicity by oral gavage in one study in mice and one study in rats. In mice, no treatment-related tumour was observed. It significantly increased the combined incidence of renal-cell adenomas and carcinomas and induced renal tubular hyperplasia in male rats.

In a two-stage experiment, oral treatment with *d*-limonene after administration of *N*-nitrosoethylhydroxyethylamine enhanced the development of renal adenomas and renal tubular hyperplasia in male Fischer 344 rats, which synthesize $\alpha_{2\mu}$-globulin, but not in male NBR rats, in which there is no evidence that $\alpha_{2\mu}$-globulin is synthesized in measurable quantities.

5.4 Other relevant data

In men, oral intake of *d*-limonene induced transient proteinuria. *d*-Limonene induced nephrotoxicity in male Fischer 344 but not NBR rats.

Table 4. Genetic and related effects of d-limonene and its metabolites

Test system	Result		Dose (LED/HID)[a]	Reference
	Without exogenous metabolic system	With exogenous metabolic system		
d-Limonene				
SA0, *Salmonella typhimurium* TA100, reverse mutation	–	–	1360.0000	Watabe *et al.* (1980)
SA0, *Salmonella typhimurium* TA100, reverse mutation	–	–	1667.0000	Haworth *et al.* (1983)
SA5, *Salmonella typhimurium* TA1535, reverse mutation	–	–	1360.0000	Watabe *et al.* (1980)
SA5, *Salmonella typhimurium* TA1535, reverse mutation	–	–	1667.0000	Haworth *et al.* (1983)
SA7, *Salmonella typhimurium* TA1537, reverse mutation	–	–	1360.0000	Watabe *et al.* (1980)
SA7, *Salmonella typhimurium* TA1537, reverse mutation	–	–	1667.0000	Haworth *et al.* (1983)
SA8, *Salmonella typhimurium* TA1538, reverse mutation	–	–	1360.0000	Watabe *et al.* (1980)
SA9, *Salmonella typhimurium* TA98, reverse mutation	–	–	1360.0000	Watabe *et al.* (1980)
SA9, *Salmonella typhimurium* TA98, reverse mutation	–	–	1667.0000	Haworth *et al.* (1983)
G51, Gene mutation, mouse lymphoma L5178Y cells, TFT resistance *in vitro*	–		60.0000	US National Toxicology Program (1990)
SIC, Sister chromatid exchange, Chinese hamster ovary cells *in vitro*	–	–	162.0000	US National Toxicology Program (1990)
CIC, Chromosomal aberrations, Chinese hamster ovary cells *in vitro*	–	–	500.0000	US National Toxicology Program (1990)
Transformation, rat tracheal cells *in vitro*[b]	–	0	3.0000	Steele *et al.* (1990)
d-Limonene-1,2-oxide				
URP, Unscheduled DNA synthesis, primary rat hepatocytes *in vitro*	–	0	13.6000	von der Hude *et al.* (1990)
Essential oils containing *d*-limonene				
BSD, *Bacillus subtilis*, rec strains, differential toxicity	–	0	4 mg/plate	Zani *et al.* (1991)
SA0, *Salmonella typhimurium* TA100, reverse mutation	–	–	200 µg/plate	Zani *et al.* (1991)
SA5, *Salmonella typhimurium* TA1535, reverse mutation	–	–	200 µg/plate	Zani *et al.* (1991)
SA7, *Salmonella typhimurium* TA1537, reverse mutation	–	–	200 µg/plate	Zani *et al.* (1991)
SA9, *Salmonella typhimurium* TA98, reverse mutation	–	–	200 µg/plate	Zani *et al.* (1991)

–, negative; 0, not tested
[a]In-vitro tests, µg/ml
[b]Not on profile

No data were available on the genetic and related effects of *d*-limonene in humans. In a small number of studies with a variety of end-points, *d*-limonene showed no evidence of genotoxic activity.

5.5 Evaluation[1]

No data were available on the carcinogenicity of *d*-limonene to humans.

There is *limited evidence* in experimental animals for the carcinogenicity of *d*-limonene.

Overall evaluation

d-Limonene *is not classifiable as to its carcinogenicity to humans (Group 3)*.

6. References

Aldrich Chemical Co. (1992) *Aldrich Catalog/Handbook of Fine Chemicals 1992- 1993*, Milwaukee, WI, p. 766

Anon. (1981) Italy: citrus oil production (Ger.). *Seifen Oele Fette Wachse*, **107**, 358

Anon. (1984) JCW spotlight on flavors and fragrances. *Jpn. chem. Week*, 3 May, pp. 4, 61

Anon. (1988a) Mexican lime oil squeeze. *Chem. Mark. Rep.*, **234**, 5, 26-27

Anon. (1988b) Annual citrus crop is strong; demand for oil even higher. *Chem. Mark. Rep.*, **234**, 30–31

Anon. (1989) *d*-Limonene's price is soft; buyers now look to Brazil. *Chem. Mark. Rep.*, **236**, 24

Anon. (1990) Florida freeze chills oils; citruses squeezed bitterly. *Chem. Mark. Rep.*, **237**, 22–23

Baetcke, K.P., Hard, G.C., Rodgers, I.S., McGaughy, R.E. & Tahan, L.M. (1991) *Alpha$_{2\mu}$-globulin: Association with Chemically Induced Renal Toxicity and Neoplasia in the Male Rat* (EPA/625/3-91/019F), Washington DC, Environmental Protection Agency

Bernard, A.M., Lauwerys, R.R., Nöel, A., Vandeleene, B. & Lambert, A. (1989) Urine protein 1: a sex-dependent marker of tubular or glomerular dysfunction. *Clin. Chem.*, **35**, 2141–2142

Bertsch, W., Chang, R.C. & Zlatkis, A. (1974) The determination of organic volatiles in air pollution studies: characterization of profiles. *J. chromatogr. Sci.*, **12**, 175–182

Boland, F.E. (1984) Fruit and fruit products. Oil (recoverable) in fruits and fruit products. Titrimetric method. Final action. In: Williams, S., ed., *Official Methods of Analysis of the Association of Official Analytical Chemists*, 14th ed., Arlington, VA, Association of Official Analytical Chemists, p. 424

Borghoff, S.J., Short, B.G. & Swenberg, J.A. (1990) Biochemical mechanisms and pathobiology of $\alpha_{2\mu}$-globulin nephropathy. *Ann. Rev. Pharmacol. Toxicol.*, **30**, 349–67

Budavari, S., ed. (1989) *The Merck Index*, 11th ed., Rahway, NJ, Merck & Co., p. 865

Clayton, G.D. & Clayton, F.E., eds (1981) *Patty's Industrial Hygiene and Toxicology*, 3rd rev. ed., New York, John Wiley & Sons, pp. 3237–3238

Connor, T.H., Theiss, J.C., Hanna, H.A., Monteith, D.K. & Matney, T.S. (1985) Genotoxicity of organic chemicals frequently found in the air of mobile homes. *Toxicol. Lett.*, **25**, 33–40

---

[1]For definition of the italicized terms, see Preamble, pp. 26–29.

Dietrich, D.R. & Swenberg, J.A. (1991) The presence of $\alpha_{2\mu}$-globulin is necessary for *d*-limonene promotion of male rat kidney tumors. *Cancer Res.*, **51**, 3512-3521

Elegbede, J.A., Elson, C.E., Qureshi, A., Tanner, M.A. & Gould, M.N. (1984) Inhibition of DMBA-induced mammary cancer by the monoterpene *d*-limonene. *Carcinogenesis*, **5**, 661-664

Elegbede, J.A., Maltzman, T.H., Verma, A.K., Tanner, M.A., Elson, C.E. & Gould, M.N. (1986) Mouse skin tumor promoting activity of orange peel oil and *d*-limonene: a re-evaluation. *Carcinogenesis*, **7**, 2047-2049

Elson, C.E., Maltzman, T.H., Boston, J.L., Tanner, M.A. & Gould, M.N. (1988) Anti-carcinogenic activity of *d*-limonene during the initiation and promotion/progression stages of DMBA-induced rat mammary carcinogenesis. *Carcinogenesis*, **9**, 331-332

Field, W.E.H. & Roe, F.J.C. (1965) Tumor promotion in the forestomach epithelium of mice by oral administration of citrus oils. *J. natl Cancer Inst.*, **35**, 771-787

Flavor and Extract Manufacturers' Association (1975) *Scientific Literature Review of Aliphatic Hydrocarbons in Flavor Usage*, Vol. I, Introduction and Summary, Tables of Data, Bibliography, Washington DC, pp. 1-9, 68-81, 136-139, 143-144

Flavor and Extract Manufacturers' Association (1991) *d-Limonene Monograph*, Washington DC, pp. 1-4

Florida Chemical Co. (1991a) *Marketing Data Sheet: d-Limonene*, Lake Alfred, FL

Florida Chemical Co. (1991b) *Product Data Sheet: d-Limonene*, Lake Alfred, FL

Florida Chemical Co. (1991c) *Material Safety Data Sheet: d-Limonene*, Lake Alfred, FL

Furia, T.E. & Bellanca, N., eds (1975) *Fenaroli's Handbook of Flavor Ingredients*, Vol. 2, 2nd ed., Cleveland, OH, CRC Press, p. 319

Gerow, G.P. (1974) Economics of *d*-limonene recovery. *Trans. Citrus Eng. Conf.*, **20**, 61-66

Haworth, S., Lawlor, T., Mortelmans, K., Speck, W. & Zeiger, E. (1983) *Salmonella* mutagenicity test results for 250 chemicals. *Environ. Mutag.*, **Suppl. 1**, 3-142

Homburger, F., Treger, A. & Boger, E. (1971) Inhibition of murine subcutaneous and intravenous benzo(*rst*)pentaphene carcinogenesis by sweet orange oils and *d*-limonene. *Oncology*, **25**, 1-10

Hooser, S.B. (1990) *d*-Limonene, linalool, and crude citrus oil extracts. *Vet. Clin. North Am. small Anim. Pract.*, **20**, 383-385

Hooser, S.B., Beasley, V.R. & Everitt, J.I. (1986) Effects of an insecticidal dip containing *d*-limonene in the cat. *J. Am. vet. Med. Assoc.*, **189**, 905-908

von der Hude, W., Mateblowski, R. & Basler, A. (1990) Induction of DNA-repair synthesis in primary rat hepatocytes by epoxides. *Mutat. Res.*, **245**, 145-150

Igimi, H., Nishimura, M., Kodama, R. & Ide, H. (1974) Studies on the metabolism of *d*-limonene (*p*-mentha-1,8-diene). I. The absorption, distribution and excretion of *d*-limonene in rats. *Xenobiotica*, **4**, 77-84

Igimi, H., Hisatsugu, T. & Nishimura, M. (1976) The use of *d*-limonene preparation as a dissolving agent of gallstones. *Digest. Dis.*, **21**, 926-939

Igimi, H., Tamura, R., Toraishi, K., Yamamoto, F., Kataoka, A., Ikejiri, Y., Hisatsugu, T. & Shimura, H. (1991) Medical dissolution of gallstones. Clinical experience of *d*-limonene as a simple, safe, and effective solvent. *Dig. Dis. Sci.*, **36**, 200-208

Jackson, P.J., Turner, R., Keen, J.N., Brooksbank, R.A. & Cooper, E.H. (1988) Purification and partial amino acid sequence of human urine protein 1. Evidence for homology with rabbit uteroglobin. *J. Chromatogr.*, **452**, 359-367

Kanerva, R.L., Ridder, G.M., Lefever, F.R. & Alden, C.L. (1987) Comparison of short-term renal effects due to oral administration of decalin or d-limonene in young adult male Fischer-344 rats. *Food chem. Toxicol.*, **25**, 345–353

Karr, L.L. & Coats, J.R. (1988) Insecticidal properties of d-limonene. *J. Pestic. Sci.*, **13**, 287–290

Kodama, R., Noda, K. & Ide, H. (1974) Studies on the metabolism of d-limonene (p-mentha-1,8-diene). II. The metabolic fate of d-limonene in rabbits. *Xenobiotica*, **4**, 85–95

Kodama, R., Yano, T., Furukawa, K., Noda, K. & Ide, H. (1976) Studies on the metabolism of d-limonene (p-mentha-1,8-diene). IV. Isolation and characterization of new metabolites and species differences in metabolism. *Xenobiotica*, **6**, 377–389

Kodama, R., Okubo, A., Araki, E., Noda, K., Ide, H. & Ikeda, T. (1977a) Studies on d-limonene as a gallstone solubilizer. VII. Effects on development of mouse fetuses and offsprings (Jpn.). *Oyo Yakuri*, **13**, 863–873

Kodama, R., Okubo, A., Sato, K., Araki, E., Noda, K., Ide, H. & Ikeda, T. (1977b) Studies on d-limonene as a gallstone solubilizer. IX. Effects on development of rabbit fetuses and offsprings (Jpn.). *Oyo Yakuri*, **13**, 885–898

Lehmann-McKeeman, L.D. & Caudill, D. (1992a) Biochemical basis for mouse resistance to hyaline droplet nephropathy: lack of relevance of the $\alpha_{2\mu}$-globulin protein superfamily in this male rat-specific syndrome. *Toxicol. appl. Pharmacol.*, **112**, 214–221

Lehman-McKeeman, L. & Caudill, D. (1992b) $\alpha_{2\mu}$-Globulin is the only member of the lipocalin protein superfamily that binds to hyaline droplet inducing agents. *Toxicol. appl. Pharmacol.*, **116**, 170–176

Lehman-McKeeman, L.D., Rodriguez, P.A., Takigiku, R., Caudill, D. & Fey, M.L. (1989) d-Limonene-induced male rat-specific nephrotoxicity: evaluation of the association between d-limonene and $\alpha_{2\mu}$-globulin. *Toxicol. appl. Pharmacol.*, **99**, 250–259

Lehman-McKeeman, L.D., Rivera-Torres, M.I. & Caudill, D. (1990) Lysosomal degradation of $\alpha_{2\mu}$-globulin and $\alpha_{2\mu}$-globulin-xenobiotic conjugates. *Toxicol. appl. Pharmacol.*, **103**, 539–548

Leung, A.Y. (1980) *Encyclopedia of Common Natural Ingredients Used in Food, Drugs and Cosmetics*, New York, John Wiley & Sons

Lide, D.R., ed. (1991) *CRC Handbook of Chemistry and Physics*, 72nd ed., Boca Raton, FL, CRC Press, p. 3–308

Lloyd, R.J. (1984) Instrumentation for automated thermal desorption—pyrolysis capillary gas chromatography. *J. Chromatogr.*, **284**, 357–371

Lockhart, L.A. (1990) The xylem resin terpene composition of *Pinus greggii* Engelm. and *Pinus pringlei* Shaw. *Silvae genet.*, **39**, 198–202

Maarse, H. & Visscher, C.A., eds (1988) *Volatile Compounds in Food. Quantitative Data*, Vol. 7, Zeist, TNO-CIVO Food Analysis Institute

Maltzman, T.H., Hurt, L.M., Elson, C.E., Tanner, M.A. & Gould, M.N. (1989) The prevention of nitrosomethylurea-induced mammary tumors by d-limonene and orange oil. *Carcinogenesis*, **10**, 781–783

Mannheim, C.H., Miltz, J. & Letzter, A. (1987) Interaction between polyethylene laminated cartons and aseptically packed citrus juice. *J. Food Sci.*, **52**, 737–740

Marsili, R. (1986) Measuring volatiles and limonene-oxidation products in orange juice by capillary GC. *LC–GC*, **4**, 358–362

Massaldi, H.A. & King, C.J. (1974) Determination of volatiles by vapor headspace analysis in a multiphase system: d-limonene in orange juice. *J. Food Sci.*, **39**, 434–437

Mosandl, A., Hener, U., Kreis, P. & Schmarr, H.-G. (1990) Enantiomeric distribution of *alpha*-pinene, *beta*-pinene and limonene in essential oils and extracts. Part 1. Rutaceae and Gramineae. *Flavour. Fragrance J.*, **5**, 193–199

Neuhaus, O.W., Flory, W., Biswas, N. & Hollerman, C.E. (1981) Urinary excretion of $\alpha_{2\mu}$-globulin and albumin by adult male rats following treatment with nephrotoxic agents. *Nephron*, **28**, 133–140

Opdyke, D.L.J. (1975) *d*-Limonene. *Food Cosmet. Toxicol.*, **13** (Suppl.), 825–826

Ranganna, S., Govindarajan, V.S. & Ramana, K.V.R. (1983) Citrus fruits—varieties, chemistry, technology, and quality evaluation. Part II. Chemistry, technology, and quality evaluation. A. Chemistry. *CRC crit. Rev. Food Sci. Nutr.*, **18**, 313–386

Roe, F.J.C. (1959) Oil of sweet orange: a possible role in carcinogenesis. *Br. J. Cancer*, **13**, 92–93

Roe, F.J.C. & Peirce, W.E.H. (1960) Tumor promotion by citrus oils: tumors of the skin and urethral orifice in mice. *J. natl Cancer Inst.*, **24**, 1389–1403

Russin, W.A., Hoesly, J.D., Elson, C.E., Tanner, M.A. & Gould, M.N. (1989) Inhibition of rat mammary carcinogenicity by monoterpenoids. *Carcinogenesis*, **10**, 2161–2164

Saleh, M.M., Zwaving, J.H., Malingré, T.M. & Bos, R. (1985) The essential oil of *Apium graveolens* var. *secalinum* and its cercaricidal activity. *Pharm. Weekbl. [Sci.]*, **7**, 277–279

Sax, N.I. & Lewis, R.J. (1987) *Hawley's Condensed Chemical Dictionary*, 11th ed., New York, Van Nostrand Reinhold, p. 701

Schulz, H.E. (1972) *d*-Limonene recovery in the Florida citrus industry. *Citrus Eng. Conf. Trans.*, **18**, 1–6

Searle, E. (1989) Determination of airborne limonene vapour by charcoal tube sampling and gas-liquid chromatographic analysis. *Analyst*, **114**, 113–114

Shaw, P.E. (1979) Review of quantitative analyses of citrus essential oils. *J. agric. Food Chem.*, **27**, 246–257

Simon, P.W., Lindsay, R.C. & Peterson, C.E. (1980) Analysis of carrot volatiles collected on porous polymer traps. *J. agric. Food Chem.*, **28**, 549–552

Steele, V.E., Kelloff, G.J., Wilkinson, B.P. & Arnold, J.T. (1990) Inhibition of transformation in cultured rat tracheal epithelial cells by potential chemopreventive agents. *Cancer Res.*, **50**, 2068–2074

STN International (1992) *HODOC Database*, Columbus, OH

Stofberg, J. & Grundschober, F. (1987) Consumption ratio and food predominance of flavoring materials. *Perfum. Flavor.*, **12**, 27–32

Stoner, G.D., Shimkin, M.B., Kniazeff, A.J., Weisburger, J.H., Weisburger, E.K. & Gori, G.B. (1973) Test for carcinogenicity of food additives and chemotherapeutic agents by the pulmonary tumor response in strain A mice. *Cancer Res.*, **33**, 3069–3085

Topfer, K. (1990) Orange oil price decreases as Brazilian harvest begins. *Chem. Mark. Rep.*, **238**, 21

Tsuji, M., Fujisaki, Y., Arikawa, Y., Masuda, S., Kinoshita, S., Okubo, A., Noda, K., Ide, H. & Iwanaga, Y. (1975a) Studies on *d*-limonene, as gallstone solubilizer. II. Acute and subacute toxicities (Jpn.). *Oyo Yakuri*, **9**, 387–401

Tsuji, M., Fujisaki, Y., Arikawa, Y., Masuda, S., Tanaka, T., Sato, K., Noda, K., Ide, H. & Kikuchi, M. (1975b) Studies on *d*-limonene, as gallstone solubilizer. IV. Chronic toxicity in dogs (Jpn.). *Oyo Yakuri*, **9**, 775–808

Tsuji, M., Fujisaki, Y., Saita, M., Noda, K. & Ide, H. (1975c) Studies on *d*-limonene as a gallstone solubilizer. VI. The pharmacological effects of *d*-limonene on the biliary and gastrointestinal system (Jpn.). *Oyo Yakuri*, **10**, 187–197

Tsuji, M., Fujisaki, Y., Okubo, A., Arikawa, Y., Noda, K., Ide, H. & Ikeda, T. (1975d) Studies on *d*-limonene as a gallstone solubilizer. V. Effects on development of rat fetuses and offspring (Jpn.). *Oyo Yakuri*, **10**, 179–186

US Department of Agriculture (1982) *Foods Commonly Eaten by Individuals: Amount per Day and per Eating Occasion* (Home Economics Research Report No. 44), Washington DC, Human Nutrition Information Service, pp. 76–77

US Department of Commerce (1991) *US Imports for Consumption* (FT247/Annual 1990), Washington DC, Bureau of the Census, p. F-49

US Food and Drug Administration (1991) Substances generally recognized as safe—synthetic flavoring substances and adjuvants. *US Code fed. Regul.*, **Title 21**, Section 182.60, p. 404

US National Toxicology Program (1990) *Toxicology and Carcinogenesis Studies of d-Limonene (CAS No. 5989-27-5) in F344/N Rats and B6C3F$_1$ Mice (Gavage Studies)* (NTP TR 347; NIH Publ. No. 90–2802), Research Triangle Park, NC

Van Duuren, B.L. & Goldschmidt, B.M. (1976) Cocarcinogenic and tumor-promoting agents in tobacco carcinogenesis. *J. natl Cancer Inst.*, **56**, 1237–1242

Verghese, J. (1968) The chemistry of limonene and of its derivatives—Part I. *Perfum. Essent. Oil. Rec.*, **59**, 439–454

Watabe, T., Hiratsuka, A., Isobe, M. & Ozawa, N. (1980) Metabolism of *d*-limonene by hepatic microsomes to non-mutagenic epoxides toward *Salmonella typhimurium*. *Biochem. Pharmacol.*, **29**, 1068–1071

Watabe, T., Hiratsuka, A., Ozawa, N. & Isobe, M. (1981) A comparative study of the metabolism of *d*-limonene and 4-vinylcyclohex-1-ene by hepatic microsomes. *Xenobiotica*, **11**, 333–344

Wattenberg, L.W. & Coccia, J.B. (1991) Inhibition of 4-(methylnitrosamino)-1-(3-pyridyl)-1-butanone carcinogenesis in mice by *d*-limonene and citrus fruit oils. *Carcinogenesis*, **12**, 115–117

Wattenberg, L.W., Sparnins, V.L. & Barany, G. (1989) Inhibition of *N*-nitrosodiethylamine carcinogenesis in mice by naturally occurring organosulfur compounds and monoterpenes. *Cancer Res.*, **49**, 2689–2692

Webb, D.R., Kanerva, R.L., Hysell, D.K., Alden, C.L. & Lehman-McKeeman, L.D. (1990) Assessment of the subchronic oral toxicity of *d*-limonene in dogs. *Food chem. Toxicol.*, **28**, 669–675

Yokoyama, K.M., Wanitprapha, K., Nakamoto, S.T., Leung, P.-S. & Roecklein, J.C. (1988) *US Import Statistics for Agricultural Commodities 1981–1986*, New Brunswick, NJ, Transaction Books, pp. 223, 250, 258, 321

Zani, F., Massimo, G., Benvenuti, S., Bianchi, A., Albasini, A., Melegari, M., Vampa, G., Bellotti, A. & Mazza, P. (1991) Studies on the genotoxic properties of essential oils with *Bacillus subtilis rec*-assay and *Salmonella*/microsome reversion assay. *Planta med.*, **57**, 237–241

Zlatkis, A. & Liebich, H.M. (1971) Profile of volatile metabolites in human urine. *Clin. Chem.*, **17**, 592–594

HETEROCYCLIC AROMATIC AMINES

IQ (2-AMINO-3-METHYLIMIDAZO[4,5-f]QUINOLINE)

This substance was considered by a previous Working Group, in October 1985 (IARC, 1986a). Since that time, new data have become available, and these have been incorporated into the monograph and taken into consideration in the present evaluation.

1. Exposure Data

1.1 Chemical and physical data

1.1.1 *Synonyms, structural and molecular data*

Chem. Abstr. Services Reg. No.: 76180-96-6
Chem. Abstr. Name: 3-Methyl-3H-imidazo[4,5-f]quinolin-2-amine
IUPAC Systematic Name: 2-Amino-3-methyl-3H-imidazo[4,5-f]quinoline

$C_{11}H_{10}N_4$ Mol. wt: 198.23

1.1.2 *Chemical and physical properties*

(a) *Description*: Crystalline solid (Kasai *et al.*, 1981)
(b) *Melting-point*: > 300 °C (Kasai *et al.*, 1981)
(c) *Spectroscopy data*: Ultraviolet (Sugimura *et al.*, 1981), proton nuclear magnetic resonance (Kasai *et al.*, 1980a), infrared absorbance (Kasai *et al.*, 1981) and mass spectral data (Spingarn *et al.*, 1980) have been reported.
(d) *Solubility*: Soluble in methanol, ethanol and dimethyl sulfoxide (Kasai *et al.*, 1980a, 1981; Lee *et al.*, 1982; Schunk *et al.*, 1984)
(e) *Stability*: Stable under moderately acidic and alkaline conditions and in cold dilute aqueous solutions protected from light (Sugimura *et al.*, 1983)
(f) *Reactivity*: Rapidly degraded by dilute hypochlorite; not deaminated by weakly acidic nitrite solutions (Tsuda *et al.*, 1985)

1.1.3 *Trade names, technical products and impurities*

No data were available to the Working Group.

1.1.4 *Analysis*

The complex matrix of cooked foods makes analysis of IQ difficult. IQ was originally isolated from broiled, sun-dried sardines, extracted with methanol and purified by Diaion HP-20 column chromatography, chloroform–methanol–water partitioning and Sephadex LH-20 column chromatography, silica-gel column chromatography and, finally, reverse-phase high-performance liquid chromatography (HPLC). The structure was deduced mainly from data obtained by proton nuclear magnetic resonance and high-resolution mass spectral analysis (Kasai *et al.*, 1980b, 1981).

IQ has been isolated from beef extract by dichloromethane extraction, column chromatography on Adsorbosil-5 and Sephadex LH-20 and HPLC, with analysis by mass spectrometry, ultraviolet spectrophotometry and/or mutagenesis assay (Hargraves & Pariza, 1983; Turesky *et al.*, 1983). IQ was detected in fried ground beef following dichloromethane extraction, chromatography on XAD-2 resin, three different HPLC separations and off-line mass spectrometry (Felton *et al.*, 1984).

IQ can be adsorbed from aqueous solutions onto cellulose or cotton to which CI Reactive Blue 21, a trisulfo-copper phthalocyanine dye, has been bound covalently (referred to as the 'blue cotton' adsorption technique). The adsorbed IQ is eluted with an ammonia–methanol solution and quantified by HPLC (Hayatsu *et al.*, 1983).

More sophisticated methods of detecting IQ in cooked foods using deuterium-labelled internal standards and HPLC-thermospray-mass spectrometry have been devised (Yamaizumi *et al.*, 1986; Turesky *et al.*, 1988). Although they differ in detail, these methods involve methanol extraction, acid–base partitioning and 'blue cotton' adsorption prior to analysis by HPLC–thermospray–mass spectrometry.

Monoclonal antibodies to IQ have been developed (Vanderlaan *et al.*, 1988), and Yanagisawa *et al.* (1990) showed that their antibodies bound to IQ and related heterocyclic amines with varying specificities. Turesky *et al.* (1989) used monoclonal antibodies immobilized on a support for selective immunoaffinity chromatography as a clean-up procedure in the analysis of beef extracts.

A practical, solid-phase extraction and HPLC (Kieselgur–Sephasorb) method for the analysis of IQ and other heterocyclic amines in foods and food extracts was devised by Gross *et al.* (1989). Improvements to the method (Gross, 1990; Gross & Grüter, 1992) allow determination of IQ and most of the other known heterocyclic amines at a level of 1 ng/g from only 10 g of food sample. Replicate samples and spiking allow accurate determination of extraction losses; chromatographic peak identities are confirmed using a diode array-ultraviolet detector.

1.2 Production and use

1.2.1 *Production*

The isolation and identification of IQ were first reported by Kasai *et al.* (1980b). Its structure was confirmed by chemical synthesis, in which 5,6-diaminoquinoline was reacted with cyanogen bromide and the resulting cyclic intermediate was converted to IQ by heating the tetramethylammonium salt under reduced pressure. Final purification was accomplished by sublimation, silica-gel column chromatography and crystallization from aqueous methanol (Kasai *et al.*, 1980a, 1981).

Improved synthetic routes were devised by Lee *et al.* (1982), Adolfsson and Olsson (1983) and Waterhouse and Rapoport (1985). Synthesis of ^{14}C-labelled IQ was reported by Adolfsson and Olsson (1983) and of tritium- and deuterium-labelled IQ by Waterhouse and Rapoport (1985).

IQ is produced commercially in small quantities for research purposes.

1.2.2 *Use*

IQ is not used commercially.

1.3 Occurrence

As cooking terms such as broiling and grilling have different meanings in different parts of the world, the authors' terminology has been retained.

IQ is formed during the cooking of various meats and fish. It was originally isolated from broiled fish and has also been quantified in fried ground beef and beef extracts. The amounts found in these and other samples are listed in Table 1. IQ was also detected but not quantified in fried ground pork (Gry *et al.*, 1986), in beef extracts used for bacteriological media and in food-grade beef extracts (Hargraves & Pariza, 1983).

Table 1. Concentrations of IQ in foods

Sample	Concentration (ng/g)	No. of samples	Reference
Sardines, sun-dried, broiled	~ 20	1	Kasai *et al.* (1980b)
	158	1	Sugimura *et al.* (1981)
Ground beef, fried			
240 °C	0.5–20^a	2	Barnes *et al.* (1983)
250 °C	0.02	1	Felton *et al.* (1984)
275 °C	0.3–1.9	3	Turesky *et al.* (1988)
Ground beef, broiled	0.5	1	Yamaizumi *et al.* (1986)
Beef, broiled	0.19	1	Wakabayashi *et al.* (1992)
Beef extract, food-grade	< 0.2	1	Takahashi *et al.* (1985)
	< 0.1–6.2	3	Turesky *et al.* (1989)
Salmon, broiled	0.3–1.8	2	Yamaizumi *et al.* (1986)
Fish, fried at 260 °C	0.16	1	Zhang *et al.* (1988)
Egg, fried at 325 °C	0.1	1	Grose *et al.* (1986)

^{a}20 ng/g in a high-fat sample, 0.5 ng/g in a low-fat sample

IQ was found in a mixture of creatine and proline heated to 180 °C (Yoshida *et al.*, 1984), in a mixture of glycine, fructose and creatinine heated to 128 °C (Grivas *et al.*, 1986), in dry mixtures of creatinine and phenylalanine or of creatinine, phenylalanine and glucose heated to 200 °C (Felton & Knize, 1990) and in a dry mixture of serine and creatinine heated to 200 °C (Knize et al., 1988). Quantities of mixtures of heterocyclic amines found in different foods are listed in Table 2.

Table 2. Representative concentrations of heterocyclic amine (ng/g) in various food samples and in cigarette smoke condensate

Sample	PhIP (ng/g)	MeIQx (ng/g)	IQ (ng/g)	MeIQ (ng/g)	Reference
Beef, fried at 190 °C	48.5	8.3	ND[a]	ND[a]	Gross (1990)
Walleye pollack, fried at 260 °C	69.2	6.44	0.16	0.03	Zhang et al. (1988)
Ground beef, fried at 250 or 300 °C	15	1.0	< 0.1	Trace	Felton et al. (1986)
Chicken, broiled	38.1	2.33	NA	NA	Hayatsu et al. (1991)
Ground beef, fried at 250°C	NA	1.0	0.02	ND	Felton et al. (1984)
Salmon, baked at 200 °C for 30'	18	4.6	ND	ND	Gross & Grüter (1992)
Cigarette smoke condensate	NA	ND	0.26 ng per cigarette	ND	Yamashita et al. (1986)

ND, not detected by method used; NA, not analysed
[a]Limit of detection 1 and 2 ng/g according to the method used

1.4 Regulations and guidelines

No data were available to the Working Group.

2. Studies of Cancer in Humans

No epidemiological study was available that addressed the carcinogenic risk to humans of IQ itself. Cancer risks associated with consumption of broiled and fried foods, which may contain IQ as well as other heterocyclic amines, have, however, been addressed in a number of case–control studies. Several of these are summarized below. IQ is also a component of tobacco smoke, which has been considered in a previous IARC monograph (IARC, 1986b).

A large number of studies on diet and cancer have been conducted, most of which addressed specific hypotheses (e.g., dietary fat intake and risk of colorectal cancer) (see Tomatis et al., 1990). Many of the investigators, however, either failed to collect data on methods of food processing, or, if they collected such data, did not analyse the findings at all or reported only summary findings. Thus, it is possible that the studies reviewed below are only a small segment of the data potentially available on the effects in humans of heterocyclic amines formed in cooking, and a positive reporting bias cannot be ruled out.

2.1 Cohort studies

A study by Ikeda et al. (1983) involved 11 203 subjects already enrolled in the Adult Health Study of the Radiation Effects Research Foundation in Hiroshima and Nagasaki between 1968 and 1970. Complete personal histories and information on current dietary and other habits were available for 7553 of these subjects (1781 men and 3341 women in Hiroshima; 965 men and 1466 women in Nagasaki). After 11 years of follow-up, there were 244 deaths from cancer, 79 of which were from stomach cancer. Intake of five foods was assessed in a multivariate analysis: the relative risks associated with consumption of broiled fish were 1.7 for gastric cancer and 1.3 for cancers at all sites, after other food variables had been taken into account (both $p < 0.05$).

The report of a large prospective study of 88 751 nurses followed up from 1980, when they answered a mailed questionnaire, to 1986 mentions that no association was found between 150 incident cases of colon cancer and the degree of cooking of red meat, but no data were shown (Willett et al., 1990).

2.2 Case–control studies

2.2.1 Cancer of the colon and rectum

A case–control study included 340 colon and rectum cancer patients and 1020 hospital controls enrolled during 1959 from seven hospitals in Kansas City, USA (Higginson, 1966). A food frequency questionnaire was used. Unadjusted proportions of subjects in different categories were presented. No difference was observed between cases and controls with respect to consumption of fried potatoes, fried meats, fried food for breakfast or method of cooking meats. [The Working Group noted that confounding by intake of other nutrients was not controlled.]

Wynder et al. (1969) carried out a hospital-based case–control study in two cancer hospitals in Tokyo, Japan, in which 157 cases of colonic or rectal cancer and 307 sex- and age-matched controls were interviewed concerning their usual adult intake of a number of foods, including fried foods and charcoal-broiled fish. No consistent difference was found. [The Working Group noted that confounding by intake of other nutrients and method of cooking was not controlled.]

Phillips (1975) interviewed 41 Seventh-day Adventists with colonic–rectal cancer and 77 with breast cancer discharged from two Adventist-operated hospitals in 1969–73, each matched by age, sex and race with three Adventist controls: two hospitalized for hernia or osteoarthritis and the third from the general population [details not given]. Consumption of beef, meat in general and several sources of saturated fat were positively associated with the incidence of colonic but not breast cancer. Fried foods were positively associated with cancers at each site; the association with fried potatoes was statistically significant at the 5% level (odds ratio (OR), 2.7 for colonic and 2.4 for breast cancer). [The Working Group noted that only a crude, unmatched analysis was presented and that confounding by intake of other nutrients and method of cooking was not controlled.]

Young and Wolf (1988) reported a case–control study of 353 cases of colonic cancer (152 proximal and 201 distal subsites) in people aged 50–89, drawn from the Wisconsin (USA) Cancer Reporting System, and 618 general population controls. The study was population-based, but 62.2% of the potential study subjects were alive and could be contacted and interviewed; 17.7% refused to participate, 13 patients were < 50 years old, thus leaving 353 cases for the study. Information on the frequency of consumption of foods in 27 groups and replies to questions about eating habits and cooking styles were requested for three periods: before the age of 18, between 18 and 35 years of age and after the age of 35. Consumption of broiled foods increased markedly with age and was consistently lower among cases than controls; on the contrary, the frequency of consumption of pan-fried foods decreased with age but was higher among cases. Consumption of processed meat and pan-fried foods was consistently associated with increased risks for cancers at each site. A significant risk for the upper *versus* the lowest quartile of consumption of pan-fried foods was observed for cancer of the proximal colon in association with diet in young adults (18–35 years old) (OR, 1.79; 95% confidence interval [CI], 1.15–2.80, adjusted for age and sex). The overall association (for both proximal and distal subsites) was not, however, significant. [The Working Group noted that confounding by intake of other nutrients was not controlled.]

Peters *et al.* (1989) reported a study of 147 men with colorectal carcinomas among 232 eligible cases, all of whom were aged 24–44 years at diagnosis and were identified through the Los Angeles County (USA) Cancer Surveillance Program, and 147 neighbourhood controls who were compared in terms of occupational exposure, tobacco and alcohol use and usual consumption of foods grouped into a few broad categories. There was no significant difference between cases who were interviewed and those not interviewed with regard to marital status, religion, birth place, social class or subsite. Elevated risks for tumours located in the right (ascending) side of the colon were associated with heavy consumption (five or more times per week) of deep-fried foods (OR, 3.9; 95% CI, 1.4–10.7), fried bacon or ham (2.6; 0.9–7.9) and barbecued or smoked meats (2.9; 1.2–7.3). The only item associated with risk for rectal cancer was deep-fried food (4.3; 1.5–12.1). These findings did not change after control for physical activity, body mass and occupational exposure to dust or fumes in a multivariate analysis. [The Working Group noted that it was not clear whether each food was adjusted for the others.]

Lyon and Mahoney (1988) identified all histologically confirmed cases of adenocarcinoma of the colon occurring in the population covered by the Utah (USA) Cancer Registry from 1976 to 1978 and aged 40–74 at diagnosis. Out of 348 eligible cases, 246 (71%) were interviewed. Controls were chosen by random-digit dialling: 484 subjects were interviewed out of 560 eligible people identified through a census of all adult householders from 92% of all residential telephone numbers selected at random. The food frequency questionnaire, which focused on diet five years before the interview, included questions on the method of food preparation, from which a score was derived for the frequency of consumption of broiled (including barbecued) foods and deep-fat and/or pan-fried foods. A slight increase in risk was seen with increasing level of ingestion of fried meats, which, however, was present only in women and was no longer significant after adjustment for total caloric intake. The age- and calorie-adjusted ORs for the upper tertile *versus* the lower tertile of intake were 1.3 (95% CI, 0.8–2.1) for women and 1.2 (0.8–1.9) for men. No association was found with eating broiled meat.

In a population-based case–control study of colorectal adenocarcinoma conducted in Stockholm, Sweden, in 1986–88 (Gerhardsson de Verdier *et al.*, 1990a,b, 1991), fairly high participation rates were obtained for both cases (559 interviewed; 78% of incident cases) and controls (505 interviewed; 81% of an age- and sex-stratified random sample of the resident population). Diet was investigated by replies to a self-administered food frequency questionnaire, supplemented, when necessary, by interview by a nurse's aide for cases and by telephone for controls. The usual portion size was estimated from a photograph to be small, moderate or large. The overall results of the study showed that protein, fat intake and body mass index increased risk, while dietary fibre and physical activity decreased risk—each factor having different effects on cancers at different colorectal subsites. The questions regarding meat (Gerhardsson de Verdier *et al.*, 1991) focused on bacon/smoked ham, beef/pork and sausages; for beef/pork and sausages, separate questions were asked for each of three cooking methods: fried, oven-roasted and boiled. A significantly increased risk, systematically higher for rectal than for colonic cancer, was observed for high consumption of each type of meat; the ORs were higher, however, for eating boiled meat than for fried or roasted meat. The risk associated with consumption of fried or roasted meat disappeared after adjustment for protein. Respondents who ever ate fried meat were also asked about browning of the meat surface during the previous five and previous 20 years, with three response alternatives: preference for meat with a light, moderately or heavily browned surface. They were also asked if they preferred fried meat to be prepared by high- or low-temperature frying. The risks for both colonic and rectal cancer were significantly higher for frequent consumption (more than once a week) of brown gravy (OR, 1.8 for colon and 2.1 for rectum), for preference for heavily browned meat surface, especially in the past five years (ORs, 2.3 and 3.7) and for high-temperature frying (ORs, 1.9 and 1.6). The association was slightly reduced when total fat was adjusted for and when browned meat was adjusted for high frying temperature and *vice versa*, indicating that these questions may partly reflect the same exposure. The adjusted risk estimates nevertheless remained elevated and statistically significant. Adjustment for other potential confounding factors, namely total energy, dietary fibre, body mass and physical activity, had little or no effect on the magnitude of the association. The authors stated that restriction of the analysis to patients without gastrointestinal symptoms at diagnosis or to patients who had not required help in filling in the questionnaire did not alter the overall findings. People with a preference for heavily browned meat surface and with high consumption of fried meat and brown gravy had higher risks than people with a preference for moderately browned meat and low consumption of fried meat and brown gravy. Fat intake but not protein was adjusted for in the analysis. [The Working Group noted that the finding of a higher risk with boiled than with other methods of cooking meat makes these results difficult to interpret with respect to the carcinogenicity of heterocyclic amines formed in cooking.]

Schiffman *et al.* (1989) and Schiffman and Felton (1990) reported the results of a case-control study of adenocarcinoma of the colon and rectum, diet and faecal mutagenicity carried out in three hospitals in Washington DC, USA, which was based on 50 cases diagnosed in 1985–87 and 96 age- and sex-matched surgical controls. The response rates were about 30%. Virtually all subjects reported frequent consumption of cooked meats, but cases were more likely than controls to report that they usually ate their red meat well done (medium to medium-well done, OR, 0.9; 95% CI, 0.4–2.5; well done, OR, 3.5; 95% CI, 1.3–9.6; compared to rare to medium rare). [The Working Group noted that only a crude analysis was reported and that details on how information was collected was not provided.]

2.2.2 Other sites

In a population-based case–control study carried out in Stockholm, Sweden, in 1985–87 (Steineck *et al.*, 1990), information was collected from 323 patients (78% response rate) with urothelial cancer and/or squamous-cell carcinoma of the lower urinary tract (94% bladder cancers) and from 392 subjects (77% response rate) randomly sampled within strata of gender and year of birth. Diet was investigated on the basis of replies to a mailed food frequency questionnaire supplemented by a telephone interview. Usual portion size was estimated from a photograph to be small, moderate or large. Information on intake of fried meat was obtained by asking questions about consumption of fried and oven-cooked meat/pork/sausages, smoked ham and bacon. An increased OR was seen for eating fried eggs (1.8; 95% CI, 1.0–3.1, for weekly *versus* less often), gravy (1.6; 1.0–2.4) and fried potatoes (1.6; 1.1–2.6); increased ORs were also suggested to be associated with consumption of fried meat and grilled foods, but not fried fish. Collating the data on fried eggs, fried potatoes, fried meat and gravy in a single variable and adjusting for age, gender, smoking habits and average daily intake of fat gave an OR for moderate intake (exposed to two or three of the fried foods) of 1.7 (95% CI, 0.9–3.0) and an OR for high intake (exposed to all four fried foods) of 2.2 (1.2–4.1).

In the study from Kansas City, described in detail above, a series of 93 gastric cancer patients were also compared to 279 controls (Higginson, 1966). Cases reported slightly greater consumption of fried potatoes, fried meats and fried food for breakfast, but the difference was not significant. No difference was suggested according to the method of cooking meat.

Norell *et al.* (1986) carried out a case–control study in Sweden on dietary habits and tobacco smoking among 99 cases of pancreatic cancer (out of 120 eligible patients) and 138 population controls (out of 162 eligible subjects). As an additional control group, 163 patients hospitalized for inguinal hernia were interviewed, out of 179 eligible subjects. Cases were drawn from all patients diagnosed in 1982–84 at the three surgical departments in Stockholm and Uppsala where suspected pancreatic cancer cases are referred. Frequent consumption of a number of meat items was associated with an increased risk, whatever control group was used for comparison, but most of the effect was confined to fried/grilled meat. Subjects who ate meat at least twice a week and grilled/fried meat at least twice a week showed an OR of 2.5 (95% CI, 1.2–5.3; population controls) in comparison with subjects who ate meat less often. Eating meat at least twice a week but grilled/fried meat less often was not associated with increased risk. Controlling for tobacco smoking and a number of other food items associated with the risk of pancreatic cancer was stated to increase slightly the association with fried/grilled meat, but data were not shown. [The Working Group noted that adjustment for nutrients was not attempted.]

A case–control study from Ankara, Turkey, included 100 cases of adenocarcinoma of the stomach, enrolled during 1987–88 from seven hospitals, 61 controls from the same hospitals with no cancer and 39 healthy controls (Demirer *et al.*, 1990). Cases and controls did not differ in their consumption of fried potatoes, fried meat or fried fish.

Kono *et al.* (1988) conducted a study in a low-risk area for stomach cancer in rural northern Kyushu, Japan. Between 1979 and 1982, 139 newly diagnosed cases of gastric cancer (85% histologically confirmed) were identified among 4729 subjects who had visited a referral centre in the area for the diagnosis of gastrointestinal diseases. Cases were compared with two sets of controls: 2574 hospital controls free of gastrointestinal disease and 278 general population controls who were similar to the cases by sex, year of birth and residence. Two different groups of people, using a standard questionnaire, interviewed patients before diagnostic procedures at the referral centre and the general population about dietary habits in the year preceding the interview or before a change in dietary habits. Consumption of broiled fish or grilled meat was not associated with an increased risk.

3. Studies of Cancer in Experimental Animals

3.1 Oral administration

3.1.1 *Mouse*

A group of 40 male and 40 female CDF_1 mice [(BALB/cAnN × DBA/2N)F_1], seven weeks of age, were fed a pelleted diet containing 300 mg/kg IQ (purity, > 99.6%) for 675 days, at which time the experiment was terminated. A group of 40 males and 40 females fed basal diet alone served as controls. The numbers of survivors on day 394, when leukaemia was found in a female control, were similar in the four groups: 39/40 treated males, 36/40 treated females, 33/40 control males and 38/40 control females. The number of mice with liver tumours (hepatocellular adenomas and hepatocellular carcinomas) was significantly higher in treated groups than in controls: 16/39 *versus* 2/33 in males; 27/36 *versus* 0/38 in females; hepatocellular carcinomas occurred in eight treated males and 22 treated females. One male and three female controls developed haemangioendotheliomas. The incidences of tumours of the lung and forestomach were also significantly higher in treated mice than in controls: combined incidences of adenoma and adenocarcinoma of the lung, 27/39 *versus* 7/33 in males and 15/36 *versus* 7/38 in females; incidences of adenocarcinoma of the lung, 14/39 *versus* 3/33 in males and 8/36 *versus* 4/38 in females; combined incidences of papilloma and squamous-cell carcinoma of the forestomach, 16/39 *versus* 1/33 in males and 11/36 *versus* 0/38 in females; and incidences of squamous-cell carcinoma of the forestomach, 5/39 *versus* 0/33 in males and 3/36 *versus* 0/38 in females (Ohgaki *et al.*, 1984, 1986).

Groups of 10 or more female CDF1 mice, 27-31 days old, were treated with IQ (purity, > 98%; dissolved in 55% ethanol in 0.9% sodium chloride solution) at 200 or 400 mg/kg bw (one-half of the LD_{50}) by gavage twice at a four-day interval. The numbers of aberrant crypts in the colon scored 21 days after the first IQ treatment were dose-related. Crypts were found more frequently in the caecal end (Tudek *et al.*, 1989).

3.1.2 *Rat*

A group of 32 female Sprague-Dawley rats, six weeks of age, received IQ hydrochloride at 0.35 mmol [70 mg]/kg bw in 5% Emulphor by gavage three times per week during weeks 1-4, twice per week during weeks 5-8 and weekly during weeks 9-31 and were maintained without further treatment until sacrifice at week 52. A group of 27 rats received 0.25 ml 5% Emulphor according to the same schedule, and a further group of nine animals served as untreated controls. Treated rats showed a 94% weight gain by the end of the experiment compared with controls. Twenty-one adenocarcinomas of the mammary gland were observed in 14/32 ($p < 0.05$) treated animals. No such tumour was observed in controls. Liver tumours were observed in 6/32 (three neoplastic nodules, two hepatocellular carcinomas and two haemangioendotheliomas) treated animals, but in none of the controls. Twelve squamous-cell carcinomas of the Zymbal gland were found in 11/32 treated animals, and no such tumour occurred in controls [$p = 0.002$; Fisher exact test]. The treated group also had altered liver-cell foci (17/32), atypical hyperplastic acinar-cell lesions in the pancreas (19/32) and altered proliferative foci in the adrenal cortex (5/32), none of which was present in the control group (Tanaka et al., 1985).

Groups of 40 male and 40 female Fischer 344 rats, eight weeks of age, were fed a pelleted diet containing 300 mg/kg IQ (purity confirmed by HPLC [percentage not indicated]) for 104 weeks. Control groups of 50 males and 50 females were fed basal diet alone. The times of appearance of the first tumours in males were day 255 in the colon, day 239 in the small intestine and day 288 in the liver. Twenty males and four females from the treated group [but no control] were killed at 300 days, because the animals were moribund owing to the occurrence of tumours. Treated animals had a significantly increased incidence of tumours of the liver, Zymbal gland, colon and small intestine during the 104-week study (Table 3). No such tumour, except one hepatocellular carcinoma in a male, occurred in concurrent controls (Takayama et al., 1984; Ohgaki et al., 1986).

Table 3. Tumour incidence in Fischer rats fed a diet containing IQ at 300 mg/kg for 104 weeks

Tumour type	Males	Females
Squamous-cell carcinoma, Zymbal gland	36/40	27/40
Adenocarcinoma, colon	25/40	9/40
Adenocarcinoma, small intestine	12/40	1/40
Hepatocellular carcinoma	27/40	18/40
Carcinoma, skin	17/40	3/40
Squamous-cell carcinoma, oral cavity	2/40	1/40
Squamous-cell carcinoma, clitoral gland	–	20/40

From Ohgaki et al. (1986)

Groups of 10 or more female Sprague-Dawley rats, 21 days old, were treated with IQ (purity, > 98%; dissolved in 55% ethanol in 0.9% sodium chloride solution) at 200 or 400 mg/kg bw (one-half of the LD_{50}) by gavage twice at a four-day interval. The numbers of aberrant crypts in the colon scored 21 days after the first IQ treatment were dose-related. Crypts were found more frequently in the caecal end (Tudek et al., 1989).

3.1.3 Monkey

Twenty cynomolgus monkeys (*Macaca fascicularis*) (14 males, 6 females), about one year old, were administered 10 mg/kg bw IQ (purity, > 99.9%) suspended in hydroxypropyl cellulose by gavage five times a week for up to 60 months. Another 20 monkeys (8 males,

12 females) were administered IQ by gavage five times a week at a dose of 20 mg/kg bw. Hepatocellular carcinomas were found in 13 monkeys which were necropsied; 10 had received 20 mg/kg (average latent period, 37 months) and three, 10 mg/kg (first tumour seen after 30 months; average latent period, 45 months). Metastases to the lung occurred in several monkeys. No such tumour occurred in colony controls (Adamson et al., 1990, 1991).

3.2 Intraperitoneal administration

Mouse

Groups [initial number unspecified] of newborn male B6C3F$_1$ mice were injected intraperitoneally with IQ (> 98% pure) at total doses of 0, 0.625 or 1.25 μmol [125–250 μg] (maximal tolerated dose) dissolved in 5, 10 or 20 μl dimethyl sulfoxide and administered on days 1, 8 and 15 after birth, respectively. Animals were sacrificed at 8 and 12 months. The incidence of hepatocellular adenomas was significantly higher in treated mice than in controls: at eight months, 1/44 in controls, 5/24 at the low dose and 5/16 at the high dose; at 12 months, 5/44 in controls, 7/19 at the low dose and 14/20 at the high dose. Two hepatocellular carcinomas were found in the high-dose group at 12 months (Dooley et al., 1992).

3.3 Administration with known carcinogens

3.2.1 *Mouse*

In a two-stage skin carcinogenesis study, a group of 20 female CD-1 mice, seven weeks of age, received topical applications on the dorsal skin of 0.75 mg IQ in 0.1 ml dimethyl sulfoxide twice weekly for five weeks, followed one week later by topical applications of 2.5 μg 12-*O*-tetradecanoylphorbol 13-acetate (TPA) twice weekly for 47 weeks. A positive control group received applications of 7,12-dimethylbenz[*a*]anthracene (DMBA; total dose, 100 μg) plus TPA. Skin tumours were found in 0/20 mice treated with IQ, 1/20 mice treated with IQ and TPA, 4/19 mice treated with DMBA, 17/18 treated with DMBA and TPA and 0/20 solvent controls (Sato et al., 1987).

3.2.2 *Rat*

Groups of 40 male Wistar rats, six weeks old, were given IQ at 10 mg/kg bw or solvent (water, acidified to pH 3.5 with citric acid) by gavage every day for two weeks. One week later, the rats were divided into two groups and received either no additional treatment or 500 ppm [mg/l] phenobarbital sodium in the drinking-water for the remainder of the study. Ten animals from each group were sacrificed at week 42, and the study was terminated after 58 weeks. Zymbal gland carcinomas were found in 2/40 rats that received IQ only and in 2/40 that received IQ plus phenobarbital but not in the untreated group or in the group that received phenobarbital only. A hepatocellular adenoma was found in one rat given IQ alone, and a hepatocellular adenoma and a tumour diagnosed as a cystic cholangiocarcinoma occurred in one rat in the group treated with IQ plus phenobarbital. Treatment with IQ and phenobarbital significantly increased ($p < 0.01$) the number of γ-glutamyl transpeptidase-positive foci of altered hepatocytes in comparison with the respective controls; some such foci were also found after administration of phenobarbital alone (Kristiansen et al., 1989).

In a short-term assay for tumour-initiating activity in the liver, groups of 20 male Fischer 344 rats, five weeks of age, each received a two-week dietary treatment with IQ at doses of 0.025, 0.05 and 0.1% (250, 500 and 1000 mg/kg of diet) and were then maintained on a diet supplemented with either 500 mg/kg phenobarbital or 24 mg/kg 3'-methyl-4-dimethyl-aminoazobenzene (3'-Me-DAB) from week 3 until final sacrifice at week 86. One group received only 0.1% IQ (1000 mg/kg of diet). All animals underwent a two-thirds partial hepatectomy at week 1. There were 14–19 effective animals per group, including both animals that survived to the end of the experiment and those that died with tumours after week 52. Administration of IQ alone or IQ at 0.025–0.1% with either phenobarbital or 3'-Me-DAB caused hyperplastic liver nodules in all animals; and subsequent administration of phenobarbital or 3'-Me-DAB caused a significant increase in the incidence of hepatocellular carcinomas: IQ alone, 0/19; 3'-Me-DAB with 0, 0.025, 0.05 and 0.1% IQ, 0/18, 1/17, 3/18 and 5/16; phenobarbital with 0, 0.025, 0.05 and 0.1% IQ, 0/18, 6/17, 7/18 and 7/14. Administration of IQ plus phenobarbital increased the incidence of thyroid adenomas and carcinomas: IQ alone, 1/19; phenobarbital plus 0, 0.025, 0.5 and 0.1% IQ, 2/18, 8/17, 8/18 and 9/14. IQ alone caused squamous-cell carcinomas or keratoacanthomas of the Zymbal gland (5/19; 26%), but the only effect of subsequent treatment with phenobarbital or 3'-Me-DAB was with 0.1% IQ plus phenobarbital (6/14; 43%). Animals that received IQ, phenobarbital or 3'-Me-DAB alone or in combination had preputial gland tumours at incidences ranging from 22 to 50%, but there did not appear to be a dose–response relationship (Tsuda et al., 1988).

In a short-term assay for tumour-initiating activity in the liver, a group of 10 male Fischer 344 rats, five weeks of age, each received a single intragastric dose of 80 mg/kg bw IQ and, two weeks later, were fed a diet containing 0.05% phenobarbital for six weeks. Rats received a two-thirds partial hepatectomy three weeks after the IQ treatment. The number and total area of foci of phenotypically altered hepatocytes in the liver were scored using expression of placental-form glutathione S-transferase (GST-P) as the marker. Treated rats had a significant, two-fold greater number of foci than five vehicle-treated control rats; IQ without subsequent phenobarbital treatment did not induce a significant increase in the number of foci in five rats. Another group of 10 rats received a two-thirds partial hepatectomy 12 h before IQ treatment, two weeks later was fed the diet containing phenobarbital as above, and then received a single intraperitoneal injection of 300 mg/kg D-galactosamine one week after the phenobarbital treatment. Administration of IQ increased the number and area of foci more than 10 fold over that in five vehicle-treated control rats. IQ without subsequent phenobarbital treatment, but with galactosamine, produced a smaller, nonsignificant increase in the number and area of foci. These results suggest that IQ has tumour-initiating activity in rat liver, especially if combined with partial hepatectomy (Tsuda et al., 1990).

As part of a medium-term carcinogenicity study on the synergistic effects of five heterocyclic amines, groups of 13–15 male Fischer 344 rats, six weeks of age, each received a single intraperitoneal injection of 200 mg/kg bw N-nitrosodiethylamine and, two weeks later, were fed a diet containing IQ at 12, 60 or 300 ppm (mg/kg). A two-thirds partial hepatectomy was performed in week 3 of the experiment; all animals were killed after eight weeks. Fifteen rats treated only with the nitrosamine served as controls. The effects were assessed by counting the numbers of GST-P-positive foci in the liver. IQ alone at the mid- and high-dose levels significantly increased the area of GST-P-positive foci (Ito et al., 1991).

4. Other Relevant Data

4.1 Absorption, distribution, metabolism and excretion

The toxicology and metabolism of heterocyclic aromatic amines have been reviewed (Övervik & Gustafsson, 1990; Aeschbacher & Turesky, 1991).

4.1.1 Humans

No data were available to the Working Group.

4.1.2 Experimental systems

The absorption and excretion of ^{14}C- and ^{3}H-IQ have been studied by fluorescence in rats and mice following gavage (Sjödin & Jägerstad, 1984; Alldrick & Rowland, 1988; Inamasu et al., 1989) and that of IQ after intraperitoneal administration (Størmer et al., 1987). IQ was absorbed rapidly, mainly from the small intestine (Alldrick & Rowland, 1988), metabolized and excreted almost quantitatively within three days; in rats, 36–49% and 46–68% of the administered dose was recovered in the urine and faeces, respectively (Sjödin & Jägerstad, 1984). The excretion pathways were essentially similar following dietary exposure (Inamasu et al., 1989).

The fate of intravenously injected ^{14}C-IQ was studied in male NMRI, pregnant NMRI and female C3H mice. Whole-body autoradiograms were characterized by an accumulation of radiolabel in metabolic and excretory organs (liver, kidney, bile, urine, gastric and intestinal contents, salivary glands, nasal mucosa and Harder's gland) and in lymphomyeloid tissues (bone marrow, thymus, spleen and lymph nodes) and endocrine and reproductive tissues (adrenal medulla, pancreatic islets, thyroid, hypophysis, testis, epididymis, seminal vesicles, ampulla and prostate). The liver and kidney cortex were identified as sites of retention of nonextractable radiolabel. IQ crossed the placenta, but no radiolabel was retained in fetal tissues (Bergman, 1985).

IQ was oxidized to N-hydroxy-IQ in the presence of rat and rabbit liver microsomal homogenates (Yamazoe et al., 1983; Kato, 1986; McManus et al., 1988a). The activity of cytochrome P450 IA2 isozyme in the liver was induced in rats by prior intraperitoneal injection of IQ and other heterocyclic amines (Degawa et al., 1989); the activity of this enzyme was induced in cultured rat hepatocytes by β-naphthoflavone and polychlorinated biphenyls (Wallin et al., 1992). IQ can also be oxidized via a prostaglandin hydroperoxidase-dependent pathway, as shown in microsomes isolated from ram seminal vesicles (Wild & Degen, 1987; Petry et al., 1989).

N-Hydroxy-IQ can be esterified by O-acetyltransferase, sulfotransferase and prolyl-tRNA synthetase but at much lower rates than aromatic amines (Kato & Yamazoe, 1987).

N-Acetylation of IQ may not be important for DNA binding (Fig. 1; Snyderwine et al., 1988a). Human liver microsomes could activate IQ into a DNA-reactive species. The isozyme involved was tentatively identified as CYP IA2 (P450 IA2) (Shimada et al., 1989). The same enzyme was shown to be responsible for the formation of N-hydroxy-IQ in human hepatic cytosols (Butler et al., 1989; McManus et al., 1990). Human liver and colon cytosols catalysed the formation of N-hydroxylated IQ into a DNA-binding form, but N-acetylation was not observed under the same conditions (Turesky et al., 1991). DNA-binding products were also found in human mammary epithelial cells cultured in the presence of IQ (Pfau et al., 1992). In human fetal liver tissue, cytochrome P450 HFLa was the main activating enzyme of IQ (Kitada et al., 1990).

Fig. 1. Schematic pathway of activation of IQ to DNA-binding products

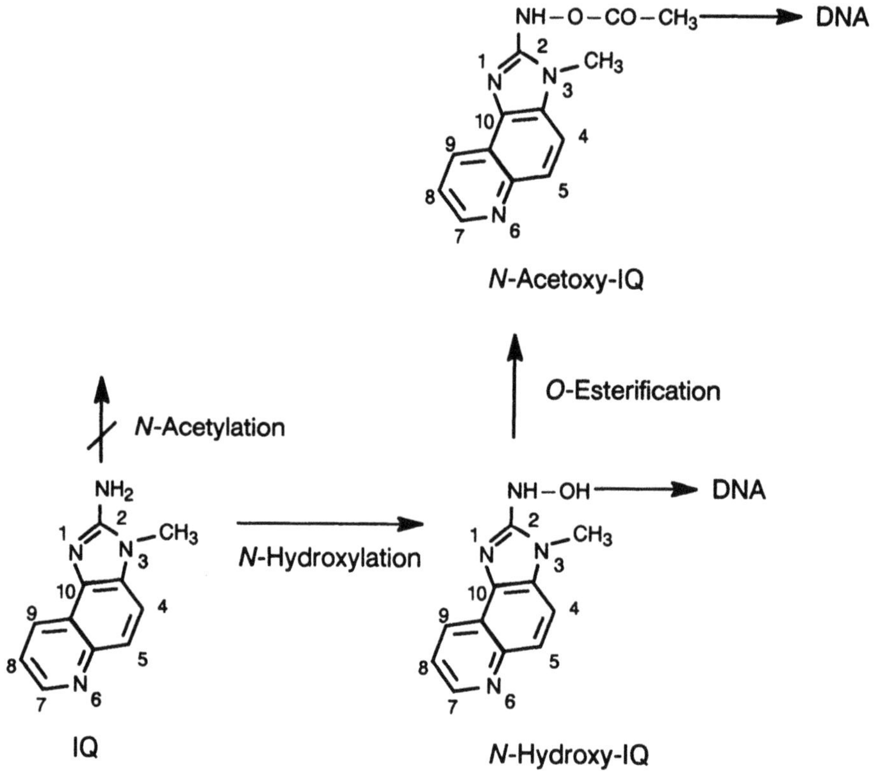

Adapted from Snyderwine et al. (1988a)

IQ binds to rat haemoglobin and albumin *in vivo*. One of the haemoglobin products was identified as a sulfinamide at a cysteine residue (Turesky et al., 1987).

Mixed and pure cultures of human intestinal anaerobic bacteria metabolized IQ to IQ-7-one (Carman et al., 1988). In rats, the routes of detoxication of IQ include cytochrome

P450-mediated ring hydroxylation at the C5 position, followed by conjugation to a sulfate or glucuronic acid (Luks *et al.*, 1989; Vavrek *et al.*, 1989). Another pathway involves conjugation of the exocyclic amine group to a glucuronic acid or sulfate (Inamasu *et al.*, 1989). Conjugated 5-hydroxy-IQ accounted for about 40% of urinary and biliary metabolites in rats; *N*-sulfamates are another major group of excretion products (Inamasu *et al.*, 1989; Turesky *et al.*, 1986; Luks *et al.*, 1989). In the urine of monkeys, sulfate and glucuronide conjugates predominated (Snyderwine *et al.*, 1991). Treatment of rats with polychlorinated biphenyls increased the excretion of sulfate conjugates into the urine (Vavrek *et al.*, 1989).

Constituents of feed may play a role in the metabolism of IQ. Dietary fibres can bind IQ *in vitro* (Sjödin *et al.*, 1985), and a high-fat diet increased the capacity of rat liver microsomes to activate IQ (Alldrick *et al.*, 1987).

4.2 Toxic effects

No data were available to the Working Group.

4.3 Reproductive and developmental toxicity

No data were available to the Working Group.

4.4 Genetic and related effects

The genetic effects of IQ have been reviewed (Sugimura, 1985; Hatch, 1986; de Meester, 1989; Sugimura *et al.*, 1989).

4.4.1 *Humans*

No data were available to the Working Group.

4.4.2 *Experimental systems* (see also Table 4 and Appendices 1 and 2)

N-Hydroxy-IQ binds nonenzymatically to DNA *in vitro* at pH 7.4. In the presence of polynucleotides, the *N*-hydroxy-IQ was bound particularly extensively with polyguanylic acid (Snyderwine *et al.*, 1988a,b). *N*-Hydroxy-IQ reacts with DNA to form up to five adducts; the major one co-chromatographs with *N*-(deoxyguanosin-8-yl)–IQ (Schut *et al.*, 1991).

IQ induced prophage, SOS repair and mutation in bacteria. Bacterial mutations were also induced in the intrasanguinous mouse host-mediated assay and following exposure to the urine of IQ-dosed rats.

IQ induced somatic and sex-linked recessive lethal mutations in *Drosophila melanogaster*. It formed DNA adducts and DNA strand breaks in cultured mammalian cells and induced unscheduled DNA synthesis in primary hepatocytes cultured from mice, rats and Syrian hamsters, but not in those from guinea-pigs. Responses in other cultured mammalian cell assays were complex. IQ induced gene mutation to diphtheria toxin resistance and at the *hprt* locus; mutations at the *hprt* locus were observed only in single studies in repair-deficient cell lines and in a repair-proficient cell line co-cultured with hepatocytes from poly-

Table 4. Genetic and related effects of IQ

Test system	Result		Dose[a] (LED/HID)	Reference
	Without exogenous metabolic system	With exogenous metabolic system		
PRB, *Escherichia coli* K12, prophage λ induction	0	+	1.0000	Nagao *et al.* (1983a)
PRB, *Salmonella typhimurium* TA1535, SOS repair	0	+	0.0300	Nakamura *et al.* (1987)
PRB, *Salmonella typhimurium*, SOS repair, with human microsomes	0	+	2.0000	Kitada *et al.* (1990)
PRB, *umu* expression, *Salmonella typhimurium* TA1535/pSK 1002	0	+	2.0000	Shimada *et al.* (1989)
ERD, *Escherichia coli rec* strains, differential toxicity	−	+	0.2600	Knasmüller *et al.* (1992)
SA0, *Salmonella typhimurium* TA100, reverse mutation	0	+	0.0500	Nagao *et al.* (1981)
SA0, *Salmonella typhimurium* TA100, reverse mutation	−	+	0.0050	Wild *et al.* (1985)
SA0, *Salmonella typhimurium* TA100, reverse mutation	0	+	0.0000	Barnes *et al.* (1985)
SA0, *Salmonella typhimurium* TA100, reverse mutation	0	+	0.0000	Felton & knize (1990)
SA0, *Salmonella typhimurium* TA100, reverse mutation	−	+	0.0000	Grivas & Jägerstad (1984)
SA2, *Salmonella typhimurium* TA102, reverse mutation	0	+	0.0000	Felton & Knize (1990)
SA4, *Salmonella typhimurium* TA104, reverse mutation	0	+	0.0000	Felton & Knize (1990)
SA5, *Salmonella typhimurium* TA1535, reverse mutation	−	−	25.0000	Wild *et al.* (1985)
SA7, *Salmonella typhimurium* TA1537, reverse mutation	+	+	0.0025	Wild *et al.* (1985)
SA8, *Salmonella typhimurium* TA1538, reverse mutation	0	+	0.0005	Thompson *et al.* (1983)
SA8, *Salmonella typhimurium* TA1538, reverse mutation	+	+	0.0003	Wild *et al.* (1985)
SA8, *Salmonella typhimurium* TA1538, reverse mutation	0	+	0.1500	Felton & Knize (1990)
SA9, *Salmonella typhimurium* TA98, reverse mutation	0	+	0.0050	Nagao *et al.* (1981)
SA9, *Salmonella typhimurium* TA98, reverse mutation	+	+	0.0003	Wild *et al.* (1985)
SA9, *Salmonella typhimurium* TA98, reverse mutation	0	+	0.0015	Barnes *et al.* (1985)
SA9, *Salmonella typhimurium* TA98, reverse mutation	0	+[b]	0.0050	Ishida *et al.* (1987)
SA9, *Salmonella typhimurium* TA98, reverse mutation	−	+	0.3125	Loprieno *et al.* (1991)
SA9, *Salmonella typhimurium* TA98, reverse mutation	−	+[c]	0.2000	Holme *et al.* (1987)
SA9, *Salmonella typhimurium* TA98, reverse mutation	+	+	0.0017	Lin *et al.* (1992)
SA9, *Salmonella typhimurium* TA98, reverse mutation	0	+	0.0000	Felton & Knize (1990)
SA9, *Salmonella typhimurium* TA98, reverse mutation	0	+	0.0010	Nagao *et al.* (1983b)
SA9, *Salmonella typhimurium* TA98, reverse mutation	0	+	0.1250	Hayashi *et al.* (1985)
SA9, *Salmonella typhimurium* TA98, reverse mutation	0	+	0.0003	Wild *et al.* (1991)
SA9, *Salmonella typhimurium* TA98, reverse mutation	0	+	0.0005	Buonarati & Felton (1990)
SA9, *Salmonella typhimurium* TA98, reverse mutation	−	+	0.0010	Grivas & Jägerstad (1984)
SAS, *Salmonella typhimurium* TA96, reverse mutation	0	+	0.0000	Felton & Knize (1990)

Table 4 (contd)

Test system	Result		Dose[a] (LED/HID)	Reference
	Without exogenous metabolic system	With exogenous metabolic system		
SAS, *Salmonella typhimurium* TA97, reverse mutation	0	+	0.0000	Felton & Knize (1990)
SAS, *Salmonella typhimurium* TA97, reverse mutation	+	+	0.0017	Lin et al. (1992)
SAS, *Salmonella typhimurium* TA97, reverse mutation[d]	+	+	0.0017	Lin et al. (1992)
SAS, *Salmonella typhimurium* TA98, reverse mutation[d]	+	+	0.0017	Lin et al. (1992)
SAS, *Salmonella typhimurium* TA98/1,8-DNP$_6$, reverse mutation	0	–	0.0250	Nagao et al. (1983b)
SAS, *Salmonella typhimurium* TA98/1,8-DNP$_6$, reverse mutation	0	–	0.0005	Buonarati & Felton (1990)
SAS, *Salmonella typhimurium* TA98/1,8-DNP$_6$, reverse mutation	0	–	0.0015	Wild et al. (1991)
SAS, *Salmonella typhimurium* TA1535/pSK 1002	0	(+)	2.0000	Ubukata et al. (1992)
SAS, *Salmonella typhimurium* TA1978, reverse mutation	0	(+)	0.0500	Thompson et al. (1983)
SAS, *Salmonella typhimurium* TA1978, reverse mutation	0		0.0050	Wild et al. (1985)
SAS, *Salmonella typhimurium* YG1024, reverse mutation	0	+	0.0001	Wild et al. (1991)
DMM, *Drosophila melanogaster*, somatic mutation and recombination	+	0	25.0000	Yoo et al. (1985)
DMM, *Drosophila melanogaster*, somatic mutation and recombination	+	0	250.0000	Graf et al. (1992)
DMX, *Drosophila melanogaster*, sex-linked recessive lethal mutation	+	0	200.0000	Wild et al. (1985)
DMX, *Drosophila melanogaster*, sex-linked recessive lethal mutation	+	0	200.0000	Graf et al. (1992)
DMM, *Drosophila melanogaster*[f], wing spot test	+	0	250.0000	Graf et al. (1992)
DIA, DNA strand breaks, radiation-induced mouse leukaemic cells *in vitro*	–	+	1.9800	Caderni et al. (1983)
DIA, DNA strand breaks, rat hepatocytes *in vitro*	+	0	0.9900	Caderni et al. (1983)
DIA, DNA strand breaks, mouse hepatocytes *in vitro*	+	0	2.0000	Hayashi et al. (1985)
DIA, DNA strand breaks, Chinese hamster V79 cells *in vitro*	–	–[c,e]	100.0000	Holme et al. (1987)
DIA, DNA strand breaks, rat hepatocytes[e] *in vitro*	+	0	20.0000	Holme et al. (1987)
URP, Unscheduled DNA synthesis, rat primary hepatocytes *in vitro*	+	0	0.0250	Barnes et al. (1985)
URP, Unscheduled DNA synthesis, rat primary hepatocytes *in vitro*	+	0	0.1000	Yoshimi et al. (1988)
UIA, Unscheduled DNA synthesis, mouse hepatocytes *in vitro*	+	0	1.0000	Yoshimi et al. (1988)
UIA, Unscheduled DNA synthesis, Syrian hamster hepatocytes *in vitro*	+	0	1.0000	Yoshimi et al. (1988)
GCL, Gene mutation, Chinese hamster lung cells, DT⁻, *in vitro*	–	+	5.0000	Nakayasu et al. (1983)
GCL, Gene mutation, Chinese hamster lung cells *in vitro*	0	+	1.0000	Sugimura et al. (1989)
GCO, Gene mutation, Chinese hamster ovary cells (*uv-5*) *hprt* locus *in vitro*	0	+	50.0000	Thompson et al. (1983)
GCO, Gene mutation, Chinese hamster ovary cells (*uv-5*) *aprt* locus *in vitro*	0	+	50.0000	Thompson et al. (1983)

Table 4 (contd)

Test system	Result		Dose[a] (LED/HID)	Reference
	Without exogenous metabolic system	With exogenous metabolic system		
GCO, Gene mutation, Chinese hamster ovary cells (AA8) hprt locus in vitro	0	–	300.0000	Thompson et al. (1983)
GCO, Gene mutation, Chinese hamster ovary cells (AA8) aprt locus in vitro	0	–	300.0000	Thompson et al. (1983)
GCO, Gene mutation, Chinese hamster ovary cells (uv-5) in vitro	0	(+)	26.0000	Brookman et al. (1985)
G9H, Gene mutation, Chinese hamster lung V79 cells, hprt locus in vitro	0	(+)[c,e]	20.0000	Holme et al. (1987)
G9H, Gene mutation, Chinese hamster lung V79 cells, hprt locus in vitro	0	–	25.6000	Loprieno et al. (1991)
G9O, Gene mutation, Chinese hamster lung V79 cells, ouabain[r] in vitro	0	–	50.0000	Takayama & Tanaka (1983)
SIC, Sister chromatid exchange, Chinese hamster lung V79 cells in vitro	–	+[c,e]	20.0000	Holme et al. (1987)
SIC, Sister chromatid exchange, Chinese hamster ovary cells (uv-5) in vitro	0	+	50.0000	Thompson et al. (1983)
SIC, Sister chromatid exchange, Chinese hamster ovary cells (AA8) in vitro	0	(+)	300.0000	Thompson et al. (1983)
CIC, Chromosome aberration, Chinese hamster ovary cells in vitro	0	+	12.8000	Loprieno et al. (1991)
CIC, Chromosome aberration, Chinese hamster ovary cells (uv-5) in vitro	0	–	80.0000	Thompson et al. (1983)
CIC, Chromosome aberration, Chinese hamster ovary cells (AA8) in vitro	0	–	300.0000	Thompson et al. (1983)
G1H, Gene mutation, human lymphocytes, hprt locus in vitro	0	–	200.0000	McManus et al. (1988b)
SHL, Sister chromatid exchange, human lymphocytes in vitro	–	+	0.2000	Aeschbacher & Ruch (1989)
MIH, Micronucleus test, human lymphocytes in vitro	0	(+)	200.0000	McManus et al. (1988b)
CHL, Chromosomal aberration, human lymphocytes in vitro	0	–	1000.0000	Aeschbacher & Ruch (1989)
CHL, Chromosomal aberration, human lymphocytes in vitro	–	+	116.2000	Loprieno et al. (1991)
BFA, Body fluids from rats, mutagenicity to S. typhimurium TA98, TA100	–	+	7.0000	Barnes & Weisburger (1985)
HMM, Host-mediated assay, intrasanguinous NMRI mouse, S. typhimurium TA98	+	0	0.198 × 1 ip, po	Wild et al. (1985)
HMM, Host-mediated assay, intrasanguinous Swiss albino mice, Escherichia coli strains M343/753, M343/765	+	0	2.3 × 1 ip, po	Knasmüller et al. (1992)
DVA, DNA strand breaks, mouse liver cells in vivo	+		10.0000	Hayashi et al. (1985)
GVA, Gene mutation, mouse melanocytes in vivo	–		400.0 × 1 in utero	Wild et al. (1985)
MST, Mouse coat colour spot test in vivo	–		20.0 × 1 ip	Wild et al. (1985)
GVA, Gene mutation, rat granuloma cells, hprt locus in vivo	+		10.0 into pouch	Radermacher et al. (1987)

Table 4 (contd)

Test system	Result		Dose[a] (LED/HID)	Reference
	Without exogenous metabolic system	With exogenous metabolic system		
SVA, Sister chromatid exchange, mouse bone-marrow cells[e] in vivo	+		20.0 × 1 ip	Minkler & Carrano (1984)
SVA, Sister chromatid exchange, rat hepatocytes in vivo	+		50.0 × 1 ip	Sawada et al. (1991)
MVM, Micronucleus test, mouse bone-marrow cells in vivo	−		594.0 × 1 ip	Wild et al. (1985)
MVM, Micronucleus test, mouse bone-marrow cells in vivo	−	0	40.0 × 1 po	Loprieno et al. (1991)
CBA, Chromosomal aberration, mouse bone-marrow cells[e] in vivo	−		160.0 × 1 ip	Minkler & Carrano (1984)
CVA, Chromosomal aberration, rat hepatocytes in vivo	+		50.0 × 1 ip	Sawada et al. (1991)
BID, Binding (covalent) to DNA in S. typhimurium[g] in vitro	+	0	0.0000	Schut et al. (1991)
BID, Binding (covalent) to DNA in vitro (bacterial DNA)[g]	0	+	100.0000	Asan et al. (1987)
BID, Binding (covalent) to DNA in vitro (rat hepatocytes)[g]	+		24.0000	Dirr et al. (1989)
BID, Binding (covalent) to DNA in vitro (Syrian hamster embryo)[g]	0	+	10.0000	Asan et al. (1987)
BID, Binding (covalent) to DNA in vitro (rat hepatocytes)	+	0	10.0000	Wallin et al. (1992)
BVD, Binding (covalent) to DNA in rats (multiple organs) in vivo[g]	+		5.00 × 1 ip	Schut et al. (1988)
BVD, Binding (covalent) to DNA in mice (multiple organs) in vivo[g]	+		25.00 × 1 po	Hall et al. (1990)
BVD, Binding (covalent) to DNA in mice (multiple organs) in vivo[g]	+		5.00 × 1 po	Zu & Schut (1991a)
BVD, Binding (covalent) to DNA in rats (multiple organs) in vivo[g]	+		5.00 × 1 po	Zu & Schut (1991b)
BVD, Binding (covalent) to DNA in rats liver and heart in vivo[g]	+		36.0000 × 4 wk diet	Övervik et al. (1991)
BVD, Binding (covalent) to DNA in liver of cynomolgus monkeys in vivo[g]	+		20.00 × 15 po	Snyderwine et al. (1988c)
BVD, Binding (covalent) to DNA in mice (multiple organs) in vivo[h]	+		40.0 × 1 po	Loprieno et al. (1991)
BVD, Binding (covalent) to DNA in rat liver in vivo[g]	+		100.0 × 1 po	Yamashita et al. (1988)
BVD, Binding (covalent) to DNA in mouse liver in vivo[g]	+		50.0 × 1 po	Schut et al. (1991)
BVD, Binding (covalent) to DNA in rat liver in vivo[g]	+		50.0 × 1 po	Schut et al. (1991)

+, positive; (+), weakly positive; −, negative; 0, not tested; ?, inconclusive (variable response in several experiments within an adequate study)
[a]In-vitro tests, μg/ml; in-vivo tests, mg/kg bw; 0.0000, not given
[b]Rhesus liver S9
[c]Hepatocytes
[d]IQ reacted with nitrite (not on profile)
[e]Polychlorinated biphenyl-treated
[f]Nitro-IQ synthesized
[g]32P-Postlabel
[h]14C-Label

chlorinated biphenyl-treated rats. No *hprt* locus or ouabain-resistance mutations were observed in other studies, in which exogenous metabolic activation systems were provided by rat liver homogenates. IQ induced sister chromatid exchange in Chinese hamster cells. Chromosomal aberrations were induced in one study with Chinese hamster ovary cells but not in studies in which a repair-deficient cell line and a repair-proficient cell line were used (Thompson *et al.*, 1983).

In cultured human lymphocytes, IQ did not induce *hprt* locus mutations but did induce sister chromatid exchange and micronucleus formation. Inconsistent results were obtained for chromosomal aberrations in metaphases.

IQ–DNA adducts were formed *in vivo* in multiple organs of rats and mice and in the liver (only organ examined) of cynomolgus monkeys given oral doses of IQ. These results obtained *in vivo* confirm that the major DNA adduct co-chromatographs with N-(deoxyguanosin-8-yl)–IQ (Schut *et al.*, 1991).

After administration *in vivo*, IQ induced DNA strand breaks in mouse liver, but it did not induce unscheduled DNA synthesis in rat stomach. Gene mutations were induced in neither the mouse coat colour spot test nor in a transplacental assay in mice, but IQ induced *hprt* locus mutations in a single granuloma pouch assay. Sister chromatid exchange was induced in mouse bone marrow and rat liver. Whereas chromosomal aberrations were induced in rat liver, neither these nor micronuclei were induced in mouse bone marrow.

(*a*) *IQ–nitrite interaction*

Reaction mixtures of IQ and nitrite were mutagenic to *Salmonella typhimurium* strains TA97 and 98 both in the absence and presence of an exogenous metabolic activation system. Nitro-IQ induced somatic mutation in *Drosophila melanogaster*.

(*b*) *Genetic changes in animal tumours*

Activated c-Ha-*ras* proto-oncogenes were found in four of seven Zymbal gland tumours induced in rats by IQ. The mutations were G to C transversions at the first base of codon 13 (two tumours), a G to T transversion at the second base of codon 13 (one tumour) and an A to T transversion (one tumour) at the second base of codon 61 (Kudo *et al.*, 1991). p53 Gene mutations were found in four of 15 Zymbal gland tumours induced in rats by IQ. These involved changes of *C*GT to *G*GT, T*G*C to T*T*C, *G*TG to *T*TG and *G*AA deletion at codons 156, 174, 214 and 256, respectively (Makino *et al.*, 1992).

5. Summary of Data Reported and Evaluation

5.1 Exposure data

IQ (2-Amino-3-methylimidazo[4,5-*f*]quinoline) has been found in cooked meat and fish. A few determinations indicated that the levels of IQ were lower than those of MeIQx (2-amino-3,8-dimethylimidazo[4,5-*f*]quinoxaline) and PhIP (2-amino-1-methyl-6-phenylimidazo[4,5-*b*]pyridine). IQ was reported in the only sample of cigarette smoke condensate tested.

5.2 Human carcinogenicity data

No data directly relevant to an evaluation of the carcinogenicity to humans of IQ were available; however, several studies that were potentially relevant were considered.

The only cohort study in which detailed results were presented showed a significantly increased risk for cancers at all sites and for gastric cancer associated with the consumption of broiled fish.

Two case–control studies, in Sweden and the USA, in which consumption of meat cooked in different ways was addressed and in which consumption of a number of nutrients was controlled did not show increased risks for colorectal cancer associated with consumption of fried meat; however, the study from Sweden showed an association with a preference for browned meat. One case–control study on gastric cancer in Japan showed no association with consumption of broiled fish or grilled meat.

The available information was insufficient to establish whether cooking methods that result in the formation of heterocyclic amines are a risk factor for cancer independent of the food item itself.

5.3 Animal carcinogenicity data

IQ was tested for carcinogenicity by oral administration in one experiment in mice, in two experiments in rats and in one study in monkeys. Hepatocellular adenomas and carcinomas, adenomas and adenocarcinomas of the lung and squamous-cell papillomas and carcinomas of the forestomach were produced in mice. In rats, hepatocellular carcinomas, adenocarcinomas of the small and large intestine, and squamous-cell carcinomas of the Zymbal gland were produced in animals of each sex. A high incidence of mammary adenocarcinomas was observed in females. In addition, squamous-cell carcinomas were found in the skin of males and in the clitoral gland of females. Hepatocellular carcinomas were produced in one study in monkeys.

Intraperitoneal injection of IQ to newborn male mice increased the incidence of hepatic adenomas.

Single dose or short-term oral treatment of rats with IQ followed by phenobarbital, with or without further modulating procedures, increased the numbers of foci of altered hepatocytes and of carcinomas in the liver. Sequential administration of IQ after *N*-nitrosodiethylamine enhanced the appearance of foci of altered hepatocytes in rats.

5.4 Other relevant data

No data were available on the genetic and related effects of IQ in humans.

IQ bound to DNA in many organs of cynomolgus monkeys and rodents dosed *in vivo*. In rodents treated *in vivo*, IQ induced DNA damage, gene mutation and chromosomal anomalies. It induced chromosomal anomalies in human cells *in vitro* and chromosomal anomalies, gene mutation and DNA damage in animal cells *in vitro*. It induced mutations in *Drosophila melanogaster* and DNA damage and mutations in bacteria. Gene mutations in c-Ha-*ras* and p53 genes were found in some Zymbal gland carcinomas induced in rats by IQ.

5.5 Evaluation[1]

There is *inadequate evidence* in humans for the carcinogenicity of IQ.

There is *sufficient evidence* in experimental animals for the carcinogenicity of IQ.

Overall evaluation

IQ (2-Amino-3-methylimidazo[4,5-*f*]quinoline) *is probably carcinogenic to humans (Group 2A)*.

In arriving at the overall evaluation, the Working Group took into consideration the following contributory information:

IQ is comprehensively genotoxic, and this activity can be expressed *in vivo* in rodents. IQ can be metabolized by human microsomes to a species that damages bacterial DNA.

6. References

Adamson, R.H., Thorgeirsson, U.P., Snyderwine, E.G., Thorgeirsson, S.S., Reeves, J., Dalgard, D.W., Takayama, S. & Sugimura, T. (1990) Carcinogenicity of 2-amino-3-methylimidazo[4,5-*f*]quinoline in nonhuman primates: induction of tumors in three macaques. *Jpn. J. Cancer Res.*, **81**, 10–14

Adamson, R.H., Snyderwine, E.G., Thorgeirsson, U.P., Schut, H.A.J., Turesky, R.J., Thorgeirsson, S.S., Takayama, S. & Sugimura, T. (1991) Metabolic processing and carcinogenicity of heterocyclic amines in nonhuman primates. In: Ernster, L., Esumi, M., Fujii, Y., Gelboin, H.Y., Kato, R. & Sugimura, T., eds, *Xenobiotics and Cancer*, Tokyo/London, Japan Scientific Societies Press/Taylor & Francis, pp. 289–301

Adolfsson, L. & Olsson, K. (1983) A convenient synthesis of mutagenic 3*H*-imidazo[4,5-*f*]quinoline-2-amines and their 2-^{14}C-labelled analogues. *Acta chem. scand.*, **B37**, 157–159

Aeschbacher, H.-U. & Ruch, E. (1989) Effect of heterocyclic amines and beef extract on chromosome aberrations and sister chromatid exchanges in cultured human lymphocytes. *Carcinogenesis*, **10**, 429–433

Aeschbacher, H.-U. & Turesky, R.J. (1991) Mammalian cell mutagenicity and metabolism of heterocyclic aromatic amines. *Mutat. Res.*, **259**, 235–250

Alldrick, A.J. & Rowland, I.R. (1988) Distribution of radiolabelled [2-^{14}C]IQ and MeIQx in the mouse. *Toxicol. Lett.*, **44**, 183–190

Alldrick, A.J., Rowland, I.R., Lake, B.G., & Flynn, J. (1987) High levels of dietary fat: alteration of hepatic promutagen activation in the rat. *J. natl Cancer Inst.*, **79**, 269–272

Asan, E., Fasshauer, I., Wild, D. & Henschler, D. (1987) Heterocyclic aromatic amine–DNA-adducts in bacteria and mammalian cells detected by ^{32}P-postlabeling analysis. *Carcinogenesis*, **8**, 1589–1593

Barnes, W.S. & Weisburger, J.H. (1985) Fate of the food mutagen 2-amino-3-methylimidazo[4,5-*f*]quinoline (IQ) in Sprague-Dawley rats. I. Mutagens in the urine. *Mutat. Res.*, **156**, 83–91

Barnes, W.S., Maher, J.C. & Weisburger, J.H. (1983) High pressure liquid chromatographic method for the analysis of 2-amino-3-methylimidazo[4,5-*f*]quinoline, a mutagen formed from the cooking of food. *J. agric. Food Chem.*, **31**, 883–886

---

[1]For definition of the italicized terms, see Preamble, pp. 26–29.

Barnes, W.S., Lovelette, C.A., Tong, C., Williams, G.M. & Weisburger, J.H. (1985) Genotoxicity of the food mutagen 2-amino-3-methylimidazo[4,5-f]quinoline (IQ) and analogs. *Carcinogenesis*, **6**, 441-444

Bergman, K. (1985) Autoradiographic distribution of ^{14}C-labeled 3H-imidazo[4,5-f]quinoline-2-amines in mice. *Cancer Res.*, **45**, 1351-1356

Brookman, K.W., Salazar, E.P. & Thompson, L.H. (1985) Comparative mutagenic efficiencies of the DNA adducts from the cooked-food-related mutagens Trp-P-2 and IQ in CHO cells. *Mutat. Res.*, **149**, 249-255

Buonarati, M.H. & Felton, J.S. (1990) Activation of 2-amino-1-methyl-6-phenylimidazo[4,5-b]-pyridine (PhIP) to mutagenic metabolites. *Carcinogenesis*, **11**, 1133-1138

Butler, M.A., Iwasaki, M., Guengerich, F.P. & Kadlubar, F.F. (1989) Human cytochrome P-450$_{PA}$ (P-450IA2), the phenacetin O-deethylase is primarily responsible for the hepatic 3-demethylation of caffeine and N-oxidation of carcinogenic arylamines. *Proc. natl Acad. Sci. USA*, **86**, 7696-7700

Caderni, G., Kreamer, B.L. & Dolara, P. (1983) DNA damage of mammalian cells by the beef extract mutagen 2-amino-3-methylimidazo[4,5-f]quinoline. *Food chem. Toxicol.*, **21**, 641-643

Carman, R.J., Van Tassell, R.L., Kingston, D.G.I., Bashir, M. & Wilkins, T.D. (1988) Conversion of IQ, a dietary pyrolysis carcinogen to a direct-acting mutagen by normal intestinal bacteria of humans. *Mutat. Res.*, **206**, 335-342

Degawa, M., Tanimura, S., Agatsuma, T. & Hashimoto, Y. (1989) Hepatocarcinogenic heterocyclic aromatic amines that induce cytochrome P-448 isozymes, mainly cytochrome P-448H (P-450IA2), responsible for mutagenic activation of the carcinogens in rat liver. *Carcinogenesis*, **10**, 1119-1122

Demirer, T., Icli, F., Uzunalimoglu, O. & Kucuk, O. (1990) Diet and stomach cancer incidence. A case-control study in Turkey. *Cancer*, **65**, 2344-2348

Dirr, A., Fasshauer, I., Wild, D. & Henschler, D. (1989) The DNA-adducts of the food mutagen and carcinogen IQ (2-amino-3-methylimidazo[4,5-f]-quinoline). *Arch. Toxicol.*, **Suppl. 13**, 224-226

Dooley, K.L., Von Tungeln, L.S., Bucci, T., Fu, P.P. & Kadlubar, F.F. (1992) Comparative carcinogenicity of 4-aminobiphenyl and the food pyrolysates, Glu-P-1, IQ, PhIP, and MeIQx in the neonatal B6C3F$_1$ male mouse. *Cancer Lett.*, **62**, 205-209

Felton, J.S. & Knize, M.G. (1990) Heterocyclic-amine mutagens/carcinogens in foods. In: Cooper, C.S. & Grover, P.L., eds, *Handbook of Experimental Pharmacology*, Vol. 94/I, *Chemical Carcinogenesis and Mutagenesis*, Berlin, Springer Verlag, pp. 471-502

Felton, J.S., Knize, M.G., Wood, C., Wuebbles, B.J., Healy, S.K., Stuermer, D.H., Bjeldanes, L.F., Kimble, B.J. & Hatch, F.T. (1984) Isolation and characterization of new mutagens from fried ground beef. *Carcinogenesis*, **5**, 95-102

Felton, J.S., Knize, M.G., Shen, N.H., Andresen, B.D., Bjeldanes, L.F. & Hatch, F.T. (1986) Identification of the mutagens in cooked beef. *Environ. Health Perspectives*, **67**, 17-24

Gerhardsson de Verdier, M., Steineck, G., Hagman, U., Rieger, Å. & Norell, S.E. (1990a) Physical activity and colon cancer: a case-referent study in Stockholm. *Int. J. Cancer*, **46**, 985-989

Gerhardsson de Verdier, M., Hagman, U., Steineck, G., Rieger, Å. & Norell, S.E. (1990b) Diet, body mass and colorectal cancer: a case-referent study in Stockholm. *Int. J. Cancer*, **46**, 832-838

Gerhardsson de Verdier, M., Hagman, U., Peters, R.K., Steineck, G. & Övervik, E. (1991) Meat, cooking methods and colorectal cancer: a case-referent study in Stockholm. *Int. J. Cancer*, **49**, 520-525

Graf, U., Wild, D. & Würgler, F.E. (1992) Genotoxicity of 2-amino-3-methylimidazo[4,5-f]quinoline (IQ) and related compounds in *Drosophila. Mutagenesis*, **7**, 145–149

Grivas, S. & Jägerstad, M. (1984) Mutagenicity of some synthetic quinolines and quinoxalines related to IQ, MeIQ and MeIQx in Ames test. *Mutat. Res.*, **137**, 29–32

Grivas, S., Nyhammar, T., Olsson, K. & Jägerstad, M. (1986) Isolation and identification of the food mutagens IQ and MeIQx from a heated model system of creatinine, glycine and fructose. *Food Chem.*, **20**, 127–136

Grose, K.R., Grant, J.L., Bjeldanes, L.F., Andresen, B.D., Healy, S.K., Lewis, P.R., Felton, J.S. & Hatch, F.T. (1986) Isolation of the carcinogen IQ from fried egg patties. *J. agric. Food Chem.*, **34**, 201–202

Gross, G.A. (1990) Simple methods for quantifying mutagenic heterocyclic aromatic amines in food products. *Carcinogenesis*, **11**, 1597–1603

Gross, G.A. & Grüter, A. (1992) Quantitation of mutagenic/carcinogenic heterocyclic aromatic amines in food products. *J. Chromatogr.*, **592**, 271–278

Gross, G.A., Philippossian, G. & Aeschbacher, H.-U. (1989) An efficient and convenient method for the purification of mutagenic heterocyclic amines in heated meat products. *Carcinogenesis*, **10**, 1175–1182

Gry, J., Vahl, M. & Nielsen, P.A. (1986) *Mutagener i Stegt Kød* [Mutagens in fried meat] (National Food Agency Publication No. 139), Copenhagen, Ministry of the Environment

Hall, M., Ni Shé, M., Wild, D., Fasshauer, I., Hewer, A. & Phillips, D.H. (1990) Tissue distribution of DNA adducts in CDF_1 mice fed 2-amino-3-methylimidazo[4,5-f]quinoline (IQ) and 2-amino-3,4-dimethylimidazo[4,5-f]quinoline (MeIQ). *Carcinogenesis*, **6**, 1005–1011

Hargraves, W.A. & Pariza, M.W. (1983) Purification and mass spectral characterization of bacterial mutagens from commercial beef extract. *Cancer Res.*, **43**, 1467–1472

Hatch, F.T. (1986) A current genotoxicity database for heterocylic thermic food mutagens. 1. Genetically relevant endpoints. *Environ. Health Perspectives*, **67**, 93–103

Hayashi, S., Møller, M.E. & Thorgeirsson, S.T. (1985) Genotoxicity of heterocyclic amines in the *Salmonella*/hepatocyte system. *Jpn. J. Cancer Res. (Gann)*, **76**, 835–845

Hayatsu, H., Matsui, Y., Ohara, Y., Oka, T. & Hayatsu, T. (1983) Characterization of mutagenic fractions in beef extract and in cooked ground beef. Use of blue-cotton for efficient extraction. *Gann*, **74**, 472–482

Hayatsu, H., Arimoto, S. & Wakabayashi, K. (1991) Methods for the separation and detection of heterocyclic amines. In: Hayatsu, H., ed., *Mutagens in Food: Detection and Prevention*, Ann Arbor, MI, CRC Press, pp. 101–112

Higginson, J. (1966) Etiological factors in gastrointestinal cancer in man. *J. natl Cancer Inst.*, **37**, 527–545

Holme, J.A., Hongslo, J.K., Søderlund, E., Brunborg, G., Christensen, T., Alexander, J. & Dybing, E. (1987) Comparative genotoxic effects of IQ and MeIQ in *Salmonella typhimurium* and cultured mammalian cells. *Mutat. Res.*, **187**, 181–190

IARC (1986a) *IARC Monographs on the Evaluation of the Carcinogenic Risk of Chemicals to Humans*, Vol. 40, *Some Naturally Occurring and Synthetic Food Components, Furocoumarins and Ultraviolet Radiation*, Lyon, pp. 261–273

IARC (1986b) *IARC Monographs on the Evaluation of the Carcinogenic Risk of Chemicals to Humans*, Vol. 38, *Tobacco Smoking*, Lyon

Ikeda, M., Yoshimoto, K., Yoshimura, T., Kono, S., Kato, H. & Kuratsune, M. (1983) A cohort study on the possible association between broiled fish intake and cancer. *Gann*, **74**, 640–648

Inamasu, T., Luks, H., Vavrek, M.T. & Weisburger, J.H. (1989) Metabolism of 2-amino-3-methylimidazo[4,5-f]quinoline in the male rat. *Food chem. Toxicol.*, **27**, 369–376

Ishida, Y., Negishi, C., Umemoto, A., Fujita, Y., Sato, S., Sugimura, T., Thorgeirsson, S.A. & Adamson, R.H. (1987) Activation of mutagenic and carcinogenic heterocyclic amines by S-9 from the liver of a rhesus monkey. *Toxicol. in vitro*, **1**, 45–48

Ito, N., Hasegawa, R., Shirai, T., Fukushima, S., Hakoi, K., Takaba, K., Iwasaki, S., Wakabayashi, K., Nagao, M. & Sugimura, T. (1991) Enhancement of GST-P positive liver cell foci development by combined treatment of rats with five heterocylic amines at low doses. *Carcinogenesis*, **12**, 767–772

Kasai, H., Nishimura, S., Wakabayashi, K., Nagao, M. & Sugimura, T. (1980a) Chemical synthesis of 2-amino-3-methylimidazo[4,5-f]quinoline (IQ), a potent mutagen isolated from broiled fish. *Proc. Jpn. Acad.*, **58** (Ser. B), 382–384

Kasai, H., Yamaizumi, Z., Wakabayashi, K., Nagao, M., Sugimura, T., Yokoyama, S., Miyazawa, T., Spingarn, N.E., Weisburger, J.H. & Nishimura, S. (1980b) Potent novel mutagens produced by broiling fish under normal conditions. *Proc. Jpn. Acad.*, **56** (Ser. B), 278–283

Kasai, H., Yamaizumi, Z., Nishimura, S., Wakabayashi, K., Nagao, M., Sugimura, T., Spingarn, N.E., Weisburger, J.H., Yokoyama, S. & Miyazawa, T. (1981) A potent mutagen in broiled fish. Part 1. 2-Amino-3-methyl-3H-imidazo[4,5-f]quinoline. *J. chem. Soc.*, 2290–2293

Kato, R. (1986) Metabolic activation of mutagenic heterocyclic amines from protein pyrolysates. *CRC crit. Rev. Toxicol.*, **16**, 307–348

Kato, R. & Yamazoe, Y. (1987) Metabolic activation and covalent binding to nucleic acids of carcinogenic heterocyclic amines from cooked foods and amino acid pyrolysates. *Jpn. J. Cancer Res. (Gann)*, **78**, 297–311

Kitada, M., Taneda, M., Ohta, K., Nagashima, K., Itahashi, K. & Kamataki, T. (1990) Metabolic activation of aflatoxin B_1 and 2-amino-3-methylimidazo[4,5-f]quinoline by human adult and fetal livers. *Cancer Res.*, **50**, 2641–2645

Knasmüller, S., Kienzl, H., Huber, W. & Hermann, R.S. (1992) Organ-specific distribution of genotoxic effects in mice exposed to cooked food mutagens. *Mutagenesis*, **7**, 235–241

Knize, M.G., Shen, N.H. & Felton, J.S. (1988) *The Production of Mutagens in Foods*. In: *81st Annual Meeting of APCA, Dallas, Texas, June 19–24, 1988*, Pittsburgh, PA, Air Pollution Control Association, paper 130.5

Kono, S., Ikeda, M., Tokudome, S. & Kuratsune, M. (1988) A case–control study of gastric cancer and diet in northern Kyushu, Japan. *Jpn. J. Cancer Res. (Gann)*, **79**, 1067–1074

Kristiansen, E., Clemmensen, S. & Olsen, P. (1989) Carcinogenic potential of cooked food mutagens (IQ and MeIQ) in Wistar rats after short-term exposure. *Pharmacol. Toxicol.*, **65**, 332–335

Kudo, M., Ogura, T., Esumi, H. & Sugimura, T. (1991) Mutational activation of c-Ha-*ras* gene in squamous cell carcinomas of rat Zymbal gland induced by carcinogenic heterocyclic amines. *Mol. Carcinog.*, **4**, 36–42

Lee, C.-S., Hashimoto, Y., Shudo, K. & Okamoto, T. (1982) Synthesis of mutagenic heteroaromatics: 2-aminoimidazo[4,5-f]quinolines. *Chem. pharm. Bull.*, **30**, 1857–1859

Lin, J.-K., Cheng, J.-T. & Lin-Shiau, S.-Y. (1992) Enhancement of the mutagenicity of IQ and MeIQ by nitrite in the *Salmonella* system. *Mutat. Res.*, **278**, 277–287

Loprieno, N., Boncristiani, G. & Loprieno, G. (1991) An experimental approach to identifying the genotoxic risk from cooked meat mutagens. *Food chem. Toxicol.*, **29**, 377–386

Luks, H.J., Spratt, T.E., Vavrek, M.T., Roland, S.F. & Weisburger, J.H. (1989) Identification of sulfate and glucuronic acid conjugates of the 5-hydroxy derivative as major metabolites of 2-amino-3-methylimidazo[4,5-f]quinoline in rats. *Cancer Res.*, **49**, 4407–4411

Lyon, J.L. & Mahoney, A.W. (1988) Fried foods and the risk of colon cancer. *Am. J. Epidemiol.*, **128**, 1000–1006

Makino, H., Ishizaka, Y., Tsujimoto, A., Nakamura, T., Onda, M., Sugimura, T. & Nagao, M. (1992) Rat p53 gene mutations in primary Zymbal gland tumors induced by 2-amino-3-methyl-imidazo[4,5-*f*]quinoline, a food mutagen. *Proc. natl Acad. Sci. USA*, **89**, 4850–4854

McManus, M.E., Burgess, W., Snyderwine, E. & Stupans, I. (1988a) Specificity of rabbit cytochrome P-450 isozymes involved in the metabolic activation of the food derived mutagen 2-amino-3-methylimidazo[4,5-*f*]quinoline. *Cancer Res.*, **48**, 4513–4519

McManus, M.E., Burgess, W., Stupans, I., Trainor, K.J., Fenech, M., Robson, R.A., Morley, A.A. & Snyderwine, E.G. (1988b) Activation of the food-derived mutagen 2-amino-3-methylimidazo-[4,5-*f*]quinoline by human-liver microsomes. *Mutat. Res.*, **204**, 185–193

McManus, M.E., Burgess, W.M., Veronese, M.E., Huggett, A., Quattrochi, L.C. & Tukey, R.H. (1990) Metabolism of 2-acetylaminofluorene and benzo(*a*)pyrene and activation of food-derived heterocyclic amine mutagens by human cytochromes P-450. *Cancer Res.*, **50**, 3367–3376

de Meester, C. (1989) Bacterial mutagenicity of heterocyclic amines found in heat-processed food. *Mutat. Res.*, **221**, 235–262

Minkler, J.L. & Carrano, A.V. (1984) In vivo cytogenetic effects of the cooked-food-related mutagens Trp-P-2 and IQ in mouse bone marrow. *Mutat. Res.*, **140**, 49–53

Nagao, M., Wakabayashi, K., Kasai, H., Nishimura, S. & Sugimura, T. (1981) Effect of methyl substitution on mutagenicity of 2-amino-3-methylimidazo[4,5-*f*]quinoline, isolated from broiled sardine. *Carcinogenesis*, **2**, 1147–1149

Nagao, M., Sato, S. & Sugimura, T. (1983a) Mutagens produced by heating foods. In: *Maillard Reactions* (ACS Monographs, 13), Washington DC, American Chemical Society, pp. 521–536

Nagao, M., Fujita, Y., Wakabayashi, K. & Sugimura, T. (1983b) Ultimate forms of mutagenic and carcinogenic heterocyclic amines produced by pyrolysis. *Biochem. biophys. Res. Commun.*, **114**, 626–631

Nakamura, S.I., Oda, Y., Shimada, T., Oki, I. & Sugimoto, K. (1987) SOS-Inducing activity of chemical carcinogens and mutagens in *Salmonella typhimurium* TA1535/pSK1002: examination with 151 chemicals. *Mutat. Res.*, **192**, 239–246

Nakayasu, M., Nakasato, F., Sakamoto, H., Terada, M. & Sugimura, T. (1983) Mutagenic activity of heterocyclic amines in Chinese hamster lung cells with diphtheria toxin resistance as a marker. *Mutat. Res.*, **118**, 91–102

Norell, S.E., Ahlbom, A., Erwald, R., Jacobson, G., Lindberg-Navier, I., Olin, R., Törnberg, B. & Wiechel, K.-L. (1986) Diet and pancreatic cancer: a case–control study. *Am. J. Epidemiol.*, **124**, 894–902

Ohgaki, H., Kusama, K., Matsukura, N., Morino, K., Hasegawa, H., Sato, S., Takayama, S. & Sugimura, T. (1984) Carcinogenicity in mice of a mutagenic compound, 2-amino-3-methyl-imidazo[4,5-*f*]quinoline, from broiled sardine, cooked beef and beef extract. *Carcinogenesis*, **5**, 921–924

Ohgaki, H., Hasegawa, H., Kato, T., Suenaga, M., Ubukata, M., Sato, S., Takayama, S. & Sugimura, T. (1986) Carcinogenicity in mice and rats of heterocyclic amines in cooked foods. *Environ. Health Perspectives*, **67**, 129–134

Övervik, E. & Gustafsson, J.-Å. (1990) Cooked-food mutagens: current knowledge of formation and biological significance. *Mutagenesis*, **5**, 437–446

Övervik, E., Ochiai, M., Hirose, M., Sugimura, T. & Nagao, M. (1991) The formation of heart DNA adducts in F344 rats following dietary administration of heterocyclic amines. *Mutat. Res.*, **256**, 37–43

Peters, R.K., Garabrant, D.H., Yu, M.C. & Mack, T.M. (1989) A case–control study of occupational and dietary factors in colorectal cancer in young men by subsite. *Cancer Res.*, **49**, 5459–5468

Petry, T.W., Josephy, P.D., Pagano, D.A., Zeiger, E., Knecht, K.T. & Eling, T.E. (1989) Prostaglandin hydroperoxidase-dependent activation of heterocyclic aromatic amines. *Carcinogenesis*, **10**, 2201–2207

Pfau, W., O'Hare, M.J., Grover, P.L. & Phillips, D.H. (1992) Metabolic activation of the food mutagens 2-amino-3-methylimidazo[4,5-*f*]quinoline (IQ) and 2-amino-3,4-dimethylimidazo[4,5-*f*]quinoline (MeIQ) to DNA binding species in human mammary epithelial cells (Short communication). *Carcinogenesis*, **13**, 907–909

Phillips, R.L. (1975) Role of life-style and dietary habits in risk of cancer among Seventh-day Adventists. *Cancer Res.*, **35**, 3513–3522

Radermacher, J., Thorne, G.M., Goldin, B.R., Sullivan, C.E. & Gorbach, S.L. (1987) Mutagenicity of 2-amino-3-methylimidazo[4,5-*f*]quinoline (IQ) in the granuloma pouch assay. *Mutat. Res.*, **187**, 99–103

Sato, H., Takahashi M., Furukawa, F., Miyakawa, Y., Hasegawa, R., Toyoda, K. & Hayashi Y. (1987) Initiating activity in a two-stage mouse skin model of nine mutagenic pyrolysates of amino acids, soybean globulin and proteinaceous food. *Carcinogenesis*, **8**, 1231–1234

Sawada, S., Yamanaka, T., Yamatsu, K., Furihata, C. & Matsushima, T. (1991) Chromosomal aberrations, micronuclei and sister-chromatid exchanges (SCEs) in rat liver induced *in vivo* by hepatocarcinogens including heterocyclic amines. *Mutat. Res.*, **251**, 59–69

Schiffman, M.H. & Felton, J.S. (1990) Re: 'Fried foods and the risk of colon cancer' (Letter to the Editor). *Am. J. Epidemiol.*, **131**, 376–378

Schiffman, M.H., Andrews, A.W., Van Tassell, R.L., Smith, L., Daniel, J., Robinson, A., Hoover, R.N., Rosenthal, J., Weil, R., Nair, P.P., Schwartz, S., Pettigrew, H., Batist, G., Shaw, R. & Wilkins, T.D. (1989) Case–control study of colorectal cancer and fecal mutagenicity. *Cancer Res.*, **49**, 3420–3424

Schunk, H., Hayashi, Y. & Shibamoto, T. (1984) Analysis of mutagenic amino acid pyrolysates with a fused silica capillary column. *J. high Resolut. Chromatogr. Chromatogr. Commun.*, **7**, 563–565

Schut, H.A.J., Putman, K.L. & Randerath, K. (1988) DNA adduct formation of the carcinogen 2-amino-3-methylimidazo[4,5-*f*]quinoline in target tissues of the F-344 rat. *Cancer Lett.*, **41**, 345–352

Schut, H.A.J., Snyderwine, E.G., Zu, H.-X. & Thorgeirsson, S.S. (1991) Similar patterns of DNA adduct formation of 2-amino-3-methylimidazo[4,5-*f*]quinoline in the Fischer 344 rat, CDF$_1$, mouse, cynomolgus monkey and *Salmonella typhimurium*. *Carcinogenesis*, **12**, 931–934

Shimada, T., Iwasaki, M., Martin, M.V. & Guengerich, F.P. (1989) Human liver microsomal cytochrome P-450 enzymes involved in the bioactivation of procarcinogens detected by *umu* gene response in *Salmonella typhimurium* TA1535/pSK1002. *Cancer Res.*, **49**, 3218–3228

Sjödin, P. & Jägerstad, M. (1984) A balance study of ^{14}C-labelled 3*H*-imidazo[4,5-*f*]quinolin-2-amines (IQ and MeIQ) in rats. *Food chem. Toxicol.*, **22**, 207–210

Sjödin, P.B., Nyman, M.E., Nilsson, L., Asp, N.-G.L. & Jägerstad, M.I. (1985) Binding of ^{14}C-labeled food mutagens (IQ, MeIQ, MeIQx) by dietary fiber *in vitro*. *J. Food Sci.*, **50**, 1680–1684

Snyderwine, E.G., Roller, P.P., Adamson, R.H., Sato, S. & Thorgeirsson, S.S. (1988a) Reaction of *N*-hydroxylamine and *N*-acetoxy derivatives of 2-amino-3-methylimidazo[4,5-*f*]quinoline with DNA. Synthesis and identification of *N*-(deoxyguanosin-8-yl)-IQ. *Carcinogenesis*, **9**, 1061–1065

Snyderwine, E.G., Wirth, P.J., Roller, P.P., Adamson, R.H., Sato, S. & Thorgeirsson, S.S. (1988b) Mutagenicity and in vitro covalent DNA binding of 2-hydroxyamino-3-methylimidazolo-[4,5-f]quinoline. *Carcinogenesis*, **9**, 411–418

Snyderwine, E.G., Yamashita, K., Adamson, R.H., Sato, S., Nagao, M., Sugimura, T. & Thorgeirsson, S.S. (1988c) Use of the ^{32}P-postlabeling method to detect DNA adducts of 2-amino-3-methylimidazo[4,5-f]quinoline (IQ) in monkeys fed IQ: identification of the N-(deoxyguanosin-8-yl)-IQ adduct. *Carcinogenesis*, **9**, 1739–1743

Snyderwine, E.G., Adamson, R.H., Welti, D.H., Richli, U., Thorgeirsson, S.S., Würzner, H.P. & Turesky, R.J. (1991) Metabolism of 2-amino-3-methylimidazo[4,5-f]quinoline (IQ) in the monkey (Abstract No. 721). *Proc. Am. Assoc. Cancer Res.*, **32**, 121

Spingarn, N.E., Kasai, H., Vuolo, L.L., Nishimura, S., Yamaizumi, Z., Sugimura, T., Matsushima, T. & Weisburger, J.H. (1980) Formation of mutagens in cooked foods. III. Isolation of a potent mutagen from beef. *Cancer Lett.*, **9**, 177–183

Steineck, G., Hagman, U., Gerhardsson, M. & Norell, S.E. (1990) Vitamin A supplements, fried foods, fat and urothelial cancer. A case–referent study in Stockholm in 1985–87. *Int. J. Cancer*, **45**, 1006–1011

Størmer, F.C., Alexander, J. & Becher, G. (1987) Fluorimetric detection of 2-amino-3-methylimidazo-[4,5-f]quinoline, 2-amino-3,4-dimethylimidazo[4,5-f]quinoline and their N-acetylated metabolites excreted by the rat. *Carcinogenesis*, **8**, 1277–1280

Sugimura, T. (1985) Carcinogenicity of mutagenic heterocyclic amines formed during the cooking process. *Mutat. Res.*, **150**, 33–41

Sugimura, T., Nagao, M. & Wakabayashi, K. (1981) Mutagenic heterocyclic amines in cooked food. In: Egan, H., Fishbein, L., Castegnaro, M., O'Neill, I.K. & Bartsch, H., eds, *Environmental Carcinogens. Selected Methods of Analysis*, Vol. 4, *Some Aromatic and Azo Dyes in the General and Industrial Environment* (IARC Scientific Publications No. 40), Lyon, IARC, pp. 251–267

Sugimura, T., Sato, S. & Takayama, S. (1983) New mutagenic heterocyclic amines found in amino acid and protein pyrolysates and in cooked food. In: Wynder, E.L., Leveille, G.A., Weisburger, J.H. & Livingston, G.E., eds, *Environmental Aspects of Cancer: The Role of Macro and Micro Components of Foods*, Westport, CT, Food and Nutrition Press, pp. 167–186

Sugimura, T., Wakabayashi, K., Nagao, M. & Ohgaki, H. (1989) Heterocyclic amines in cooked food. In: Taylor & Scanlan, eds, *Food Toxicology. A Perspective on the Relative Risks* (IFT Basic Symposium Series), New York, Marcel Dekker, pp. 31–55

Takahashi, M., Wakabayashi, K., Nagao, M., Yamamoto, M., Masui, T., Goto, T., Kinae, N., Tomita, I. & Sugimura, T. (1985) Quantification of 2-amino-3-methylimidazo[4,5-f]quinoline (IQ) and 2-amino-3,8-dimethylimidazo[4,5-f]quinoxaline (MeIQx) in beef extracts by liquid chromatography with electrochemical detection (LCEC). *Carcinogenesis*, **6**, 1195–1199

Takayama, S. & Tanaka, M. (1983) Mutagenesis of amino acid pyrolysis products in Chinese hamster V79 cells. *Toxicol. Lett.*, **17**, 23–28

Takayama, S., Nakatsuru, Y., Masuda, M., Ohgaki, H., Sato, S. & Sugimura, T. (1984) Demonstration of carcinogenicity in F344 rats of 2-amino-3-methylimidazo[4,5-f]quinoline from broiled sardine, fried beef and beef extract. *Gann*, **75**, 467–470

Tanaka, T., Barnes, W.S., Williams, G.M. & Weisburger, J.H. (1985) Multipotential carcinogenicity of the fried food mutagen 2-amino-3-methylimidazo[4,5-f]quinoline in rats. *Jpn. J. Cancer Res. (Gann)*, **76**, 570–576

Thompson, L.H., Carrano, A.V., Salazar, E., Felton, J.S. & Hatch, F.T. (1983) Comparative genotoxic effects of the cooked-food-related mutagens Trp-P-2 and IQ in bacteria and cultured mammalian cells. *Mutat. Res.*, **117**, 243–257

Tomatis, L., Aitio, A., Day, N.E., Heseltine, E., Kaldor, J., Miller, A.B., Parkin, D.M. & Riboli, E., eds (1990) *Cancer: Causes, Occurrence and Control* (IARC Scientific Publications No. 100), Lyon, IARC

Tsuda, M., Negishi, C., Makino, R., Sato, S., Yamaizumi, Z., Hirayama, T. & Sugimura, T. (1985) Use of nitrite and hypochlorite treatments in determination of the contributions of IQ-type and non-IQ type heterocyclic amines to the mutagenicities in crude pyrolized materials. *Mutat. Res.*, 147, 335-341

Tsuda, H., Asamoto, M., Ogiso, T., Inoue, T., Ito, N. & Nagao, M. (1988) Dose-dependent induction of liver and thyroid neoplastic lesions by short-term administration of 2-amino-3-methylimidazo-[4,5-f]quinoline combined with partial hepatectomy followed by phenobarbital or low dose 3'-methyl-4-dimethylaminoazobenzene promotion. *Jpn. J. Cancer Res. (Gann)*, 79, 691-697

Tsuda, H., Takahashi, S., Yamaguchi, S., Ozaki, K. & Ito, N. (1990) Comparison of initiation potential of 2-amino-3-methylimidazo[4,5-f]quinoline and 2-amino-3,8-dimethylimidazo[4,5-f]quinoxaline in an in vivo carcinogen bioassay system. *Carcinogenesis*, 11, 549-552

Tudek, B., Bird, R.P. & Bruce, W.R. (1989) Foci of aberrant crypts in the colons of mice and rats exposed to carcinogens associated with foods. *Cancer Res.*, 49, 1236-1240

Turesky, R.J., Wishnok, J.S., Tannenbaum, S.R., Pfund, R.A. & Buchi, G.H. (1983) Qualitative and quantitative characterization of mutagens in commercial beef extract. *Carcinogenesis*, 4, 863-866

Turesky, R.J., Skipper, P.L., Tannenbaum, S.R., Coles, B. & Ketterer, B. (1986) Sulfamate formation is a major route for detoxification of 2-amino-3-methylimidazo[4,5-f]quinoline in the rat. *Carcinogenesis*, 7, 1483-1485

Turesky, R.J., Skipper, P.L. & Tannenbaum, S.R. (1987) Binding of 2-amino-3-methylimidazo[4,5-f]-quinoline to hemoglobin and albumin in vivo in the rat. Identification of an adduct suitable for dosimetry. *Carcinogenesis*, 8, 1537-1542

Turesky, R.J., Bur, H., Huynh-Ba, T., Aeschbacher, H.U. & Milon, H. (1988) Analysis of mutagenic heterocyclic amines in cooked beef products by high-performance liquid chromatography in combination with mass spectrometry. *Food chem. Toxicol.*, 26, 501-509

Turesky, R.J., Forster, C.M., Aeschbacher, H.U., Würzner, H.P., Skipper, P.L., Trudel, L.J. & Tannenbaum, S.R. (1989) Purification of the food-borne carcinogens 2-amino-3-methylimidazo-[4,5-f]quinoline and 2-amino-3,8-dimethylimidazo[4,5-f]quinoline in heated meat products by immunoaffinity chromatography. *Carcinogenesis*, 10, 151-156

Turesky, R.J., Lang, N.P., Butler, M.A., Teitel, C.H. & Kadlubar, F.F. (1991) Metabolic activation of carcinogenic heterocyclic aromatic amines by human liver and colon. *Carcinogenesis*, 10, 1839-1845

Ubukata, K., Ohi, H., Kitada, M. & Kamataki, T. (1992) A new form of cytochrome P-450 responsible for mutagenic activation of 2-amino-3-methylimidazo[4,5-f]quinoline in human livers. *Cancer Res.*, 52, 758-763

Vanderlaan, M., Watkins, B.E., Hwang, M., Knize, M.G. & Felton, J.S. (1988) Monoclonal antibodies for the immunoassay of mutagenic compounds produced by cooking beef. *Carcinogenesis*, 9, 153-160

Vavrek, M.T., Sidoti, P., Reinhardt, J. & Weisburger, J.H. (1989) Effect of enzyme inducers on the metabolism of 2-amino-3-methylimidazo[4,5-f]quinoline (IQ) in the rat. *Cancer Lett.*, 48, 183-188

Wakabayashi, K., Nagao, M., Esumi, H. & Sugimura, T. (1992) Food-derived mutagens and carcinogens. *Cancer Res.*, 52(Suppl.), 2092s-2098s

Wallin, H., Holme, J.A. & Alexander, J. (1992) Covalent binding of food carcinogens MeIQx, MeIQ and IQ to DNA and protein in microsomal incubations and isolated rat hepatocytes. *Pharmacol. Toxicol.*, **70**, 220–225

Waterhouse, A.L. & Rapoport, H. (1985) Synthesis and tritium labeling of the food mutagens IQ and methyl-IQ. *J. labeled Comp. Radiopharm.*, **22**, 201–216

Wild, D. & Degen, G.H. (1987) Prostaglandin H synthase-dependent mutagenic activation of heterocyclic aromatic amines of the IQ-type. *Carcinogenesis*, **8**, 541–545

Wild, D., Gocke, E., Harnasch, D., Kaiser, G. & King, M.-T. (1985) Differential mutagenic activity of IQ (2-amino-3-methylimidazo[4,5-f]quinoline) in *Salmonella typhimurium* strains *in vitro* and *in vivo*, in *Drosophila*, and in mice. *Mutat. Res.*, **156**, 93–102

Wild, D., Watkins, B.E. & Vanderlaan, M. (1991) Azido- and nitro-PhIP, relatives of the heterocyclic arylamine and food mutagen PhIP—mechanism of their mutagenicity in *Salmonella*. *Carcinogenesis*, **12**, 1091–1096

Willett, W.C., Stampfer, M.J., Colditz, G.A., Rosner, B.A. & Speizer, F.E. (1990) Relation of meat, fat, and fiber intake to the risk of colon cancer in a prospective study among women. *New Engl. J. Med.*, **323**, 1664–1672

Wynder, E.L., Kajitani, T., Ishikawa, S., Dodo, H. & Takano, A. (1969) Environmental factors of cancer of the colon and rectum. II. Japanese epidemiological data. *Cancer*, **23**, 1210–1220

Yamaizumi, Z., Kasai, H., Nishimura, S., Edmonds, C.G. & McCloskey, J.A. (1986) Stable isotope dilution quantification of mutagens in cooked foods by combined liquid chromatography-thermospray mass spectrometry. *Mutat. Res.*, **173**, 1–7

Yamashita, M., Wakabayashi, J., Nagao, M., Sato, S., Yamaizumi, Z., Takahashi, M., Kinae, N., Tomita, I. & Sugimura, T. (1986) Detection of 2-amino-3-methylimidazo[4,5-f]quinoline in cigarette smoke condensate. *Jpn. J. Cancer Res. (Gann)*, **77**, 419–422

Yamashita, K., Umemoto, A., Grivas, S., Kato, S., Sato, S. & Sugimura, T. (1988) Heterocyclic amine–DNA adducts analyzed by ^{32}P-postlabeling method. *Nucleic Acids Res.*, **19**, 111–114

Yamazoe, Y., Shimada, M., Kamataki, T. & Kato, R. (1983) Microsomal activation of 2-amino-3-methylimidazo[4,5-f]quinoline, a pyrolysate of sardine and beef extracts, to a mutagenic intermediate. *Cancer Res.*, **43**, 5768–5774

Yanagisawa, H., Tachikawa, T., Shin, S. & Wada, O. (1990) Monoclonal antibody to 2-amino-3-methylimidazo[4,5-f]quinoline, a dietary carcinogen. *Appl. Biochem. Biotechnol.*, **23**, 1–13

Yoo, M.A., Ryo, H., Todo, T. & Kondo, S. (1985) Mutagenic potency of heterocyclic amines in the *Drosophila* wing spot test and its correlation to carcinogenic potency. *Jpn. J. Cancer Res. (Gann)*, **76**, 468–473

Yoshida, D., Saito, Y. & Mizusaki, S. (1984) Isolation of 2-amino-3-methylimidazo[4,5-f]quinoline as mutagen from the heated product of a mixture of creatine and proline. *Agric. Biol. Chem.*, **48**, 241–243

Yoshimi, N., Sugie, S., Iwata, H., Mori, H. & Williams, G.M. (1988) Species and sex differences in genotoxicity of heterocyclic amine pyrolysis and cooking products in the hepatocyte primary culture/DNA repair test using rat, mouse, and hamster hepatocytes. *Environ. mol. Mutag.*, **12**, 53–64

Young, T.B. & Wolf, D.A. (1988) Case-control study of proximal and distal colon cancer and diet in Wisconsin. *Int. J. Cancer*, **42**, 167–175

Zhang, X.-M., Wakabayashi, K., Liu, Z.-C., Sugimura, T. & Nagao, M. (1988) Mutagenic and carcinogenic heterocyclic amines in Chinese cooked foods. *Mutat. Res.*, **201**, 181–188

Zu, H.-X. & Schut, H.A.J. (1991a) Sex differences in the formation and persistence of DNA adducts of 2-amino-3-methylimidazo[4,5-*f*]quinoline (IQ) in CDF1 mice. *Carcinogenesis*, **12**, 2163–2168

Zu, H.-X. & Schut, H.A.J. (1991b) Formation and persistence of DNA adducts of 2-amino-3-methylimidazo[4,5-*f*]quinoline in male Fischer-344 rats. *Cancer Res.*, **51**, 5636–5641

MeIQ (2-AMINO-3,4-DIMETHYLIMIDAZO[4,5-*f*]QUINOLINE)

This substance was considered by a previous Working Group, in October 1985 (IARC, 1986). Since that time, new data have become available, and these have been incorporated into the monograph and taken into consideration in the present evaluation.

1. Exposure Data

1.1 Chemical and physical data

1.1.1 *Synonyms, structural and molecular data*

Chem. Abstr. Services Reg. No.: 77094-11-2
Chem. Abstr. Name: 3,4-Dimethyl-3*H*-imidazo[4,5-*f*]quinolin-2-amine
IUPAC Systematic Name: 2-Amino-3,4-dimethyl-3*H*-imidazo[4,5-*f*]quinoline

$C_{12}H_{12}N_4$ 	Mol. wt: 212.25

1.1.2 *Chemical and physical properties*

(a) *Description*: Brown crystalline solid (Lee *et al.*, 1982)
(b) *Melting-point*: 296–298 °C (Adolfsson & Olsson, 1983)
(c) *Spectroscopy data*: Ultraviolet, proton nuclear magnetic resonance (Kasai *et al.*, 1980a) and mass spectral data (Hargraves & Pariza, 1983) have been reported.
(d) *Solubility*: Soluble in methanol, ethanol and dimethyl sulfoxide (Lee *et al.*, 1982; Adolfsson & Olsson, 1983; Schunk *et al.*, 1984)
(e) *Stability*: Stable under moderately acidic and alkaline conditions and in cold dilute aqueous solutions protected from light (Sugimura *et al.*, 1983). Exposure of an acetone solution of MeIQ to sunlight for 1 h resulted in conversion of the 2-amino group to a 2-nitro group (Hirose *et al.*, 1990).
(f) *Reactivity*: Rapidly degraded by dilute hypochlorite; not deaminated by weakly acidic nitrite solutions (Tsuda *et al.*, 1985)

1.1.3 *Trade names, technical products and impurities*

No data were available to the Working Group.

1.1.4 *Analysis*

MeIQ was originally isolated from broiled, sun-dried sardines extracted with methanol. The neutral fraction obtained from the extract was subjected to Diaion HP-20 column chromatography, to chloroform–methanol–water partitioning and to Sephadex LH-20 column and silica-gel column chromatography. It was further purified by reverse-phase high-performance liquid chromatography. The structure was deduced mainly from data obtained by proton nuclear magnetic resonance and high-resolution mass spectral analysis (Kasai *et al.*, 1980b).

MeIQ was quantified in broiled fish (Yamaizumi *et al.*, 1986) after methanol extraction, acid/base partition and 'blue cotton' adsorption prior to analysis by high-performance liquid chromatography–thermospray–mass spectrometry. Extraction recoveries were determined using deuterium-labelled MeIQ.

1.2 Production and use

1.2.1 *Production*

The isolation and identification of MeIQ were first reported by Kasai *et al.* (1980b). Its structure was confirmed by chemical synthesis (Kasai *et al.*, 1980a).

Improved syntheses of MeIQ were devised by Lee *et al.* (1982), Adolfsson and Olsson (1983) and Waterhouse and Rapoport (1985). ^{14}C-Labelled MeIQ was synthesized by Adolfsson and Olssen (1983) and tritium-labelled MeIQ by Waterhouse and Rapoport (1985).

MeIQ is produced commercially in small quantities for research purposes.

1.2.2 *Use*

MeIQ is not used commercially.

1.3 Occurrence

MeIQ has been detected in grilled, sun-dried sardines at 20–72 ng/g (Kasai *et al.*, 1980b; Sugimura *et al.*, 1981), in broiled salmon at 0.6–3.1 ng/g, in broiled sardine at 16.6 ng/g (Yamaizumi *et al.*, 1986) and in fried cod at 0.03 ng/g (Wakabayashi *et al.*, 1992). MeIQ was found in fried ground beef cooked at 250 °C but at levels of less than 0.1 ng/g (Felton & Knize, 1990). As described in the monograph on IQ (p. 168), MeIQ is present in foods at lower levels than IQ, MeIQx and PhIP.

MeIQ was also detected but not quantified in fried ground pork (Gry *et al.*, 1986) and in beef extract used for bacteriological media (Hargraves & Pariza, 1983). It was reported in high-temperature-roasted coffee (Kikugawa *et al.*, 1989), but another study of commercial instant and roasted coffees did not demonstrate the presence of any MeIQ (Gross & Wolleb, 1991). A trace amount was found in a refluxed mixture of alanine, fructose and creatinine (Grivas *et al.*, 1985).

1.4 Regulations and guidelines

No data were available to the Working Group.

2. Studies of Cancer in Humans

No epidemiological study was available that addressed the carcinogenic risk to humans of MeIQ itself. Cancer risks associated with consumption of broiled and fried foods, which may contain MeIQ as well as other heterocyclic amines, have, however, been addressed in a number of case-control studies. Several of these are summarized in the monograph on IQ.

3. Studies of Cancer in Experimental Animals

3.1 Oral administration

3.1.1 *Mouse*

Groups of 40 male and 40 female CDF_1 (BALB/cAnN × DBA/2N) F_1 mice, six weeks old, were fed a diet containing 0, 100 or 400 mg/kg of diet MeIQ ('analytical grade') for 91 weeks. Animals that became moribund were killed and autopsied during the experiment. The first forestomach papilloma was detected on day 324 in a male mouse given the highest dose, and the numbers of mice that survived after that time were taken as the effective numbers: males—38/40 high-dose, 38/40 low-dose and 29/40 control; females—38/40 high-dose, 36/40 low-dose and 40/40 control. Significantly more treated female mice than controls had hepatocellular tumours (adenomas and carcinomas combined): 0/40, 4/36 and 27/38 in the control, low-dose and high-dose groups, respectively; there was no significant increase in the incidence of liver tumours in males: 4/29 in controls, 11/38 in low-dose and 7/38 in high-dose animals. The incidence of forestomach tumours (papillomas, carcinomas and one sarcoma) was significantly elevated in both treated males and females compared to controls: males, 0/29, 7/38 and 35/38; females—0/40, 19/36 and 34/38, in control, low-dose and high-dose animals, respectively. The incidence of squamous-cell carcinomas was also increased: males, 0/29, 3/38, 30/38; females, 0/40, 11/36, 24/38, in controls, low-dose and high-dose animals, respectively; one high-dose female had a sarcoma. Squamous-cell carcinomas of the forestomach, many of which metastasized to the liver, were observed in 30 males and 24 females in the high-dose group, in three males and 11 females in the low-dose group and in none of the control animals (Ohgaki *et al.*, 1986).

3.1.2 *Rat*

Groups of 20 male and 20 female Fischer 344 rats, seven weeks old, were fed a diet containing 0 or 300 mg/kg of diet MeIQ (purity, > 99%) for 286 days. Animals that became moribund were killed and autopsied during the experiment. The first tumours of the Zymbal gland, oral cavity and skin were seen independently in three rats on day 139; the numbers of rats that lived after that time being taken as the effective numbers. At termination of the experiment, none of the treated males and 4/20 treated females were still alive. Zymbal gland tumours were found in 19/20 ($p < 0.001$) treated males and 17/20 ($p < 0.001$) treated females; most of these tumours were squamous-cell carcinomas, one of which (in a female) metastasized to the lung. Oral cavity tumours were observed in 7/20 ($p < 0.005$) treated males and 7/20 ($p < 0.005$) treated females and were diagnosed histologically as squamous-cell carcinomas or sebaceous squamous-cell carcinomas. Colonic tumours (adenomas or adenocarcinomas) were found in 7/20 ($p < 0.005$) treated males and 5/20 ($p < 0.025$) treated females. Skin tumours, mainly squamous-cell carcinomas, were found in 10/20 ($p < 0.001$) treated males and in 1/20 treated female. Mammary gland tumours (mostly adenocarcinomas) were found in 5/20 ($p < 0.025$) treated females. A few additional tumours were observed at different sites (males: one adenoma and two adenocarcinomas of the small intestine, one neoplastic nodule of the liver and one papilloma of the forestomach; females: two adenocarcinomas of the small intestine and one squamous-cell carcinoma of the clitoral gland). No tumour of the Zymbal gland, oral cavity, colon, skin or mammary gland was found in the controls (Kato et al., 1989).

3.2 Administration with known carcinogens

3.2.1 Sequential exposure

Rat

As part of a mid-term carcinogenicity study on the synergistic effects of five heterocyclic amines, groups of 14–15 male Fischer 344 rats, six weeks of age, each received a single intraperitoneal injection of 200 mg/kg N-nitrosodiethylamine and, two weeks later, were fed a diet containing MeIQ at 12, 60 or 300 mg/kg. A two-thirds partial hepatectomy was performed in week 3 of the experiment; all animals were killed after eight weeks. Fifteen rats treated only with the nitrosamine served as controls. The effects were assessed by counting placental-form glutathione S-transferase-positive foci in the liver. MeIQ alone at the mid- and high-dose levels significantly increased the area of positive foci (Ito et al., 1991).

3.2.2 Prior exposure

Rat

Groups of 50 male Wistar rats, six weeks old, were given MeIQ at 10 mg/kg bw or solvent (water, acidified to pH 3.5 with citric acid) by gavage every day for two weeks. One week later, the rats were divided into two groups and received either no additional treatment or 500 ppm [mg/l] phenobarbital sodium in the drinking-water until the end of the study at week 58. Interim sacrifice was performed on 10 animals from each group at week 42. Zymbal gland tumours were found in 5/40 rats that received MeIQ alone ($p < 0.05$) and in 3/40 that received MeIQ plus phenobarbital. One animal given MeIQ alone also had a lymphatic leukaemia and another a squamous-cell carcinoma of the skin. Two rats given MeIQ plus phenobarbital developed hepatocellular carcinomas, one, a sarcoma of unclassified histogenesis in the liver, one, lymphatic leukaemia and one, a cutaneous papilloma. One lipoma of the mesentery was found in untreated animals (Kristiansen et al., 1989)

4. Other Relevant Data

4.1 Absorption, distribution, metabolism and excretion

The toxicology and metabolism of heterocyclic aromatic amines have been reviewed (Övervik & Gustafsson, 1990; Aeschbacher & Turesky, 1991).

4.1.1 *Humans*

No data were available to the Working Group.

4.1.2 *Experimental systems*

^{14}C-MeIQ was rapidly absorbed, metabolized and excreted following its oral administration to rats (Sjödin & Jägerstad, 1984; Størmer *et al.*, 1987); excretion occurred through the urine, bile and faeces, mostly within the first 24 h, regardless of the exposure route. The proportion of radiolabel excreted in the urine and faeces was about the same (Sjödin & Jägerstad, 1984).

In mice, intravenously administered ^{14}C-MeIQ was distributed rapidly to the liver, kidney, stomach, lymphomyeloid tissues and endocrine tissues. MeIQ crossed the placenta to reach the fetus in pregnant NMRI mice, but no radiobel was retained in fetal tissues after 24 h (Bergman, 1985). MeIQ is metabolized along a number of pathways, including *N*-hydroxylation, aromatic hydroxylation and conjugation reactions of acetylation, sulfation and glucuronidation, to produce a complex array of metabolites both *in vivo* and *in vitro* (Størmer *et al.*, 1987; Alexander *et al.*, 1989).

Once absorbed, MeIQ is activated through *N*-hydroxylation to its mutagenic and reactive form, mainly by the human hepatic cytochrome P450 isozyme P450 IA2 and to some extent by P450 IA1 (McManus *et al.*, 1990). Human liver microsomes can activate MeIQ into a DNA-reactive species. (See Fig. 1, in the monograph on IQ, p. 178.) The cytochrome P450 isozyme responsible has been identified tentatively as CYP IA2 (P450 IA2) (Shimada *et al.*, 1989). Other metabolic pathways appeared to result in detoxication (Alldrick *et al.*, 1986). MeIQ can also be activated by a prostaglandin hydroperoxidase-dependent pathway, as shown in microsomes isolated from ram seminal vesicles (Wild & Degen, 1987; Petry *et al.*, 1989).

In a study designed to provide information on the metabolic steps involved in the formation of the active mutagenic form, MeIQ was not mutagenic to *Salmonella typhimurium* TA98/1,8-DNP6 (defective in esterifying activity) in the presence of an exogenous metabolic system (S9 mix), but it was strongly mutagenic to the original TA98 with S9 mix, suggesting that the ultimate form of MeIQ is a reactive ester of its *N*-hydroxy derivative (Nagao *et al.*, 1983a). In the presence of a microsomal activation system, MeIQ bound to protein and DNA. Binding of MeIQ to protein was enhanced in the presence of microsomes from animals pretreated with Aroclor 1254 and β-naphthoflavone (Wallin & Alexander, 1988).

Certain species of colonic bacteria (i.e., *Eubacterium* and *Clostridium*) have been reported to activate MeIQ to 7-hydroxy-MeIQ, a highly mutagenic metabolite for *Salmonella typhimurium* strain TA98 (Van Tassell *et al.*, 1990). Diet may influence the metabolism and activation of MeIQ. Thus, hepatic postmitochondrial fractions from Ola:Sprague-Dawley rats fed high-fat diets showed increased metabolic activation of MeIQ (Alldrick *et al.*, 1987). MeIQ bound to various dietary fibres *in vitro* (Sjödin *et al.*, 1985).

4.2 Toxic effects

No data were available to the Working Group.

4.3 Reproductive and developmental toxicity

No data were available to the Working Group.

4.4 Genetic and related effects

The genetic effects of MeIQ have been reviewed (Sugimura, 1985; Hatch, 1986; de Meester, 1989).

4.4.1 *Humans*

No data were available to the Working Group.

4.4.2 *Experimental systems* (see also Table 1 and Appendices 1 and 2)

MeIQ induced prophage, SOS repair and mutation in bacteria and somatic mutations in *Drosophila melanogaster*. In cultured hepatocytes of rat, mouse, Syrian hamster and guinea-pig, MeIQ induced DNA strand breaks and unscheduled DNA synthesis. In mammalian cell lines, it induced mutations at the *hprt* locus as well as diphtheria toxin-resistant mutants; it induced sister chromatid exchange and micronucleus formation in one experiment but not in others.

After administration *in vivo*, MeIQ formed DNA adducts in multiple mouse organs and induced DNA strand breaks in several rat organs. It also induced sister chromatid exchange in the colonic mucosa of mice treated orally.

4.4.3 *Genetic changes in animal tumours*

Activated c-Ha-*ras* proto-oncogenes were found in four of seven squamous-cell carcinomas of the forestomach induced by MeIQ in mice as a result of G to T transversion at the second base of codon 13 (Makino *et al.*, 1992). Activated c-Ha-*ras* was also found in all of 15 Zymbal gland tumours induced in rats by MeIQ. The mutations were G to T transversions at the second base of codon 13 (10 tumours), a G to T transversion at the second base of codon 12 (one tumour) and a G to A transition at the second base of codon 12 (one tumour) (Kudo *et al.*, 1991).

5. Summary of Data Reported and Evaluation

5.1 Exposure data

MeIQ (2-Amino-3,4-dimethylimidazo[4,5-*f*]quinoline) has been found in cooked meat and fish. A few determinations indicated that the levels of MeIQ were lower than those of IQ, MeIQx and PhIP.

5.2 Human carcinogenicity data

No data directly relevant to an evaluation of the carcinogenicity to humans of MeIQ were available. Studies on the consumption of cooked meat and fish are summarized in the monograph on IQ.

Table 1. Genetic and related effects of MeIQ

Test system	Result without exogenous metabolic system	Result with exogenous metabolic system	Dose (LED/HID)[a]	Reference
PRB, Prophage induction, *Escherichia coli* K12	0	+	0.1000	Nagao et al. (1983b)
PRB, SOS repair, *Salmonella typhimurium* TA1535	0	+	0.5700	Nakamura et al. (1987)
PRB, *umu* expression, *Salmonella typhimurium* TA1535/pSK1002	0	+	2.0000	Shimada et al. (1989)
PRB, SOS repair, *Salmonella typhimurium* with human adult and fetal microsomes	0	+	2.0000	Kitada et al. (1990)
ERD, *Escherichia coli* rec strains, differential toxicity	–	+	0.0700	Knasmüller et al. (1992)
SA0, *Salmonella typhimurium* TA100, reverse mutation	0	+	0.0250	Nagao et al. (1981)
SA0, *Salmonella typhimurium* TA100, reverse mutation	–	+	0.0000	Grivas & Jägerstad (1984)
SA0, *Salmonella typhimurium* TA100, reverse mutation	0	+	0.0000	Felton & Knize (1990)
SA0, *Salmonella typhimurium* TA100, reverse mutation	+[b]	+	0.0500	Lin et al. (1992)
SA0, *Salmonella typhimurium* TA100, reverse mutation	+[b]	+	0.0080	Lin et al. (1992)
SA2, *Salmonella typhimurium* TA102, reverse mutation	0	–	0.0000	Felton & Knize (1990)
SA4, *Salmonella typhimurium* TA104, reverse mutation	0	+	0.0000	Felton & Knize (1990)
SA5, *Salmonella typhimurium* TA1535, reverse mutation	(+)[b]	–	0.0080	Lin et al. (1992)
SA8, *Salmonella typhimurium* TA1538, reverse mutation	0	+	0.0400	Felton & Knize (1990)
SA9, *Salmonella typhimurium* TA98, reverse mutation	–	+	0.0000	Kasai et al. (1980b)
SA9, *Salmonella typhimurium* TA98, reverse mutation	0	+	0.0025	Nagao et al. (1981)
SA9, *Salmonella typhimurium* TA98, reverse mutation	0	+	0.0012	Nagao et al. (1983a)
SA9, *Salmonella typhimurium* TA98, reverse mutation	–	+	0.0000	Grivas & Jägerstad (1984)
SA9, *Salmonella typhimurium* TA98, reverse mutation	0	+	0.0020	Loury et al. (1985)
SA9, *Salmonella typhimurium* TA98, reverse mutation	0	+	0.0050	Howes et al. (1986)
SA9, *Salmonella typhimurium* TA98, reverse mutation	–	+[d]	0.0200	Holme et al. (1987)
SA9, *Salmonella typhimurium* TA98, reverse mutation[c]	0	+	0.0005	Buonarati & Felton (1990)
SA9, *Salmonella typhimurium* TA98, reverse mutation	0	+	0.0000	Ishida et al. (1987)
SA9, *Salmonella typhimurium* TA98, reverse mutation	0	+	0.0000	Felton & Knize (1990)
SA9, *Salmonella typhimurium* TA98, reverse mutation	+[b]	+	0.0001	Lin et al. (1992)
SA9, *Salmonella typhimurium* TA98, reverse mutation	+[b]	+	0.0001	Lin et al. (1992)
SAS, *Salmonella typhimurium* TA98/1,8-DNP$_6$, reverse mutation	0	–	0.0050	Nagao et al. (1983a)

Table 1 (contd)

Test system	Result		Dose (LED/HID)[a]	Reference
	Without exogenous metabolic system	With exogenous metabolic system		
SAS, *Salmonella typhimurium* TA98/1,8-DNP$_6$, reverse mutation	0	–	0.0005	Buonarati & Felton (1990)
SAS, *Salmonella typhimurium* TA98/1,8-DNP$_6$, reverse mutation	0	–[d,e]	100.0000	Holme et al. (1988)
SAS, *Salmonella typhimurium* TA98NR, reverse mutation	0	+[d,e]	100.0000	Holme et al. (1988)
SAS, *Salmonella typhimurium* TA96, reverse mutation	0	+	0.0000	Felton & Knize (1990)
SAS, *Salmonella typhimurium* TA97, reverse mutation	0	+	0.0000	Felton & Knize (1990)
SAS, *Salmonella typhimurium* TA97, reverse mutation	+[b]	+	0.0500	Lin et al. (1992)
SAS, *Salmonella typhimurium* TA97, reverse mutation		+	0.0008	Lin et al. (1992)
DMM, *Drosophila melanogaster*, somatic mutation and recombination	+	0	25.0000	Yoo et al. (1985)
DIA, DNA strand breaks, rat hepatocytes[e] *in vitro*	+	0	20.0000	Holme et al. (1987)
DIA, DNA strand breaks, Chinese hamster lung V79 cells *in vitro*	–	–[d,e]	100.0000	Holme et al. (1987)
URP, Unscheduled DNA synthesis, rat primary hepatocytes *in vitro*	+	0	0.1000	Yoshimi et al. (1988)
URP, Unscheduled DNA synthesis, rat hepatocytes *in vitro*	+	0	1.0000	Liu et al. (1990)
UIA, Unscheduled DNA synthesis, Syrian hamster primary hepatocytes *in vitro*	+	0	1.0000	Yoshimi et al. (1988)
UIA, Unscheduled DNA synthesis, mouse primary hepatocytes (female) *in vitro*	+	0	1.0000	Yoshimi et al. (1988)
UIA, Unscheduled DNA synthesis, mouse primary hepatocytes (male) *in vitro*	+	0	10.0000	Yoshimi et al. (1988)
GCL, Gene mutation, Chinese hamster lung cells, DT^r *in vitro*	–	+	7.5000	Nakayasu et al. (1983)
GCL, Gene mutation, Chinese hamster lung cells *in vitro*	0	+	1.0000	Sugimura et al. (1989)
GCO, Gene mutation, Chinese hamster ovary cells (*uv5*) *in vitro*	0	+	75.0000	Thompson et al. (1987)
G9H, Gene mutation, Chinese hamster V79 cells *hprt* locus *in vitro*	–	(+)[d,e]	20.0000	Holme et al. (1987)
G90, Gene mutation, Chinese hamster lung V79 cells ouabain^r *in vitro*	0	–	50.0000	Takayama & Tanaka (1983)
SIC, Sister chromatid exchange, Chinese hamster V79 cells *in vitro*	–	+[d,e]	20.0000	Holme et al. (1987)

Table 1 (contd)

Test system	Result Without exogenous metabolic system	Result With exogenous metabolic system	Dose (LED/HID)[a]	Reference
SIC, Sister chromatid exchange, Chinese hamster ovary (uv5) cells *in vitro*	0	(+)	200.0000	Thompson *et al.* (1987)
SIC, Sister chromatid exchange, IAR 20 cells *in vitro*	+	0	0.1000	Liu *et al.* (1990)
SIC, Sister chromatid exchange, Chinese hamster V79 cells *in vitro*	0	+	0.0035	Liu *et al.* (1990)
MIA, Micronucleus test, Chinese hamster V79 cells *in vitro*	0	+	0.0020	Liu *et al.* (1990)
CIC, Chromosomal aberrations, Chinese hamster ovary (uv5) cells *in vitro*	0	(+)	400.0000	Thompson *et al.* (1987)
HMM, Host-mediated assay, *Escherichia coli* in intrasanguinous mice *in vivo*	+		2.3×1 ip	Knasmüller *et al.* (1992)
DVA, DNA strand breaks, rat[e] liver, large intestine, kidney *in vivo*	+		5.00×1 ip (po)	Holme *et al.* (1991)
SVA, Sister chromatid exchange, mouse colonic epithelial cells *in vivo*	+		50.00×1 ip	Couch *et al.* (1987)
BID, Binding (covalent) to DNA in rat hepatocytes *in vitro*[f]	+	0	10.0000	Wallin *et al.* (1992)
BVD, Binding (covalent) to DNA in mice (multiple organs) *in vivo*[g]	+		25.00×1 po	Hall *et al.* (1990)

+, positive; (+), weakly positive; −, negative; 0, not tested; ?, inconclusive (variable response in several experiments within an adequate study)

[a]In-vitro tests, μg/ml; in-vivo tests, mg/kg bw
[b]MeIQ reacted with nitrite (not on profile)
[c]Rhesus liver S9
[d]Hepatocytes
[e]Polychlorinated biphenyl-treated
[f]¹⁴C-Label
[g]³²P-Postlabel

5.3 Animal carcinogenicity data

MeIQ was tested for carcinogenicity by dietary administration in one study in mice and in one study in rats. In mice, hepatocellular adenomas and carcinomas were induced in females and papillomas and squamous-cell carcinomas of the forestomach in animals of each sex in a dose-dependent manner. In rats, oral administration of MeIQ produced squamous-cell carcinomas of the Zymbal gland and oral cavity and adenomas and adenocarcinomas of the colon in animals of each sex, squamous-cell carcinomas of the skin in male rats and mammary adenocarcinomas in female rats.

Sequential administration of MeIQ after N-nitrosodiethylamine enhanced the appearance of foci of altered hepatocytes in rat liver.

5.4 Other relevant data

No data were available on the genetic and related effects of MeIQ in humans.

MeIQ bound to DNA and induced DNA damage and sister chromatid exchange in rodents treated *in vivo*. It induced DNA damage and gene mutation in rodent cells *in vitro* and gene mutation in insects. It induced DNA damage and mutation in bacteria.

MeIQ can be metabolized by human liver microsomes to a species that damages bacterial DNA.

5.5 Evaluation[1]

There is *inadequate evidence* in humans for the carcinogenicity of MeIQ.

There is *sufficient evidence* in experimental animals for the carcinogenicity of MeIQ.

Overall evaluation

MeIQ (2-Amino-3,4-dimethylimidazo[4,5-*f*]quinoline) *is possibly carcinogenic to humans (Group 2B)*.

6. References

Adolfsson, L. & Olsson, K. (1983) A convenient synthesis of mutagenic 3*H*-imidazo[4,5-*f*]quinoline-2-amines and their 2-^{14}C-labelled analogues. *Acta chem. scand.*, **B37**, 157–159

Aeschbacher, H.-U. & Turesky, R.J. (1991) Mammalian cell mutagenicity and metabolism of heterocyclic aromatic amines. *Mutat. Res.*, **259**, 235–250

Alexander, J., Holme, J.A., Wallin, H. & Becher, G. (1989) Characterisation of metabolites of the food mutagens 2-amino-3-methylimidazo[4,5-*f*]quinoline and 2-amino-3,4-dimethylimidazo[4,5-*f*]quinoline formed after incubation with isolated rat liver cells. *Chem.-Biol. Interactions*, **72**, 125–142

---

[1]For definition of the italicized terms, see Preamble, pp. 26-29.

Alldrick, A.J., Lake, B.G., Flynn, J. & Rowland, I.R. (1986) Metabolic conversion of IQ and MeIQ to bacterial mutagens. *Mutat. Res.*, **163**, 109–114

Alldrick, A.J., Rowland, I.R., Lake, B.G. & Flynn, J. (1987) High levels of dietary fat: alteration of hepatic promutagen activation in the rat. *J. natl Cancer Inst.*, **79**, 269–272

Bergman, K. (1985) Autoradiographic distribution of ^{14}C-labeled 3H-imidazo[4,5-f]quinoline-2-amines in mice. *Cancer Res.*, **45**, 1351–1356

Buonarati, M.H. & Felton, J.S. (1990) Activation of 2-amino-1-methyl-6-phenylimidazo[4,5-b]-pyridine (PhIP) to mutagenic metabolites. *Carcinogenesis*, **11**, 1133–1138

Couch, D.B., Stuart, E. & Heddle, J.A. (1987) Effect of oral administration of mutagens found in food on the frequency of sister chromatid exchanges in the colonic epithelium of mice. *Environ. mol. Mutag.*, **10**, 205–209

Felton, J.S. & Knize, M.G. (1990) Heterocyclic-amine mutagens/carcinogens in foods. In: Cooper, C.J. & Grover, P.L., eds, *Handbook of Experimental Pharmacology*, Vol. 94/I, *Chemical Carcinogenesis and Mutagenesis*, Berlin, Springer Verlag, pp. 471–502

Grivas, S. & Jägerstad, M. (1984) Mutagenicity of some synthetic quinolines and quinoxalines related to IQ, MeIQ or MeIQx in Ames test. *Mutat. Res.*, **137**, 29–32

Grivas, S., Nyhammar, T., Olsson, K. & Jägerstad, M. (1985) Formation of a new mutagenic DiMeIQx compound in a model system by heating creatinine, alanine and fructose. *Mutat. Res.*, *151*, 177–183

Gross, G.A. & Wolleb, U. (1991) 2-Amino-3,4-dimethylimidazo[4,5-f]quinoline is not detectable in commercial instant and roasted coffees. *J. agric. Food Chem.*, **39**, 2231–2236

Gry, J., Vahl, M. & Nielsen, P.A. (1986) *Mutagener i Stegt Kød* [Mutagens in fried meat] (National Food Agency Publication No. 139), Copenhagen, Ministry of the Environment

Hall, M., Shé, M.N., Wild, D., Fasshauer, I., Hewer, A. & Phillips, D.H. (1990) Tissue distribution of DNA adducts in CDF$_1$ mice fed 2-amino-3-methylimidazo[4,5-f]quinoline (IQ) and 2-amino-3,4-dimethylimidazo[4,5-f]quinoline (MeIQ). *Carcinogenesis*, **11**, 1005–1011

Hargraves, W.A. & Pariza, M.W. (1983) Purification and mass spectral characterization of bacterial mutagens from commercial beef extract. *Cancer Res.*, **43**, 1467–1472

Hatch, F.T. (1986) A current genotoxicity database for heterocyclic thermic food mutagens. 1. Genetically relevant endpoints. *Environ. Health Perspectives*, **67**, 93–103

Hirose, M., Wakabayashi, K., Grivas, S., De Flora, S., Arakawa, N., Nagao, M. & Sugimura, T. (1990) Formation of a nitro derivative of 2-amino-3,4-dimethylimidazo[4,5-f]quinoline by photoirradiation. *Carcinogenesis*, **11**, 869–871

Holme, J.A., Hongslo, J.K., Søderlund, E., Brunborg, G., Christensen, T., Alexander, J. & Dybing, E. (1987) Comparative genotoxic effects of IQ and MeIQ in *Salmonella typhimurium* and cultured mammalian cells. *Mutat. Res.*, **187**, 181–190

Holme, J.A., Brunborg, G., Alexander, J., Trygg, B. & Bjørnstad, C. (1988) Modulation of the mutagenic effects of 2-amino-3-methylimidazo[4,5-f]quinoline (IQ) and 2-amino-3,4-dimethylimidazo[4,5-f]quinoline (MeIQ) in bacteria with rat-liver 9000 × g supernatant or monolayers of rat hepatocytes as an activation system. *Mutat. Res.*, **197**, 39–49

Holme, J.A., Brunborg, G., Alexander, J., Trygg, B. & Søderlund, E.J. (1991) Genotoxic effects of 2-amino-3,4-dimethylimidazo[4,5-f]quinoline (MeIQ) in rats measured by alkaline elution. *Mutat. Res.*, **251**, 1–6

Howes, A.J., Beamand, J.A. & Rowland, I.R. (1986) Induction of unscheduled DNA synthesis in rat and hamster hepatocytes by cooked food mutagens. *Food chem. Toxicol.*, **24**, 383–387

IARC (1986) *IARC Monographs on the Evaluation of the Carcinogenic Risk of Chemicals to Humans*, Vol. 40, *Some Naturally Occurring and Synthetic Food Components, Furocoumarins and Ultraviolet Radiation*, Lyon, pp. 275-281

Ishida, Y., Negishi, C., Umemoto, A., Fujita, Y., Sato, S., Sugimura, T., Thorgeirsson, S.A. & Adamson, R.H. (1987) Activation of mutagenic and carcinogenic heterocyclic amines by S-9 from the liver of a rhesus monkey. *Toxicol. in vitro*, **1**, 45–48

Ito, N., Hasegawa, R., Shirai, T., Fukushima, S., Hakoi, K., Takaba, K., Iwasaki, S., Wakabayashi, K., Nagao, M. & Sugimura, T. (1991) Enhancement of GST-P positive liver cell foci development by combined treatment of rats with five heterocyclic amines at low doses. *Carcinogenesis*, **12**, 767–772

Kasai, H., Yamaizumi, Z., Wakabayashi, K., Nagao, M., Sugimura, T., Yokoyama, S., Miyazawa, T. & Nishimura, S. (1980a) Structure and chemical synthesis of ME-IQ, a potent mutagen isolated from broiled fish. *Chem. Lett.*, **11**, 1391–1394

Kasai, H., Yamaizumi, Z., Wakabayashi, K., Nagao, M., Sugimura, T., Yokoyama, S., Miyazawa, T., Spingarn, N.E., Weisburger, J.H. & Nishimura, S. (1980b) Potent novel mutagens produced by broiling fish under normal conditions. *Proc. Jpn. Acad.*, **56** (Serie B), 278–283

Kato, T., Migita, H., Ohgaki, H., Sato, S., Takayama, S. & Sugimura, T. (1989) Induction of tumors in the Zymbal gland, oral cavity, colon, skin and mammary gland of F344 rats by a mutagenic compound, 2-amino-3,4-dimethylimidazo[4,5-*f*]quinoline. *Carcinogenesis*, **10**, 601–603

Kikugawa, K., Kato, T. & Takahashi, S. (1989) Possible presence of 2-amino-3,4-dimethylimidazo-[4,5-*f*]quinoline and other heterocyclic amine-like mutagens in roasted coffee beans. *J. agric. Food Chem.*, **37**, 881–886

Kitada, M., Taneda, M., Ohta, K., Nagashima, K., Itahashi, K. & Kamataki, T. (1990) Metabolic activation of aflatoxin B_1 and 2-amino-3-methylimidazo[4,5-*f*]quinoline by human adult and fetal livers. *Cancer Res.*, **50**, 2641–2645

Knasmüller, S., Kienzl, H., Huber, W. & Hermann, R.S. (1992) Organ-specific distribution of genotoxic effects in mice exposed to cooked food mutagens. *Mutagenesis*, **7**, 235–241

Kristiansen, E., Clemmensen, S. & Olsen, P. (1989) Carcinogenic potential of cooked food mutagens (IQ and MeIQ) in Wistar rats after short-term exposure. *Pharmacol. Toxicol.*, **65**, 332–335

Kudo, M., Ogura, T., Esumi, H. & Sugimura, T. (1991) Mutational activation of c-Ha-*ras* gene in squamous cell carcinoma of rat Zymbal gland induced by carcinogenic heterocyclic amines. *Mol. Carcinog.*, **4**, 36–42

Lee, C.-S., Hashimoto, Y., Shudo, K. & Okamoto, T. (1982) Synthesis of mutagenic heteroaromatics: 2-aminoimidazo[4,5-*f*]quinolines. *Chem. pharm. Bull.*, **30**, 1857–1859

Lin, J.-K., Cheng, J.-T. & Lin-Shiau, S.-Y. (1992) Enhancement of the mutagenicity of IQ and MeIQ by nitrite in the *Salmonella* system. *Mutat. Res.*, **278**, 277–287

Liu, X.-L., Chen, S.-J. & Li, M.-X. (1990) Genotoxicity of fried fish extract, MeIQ and inhibition by green tea oxidant (Chin.). *Chin. J. Oncol.*, **12**, 170–173

Loury, D.J., Kado, N.Y. & Byard, J.L. (1985) Enhancement of hepatocellular genotoxicity of several mutagens from amino acid pyrolysates and broiled foods following ethanol pretreatment. *Food chem. Toxicol.*, **23**, 661–667

Makino, H., Ochiai, M., Caignard, A., Ishizaka, Y., Onda, M., Sugimura, T. & Nagao, M. (1992) Detection of a Ha-*ras* point mutation by polymerase chain reaction–single strand conformation polymorphism analysis in 2-amino-3,4-dimethylimidazo[4,5-*f*]quinoline-induced mouse forestomach tumors. *Cancer Lett.*, **62**, 115–121

McManus, M.E., Burgess, W.M., Veronese, M.E., Huggett, A., Quattrochi, L.C. & Tukey, R.H. (1990) Metabolism of 2-acetylaminofluorene and benzo(a)pyrene and activation of food-derived heterocyclic amino mutagens by human cytochromes P-450. *Cancer Res.*, **50**, 3367–3376

de Meester, C. (1989) Bacterial mutagenicity of heterocyclic amines found in heat-processed food. *Mutat. Res.*, **221**, 235–262

Nagao, M., Wakabayashi, K., Kasai, H., Nishimura, S. & Sugimura, T. (1981) Effect of methyl substitution on mutagenicity of 2-amino-3-methylimidazo[4,5-f]quinoline, isolated from broiled sardine. *Carcinogenesis*, **2**, 1147–1149

Nagao, M., Fujita, Y., Wakabayashi, K. & Sugimura, T. (1983a) Ultimate forms of mutagenic and carcinogenic heterocyclic amines produced by pyrolysis. *Biochem. biophys. Res. Commun.*, **114**, 626–631

Nagao, M., Sato, S. & Sugimura, T. (1983b) Mutagens produced by heating foods. In: *Maillard Reactions* (ACS Monographs, 13), Washington DC, American Chemical Society, pp. 521–536

Nakamura, S.-I., Oda, Y., Shimada, T., Oki, I. & Sugimoto, K. (1987) SOS-Inducing activity of chemical carcinogens and mutagens in *Salmonella typhimurium* TA1535/pSK1002: examination with 151 chemicals. *Mutat. Res.*, **192**, 239–246

Nakayasu, M., Nakasato, F., Sakamoto, H., Terada, M. & Sugimura, T. (1983) Mutagenic activity of heterocyclic amines in Chinese hamster lung cells with diphtheria toxin resistance as a marker. *Mutat. Res.*, **118**, 91–102

Ohgaki, H., Hasegawa, H., Suenaga, M., Kato, T., Sato, S., Takayama, S. & Sugimura, T. (1986) Induction of hepatocellular carcinoma and highly metastatic squamous cell carcinomas in the forestomach of mice by feeding 2-amino-3,4-dimethylimidazo[4,5-f]quinoline. *Carcinogenesis*, **7**, 1889–1893

Övervik, E. & Gustafsson, J.-Å. (1990) Cooked-food mutagens: current knowledge of formation and biological significance. *Mutagenesis*, **5**, 437–446

Petry, T.W., Josephy, P.D., Pagano, D.A., Zeiger, E., Knecht, K.T. & Eling, T.E. (1989) Prostaglandin hydroperoxidase-dependent activation of heterocyclic aromatic amines. *Carcinogenesis*, **10**, 2201–2207

Schunk, H., Hayashi, T. & Shibamoto, T. (1984) Analysis of mutagenic amino acid pyrolyzates with a fused silica capillary column. *J. high Resolut. Chromatogr. Chromatogr. Commun.*, **7**, 563–565

Shimada, T., Iwasaki, M., Martin, M.V. & Guengerich, F.P. (1989) Human liver microsomal cytochrome P-450 enzymes involved in the bioactivation of procarcinogens detected by *umu* gene response in *Salmonella typhimurium* TA1535/pSK1002. *Cancer Res.*, **49**, 3218–3228

Sjödin, P.B. & Jägerstad, M. (1984) A balance study of ^{14}C-labelled 3H-imidazo[4,5-f]quinolin-2-amines (IQ and MeIQ) in rats. *Food Chem. Toxicol.*, **22**, 207–210

Sjödin, P.B., Nyman, M.E., Nilsson, L., Asp, N.G.-L. & Jägerstad, M.I. (1985) Binding of ^{14}C-labelled food mutagens (IQ, MeIQ, MeIQx) by dietary fiber *in vitro*. *J. Food Sci.*, **50**, 1680–1684

Størmer, F.C., Alexander, J. & Becher, G. (1987) Fluorimetric detection of 2-amino-3-methylimidazo-[4,5-f]quinoline, 2-amino-3,4-dimethylimidazo[4,5-f]quinoline and their N-acetylated metabolites excreted by the rat. *Carcinogenesis*, **8**, 1277–1280

Sugimura, T. (1985) Carcinogenicity of mutagenic heterocyclic amines formed during the cooking process. *Mutat. Res.*, **150**, 33–41

Sugimura, T., Nagao, M. & Wakabayashi, K. (1981) Mutagenic heterocyclic amines in cooked food. In: Egan, H., Fishbein, L., Castegnaro, M., O'Neill, I.K. & Bartsch, H., eds, *Environmental Carcinogens: Selected Methods of Analysis*, Vol. 4, *Some Aromatic Amines and Azo Dyes in the General and Industrial Environment* (IARC Scientific Publications No. 40), Lyon, IARC, pp. 251–267

Sugimura, T., Sato, S. & Takayama, S. (1983) New mutagenic heterocyclic amines found in amino acid and protein pyrolysates in cooked food. In: Wynder, E.L., Leveille, G.A., Weisburger, J.H. & Livingston, G.E., eds, *Environmental Aspects of Cancer: The Role of Macro and Micro Components of Foods*, Westport, CT, Food and Nutrition Press, pp. 167–186

Sugimura, T., Wakabayashi, K., Nagao, M. & Ohgaki, H. (1989) Heterocyclic amines in cooked food. In: Taylor & Scanlan, *Food Toxicology. A Perspective on the Relative Risks* (IFT Basic Symposium Ser.), New York, Marcel Dekker, pp. 31–55

Takayama, S. & Tanaka, M. (1983) Mutagenesis of amino acid pyrolysis products in Chinese hamster V79 cells. *Toxicol. Lett.*, **17**, 23–28

Thompson, L.H., Tucker, J.D., Stewart, S.A., Christensen, M.L., Salazar, E.P., Carrano, A.V. & Felton, J.S. (1987) Genotoxicity of compounds from cooked beef in repair-deficient CHO cells versus *Salmonella* mutagenicity. *Mutagenesis*, **2**, 483–487

Tsuda, M., Negishi, C., Makino, R., Sato, S., Yamaizumi, Z., Hirayama, T. & Sugimura, T. (1985) Use of nitrite and hypochlorite treatments in determination of the contributions of IQ-type and non-IQ type heterocyclic amines to the mutagenicities in crude pyrolized materials. *Mutat. Res.*, **147**, 335–341

Van Tassell, R.L., Kingston, D.G.I. & Wilkins, T.D. (1990) Metabolism of dietary genotoxins by the human colonic microflora; the fecapentaenes and heterocyclic amines. *Mutat. Res.*, **238**, 209–221

Wakabayashi, K., Nagao, M., Esumi, H. & Sugimura, T. (1992) Food-derived mutagens and carcinogens. *Cancer Res.*, **52**(Suppl.), 2092s–2098s

Wallin, H. & Alexander, J. (1988) A model system for studying covalent binding of food carcinogens MeIQx, MeIQ and IQ to DNA and protein. In: Bartsch, H., Hemminki, K. & O'Neill, I.K., eds, *Methods for Detecting DNA Damaging Agents in Humans: Applications in Cancer Epidemiology and Prevention* (IARC Scientific Publications No. 89), Lyon, IARC, pp. 113–117

Wallin, H., Holme, J.A. & Alexander, J. (1992) Covalent binding of food carcinogens MeIQx, MeIQ and IQ to DNA and protein in microsomal incubations and isolated rat hepatocytes. *Pharmacol. Toxicol.*, **70**, 220–225

Waterhouse, A.L. & Rapoport, H. (1985) Synthesis and tritium labeling of the food mutagens IQ and methyl-IQ. *J. labelled Compd Radiopharmacol.*, **22**, 201–216

Wild, D. & Degen, G.H. (1987) Prostaglandin H synthase-dependent mutagenic activation of heterocyclic aromatic amines of the IQ-type. *Carcinogenesis*, **8**, 541–545

Yamaizumi, Z., Kasai, H., Nishimura, S., Edmonds, C.G. & McCloskey, J.A. (1986) Stable isotope dilution quantification of mutagens in cooked foods by combined liquid chromatography-thermospray mass spectrometry. *Mutat. Res.*, **173**, 1–7

Yoo, M.A., Ryo, H., Todo, T. & Kondo, S. (1985) Mutagenic potency of heterocyclic amines in the *Drosophila* wing spot test and its correlation to carcinogenic potency. *Jpn. J. Cancer Res. (Gann)*, **76**, 468–473

Yoshimi, N., Sugie, S., Iwata, H., Mori, H. & Williams, G.M. (1988) Species and sex differences in genotoxicity of heterocyclic amine pyrolysis and cooking products in the hepatocyte primary culture/DNA repair test using rat, mouse, and hamster hepatocytes. *Environ. mol. Mutag.*, **12**, 53–64

MeIQx (2-AMINO-3,8-DIMETHYLIMIDAZO[4,5-*f*]QUINOXALINE)

This substance was considered by a previous Working Group, in October 1985 (IARC, 1986). Since that time, new data have become available, and these have been incorporated into the monograph and taken into consideration in the present evaluation.

1. Exposure Data

1.1 Chemical and physical data

1.1.1 Synonyms, structural and molecular data

Chem. Abstr. Services Reg. No.: 77500-04-0
Chem. Abstr. Name: 3,8-Dimethyl-3*H*-imidazo[4,5-*f*]quinoxalin-2-amine
IUPAC Systematic Name: 2-Amino-3,8-dimethyl-3*H*-imidazo[4,5-*f*]quinoxaline

$C_{11}H_{11}N_5$ Mol. wt: 213.24

1.1.2 Chemical and physical properties

(a) *Melting-point*: 295–300 °C (with slight decomposition) (Grivas & Olsson, 1985)
(b) *Spectroscopy data*: Ultraviolet (Kasai *et al.*, 1981a,b), proton nuclear magnetic resonance (Kasai *et al.*, 1981b; Knize *et al.*, 1987) and mass spectral data (Kasai *et al.*, 1981b; Hargraves & Pariza, 1983; Felton *et al.*, 1984) have been reported.
(c) *Solubility*: Soluble in methanol (Kasai *et al.*, 1981a) and dimethyl sulfoxide (Dooley *et al.*, 1992)
(d) *Stability*: Stable under moderately acidic and alkaline conditions and in cold dilute aqueous solutions protected from light (Sugimura *et al.*, 1983)
(e) *Reactivity*: Rapidly degraded by dilute hypochlorite; not deaminated by weakly acidic nitrite solutions (Tsuda *et al.*, 1985)

1.1.3 Trade names, technical products and impurities

No data were available to the Working Group.

1.1.4 *Analysis*

MeIQx was initially isolated and purified by acid–base partitioning, Sephadex LH-20 column chromatography and reverse-phase high-performance liquid chromatography (HPLC). The structure was deduced mainly from data obtained by proton nuclear magnetic resonance and high-resolution mass spectral analysis and by comparison with synthetic MeIQx (Kasai *et al.*, 1981b).

MeIQx was isolated from beef extract by dichloromethane extraction, column chromatography on Adsorbosil-5 and Sephadex LH20 and HPLC, with analysis by mass spectrometry and ultraviolet spectrophotometry (Hargraves & Pariza, 1983; Turesky *et al.*, 1983). MeIQx was detected in fried ground beef at 1 ng/g by acid extraction, absorption on XAD-2 resin, preparative and analytical HPLC and off-line mass spectrometry (Felton *et al.*, 1984).

MeIQx can be adsorbed from aqueous solutions onto cellulose or cotton to which CI Reactive Blue 21, a trisulfo-copper phthalocyanine dye, has been bound covalently. The adsorbed MeIQx is eluted with an ammonia–methanol and quantified by assay for mutagenic activity (Hayatsu *et al.*, 1983). This 'blue cotton' adsorption technique has been used in several procedures for the detection of MeIQx.

MeIQx has also been quantified in cooked beef products using a deuterium-labelled internal standard and liquid chromatography–thermospray–mass spectrometry (Turesky *et al.*, 1988a). This method involves methanol extraction, acid–base partition and 'blue cotton' adsorption prior to analysis.

Turesky *et al.* (1989) used monoclonal antibodies immobilized on a support for selective immunoaffinity chromatography as a clean-up procedure in the analysis of beef extracts used for bacteriological media and of food-grade beef extracts. Vanderlaan *et al.* (1988) produced monoclonal antibodies which bound to MeIQx and related heterocyclic amines with varying specificities.

Derivatization and capillary gas chromatography–mass spectrometry analysis of fried beef with picogram sensitivity for MeIQx has been reported (Murray *et al.*, 1988). Cooked beef was analysed using only a 10-g sample, and isotope-labelled MeIQx internal standard was used to determine recovery accurately.

A practical, solid-phase extraction and medium-pressure liquid chromatographic method for the analysis of MeIQx and other heterocyclic amines was devised by Gross *et al.* (1989) and used with food-grade and bacterial beef extracts as well as fried beef. Improvements to the method (Gross, 1990; Gross & Grüter, 1992) allow determination of MeIQx and most of the other known heterocyclic amines at a level of 1 ng/g from only 3 g of meat and 10 g of fish. Replicate samples and spiking allow accurate determination of extraction losses; chromatographic peak identities are confirmed using a diode array–ultraviolet detector.

1.2 Production and use

1.2.1 *Production*

The isolation and identification of MeIQx were first reported by Kasai *et al.* (1981b). Its structure was confirmed by chemical synthesis, in which 6-amino-3-methyl-5-nitroquinoxaline was methylated, reduced to the 5-amino derivative and cyclized with cyanogen bromide to form MeIQx (Kasai *et al.*, 1981a; Sugimura *et al.*, 1983). Subsequent purification from fried beef confirmed the original structure (Knize *et al.*, 1987).

Improved synthetic routes have been reported from 4-fluoro-*ortho*-phenylenediamine (Grivas & Olsson, 1985), through benzoselenadiazole (Grivas, 1986) and through copper(I)-promoted quinoxaline formation (Knapp *et al.*, 1989).

Synthesis of isotopically-labelled [^{13}C, ^{15}N$_2$] MeIQx was reported by Murray *et al.* (1988).

MeIQx is produced commercially in small quantities for research purposes.

1.2.2 *Use*

MeIQx is not used commercially.

1.3 Occurrence

MeIQx has been detected in cooked beef, fish, chicken and mutton in varying amounts (Table 1).

Table 1. Concentrations of MeIQx in foods

Sample	Concentration (ng/g)	No. of samples	Reference
Ground beef, uncooked	ND	1	Murray *et al.* (1988)
Ground beef, fried			
190 °C	0.45	1	Hargraves & Pariza (1983)
	5.1–8.3	2	Gross *et al.* (1990)
200 °C	1.3–2.4	2	Murray *et al.* (1988)
250 °C	0.5–1.5	2	Turesky *et al.* (1989)
	1.0	1	Felton *et al.* (1984)
	1.1–1.4[a]	1	Gross *et al.* (1989); Turesky *et al.* (1989)
275 °C	2.7–12.3	3	Turesky *et al.* (1988a)
300 °C	1.0	1	Felton *et al.* (1986)
No temperature given	0.64	1	Wakabayashi *et al.* (1992)
Beef, broiled	2.11	1	Wakabayashi *et al.* (1992)
Beef extract, food-grade	3.1	1	Takahashi *et al.* (1985)
	8.5–30	4	Gross *et al.* (1989)
	28	1	Hargraves & Pariza (1983)
Mutton, broiled	1.01	1	Wakabayashi *et al.* (1992)
Fish, fried	6.44	1	Zhang *et al.* (1988)
Salmon, pan-broiled, 200 °C	1.4–5	4	Gross & Grüter (1992)
Salmon, oven-cooked, 200 °C	<1–4.6	3	Gross & Grüter (1992)
Salmon, barbecued, 270 °C	<1	4	Gross & Grüter (1992)
Mackerel, smoked, dried	0.8	1	Kato *et al.* (1986)
Chicken, broiled	2.33	1	Wakabayashi *et al.* (1992)
Eel, roasted, tinned	1.1	1	Lee & Tsai (1991)

ND, not detected
[a]Depending on the method used

Amounts of MeIQx relative to the three other heterocyclic amines included in this volume are listed in the monograph on IQ (p. 168).

MeIQx has also been found in refluxing diethylene glycol mixtures of glycine, glucose and creatinine (Jägerstad *et al.*, 1984) and has been formed from mixtures of glycine, fructose and creatinine (Grivas *et al.*, 1986) and glycine, glucose and creatinine (Skog & Jägerstad, 1990). Dry mixtures, heated at 200 °C, of creatine and either serine, alanine or tyrosine formed MeIQx (Övervik *et al.*, 1989).

Since many foods in the human diet contain heterocyclic amines, methods have been developed to estimate the dietary dose from urinary excretion after food ingestion and metabolism. Murray *et al.* (1989) determined the recovery of MeIQx in urine to be 1.8–4.9% of the ingested amount after 12 h. The amount of MeIQx in the urine of 10 subjects on a normal diet was 11–47 ng/24 h. After adjustment for metabolic transformation, the daily dose of MeIQx was estimated to range from 0.2 to 2.6 µg for the individuals who participated in the study. MeIQx was not detected (< 1 ng/24-h urine sample) in the urine of three patients who were receiving parenteral nutrients (Ushiyama *et al.*, 1991). Total amounts in the diet thus appear to range from virtually none for people eating a diet contained few cooked foods (especially meat) to perhaps a few micrograms per day.

1.4 Regulations and guidelines

No data were available to the Working Group.

2. Studies of Cancer in Humans

No epidemiological study was available that addressed the carcinogenic risk to humans of MeIQx itself. Cancer risks associated with consumption of broiled and fried foods, which may contain MeIQx as well as other heterocyclic amines, have, however, been addressed in a number of case–control studies. Several of these are summarized in the monograph on IQ.

3. Studies of Cancer in Experimental Animals

3.1 Oral administration[1]

3.1.1 *Mouse*

Groups of 40 male and 40 female CDF_1 [(BALB/cAnN × DBA/2N) F_1] mice, six weeks old, were fed a diet containing 0 or 600 mg/kg MeIQx [purity unspecified] for 84 weeks. Animals that became moribund were killed and autopsied. The first leukaemia, in a treated

---

[1]The Working Group was aware of a study in progress on MeIQx given in the diet of mice (Ghess *et al.*, 1992) and of a study by gavage to non-human primates carried out at the US National Cancer Institute, Division of Cancer Etiology.

male mouse, was detected in week 48, and the numbers of mice that survived after that time were taken as the effective numbers. The average survival times in weeks (mean ± SD) of treated males, treated females, control males and control females were 76 ± 11, 75 ± 7, 77 ± 11 and 82 ± 4, respectively. A significantly larger number of treated mice than controls had liver tumours (adenomas and carcinomas combined): males, 16/37 versus 6/36; females, 32/35 versus 0/39; 10 males and 25 females in the treated group developed hepatocellular carcinomas. The incidence of lung tumours (adenomas and adenocarcinomas) was significantly greater in treated female mice than in controls: 15/35 versus 4/39; adenocarcinoma of the lung was observed in 11 treated males, six treated females, seven control males and two control females. The incidence of lymphoma and leukaemia in treated male mice was significantly greater than in controls: 11/37 versus 2/36. In mice of each sex, the average time to appearance of lymphomas or leukaemias was significantly shorter among treated animals than controls: males, 69 versus 84 weeks; females, 70 versus 79 weeks (Ohgaki et al., 1987).

3.1.2 *Rat*

Groups of 20 male and 20 female Fischer 344 rats, seven weeks old, were fed a diet containing 0 or 400 mg/kg of diet MeIQx (purity, > 99%) for 429 days. Animals that became moribund were killed and autopsied. The first tumour of the Zymbal gland was seen in a treated female on day 177, and the numbers of rats that lived after that time were taken as the effective numbers. The average survival times in days (mean ± SD) of treated males, treated females, control males and control females were 326 ± 36, 364 ± 61, 427 ± 8 and 429 ± 0, respectively. Treated animals had significantly more liver tumours than controls: males, 20/20 (19 hepatocellular carcinomas, 1 neoplastic nodule) versus 0/19; females, 10/19 (all neoplastic nodules) versus 0/20. In six of the treated males, hepatocellular carcinomas metastasized to the lung. A significant increase in the incidence of tumours of the Zymbal gland was found in treated animals over that in controls: males, 15/20 (13 squamous-cell carcinomas, 2 squamous-cell papillomas) versus 0/19; females, 10/19 (all squamous-cell carcinomas) versus 0/20. Clitoral gland tumours (squamous-cell carcinomas) were found in 12/19 treated females and in no control. Skin tumours were found in 7/20 treated males and in 1/19 treated females; the skin tumours in males were diagnosed as five squamous-cell carcinomas, one basal-cell carcinoma and one squamous-cell papilloma, and the tumour in the female as a squamous-cell carcinoma (Kato et al., 1988).

3.2 Intraperitoneal administration

Mouse

Groups [initial numbers unspecified] of newborn male $B6C3F_1$ mice were injected intraperitoneally with MeIQx (> 98% pure) at total doses of 0, 0.625 or 1.25 μmol [133.3 or 266.5 μg] (maximal tolerated dose) dissolved in 5, 10 or 20 μl dimethyl sulfoxide and administered on days 1, 8 and 15 after birth, respectively. The incidence of hepatocellular adenomas was significantly higher in treated mice than in controls at 12 months: controls, 5/44; low-dose, 8/24; and high-dose, 17/20 (Dooley et al., 1992).

3.3 Administration with known carcinogens

Rat

In a short-term assay for tumour-initiating activity in the liver, a group of 10 male Fischer 344 rats, five weeks of age, each received a single intragastric dose of 80 mg/kg bw MeIQx [source and purity unspecified] in corn oil, and, two weeks later, were fed a diet containing 0.05% phenobarbital for six weeks. The rats received a two-thirds partial hepatectomy three weeks after the MeIQx treatment. The number and total area of foci of phenotypically altered hepatocytes in the liver were scored using expression of placental-form glutathione *S*-transferase (GST-P) as the marker. The number of foci was not significantly increased over that in five vehicle-treated control rats, and, in the absence of subsequent phenobarbital treatment, MeIQx did not induce a significant increase in the number of foci in five rats. Another group of 10 rats received a two-thirds partial hepatectomy 12 h before MeIQx treatment, were placed on phenobarbital as above and one week after phenobarbital treatment received a single intraperitoneal injection of 300 mg/kg D-galactosamine. Administration of MeIQx significantly increased the number of foci by six fold and the area by five fold over that in five vehicle-treated control rats. MeIQx without subsequent phenobarbital treatment, but with galactosamine, produced a smaller increase in the number of foci in five rats. These results suggest that MeIQx has tumour-initiating activity in rat liver if combined with prior partial hepatectomy (Tsuda *et al.*, 1990).

As part of a medium-term carcinogenicity study on the synergistic effects of five heterocyclic amines, groups of 14–18 male Fischer 344 rats, six weeks of age, each received a single intraperitoneal injection of 200 mg/kg *N*-nitrosodiethylamine and, two weeks later, were fed a diet containing MeIQx at 16, 80 or 400 ppm (mg/kg diet). A two-thirds partial hepatectomy was performed in week 3 of the experiment; all animals were killed after eight weeks. Fifteen rats treated with the nitrosamine alone and groups of three rats given MeIQx at the three dose levels and saline instead of nitrosamine served as controls. The effects were assessed by counting the numbers of GST-P-positive foci in the liver. MeIQx at the high dose level significantly increased the area and the number of GST-P-positive foci (Ito *et al.*, 1991).

A group of 15 male Wistar rats, six weeks old, underwent a two-thirds partial hepatectomy, followed 17 h later by a single intraperitoneal injection of 50 mg/kg bw MeIQx (purity, > 99%) dissolved in acid water (pH 5.0); animals were then fed a diet containing 200 mg/kg 2-acetylaminofluorene during weeks 2 and 3 and received 2 ml[3.2 mg]/kg bw carbon tetrachloride by gavage at the beginning of week 3. All animals were killed at the end of week 6. Significantly more γ-glutamyl transpeptidase-positive liver-cell foci were found in the group treated with MeIQx (1.7 foci/cm^2) than in 17 controls given 0.9% saline instead of MeIQx (0.48 foci/cm^2) (Kleman *et al.*, 1989).

4. Other Relevant Data

4.1 Absorption, distribution, metabolism and excretion

The toxicology and metabolism of heterocyclic aromatic amines have been reviewed (Övervik & Gustafsson, 1990; Aeschbacher & Turesky, 1991).

4.1.1 Humans

Six male volunteers had a test meal of 320 g of ground beef fried at normal cooking temperature. In all six, MeIQx was detected in 12-h urine collected after the test meal, while no detectable amount was found before the meal. Two to five percent of the ingested amount of MeIQx was excreted unchanged in the urine (Murray et al., 1989).

4.1.2 Experimental systems

After intravenous administration of ^{14}C-MeIQx to mice, radiolabel, measured by autoradiography, was distributed throughout the body after 10 min. The label persisted in the liver and in the gut contents four days after exposure (Gooderham et al., 1991). In rats administered ^{14}C-MeIQx by gavage, radiolabel was widely distributed in the body after 1 h. Non-extractable radiolabel persisted in the liver and kidneys six days after administration (Tjøtta et al., 1992).

After oral administration to mice of [2-^{14}C]-MeIQx, MeIQx was rapidly absorbed and metabolized, 20–25% of the radiolabel being excreted in the urine within 6 h. Radiolabel was found in all organs studied: stomach, small intestine, large intestine, blood, liver, spleen, lung, kidney (Alldrick & Rowland, 1988).

When male Sprague-Dawley rats were given [2-^{14}C]-MeIQx at 0.01, 0.2 or 20 mg/kg bw by gavage, the recovery of radiolabel was about 90% of the administered dose, and higher in faeces than in urine. Small amounts of radiolabel were found after 72 h, especially in liver and kidney. At the two higher doses, excretion occurred mainly via sulfamate formation, while at the low dose conjugation with glucuronic acid was the most important pathway. Induction of cytochrome P450 enzymes by polychlorinated biphenyls led to increased levels of glucuronic and sulfuric acid conjugates and a decreased level of sulfamate formation (Turesky et al., 1991a).

Human liver microsomes could activate MeIQx to a DNA-reactive species. The isozyme involved was tentatively identified as CYP IA2 (P450 IA2) (Shimada et al., 1989). Human liver and colon cytosols contained O-acetyltransferase activity that metabolized N-hydroxy-MeIQx into a DNA-binding form (Turesky et al., 1991b). (See Fig. 1, in the monograph on IQ, p. 178).

In the presence of liver microsomes from rats induced with 3-methylcholanthrene or polychlorinated biphenyls, MeIQx was metabolized to an N-hydroxylated derivative (Yamazoe et al., 1988; Turesky et al., 1991a). Metabolism occurred mainly by two forms of cytochrome P450, P448-H and P448-L (Yamazoe et al., 1988). Formation of the 5-hydroxylated derivatives was strongly induced by polychlorinated biphenyls (Turesky et al., 1991a).

In isolated rat liver cells, MeIQx was transformed into 10 metabolites, including N-hydroxy, 4- or 5-hydroxy, 8-hydroxymethyl and N3-demethylated derivatives, sulfate and glucuronide derivatives of these compounds, and sulfamate and glucuronide derivatives of MeIQx (Wallin et al., 1989) (Fig. 1). After dietary administration of MeIQx to rats in vivo, N3-demethylated, C8-hydroxymethylated and N-acetylated derivatives of MeIQx were detected (Hayatsu et al., 1987).

Similar metabolites were found in another study in rats. After intragastric administration of [2-^{14}C]-MeIQx, 40 and 49% of the radiolabel was recovered in the urine

Fig. 1. Oxidative metabolic pathways of MeIQx

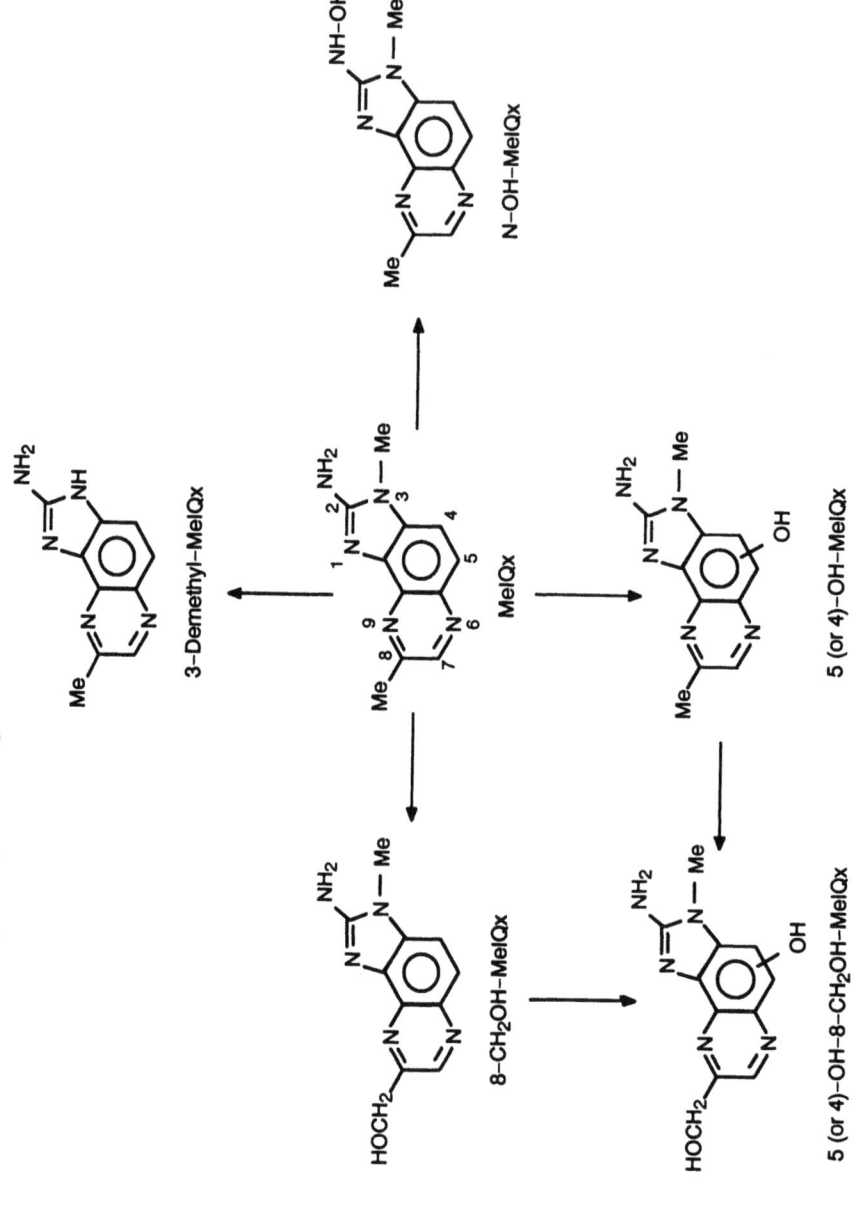

From Wallin *et al.* (1989)

and faeces, respectively; 25–50% of the dose was recovered in the bile within 24 h. Five metabolites—two sulfamates, two glucuronides and an acetylated derivative—accounted for 96% of the radiolabel excreted in bile (Turesky et al., 1988b,c). The importance of the acetylation reaction was also seen in bacterial mutagenicity studies in which inhibitors of O-acetyltransferase were applied (Negishi et al., 1989).

Prior treatment of rats with MeIQx induced the metabolism of MeIQx administered subsequently (Degawa et al., 1989).

In a study designed to provide information on the metabolic steps involved in the formation of the active mutagenic form, MeIQx was not mutagenic to *Salmonella typhimurium* TA98/1,8-DNP$_6$ (defective in esterifying activity) in the presence of an exogenous metabolic system (S9 mix), but it was strongly mutagenic to the original strain TA98 with S9 mix, suggesting that the ultimate mutagenic form of MeIQx is a reactive ester of its N-hydroxy derivative (Nagao et al., 1983).

Covalent binding of MeIQx to albumin and DNA in microsomes and rat hepatocytes *in vitro* was dependent on cytochrome P450 enzymes. The efficiency of binding of MeIQx was similar to that of MeIQ and less than that of IQ (Wallin et al., 1992).

4.2 Toxic effects

No data were available to the Working Group.

4.3 Reproductive and developmental toxicity

No data were available to the Working Group.

4.4 Genetic and related effects

The genetic effects of MeIQx have been reviewed (Sugimura, 1985; de Meester, 1989).

4.4.1 *Humans*

No data were available to the Working Group.

4.4.2 *Experimental systems* (see also Table 2 and Appendices 1 and 2)

MeIQx induced SOS repair and mutation in bacteria and somatic mutations in *Drosophila melanogaster*. It induced unscheduled DNA synthesis in primary cultures of hepatocytes from various species but, in hepatocytes from mice, only in cells derived from females. In mammalian cell lines, it induced chromosomal anomalies and diphtheria toxin-resistant mutants. MeIQx induced *hprt* locus mutations in a repair-deficient but not in a repair-proficient cell line. It induced sister chromatid exchange but not chromosomal aberrations in cultured human lymphocytes.

After administration *in vivo*, MeIQx formed DNA adducts in various rat and mouse organs. It did not induce mutation in the Dlb-1 locus (which determines the tissue-specific pattern of expression of the binding site for the lectin *Dolichos biflorus* agglutinin) in mouse

Table 2. Genetic and related effects of MeIQx

Test system	Result		Dose (LED/HID)[a]	Reference
	Without exogenous metabolic system	With exogenous metabolic system		
PRB, SOS repair test, *Salmonella typhimurium* TA1535	0	+	0.0700	Nakamura et al. (1987)
PRB, *umu* expression, *Salmonella typhimurium* TA1535/psk 1002	0	+	2.0000	Shimada et al. (1989)
PRB, SOS repair, *Salmonella typhimurium* with human adult and fetal microsomes	0	+	2.0000	Kitada et al. (1990)
ERD, *Escherichia coli* K12, differential toxicity	–	+	0.2600	Knasmüller et al. (1992)
SA0, *Salmonella typhimurium* TA100, reverse mutation	0	+	0.0100	Kasai et al. (1981b)
SA0, *Salmonella typhimurium* TA100, reverse mutation	–	+	0.0000	Grivas & Jägerstad (1984)
SA0, *Salmonella typhimurium* TA100, reverse mutation	0	+	0.0000	Felton & Knize (1990)
SA2, *Salmonella typhimurium* TA102, reverse mutation	0	+	0.0000	Felton & Knize (1990)
SA4, *Salmonella typhimurium* TA104, reverse mutation	0	–	0.0000	Felton & Knize (1990)
SA8, *Salmonella typhimurium* TA1538, reverse mutation	0	+	0.0000	Felton et al. (1986)
SA8, *Salmonella typhimurium* TA1538, reverse mutation	0	+	0.4000	Felton & Knize (1990)
SA9, *Salmonella typhimurium* TA98, reverse mutation	0	+	0.0050	Kasai et al. (1981a)
SA9, *Salmonella typhimurium* TA98, reverse mutation[b]	0	+	0.0000	Ishida et al. (1987)
SA9, *Salmonella typhimurium* TA98, reverse mutation	0	+	0.0000	Felton & Knize (1990)
SA9, *Salmonella typhimurium* TA98, reverse mutation	–	+	0.0025	Winton et al. (1990)
SA9, *Salmonella typhimurium* TA98, reverse mutation	0	+	0.3125	Loprieno et al. (1991)
SAS, *Salmonella typhimurium* TA98/1,8-DNP$_6$, reverse mutation	0	–	0.0250	Nagao et al. (1983)
SAS, *Salmonella typhimurium* TA96, reverse mutation	–	–	0.0000	Felton & Knize (1990)
SAS, *Salmonella typhimurium* TA97, reverse mutation	0	+	0.0000	Felton & Knize (1990)
DMM, *Drosophila melanogaster*, somatic mutation and recombination	+	0	25.0000	Yoo et al. (1985)
URP, Unscheduled DNA synthesis, rat primary hepatocytes *in vitro*	+	0	1.0000	Yoshimi et al. (1988)
UIA, Unscheduled DNA synthesis, mouse primary hepatocytes (female) *in vitro*	+	0	10.0000	Yoshimi et al. (1988)
UIA, Unscheduled DNA synthesis, mouse primary hepatocytes (male) *in vitro*	–	0	10.0000	Yoshimi et al. (1988)
UIA, Unscheduled DNA synthesis, Syrian hamster primary hepatocytes *in vitro*	+	0	1.0600	Yoshimi et al. (1988)
GCL, Gene mutation, Chinese hamster lung cells DT *in vitro*	–	+	10.0000	Nakayasu et al. (1983)
GCL, Gene mutation, Chinese hamster lung cells *in vitro*	0	(+)	1.0000	Sugimura et al. (1989)

Table 2 (contd)

Test system	Result		Dose (LED/HID)[a]	Reference
	Without exogenous metabolic system	With exogenous metabolic system		
GCO, Gene mutation, Chinese hamster ovary cells (uvS) hprt locus in vitro	0	+	300.0000	Thompson et al. (1987)
G9H, Gene mutation, Chinese hamster lung V79 cells, hprt locus in vitro	0	–	51.2000	Loprieno et al. (1991)
SIC, Sister chromatid exchange, Chinese hamster ovary (uvS) cells in vitro	0	(+)	200.0000	Thompson et al. (1987)
CIC, Chromosomal aberrations, Chinese hamster ovary (uvS) cells in vitro	0	(+)	600.0000	Thompson et al. (1987)
CIC, Chromosomal aberrations, Chinese hamster ovary cells in vitro	0	–	80.0000	Loprieno et al. (1991)
SHL, Sister chromatid exchange, human lymphocytes in vitro	(+)	+	0.1000	Aeschbacher & Ruch (1989)
CHL, Chromosomal aberrations, human lymphocytes in vitro	–	–	130.0000	Loprieno et al. (1991)
CHL, Chromosomal aberrations, human lymphocytes in vitro	–	–	1000.0000	Aeschbacher & Ruch (1989)
HMM, Host-mediated assay, Escherichia coli in intrasanguinous mice	+		2.30×1 ip, po	Knasmüller et al. (1992)
GVA, Gene mutation, mouse small intestinal cells, Dlb-1 expression in vivo (females)	–		20.00×10 po	Winton et al. (1990)
SVA, Sister chromatid exchange, rat hepatocytes in vivo	+		200.00×1 ip	Sawada et al. (1991)
SVA, Sister chromatid exchange, mouse (PCB-treated) bone-marrow cells in vivo (males)	–		50.00×1 ip	Tucker et al. (1989)
MVM, Micronucleus test, mouse bone-marrow cells in vivo (males)	–		40.00×1 po	Loprieno et al. (1991)
CVA, Chromosomal aberrations, rat hepatocytes in vivo	+		200.00×1 ip	Sawada et al. (1991)
BID, Binding (covalent) to DNA, rat hepatocytes in vitro	+	0	10.0000	Wallin et al. (1992)
BVD, Binding (covalent) to DNA, rat liver in vivo[c]	+		50.00×1 ip	Yamashita et al. (1988)
BVD, Binding (covalent) to DNA, mouse (various tissues) in vivo[d]	+		0.01065×1 po	Alldrick & Lutz (1989)
BVD, Binding (covalent) to DNA, rat liver in vivo[c]	+		0.48 diet $\times 1$ wk	Yamashita et al. (1990)
BVD, Binding (covalent) to DNA, rat heart and liver in vivo[c]	+		48 diet $\times 4$ wk	Övervik et al. (1991)
BVP, Binding (covalent) to DNA, mouse bone marrow in vivo (males)[d]	+		40.00×1 po	Loprieno et al. (1991)

+, positive; (+), weakly positive; –, negative; 0, not tested
[a]In-vitro tests, μg/ml; in-vivo tests, mg/kg bw; 0.0000, dose not given
[b]Rhesus liver S9
[c]32P-Postlabel
[d]14C-Label

small intestinal cells. Sister chromatid exchange was induced in rat hepatocytes, but not in mouse bone-marrow cells, following intraperitoneal administration of MeIQx.

N-Hydroxy-MeIQx binds nonenzymatically to DNA *in vitro* (Negishi et al., 1989).

4.4.3 *Genetic changes in animal tumours*

Two of six Zymbal gland tumours induced in rats by MeIQx carried mutated c-Ha-*ras* genes, one being a G to T transversion at the second base of codon 13 and the other an A to T transversion at the second base of codon 61 (Kudo et al., 1991).

5. Summary of Data Reported and Evaluation

5.1 Exposure data

MeIQx (2-Amino-3,8-dimethylimidazo[4,5-*f*]quinoxaline) has been found in cooked meat and fish at concentrations of up to 12 ng/g. A few determinations indicated that the levels of MeIQx were lower than those of PhIP and higher than those of IQ and MeIQ.

5.2 Human carcinogenicity data

No data directly relevant to an evaluation of the carcinogenicity to humans of MeIQx were available. Studies on the consumption of cooked meat and fish are summarized in the monograph on IQ.

5.3 Animal carcinogenicity data

MeIQx was tested for carcinogenicity by oral administration in the diet in one experiment in mice and in one experiment in rats. In mice, it produced hepatocellular carcinomas in animals of each sex, lymphomas and leukaemias in males and lung tumours in females. In rats, it produced hepatocellular carcinomas in males, squamous-cell carcinomas of the Zymbal gland in animals of each sex, squamous-cell carcinomas of the skin in males and squamous-cell carcinomas of the clitoral gland in females.

Intraperitoneal injection of MeIQx to newborn male mice increased the incidence of hepatic adenomas.

A single oral treatment of rats with MeIQx followed by phenobarbital, combined with further modulating procedures, stimulated development of foci of altered hepatocytes. Sequential administration of MeIQx after N-nitrosodiethylamine enhanced the appearance of foci of altered hepatocytes in rats.

5.4 Other relevant data

No data were available on the genetic and related effects of MeIQx in humans.

MeIQx bound to DNA in several tissues of rodents dosed *in vivo*, and, in single studies, it induced chromosomal anomalies. It induced sister chromatid exchange in human cells *in vitro* and DNA damage, gene mutation and sister chromatid exchange in rodent cells *in vitro*. It induced gene mutation in insects and gene mutation and DNA damage in bacteria.

MeIQx can be metabolized by human microsomes to a species that damages bacterial DNA.

5.5 Evaluation[1]

There is *inadequate evidence* in humans for the carcinogenicity of MeIQx.

There is *sufficient evidence* in experimental animals for the carcinogenicity of MeIQx.

Overall evaluation

MeIQx (2-Amino-3,8-dimethylimidazo[4,5-*f*]quinoxaline) *is possibly carcinogenic to humans (Group 2B)*.

6. References

Aeschbacher, H.-U. & Ruch, E. (1989) Effect of heterocyclic amines and beef extract on chromosome aberrations and sister chromatid exchanges in cultured human lymphocytes. *Carcinogenesis*, **10**, 429-433

Aeschbacher, H.-U. & Turesky, R. (1991) Mammalian cell mutagenicity and metabolism of heterocyclic aromatic amines. *Mutat. Res.*, **259**, 235-250

Alldrick, A.J. & Lutz, W.K. (1989) Covalent binding of [2-^{14}C]2-amino-3,8-dimethylimidazo[4,5-*f*]quinoxaline (MeIQx) to mouse DNA *in vivo*. *Carcinogenesis*, **10**, 1419-1423

Alldrick, A.J. & Rowland, I.R. (1988) Distribution of radiolabelled [2-^{14}C]IQ and MeIQx in the mouse. *Toxicol. Lett.*, **44**, 183-190

Degawa, M., Tanimura, S., Agatsuma, T. & Hashimoto, Y. (1989) Hepatocarcinogenic heterocyclic aromatic amines that induce cytochrome P-448 isozymes, mainly cytochrome P-448H (P-450IA2), responsible for mutagenic activation of the carcinogens in rat liver. *Carcinogenesis*, **10**, 1119-1122

Dooley, K.L., Von Tungeln, L.S., Bucci, T., Fu, P.P. & Kadlubar, F.F. (1992) Comparative carcinogenicity of 4-aminobiphenyl and the food pyrolysates, Glu-P-1, IQ, PhIP, and MeIQx in the neonatal B6C3F$_1$ male mouse. *Cancer Lett.*, **62**, 205-209

Felton, J.S. & Knize, M.G. (1990) Heterocyclic-amine mutagens/carcinogens in foods. In: Cooper, C.S. & Grover, P.L., eds, *Handbook of Experimental Pharmacology*, Vol. 94/I, *Chemical Carcinogenesis and Mutagenesis*, Berlin, Springer Verlag, pp. 471-502

Felton, J.S., Knize, M.G., Wood, C., Wuebbles, B.J., Healy, S.K., Stuermer, D.H., Bjeldanes, L.F., Kimble, B.J. & Hatch, F.T. (1984) Isolation and characterization of new mutagens from fried ground beef. *Carcinogenesis*, **5**, 95-102

Felton, J.S., Knize, M.G., Shen, N.H., Andresen, D.B., Bjeldanes, L.F. & Hatch, F.T. (1986) Identification of the mutagens in cooked beef. *Environ. Health Perspectives*, **67**, 17-24

---

[1]For definition of the italicized terms, see Preamble, pp. 26-29.

Ghess, M.-J., Wilbourn, J.D. & Vainio, H., eds (1992) *Directory of Agents Being Tested for Carcinogenicity*, No. 15, Lyon, IARC, p. 93

Gooderham, N.J., Soames, A., Rice, J.C., Boobis, A.R. & Davies, D.S. (1991) Distribution and elimination of [2-^{14}C]amino-3,8-dimethylimidazo[4,5-f]quinoxaline in mice. Analysis by whole animal autoradiography. *Hum. exp. Toxicol.*, **10**, 337–345

Grivas, S. (1986) Efficient synthesis of mutagenic imidazo[4,5-f]quinoxalin-2-amines via readily accessible 2,1,3-benzoselenadiazoles. *Acta chem. scand.*, **B40**, 404–406

Grivas, S. & Olsson, K. (1985) An improved synthesis of 3,8-dimethyl-3H-imidazo[4,5-f]quinoxaline-2-amine (MeIQx) and its 2-^{14}C-labelled analogue. *Acta chem. scand.*, **B39**, 31–34

Grivas, S., Nyhammar, T., Olsson, K. & Jägerstad, M. (1986) Isolation and identification of the food mutagens IQ and MeIQx from a heated model system of creatinine, glycine and fructose. *Food Chem.*, **20**, 127–136

Gross, G.A. (1990) Simple methods for quantifying mutagenic heterocyclic aromatic amines in food products. *Carcinogenesis*, **11**, 1597–1603

Gross, G.A. & Grüter, A. (1992) Quantification of mutagenic/carcinogenic heterocyclic aromatic amines in food products. *J. Chromatogr.*, **592**, 271–278

Gross, G.A., Philippossian, G. & Aeschbacher, H.-U. (1989) An efficient and convenient method for the purification of mutagenic heterocyclic amines in heated meat products. *Carcinogenesis*, **10**, 1175–1182

Hargraves, W.A. & Pariza, M.W. (1983) Purification and mass spectral characterization of bacterial mutagens from commercial beef extract. *Cancer Res.*, **43**, 1467–1472

Hayatsu, H., Matsui, Y., Ohara, Y., Oka, T. & Hayatsu, T. (1983) Characterization of mutagenic fractions in beef extract and in cooked ground beef. Use of blue-cotton for efficient extraction. *Gann*, **74**, 472–482

Hayatsu, H., Kasai, H., Yokoyama, S., Miyazawa, T., Yamaizumi, Z., Sato, S., Nishimura, S., Arimoto, S., Hayatsu, T. & Ohara, Y. (1987) Mutagenic metabolites in urine and feces of rats fed with 2-amino-3,8-dimethylimidazo[4,5-f]quinoxaline, a carcinogenic mutagen present in cooked meat. *Cancer Res.*, **47**, 791–794

IARC (1986) *IARC Monographs on the Evaluation of the Carcinogenic Risk of Chemicals to Humans, Vol. 40, Some Naturally Occurring and Synthetic Food Components, Furocoumarins and Ultraviolet Radiation*, Lyon, pp. 283–288

Ishida, Y., Negishi, C., Umemoto, A., Fujita, Y., Sato, S., Sugimura, T., Thorgeirsson, S.S. & Adamson, R.H. (1987) Activation of mutagenic and carcinogenic heterocyclic amines by S-9 from the liver of a rhesus monkey. *Toxicol. in vitro*, **1**, 45–48

Ito, N., Hasegawa, R., Shirai, T., Fukushima, S., Hakoi, K., Takaba, K., Iwasaki, S., Wakabayashi, K., Nagao, M. & Sugimura, T. (1991) Enhancement of GST-P positive liver cell foci development by combined treatment of rats with five heterocyclic amines at low doses. *Carcinogenesis*, **12**, 767–772

Jägerstad, M., Olsson, K., Grivas, S., Negishi, C., Wakabayashi, K., Tsuda, M., Sato, S. & Sugimura, T. (1984) Formation of 2-amino-3,8-dimethylimidazo[4,5-f]quinoxaline in a model system by heating creatinine, glycine and glucose. *Mutat. Res.*, **126**, 239–244

Kasai, H., Shiomi, T., Sugimura, T. & Nishimura, S. (1981a) Synthesis of 2-amino-3,8-dimethyl-imidazo[4,5-f]quinoxaline (Me-IQx), a potent mutagen isolated from fried beef. *Chem. Lett.*, 675–678

Kasai, H., Yamaizumi, Z., Shiomi, T., Yokoyama, S., Miyazawa, T., Wakabayashi, K., Nagao, M., Sugimura, T. & Nishimura, S. (1981b) Structure of a potent mutagen isolated from fried beef. *Chem. Lett.*, 485–488

Kato, T., Kikugawa, K. & Hayatsu, H. (1986) Occurrence of the mutagens 2-amino-3,8-dimethylimidazo[4,5-*f*]quinoxaline (MeIQx) and 2-amino-3,4,8-trimethylimidazo[4,5-*f*]quinoxaline (4,8-Me$_2$IQx) in some Japanese smoked, dried fish products. *J. agric. Food Chem.*, **34**, 810–814

Kato, T., Ohgaki, H., Hasegawa, H., Sato, S., Takayama, S. & Sugimura, T. (1988) Carcinogenicity in rats of a mutagenic compound, 2-amino-3,8-dimethylimidazo[4,5-*f*]quinoxaline. *Carcinogenesis*, **9**, 71–73

Kleman, M., Övervik, E., Blanck, A. & Gustafsson, J.-Å. (1989) The food-mutagens 2-amino-1-methyl-6-phenylimidazo[4,5-*b*]pyridine (PhIP) and 2-amino-3,8-dimethylimidazo-[4,5-*f*]quinoxaline (MeIQx) initiate enzyme-altered hepatic foci in the resistant hepatocyte model. *Carcinogenesis*, **10**, 1697–1700

Knapp, S., Ziv, J. & Rosen, J.D. (1989) Synthesis of the food mutagens MeIQx and 4,8-DiMeIQx by copper(i) promoted quinoxaline formation. *Tetrahedron*, **45**, 1293–1298

Knasmüller, S., Kienzl, H., Huber, W. & Hermann, R.S. (1992) Organ-specific distribution of genotoxic effects in mice exposed to cooked food mutagens. *Mutagenesis*, **7**, 235–241

Knize, M.G., Happe, J.A., Healy, S.K. & Felton, J.S. (1987) Identification of the mutagenic quinoxaline isomers from fried ground beef. *Mutat. Res.*, **178**, 25–32

Kudo, M., Ogura, T., Esumi, H. & Sugimura, T. (1991) Mutational activation of c-Ha-*ras* gene in squamous-cell carcinomas of rat Zymbal gland induced by carcinogenic heterocyclic amines. *Mol. Carcinog.*, **4**, 36–42

Lee, H. & Tsai, S.-J. (1991) Detection of IQ-type mutagens in canned roasted eel. *Food chem. Toxicol.*, **29**, 517–522

Loprieno, N., Boncristiani, G. & Loprieno, G. (1991) An experimental approach to identifying the genotoxic risk from cooked meat mutagens. *Food Chem. Toxicol.*, **29**, 377–386

de Meester, C. (1989) Bacterial mutagenicity of heterocyclic amines found in heat-processed food. *Mutat. Res.*, **221**, 235–262

Murray, S., Gooderham, N.J., Boobis, A.R. & Davies, D.S. (1988) Measurement of MeIQx and DiMeIQx in fried beef by capillary column gas chromatography electron capture negative ion chemical ionisation mass spectrometry. *Carcinogenesis*, **9**, 321–325

Murray, S., Gooderham, N.J., Boobis, A.R. & Davies, D.S. (1989) Detection and measurement of MeIQx in human urine after ingestion of a cooked meat meal. *Carcinogenesis*, **10**, 763–765

Nagao, M., Fujita, Y., Wakabayashi, K. & Sugimura, T. (1983) Ultimate forms of mutagenic and carcinogenic heterocyclic amines produced by pyrolysis. *Biochem. biophys. Res. Commun.*, **114**, 626–631

Nakamura, S.-I., Oda, Y., Shimada, T., Oki, I. & Sugimoto, K. (1987) SOS-inducing activity of chemical carcinogens and mutagens in *Salmonella typhimurium* TA1535/pSK1002: examination with 151 chemicals. *Mutat. Res.*, **192**, 239–246

Nakayasu, M., Nakasato, F., Sakamoto, H., Terada, M. & Sugimura, T. (1983) Mutagenic activity of heterocyclic amines in Chinese hamster lung cells with diphtheria toxin resistance as a marker. *Mutat. Res.*, **118**, 91–102

Negishi, C., Yamaizumi, Z. & Sato, S. (1989) Nucleic acid binding and mutagenicity of active metabolites of 2-amino-3,8-dimethylimidazo[4,5-*f*]quinoxaline. *Mutat. Res.*, **210**, 127–134

Ohgaki, H., Hasegawa, H., Suenaga, M., Sato, S., Yakayama, S. & Sugimura, T. (1987) Carcinogenicity in mice of a mutagenic compound, 2-amino-3,8-dimethylimidazo[4,5-f]quinoxaline (MeIQx) from cooked foods. *Carcinogenesis*, **8**, 665–668

Övervik, E. & Gustafsson, J.-Å. (1990) Cooked-food mutagens: current knowledge of formation and biological significance. *Mutagenesis*, **5**, 437–446

Övervik, E., Kleman, M., Berg, I. & Gustafsson, J.-Å. (1989) Influence of creatine, amino acids and water on the formation of the mutagenic heterocyclic amines found in cooked meat. *Carcinogenesis*, **10**, 2293–2301

Övervik, E., Ochiai, M., Hirose, M., Sugimura, T. & Nagao, M. (1991) The formation of heart DNA adducts in F344 rats following dietary administration of heterocyclic amines. *Mutat. Res.*, **256**, 37–43

Sawada, S., Yamanaka, T., Yamatsu, K., Furihata, C. & Matsushima, T. (1991) Chromosome aberrations, micronuclei and sister-chromatid exchanges (SCEs) in rat liver induced *in vivo* by hepatocarcinogens including heterocyclic amines. *Mutat. Res.*, **251**, 59–69

Shimada, T., Iwasaki, M., Martin, M.V. & Guengerich, F.P. (1989) Human liver microsomal cytochrome P-450 enzymes involved in the bioactivation of procarcinogens detected by *umu* gene response in *Salmonella typhimurium* TA1535/pSK1002. *Cancer Res.*, **49**, 3218–3228

Skog, K. & Jägerstad, M. (1990) Effects of monosaccharides and disaccharides on the formation of food mutagens in model systems. *Mutat. Res.*, **230**, 263–272

Sugimura, T. (1985) Carcinogenicity of mutagenic heterocyclic amines formed during the cooking process. *Mutat. Res.*, **150**, 33–41

Sugimura, T., Sato, S. & Takayama, S. (1983) New mutagenic heterocyclic amines found in amino acid and protein pyrolysates and in cooked food. In: Wynder, E.L., Leveille, G.A., Weisburger, J.H. & Livingston, G.E., eds, *Environmental Aspects of Cancer: The Role of Macro and Micro Components of Foods*, Westport, CT, Food and Nutrition Press, pp. 167–186

Takahashi, M., Wakabayashi, K., Nagao, M., Yamamoto, M., Masui, T., Goto, T., Kinae, N., Tomita, I. & Sugimura, T. (1985) Quantification of 2-amino-3-methylimidazo[4,5-f]quinoline (IQ) and 2-amino-3,8-dimethylimidazo[4,5-f]quinoxaline (MeIQx) in beef extracts by liquid chromatography with electrochemical detection (LCEC). *Carcinogenesis*, **6**, 1195–1199

Thompson, L.H., Tucker, J.D., Stewart, S.A., Christensen, M.L., Salazar, E.P., Carrano, A.V. & Felton, J.S. (1987) Genotoxicity of compounds from cooked beef in repair-deficient CHO cells *versus Salmonella* mutagenicity. *Mutagenesis*, **2**, 483–487

Tjøtta, K., Ingebrigtsen, K. & Aune, T. (1992) Tissue distribution of 2-amino-3,8-dimethylimidazo-[4,5-f]quinoxaline (MeIQx) in two strains of male rats. *Carcinogenesis*, **13**, 779–782

Tsuda, M., Negishi, C., Makino, R., Sato, S., Yamaizumi, Z., Hirayama, T. & Sugimura, T. (1985) Use of nitrite and hypochlorite treatments in determination of the contributions of IQ-type and non-IQ type heterocyclic amines to the mutagenicities in crude pyrolyzed materials. *Mutat. Res.*, **147**, 335–341

Tsuda, H., Takahashi, S., Yamaguchi, S., Ozaki, K. & Ito, N. (1990) Comparison of initiation potential of 2-amino-3-methylimidazo[4,5-f]quinoline and 2-amino-3,8-dimethylimidazo[4,5-f]quinoxaline in an in vivo carcinogen bioassay system. *Carcinogenesis*, **11**, 549–552

Tucker, J.D., Carrano, A.V., Allen, N.A., Christensen, M.L., Knize, M.G., Strout, C.L. & Felton, J.S. (1989) In vivo cytogenetic effects of cooked food mutagens. *Mutat. Res.*, **224**, 105–113

Turesky, R.J., Wishnok, J.S., Tannenbaum, S.R., Pfund, R.A. & Buchi, G.H. (1983) Qualitative and quantitative characterization of mutagens in commercial beef extract. *Carcinogenesis*, **4**, 863–866

Turesky, R.J., Bur, H., Huynh-Ba, T., Aeschbacher, H.-U. & Milon, H. (1988a) Analysis of mutagenic heterocyclic amines in cooked beef products by high-performance liquid chromatography in combination with mass spectrometry. *Food chem. Toxicol.*, **26**, 501–509

Turesky, R.J., Aeschbacher, H.-U., Malnoë, A. & Würzner, H.P. (1988b) Metabolism of the food-borne mutagen/carcinogen 2-amino-3,8-dimethylimidazo[4,5-*f*]quinoxaline in the rat: assessment of biliary metabolites for genotoxicity. *Food chem. Toxicol.*, **26**, 105–110

Turesky, R.J., Aeschbacher, H.-U., Würzner, H.P., Skipper, P.L. & Tannenbaum, S.R. (1988c) Major routes of metabolism of the food-borne carcinogen 2-amino-3,8-dimethylimidazo[4,5-*f*]-quinoxaline in the rat. *Carcinogenesis*, **9**, 1043–1048

Turesky, R.J., Forster, C.M., Aeschbacher, H.-U., Würzner, H.P., Skipper, P.L., Trudel, L.J. & Tannenbaum, S.R. (1989) Purification of the food-borne carcinogens 2-amino-3-methylimidazo-[4,5-*f*]quinoxaline and 2-amino-3,8-dimethylimidazo[4,5-*f*]quinoline in heated meat products by immonoaffinity chromatography. *Carcinogenesis*, **10**, 151–156

Turesky, R.J., Markovic, J., Bracco-Hammer, I. & Fay, L.B. (1991a) The effect of dose and cytochrome P450 induction on the metabolism and disposition of the food-borne carcinogen 2-amino-3,8-dimethylimidazo[4,5-*f*]-quinoxaline (MeIQx) in the rat. *Carcinogenesis*, **10**, 1847–1855

Turesky, R.J., Lang, N.P., Butler, M.A., Teitel, C.H. & Kadlubar, F.F. (1991b) Metabolic activation of carcinogenic heterocyclic aromatic amines by human liver and colon. *Carcinogenesis*, **10**, 1839–1845

Ushiyama, H., Wakabayashi, K., Hirose, M., Itoh, H., Sugimura, T. & Nagao, M. (1991) Presence of carcinogenic heterocyclic amines in urine of healthy volunteers eating normal diet, but not of patients receiving parenteral alimentation. *Carcinogenesis*, **12**, 1417–1422

Vanderlaan, M., Watkins, B.E., Hwang, M., Knize, M.G. & Felton, J.S. (1988) Monoclonal antibodies for the immunoassay of mutagenic compounds produced by cooking beef. *Carcinogenesis*, **9**, 153–160

Wakabayashi, K., Nagao, M., Esumi, H. & Sugimura, T. (1992) Food-derived mutagens and carcinogens. *Cancer Res.*, **52**(Suppl.), 2092s–2098s

Wallin, H., Holme, J.A., Becher, G. & Alexander, J. (1989) Metabolism of the food carcinogen 2-amino-3,8-dimethylimidazo[4,5-*f*]quinoxaline in isolated rat liver cells. *Carcinogenesis*, **10**, 1277–1283

Wallin, H., Holme, J.A. & Alexander, J. (1992) Covalent binding of food carcinogens MeIQx, MeIQ and IQ to DNA and protein in microsomal incubations and isolated rat hepatocytes. *Pharmacol. Toxicol.*, **70**, 220–225

Winton, D.J., Gooderham, N.J., Boobis, A.R., Davies, D.S. & Ponder, B.A.J. (1990) Mutagenesis of mouse intestine *in vivo* using the *Dlb-1* specific locus test: studies with 1,2-dimethylhydrazine, dimethylnitrosamine, and the dietary mutagen 2-amino-3,8-dimethylimidazo[4,5-*f*]quinoxaline. *Cancer Res.*, **50**, 7992–7996

Yamashita, K., Umemoto, A., Grivas, S., Kato, S. & Sugimura, T. (1988) In vitro reaction of hydroxyamino derivatives of MeIQx, Glu-P-1 and Trp-P-1 with DNA: ^{32}P-postlabelling analysis of DNA adducts formed *in vivo* by the parent amines and *in vitro* by their hydroxylamino derivatives. *Mutagenesis*, **3**, 515–520

Yamashita, K., Adachi, M., Kato, S., Nakagama, H., Ochiai, M., Wakabayashi, K., Sato, S., Nagao, M. & Sugimura, T. (1990) DNA adducts formed by 2-amino-3,8-dimethylimidazo[4,5-*f*]quinoxaline in rat liver: dose-response on chronic administration. *Jpn. J. Cancer Res.*, **81**, 470–476

Yamazoe, Y., Abu-Zeid, M., Manabe, S., Toyama, S. & Kato, R. (1988) Metabolic activation of a protein pyrolysate promutagen 2-amino-3,8-methylimidazo[4,5-*f*]quinoxaline by rat liver microsomes and purified cytochrome P450. *Carcinogenesis*, **9**, 105–109

Yoo, M.A., Ryo, H., Todo, T. & Kondo, S. (1985) Mutagenic potency of heterocyclic amines in the *Drosophila* wing spot test and its correlation to carcinogenic potency. *Jpn. J. Cancer Res. (Gann)*, **76**, 468–473

Yoshimi, N., Sugie, S., Iwata, H., Mori, H. & Williams, G.M. (1988) Species and sex differences in genotoxicity of heterocyclic amine pyrolysis and cooking products in the hepatocyte primary culture/DNA repair test using rat, mouse, and hamster hepatocytes. *Environ. mol. Mutag.*, **12**, 53–64

Zhang, X.-M., Wakabayashi, K., Liu, Z.-C., Sugimura, T. & Nagao, N. (1988) Mutagenic and carcinogenic heterocyclic amines in Chinese cooked foods. *Mutat. Res.*, **201**, 181–188

PhIP (2-AMINO-1-METHYL-6-PHENYLIMIDAZO[4,5-b]PYRIDINE)

1. Exposure Data

1.1 Chemical and physical data

1.1.1 Synonyms, structural and molecular data

Chem. Abstr. Services Reg. No.: 105650–23–5
Chem. Abstr. Name: 1-Methyl-6-phenyl-1H-imidazo[4,5-b]pyridin-2-amine
IUPAC Systematic Name: 2-Amino-1-methyl-6-phenylimidazo[4,5-b]pyridine

$C_{13}H_{12}N_4$ Mol. wt: 224.11

1.1.2 Chemical and physical properties

(a) *Description*: Grey-white crystals
(b) *Melting-point*: 327–328 °C (Knize & Felton, 1986)
(c) *Spectroscopy data*: Ultraviolet, mass, proton nuclear magnetic resonance, carbon-13 nuclear magnetic resonance and infrared absorbance spectra have been reported (Knize & Felton, 1986).
(d) *Solubility*: Soluble in methanol (Zhang *et al.*, 1988) and in dimethyl sulfoxide (Dooley *et al.*, 1992)
(e) *Stability*: Stable under moderately acidic and alkaline conditions (Sugimura *et al.*, 1983)

1.1.3 Trade names, technical products and impurities

No data were available to the Working Group.

1.1.4 Analysis

PhIP was originally isolated from fried beef by acid extraction, XAD-2 resin absorption and a series of preparative and analytical high-performance liquid chromatography (HPLC) purifications. The structure of PhIP was determined on the basis of data obtained by mass and proton nuclear magnetic resonance spectral analysis (Felton *et al.*, 1986a).

PhIP was isolated and identified using methanol extraction, 'blue cotton' adsorption and a series of HPLC purifications (Zhang et al., 1988).

A practical, solid-phase extraction and HPLC method for the analysis of PhIP and other heterocyclic amines was devised by Gross et al. (1989) and used on foods and food extracts. Improvements to the method (Gross, 1990; Gross & Grüter, 1992) allow determination of PhIP and most of the other known heterocyclic amines at a level of 1 ng/g from only 10 g of food sample. Replicate samples and spiking allow accurate determination of extraction losses; chromatographic peak identities are confirmed using a diode array–ultraviolet fluorescence detector.

1.2 Production and use

1.2.1 *Production*

The isolation and identification of PhIP were first reported by Felton et al. (1986a). Its structure was confirmed by chemical synthesis (Knize & Felton, 1986). Synthesis of [2-^{14}C]-PhIP from 2-amino-3-bromo-5-phenylpyridine was reported by Turteltaub et al. (1989). [2'-^{3}H]-PhIP was made by catalytic tritiation of 2'-bromo-PhIP. 2'-Bromo-PhIP was made by diazotization of bromobenzene and 2,5-diaminopyridine, the product of which was brominated at the 3-position of pyridine and then displaced with methylamine and cyclized with cyanogen bromide in analogy with the original synthesis of PhIP. Pentadeutero-PhIP was made from 5-amino-2-chloropyridine by a four-step synthetic route (Lynch et al., 1992).

PhIP is produced commercially in small quantities for research purposes.

1.2.2 *Use*

PhIP is not used commercially.

1.3 Occurrence

PhIP was originally isolated from fried ground beef cooked at 300 °C (Felton et al., 1986a). It has been reported in cooked beef, chicken, fish and pork (Table 1). In investigations of foods for the presence of multiple heterocyclic amines, PhIP is usually found to be the most abundant (see monograph on IQ, p. 168). PhIP was also isolated from a complete human diet cooked simulating 'household' conditions (Alink et al., 1988).

PhIP has been produced in several model systems, including refluxed phenylalanine, glucose and creatinine (Shioya et al., 1987; Skog & Jägerstad, 1991), and dry-heated reactions of phenylalanine and creatine, of phenylalanine, creatine and glucose (Taylor et al., 1987; Felton & Knize, 1991) and of leucine and creatine (Övervik et al., 1989).

The amount of PhIP in the urine of 10 subjects on a normal diet ranged from 0.1 to 2.0 ng/24-h urine sample. PhIP was not detected (< 0.01 ng/24-h urine sample) in the urine of three patients who were receiving parenteral nutrients (Ushiyama et al. 1991). Peluso et al. (1991) inferred that PhIP was present in the urine of smokers of black tobacco. PhIP has also been found in mainstream cigarette smoke condensate at 11–23 ng/cigarette, with an average of 16.4 ng/cigarette in six samples (Manabe et al., 1991).

Table 1. Concentrations of PhIP in foods

Sample	Concentration (ng/g)	No. of samples	Reference
Ground beef, fried			
190 °C	23.5–48.5	2	Gross (1990)
250 °C	1.2	1	Gross et al. (1989); Turesky et al. (1989)
300 °C	15	1	Felton et al. (1986b)
No temperature given	0.56	1	Wakabayashi et al. (1992)
Beef, broiled	15.7	1	Wakabayashi et al. (1992)
Beef extract, food-grade	3.62	1	Wakabayashi et al. (1992)
Chicken, broiled	38.1	1	Wakabayashi et al. (1992)
Mutton, broiled	42.5	1	Wakabayashi et al. (1992)
Fish, walleye pollack, fried at 260 °C	69.2	1	Zhang et al. (1988)
Salmon			Gross & Grüter (1992)
Pan-broiled at 200 °C	1.7–23	4	
Oven-cooked at 200 °C	ND–18	3	
Barbecued at 270 °C	2–73	4	

ND, not detected (< 1 ng/g)

1.4 Regulations and guidelines

No data were available to the Working Group.

2. Studies of Cancer in Humans

No epidemiological study was available that addressed the carcinogenic risk to humans of PhIP itself. Cancer risks associated with consumption of broiled and fried foods, which have a high content of PhIP as well as other heterocyclic amines have, however, been addressed in a number of case–control studies. Several of these are summarized in the monograph on IQ. PhIP is also a component of tobacco smoke, which has been covered in a previous IARC monograph (IARC, 1986).

3. Studies of Cancer in Experimental Animals

3.1 Oral administration[1]

3.1.1 *Mouse*

Groups of 40 male and 40 female CDF_1 (BALB/cAnN × DBA/2N F_1) mice, six weeks old, were fed a diet containing 0 or 400 mg/kg PhIP [purity unspecified] for 579 days. Animals

---

[1]The Working Group was aware of studies in progress on PhIP given in the diet of rats (Ghess et al., 1992) and of a study by gavage to non-human primates being carried out at the US National Cancer Institute, Division of Cancer Etiology.

that became moribund were killed and autopsied. The first lymphoma was detected on day 236 in a female mouse given PhIP, and the numbers of mice that survived after that time were taken as the effective numbers. Higher incidences of lymphomas were found in treated than control animals: males, 11/35 *versus* 2/36; females, 26/38 *versus* 6/40. Lung tumours [histological type unspecified] were observed in two controls and in nine treated animals [sex unspecified] (Esumi *et al.*, 1989).

Groups of 10 or more female CF_1 (Charles River) mice, 27–31 days old, were treated with 35, 70 or 150 mg/kg bw PhIP ($< LD_{50}$; dissolved in medium-chain triglycerides) by gavage twice at a four-day interval. Aberrant crypts of colonic mucosa (as defined by light microscopy in methylene blue-stained mucosa; McLellan & Bird, 1988) were scored 21 days after the first treatment. Crypts (0.5 ± 0.3) were observed only after the highest dose (Tudek *et al.*, 1989). [The Working Group noted the lack of concurrent controls.]

3.1.2 *Rat*

A group of 30 male and 30 female Fischer 344 rats, six weeks old, were fed a diet containing 400 mg/kg PhIP (purity, > 99.9%) for 52 weeks. A control group of 40 males and 40 females was available. Animals that became moribund were killed. The first tumour [unspecified] was detected at week 34 in a female rat given PhIP, and the numbers of rats that survived after that time were taken as the effective numbers. At the end of the experiment, 24/30 treated males, 18/30 treated females, 40/40 control males and 40/40 control females were still alive. A total of 38 colon adenocarcinomas were found in 16/29 treated males, four in 2/30 treated females, none in 40 control males and none in 40 control females. Mammary adenocarcinomas were observed in 14/30 treated females and 0/40 controls (Ito *et al.*, 1991).

A group of 13 male Nagase analbuminaemic rats [known to be particularly sensitive to chemical carcinogens (Kakizoe & Sugimura, 1988)], eight weeks old, were fed a diet containing 400 mg/kg PhIP as the hydrochloric acid salt [purity not given] for the first 108 days. The dose was reduced to 300 mg/kg and finally to 100 mg/kg between day 144 and day 311, when the two surviving rats were sacrificed. The presence of at least 90% of the added PhIP in the diet was confirmed by HPLC analysis. Animals that became moribund were killed. The first tumours were detected in the small and large intestine of a rat autopsied on day 136. A total of 36 intestinal tumours were found in 10/13 animals; 22 were adenocarcinomas of the small intestine, two were adenomas at this site, four were carcinomas of the caecum and eight were carcinomas of the large intestine (Ochiai *et al.*, 1991). [The Working Group noted the small number of animals and that no concurrent controls were used but that historical controls were reported to have low rates of intestinal tumours.]

Male Fischer 344 rats, six weeks old, were fed a diet containing 500 mg/kg PhIP for 16 weeks. Animals were killed at weeks 2 and 4 in one experiment and at weeks 4, 8 and 12 and up to 16 weeks in the second. Aberrant crypt foci were induced at a rate of 0.3 ± 0.4 in one out of four rats at week 2 and at 1.3 ± 0.8 in four out of four rats at week 4. In the second experiment, the numbers of aberrant crypt foci induced were 1.3 ± 0.6 (3/3 rats) at week 4, 0.7 ± 0.6 (2/3 rats) at week 8, 3.0 ± 0 (3/3 rats) at week 12 and 11.3 ± 7.0 (3/3 rats) in those killed up to week 16. In age-matched controls, no aberrant crypt foci was found. The numbers of foci per colon found were almost half those induced by Glu-P-1, a known colonic carcinogen (Takahashi *et al.*, 1991).

3.2 Intraperitoneal administration

Mouse

Groups [initial numbers unspecified] of newborn male $B6C3F_1$ mice were injected intraperitoneally with PhIP (purity, 98%) at total doses of 0, 0.625 or 1.25 µmol [140 or 280 µg] (maximal tolerated dose) dissolved in 5, 10 or 20 µl dimethyl sulfoxide and administered on days 1, 8 and 15 after birth, respectively. The incidence of hepatocellular adenomas was significantly higher in treated mice than in controls: at eight months, 1/44 in controls, 2/19 at the low dose and 7/24 at the high dose; at 12 months, 5/44 in controls, 8/16 at the low dose and 14/21 at the high dose. One hepatocellular carcinoma was found in the high-dose group at 12 months (Dooley *et al.*, 1992).

3.3 Administration with known carcinogens

Rat

Groups of 15 male Wistar rats, six weeks old, underwent a two-thirds partial hepatectomy, followed 16–19 h later by a single intraperitoneal injection of 50 or 75 mg/kg bw PhIP (purity, > 99%) dissolved in acid water (pH 5.0.); animals were then fed a diet containing 200 mg/kg 2-acetylaminofluorene during weeks 2 and 3 and received 2 ml [3.2 mg]/kg bw carbon tetrachloride by gavage at the beginning of week 3. All animals were killed at the end of week 6. Significantly more γ-glutamyl transpeptidase-positive liver-cell foci were found in the group treated with the higher dose of PhIP (1.4 foci/cm^2) than in 17 controls given 0.9% saline instead of PhIP (0.48 foci/cm^2). The lower dose of PhIP did not induce significant development of lesions (Kleman *et al.*, 1989).

4. Other Relevant Data

4.1 Absorption, distribution, metabolism and excretion

The toxicology and metabolism of heterocyclic aromatic amines have been reviewed (Övervik & Gustafsson, 1990; Aeschbacher & Turesky, 1991).

4.1.1 *Humans*

No data were available to the Working Group.

4.1.2 *Experimental systems*

In rats and mice, absorbed PhIP was rapidly distributed to most tissues, the highest concentrations being found in the liver and intestines. Over 90% of the radiolabel was eliminated from the body in the two species within 24 h. At that time, much of the retained radiolabel was in ethanol-insoluble, perhaps covalently bound, form. In mice, 11 different radioactive products were identified in the urine and two in the faeces. The profile of urinary and fecal metabolites of PhIP in rats depended on pretreatment (Turteltaub *et al.*, 1989; Watkins *et al.*, 1991).

Human liver and colon cytosols contained O-acetyltransferase activity that metabolized N-hydroxy-PhIP into a DNA-binding form. PhIP was not a substrate to liver cytosolic N-acetyltransferase (Turesky et al., 1991). (See Fig. 1, in the monograph on IQ, p. 178).

Hepatocytes isolated from Aroclor 1254-pretreated rats activated PhIP into a mutagen, as detected in a number of systems. The activation was inhibited by α-napththoflavone, indicating involvement of the cytochrome P450 system. The active metabolite, out of at least eight different species, was concluded to be N-hydroxy-PhIP (Holme et al., 1989).

Rabbit and human liver microsomes could activate PhIP to a mutagenic form. Mutagenic activity was increased when 2,3,7,8-tetrachlorodibenzo-*para*-dioxin (TCDD)-induced rabbit liver and lung microsomes were used. Rabbit cytochrome P450 forms 4 and 6 were mainly responsible for the activation, suggesting involvement of polycyclic hydrocarbon-inducible enzymes (McManus et al., 1989). Cytochrome P450 IA enzymes are responsible for the conversion of PhIP to mutagenic metabolites (Shimada & Guengerich, 1991). Use of purified rat and rabbit cytochrome P450 preparations led to the conclusion that the activation pathway to N-hydroxy-PhIP is catalysed mainly by CYP IA2 (P450 IA2) and to a lesser extent by CYP IA1 (P450 IA1) (Wallin et al., 1990).

4'-Hydroxy-PhIP, thought to be a detoxified metabolite in rats, was formed five times more rapidly by CYP IA1 than by CYP IA2 (Wallin et al., 1990). A sulfate ester of 4'-hydroxy-PhIP was found to be the main metabolite of PhIP in cultured rat hepatocytes and in rat urine but not in bile *in vivo*. N^2-Sulfamate and N^2-acetyl derivatives of PhIP were not present in urine or bile (Alexander et al., 1989); the β-glucuronide conjugate of N-hydroxy-PhIP is the major PhIP metabolite excreted in bile and to a smaller extent in the urine. A glutathione conjugate of N-hydroxy-PhIP was also identified (Alexander et al., 1991).

In mice, 31% of the radiolabel was found in urine and 30% in faeces 24 h after gavage with radioactive PhIP. Following intraperitoneal exposure, the respective amounts were 39 and 12% (Turteltaub et al., 1989). In rats, 51% of the total dose was recovered as unmetabolized PhIP in the faeces 24 h after gavage, representing 66% of the faecal radiolabel (Watkins et al., 1991).

4.2 Toxic effects

No data were available to the Working Group.

4.3 Reproductive and developmental toxicity

No data were available to the Working Group.

4.4 Genetic and related effects

The genetic effects of PhIP have been reviewed (de Meester, 1989; Sugimura et al., 1989; Felton & Knize, 1990).

4.4.1 *Humans*

No data were available to the Working Group.

4.4.2 *Experimental systems* (see also Table 2 and Appendices 1 and 2)

PhIP was mutagenic to *Salmonella typhimurium* strains. In cultured mammalian cells, it induced DNA strand breaks; unscheduled DNA synthesis was found in rat primary hepatocytes, provided that they were derived from polychlorinated biphenyl-treated animals. In cultured repair-deficient Chinese hamster ovary cells, PhIP induced mutation at the *hprt* locus, sister chromatid exchange and chromosomal aberrations.

PhIP was reported to have induced sister chromatid exchange in bone-marrow cells from mice treated *in vivo*. [The Working Group noted that the control values in different experiments were variable.] Chromosomal aberrations were not induced in bone-marrow cells, but the frequency of aberrations was slightly increased in circulating blood cells. [The Working Group noted that a single sampling time, 50 h, was used to obtain the observations in bone marrow.]

PhIP binds covalently to the DNA of various tissues in rats and cynomolgus monkeys following oral administration. Hydrolysis of DNA from the livers of rats dosed with ^{3}H-PhIP, followed by HPLC separation, showed that the radiolabel co-chromatographed with the acid hydrolysis product of N^2-(2'-deoxyguanosin-8-yl)–PhIP (Frandsen *et al.*, 1992).

N-Hydroxy-PhIP does not react with DNA *in vitro* unless it is esterified, particularly by sulfotransferase or *O*-acetyltransferase reactions (Buonarati *et al.*, 1990). Chemical acetylation of N^2-hydroxy-PhIP produced two products, one of which reacted with DNA and with 2'-deoxyguanosine but not with 2'-deoxycytidine, 2'-deoxyadenosine or 2'-deoxythymidine. The adduct was identified as N^2-(2'-deoxyguanosin-8-yl)–PhIP (Frandsen *et al.*, 1992).

4.4.3 *Genetic changes in animal tumours*

As reported in an abstract (Ushijima *et al.*, 1992), colonic carcinomas induced in rats by PhIP contained no mutation in codons 12, 13 or 61 of Ki-*ras* or Ha-*ras*, and polymerase chain reaction and single-strand conformation polymorphism analysis showed no mutation or mutational hot spot in p53. In breast carcinomas induced in rats by PhIP, Ha-*ras* was mutated at a rate of 20%, but no p53 mutation was found.

5. Summary of Data Reported and Evaluation

5.1 Exposure data

PhIP (2-Amino-1-methyl-6-phenylimidazo[4,5-*b*]pyridine) has been found in cooked meat and fish at concentrations of up to 70 ng/g. A few determinations indicated that the levels of PhIP were higher than those of IQ, MeIQ and MeIQx.

5.2 Human carcinogenicity data

No data directly relevant to an evaluation of the carcinogenicity to humans of PhIP were available. Studies on the consumption of cooked meat and fish are summarized in the monograph on IQ.

Table 2. Genetic and related effects of PhIP

Test system	Result without exogenous metabolic system	Result with exogenous metabolic system	Dose (LED/HID)[a]	Reference
ERD, *Escherichia coli* differential toxicity	−	+	3.1 μg/ml	Knasmüller et al. (1992)
HMM, Host-mediated assay, *Escherichia coli* in intrasanguinous mice	+	0	2.3 mg/kg	Knasmüller et al. (1992)
SA0, *Salmonella typhimurium* TA100, reverse mutation	0	+	0.0000	Sugimura et al. (1989)
SA0, *Salmonella typhimurium* TA100, reverse mutation	0	+	0.0000	Felton & Knize (1990)
SA2, *Salmonella typhimurium* TA102, reverse mutation	0	−	0.0000	Felton & Knize (1990)
SA4, *Salmonella typhimurium* TA104, reverse mutation	0	+	0.0000	Felton & Knize (1990)
SA8, *Salmonella typhimurium* TA1538, reverse mutation	0	+	0.0000	Felton et al. (1986a)
SA8, *Salmonella typhimurium* TA1538, reverse mutation	0	+	15	Felton & Knize (1990)
SA9, *Salmonella typhimurium* TA98, reverse mutation	0	+	0.5000	Shioya et al. (1987)
SA9, *Salmonella typhimurium* TA98, reverse mutation	−	+[b]	2.2400	Holme et al. (1989)
SA9, *Salmonella typhimurium* TA98, reverse mutation	0	+	0.0000	Sugimura et al. (1989)
SA9, *Salmonella typhimurium* TA98, reverse mutation	0	+	0.5000	Buonarati & Felton (1990)
SA9, *Salmonella typhimurium* TA98, reverse mutation	0	+	0.0000	Felton & Knize (1990)
SA9, *Salmonella typhimurium* TA98, reverse mutation	0	+	0.05	Wild et al. (1991)
SAS, *Salmonella typhimurium* TA98/1,8-DNP$_6$, reverse mutation	0	+[b]	112.0000	Holme et al. (1989)
SAS, *Salmonella typhimurium* TA98/1,8-DNP$_6$, reverse mutation	0	+	0.5000	Buonarati & Felton (1990)
SAS, *Salmonella typhimurium* TA98/1,8-DNP$_6$, reverse mutation	0	+	0.0050	Wild et al. (1991)
SAS, *Salmonella typhimurium* TA98NR, reverse mutation	0	+[b]	112.0000	Holme et al. (1989)
SAS, *Salmonella typhimurium* TA97, reverse mutation	0	+	0.0000	Felton & Knize (1990)
SAS, *Salmonella typhimurium* YG1024, reverse mutation	0	+	0.0500	Wild et al. (1991)
SAS, *Salmonella typhimurium* TA96, reverse mutation	0	(+)	0.0000	Felton & Knize (1990)
DIA, DNA strand breaks, rat hepatocytes *in vitro*	+	0	11.2000	Holme et al. (1989)
DIA, DNA strand breaks, Chinese hamster V79 cells *in vitro*	−	+[b]	11.2000	Holme et al. (1989)
GCO, Gene mutation, Chinese hamster ovary cells (*uvr5*) *in vitro*	0	+	2.0000	Thompson et al. (1987)

Table 2 (contd)

Test system	Result		Dose (LED/HID)[a]	Reference
	Without exogenous metabolic system	With exogenous metabolic system		
SIC, Sister chromatid exchange, Chinese hamster ovary cells (uv5) in vitro	0	+	1.0000	Thompson et al. (1987)
SIA, Sister chromatid exchange, Chinese hamster lung V79 cells in vitro	–	+[b,c]	5.6000	Holme et al. (1989)
CIC, Chromosome aberrations, Chinese hamster ovary cells (uv5) in vitro	0	+	2.0000	Thompson et al. (1987)
SVA, Sister chromatid exchange, mouse[c] bone-marrow cells in vivo	+		3.1 × 1 ip	Tucker et al. (1989)
MVM, Micronucleus test, mouse[c] bone-marrow cells in vivo	?		50.0 × 1 ip	Tucker et al. (1989)
MVM, Micronucleus test, mouse[c] peripheral blood cells in vivo	?		100.0 × 1 ip	Tucker et al. (1989)
CBA, Chromosomal aberration, mouse[c] bone-marrow cells in vivo	–		100.0 × 1 ip	Tucker et al. (1989)
CLA, Chromosomal aberration, mouse[c] peripheral blood cells in vivo	(+)		100.0 × 1 ip	Tucker et al. (1989)
BVD, Binding (covalent) to DNA in rats (multiple organs) in vivo[d]	+		60 ppm diet × 2/4 wk	Takayama et al. (1989)
BVD, Binding (covalent) to DNA in cynomolgus monkey (multiple organs) in vivo[d]	+		20.00 × 9 po	Adamson et al. (1991)
BID, Binding (covalent) to calf thymus DNA in vitro	0	+	0	Peluso et al. (1991)

+, positive; (+), weakly positive; –, negative; 0, not tested; ?, inconclusive (e.g., variable response in several experiments within an adequate study)
[a]In-vitro tests, μg/ml; in-vivo tests, mg/kg bw
[b]Hepatocytes
[c]Polychlorinated biphenyl-treated
[d]^{32}P-Postlabel

5.3 Animal carcinogenicity data

PhIP was tested for carcinogenicity in one experiment in mice and in two experiments in rats by oral administration in the diet. It increased the incidence of lymphomas in mice of each sex. In rats, it produced adenocarcinomas of the small and large intestine in males and mammary adenocarcinomas in females.

Intraperitoneal injection of PhIP to newborn male mice increased the incidence of hepatic adenomas.

A single intraperitoneal dose of PhIP after a two-thirds hepatectomy, followed by further modulating treatment, enhanced development of foci of altered hepatocytes in the livers of rats.

5.4 Other relevant data

PhIP formed DNA adducts *in vivo* in rats and monkeys. In rodent cells *in vitro*, it induced DNA damage, gene mutation and chromosomal anomalies. It induced DNA damage and mutation in bacteria.

PhIP can be metabolized by human microsomes isolated from liver and colon to a species that damages bacterial DNA.

5.5 Evaluation[1]

There is *inadequate evidence* in humans for the carcinogenicity of PhIP.

There is *sufficient evidence* in experimental animals for the carcinogenicity of PhIP.

Overall evaluation

PhIP (2-Amino-1-methyl-6-phenylimidazo[4,5-*b*]pyridine) *is possibly carcinogenic to humans (Group 2B)*.

6. References

Adamson, R.H., Snyderwine, E.G., Thorgeirsson, U.P., Schut, H.A.J., Turesky, R.J., Thorgeirsson, S.S., Takayama, S. & Sugimura, T. (1991) Metabolic processing and carcinogenicity of heterocyclic amines in nonhuman primates. In: Ernster, L., Esumi, H,. Fujii, Y., Gelboin, H.Y., Kato, R. & Sugimura, T., eds, *Xenobiotics and Cancer*, Tokyo/London, Japan Scientific Societies Press/Taylor & Francis, pp. 289–301

Aeschbacher, H.-U. & Turesky, R.J. (1991) Mammalian cell mutagenicity and metabolism of heterocyclic aromatic amines. *Mutat. Res.*, **259**, 235–250

Alexander, J., Wallin, H., Holme, J.A. & Becher, G. (1989) 4-(2-Amino-1-methylimidazo[4,5-*b*]pyrid-6-yl)phenyl sulfate—a major metabolite of the food mutagen 2-amino-1-methyl-6-phenylimidazo[4,5-*b*]pyridine in the rat. *Carcinogenesis*, **10**, 1543–1547

---

[1]For definition of the italicized terms, see Preamble, pp. 26–29.

Alexander, J., Wallin, H., Rossland, O.J., Solberg, K.E., Holmes, J.A., Becher, G., Andersson, R. & Grivas, S. (1991) Formation of a glutathione conjugate and a semistable transportable glucuronide conjugate of N^2-oxidized species of 2-amino-1-methyl-6-phenylimidazo[4,5-*b*]pyridine (PhIP) in rat liver. *Carcinogenesis*, **12**, 2239-2245

Alink, G.M., Knize, M.G., Shen, N.H., Hesse, S.P. & Felton, J.S. (1988) Mutagenicity of food pellets from human diets in the Netherlands. *Mutat. Res.*, **206**, 387-393

Buonarati, M.H. & Felton, J.S. (1990) Activation of 2-amino-1-methyl-6-phenylimidazo[4,5-*b*]pyridine (PhIP) to mutagenic metabolites. *Carcinogenesis*, **11**, 1133-1138

Buonarati, M.H., Turteltaub, K.W., Shen, N.H. & Felton, J.S. (1990) Role of sulfation and acetylation in the activation of 2-hydroxyamino-1-methyl-6-phenylimidazo[4,5-*b*]pyridine to intermediates which bind DNA. *Mutat. Res.*, **245**, 185-190

Dooley, K.L., Von Tungeln, L.S., Bucci, T., Fu, P.P. & Kadlubar, F.F. (1992) Comparative carcinogenicity of 4-aminobiphenyl and the food pyrolysates, Glu-P-1, IQ, PhIP, and MeIQx in the neonatal B6C3F$_1$ male mouse. *Cancer Lett.*, **62**, 205-209

Esumi, H., Ohgaki, H., Kohzen, E., Takayama, S. & Sugimura, T. (1989) Induction of lymphoma in CDF$_1$ mice by the food mutagen, 2-amino-1-methyl-6-phenylimidazo[4,5-*b*]pyridine. *Jpn. J. Cancer Res.*, **80**, 1176-1178

Felton, J.S. & Knize, M.G. (1990) Heterocyclic-amine mutagens/carcinogens in foods. *Handb. exp. Pharmacol.*, **94**, 471-502

Felton, J.S. & Knize, M.G. (1991) Mutagen formation in muscle meats and model heating systems. In: Hayatsu, H., ed., *Mutagens in Food: Detection and Prevention*, Boca Raton, FL, CRC Press, pp. 57-66

Felton, J.S., Knize, M.G., Shen, N.H., Lewis, P.R., Andresen, B.D., Happe, J. & Hatch, F.T. (1986a) The isolation and identification of a new mutagen from fried ground beef: 2-amino-1-methyl-6-phenylimidazo[4,5-*b*]pyridine (PhIP). *Carcinogenesis*, **7**, 1081-1086

Felton, J.S., Knize, M.G., Shen, N.H., Andresen, B.D., Bjeldanes, L.F. & Hatch, F.T. (1986b) Identification of the mutagens in cooked beef. *Environ. Health Perspectives*, **67**, 17-24

Frandsen, H., Grivas, S., Andersson, R., Dragsted, L. & Larsen, J.C. (1992) Reaction of the N^2-acetoxy derivative of 2-amino-1-methyl-6-phenylimidazo[4,5-*b*]pyridine (PhIP) with 2'-deoxyguanosine and DNA. Synthesis and identification of N^2-(2'-deoxyguanosin-8-yl)-PhIP. *Carcinogenesis*, **13**, 629-635

Ghess, M.-J., Wilbourn, J.D. & Vainio, H., eds (1992) *Directory of Agents Being Tested for Carcinogenicity*, No. 15, Lyon, IARC, pp. 87, 209

Gross, G.A. (1990) Simple methods for quantifying mutagenic heterocyclic aromatic amines in food products. *Carcinogenesis*, **11**, 1597-1603

Gross, G.A. & Grüter, A. (1992) Quantitation of mutagenic/carcinogenic heterocyclic aromatic amines in food products. *J. Chromatogr.*, **592**, 271-278

Gross, G.A., Philippossian, G. & Aeschbacher, H.-U. (1989) An efficient and convenient method for the purification of mutagenic heterocyclic amines in heated meat products. *Carcinogenesis*, **10**, 1175-1182

Holme, J.A., Wallin, H., Brunborg, G., Søderlund, E.J., Hongslo, J.K. & Alexander, J. (1989) Genotoxicity of the food mutagen 2-amino-1-methyl-6-phenylimidazo[4,5-*b*]pyridine (PhIP): formation of 2-hydroxamino-PhIP, a directly acting genotoxic metabolite. *Carcinogenesis*, **10**, 1389-1396

IARC (1986) *IARC Monographs on the Evaluation of the Carcinogenic Risk of Chemicals to Humans*, Vol. 38, *Tobacco Smoking*, Lyon

Ito, N., Hasegawa, R., Sano, M., Tamano, S., Esumi, H., Takayama, S. & Sugimura, T. (1991) A new colon and mammary carcinogen in cooked food, 2-amino-1-methyl-6-phenylimidazo[4,5-b]pyridine (PhIP). *Carcinogenesis*, **12**, 1503–1506

Kakizoe, T. & Sugimura, T. (1988) Chemical carcinogenesis in analbuminemic rats. *Jpn. J. Cancer Res. (Gann)*, **79**, 775–784

Kleman, M., Övervik, E., Blanck, A. & Gustafsson, J.Å. (1989) The food-mutagens 2-amino-1-methyl-6-phenylimidazo[4,5-b]pyridine (PhIP) and 2-amino-3,8-dimethylimidazo-[4,5-f]quinoxaline (MeIQx) initiate enzyme-altered hepatic foci in the resistant hepatocyte model. *Carcinogenesis*, **10**, 1697–1700

Knasmüller, S., Kienzl, H., Huber, W. & Hermann, R.S. (1992) Organ-specific distribution of genotoxic effects in mice exposed to cooked food mutagens. *Mutagenesis*, **7**, 235–241

Knize, M.G. & Felton, J.S. (1986) The synthesis of the cooked-beef mutagen 2-amino-1-methyl-6-phenylimidazo[4,5-b]pyridine and its 3-methyl isomer. *Heterocycles*, **24**, 1815–1819

Lynch, A.M., Knize, M.G., Boobis, A.R., Gooderham, N.J., Davies, D.S. & Murray, S. (1992) Intra- and interindividual variability in systemic exposure in humans to 2-amino-3,8-dimethylimidazo[4,5-f]quinoxaline and 2-amino-1-methyl-6-phenylimidazo[4,5-b]pyridine, carcinogens present in cooked beef. *Cancer Res.*, **52**, 6216–6623

Manabe, S., Tohyama, K., Wada, O. & Aramaki, T. (1991) Detection of a carcinogen, 2-amino-1-methyl-6-phenylimidazo[4,5-b]pyridine (PhIP), in cigarette smoke condensate. *Carcinogenesis*, **12**, 1945–1947

McLellan, E.A. & Bird, R.P. (1988) Specificity study to evaluate induction of aberrant crypts in murine colons. *Cancer Res.*, **48**, 6183–6186

McManus, M.E., Felton, J.S., Knize, M.G., Burgess, W.M., Roberts-Thomson, S., Pond, S.M., Stupans, I. & Veronese, M.E. (1989) Activation of the food-derived mutagen 2-amino-1-methyl-6-phenylimidazo[4,5-b]pyridine by rabbit and human liver microsomes and purified forms of cytochrome P-450. *Carcinogenesis*, **10**, 357–363

de Meester, C. (1989) Bacterial mutagenicity of heterocyclic amines found in heat-processed food. *Mutat. Res.*, **221**, 235–262

Ochiai, M., Ogawa, K., Wakabayashi, K., Sugimura, T., Nagase, S., Esumi, H. & Nagao, M. (1991) Induction of intestinal adenocarcinomas by 2-amino-1-methyl-6-phenylimidazo[4,5-b]pyridine in Nagase analbuminemic rats. *Jpn. J. Cancer Res.*, **82**, 363–366

Övervik, E. & Gustafsson, J.-Å. (1990) Cooked-food mutagens: current knowledge of formation and biological significance. *Mutagenesis*, **5**, 437–446

Övervik, E., Kleman, M., Berg, I. & Gustafsson, J.-Å. (1989) Influence of creatine, amino acids and water on the formation of the mutagenic heterocyclic amines found in cooked meat. *Carcinogenesis*, **10**, 2293–2301

Peluso, M., Castegnaro, M., Malaveille, C., Friesen, M., Garren, L., Hautefeuille, A., Vineis, P., Kadlubar, F. & Bartsch, H. (1991) ^{32}P-Postlabelling analysis of urinary mutagens from smokers of black tobacco implicates 2-amino-1-methyl-6-phenylimidazo[4,5-b]pyridine (PhIP) as a major DNA-damaging agent. *Carcinogenesis*, **12**, 713–717

Shimada, T. & Guengerich, F.P. (1991) Activation of amino-α-carboline, 2-amino-1-methyl-6-phenylimidazo[4,5-b]pyridine, and a copper phthalocyanine cellulose extract of cigarette smoke condensate by cytochrome P-450 enzymes in rat and human liver microsomes. *Cancer Res.*, **51**, 5284–5291

Shioya, M., Wakabayashi, K., Sato, S., Nagao, M. & Sugimura, T. (1987) Formation of a mutagen, 2-amino-1-methyl-6-phenylimidazo[4,5-*b*]pyridine (PhIP) in cooked beef, by heating a mixture containing creatinine, phenylalanine and glucose. *Mutat. Res.*, **191**, 133–138

Skog, K. & Jägerstad, M. (1991) Effects of glucose on the formation of PhIP in a model system. *Carcinogenesis*, **12**, 2297–2300

Sugimura, T., Sako, S. & Takayama, S. (1983) New mutagenic heterocyclic amines found in amino acid and protein pyrolysates in cooked food. In: Wynder, E.L., Leveille, G.A., Weisburger, J.H. & Livingstone, G.E., eds, *Environmental Aspects of Cancer: The Role of Macro and Micro Components of Foods*, Westport, CT, Food and Nutrition Press, pp. 167–186

Sugimura, T., Wakabayashi, K., Nagao, M. & Ohgaki, H. (1989) Heterocyclic amines in cooked food. In: Taylor & Scanlan, eds, *Food Toxicology. A Perspective on the Relative Risks*, New York, Marcel Dekker, pp. 31–55

Takahashi, S., Ogawa, K., Ohshima, H., Esumi, H., Ito, N. & Sugimura, T. (1991) Induction of aberrant crypt foci in the large intestine of F344 rats by oral administration of 2-amino-1-methyl-6-phenylimidazo[4,5-*b*]pyridine. *Jpn. J. Cancer Res. (Gann)*, **82**, 135–137

Takayama, K., Yamashita, K., Wakabayashi, K., Sugimura, T. & Nagao, M. (1989) DNA modification by 2-amino-1-methyl-6-phenylimidazo[4,5-*b*]pyridine in rats. *Jpn. J. Cancer Res.*, **80**, 1145–1148

Taylor, R.T., Fultz, E., Knize, M.G. & Felton, J.S. (1987) Formation of the fried ground beef mutagens 2-amino-3-methylimidazo[4,5-*f*]quinoline (IQ) and 2-amino-1-methyl-6-phenylimidazo[4,5-*b*]pyridine (PhIP) from L-phenylalanine (Phe) + creatinine (Cre) (or creatine) (Abstract). *Environ. Mutag.*, **9** (Suppl. 8), 106

Thompson, L.H., Tucker, J.D., Stewart, S.A., Christensen, M.L., Salazar, E.P., Carrano, A.V. & Felton, J.S. (1987) Genotoxicity of compounds from cooked beef in repair-deficient CHO cells *versus Salmonella* mutagenicity. *Mutagenesis*, **2**, 483–487

Tucker, J.D., Carrano, A.V., Allen, N.A., Christensen, M.L., Knize, M.G., Strout, C.L. & Felton, J.S. (1989) In vivo cytogenetic effects of cooked food mutagens. *Mutat. Res.*, **224**, 105–113

Tudek, B., Bird, R. & Bruce, W.R. (1989) Foci of aberrant crypts in the colons of mice and rats exposed to carcinogens associated with foods. *Cancer Res.*, **49**, 1236–1240

Turesky, R.J., Forster, C.M., Aeschbacher, H.-U., Würzner, H.P., Skipper, P.L., Trudel, L.J. & Tannenbaum, S.R. (1989) Purification of the food-borne carcinogens 2-amino-3-methylimidazo-[4,5-*f*]quinoline and 2-amino-3,8-dimethylimidazo[4,5-*f*]quinoxaline in heated meat products by immunoaffinity chromatography. *Carcinogenesis*, **10**, 151–156

Turesky, R.J., Lang, N.P., Butler, M.A., Teitel, C.H. & Kadlubar, F.F. (1991) Metabolic activation of carcinogenic heterocyclic aromatic amines by human liver and colon. *Carcinogenesis*, **12**, 1839–1845

Turteltaub, K.W., Knize, M.G., Healy, S.K., Tucker, J.D. & Felton, J.S. (1989) The metabolic disposition of 2-amino-1-methyl-6-phenylimidazo[4,5-*b*]pyridine in the induced mouse. *Food chem. Toxicol.*, **27**, 667–673

Ushijima, T., Kakiuchi, H., Makino, H., Tsujimoto, A., Ishizaka, Y., Sugimura, T. & Nagao, M. (1992) *ras* and *p53* Mutations in rat carcinomas induced by 2-amino-1-methyl-6-phenylimidazo-[4,5-*b*]pyridine (PhIP) (Abstract No. 665). *Proc. Am. Assoc. Cancer Res.*, **33**, 111

Ushiyama, H., Wakabayashi, K., Hirose, M., Itoh, H., Sugimura, T. & Nagao, M. (1991) Presence of carcinogenic heterocyclic amines in urine of healthy volunteers eating normal diet, but not in patients receiving parenteral alimentation. *Carcinogenesis*, **12**, 1417–1422

Wakabayashi, K., Nagao, M., Esumi, H. & Sugimura, T. (1992) Food-derived mutagens and carcinogens. *Cancer Res.*, **52** (Suppl.), 2092s–2098s

Wallin, H., Mikalsen, A., Guengerich, F.P., Ingelman-Sundberg, M., Solberg, K.E., Rossland, O.J. & Alexander, J. (1990) Differential rates of metabolic activation and detoxication of the food mutagen 2-amino-1-methyl-6-phenylimidazo[4,5-b]pyridine by different cytochrome P450 enzymes. *Carcinogenesis*, **11**, 489–492

Watkins, B.E., Esumi, H., Wakabayashi, K., Nagao, M. & Sugimura, T. (1991) Fate and distribution of 2-amino-1-methyl-6-phenylimidazo[4,5-b]pyridine (PhIP) in rats. *Carcinogenesis*, **12**, 1073–1078

Wild, D., Watkins, B.E. & Vanderlaan, M. (1991) Azido- and nitro-PhIP, relatives of the heterocyclic arylamine and food mutagen PhIP—mechanism of their mutagenicity in *Salmonella*. *Carcinogenesis*, **12**, 1091–1096

Zhang, X.-M., Wakabayashi, K., Liu, Z.-C., Sugimura, T. & Nagao, M. (1988) Mutagenic and carcinogenic heterocyclic amines in Chinese cooked foods. *Mutat. Res.*, **201**, 181–188

MYCOTOXINS

AFLATOXINS

These substances were considered by previous working groups, in December 1971 (IARC, 1972), October 1975 (IARC, 1976) and March 1987 (IARC, 1987a). Since that time, new data have become available, and these have been incorporated into the monograph and taken into account in the present evaluation.

1. Exposure Data

1.1 Chemical and physical data

1.1.1 *Synonyms, structural and molecular data*

Aflatoxin B$_1$

Chem. Abstr. Services Reg. No.: 1162-65-8
Chem. Abstr. Name: (6aR-*cis*)(2,3,6a,9a)Tetrahydro-4-methoxycyclopenta[*c*]furo[3',2': 4,5]furo[2,3-*h*][*l*]benzopyran-1,11-dione
Synonyms: 6-Methoxydifurocoumarone; 2,3,6aα,9aα-tetrahydro-4-methoxycyclopenta-[*c*]furo[3',2':4,5]furo[2,3-*h*][*l*]benzopyran-1,11-dione

Aflatoxin B$_2$

Chem. Abstr. Services Reg. No.: 7220-81-7
Chem. Abstr. Name: (6aR-*cis*)(2,3,6a,8,9,9a)Hexahydro-4-methoxycyclopenta[*c*]furo-[3',2':4,5]furo[2,3-*h*][*l*]benzopyran-1,11-dione
Synonyms: Dihydroaflatoxin B$_1$; 2,3,6aα,8,9,9aα-hexahydro-4-methoxycyclopenta[*c*]-furo[3',2':4,5]furo[2,3-*h*][*l*]benzopyran-1,11-dione

Aflatoxin G$_1$

Chem. Abstr. Services Reg. No.: 1165-39-5
Chem. Abstr. Name: (7aR-*cis*)(3,4,7a,10a)Tetrahydro-5-methoxy-1*H*,12*H*-furo[3',2': 4,5]furo[2,3-*h*]pyrano[3,4-*c*][*l*]benzopyran-1,12-dione
Synonym: 3,4,7aα,10aα-Tetrahydro-5-methoxy-1*H*,12*H*-furo[3',2':4,5]furo[2,3-*h*]pyrano[3,4-*c*][*l*]benzopyran-1,12-dione

Aflatoxin G$_2$

Chem. Abstr. Services Reg. No.: 7241-98-7
Chem. Abstr. Name: (7aR-*cis*)(3,4,7a,9,10,10a)Hexahydro-5-methoxy-1*H*,12*H*-furo-[3',2':4,5]furo[2,3-*h*]pyrano[3,4-*c*][*l*]benzopyran-1,12-dione

Synonyms: Dihydroaflatoxin G$_1$; 3,4,7aα,9,10,10aα-hexahydro-5-methoxy-1*H*,12*H*-furo[3',2':4,5]furo[2,3-*h*]pyrano[3,4-*c*][*l*]benzopyran-1,12-dione

Aflatoxin M$_1$

Chem. Abstr. Services Reg. No.: 6795-23-9
Chem. Abstr. Name: 2,3,6a,9a-Tetrahydro-9a-hydroxy-4-methoxycyclopenta[*c*]furo-[3',2':4,5]furo[2,3-*h*][*l*]benzopyran-1,11-dione
Synonym: 4-Hydroxyaflatoxin B$_1$

B$_1$: C$_{17}$H$_{12}$O$_6$ Mol. wt: 312.3

B$_2$: C$_{17}$H$_{14}$O$_6$ Mol. wt: 314.3

G$_1$: C$_{17}$H$_{12}$O$_7$ Mol. wt: 328.3

G$_2$: C$_{17}$H$_{14}$O$_7$ Mol. wt: 330.3

M$_1$: C$_{17}$H$_{12}$O$_7$ Mol. wt: 328.3

The structures of other metabolites and degradation products mentioned in the monographs are shown in Figure 1 and given by Castegnaro *et al.* (1980).

1.1.2 *Chemical and physical properties of aflatoxins* (from Castegnaro *et al.*, 1980, 1991; Budavari, 1989, unless otherwise stated)

(*a*) *Description*: Colourless to pale-yellow crystals. Intensely fluorescent in ultraviolet light, emitting blue (aflatoxins B$_1$ and B$_2$) or yellow-green (aflatoxins G$_1$, G$_2$) fluorescence, from which the designations B and G were derived, or blue-violet fluorescence (aflatoxin M$_1$)

Fig. 1. Structures of other metabolites and degradation products of aflatoxins

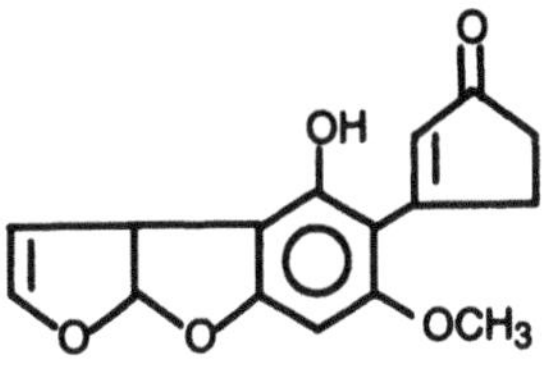

Aflatoxin D_1

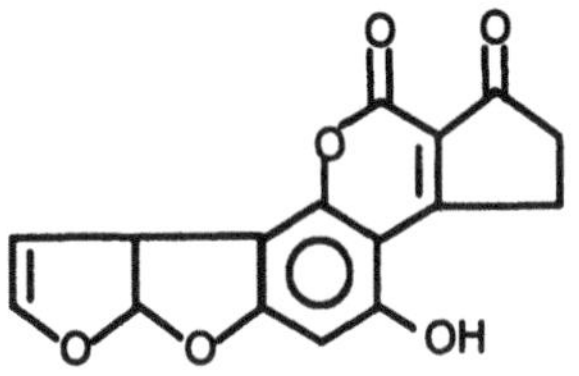

Aflatoxin P_1

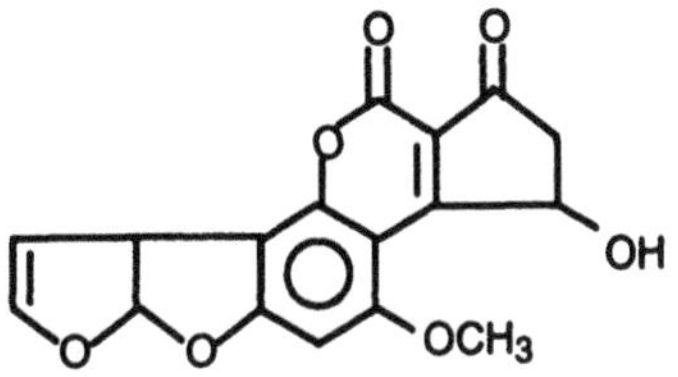

Aflatoxin Q_1

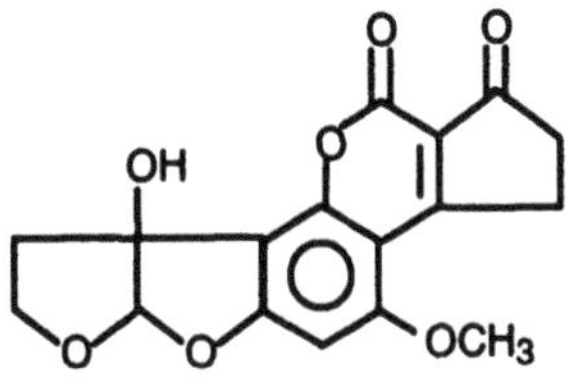

Aflatoxin M_2

(b) Melting-point and (c) Absorption spectroscopy:

Table 1. Melting-points and ultraviolet absorption of aflatoxins

Aflatoxin	Melting-point (°C)	Ultraviolet absorption (ethanol)	
		λ_{max} (nm)	ϵ
B_1	268–269 (decomposition) (crystals from chloroform)	223 265 362	25 600 13 400 21 800
B_2	287–289 (decomposition) (crystals from chloroform–pentane)	222 265 363	17 000 11 700 23 400
G_1	244–246 (decomposition) (crystals from chloroform–methanol)	243 257 264 362	11 500 9 900 10 000 16 100
G_2	237–239 (decomposition) (crystals from ethyl acetate)	214 265 363	28 100 11 600 21 000
M_1	299 (decomposition) (crystals from methanol)	226 265 357	23 100 11 600 19 000

From IARC (1976); Castegnaro *et al.* (1980); Cole & Cox (1981); Budavari (1989)

(d) *Solubility*: Very slightly soluble in water (10–30 µg/ml); insoluble in non-polar solvents; freely soluble in moderately polar organic solvents (e.g., chloroform and methanol) and especially in dimethyl sulfoxide (Cole & Cox, 1981)

(e) *Stability*: Unstable to ultraviolet light in the presence of oxygen, to extremes of pH (< 3, > 10) and to oxidizing agents

(f) *Reactivity*: The lactone ring is susceptible to alkaline hydrolysis. Aflatoxins are also degraded by reaction with ammonia or sodium hypochlorite.

1.1.3 *Trade names, technical products and impurities*

No data were available to the Working Group.

1.1.4 *Analysis*

The numerous methods for the determination of aflatoxins B and G in maize (known as 'corn' in the USA), groundnuts (peanuts) and cottonseed meal (Egan *et al.*, 1982) and of aflatoxin M_1 in milk products (Scott, 1989) have been reviewed. Those that have been verified in collaborative studies and have been proposed as official methods by the Association of Official Analytical Chemists (Scott, 1990) are shown in Table 2.

Table 2. Analytical methods validated by the Association of Official Analytical Chemists

Method no.	Aflatoxin	Food	Method[a]	Detection limit (µg/kg)
975.36	All	Foods and feeds (screening)	MC	10
979.18	All	Maize and groundnuts (screening)	MC	10
990.31	All	Maize and groundnuts (Aflatest screening)	IC	10
990.34	All	Maize, cottonseed, groundnuts (screening)	ELISA	20–30
989.06	B_1	Cottonseed products and mixed feed (screening)	ELISA	15
990.32	B_1	Maize and groundnuts (screening)	ELISA	20
968.22	B_1, B_2, G_1, G_2	Groundnuts and groundnut products	TLC	5
970.45	B_1, B_2, G_1, G_2	Groundnuts and groundnut products	TLC	10
971.23	B_1, B_2, G_1, G_2	Cocoa beans	TLC	10
971.24	B_1, B_2, G_1, G_2	Coconut, copra and copra meal	TLC	50
972.26	B_1, B_2, G_1, G_2	Maize	TLC	5
980.20	B_1, B_2, G_1, G_2	Cottonseed products	TLC, HPLC	10, 5
970.46	B_1, B_2, G_1, G_2	Green coffee	TLC	25
974.16	B_1, B_2, G_1, G_2	Pistachio nuts	TLC	15
972.27	B_1, B_2, G_1, G_2	Soya beans	TLC	10
990.33	B_1, B_2, G_1, G_2	Maize and groundnut butter	HPLC	5
978.15	B_1	Eggs	TLC	0.1
982.24	B_1 and M_1	Liver	TLC	0.1
974.17	M_1	Dairy products	TLC	0.1
980.21	M_1	Milk and cheese	TLC	0.1
986.16	M_1 and M_2	Fluid milk	HPLC	0.1

From Scott (1990)

[a]MC, minicolumn; IC, immunoaffinity column, ELISA, enzyme-linked immunosorbent assay; TLC, thin-layer chromatography; HPLC, high-performance liquid chromatography

Analytical quality assurance for the analysis of aflatoxins B_1, B_2, G_1, G_2 and M_1 in foods is available for laboratories through the International Mycotoxin Check Sample Programme organized by IARC (Friesen, 1989).

As contamination may not occur in a homogeneous way throughout a sample of maize or groundnuts, good sampling and sample preparation procedures must be used to obtain accurate quantitative results. Sampling and sample preparation procedures for aflatoxin analysis have been reviewed (Egan *et al.*, 1982; van Egmond, 1984), and the Association of Official Analytical Chemists' method (977.16) for sampling aflatoxin-containing commodities has been published (Scott, 1990).

A number of approaches have been used to analyse for aflatoxins and their metabolites in human tissues and body fluids. These include immunoaffinity purification, immunoassay (Wild *et al.*, 1987), high-performance liquid chromatography with fluorescence or ultraviolet detection and synchronous fluorescence spectroscopy (Groopman & Sabbioni, 1991). Molecular biomarkers, such as urinary markers, metabolites in milk and parent compounds in blood, are used for determining exposure to aflatoxins (Groopman, 1993).

1.2 Production and use

1.2.1 *Production*

Aflatoxins were first identified in 1961 in animal feed contaminated by *Aspergillus parasiticus* (Sargeant *et al.*, 1961). They are known to be produced by three species: *A. flavus*, *A. parasiticus* and the rare species *A. nomius* (Kurtzman *et al.*, 1987). It is generally considered that *A. flavus* produces aflatoxins B_1 and B_2, whereas *A. parasiticus* produces aflatoxins B_1, B_2, G_1 and G_2 (Dorner *et al.*, 1984). In the USA and Africa, *A. flavus* and *A. parasiticus* are widely distributed; in southeast Asia, *A. flavus* occurs to the virtual exclusion of the other species (Pitt *et al.*, 1993).

Aflatoxins are produced, in small quantities for use in research only, by large-scale fermentation on solid substrates or liquid media, from which the aflatoxins are extracted and purified by chromatography. Total annual production probably does not exceed 100 g.

1.2.2 *Use*

Aflatoxins are not used commercially other than in research.

1.3 Occurrence

The aflatoxin-producing *Aspergillus* species, and consequently dietary aflatoxin contamination, are ubiquitous in areas of the world with hot, humid climates, including sub-Saharan Africa and Southeast Asia. Exposure in those countries results from contamination of dietary staples and is therefore likely to be chronic. Since countries in colder climatic areas import foods from areas where aflatoxin levels are high, however, aflatoxins are of worldwide concern. Data on occurrence are available predominantly from importing countries, where regulation of contamination should contribute to ensuring relatively low exposure in those countries.

The relative proportions of aflatoxin B_1, aflatoxin G_1, aflatoxin B_2 and aflatoxin G_2 on crops depend on the particular *Aspergillus* species present. Aflatoxin B_1 is the most frequent type present in contaminated samples, and aflatoxin G_1 has never been reported in the absence of aflatoxin B_1. Aflatoxin B_2 and aflatoxin G_2 are typically present in much lower quantities. Aflatoxin M_1 is a metabolic hydroxylation product of aflatoxin B_1; it can occur in the absence of other aflatoxins.

1.3.1 *Diet*

Samples of most dietary staples have been shown to be contaminated with aflatoxins at one time or another; however, the frequency with which different dietary items are contaminated and the level of contamination differs. High levels have been found notably in groundnuts and maize in African regions, Southeast Asia and southern China, where these foods are dietary staples. Contamination of raw commodities in the USA also occurs at high levels in some years. Foodstuffs may be contaminated with aflatoxin both pre- and post-harvest.

An extensive review of the levels of aflatoxins in foods and feeds from 16 countries in North America, South America, Europe, Asia and Africa was made for the period 1976–83 (Jelinek *et al.*, 1989). Data were presented as total aflatoxins. In maize and maize products from nine countries, median levels ranged from < 0.1 to 80 µg/kg; in several countries, more than 10% of grain samples contained levels above 5 µg/kg (Table 3). Data from the USA demonstrated annual and geographical fluctuations in aflatoxin levels and increased levels in maize in conditions of drought. Levels in other grains and cereal products were lower than those in maize in all the countries surveyed. Extensive surveys of groundnuts in North America, Central America, South America, Europe and Australia indicated that both median and 90% levels were almost always below 20 µg/kg. Groundnuts imported into the USA in 1981 from India, the Sudan and Brazil contained much higher levels, more than 50% of samples containing over 26 µg/kg (Table 4). Other nuts, including almonds, cashews, filberts, hazel-nuts, mixed nuts, pecans and walnuts, were contaminated to a lower extent, but higher levels were occasionally seen in surveys of pistachios, pumpkin seeds and Brazil-nuts.

Table 3. FAO/WHO/UNEP Monitoring Program: aflatoxins in maize and maize products

Country	Year	No. of samples	Median[a] (µg/kg)	90th percentile[b] (µg/kg)
Maize				
Brazil	1981	228	< 8.0	< 8.0
Canada	1976	25	< 4.0	< 4.0
Guatemala	1976–79	231	< 4.0	4–360
Kenya	1978–79	78	< 0.1–70	30–1920
Mexico	1979–80	96	< 2.5–< 10	< 2.5–30
United Kingdom	1978	29	5.0	8.0
USA	1978–83	2633	< 1–80	10–700
USSR	1981–82	219	< 1.0	< 1–662

Table 3 (contd)

Country	Year	No. of samples	Median[a] (μg/kg)	90th percentile[b] (μg/kg)
Maize products				
Canada[c]	1983	20	4.0	6.0
Guatemala[c]	1977-79	22	< 4.0	10-20
Kenya[d]	1978-79	283	< 10.0	< 10.0
Mexico[e]	1978	217	< 2.0	Not reported
Switzerland[c]	1978	40	< 0.5	2.0
United Kingdom[d]	1978	13	< 0.1	< 0.1
USA[f]	1978-83	1174	< 1.0	< 1-56
USSR[c]	1976	87	< 1.0	5.0

From Jelinek et al. (1989)
[a]Many of the median values were below the detection limits of the assay.
[b]The level below which 90% of the findings occurred in a given survey; when more than one value is given, they refer to separate surveys in different years.
[c]Maize meal
[d]Maize flour
[e]Tortillas
[f]Ground or dry-milled maize

Table 4. Aflatoxins in raw, shelled groundnuts imported into the USA, 1981

Origin	No. of lots	Lots (%)	
		Determinable	> 26 μg/kg
China	2 585	15	2.5
India	1 453	92	58.0
Sudan	932	94	78.0
Argentina	446	40	4.5
South Africa	112	41	21.0
Malawi	80	60	10.0
Australia	52	10	4.0
Brazil	44	100	95.0
Egypt	41	14	2.0
Taiwan	37	27	0.0
USA[a]	172 669	20	3.0

From Jelinek et al. (1989)
[a]For comparison, levels in US domestic groundnuts for the combined crop years 1973-79

Rice exposed during cyclones in India had levels up to 1130 µg/kg. This commodity is otherwise rarely contaminated at levels > 20 µg/kg (Choke, 1990). High levels of aflatoxins (> 100 µg/kg) occurred in cottonseed and groundnut and in sunflower seeds. Data from other food surveys similarly showed high levels of aflatoxins in groundnuts and maize (Stoloff, 1982; Choke, 1991).

Exposure can occur due to the presence of aflatoxin and aflatoxin metabolites in milk and milk products from animals that have consumed contaminated feed (Applebaum et al., 1982). Some data on the occurrence of aflatoxin M_1 in the late 1960s, 1970s and 1980s in several countries are reported in Table 5. High proportions of positive samples were found in some surveys, usually at levels of less than 0.5 µg/kg.

Table 5. Occurrence of aflatoxin M_1 in milk

Country or region	Period of sampling	No. positive/ no. samples	Proportion positive (%)	Range of concentrations (µg/kg)[a]
Austria	1983–86	0/65	0	< 0.03
Belgium	1960s, 1970s	42/68	62	0.02–0.2
	1980–86	135/809	16.7	< 0.01–0.5
China	1981, 1983	173/319	54.2	0.02–0.5
Finland	1982, 1986	0/17	0	< 0.1
France	1981–85	580/3634	16.4	0.02–0.5
Germany	1960s, 1970s	229/788	29.1	0.04–6.5[b]
India	1960s, 1970s	3/21	14	up to 13.3
Ireland	1981–82	0/36	0	< 0.015
Italy	1982–84	213/537	39.7	0.001–0.15
Netherlands	1960s, 1970s	74/95	82	0.03–0.5
	1985–86	964/1241	77.7	0.01–0.09
South Africa	1960s, 1970s	5/21	24	0.02–0.2
Spain	1983	7/95	7	0.02–0.04
Sweden	1983–86	384/647	59.4	0.005–0.3
Taiwan	1986	0/217	0	< 0.1
United Kingdom	1960s, 1970s	85/278	31	0.03–0.52
	1981–83	59/686	8.6	0.01–0.78
USA	1960s, 1970s	191/302	63	Trace–> 0.7

Compiled from van Egmond (1989a)
[a]For milk powder, calculated on the basis of reconstituted milk (dilution factor, 10 ×)
[b]Western part, 0.04–0.54; eastern part, < 0.1–6.5

The dietary exposure resulting from intake of aflatoxin-contaminated food has been estimated for a number of populations by integrating the level of aflatoxins in the dietary constituents and estimating consumption (see Table 6). High levels of exposure in the USA were calculated by Stoloff (1983) for the years prior to the surveys included in Table 6. Exposures to aflatoxin B_1 in Africa and Southeast Asia are usually in the range 3–200 ng/kg bw per day (Hall & Wild, 1993).

Table 6. Population exposures estimated on the basis of analyses of aflatoxins in food

Country or region	Period of sampling	Food source[a]	Estimated exposure to aflatoxin B_1 (ng/kg bw/day)	Reference
Kenya	1969	P (diet and beer)	3.5–14.8[b]	Peers & Linsell (1973)
Swaziland	1972–73	H (wet diet)	5.1–43.1[c]	Peers et al. (1976)
	1982–83	H	[11.4–158.6]	Peers et al. (1987)
Mozambique	1969–74	P	38.6–183.7	van Rensburg et al. (1985)
Transkei	1976–77	P	16.5	van Rensburg et al. (1985)
Gambia	1988	P	[4–115][d]	Wild et al. (1992a)
Southern Guangxi, China	1978–84	M	[11.7–2027]	Yeh et al. (1989)
Thailand	1969–70	P	5–55	Shank et al. (1972a)
USA	1960–79	M	2.7	Bruce (1990)

[ ]Calculated by Working Group, assuming 70-kg weight per person
[a]P, samples of cooked food from the plate; H, uncooked food samples from the home; M, samples from the market
[b]B_1 and B_2
[c]Aflatoxin, unspecified
[d]B_1, B_2, G_1, G_2

Aflatoxins in foods are not readily degraded under normal cooking conditions (Goldblatt, 1969; Müller, 1982).

Methods for the decontamination and detoxification of agricultural commodities contaminated with mycotoxins, including aflatoxins, have been reviewed (Jemmali, 1990). Ammoniation of aflatoxin B_1-contaminated animal feeds has been studied extensively as a potential method of detoxification. This process, when applied in combination with heat, leads to conversion of aflatoxin B_1 to aflatoxin D_1 (see Fig. 1) and a further degradation product, both reported to be non-toxic, whereas without heat, this process may be reversible and regenerate aflatoxin B_1 upon acidification. The process has been used commercially on a limited basis; however, there is still some controversy about the safety of the detoxified feedstuffs (Müller, 1983).

1.3.2 Occupation

Occupational exposures can occur through the handling and processing of aflatoxin-contaminated crops. Low-level respiratory exposure to aflatoxin-contaminated dust particles was reported in workers processing imported maize (Silas et al., 1987) and in workers extracting oil from linseeds and groundnuts (van Nieuwenhuize et al., 1973). A study of occupational exposure of Danish workers in animal feed production showed that seven of 45 workers exposed to aflatoxin B_1-contaminated feeds (0–26 μg/kg) had detectable levels of aflatoxin B_1 bound to serum albumin (none detected [< 5 pg/mg] to 100 pg/mg albumin). Dust samples collected at different sites showed none detectable to 8 μg/kg dust (Olsen et al., 1988; Autrup et al., 1991).

1.3.3 Human biological fluids

Aflatoxins have been detected in human urine, milk and blood samples. An initial study reported the presence of aflatoxin M_1 in human urine from the Philippines (Campbell et al., 1970); early observations from a number of other laboratories have been reviewed (Garner et al., 1985).

In Zimbabwe, 4.3% of 1228 urine samples collected throughout the country contained aflatoxins; the most common metabolite detected was aflatoxin M_1 (Nyathi et al., 1987). Aflatoxin metabolites were detected in urine samples from the Gambia, the Philippines, Singapore and France (Wild et al., 1988). Aflatoxin B_1-$N7$-guanine adducts were found in 12.6% of 983 urine samples from Kenya (Autrup et al., 1987).

In a study of 252 urine samples from 32 households in Guangxi Autonomous Region in China, a good correlation was seen between urinary excretion of aflatoxin M_1 over a three-day period and dietary intake of aflatoxin B_1. Between 1.2 and 2.2% of dietary aflatoxin B_1 was found as urinary aflatoxin M_1 (Zhu et al., 1987). A more detailed study (Groopman et al., 1992a) of the pattern of urinary metabolites in the same urine samples confirmed the findings on aflatoxin M_1, revealed the presence of aflatoxin P_1 (see Fig. 1) as a major metabolite and demonstrated the presence of aflatoxin B_1, aflatoxin Q_1 (see Fig. 1; rarely found) and aflatoxin B_1-$N7$-guanine adducts. The total mean percentage of dietary aflatoxin B_1 excreted as the above metabolites was calculated to be 4.4% for women and 7.6% for men. This study also established that the levels of aflatoxin M_1 and of aflatoxin B_1-$N7$-guanine adducts were correlated with intake, while those of aflatoxin P_1 and aflatoxin B_1 were not.

In the Gambia, urinary excretion of total aflatoxin metabolites and of aflatoxin B_1-$N7$-guanine adducts over a four-day period in 20 individuals was correlated with their dietary intake of aflatoxins. In contrast to the study in China, where no aflatoxin G_1 was detected, aflatoxin G_1 was present in the diet and was also found to be the major urinary aflatoxin in this study (Groopman et al., 1992b).

Several studies have demonstrated the presence of aflatoxins in human milk: aflatoxins M_1, M_2, B_1, B_2, G_1 and G_2 have been reported in the Sudan, Zimbabwe and Ghana (Coulter et al., 1984; Wild et al, 1987; Lamplugh et al., 1988). Aflatoxin M_1 was the type most frequently present, at levels up to 64 pg/ml in the Sudan (Coulter et al., 1984), up to 50.5 pg/ml in Zimbabwe (Wild et al., 1987) and 1.8 pg/ml in Ghana (Lamplugh et al., 1988). In one report on five lactating women in the Gambia, 0.09–0.43% of dietary intake was excreted in milk as aflatoxin M_1 (Zarba et al., 1992).

Aflatoxins have been detected in blood samples in Nigeria (Denning et al., 1988), the Sudan (Hendrickse et al., 1982) and Japan (Tsuboi et al., 1984). Sera from the United Kingdom were found to contain less than 20 pg/ml (Wilkinson et al., 1988). Aflatoxins were also detected in 17 of 35 umbilical cord blood samples in southern Thailand (Denning et al., 1990), in nine of 78 samples in Nigeria and in 63 of 188 samples in Ghana (Lamplugh et al., 1988).

The biologically effective dose of aflatoxin for individuals can be determined in assays for covalent binding to albumin in peripheral blood (Gan et al., 1988; Wild et al., 1990a) and for aflatoxin B_1-$N7$-guanine adducts (Groopman et al., 1992b,c). The level of urinary adducts reflects dietary exposure to aflatoxin over the previous 24–48 h (Groopman et al.,

1992b,c), while the level of aflatoxin–albumin adducts is assumed to reflect exposure over the previous two to three months (Wild et al., 1992a). The urinary aflatoxin B_1-$N7$-guanine adduct is by definition a specific indicator of exposure to aflatoxin B_1, whereas the aflatoxin–albumin adduct, determined by immunoassay, is only predominantly a measure of such exposure. An aflatoxin G_1–albumin adduct has also been identified (Sabbioni & Wild, 1991).

In a study of 42 residents of Guangxi, China, the level of aflatoxin–albumin adducts was shown to be correlated with both dietary aflatoxin B_1 intake and urinary aflatoxin M_1 excretion at the individual level (Gan et al., 1988). A later report on the same subjects demonstrated a correlation between the level of urinary aflatoxin B_1-$N7$-guanine adducts and serum aflatoxin–albumin adducts (Groopman et al., 1992a).

The levels of aflatoxin–albumin adducts in subjects from several countries in Africa, Southeast Asia and Europe have been reviewed (Wild et al., 1992b; Table 7). Although the subjects in the different studies were not matched, the data illustrate the quantitative differences in exposure that can occur worldwide. Exposure was significantly higher in the Gambia, Kenya and the Guangxi region of China than in Thailand, while in Europe the levels were below the detection limit. In the Gambia, only seven of 390 individuals (< 2%) had no detectable adduct. The same adduct was detected in 21 of 30 samples of cord blood from the Gambia, and there was a good correlation with the level of adduct in matched maternal venous blood (Wild et al., 1991). Exposure levels did not vary with age or sex in the Gambian population (Allen et al., 1992; Wild et al, 1992a), but seasonal variations in exposure occurred (May compared to November).

Table 7. Aflatoxin–albumin adducts in human sera

Country	No. of subjects	No. of subjects with different adduct levels (pg aflatoxin B_1-lysine eq./mg albumin)					
		< 5	5–25	26–50	51–75	76–100	> 100
Gambia							
May	323	7	53	76	49	40	98
November	67	0	39	13	7	3	5
Senegal	29	0	20	6	2	1	0
Kenya	91	48	26	5	1	5	6
China							
Guangxi	93	28	35	13	6	2	9
Shandong	69	69	0	0	0	0	0
Thailand	84	73	10	1	0	0	0
France	44	44	0	0	0	0	0
Poland	30	30	0	0	0	0	0

From Wild et al. (1992b); limit of detection, 5 pg aflatoxin B_1-lysine eq./mg albumin

Aflatoxins have also been reported in human tissues, either in the free form, e.g., in liver (Wray & Hayes, 1980; Stora et al., 1981; Lamplugh & Hendrickse, 1982; Onyemelukwe et al., 1982; Stora & Dvořáčková, 1986, 1987), or bound to DNA in liver (Hsieh et al., 1988; Zhang *et al.*, 1991). Aflatoxins were also reported in a case of colon adenocarcinoma (Deger, 1976) and in two cases of lung cancer (Dvořáčková *et al.*, 1981).

1.4 Regulations and guidelines

It is probably not possible to eliminate completely exposure of humans to aflatoxins. In 1987, at least 50 countries had existing or proposed regulations for aflatoxins in foodstuffs (van Egmond, 1989b, 1992; Stoloff *et al.*, 1991). The maximum limits range from none detectable to 50 µg/kg of food for either the sum of aflatoxins B_1, B_2, G_1 and G_2 or for aflatoxin B_1 alone; 5 µg/kg is the commonest maximal limit.

In 1987, aflatoxin M_1 levels in dairy products were regulated in 14 countries (van Egmond, 1989b). The tolerances in infants' and children's food were 0.05–0.5 µg/kg milk.

Aflatoxins were reviewed by a joint FAO/WHO Expert Committee on Food Additives in 1987 (WHO, 1987). No acceptable daily intake was given; it was recommended that human intake be reduced to the lowest practicable level.

2. Studies of Cancer in Humans

2.1 Descriptive studies

In a number of studies, estimates of the incidence of primary liver cancer were correlated with the intake of aflatoxins in the same population groups. Apart from methodological limitations of correlation studies, two specific problems related to liver cancer hinder the interpretation of the results of these studies. Firstly, diagnosis and registration of liver cancer are likely to be incomplete, especially in some developing areas of the world, and inaccurate. Secondly, most of the studies have not evaluated the rate of chronic infection with hepatitis B or hepatitis C virus, which are potential confounders for aflatoxins. Studies have been conducted in Swaziland (Keen & Martin, 1971; Peers *et al.*, 1976), Uganda (Alpert *et al.*, 1971), Thailand (Shank *et al.*, 1972a,b,c), Kenya (Peers & Linsell, 1973), Mozambique (van Rensburg *et al.*, 1974), China (Armstrong, 1980; Wang *et al.*, 1983; Sun & Chu, 1984; Wu *et al.*, 1984; Yu *et al.*, 1989), the USA (Stoloff, 1983) and the Transkei, South Africa (van Rensburg *et al.*, 1985, 1990). The substantial differences in the incidence of liver cancer between as well as within these areas generally are strongly correlated with differences in the estimated intake of aflatoxin B_1 within the same population groups. The intra-country correlation was first demonstrated by Linsell and Peers (1977) and was corroborated by data from the Transkei (van Rensburg *et al.*, 1985) and from Swaziland (Peers *et al.*, 1987; Bosch & Muñoz, 1988). In an early study in Kenya, there appeared to be little correlation between liver tumour incidence and an estimate of the prevalence of hepatitis B surface antigen (HBsAg), analysed using a radioimmunoassay method (Bagshawe *et al.*, 1975); in contrast, the estimated consumption of aflatoxin B_1-contaminated food in the same areas was correlated with liver tumour incidence (Peers & Linsell, 1973).

Two correlation studies have been reported in which attempts were made to evaluate differences in liver tumour incidence in areas of Africa simultaneously with data on exposure to aflatoxins and hepatitis B infection. A study in Swaziland was based on estimates of aflatoxin consumption from a defined sampling of foods in 11 designated areas in the country, subsequently analysed for aflatoxin content, and surveys of locally grown crops between harvesting and consumption (Peers *et al.*, 1987); hepatitis B virus markers were assessed in blood donors and primary liver cancer incidence from a cancer registration system that operated from 1979 to 1983. In this study, 120 of 2583 food samples were contaminated with aflatoxin, and groundnuts were the food item most commonly contaminated. Of the 120 contaminated samples, 97% contained aflatoxin B_1, 73% aflatoxin G_1 and 33% contained either B_2 or G_2. Aflatoxin B_1 constituted 19.5–57% of the estimated average daily exposure to total aflatoxin in the 11 areas. The correlation between liver cancer incidence and aflatoxin exposure was reported to be significant for both total aflatoxin and aflatoxin B_1. [The Working Group noted that the relatively high proportion of aflatoxin G_1, G_2 and B_2 in these samples could probably have contributed to the reported correlation with total aflatoxins.] In the analysis, a model incorporating daily consumption of total aflatoxins and the proportion of people positive for HBsAg provided the best fit to the variation in liver cancer incidence. It was not significantly better than a model that included only total aflatoxin consumption; however, it provided a significantly better fit to the liver cancer rates than a model that included only the proportion of people with HBsAg. Thus, at an area or ecological level, evidence of hepatitis B infection did not confound the association of liver cancer incidence with aflatoxin consumption, and although both were associated with liver cancer incidence, aflatoxin exposure appeared to be more important in explaining the variation in liver cancer incidence.

In a second study, conducted in Kenya (Autrup *et al.*, 1987), the incidence of liver cancer by area was determined from the records of one large hospital. Specimens of urine and blood were collected from patients attending the out-patient departments of various hospitals; urinary excretion of aflatoxin B_1–guanine adducts and serological markers of hepatitis B infection were determined. Of the 983 tested individuals, 12.4% excreted aflatoxin B_1–guanine adducts and 10.6% had chronic infection with hepatitis B virus. There were wide variations in the prevalences of these exposures by area: Among Bantus, there was a significant correlation between liver cancer 'incidence' [as estimated from hospital records] and the prevalence of exposure to aflatoxin (r, 0.75) but not with evidence of hepatitis B infection (r, 0.19). There was no evidence of interaction between the two exposures. [The Working Group noted that the associations were 'limited to certain ethnic groups', but no information was provided on the size of those groups in relation to the total number of subjects in the study.]

Stoloff (1983), in evaluating the role of aflatoxin ingestion in the prevalence of primary liver cancer in the USA, also did not perform a formal correlation analysis. Data from many published sources were used to estimate levels of exposure in the USA, which was found to be much higher in southeastern than in northern or western USA. [The Working Group noted that these estimates included extrapolations to time periods before the early 1960s that preceded structural identification of aflatoxins.]

In China, a number of studies have been performed to correlate liver cancer incidence or mortality with estimates of aflatoxin exposure by area, and several have attempted to take hepatitis B infection into account. Armstrong (1980) reported on data from Chinese investigators that had been presented at an international workshop. In an analysis which included a number of different factors suspected of being associated with liver cancer, including the prevalence of hepatitis B infection, only the product of proportion of mould-producing days in the year and proportion of maize in food was correlated significantly with liver cancer mortality (r, 0.71).

In a factor analysis that included data on climate, grain-oil contamination by aflatoxins and HBsAg positivity rates in relation to liver cancer mortality in China, Wang et al. (1983) found no association with HBsAg positivity rates (although they noted the absence of data from some areas) but did find an association with aflatoxin exposure, which accounted for 60.7% of the variation in liver cancer mortality in males and 44.0% in females.

Sun and Wang (1983) and Sun and Chu (1984) reviewed previous studies that showed differences in liver cancer rates which do not seem to be explained by differences in hepatitis B carrier status. The additional factor appeared to be differences in aflatoxin exposure.

Yeh et al. (1989) performed a geographical analysis within a cohort study of 7917 men aged 25–64 in Guangxi, China. The results showed a 3.5-fold difference in liver cancer mortality according to place of residence, which was linearly associated with dietary aflatoxin contamination (r, 1.00; p = 0.004), estimated for each commune after twice yearly sampling of foods. The four communes that contributed to the correlation with aflatoxins had a range of mean aflatoxin B_1 intakes of 0.3–51.8 mg/person per year. In the same four communes, the range in age-adjusted prevalence of HBsAg was 21.6–24.8%, and there was little relationship between prevalence of HBsAg positivity by area and liver cancer mortality (r, 0.28; p = 0.65). At the individual level, however, the cohort study showed a strong association between HBsAg positivity at the time of enrolment and liver cancer mortality.

In a large cross-sectional survey performed in 48 sites in China with an estimated 600-fold variation in aflatoxin exposure, a 28-fold range in HBsAg prevalence and a 39-fold range in mortality from primary liver cancer, no association with aflatoxin exposure but a strong correlation with HBsAg prevalence was noted (Campbell et al., 1990). Liver cancer mortality was determined in 1973–75, and biochemical data were obtained in 1983. Aflatoxin exposure was estimated from measures of aflatoxin metabolites in age-specific pooled urine samples from each region. [The Working Group noted that the method provided a mean intake but no estimate of standard deviation. Furthermore, the time difference between estimates of aflatoxin exposure and liver cancer incidence and mortality was substantial.]

Another correlation study was performed in Thailand, which was based on more sensitive markers of aflatoxins in serum and urine than those used in most previous studies (Srivatanakul et al., 1991a). Hepatocarcinoma incidence in five areas correlated poorly with these markers and with markers of hepatitis B infection.

2.2 Cohort studies

Hayes et al. (1984) reported on mortality between 1963 and 1980 in a cohort of 71 Dutch workers involved in extracting oil from linseeds and groundnuts, who had been exposed by inhalation to aflatoxin-containing dust during the period 1961–69 and who had been employed in the study plant for at least two years. A comparison group of 67 plant workers not exposed to aflatoxin-contaminated dust was also studied, and for each group standardized mortality ratios (SMR) with the Dutch male population as standard were computed. In the aflatoxin-contaminated group, but not in the comparison group, the observed numbers of all cancer deaths were above expectation (16 in the exposed and 7 in the controls; SMR (95% confidence intervals [CI]), 2.50 (1.40–4.00) and 1.13 (0.40–2.30), respectively). In particular, there were seven cases of respiratory cancer among aflatoxin-exposed workers (SMR, 2.53; 95% CI, 1.00–5.00). There was no death in either group from primary liver cancer. This study was an extension of a follow-up, with inclusion of younger workers from the same plant, documented in a previous report (van Nieuwenhuize et al., 1973), which gave levels of aflatoxins in residues from groundnut cakes of 75–1320 µg/kg and reported one case of primary liver cancer in the aflatoxin-exposed workers. Hayes et al. (1984) reported, however, that re-examination of the pathological material had shown the case to have been a metastasis in the liver from a tumour of undefined endocrine origin.

Yeh et al. (1985) summarized a study in China in which HBsAg-positive and -negative individuals were stratified into those who lived in villages with 'light' or 'heavy' aflatoxin contamination of foods and were followed up to 1982 (total years of follow-up, five to eight). Maize was the major source of aflatoxin contamination. The average annual intake of aflatoxins in the heavily contaminated villages was 6.016 mg per person, and that in the lightly contaminated villages, 0.638 mg per person. The mortality rate for liver cancer was [372 per 100 000] in the heavily contaminated villages, based on 15 deaths, and [32.8 per 100 000] in the lightly contaminated villages, based on one death [$p < 0.05$]. There was a multiplicative effect between contamination with aflatoxins and HBsAg status, but misclassification of HBsAg status could have affected the findings.

Olsen et al. (1988) reported on cancer risk among employees of 241 livestock feed processing companies, using a system based on record linkage between the Danish Supplementary Pension Fund and the Danish Cancer Registry, with occupational data dating back to 1964. Exposure to aflatoxins from imported crops intended for animal consumption had been documented by government surveys; average concentrations in imported groundnut oilcakes were around 1000 µg/kg in 1976–78. When mixed with other products, the concentrations of aflatoxins in prepared cattle feed were ~ 140 µg/kg. Because dust levels in the industry were high, it was estimated that workers may have inhaled 170 ng aflatoxin B_1 per working day. Aflatoxin G_1 was not measured. As population denominators were not available, the analysis was based on standardized proportional incidence ratios (SPIRs), all employees with cancer in the data base serving as the referent. All males with cancer diagnosed in 1970–84, whose longest work experience had been at one of the 241 feed processing companies in the period 1 April 1964 to the date of diagnosis, were included in the analysis. There was no excess of respiratory cancer. For liver cancer, the SPIR was 1.41 (95% CI, 0.57–2.93; six observed cases); for gall-bladder and extrahepatic bile-duct cancer, the SPIR was 2.19 (0.89–4.55; six observed cases, including one incorrectly classified as an intrahepatic cholangiocarcinoma). In an analysis restricted to the period 10 or more years before diagnosis, workers who had had major exposure to aflatoxin-contaminated feed had an SPIR for liver cancer of 2.46 (1.08–4.86; seven observed cases), while that for gall-bladder and extrahepatic bile-duct cancers was 2.98 (1.09–6.59; five cases). [The Working Group noted that although possible confounding factors were not evaluated in this study, they were not likely to have affected the results.]

Ross et al. (1992) conducted a cohort study of 18 244 male residents of Shanghai, China, 96% of whom were aged 45-64 years on entry to the study. The men were recruited by invitation from four geographically defined areas and responded to questionnaires administered by nurses, usually in their homes, on life style (including smoking and alcohol consumption) and on food frequency. Blood and urine specimens were collected. The men were followed up by identification of death records in the same or neighbouring areas and through linkage with the Shanghai Cancer Registry (estimated to be 85% complete). An attempt was also made to contact each cohort member annually. The cohort was established between January 1986 and September 1989 and was followed to 1 March 1990 for the current analysis, resulting in 35 299 person-years of follow-up. Only 24 members were lost to follow-up. Of 176 cancer cases identified, 22 were primary liver cancers, although only five were confirmed by biopsy. The first six cases were each matched for age, sample collection date and area of residence to 10 controls randomly selected from cohort members who had not been diagnosed with liver cancer at the time of diagnosis of the case. The remaining 16 cases were similarly matched to five controls each. For each case and control, the urine samples were analysed for aflatoxin B_1 and its principal metabolites, aflatoxins P_1 and M_1, as well as for aflatoxin B_1–$N7$-guanine adducts. HBsAg was measured by radioimmunoassay. The study was analysed as a nested case–control study within the cohort. Each of the four markers of exposure was more frequently present among cases than controls. For 13 of the 22 liver cancer cases and 53 of 140 controls, results were positive in at least one of the four assays (relative risk (RR), 2.4; 95% CI, 1.0–5.9). The RR for HBsAg positivity was 7.8 (95% CI, 3.0–20.6). When this and other risk factors for liver cancer (including smoking and alcohol use) were included in a multivariate logistic regression analysis, the RR for detectable urinary aflatoxins or aflatoxin–guanine adducts became 3.8 (95% CI, 1.2–12.2). Among individuals who did not have HBsAg, the RR for urinary aflatoxins was 1.9 (0.5–7.5); however, among those who had HBsAg, the RR was 60.1 (6.4–561.8), although the latter estimate was based on only seven cases and two controls who had both factors. Urinary aflatoxin metabolites and guanine adducts reflect recent dietary exposure. It was considered unlikely, however, that individuals with early symptoms could have changed their diet and somehow increased consumption of aflatoxin-contaminated food. Given the short follow-up period (less than two years as a mean), some of the observed tumours might already have been present, in a clinically silent phase, at the time of urine collection. Moreover, chronic liver disease, which frequently underlies hepatocellular carcinoma, might affect metabolism of aflatoxin. [The Working Group noted that no analysis was presented by interval between urine collection and liver cancer diagnosis and that the proportion of liver cancer cases known to be accompanied by cirrhosis was not given. It was considered, however, that to explain the observed case–control differences in terms of some metabolic effect of preclinical liver conditions would imply a substantial enhancement of all the different enzymatic metabolic pathways leading to, respectively, aflatoxin P_1, aflatoxin M_1 and aflatoxin B_1–$N7$-guanine adducts, and that the data do not support such effects. (See also Hall & Wild, 1992.)]

In an extension of the analysis (Qian et al., 1993), the risk for hepatocellular carcinoma was higher among people who excreted aflatoxin B_1–guanine adducts in their urine: the RR (adjusted for hepatitis B, smoking and alcohol use) was 7.8 (95% CI, 1.5–40.0). No association was found with dietary aflatoxin levels, as determined by an in-person interview and a survey of market foods in the survey region. There was no association between dietary and urinary aflatoxin levels among the subjects for whom both levels were determined.

2.3 Case–control studies

Bulatao-Jayme et al. (1982) reported a case–control study based on 90 hospital cases (66% non-paying) of confirmed primary liver cancer from three hospitals in the Philippines. The controls were non-paying patients from surgical and medical wards who showed normal liver function after a battery of biochemical tests and who were individually matched by age and sex to the cases [response rates for cases and controls not reported]. Recent aflatoxin exposure was estimated from 24-h urine samples collected on or before the first day of admission and analysed by thin-layer chromatography for aflatoxin B_1 and aflatoxin M_1. Of the samples from 76 cases, 51% were positive for aflatoxin B_1 and/or M_1; this result was significantly different from the 35% in 89 controls. Aflatoxin contamination of the diet was estimated by measuring the aflatoxin B_1 and M_1 content of food samples, mostly from the greater Manila area but some (maize) from 'practically' all regions of the country. These measurements (usually covering at least 20 samples cooked in various ways) were applied to a usual dietary food intake history collected from each subject by a professional nutritionist, who was given only the patient's name and bed number before the interview. A separate dietary record was taken for each period of life in which there had been a different residence for one year or more; the overall mean aflatoxin load per day from these separate periods was used in the analysis. Aflatoxin intake was generally categorized as heavy or light, heavy being defined as a total estimated aflatoxin load of 4 µg or more per day, although for some analyses the heavy category was divided into moderately heavy (4–6 µg) and very heavy (≥ 7 µg). For all aflatoxin-containing food items except rice, the aflatoxin load was significantly higher in the cases than in the controls, with correspondingly elevated odds ratios (ORs), including 4.2 for maize, 7.7 for groundnuts, 11.0 for sweet potatoes and 22.5 for cassava. When aflatoxin intake was categorized into three levels, the OR for very heavy exposure was 17.0 and that for moderately heavy, 13.9 (both significant). The OR for heavy aflatoxin and heavy alcohol consumption relative to light aflatoxin and light alcohol consumption was 35.0, that for heavy aflatoxin and light alcohol was 17.5, and that for light aflatoxin and heavy alcohol was 3.9. A similar finding was reported when the analysis was restricted to 35 pairs of subjects who consumed only aflatoxin-free alcoholic beverages. [The Working Group noted that no data were available on hepatitis B infection; however, there would have to be an unusually close correlation between estimates of aflatoxin consumption and hepatitis B infection status for the ORs reported to be fully explained by confounding. The dietary data on which the estimates of aflatoxin intake were based presumably involve substantial misclassification, and generally it would be expected that this would bias the ORs towards the null.]

In a hospital-based case–control study in Hong Kong of primary liver cancer, involving 107 pairs of Chinese subjects, Lam et al. (1982) evaluated a number of risk factors, including hepatitis B virus infection and ingestion of aflatoxin-containing foods. The cases interviewed comprised 72% of those initially ascertained. The controls were selected from among patients admitted with trauma to the orthopaedic ward of the same hospital. Frequency of consumption of a number of dietary food items was ascertained currently and 20 years previously; there was little difference in these estimates, and current frequency of consumption was used in the analysis. Aflatoxin contamination of food was based on a market survey that had been performed 10 years before the study. Maize was the most frequently contaminated food. The analysis of aflatoxin intake was based on a comparison of frequency of consumption of different foods, with no significant difference between cases and controls in these consumption frequencies. The results were similar when adjusted for HBsAg status. [The Working Group noted that no quantitative estimate of aflatoxin consumption at the individual level was made and that, given the low reported frequency of consumption of presumed aflatoxin-contaminated foods, the power of the study to detect an association was low.]

Two hospital-based case–control studies of liver cancer were carried out in patients aged less than 75 when recruited during 1987–88 from three hospitals in north-east Thailand — one on cholangiocarcinoma (Parkin et al., 1991) and one on hepatocellular carcinoma (Srivatanakul et al., 1991b). The studies included assessment of both hepatitis B virus infection status and consumption of foods possibly contaminated with aflatoxins as well as measurement in blood of aflatoxin–albumin adducts by the method of Wild et al. (1990a), which is believed to provide an indication of integrated aflatoxin exposure over recent weeks or months. Controls were selected from among individuals who were visiting clinics or who had been admitted to the same hospital as the cases and were matched to the cases on age and sex. Dietary data were obtained from responses on a food-frequency questionnaire directed to the period one year prior to interview. Several food groups were characterized, one of which was foods liable to aflatoxin contamination on the basis of previous studies in northern Thailand. Frequency of consumption was dichotomized on the basis of consumption frequency among controls. In the study of cholangiocarcinoma (Parkin et al., 1991), 113 patients were interviewed, but 10 cases that did not fulfil the diagnostic criteria and their matched controls were excluded. In the analysis of aflatoxin–albumin adducts, 21 pairs were considered, with only one discordant pair in each category and an OR of 1.0 (95% CI, 0.1–16.0). The OR for consumption of aflatoxin-contaminated foods was 1.4 (0.8–2.7). In the study of hepatocellular carcinoma, 73 cases were interviewed, and 65 pairs that fulfilled the diagnostic criteria were included in the analysis (Srivatanakul et al., 1991b). Forty-five pairs were tested for albumin-bound aflatoxins, with 16 discordant pairs — eight in each category — for an OR of 1.0 (0.4–2.7). The prevalence of detectable aflatoxins in the 99 specimens examined was 22.2%. Of the food groups considered, those potentially associated with aflatoxin contamination gave the highest OR (1.9), but it was not significant [confidence interval not given].

Sera from 100 patients with lung cancer in Italy were tested for the presence of aflatoxins, as were sera from 150 healthy donors (Cusumano, 1991). Positive results were obtained in seven and one, respectively. Only one of the seven cases was a smoker. Of the seven, five were positive for aflatoxin B_1 and two for aflatoxin B_2; the one in the control group was positive for aflatoxin G_1. [The Working Group noted that the controls were not strictly comparable to the cases.]

3. Studies of Cancer in Experimental Animals

3.1 Mixtures of aflatoxins

3.1.1 *Oral administration*

Rat: Since the first report of the induction of hepatocellular tumours in rats by groundnut meal (Lancaster *et al.*, 1961), many studies in rats have demonstrated the carcinogenic potency of aflatoxins (IARC, 1976, 1987a).

Groups of rats [strain, sex, age and initial number unspecified] were fed diets containing 5, 200, 1000, 3500 or 5000 µg/kg of diet aflatoxins for 294–384 days. A linear dose–response relationship was observed in the incidence of hepatomas: 0/10, 3/20, 13/24, 18/25 and 14/15, respectively (Newberne, 1965).

A group of six young rats [strain and sex unspecified] was fed a diet containing groundnut meal providing aflatoxins at a concentration of 3000–4000 µg/kg of diet for three weeks and then returned to a normal diet. Another group of six male rats [strain unspecified], one year old, was fed the same diet for more than 39 weeks. Of the six rats treated with aflatoxins for three weeks early in life, one animal, which lived for 106 weeks after return to the normal diet, developed a carcinosarcoma of the stomach and another animal a hepatocellular carcinoma. Of the six rats fed the diet containing aflatoxins after one year of age, five survived more than 39 weeks; anaplastic hepatocellular carcinomas were found in 3/5 animals and adenocarcinomas of the stomach and rectum in 1/5 animals each (Butler & Barnes, 1966).

Groups of six male rats [strain and age unspecified] were administered 10 or 100 µg of a mixture of aflatoxins (containing 37.7% aflatoxin B_1, 56.4% aflatoxin G_1 and traces of aflatoxins B_2 and G_2) in 100 ml drinking-water for 64 weeks. At the high dose of aflatoxins (approximately 300 µg/week), 6/6 rats treated for more than 39 weeks developed hepatocellular tumours. At the low dose (approximately 35 µg/week), 1/5 rats developed a hepatocellular tumour, which appeared at 66 weeks (Dickens *et al.*, 1966).

Groups of 16–19 female Wistar rats, weighing 50–96 g, were administered 0, or a single dose of 0.5 mg/animal aflatoxin B_1 (approximately 7650 µg/kg bw) in 0.1 ml dimethylformamide, or a mixture of aflatoxins (40% aflatoxin B_1, 60% aflatoxin G_1; approximately 5.1 mg/kg bw in terms of aflatoxin B_1) in 0.1 ml dimethylformamide by stomach tube. Survival after 10 months was 18/19 vehicle controls, 15/16 rats treated with aflatoxin B_1 and 16/18 rats receiving the mixture. The remaining animals died or were killed between 15 and 32 months after aflatoxin administration. Increased incidences of hepatic tumours (mainly hepatocellular carcinomas) were found in 7/15 rats treated with aflatoxin B_1, in 7/16 rats that received the mixture and in 0/18 vehicle controls. The first hepatic tumours were found after 21 months, and the mean time to tumour induction was approximately 26 months. In addition, nodular hepatic hyperplasia was observed in 3/15 rats treated with aflatoxin B_1, in 5/16 rats that received the mixture and in 0/18 vehicle controls. In some animals treated with aflatoxin B_1 and G_1, cystic bile-duct hyperplasia was seen (Carnaghan, 1967).

Groups of 6–36 male and female Porton rats, 40–90 days of age, were fed diets (prepared from a basic diet containing 10 000 µg/kg aflatoxin B_1 and 200 µg/kg aflatoxin B_2) containing aflatoxin at approximately 0, 100, 500 and 5000 μg/kg of diet for one to nine weeks or for lifetime. Animals were necropsied when they became moribund or died. In male rats fed 5000 μg/kg of diet aflatoxin B_1 for one to nine weeks, the incidence of hepatocarcinomas (based on the number of animals alive when the first tumour appeared) was related to the time of treatment (control, 0/46; one week, 0/13; three weeks, 3/20; six weeks, 12/19; nine weeks, 6/6). In male rats fed 0, 100 or 500 μg/kg of diet aflatoxin for lifetime, the incidences of hepatocarcinomas were 0/46 controls, 17/34 low-dose and 25/25 high-dose animals. The incidences of hepatocarcinomas in female rats were: control, 0/34; low-dose, 5/30; high-dose, 26/33. A small number of neoplasms at other sites, including kidney (adenomas and carcinomas of the renal pelvis in 5/53 rats), stomach (carcinomas of the glandular stomach in 2/53 rats after 86 weeks following feeding for 1–9 weeks), lung and salivary glands, was found in some rats given aflatoxin but not in the control rats (Butler & Barnes, 1968). In a similar experiment in male Fischer rats, histochemical findings indicated that all of the carcinomas had a hepatocellular origin (Butler & Hempsall, 1981). Various types of hepatic nodules were found which were composed of vacuolated (lipid-rich), basophilic (ribosome-rich) or eosinophilic cells (containing abundant smooth endoplasmic reticulum, frequently arranged in whorls) (Pritchard & Butler, 1988). In male Fischer rats treated with the same regimen, foci of altered hepatocytes (as defined by increased basophilia, starvation-resistant glycogen and a decrease in the activity of various enzymes) were seen in the liver parenchyma, for the first time at three weeks (Butler *et al.*, 1981).

Groups of 60 male and 30 female Sprague Dawley rats, each weighing about 40 g, were fed a diet containing aflatoxins (aflatoxin B_1 was present at a level of 5–7.5 μg/kg of diet) derived from contaminated groundnut oil or a diet prepared with aflatoxin-free corn oil for 22 months and sacrificed thereafter. Survival at that time was 90/90 controls and 76/90 animals treated with aflatoxins. Sarcomas were found in 3/76 treated animals (one sarcoma each in the liver, colon and subcutaneous tissue). In addition, 18/76 rats had parenchymal liver damage (swelling of cytoplasm, fatty infiltration) and 1/76 had a premalignant liver lesion. Five benign mammary tumours were reported to have occurred in 90 controls (Fong & Chan, 1981).

Duck: A group of 37 Khaki Campbell ducklings, seven days of age, were fed a diet containing aflatoxins derived from contaminated groundnut meal at a level of 35 μg/kg diet aflatoxin B_1. Sixteen control ducklings were fed the same diet without added groundnut meal. The survivors of both groups were killed 14 months after commencement of the experiment. Liver tumours classified as hepatomas were found in 6/11 and liver tumours classified as cholangiomas in 2/11 treated animals. No liver tumour occurred in 10 control animals (Carnaghan, 1965).

Trout: Groups of 300 hatched rainbow trout were fed various diets containing aflatoxins (values were calculated as aflatoxin B_1) at levels of 3.7–42 μg/kg of diet or a purified test diet to which aflatoxin B_1 had been added. Subsamples of fish were selected randomly at multiple time periods for evaluation. The diet containing the highest dose level of aflatoxins (42 μg/kg of diet) produced a high incidence of hepatomas (six months, 10/20; nine months, 20/20; 12 months, 20/20), and the level of 3.7 μg/kg of diet produced hepatomas in 5/10 fish by 12 months. A lower incidence of hepatomas developed in trout that received the purified test diet to which aflatoxin B_1 had been added: after 12 months, the incidence at 7.9 μg/kg of diet was 5/12, that at 4.0 μg/kg, 2/13, and that at 0.8 μg/kg, 0/12; after 15 months, the incidence was 24/30 in those given the high dose; after 20 months, the incidences were 3/21 in those given the medium dose and 3/31 in those receiving the low dose. No liver tumour was observed in a group that received the control diet (Sinnhuber *et al.*, 1968a).

Groups of 20 rainbow trout fry, 60 days old, were fed a commercial diet containing 42 μg/kg aflatoxins for two, four, six and eight weeks and then switched to control diet until the end of the experiment at nine months. There was a high incidence of hepatomas even after the shortest treatment period (two weeks, 12/20; four weeks, 15/20; six weeks, 14/20; eight weeks, 15/20) (Sinnhuber *et al.*, 1968b).

3.1.2 *Oral administration of ammoniated samples* (see also pp. 253 and 307)

Rat: Groups of 12 male and 12 female Fischer 344 rats, three weeks old, were fed diets derived from maize with no detectable aflatoxin and from maize naturally contaminated with aflatoxins, in both cases with and without ammoniation. The diets contained 0 and 176 μg/kg of a mixture of aflatoxins (150 μg B_1, 18 μg B_2 and 8 μg G_1) or the corresponding ammoniated aflatoxin by-products. The experiment was terminated at 92 weeks. The first liver tumours were found in rats treated with aflatoxin-contaminated diet after 80 weeks, when 6/12 males and 2/12 females had died. The overall incidence of hepatocellular carcinomas in rats receiving the aflatoxin-contaminated diet was 12/12 males and 11/12 females; the incidence of neoplastic hepatic nodules was 6/12 males and 10/12 females. In addition, preneoplastic foci of altered hepatocytes (clear-cell foci) were observed in this group. No neoplastic or preneoplastic liver lesion was found in rats fed diets free of aflatoxins or diets containing aflatoxins after ammoniation (Norred & Morrissey, 1983).

Groups of 20 male and 20 female Fischer 344 rats, three to five weeks of age, were fed diets naturally contaminated with aflatoxins (derived from groundnut meal) with and without ammoniation. The unammoniated diets contained 0 or 590.7 (375.0 μg B_1, 36.2 μg B_2, 156.5 μg G_1 and 23.0 μg G_2) and the ammoniated diets 0 or 37.6 (26.4 μg B_1, 6.5 μg B_2, 3.5 μg G_1 and 1.2 μg G_2) μg/kg aflatoxins. The diets were fed for up to 90 days, when the experiment was terminated. Preneoplastic foci of altered hepatocytes (as defined by γ-glutamyltransferase-positive [GGT^+] staining) were found in rats receiving the unammoniated aflatoxin-contaminated diet but not in rats receiving the ammoniated diet or an uncontaminated control diet (Manson & Neal, 1987).

Groups of 15–31 male and 10–20 female Fischer 344 or Wistar WAG rats, four to five weeks old, were fed diets containing mixtures of aflatoxins B_1 and G_1: 60, 140 or 1000 μg/kg of diet aflatoxin B_1 and 10, 20 or 170 μg/kg of diet aflatoxin G_1; the two lower dose levels of aflatoxin B_1 and G_1 were obtained by treating the diet containing the highest dose with ammonia gas at pressures of 3 and 2 bar (300 and 200 kPa), respectively, for 15 min at 95 °C. Control groups of 20 male and 10 female rats were fed a diet which was found to be contaminated with 50 μg/kg aflatoxin B_1. The experiment was terminated after 18 months. In males, 40/40 controls, 46/46 low-dose, 40/40 medium-dose, and 36/40 high-dose animals survived to 18 months; in females, survival was 20/20 controls, 39/40 low-dose, 20/20 medium-dose and 22/30 high-dose animals. The incidence of hepatocellular carcinomas (which were associated with cholangiocellular carcinomas in some animals) showed a dose-dependent increase in animals of each sex: males, 0/40 controls, 0/46 low-dose, 2/40 medium-dose and 35/36 high-dose; females, 0/20 controls, 0/40 low-dose, 1/20 medium-dose and 14/22 high-dose (Frayssinet & Lafarge-Frayssinet, 1990).

3.1.3 *Intratracheal administration*

Rat: Groups of six male rats [strain and age unspecified] were administered a mixture of aflatoxins (37.7% B_1, 56.4% G_1 and traces of B_2 and G_2) at doses of 0 or 300 µg suspended in 30 µl arachis oil intratracheally by intubation twice weekly for 30 weeks. The rats were then held without further treatment for up to 100 weeks. Three of the six treated animals developed squamous-cell carcinomas of the trachea, four of six developed hepatomas and one had a renal-cell adenoma and a carcinoma of the pylorus. No tumour was found in the vehicle controls (Dickens *et al.*, 1966).

3.1.4 *Subcutaneous administration*

Mouse: A group of 20 Tuck No. 1 strain mice [sex unspecified], weighing 20–25 g, were injected subcutaneously with 10 µg of a mixture of aflatoxins (37.7% B_1, 56.4% G_1 and traces of B_2 and G_2) suspended in arachis oil twice weekly for up to 65 weeks. Surviving animals which did not develop tumours were killed 106 weeks after the first injection. Fifteen of 17 animals which were alive when the first tumour appeared developed subcutaneous sarcomas at 23 and 76 weeks (Dickens & Jones, 1965).

Rat: Groups of six male rats [strain unspecified], weighing about 100 g, were injected subcutaneously with 0, 2, 10, 50 or 500 µg of a mixture of aflatoxins (37.7% B_1, 56.4% G_1 and traces of B_2 and G_2) suspended in arachis oil twice weekly for up to 65, 65, 60 and eight weeks, respectively. After injections at the highest dose level had ceased, the animals were kept under observation for a total of 30 weeks. A high incidence of subcutaneous sarcomas developed in all treated groups: 0 µg, 0/6; 2 µg, 5/6; 10 µg, 6/6; 50 µg, 6/6; 500 µg, 5/5. The time at which tumours first appeared was dose-dependent: 2 µg, 44 weeks; 10 µg, 24 weeks; 50 µg, 21 weeks; 500 µg, 20 weeks (Dickens & Jones, 1963, 1965).

3.1.5 *Intramuscular and oral administration*

Monkey: A male rhesus monkey was injected intramuscularly on five days a week with mixed aflatoxins (44% B_1, 44% G_1, 2% B_2 and G_2) at 50 µg for one month followed by 100 µg for 11 months, and then received them orally by gavage at 200 µg per day for 4.5 years. A hepatocellular carcinoma was found 2.5 years after the end of treatment (Gopalan *et al.*, 1972). A female rhesus monkey treated identically, except that the oral dose was 100 µg per day, developed a metastasizing intrahepatic bile-duct carcinoma 5.25 years after the end of treatment (Tilak, 1974).

3.1.6 *Pre- and postnatal exposure*

Rat: Six groups of 10 female Wistar rats [age unspecified] were fed a diet containing 25 or 50% groundnut meal containing 10 000 µg/kg aflatoxin B_1 and 200 µg/kg aflatoxin B_2 from day 10 of pregnancy to parturition, from day 1 to day 10 post-partum, or from day 10 of pregnancy to day 10 post-partum. Among 113 male and 95 female offspring, observed for up to 36 months, one male exposed *in utero* from day 10 of pregnancy and one female exposed *via* the milk for 10 days post-partum developed cholangiocarcinomas, and two females exposed *in utero* from day 10 of pregnancy and *via* the milk for 10 days post-partum developed liver-cell carcinomas. No liver tumour was reported in 50 male or 50 female controls obtained from mothers fed 25–50% soya bean meal in the diet during pregnancy (Grice *et al.*, 1973).

3.2 Aflatoxin B_1

3.2.1 *Oral administration*

Mouse: Wogan (1969a) reported in a review a series of experiments in which aflatoxin B_1 was fed at a level of 1000 μg/kg of diet to three strains of mice (Swiss, C3HfB/HEN, C57Bl/6NB) [number and sex unspecified] for 70 weeks. No tumour was reported.

Rat: Groups of 25 male and 25 female Fischer rats, weighing 80 and 70 g, respectively, were fed a semi-synthetic diet containing 0, 15, 300 or 1000 μg/kg of diet aflatoxin B_1 (purity, > 99.5%) for 52 weeks or until tumours developed. Additional groups of 25 male and 25 female rats were fed 1000 μg/kg of diet aflatoxin B_1 for 14 days and then returned to control diet for the remainder of the experiment. The incidence of hepatocellular carcinomas was increased at all dose levels of aflatoxin B_1 in rats of each sex; in addition, time to tumour decreased with increasing dose: males, 0/25 controls, 12/12 low-dose after 68 weeks, 6/20 medium-dose between 35 and 52 weeks, 18/22 high-dose between 35 and 41 weeks; females, 0/25 controls, 13/13 low-dose after 80 weeks, 11/11 medium-dose between 60 and 70 weeks and 4/4 high-dose after 64 weeks. In addition to hepatocellular carcinomas, hepatocellular adenomas and/or preneoplastic liver lesions (transitional-cell foci) were frequently found. One male animal in the low-dose group had a colonic mucinous adenocarcinoma at 68 weeks. Administration of aflatoxin B_1 at 1000 μg/kg of diet for only 14 days resulted in very low incidences of hepatocellular carcinomas in animals of each sex at 82 weeks: males, 1/16; females, 1/13 (Wogan & Newberne, 1967).

Groups of 30 male and 30 female Fischer rats, weighing 80 and 70 g, respectively, were administered aflatoxin B_1 in dimethyl sulfoxide (DMSO) at doses of 0 or 80 μg/day orally by gavage for five days (total dose, 400 μg/rat). All treated male animals died within 14 days after the last dose; mortality at 35 weeks in treated females was 11/30. Among the 16 animals surviving to 82 weeks, two had hepatocellular adenomas and three had preneoplastic liver lesions (transitional-cell foci). When the total dose of aflatoxin B_1 (400 μg per rat) was given in 10 consecutive doses of 40 μg per day, no mortality attributable to acute toxicity occurred. Following this treatment, an increase in the incidence of hepatocellular carcinomas was observed in males (4/20) but not in females (0/20) that survived for 35 or 82 weeks; in addition, hepatocellular adenomas were found in 1/19 males and 6/17 females by 82 weeks. Preneoplastic lesions (transitional-cell foci) were observed after this treatment in animals of each sex. In addition, groups of 20 and 22 male rats received a single dose of 5 mg/kg bw aflatoxin B_1 (the LD_{50}). Of five animals that survived to 69 weeks, one developed a hepatocellular adenoma and three had preneoplastic lesions (transitional-cell foci). No tumour or preneoplastic liver lesion was found in 60 vehicle controls (Wogan & Newberne, 1967).

Groups of 10–15 male and 15 female MRC rats, eight to nine weeks old, were administered 0 or 20 μg/animal aflatoxin B_1 in the drinking-water (using dark bottles to avoid photolysis) on five nights a week for 10 weeks (total dose of aflatoxin B_1, 1000 μg/animal) or 20 weeks (total dose, 2 mg/animal) and were followed until they were moribund or dead. Survival at 90 weeks was 26/30 (both sexes combined) controls, 4/10 animals (males only) treated with aflatoxin B_1 for 10 weeks and 12/30 animals (both sexes combined) treated for 20 weeks. Aflatoxin B_1 produced liver neoplasms (predominantly hepatocellular tumours, two cholangiocarcinomas) in 8/15 males and 11/15 females treated for 20 weeks and in 3/10 males treated for 10 weeks. In addition, hyperplastic hepatic nodules and cystadenomas were frequently found in the livers of treated animals. A variety of neoplasms, including two renal-cell tumours, were observed in other organs of treated rats but not in untreated controls (Butler et al., 1969).

Groups of 10–20 male Fischer rats [age unspecified] were administered 0, 25, 37.5 or 70 μg/animal aflatoxin B_1 (purified by chromatography and recrystallization) in 0.05 ml DMSO by gastric intubation four or five times a week for two to eight weeks (total doses of aflatoxin B_1, 0, 500, 630, 1000 and 1500 μg/animal) and were sacrificed between 25 and 78 weeks. A high incidence of hepatocellular carcinomas was found at all dose levels: vehicle control, 0/10 at 59 weeks; total dose of 500 μg, 7/7 at 74 weeks; 630 μg, 2/4 at 75 weeks; 1000 μg, 18/18 between 42 and 58 weeks; 1500 μg, 17/17 between 42 and 46 weeks. In addition, preneoplastic liver lesions (foci of hyperplasia and transitional cells) were often observed (Wogan et al., 1971).

Groups of male Fischer rats [initial number unspecified], weighing approximately 80 g, were fed a semi-synthetic diet containing 0, 1, 5, 15, 50 or 100 μg/kg of diet aflatoxin B_1 (purity, > 99.5%) until clinical deterioration of animals was observed, at which time all survivors in that treatment group were killed. A dose- and time-related increase in the incidence of hepatocellular carcinomas and of preneoplastic liver lesions (foci of hyperplasia and transitional cells) was observed (Table 8) (Wogan et al., 1974).

Table 8. Incidences of hyperplastic foci and hepatocellular carcinomas in rats fed aflatoxin B_1

Dietary aflatoxin B_1 (μg/kg of diet)	Time of earliest tumour (weeks)	Numbers of rats with	
		Hyperplastic foci	Hepatocellular carcinomas
0	–	1/18	0/18
1	104	7/22	2/22
5	93	5/22	1/22
15	96	13/21	4/21
50	82	15/25	20/25
100	54	12/28	28/28

From Wogan et al. (1974)

Groups of 16–26 male Wistar rats, weighing 150–200 g, were fed a diet containing aflatoxin B_1 (highly purified) at 0, 250 500 or 1000 μg/kg for 147 days and thereafter maintained on basal diet until death. Survival beyond 100 days was 24/26 control, 13/16 low-dose, 18/18 medium-dose and 14/17 high-dose animals. An increased incidence of hepatocellular carcinomas (control, 0/24; low-dose, 8/13; medium-dose, 13/18; high-dose, 12/14) was seen, which were diagnosed after mean induction periods of 742, 622 and 611 days, respectively. In addition, hyperplastic liver nodules were often observed in animals treated with aflatoxin B_1 that survived 100 days. There was also an increased incidence of renal-cell tumours (control, 0/24; low-dose, 3/13; medium-dose, 5/18; high-dose, 8/14), which were diagnosed after mean induction periods of 783, 696, and 603 days, respectively. The renal-cell tumours showed papillary, tubular and acinar formations consisting of various cell types, including clear and granular cells. Basophilic hyperplastic tubules were seen in some kidneys (Epstein et al., 1969).

Groups of 18 and 36 male Wistar R rats, weighing 80–120 g, were administered 0 and 50 μg/animal aflatoxin B_1 in 0.1 ml DMSO by gastric intubation twice a week for four weeks and then 0 and 75 μg/animal aflatoxin B_1 in 0.15 ml DMSO twice a week for 10 weeks. Groups of one to three animals were killed at various times between three and 86 weeks after the beginning of treatment. Administration of aflatoxin B_1 induced hepatocellular and hepatocholangiocellular carcinomas in 70% of rats [number unspecified] that received a total dose of 1900 μg aflatoxin B_1, the first carcinoma appearing after 44 weeks. Foci of altered hepatocytes (clear, acidophilic and hyperbasophilic cell GGT^+ foci) appeared after 15 weeks and increased in number and size with time, to form hyperplastic nodules (Kalengayi et al., 1975).

Groups of about 30 male Charles River rats, 63 days old, were fed aflatoxin B_1 at 0 or 1000 μg/kg of diet for 15 weeks, and unspecified numbers were killed 8 and 16 weeks after beginning of exposure. The 16 treated and 16 control surviving animals were kept on control diet until week 88. Small foci of vacuolated hepatocytes were observed in treated animals for the first time after 16 weeks, and hepatocellular carcinomas after 68 weeks; the cumulative incidence of hepatocellular carcinomas reached 40% at 88 weeks (Nishizumi et al., 1977).

Groups of 25 male and 25 female Wistar rats, seven weeks old, were administered 0, 100 (males) or 75 (females) μg/animal aflatoxin B_1 in DMSO intragastrically twice a week for five weeks, followed by 0, 20 (males) or 15 (females) μg/animal aflatoxin B_1 in DMSO twice a week for 10 weeks. Treated rats died or were killed between 184 and 486 days after cessation of aflatoxin B_1 treatment, and vehicle controls were killed between 250 and 500 days. Malignant hepatomas, sometimes accompanied by cholangiocellular adenomas, appeared for the first time after 386 days in males and after 417 days in females treated with aflatoxin B_1. The incidences of malignant hepatomas at those times were 8/8 males and 5/8 females. Benign hepatomas appeared for the first time after 265 days in males and 295 days in females, and their incidences at those times were 14/22 males and 10/26 females. Precancerous focal lesions (designated as hyperplastic) were observed in all treated rats starting from 184 days after the cessation of aflatoxin B_1 administration. No neoplastic or preneoplastic lesion was observed in vehicle controls. The progress of hepatocarcinogenesis induced by aflatoxin B_1 was accompanied by an increased level of porphyrins in liver and faeces and by an increased concentration of GGT in liver, especially in hepatomas, and in serum (Zawirska & Bednarz, 1981).

Groups of 66 and 120 female Wistar rats, weighing about 200 g, were administered a single dose of 0 or 5000 µg/kg bw aflatoxin B_1 in olive oil by gavage. Of the treated rats, 29 died in the first few days and eight between 52 and 104 weeks after dosing. Groups of five aflatoxin B_1-treated rats and two to five untreated control rats were killed at two days and at various times between one and 104 weeks after the beginning of the experiment. Thirteen additional aflatoxin B_1-treated rats which had survived 52 weeks were killed when moribund. Hepatocellular adenomas (designated as neoplastic nodules) were found in 10/26 animals that survived 78 weeks but not in controls. Foci of altered hepatocytes (predominantly tigroid-cell foci) were detected for the first time eight weeks after administration of aflatoxin B_1; these increased in number and size up to 104 weeks. Rarely, foci of altered hepatocytes were also observed in controls that survived 52 weeks (Bannasch et al., 1985).

Groups of 15 and 30 male Wistar rats, weighing 100 g, were administered 0 or 50 µg/animal aflatoxin B_1 in 0.1 ml DMSO by gastric intubation twice a week for four weeks and then 0 and 75 µg/animal in 0.15 ml DMSO for 10 weeks [number of times per week unspecified]. Randomly selected animals were sacrificed 4–78 weeks after the beginning of treatment. Preneoplastic hepatic foci (clear, mixed, diffusely basophilic and tigroid-cell foci) appeared in the livers of aflatoxin B_1-treated rats from 22 weeks onward but were not observed in untreated controls (Gil et al., 1988).

Groups of 56 male Fischer 344/CDF rats, six weeks old, were exposed [route not clearly specified but probably gavage] to 0 or 25 µg/animal aflatoxin B_1 in 0.1 ml DMSO five times a week for eight weeks, or to 0 or 70 µg/animal in 0.15 ml DMSO nine times in two weeks. Six animals per group were sacrificed 2, 6, 10, 14, 22, 32 and 47 weeks after the last treatment with aflatoxin B_1, and all surviving rats were killed after 66 weeks. After treatment with 25 µg/animal aflatoxin B_1, hepatocellular carcinomas were observed in 1/6 rats after 47 weeks and in 3/10 rats after 66 weeks; neoplastic hepatic nodules were found in 3/6 rats after 32, 3/6 rats after 47 and in 6/10 rats after 66 weeks; preneoplastic hepatic foci (as defined by glycogen storage, cytoplasmic basophilia or GGT^+ staining) were found for the first time two weeks after the last exposure to aflatoxin B_1, and their number and size increased with time. In rats exposed to 70 µg/animal aflatoxin B_1 for two weeks only, hepatocellular carcinomas occurred in 1/13 and neoplastic hepatic nodules in 3/13 rats after 66 weeks but not earlier; preneoplastic hepatic foci were first observed 6–14 weeks after the last administration of aflatoxin B_1 and increased in number and size with time. No neoplastic or preneoplastic hepatic lesion was observed in vehicle controls (Soffritti & McConnell, 1988).

Hamster: Male Syrian golden hamsters [initial number not specified], four weeks old, were given 2000 µg/kg bw aflatoxin B_1 (dissolved in DMSO and diluted with an equal volume of trioctanoin) by gavage on five days a week for six consecutive weeks. Starting 24 h after the last dose of aflatoxin B_1, some animals [number unspecified] were given 0.1% phenobarbital in the drinking-water. Control animals received only the vehicle (2 ml/kg bw DMSO–trioctanoin per day). At least six hamsters per group were killed at 6, 14 and 46 weeks, and groups of 30 hamsters were observed for tumour development; the 15 surviving animals were killed at 78 weeks. The hamsters killed at or after 23 weeks included those given phenobarbital in the drinking-water and groups given only tap-water. In hamsters that were treated with aflatoxin B_1 and were killed at or survived to 46 weeks, a high incidence of cholangiocellular tumours was observed: cholangiocarcinomas in 8/30 hamsters given aflatoxin B_1 and phenobarbital and in 9/33 without phenobarbital, and cystic cholangiomas in 19/30 given aflatoxin B_1 and phenobarbital and 21/33 without phenobarbital. Focal bile-duct proliferation and foci of altered hepatocytes were also frequently observed in aflatoxin-treated hamsters. Hepatocellular carcinomas were found in two aflatoxin-treated hamsters killed at 78 weeks, one of which had been given phenobarbital. No notable change was seen in control livers (Moore et al., 1982).

Monkey: A total of 47 monkeys (rhesus, cynomolgus and African green) [sex and age unspecified] received aflatoxin B_1 dissolved in > 0.2 ml/kg DMSO by intraperitoneal injection at 125–250 μg/kg bw and/or orally at 100–800 μg/kg bw for two months or longer. Thirteen of 35 necropsied monkeys had developed one or more malignant neoplasms: two hepatocellular carcinomas, three liver angiosarcomas, two osteogenic sarcomas, six adenocarcinomas of the gall-bladder or bile duct, two pancreatic adenocarcinomas, one undifferentiated pancreatic tumour and one papillary grade-I carcinoma of the urinary bladder. The tumours developed at an average total dose of 709 mg (range, 99–1354 mg) aflatoxin B_1 after an average period of 114 months (range, 47–147 months). Fifteen of the 22 necropsied monkeys without tumours showed histological evidence of liver damage, including toxic hepatitis, cirrhosis and hyperplastic nodules. These animals had received an average total dose of 363 mg (range, 0.35–1368 mg) for an average of 55 months (range, 2–141 months). No liver tumour, but one gall-bladder carcinoma and three malignant lymphomas, were found in 68 necropsied, untreated breeder and vehicle-treated (0.2 ml/kg DMSO) monkeys which had died in the colony during the observation period (Sieber et al., 1979).

Tree shrew: Groups of 10 female and 8 male tree shrews (*Tupaia glis*), weighing 95–140 g, were intermittently given aflatoxin B_1 at a dietary concentration of 2000 μg/kg. Groups of three female and five male animals were fed control diet. The experiment was terminated after 172 weeks. Of 12 animals that survived, 6/6 female and 3/6 male tree shrews developed hepatocellular carcinomas after 74–172 weeks of treatment (total dose of aflatoxin B_1, 24–66 mg). None of the eight control animals developed liver cancers. In two tree shrews, the liver tumours were associated with severe post-necrotic scarring; in the other seven tumour-bearing livers, only mild to moderate portal fibrosis was seen (Reddy et al., 1976).

Guppy: Groups of 50 male and female guppies (*Lebistes reticularis*), four weeks old, were fed 0 or 6000 μg/kg of diet aflatoxin B_1 and were killed at intervals of 2–4 months between 4 and 13 months or were followed until death. Aflatoxin B_1 induced hepatic tumours in 9/16 fish between nine and 11 months. No liver tumour was observed in five untreated fish that survived nine months (Sato et al., 1973).

Salmon and trout: Groups of hatched coho salmon and rainbow trout [initial numbers unspecified] were fed aflatoxin B_1 at 0, 20, 40 or 5000 μg/kg of diet for four (20 and 40 μg) or three (5000 μg) weeks and were sacrificed at 12 months of age. The incidence of liver lesions was much lower in salmon (control, 0/118; 40 μg aflatoxin B_1, 0/116; 5000 μg aflatoxin B_1, 9/176—one basophilic focus and eight hepatic adenomas) than in rainbow trout (control, 1/121; 20 μg aflatoxin B_1, 74/120—predominantly hepatocellular carcinomas). Higher doses were lethal to the trout (Bailey et al., 1988).

Trout: Groups of rainbow trout [number and age unspecified] were fed aflatoxin B_1 at 0 or 8 μg/kg of diet for up to 12 months, when the experiment was terminated. A total of

95 liver lesions, ranging from small preneoplastic foci to hepatocellular, cholangiocellular and mixed neoplasms, were observed in treated trout (Kirby et al., 1990).

3.2.2 Subcutaneous administration

Rat: Groups of six male rats [strain unspecified], weighing about 100 g, were injected subcutaneously with 0 or 20 µg aflatoxin B_1 suspended in 0.5 ml arachis oil twice weekly for up to 65 weeks. Subcutaneous sarcomas were found in 6/6 treated rats within 18–37 weeks and in 0/6 controls (Dickens & Jones, 1965).

Groups of male Fischer rats [initial number and age not clearly specified] were administered 0 or 10 µg/rat aflatoxin B_1 suspended in 0.2 ml trioctanoin subcutaneously twice weekly for 20 weeks (total dose of aflatoxin B_1, 400 µg/rat). Subcutaneous sarcomas were found by 58 weeks at the injection site in 9/9 treated rats but not in vehicle controls (Wogan et al., 1971).

3.2.3 Intraperitoneal administration

Mouse: Groups of 16 female strain A/He mice, two months old, were administered 0 or 2000 µg/kg bw aflatoxin B_1 (purity, > 99.5%) in 0.1 ml DMSO intraperitoneally three times a week for four weeks (total average dose of aflatoxin B_1, 5600 µg/animal). A group of 16 females served as untreated controls. The experiment was terminated at 24 weeks after the start of treatment. Survival was similar in all three groups: untreated controls, 14/16; vehicle control, 15/16; aflatoxin B_1, 14/16. Aflatoxin B_1 produced an average of 5.6 pulmonary adenomas/mouse in 14/14 animals. No tumour occurred in untreated controls. Of the vehicle controls, 4/15 had lung adenomas, with an average of 0.3 tumours/mouse (Wieder et al., 1968).

Groups of eight male and eight female strain A/J mice, six to eight weeks old, were administered 5000, 12 500 or 25 000 µg/kg bw aflatoxin B_1 in 0.1 ml DMSO intraperitoneally once a week for six weeks. A group of 16 male and 16 female DMSO controls was available, and groups of 136 males and 131 females served as untreated controls. The experiment was terminated 24 weeks after the start of treatment, at which time survival was 135/136 male and 131/131 female untreated controls; survival was reduced in vehicle controls and in animals treated with aflatoxin B_1 at the highest dose level: males, 12/16 vehicle controls, 9/9 low-dose, 8/8 medium-dose, 5/8 high-dose; females, 14/16 vehicle controls, 7/7 low-dose, 8/8 medium-dose and 5/8 high-dose animals. Treatment with aflatoxin B_1 produced lung adenomas in 100% of the surviving animals at all three dose levels, and the average number of lung adenomas per mouse was dose-related: males, low-dose, 6.56; medium-dose, 15.75; high-dose, 20.20; females, low-dose, 11.57; medium-dose, 16.13; high-dose, 28.80. The proportion of untreated controls with lung tumours was 38% in males and 25% in females, and that in vehicle controls was 17% in males and 50% in females. The average number of lung tumours in untreated and vehicle controls of each sex ranged between 0.29 and 0.57 (Stoner et al., 1986).

Groups of newborn male and female inbred (C57Bl × C3H)F_1 mice [initial numbers unspecified], one to seven days old, were administered 0, 250, 1000, 2000 or 6000 µg/kg bw aflatoxin B_1 in 10 ml/kg bw trioctanoin intraperitoneally once, three or five times at three-day intervals between one and 16 days after birth (total doses of aflatoxin B_1, 1250, 2000, 3000 or 6000 µg/kg bw. Of the mice treated at day 1 with a single dose of 2000 µg/kg bw aflatoxin B_1, 55% died within four days. Seventy animals [sex unspecified] selected at random (10–20 per group) were sacrificed at 52 weeks, and 105 animals (16–29 per group) at 82 weeks of age. An increased incidence of hepatomas (total, 31/70) was found at 52 weeks in all aflatoxin B_1-treated groups except that given a single dose of 2000 µg/kg bw. No hepatoma was observed in 100 vehicle controls at that time. By 82 weeks, all aflatoxin-treated groups, including the group that received a total dose of 1250 µg/kg bw, showed increased incidences of hepatomas (total, 82/105). The incidence of hepatomas in the control group at 82 weeks was 3/100 (Vesselinovitch et al., 1972).

Transgenic mouse: Groups of 10 female transgenic mice of C57Bl/6 background, which overexpress the hepatitis B virus large envelope polypeptide, two to three months of age, and 9–10 female, age-matched nontransgenic littermates were given intraperitoneal injections of aflatoxin B_1 in tricaprylin at doses of 0 or 250 µg/kg bw as a single dose or five doses at approximately monthly intervals or 2000 µg/kg bw in three weekly doses. At 15 months, when the experiment was terminated, seven to nine animals in each group were still alive. After three doses of 2000 µg/kg bw, two hepatocellular carcinomas and 10 hepatocellular adenomas were found in transgenic mice. At that time, four hepatocellular adenomas and no carcinoma were observed in transgenic mice that received 250 µg/kg bw as five doses and six adenomas but no carcinoma in those that received 250 µg/kg bw as a single dose. No liver neoplasm was described in nontransgenic mice with or without aflatoxin B_1 treatment; however, multiple liver nodules of different sizes (which were not regarded as neoplasms) were observed in control transgenic mice, and four such nodules of < 2 mm per group were seen in nontransgenic mice exposed to aflatoxin B_1 (Sell et al., 1991).

Rat: Groups of male Fischer rats [initial number and age not clearly specified] were administered 0 or 32.5 µg aflatoxin B_1 in 0.05 ml DMSO intraperitoneally five times a week for eight weeks (total dose of aflatoxin B_1, 1300 µg/animal). By 46 weeks, the incidence of hepatocellular carcinomas was 9/9 in rats treated with aflatoxin B_1 and 0/10 in vehicle controls (Wogan et al., 1971).

Toad: Groups of 50 male and 50 female Egyptian toads (*Bufo regularis*) [age unspecified] were injected into the dorsal lymph sac with 0 or 200 µg/kg bw aflatoxin B_1 in 1 ml of commercial corn oil once a week for 15 weeks. Experimental and control animals were autopsied at various times between two and 15 weeks; four males and seven females died spontaneously but were also autopsied. Hepatocellular carcinomas were found in 12 male and seven female treated toads. Tumours observed in the kidneys of three males and one female were diagnosed by the authors as metastases of the hepatic tumours. No tumour was detected in the controls (El-Mofty & Sakr, 1988).

3.2.4 Intraperitoneal administration after partial hepatectomy

Mouse: Two groups of 16 male $STCF_1$ mice, nine weeks of age, were administered 0 or 6000 µg/kg bw aflatoxin B_1 in 10 ml/kg bw tricaprylin intraperitoneally 46 h after a two-thirds partial hepatectomy. A group of 24 mice received the same single dose of aflatoxin B_1 without partial hepatectomy, and 38 intact mice served as untreated controls. The experiment was terminated at 104 weeks. At that time, survival was 13/38 in intact, untreated controls, 4/16 in controls that had a partial hepatectomy, 8/24 in the group treated with aflatoxin B_1 without partial hepatectomy and 4/16 in those given aflatoxin B_1 and a partial hepatectomy. In mice exposed to aflatoxin B_1, hepatocellular carcinomas were observed in 9/16 animals with and in 0/24 animals without preceding partial hepatectomy. In untreated controls, one hepatocellular carcinoma each was observed in the groups with and without partial hepatectomy (in animals that died at 98 and 103 weeks, respectively). In the mice that received aflatoxin B_1 after partial hepatectomy, the first hepatocellular carcinoma was observed after 55 weeks. Hepatocellular adenomas were observed in 4/38 controls without partial hepatectomy, in 3/16 controls with partial hepatectomy, in 8/16 mice exposed to aflatoxin B_1 with preceding partial hepatectomy and in 2/24 treated mice without partial hepatectomy. Statistical analyses revealed a significantly increased risk for the development of hepatocellular carcinomas ($p < 0.001$) and hepatocellular adenomas ($p < 0.01$) in mice exposed to aflatoxin B_1 with but not without partial hepatectomy. There was no significant increase in risk for lung tumour development in mice following a single exposure to aflatoxin B_1 (Dix, 1984).

Rat: Groups of 10–15 male Fischer 344 rats, six to eight weeks old, were administered a single intraperitoneal dose of 0, 75 or 250 μg/kg bw aflatoxin B_1 in DMSO 6, 24 and 31 h after partial hepatectomy, sham operation or no operation. Survival of rats that received the high dose of aflatoxin B_1 after partial hepatectomy was markedly reduced at 62 weeks: untreated control, 10/12; high-dose aflatoxin B_1 without prior operation, 12/12; high-dose 24 h after sham operation, 8/10; high-dose 6 h after partial hepatectomy, 2/12; high-dose 24 h after partial hepatectomy, 0/12; high-dose 31 h after partial hepatectomy, 2/12. All surviving animals were killed at 62 weeks. No hepatocellular neoplasm was found, but there was an increased incidence of preneoplastic foci of altered hepatocytes (basophilic and clear-cell foci) at both dose levels of aflatoxin B_1 following partial hepatectomy, and an increased incidence of clear-cell foci after the high dose of aflatoxin B_1 following sham operation (Neal & Cabral, 1980).

Groups of 20–40 male inbred AS_2 rats, weighing 120–150 g, were administered a single intraperitoneal dose of 250 μg/kg bw aflatoxin B_1 in DMSO 24 h after two-thirds partial hepatectomy or sham operation. At 40 weeks, 11/40 hepatectomized and 12/20 non-hepatectomized rats were still alive; those that had undergone hepatectomy [number unspecified], but not non-hepatectomized animals, had preneoplastic hepatic foci consisting of clear, vacuolated, acidophilic or basophilic cells. A high incidence of hepatocellular carcinomas (9/10) and neoplastic hepatic nodules (10/10) was found in hepatectomized rats 55–65 weeks after aflatoxin B_1 administration; none of the 10 non-hepatectomized rats killed at that time had hepatocellular carcinomas or neoplastic hepatic nodules, although there were small preneoplastic hepatic foci. One hepatectomized and 1/2 non-hepatectomized rats followed until 90 weeks after administration of aflatoxin B_1 had both hepatocellular carcinomas and neoplastic hepatic nodules (Rizvi *et al.*, 1989).

3.2.5 *Pre- and postnatal exposure*

Rat: Female Fischer 344 rats were fed aflatoxin B_1 at 2000 μg/kg of diet throughout pregnancy and lactation, and then 26 males and 21 females of their weaned offspring were fed the same diet until they died. An additional six male and 19 female weaned offspring of untreated Fischer 344 rats were fed aflatoxin B_1 at the same dose from six to seven weeks of age until death. A control group of male and female offspring of untreated Fischer 344 rats [numbers unspecified] remained untreated until death or were killed at 75 weeks, at which time the majority [numbers unspecified] had survived. Mean survival of groups exposed to aflatoxin B_1 both *in utero* and postnatally (males, 45.6 weeks of age; females, 61.6) and groups exclusively exposed postnatally (males, 50.2 weeks of age; females, 63.1) did not differ significantly ($p > 0.1$), but mean survival was significantly ($p < 0.01$) shortened in males of both treatment groups as compared to females. More than 75% [detailed data not given] of animals in both treatment groups [groups exposed *in utero* and postnatally and exclusively postnatally were not separated; sex unspecified] developed malignant liver neoplasms (haemangiosarcomas, hepatocellular carcinomas), which were the main cause of death. The numbers of rats still alive after the first malignant liver neoplasms were detected in males (34 weeks) and females (49 weeks) were 20/26 males and 14/21 females exposed to aflatoxin B_1 from conception and 5/6 males and 14/19 females fed aflatoxin B_1 from six to seven weeks of age. In addition to malignant liver neoplasms, hyperplastic hepatic nodules and two types of areas of hyperplasia (consisting either of small dark-staining or large pale-staining liver cells) were found frequently. An increased incidence of colonic tumours (predominantly small polypoid tumours without metastases, one adenocarcinoma) was observed in rats of each sex which had been exposed to aflatoxin B_1 from conception (males 2/20; females, 5/14) and in rats of each sex that had been fed aflatoxin B_1 from weeks six to seven of age (males, 2/5; females, 3/14). A few treated rats had tumours at other sites, including two renal-cell tumours. No tumour was reported in controls (Ward *et al.*, 1975). [The Working Group noted the limited reporting of the data.]

Groups of 15 female Sprague-Dawley rats, 12 weeks old, were administered 500 µg/kg bw aflatoxin B_1 intraperitoneally from days 15 to 18 or 18 to 21 of pregnancy, and then 872 offspring of each sex received additional intraperitoneal injections of 0 or 500 µg/kg bw on days 2-5 or 14-17 after birth. A control group of 10 pregnant rats received daily intraperitoneal injections of saline on days 15-21 of pregnancy, and their neonatal offspring were given intraperitoneal injections of saline on days 2-5 or 14-17 after birth. The mean survival time of dams was 834 days in controls, 791 days in those treated with aflatoxin B_1 on days 15-18 and 792 days in those treated on days 18-21 of pregnancy. Of the offspring treated with aflatoxin B_1, 483/717 treated and 113/113 control animals survived the lactation period; the median survival times were 834 in male and 839 days in female controls and 728-844 days in male and 744-842 days in female animals treated with aflatoxin B_1. In dams treated with aflatoxin B_1, increased incidences of benign and malignant neoplasms (other than leukaemia) were observed in various organs: benign tumours, liver and gastrointestinal tract, 34/90; endocrine organs, 49/90; mammary gland, 30/90; other organs, 11/90; malignant tumours, liver and gastrointestinal tract, 6/90; endocrine organs, 6/90; mammary gland, 5/90; other organs, 10/90. No malignant neoplasm was detected in controls; the incidences of benign tumours in controls were 0/10 in liver and gastrointestinal tract, 3/10 in endocrine organs, 6/10 in mammary gland and 0/10 in other organs. In offspring that had been exposed to aflatoxin B_1 *in utero* (with and without additional aflatoxin B_1 treatment after birth), a significant increase was seen in the total number of malignant tumours, except for mammary gland tumours and leukaemias. The location and number of benign and malignant tumours in offspring that had been exposed to aflatoxin B_1 *in utero* (with and without additional aflatoxin B_1 treatment after birth) are given in Table 9. Significant differences in the incidences of benign tumours between control and treated offspring were found in the trachea (control, 1/113; treated, 17/483), Harderian gland (control, 2/113; treated, 27/483) and liver (control, 14/113; treated, 207/483). Benign and malignant tumours were also observed in the central and peripheral nervous systems of offspring exposed to aflatoxin B_1 but not in controls; prenatal exposure alone was sufficient to produce these tumours (Goerttler *et al.*, 1980).

Trout: Groups of 200 rainbow trout eggs, 14 days after fertilization, were exposed to 0 or 500 μg/l aflatoxin B_1 in water for 1 or 2 h. After birth, the trout were fed a semipurified diet for up to 12 months, when the experiment was terminated. Aflatoxin exposure for 1 h resulted in hepatocellular carcinomas in 42/100 fish, and exposure for 2 h resulted in hepatocellular carcinomas in 45/100 fish at 12 months. No control fish kept in well water developed a hepatocellular carcinoma (Wales *et al.*, 1978).

In another experiment, groups of 200 rainbow trout eggs were exposed to aflatoxin B_1 at 1000 μg/l of water for 1 h/day on either day 1, 3, 4, 7, 9, 11, 13, 15, 17, 19, 21 or 23 after fertilization. A control group was not exposed to aflatoxin B_1. Random samples of fish were taken at 10 and 12 months. The incidence of hepatocellular carcinomas increased with increasing age of the embryo at the time of exposure to aflatoxin B_1 and was 20/35 at 10 months and 35/60 at 12 months in fish exposed on day 23 after fertilization. None of the control fish developed tumours. The earliest indication of neoplastic development was occasional foci of basophilic cells in the livers of fish at four months, and the first hepatocellular carcinomas were seen at six months after birth (Wales *et al.*, 1978).

Salmon and trout: Ninety-nine coho salmon eggs, 20 days before hatching, were administered 90 ng/egg (400 μg/kg) aflatoxin B_1 in DMSO by microinjection, and hatched fish were killed eight months after exposure to aflatoxin B_1. A further 199 rainbow trout eggs, < 3 days before hatching, were injected similarly with 100 ng/egg (1400 μg/kg), and hatched fish were killed nine months after exposure. The mortality of trout eggs was: uninjected control, 13/199; vehicle control, 78/200; and 100 ng aflatoxin B_1, 145/199; and that of salmon eggs was: uninjected control, 14/117; vehicle control, 107/153; and 90 ng aflatoxin B_1, 53/99. Hepatocellular neoplasms (carcinomas and neoplastic nodules combined) occurred in 5/20 rainbow trout and 3/18 coho salmon exposed as embryos to aflatoxin B_1 and in none of 48 rainbow trout or 21 coho salmon exposed as embryos to the vehicle only (Black *et al.*, 1985).

Salmon: Groups of 300 coho salmon eggs were administered 50 or 100 ng/egg (217 or 434 μg/kg) aflatoxin B_1 in 1 μl DMSO by microinjection, and all surviving hatched fish were killed 14 months later. None of the fish that had been exposed as embryos to the high dose of aflatoxin B_1 survived to the end of the experiment; the effective numbers at 14 months were 38 in the low-dose group and 64 vehicle controls. Neoplastic hepatic nodules were found in 4/38 and focal bile-duct proliferation in 16/38 fish at the low dose; no liver lesion was observed in vehicle controls (Black *et al.*, 1988).

Table 9. Location and numbers of malignant and benign tumours in offspring treated with aflatoxin B_1 *in utero* and in control rats

Group	Sex	No.	Liver, stomach and intestines		Endocrine organs		Other organs		Mammary gland		Leukaemias (malignant)
			Benign[a]	Malignant[b]	Benign[c]	Malignant[d]	Benign[e]	Malignant[f]	Benign	Malignant	
Treated	M	228	68	16	143	12	48	49	13	–	22
	F	255	129	21	198	15	34	29	134	14	19
Control	M	53	–	–	18	1	–	–	3	–	2
	F	60	–	–	41	–	1	–	30	4	3

From Goerttler *et al.* (1980)

[a]Mainly hepatocellular and cholangiocellular adenomas, some papillomas and adenomas of the stomach and intestines

[b]21 tumours in liver; 16 in stomach and intestines

[c]Pituitary gland, adrenal glands, ovaries, testes, thyroid gland, parathyroid glands and endocrine pancreas: treated, 170 pituitary gland tumours (129 F, 41 M), 82 adrenal gland tumours (22 F, 60 M), 65 thyroid gland tumours (39 F, 26 M) and 24 others; control, 39 pituitary gland tumours (31 F, 8 M), 10 adrenal gland tumours (4 F, 6M), nine thyroid gland tumours (6 F, 3M) and two others

[d]Treated, 11 tumours of the adrenal gland, six tumours of the thyroid gland, six tumours of the pituitary gland and four others; control, one papilloma of the thyroid gland

[e]Treated, 30 central and peripheral nervous system tumours, 21 skin tumours, nine urogenital tract tumours (except ovaries and testes), four skeletal tumours, three nasal cavity tumours and 15 others; control, one adenoma of the exocrine pancreas

[f]16 skin tumours (3 F, 13 M), 11 skeletal tumours (3 F, 8 M), 12 tumours of uterus or vagina, 14 central or peripheral nervous system tumours (5 F, 9 M), eight nasal cavity tumours (2 F, 6 M) and 17 others

3.3 Carcinogenicity of metabolites

3.3.1 Aflatoxin M_1

Oral administration

Rat: Groups of 10–30 male weanling Fischer rats were administered 0 or 25 μg/rat of synthetic aflatoxin M_1 (purity, > 95%) or purified aflatoxin B_1 suspended in 0.5 ml distilled water by gastric intubation on five days a week for eight consecutive weeks and held thereafter without treatment. Only 1/29 animals treated with aflatoxin M_1 had developed a hepatocellular carcinoma by 96 weeks. The remaining animals in that group were killed after 100 weeks and, of these, eight had preneoplastic liver lesions (hyperplasia or transitional cells). Nine of nine rats treated with aflatoxin B_1 developed hepatocellular carcinomas between 47 and 53 weeks. No liver tumour was observed in untreated controls (Wogan & Paglialunga, 1974).

Groups of 42 and 62–63 male Fischer rats, seven weeks old, were fed 50 μg/kg of diet aflatoxin B_1 and 0, 0.5, 5 or 50 μg/kg of diet of naturally occurring aflatoxin M_1 for up to 21 months. Serial sacrifices were carried out between three and 21 months. Liver neoplasms were found only after 16 months in rats administered aflatoxin B_1 and aflatoxin M_1 at dose levels of 50 μg/kg of diet and in one control animal. The incidence of liver tumours (diameter > 2 mm; hepatocellular carcinomas and neoplastic nodules combined) was lower after administration of aflatoxin M_1 (16 months, 1/6; 17 months, 0/6; 19 months, 2/19; 21 months, 6/18) than after treatment with aflatoxin B_1 (16 months, 9/9; 17 months, 19/20; no surviving animals later). After administration of aflatoxin M_1, 2/6 tumours found after 21 months were hepatocellular carcinomas; in contrast, in animals treated with aflatoxin B_1 and killed after 16 and 17 months all liver tumours (28/29) were hepatocellular carcinomas (Hsieh *et al.*, 1984).

Groups of 42 and 62 male Fischer 344 rats, seven weeks of age, were fed a semisynthetic diet containing 0, 0.5, 5 or 50 μg/kg of naturally occurring aflatoxin M_1 or 50 μg/kg of diet aflatoxin B_1 up to 21 months of age. Animals were killed at different times between 3 and 22 months. After 21 months, the incidence of hepatocellular carcinomas induced by the high dose of aflatoxin M_1 was 2/18, which was significantly ($p < 0.01$) lower that that (19/20) in rats administered aflatoxin B_1 at the same dose level for 17 months. As in rats administered aflatoxin B_1, the induction of hepatocellular carcinomas by aflatoxin M_1 was preceded by the appearance of foci of altered hepatocytes (predominantly eosinophilic-cell foci) and neoplastic hepatic nodules (composed of basophilic, eosinophilic and clear cells). The number and size of foci of altered hepatocytes induced by the high dose of aflatoxin M_1 were significantly ($p < 0.05$) less than those produced by the same dose of aflatoxin B_1. Intestinal adenocarcinomas were found in three animals treated with the high dose of aflatoxin M_1 after 9 and 18 months, but were not observed in any other group. No hepatocellular carcinoma or neoplastic hepatic nodule (but a very low incidence of foci of altered hepatocytes at 21 months) was found in rats that received the lower doses of aflatoxin M_1 or in controls (Cullen *et al.*, 1987).

Trout: Groups of 250 rainbow trout, four months old, were fed diets containing 0, 5.9 or 27.3 μg/kg of naturally occurring aflatoxin M_1 (the diet also contained aflatoxin M_2) or 5.8 μg/kg of diet aflatoxin B_1 for 16 months. Groups of fish were killed 5, 9, 12 and 16 months after the beginning of treatment. The hepatocarcinogenic effect of aflatoxins M_1 and M_2 was much weaker than that of aflatoxin B_1, the incidence of hepatocellular carcinomas at 16 months being 0/49 low-dose aflatoxins M_1 and M_2-treated animals, 1/48 high-dose aflatoxins M_1 and M_2-treated animals and 6/48 aflatoxin B_1-treated animals. A similar trend was observed in the incidence of hyperplastic (eosinophilic, basophilic and mixed cell) nodules (Canton *et al.*, 1975).

3.3.2 *Aflatoxin Q_1*

Oral administration

Trout: Two duplicate groups of 80 rainbow trout, four months old, were fed diets containing 0 or 100 µg/kg aflatoxin Q_1 (produced by incubation of aflatoxin B_1 with monkey liver microsomes) for 10 months or 4 µg/kg of diet aflatoxin B_1 for 12 months. Groups of fish were taken at six and nine months, and the remaining animals were killed at 12 months. Aflatoxin Q_1 induced hepatocellular carcinomas in 12/113 fish by 12 months; aflatoxin B_1 at a dose of 4 µg/kg of diet induced hepatocellular carcinomas in 55/115 fish by 12 months. No hepatocellular tumour was observed in untreated controls (Hendricks *et al.*, 1980a).

3.3.3 *Aflatoxicol*

Oral administration

Rat: Groups of 20 male Fischer 344 rats, four weeks of age, were fed diets containing 0, 50 or 200 µg/kg aflatoxicol or 50 µg/kg aflatoxin B_1 for 12 months and were observed up to 24 months, when the experiment was terminated. Survival at 24 months was 11/20 controls, 9/20 aflatoxin B_1-treated rats, 5/20 low-dose aflatoxicol-treated animals and 0/20 high-dose aflatoxicol-treated animals. The incidence of hepatocellular carcinomas in animals treated with aflatoxicol was about one-half of that in rats treated with aflatoxin B_1 at the same dose level: controls, 0/20; aflatoxin B_1, 8/20; low-dose aflatoxicol, 4/20; high-dose aflatoxicol, 14/20. The incidence of neoplastic hepatic nodules showed a similar trend: controls, 0/20; aflatoxin B_1, 7/20; low-dose aflatoxicol, 4/20; high-dose aflatoxicol, 2/20 (Nixon *et al.*, 1981). [The Working Group noted the poor survival of animals treated with aflatoxicol.]

Trout: Two duplicate groups of 120 rainbow trout fingerlings [sex unspecified] were fed a semi-synthetic diet alone or containing (per kg of diet) 20 µg aflatoxin B_1, 29 µg aflatoxicol or 61 µg of the diastereomer of aflatoxicol. Groups of 40 trout were selected randomly and killed at four, eight and 12 months. The incidences of hepatocellular carcinomas were 0/80 controls, 45/80 aflatoxin B_1-treated, 20/80 aflatoxicol-treated and 0/80 animals treated with the diastereomer at eight months after initiation of the experiment, and 0/76 controls, 62/75 aflatoxin B_1-treated, 46/57 aflatoxicol-treated and 17/70 diastereomer-treated animals 12 months after beginning of treatment (Schoenhard *et al.*, 1981).

3.4 Aflatoxin B_2

3.4.1 *Oral administration*

Rat: Groups of 30 and 10 male MRC rats, eight to nine weeks old, were administered 0 or 20 µg/animal aflatoxin B_2 in the drinking-water (using dark bottles to avoid photolysis) on five nights a week for 10 weeks (total dose of aflatoxin B_2, 1 mg/animal). Eight treated and 26 control rats were still alive at week 90; no treated rat survived after week 100. Hepatic nodules (designated as atypical hyperplastic) [incidence unspecified] were found, but neither hepatocarcinomas nor renal-cell tumours were observed in rats treated with aflatoxin B_2. No neoplasm was seen in untreated controls (Butler *et al.*, 1969). [The Working Group noted the lack of detailed reporting.]

Groups of 10 male Fischer rats [age not clearly specified] were administered 0, 50 or 100 μg/animal aflatoxin B_2 (purified by column chromatography and recrystallization) in 0.05 ml DMSO by gastric intubation on five days a week for 10 weeks (total doses of aflatoxin B_2, 0, 500 and 1000 μg/animal) and were sacrificed between 62 and 78 weeks. No hepatocellular carcinoma was observed after treatment with aflatoxin B_2 or in vehicle controls, but there was an increased incidence of preneoplastic liver lesions (foci of hyperplasia) in treated animals of each sex at 78 weeks: vehicle control, 0/10; low-dose, 6/9; high-dose, 5/7 (Wogan *et al.*, 1971).

Trout: Duplicate groups of 100 rainbow trout, 60 days old, were fed a semi-synthetic diet containing 0 or 20 μg/kg aflatoxin B_2 and 0, 4, 8 or 20 μg/kg aflatoxin B_1. Groups of fish were taken 9, 12 and 16 months after beginning of treatment for evaluation. Aflatoxin B_2 had little hepatocarcinogenic effect at the dose level used. Hepatoma incidence at 12 months was: control, 0/20; 4 μg aflatoxin B_1, 10/40; 8 μg aflatoxin B_1, 40/57; 20 μg aflatoxin B_1, 62/80; 20 μg aflatoxin B_2, 1/20. Hepatoma incidence at 16 months was: control, 0/40; 4 μg aflatoxin B_1, 14/40; 8 μg aflatoxin B_1, 32/40; 20 μg aflatoxin B_2, 0/40 (Ayres *et al.*, 1971).

3.4.2 *Subcutaneous administration*

Rat: Groups of male Fischer rats [initial number and age not clearly specified] were administered 0 or 300 μg/rat aflatoxin B_2 suspended in 0.2 ml trioctanoin subcutaneously twice weekly for 20 weeks (total dose of aflatoxin B_2, 12 000 μg/rat). Neither aflatoxin B_2 or the vehicle alone induced tumours in 20 rats that survived to 78 or 86 weeks (Wogan *et al.*, 1971).

3.4.3 *Intraperitoneal administration*

Rat: Groups of male Fischer rats [initial number and age not clearly specified] were administered 0 or 3750 μg aflatoxin B_2 in 0.05 ml DMSO intraperitoneally five times a week for eight weeks (total dose of aflatoxin B_2, 150 mg/animal). By 57–59 weeks, the incidence of hepatocellular carcinomas was 2/9 in rats treated with aflatoxin B_2 and 0/10 in vehicle controls (Wogan *et al.*, 1971).

3.5 Aflatoxins G_1 and G_2

3.5.1 *Oral administration*

Rat: Groups of 10–15 male and 15 female MRC rats, eight to nine weeks old, were administered 0, 20 or 60 μg/animal aflatoxin G_1 in the drinking-water (using dark bottles to avoid photolysis) on five nights a week for 10 weeks (low-dose only; total dose of aflatoxin G_1, 1000 μg/animal) or 20 weeks (total doses, 2000 and 6000 μg/animal), and followed until they were in poor condition or died. Survival at 90 weeks was 26/30 (the two sexes combined) controls, 6/10 low-dose (males only) treated for 10 weeks, 17/30 (the two sexes combined) low-dose groups treated for 20 weeks and 9/28 (the two sexes combined) high-dose groups treated for 20 weeks. Aflatoxin G_1 produced a dose-dependent increase in the incidence of hepatic neoplasms in animals of each sex: males, controls, 0/15; low-dose, 10 weeks' treatment, 1/10; low-dose, 20 weeks' treatment, 2/15; high-dose, 9/11; females, control, 0/15; low-dose, 1/15; high-dose, 12/15. In males treated with aflatoxin G_1, there was also a dose-dependent increase in the incidence of renal-cell tumours: control, 0/15; low-dose treated for 20 weeks, 5/15; high-dose, 6/11. A variety of neoplasms was observed in other organs of treated rats but not in untreated controls (Butler et al., 1969).

Groups of up to 30 male Fischer rats [age not clearly specified] were administered 0, 50 or 100 µg/animal aflatoxin G_1 (purified by column chromatography and recrystallization) in 0.05 ml DMSO by gastric intubation on four days a week for 2.5 or 8 weeks (total doses of aflatoxin G_1, 0, 700, 1400 and 2000 µg/animal) and were sacrificed between 4 and 68 weeks. Increased incidences of hepatocellular carcinomas were observed after treatment with aflatoxin G_1 with the two higher total doses: vehicle control, 0/10 at 59 weeks; low-dose, 0/3 at 68 weeks; medium-dose, 3/5 at 68 weeks; high-dose, 18/18 between 45 and 64 weeks. Preneoplastic liver lesions (foci of hyperplasia and transitional cells) were observed in the majority of animals treated with aflatoxin G_1 at all dose levels and sacrificed between 4 and 20 weeks. A total of 4/26 animals dosed with aflatoxin G_1 developed renal adenocarcinomas within 68 weeks. No neoplasm or preneoplastic lesion was observed in untreated controls (Wogan et al., 1971).

Trout: Duplicate groups of 100 rainbow trout, 60 days old, were fed a semi-synthetic diet containing 0 or 20 µg/kg aflatoxin G_1 or 20 µg/kg aflatoxin G_2 [purity unspecified] and 0, 4, 8 or 20 µg/kg aflatoxin B_1 [purity unspecified]. Groups of fish were taken 9, 12 and 16 months after beginning of treatment for evaluation. Aflatoxin G_1 had a weaker hepatocarcinogenic effect than aflatoxin B_1, and aflatoxin G_2 had no hepatocarcinogenic effect at the dose level used. Hepatoma incidence at 12 months was: control, 0/20; 4 µg aflatoxin B_1, 10/40; 8 µg aflatoxin B_1, 40/57; 20 µg aflatoxin B_1, 62/80; 20 µg aflatoxin G_1, 1/20; 20 µg aflatoxin G_2, 0/20. Hepatoma incidence at 16 months was: control, 0/40; 4 µg aflatoxin B_1, 14/40; 8 µg aflatoxin B_1, 32/40; 20 µg aflatoxin G_1, 7/40; 20 µg aflatoxin G_2, 0/40 (Ayres et al., 1971).

3.5.2 Subcutaneous administration

Rat: A group of six male rats [strain unspecified], weighing about 100 g, were injected subcutaneously with 20 µg aflatoxin G_1 suspended in arachis oil twice weekly for up to 65 weeks. Subcutaneous sarcomas were found in 4/6 rats within 30–50 weeks (Dickens & Jones, 1965).

3.6 Administration with known carcinogens and other modifying factors

3.6.1 Mixtures of aflatoxins

Rat: Groups of 6–10 rats of each sex [strain unspecified], 28 days old, were fed diets containing 5, 10 or 100 µg/kg aflatoxins [type unspecified] for 232, 225 and 10 days, respectively, and received either a high-protein (20% casein) or a low-protein diet (5% casein) for up to two years, when the experiment was terminated. Two control groups of six male rats each received the high- and low-protein diet but no aflatoxins. Survival in rats treated with aflatoxins at the different dose levels was 12/22 in those on the low-protein diet and 20/22 in those on the high-protein diet at 52 weeks; the majority of the remaining animals died between 52 and 104 weeks due to massive pneumonia. there was a high incidence of hepatomas (6/20) and precancerous liver lesions (10/20; described as generalized cytoplasmic vacuolization, abundant cytoplasmic eosinophilia or increase in cytoplasmic ribonucleoprotein) in aflatoxin-treated rats on the high-protein diet. No hepatoma or precancerous liver lesion but one renal-cell carcinoma was observed in 12 aflatoxin-treated rats on the low-protein diet (Madhavan & Gopalan, 1968).

Groups of 12–25 male Fischer rats, weighing about 80 g, were fed synthetic diets containing 0 or 1 mg/kg of diet aflatoxins (approximately 1:1 mixture of aflatoxin B_1 and G_1; traces of aflatoxin B_2 and G_2) plus low (5%) or high (20%) casein with and without addition of 0.5 µg/kg of diet vitamin B_{12} for up to 33 weeks, when the experiment was terminated. In rats treated with aflatoxins, the incidence of hepatocellular carcinomas was 2/23 when the aflatoxins were combined with the low-protein diet without vitamin B_{12}, 1/24 when combined with the low-protein diet plus vitamin B_{12}, 1/24 when combined with the high-protein diet without vitamin B_{12} and 6/25 when combined with high-protein diet plus vitamin B_{12}. Administration of aflatoxins at a concentration of 1 mg/kg of low-protein diet with and without vitamin B_{12} produced a higher incidence of various other types of liver lesions (designated as hyperplastic nodules, cirrhosis, cholangiofibrosis and cysts) than treatment with the same dose of aflatoxins in the high-protein diet with and without vitamin B_{12} (Temcharoen et al., 1978).

Groups of 10–15 male Long-Evans rats, weighing 50–60 g, were fed diets containing aflatoxin B_1 at concentrations of 0, 10, 100, 300 and 400 µg/kg of diet plus 0, 8.4, 84, 252 and 336 µg/kg of diet aflatoxin G_1. Similar groups received aflatoxins B_1 and G_1 plus 220 µg/kg cyclopropenoid fatty acids. In the combined experiment, another group received 100 µg/kg of diet aflatoxin B_1 and 84 µg/kg of diet aflatoxin G_1 together with 400 µg/kg of diet cyclopropenoid fatty acids. All the groups were treated for up to 18 months. Some animals were killed after 12 months or when ill, and the remaining animals were sacrificed after 18 months. Administration of aflatoxins resulted in an increased total incidence of renal-cell adenomas (control, 0/7; all aflatoxin treatments, 18/41; cyclopropenoid fatty acid, 1/10; all aflatoxins plus cyclopropenoid fatty acid treatment, 11/43). The incidences of liver-cell tumours were 0/7, 6/41, 0/10 and 2/43, respectively (Lee et al., 1969).

Groups of 20–40 male Charles River rats, three weeks of age, were fed diets containing 0 or 200 µg/kg aflatoxins (added as groundnut meal containing 600 µg/kg) and 0 or 4000 µg/kg of diet diethylstilboestrol [duration unspecified]. Survival time was not appreciably affected by any of the treatments. Diethylstilboestrol alone or in combination with aflatoxins, but not aflatoxins alone, markedly decreased body weights at all recorded times. Rats fed diethylstilboestrol in addition to aflatoxins had a lower incidence of liver carcinomas than aflatoxin-treated animals (control, 0/20; diethylstilboestrol alone, 1/28; aflatoxins alone, 25/35; aflatoxins plus diethylstilboestrol, 8/40) and a lower incidence of liver nodules (control, 0/20; diethylstilboestrol alone, 3/28; aflatoxins alone, 10/35; aflatoxins plus diethylstilboestrol,

19/40). Paired-feeding and feeding-to-weight indicated that the decreased tumour incidence associated with diethylstilboestrol was not the result of decreased food intake (Newberne & Williams, 1969).

Groups of 20 intact and 37 hypophysectomized male MRC rats, eight weeks old, were fed aflatoxins (containing about 45% aflatoxin B_1, 45% aflatoxin G_1, 5% aflatoxin B_2 and 3% aflatoxin G_2) in the diet at approximately 4000 μg/kg and were killed when moribund. The mean quantity of diet consumed by the intact rats was calculated to have been approximately 10 g/rat per day and that by the hypophysectomized rats, 8 g/rat per day. The mean body weight of the hypophysectomized animals was only 30% that of intact rats. Mortality was significantly greater in hypophysectomized than in intact rats treated with aflatoxins at every time up to 60 weeks. Survival at 49 weeks was 14/37 hypophysectomized and 14/20 intact rats treated with aflatoxins. Liver carcinomas developed after treatment with aflatoxins in 14/14 intact rats and in 0/14 hypophysectomized rats (Goodall & Butler, 1969). [The Working Group noted the poor survival and the lack of exact food consumption data.]

Groups of 10 and 20 male Porton rats [age unspecified] were fed 0 or approximately 5000 μg/kg of diet (as calculated from aflatoxin B_1 content) of a mixture of aflatoxins from weeks 1–10 and received 0 or 1 g/l of sodium phenobarbital in the drinking-water during weeks 0–12. After interim sacrifices, the remaining animals were observed for up to 24 months, when the experiment was terminated. Survival after 11 months was the same in animals treated with aflatoxin B_1 alone and with aflatoxin B_1 plus phenobarbital. At 24 months, the incidence of liver tumours (predominantly carcinomas with mixed hepatocellular and cholangiocellular appearance) was significantly higher (17/18) in animals treated with aflatoxin B_1 alone than in animals administered aflatoxin B_1 plus phenobarbital (12/18). In the groups killed at earlier times, the incidences of liver tumours induced by the combined treatment with aflatoxin B_1 and phenobarbital were also lower than those with aflatoxin B_1 alone, and the tumours had a higher degree of differentiation (McLean & Marshall, 1971).

Groups of 15 male Fischer 344 rats, eight weeks old, were fed a diet free of or naturally contaminated with aflatoxins (4 mg/kg of diet aflatoxin B_1) for six weeks, together with 0, 500 or 5000 mg/kg of diet ethoxyquin [stability not measured] either two weeks before, six weeks during or two weeks after aflatoxin. The rats were kept for 48 weeks after the start of treatment. Ethoxyquin produced a dose-related inhibition of aflatoxin-induced liver carcinogenesis. The most effective reduction in the incidence of liver-cell tumours was found after simultaneous administration of aflatoxin and ethoxyquin (aflatoxin B_1 alone, 3/11; aflatoxin plus ethoxyquin, 0/9). Similar trends were observed in the incidence of neoplastic hepatic nodules (aflatoxin B_1 alone, 7/11; aflatoxin plus ethoxyquin, 0/9) and of preneoplastic hepatic foci (aflatoxin B_1 alone, 8/11; aflatoxin plus ethoxyquin, 1/9). No liver lesion was found in rats given control diet or ethoxyquin alone. In another experiment, groups of 21 rats of the same strain, sex and age were fed the same aflatoxin-contaminated diet for six weeks after administration of 0 or 5000 mg/kg of diet ethoxyquin for two weeks and a treatment-free interval of four weeks. The authors stated that there was no difference between the two groups in the incidence of liver lesions at 52 weeks (Cabral & Neal, 1983).

3.6.2 Aflatoxin B_1

(a) Viruses

Marmoset: Groups of 12 young marmosets (*Saguinus oedipomidas*) of each sex, weighing 370–430 g, were given 2000 µg/kg aflatoxin B_1 in chow or fruit diet (200 µg/kg bw), either alone or in combination with hepatitis virus (0.25 ml of a 1/50 dilution containing 100 or greater infectious doses of the G. Baker strain of hepatitis agent), which was injected six weeks after the start of aflatoxin diet. Groups of seven marmosets receiving the hepatitis virus alone and five marmosets on chow and fruit diet served as controls. Nine of 12 marmosets treated with aflatoxin B_1 alone died between 9 and 55 weeks after the beginning of the experiment; 5/12 marmosets that received both aflatoxin B_1 and the hepatitis virus and 2/7 marmosets injected with the hepatitis virus alone were still alive at the time of reporting; 3/5 animals in the untreated control group died. Hepatocellular carcinomas developed in 1/9 marmosets fed aflatoxin B_1 alone and in 2/7 animals injected with hepatitis virus during aflatoxin exposure and which survived 3–94 weeks of treatment. No tumour was found in three animals infected with hepatitis virus alone which died 28–67 weeks after the start of the experiment (Lin *et al.*, 1974). [The Working Group noted the high mortality and that some animals were still alive at the time of reporting.]

Duck: Groups of 2–22 newly hatched ducklings of each sex, free of infection or infected with duck hepatitis virus (DHBV), were administered 0 or 100 µg/kg bw aflatoxin B_1 in DMSO by gastric intubation twice a week for 54 weeks, continuously for five weeks followed by 49 weeks without aflatoxin B_1, or for 25 weeks after an interval of 16 weeks without aflatoxin B_1. After continuous administration of aflatoxin B_1, 16/22 ducks inoculated with the virus and 8/16 ducks without virus died within 10 weeks of hatching due to extensive liver-cell necrosis. After administration of aflatoxin B_1 following an interval of 16 weeks without aflatoxin treatment, 3/5 DHBV-positive ducks died between 20 and 25 weeks. The remaining ducks were sacrificed when ill or 54 weeks after the beginning of the experiment. All ducks (except one) that had been inoculated with the virus showed the presence of DHBV, and all those that were not inoculated did not show the virus, as demonstrated by analysis for DHBV DNA in the sera or tissues and/or immunohistochemistry. After continuous administration of aflatoxin B_1, hepatocellular carcinomas were found in 2/8 DHBV-negative ducks and in 0/8 DHBV-positive ducks that survived 10 weeks. In addition, long-term administration of aflatoxin B_1 often produced cirrhotic changes and nodular liver lesions in ducks both positive and negative for DHBV. Administration of aflatoxin B_1 for five weeks induced one hepatocellular carcinoma in five DHBV-positive ducks by 52 weeks and in 0/3 DHBV-negative ducks kept for 54 weeks. DHBV did not significantly accelerate the induction of hepatic disorders by aflatoxin B_1 (Uchida *et al.*, 1988). [The Working Group noted the high mortality among aflatoxin-treated ducks and the limited reporting.]

Groups of 8–12 Pekin ducks [sex unspecified], three days of age, congenitally infected or free of infection with DHBV, were administered 0 or 200 µg/kg bw aflatoxin B_1 (dissolved in chloroform and added to corn oil) by gavage for 60 days (total dose of aflatoxin B_1, 1740–2040 µg/animal) and kept up to 28 months. Survival at 12 months was 13/22 (DHBV-positive, 9/12; DHBV-negative, 4/10) in the group treated with aflatoxin B_1 and 13/17 (DHBV-positive, 6/8; DHBV-negative, 7/9) in vehicle controls. DHBV core antigen was demonstrated by immunohistochemistry in the liver and several other tissues of most virus-infected ducks but in none of the uninfected ducks. Hepatic neoplasia occurred only in ducks treated with aflatoxin B_1; the incidence of hepatocellular and cholangiocellular carcinomas was 3/8 DHBV-positive ducks and 3/4 DHBV-negative ducks; hepatocellular adenomas occurred in 3/8 DHBV-positive and 0/4 DHBV-negative birds. In addition, foci of altered hepatocytes (as defined by morphology and changes in glycogen content) were often observed in ducks treated with aflatoxin B_1 but not in congenitally DHBV-infected or uninfected ducks that did not receive aflatoxin B_1. The authors concluded that persistent congenital DHBV infection does not contribute significantly to the emergence of hepatic neoplasia induced in ducks by aflatoxin B_1 (Cullen et al., 1990). [The Working Group noted the high mortality.]

Groups of 13–16 male and female Peking ducklings congenitally infected or free of infection with DHBV were administered 0, 20 or 80 µg/kg bw ^{3}H-aflatoxin B_1 in DMSO intraperitoneally once a week every week from the third month after hatching until they were sacrificed 2.3 years later (total doses of aflatoxin B_1/year, 1050 and 4150 µg/kg bw. Of the control ducks, which did not receive aflatoxin B_1, approximately 80% survived to 17 months, whether infected or not. Aflatoxin B_1 treatment resulted in a lower survival by 17 months (high-dose—with infection, approximately 10%; without infection, 40%; low-dose—with infection, 70%; without infection, 50%). Seven hepatocellular carcinomas and two hepatocellular adenomas were observed in eight ducks that were treated with aflatoxin B_1, regardless of the presence of DHBV infection, but not in infected ducks treated with the lower dose of aflatoxin B_1: high-dose with infection, 3/6; without infection, 3/10; low-dose with infection, 0/13; without infection, 2/10. No liver tumour was observed in ducks that were not given aflatoxin B_1 (Cova et al., 1990). [The Working Group noted the poor survival.]

(b) Parasites

Rat: Groups of 72–83 male Buffalo rats, weighing 80–100 g, were fed diets containing 0 or 2000 µg/kg of diet aflatoxin B_1 for 10 weeks, beginning 12 days after intraperitoneal inoculation of 10^6 erythrocytes infected with *Plasmodium berghei* or 0.5 ml of heparinized blood. Three animals in each group were killed at 3, 6 and 12 days after inoculation with parasite or injection of heparinized blood, and 4–25 animals in each group 1, 5, 10, 20, 40, 60 and 82 weeks after beginning of the treatment with aflatoxin B_1. Survival at 11 weeks was 72/80 rats injected with heparinized blood, 79/83 rats given infected erythrocytes, 62/73 rats given infected erythrocytes and treated with aflatoxin B_1 and 67/72 rats treated with aflatoxin B_1 alone. In rats treated with aflatoxin B_1 alone, the incidence of hepatocellular carcinomas was 13/19 after 60–82 weeks, and that of neoplastic hepatic nodules was 5/12 after 20 weeks, 12/23 after 40 weeks and 13/19 after 60–82 weeks. In rats that received infected erythrocytes and aflatoxin B_1, the incidence of hepatocellular carcinomas was 5/14 after 60–82 weeks, and that of neoplastic hepatic nodules was 0/15 after 20 weeks, 5/24 after 40 weeks and 5/14 after 60–82 weeks. Only 1/22 rats inoculated with infected erythrocytes developed a neoplastic hepatic nodule at week 82. No animal in the control group developed a hepatocellular carcinoma or a neoplastic nodule (Angsubhakorn et al., 1988).

Groups of 25 male Wistar rats, 8–10 weeks old, were fed diets containing 0 or 1000 µg/kg aflatoxin B_1 (pure) for 12 weeks and were given a single dose of 0 or 50 metacercariae of

Clonorchis sinensis by intubation at the beginning of the experiment. Three rats from each group were killed at four-week interval up to 28 weeks. Two hepatocellular carcinomas were found at 28 weeks in rats treated with aflatoxin B_1 and infected with *C. sinensis*, and mild to moderate cirrhotic changes were seen after 24 weeks. No liver tumour or cirrhotic change was observed in animals treated with aflatoxin B_1 or *C. sinensis* alone (Park, 1989).

(c) *Chemicals and dietary factors*

(i) *Oral administration*

Rat: Groups of 10–40 male CD rats, three weeks of age, were subjected to a number of additional liver insults (ethionine treatment, choline deficiency, partial hepatectomy, repeated biopsies) before, during or after feeding of 0, 70, 400, 1000 or 1500 μg/kg of diet aflatoxin B_1 (purified). Aflatoxin and ethionine appeared to have syncarcinogenic effect on the liver: The incidences of liver tumours in the various groups suggested a potentiating effect of cirrhosis on the hepatocarcinogenic effect of aflatoxin B_1, particularly when the animals were returned to an adequate diet after the induction of cirrhosis (Newberne *et al.*, 1964, 1966).

Groups of 20 male and 20 female weanling Wistar rats and 16 male and 16 female weanling Fischer rats were fed diets containing 0, 20 or 100 μg/kg aflatoxin B_1 (spectrally pure), with and without 0.02 and 0.035% cyclopropenoid fatty acids for 24 months, and were killed when moribund or showing tumour development. The fatty acids had little, if any, modifying effect on the response to aflatoxin B_1 (Nixon *et al.*, 1974).

Groups of male Sprague-Dawley and male Fischer rats [initial numbers unspecified], about six weeks of age, were fed marginal lipotrope diets (3% casein, 2% corn oil) or a control diet (23% casein, 16% corn oil) and were given 25 μg/animal aflatoxin B_1 in 0.1 ml DMSO intragastrically for 15 days. Hepatocarcinomas [number not given] were found from six months onwards in the deficient rats (6 months, 25%; 9 months, 50%; 12 months, 60%; 18 months, 85%) but not until 12 months in rats on a normal diet (12 months, 30%; 18 months, 60%). Hyperplastic hepatic nodules were present in significant numbers in deficient rats immediately after treatment with aflatoxin B_1 but not in rats fed the normal diet until six months after administration of aflatoxin B_1. The early appearance of preneoplastic focal liver lesions in animals fed the marginal lipotrope diet correlated well with the short tumour induction time (Rogers & Newberne, 1971).

Groups of 50 male weanling Sprague-Dawley rats were fed a purified basal diet containing 50 μg vitamin A palmitate (controls), a purified diet with 100 μg/kg of diet aflatoxin B_1 (purity, > 99.5%) and 5 μg/day vitamin A palmitate (low dose), 100 μg/kg aflatoxin B_1 with 50 μg/day vitamin A (medium dose) or 100 μg/kg aflatoxin B_1 plus 5000 μg/day vitamin A palmitate (high dose). Survival of rats at 24 months was 35/50 controls, 23/50 low-dose, 34/50 medium-dose and 31/50 high-dose animals. The incidence of liver-cell carcinomas was increased by aflatoxin B_1, and the dietary concentration of vitamin A had no significant effect: control, 0/50; low-dose, 11/50; medium-dose, 24/50; high-dose, 19/50. In addition, colon carcinomas were observed in rats treated with aflatoxin B_1 plus low dietary vitamin A: control, 0/50; low-dose, 6/50; medium-dose, 0/50; high-dose, 0/50. In another experiment, 49 male Sprague-Dawley rats were fed diets containing enough vitamin A (added as acetate) to provide 50 μg/day and were administered aflatoxin B_1 (in DMSO) by gastric intubation at a dose of 25 µg for 15 days. The incidence of hepatocellular carcinomas was 27/49 at 24 months (Newberne & Rogers, 1973).

Groups of 60 male Sprague-Dawley rats, about four weeks old, were each given 25 µg aflatoxin B_1 in DMSO by gastric intubation for 15 days and simultaneously or subsequently received diets containing either 28% beef fat and 2% corn oil or no beef fat and 30% corn oil. Animals fed beef fat had a lower incidence of hepatocellular carcinomas (32/60) than those fed the corn oil diet during and after exposure to aflatoxin B_1 (60/60) at 96 weeks. No difference in the incidence of liver tumours was observed between groups fed beef fat throughout exposure to aflatoxin B_1 and afterwards (32/60) and those fed beef fat diet only after exposure to aflatoxin B_1 (28/60) (Newberne et al., 1979).

Groups of 28–35 male Sprague-Dawley CD rats, weighing 40–50 g, were each administered 15 µg aflatoxin B_1 in DMSO by gavage on three to five days per week for seven weeks (total dose, 375 µg/animal) and received semi-synthetic diets containing different amounts of lipids and lipotropes (complete or deficient) throughout the experiment. The animals were killed when moribund or when they reached 90 weeks. All rats fed lipotrope-deficient diets (lipotrope-deficient, deficient plus control vitamin mix, deficient plus lipotropes, deficient plus amino acids) were dead by 82 weeks, while 4/27 of rats fed the lipotrope-complete diet and 10/29 of rats receiving complete diet with high fat content were still alive at that time. The incidences of hepatocellular carcinomas were significantly higher in rats treated with aflatoxin B_1 and fed the various deficient diets: control, 4/27; control plus high fat, 2/29; lipotrope-deficient, 12/31; deficient plus control vitamin mix, 28/34; deficient plus lipotropes, 19/33; deficient plus amino acids, 16/33 (Rogers et al., 1980).

Groups of 10 male Sprague-Dawley rats, weighing 100 g, were administered 0 or 380 µg/kg bw aflatoxin B_1 in tricaprylin by gastric intubation five times a week for up to two weeks and, after an interval of five days, were fed low-fat (11% of calories) or high-fat (35% of calories) diets without or with ethanol (35% of calories) for 12 weeks, when the experiment was terminated. Preneoplastic foci of altered hepatocytes (as defined by morphology and GGT^+ staining) were found in livers of all rats pretreated with aflatoxin B_1; their development was not affected by ethanol consumption, but their number and size were increased significantly ($p < 0.01$) by the high-fat diet as compared with the low-fat diet (Misslbeck et al., 1984).

Groups of 10–18 male Fischer rats, weighing 80 g, were each administered 0 or 25 µg aflatoxin B_1 (in 0.5 ml saline emulsion containing 2.5 µl each of DMSO and propylene glycol) by intubation on five days a week for eight weeks and were fed 0 or 0.015% β-naphthoflavone one week before, during and one week after exposure to aflatoxin B_1. The experiment was terminated 42 weeks after the beginning of aflatoxin B_1 administration. Survival was 10/10 vehicle controls without β-naphthoflavone, 18/18 vehicle controls with β-naphthoflavone, 15/17 rats receiving aflatoxin B_1 without β-naphthoflavone and 16/18 rats receiving aflatoxin B_1 with β-naphthoflavone. Administration of aflatoxin B_1 alone resulted in hepatomas in 15/15 rats and in neoplastic hepatic nodules in 11/15 rats. Combined treatment with aflatoxin B_1 and β-naphthoflavone markedly reduced the incidence of hepatomas (5/16) and neoplastic hepatic nodules (3/16). No neoplastic or preneoplastic liver lesion was found in rats receiving vehicle with or without β-naphthoflavone (Gurtoo et al., 1985).

A total of 267 male Fischer rats, eight weeks of age, were divided into nine groups [numbers of animals per group unspecified, except that there were 21 untreated controls] and were administered 0 or 25 μg/kg bw aflatoxin B_1 in 0.05 ml DMSO by gavage three times a week for 20 weeks (total dose of aflatoxin B_1, 1500 μg/kg bw) and a diet containing 0, 1000 or 6000 mg/kg of either butylated hydroxyanisole or butylated hydroxytoluene one week before, during and one week after administration of aflatoxin B_1. Rats that received aflatoxin B_1 alone were killed every two weeks from weeks 4 to 20 after the beginning of treatment; the four remaining rats in that group and animals in all other groups were killed 12 and 24 weeks after cessation of treatment with aflatoxin B_1. The additional administration of butylated hydroxytoluene and butylated hydroxyanisole resulted in a dose-related reduction in the combined incidence of hepatocellular carcinomas and neoplastic hepatic nodules: aflatoxin B_1 alone, 17/27; plus 1000 mg/kg of diet butylated hydroxytoluene, 3/25; plus 6000 mg/kg of diet butylated hydroxytoluene, 0/26; plus 1000 mg/kg of diet butylated hydroxyanisole, 3/23; plus 6000 mg/kg of diet butylated hydroxyanisole, 2/25. No hepatocellular neoplasm was observed in rats receiving basal diet, DMSO (vehicle control) or 6000 mg/kg of diet butylated hydroxyanisole or butylated hydroxytoluene alone. The number of foci of altered hepatocytes (as defined by iron exclusion and GGT^+ staining) in rats given aflatoxin B_1 alone increased with time from week 4 to week 20. In groups given butylated hydroxytoluene or butylated hydroxyanisole, there was a significant dose-related reduction in the number of foci of altered hepatocytes at the end of exposure to aflatoxin B_1 and 12 and 24 weeks thereafter (Williams *et al.*, 1986).

Groups of 6 and 12 male Fischer 344 rats, weighing 80 g, were administered 0 or 250 μg/kg bw aflatoxin B_1 in tricaprylin by gastric intubation on five days a week for two weeks and were then fed diets containing high (20%) or low (5%) levels of casein one week prior, during and one week after treatment with aflatoxin B_1; they then remained on the same diet or were switched from high to low or low to high casein for another 12 weeks. There was a significant ($p < 0.05$) increase in the number and volume of aflatoxin B_1-induced foci of altered hepatocytes (as defined by GGT^+ staining) in rats given the high-casein diet compared to those given the low-casein diet in the period after aflatoxin B_1 treatment (Appleton & Campbell, 1983a). In a similar experiment, the same treatment schedule was used up to one week after administration of aflatoxin B_1, but in addition to groups receiving diets containing high (20%) and low (5%) levels of casein throughout the subsequent 12 weeks, two groups received the high- or low-casein diet for six weeks and were then switched to low- or high-casein diet for another six weeks. As in the previous experiment, animals fed 5% casein in the diet throughout the 12-week period after treatment with aflatoxin B_1 had a marked reduction in the development of GGT^+ foci. Animals fed 20% casein in the diet throughout the same period had the greatest response. Groups fed 5% casein in the diet for half of the period after treatment and 20% for the other half had intermediate responses (Appleton & Campbell, 1983b).

Groups of 8 and 14 male Fischer 344 rats, weighing 40 g, were administered 0 or 250 μg/kg aflatoxin B_1 in tricaprylin by gastric intubation five times per week for two weeks. One week after completion of the aflatoxin treatment, the animals were given diets containing 4, 8, 10, 12, 15, 20 or 30% casein for 12 weeks and then sacrificed. Preneoplastic hepatic foci were observed in all animals treated with aflatoxin B_1; their number and volume were small in rats fed low levels of casein (4, 6, 8 and 10%) but larger in those receiving higher levels of casein (12, 20 and 30%) (Dunaif & Campbell, 1987a).

Groups of 12 male Fischer 344 rats, weighing about 40 g, were administered 0, 40, 100, 150, 200, 250, 300, 350 or 400 µg/kg bw aflatoxin B_1 in tricaprylin by gavage five days a week for two weeks and then received a diet containing 20% casein. The experiment was terminated 15 weeks after beginning of treatment with aflatoxin B_1. The incidence of preneoplastic hepatic foci (as defined by GGT^+ staining) was increased after treatment with aflatoxin B_1 at the higher doses (control, 0/12; 40, 0/12; 100, 0/12; 150, 8/12; 200, 11/12; 250, 9/12; 300, 11/11; 350, 8/8; 400 µg/kg bw, 5/5) (Dunaif & Campbell, 1987b). In another experiment, groups of 18–19 male Fischer 344 rats were administered 0, 200, 235, 270, 300 or 350 µg/kg bw aflatoxin B_1 in tricaprylin by gavage five days a week for two weeks; and, one week after the last treatment, the rats were placed on diets containing 4, 8, 12, 16 or 20% casein for 12 weeks and then sacrificed. Increasing the dose of aflatoxin B_1 from 200 to 350 µg increased the incidence and volume fraction of preneoplastic hepatic foci (as defined by GGT^+ staining) when the diet contained 20% casein but not when the diet contained 4, 8, 12 or 16% casein (Dunaif & Campbell, 1987b).

Groups of 46 and 51 male and 40 and 51 female Sprague-Dawley rats, 12 weeks old, were fed diets containing 0 or 2000 µg/kg of diet aflatoxin B_1, and a low level of protein (10.32%) and riboflavin or a normal content of proteins (24.2%) and vitamins. Four animals from each group were sacrificed at different times between 15 and 400 days after the beginning of treatment, when the experiment was terminated. Hepatocellular carcinomas were found in 18/23 males and 8/23 females fed aflatoxin B_1 in the diet with normal protein and vitamin content and which survived more than 320 days. No liver tumour was observed by 400 days in rats fed aflatoxin B_1 in a diet deficient in protein and riboflavin, or in rats on high- or low-protein diets with or without administration of aflatoxin B_1 (Stora et al., 1987).

Groups of 11 weanling male Fischer 344 rats were administered 250 µg/kg bw aflatoxin B_1 in tricaprylin intragastrically five times per week for two weeks and were then fed diets containing 20% casein or 20% gluten (low-quality protein) one week prior, during and one week after administration of aflatoxin B_1, followed by diets containing either 20% gluten, 20% casein, 5% casein or 20% gluten plus lysine for 11 weeks, when the experiment was terminated. Preneoplastic foci of altered hepatocytes (as defined by GGT^+ staining) were observed in all rats treated with aflatoxin B_1; the number and size of the foci was significantly greater ($p < 0.05$) in rats fed high-protein diets (20% casein; 20% casein followed by 20% gluten plus lysine) than in rats fed high-protein diets during aflatoxin B_1 treatment but afterwards switched to low-protein diet (20% gluten; 5% casein) or in rats fed low-protein diet throughout the experiment (20% gluten) (Schulsinger et al., 1989).

Groups of 10 weanling male Fischer rats were fed semi-purified diets containing 0 or 1000 µg/kg aflatoxin B_1 with or without 25% freeze-dried green beans, beetroot or squash for eight weeks. All animals were killed after eight weeks, and their livers were examined for the presence of preneoplastic hepatic foci (as defined by γ-glutamyltransferase positivity). Among aflatoxin B_1-treated groups, both the percentage of the liver area that was GGT^+ (aflatoxin B_1 alone, 17.9 ± 3.9; plus bean, 33.0 ± 3.8; plus beetroot, 28.5 ± 2.8; plus squash, 30.0 ± 5.0) and the number of foci per square centimetre (aflatoxin B_1 alone, 3.6 ± 3.0; plus bean, 12.8 ± 1.1; plus beetroot, 16.9 ± 6.5; plus squash, 12.8 ± 6.6) were significantly

($p \leq 0.05$) higher in each of the groups given vegetables than in the group given aflatoxin B_1 alone (Boyd et al., 1983).

Groups of 33–42 male Buffalo rats, weighing about 50 g, were fed a semi-synthetic diet containing 1000 μg/kg aflatoxin B_1, 25 mg/kg N-nitrosodimethylamine (NDMA) or a combination of the two for six months and were killed 3, 6, 9 and 12 months after the onset of the experiment. The number of effective animals ranged between 26 and 34. At 12 months, hepatocellular carcinomas were found in 9/20 rats treated with aflatoxin B_1, in 1/21 rats administered NDMA and in 11/14 rats given the combined treatment. The incidence of neoplastic hepatic nodules at this time was 7/20 in rats treated with aflatoxin B_1, 0/21 in animals administered NDMA and 11/14 after the combined treatment. Preneoplastic hepatic foci (clear-, acidophilic- and basophilic-cell foci) were observed in all treatment groups, but their incidence was higher in rats that received the combination of aflatoxin B_1 and NDMA then in those treated with aflatoxin B_1 or NDMA separately. In addition, multiple liver cysts were found at all times in animals fed aflatoxin B_1 plus NDMA, whereas rats fed aflatoxin B_1 and NDMA separately had developed only a few cysts by the end of the experiment. Neither neoplastic nor preneoplastic lesions were found in untreated controls (Angsubhakorn et al., 1981a).

Groups of about 25–28 male Buffalo rats, weighing 40–50 g, were fed diets containing 0 or 1000 μg/kg aflatoxin B_1, 500 mg/kg α-benzene hexachloride or a combination of aflatoxin B_1 and α-benzene hexachloride for 35 weeks and then placed on basal diet for 30 weeks. Three to 10 animals from each group were killed at intervals of 5, 10, 15, 35 and 65 weeks after the beginning of the experiment. Administration of aflatoxin B_1 resulted in both hepatocellular carcinomas and neoplastic hepatic nodules in 6/6 rats by 65 weeks. After additional administration of α-benzene hexachloride, no liver-cell carcinoma and only one neoplastic hepatic nodule was observed in 10 animals by 65 weeks. No change in the livers of rats fed either α-benzene hexachloride or the basal diet was noted. Foci of altered hepatocytes (predominantly acidophilic-cell foci) were found for the first time at 10 weeks (1/4 rats) and regularly (10/10 rats) at 35 and 65 weeks after treatment with aflatoxin B_1; they did not appear in rats that received aflatoxin B_1 plus α-benzene hexachloride up to 35 weeks, but were observed in 2/10 animals after the combined treatment by 65 weeks (Angsubhakorn et al., 1981b).

Groups of 10–12 male Fischer 344 rats, weighing 75–100 g, were administered 0 or 250 μg/kg bw aflatoxin B_1 in 0.1 ml tricaprylin by gavage five times a week for two weeks and then fed a diet containing 4000 mg/kg ethoxyquin one week prior to, during and one week after treatment with aflatoxin B_1. The experiment was terminated 16 weeks after the beginning of ethoxyquin administration. Preneoplastic foci of altered hepatocytes (as defined by GGT^+ staining) were found in all 12 rats treated with aflatoxin B_1 alone. Only 3/10 rats that also received ethoxyquin had preneoplastic hepatic foci, but a significantly lower number per square centimetre of liver tissue and volume fraction of total liver parenchyma ($p < 0.01$) as compared to those in rats that received aflatoxin B_1 alone (Kensler et al., 1986).

Groups of 8–10 male Fischer 344 rats, seven to eight weeks of age, were given a single intraperitoneal injection of 250 μg/kg bw aflatoxin B_1 and were then fed diets containing 0 or 5000 mg/kg of diet ethoxyquin four weeks prior to and two weeks after administration of aflatoxin B_1; they were then placed on diets containing 0 or 1000 µg/kg aflatoxin B_1 plus 0 or 5000 mg/kg ethoxyquin for another 17 weeks. Twenty-three weeks after the beginning of the experiment, all animals were killed. Preneoplastic foci of altered hepatocytes (as defined by morphology and several histochemical parameters) were often found in animals receiving aflatoxin alone but were not observed in animals additionally treated with ethoxyquin or in controls. In rats that received ethoxyquin, with or without aflatoxin B_1, many altered renal tubules (designated as hyperplastic and putatively preneoplastic) were found (Manson et al., 1987).

An unspecified number of male Fischer 344 rats, weighing 75–100 g, were administered 250 µg/kg bw aflatoxin B_1 by gavage on five days a week for two weeks and were then fed 0, 100, 200, 400 or 750 mg/kg of the schistosomicidal drug 5-(2-pyrazinyl)-4-methyl-1,2-dithiol-3-thione (oltipraz) one week prior to, during and one week after treatment with aflatoxin B_1. The experiment was terminated at the end of the fourth month after the beginning of treatment with oltipraz. All rats administered aflatoxin B_1 alone developed preneoplastic hepatic foci (as defined by GGT^+ staining); the average number of foci per square centimetre of liver tissue was 1.41 ± 0.29. Oltipraz induced a significant ($p < 0.01$), dose-dependent reduction in the number and size of preneoplastic hepatic foci (0, 1.41 ± 0.29; 100, 0.44 ± 0.11; 200, 0.21 ± 0.10; 400, 0.17 ± 0.06; 750, 0.03 ± 0.01). Preneoplastic hepatic foci were not found in 40, 20 and 30% of rats receiving 750, 400 and 200 mg oltipraz, respectively (Kensler et al., 1987).

Groups of 55 and 56 male Fischer 344 rats, five weeks of age, were each administered 25 µg aflatoxin B_1 (purity, > 97%) by gavage in 0.1 ml tricaprylin on five days a week for two weeks (total dose of aflatoxin B_1, approximately 2000 µg/kg bw) and then received a purified diet containing 0 or 0.075% [750 mg/kg of diet] of oltipraz one week before, during and one week after the beginning of treatment with aflatoxin B_1. Ten rats from each group were killed 15 weeks after the first dose of aflatoxin B_1. At 23 months after the first dose of aflatoxin B_1, survival of the remaining animals was 10/45 in the group treated with aflatoxin B_1 alone and 24/46 in the group receiving aflatoxin B_1 plus oltipraz. Oltipraz significantly decreased the incidence of hepatocellular tumours (aflatoxin B_1 alone: 9/45 (five carcinomas, four adenomas); aflatoxin B_1 plus oltipraz: 0/45). The difference in the incidence of hepatocellular carcinomas and adenomas between the two groups was statistically significant ($p < 0.05$, χ^2-test). Foci of altered hepatocytes (detected by GGT^+ staining) were found in both groups, but their number and size were significantly smaller in rats treated with aflatoxin B_1 plus oltipraz than in those treated with aflatoxin B_1 alone (Roebuck et al., 1991).

Groups of 24–34 female Wistar rats, weighing 100 g, were each administered 0.25 µg aflatoxin B_1 in 1.0 ml water (diluted after dissolution in DMSO) by gavage five times a week for eight weeks, followed, after an interval of 16 months, by water (control) or by 100 mg/animal reduced glutathione, 5 mg/animal butylated hydroxytoluene, 80 mg/animal methionine or 50 mg/animal ascorbic acid in 2.5 ml water for an additional eight months, when all remaining animals were sacrificed. None of the additional treatments increased the percentage of aflatoxin-treated animals alive at the terminal sacrifice date, but after the methionine and ascorbic acid treatments significantly ($p < 0.05$) fewer animals were alive at termination of the experiment. No treatment had any effect on the number or size of hepatic nodules, but the incidence of cystic cholangiomas was reduced by ascorbic acid treatment (Iverson *et al.*, 1987).

Groups of 25 female Buffalo rats, weighing 100 g, were fed a semi-synthetic diet, a diet containing 1000 µg/kg aflatoxin B_1 for 15 weeks, a diet containing 100 mg/kg lindane for 15 weeks or a diet containing aflatoxin B_1 and lindane for 1, 3, 5, 10 or 15 weeks and then the semi-synthetic diet until week 15. All animals then received basal diet for 67 weeks. Survival at the end of the experiment (82 weeks) ranged from 13/25 in the group that received both aflatoxin B_1 and lindane for 15 weeks to 23/25 in the untreated control group. Administration of aflatoxin B_1 alone for 15 weeks resulted in hepatocellular carcinomas in 6/19 rats by the end of the experiment. Administration of lindane in the diet appeared to prevent the induction of liver tumours by aflatoxin B_1 at 82 weeks: one hepatocellular carcinoma was seen in the 19 animals that received aflatoxin B_1 and lindane for one week, and no liver tumour occurred in any other treatment group (Angsubhakorn *et al.*, 1989).

Groups of 28–34 Fischer 344 rats of each sex, weighing 75–100 g, were each given 25 µg aflatoxin B_1 in the drinking-water five times a week for 20 weeks and administered either 2 units of adrenocorticotropin subcutaneously twice a week for 20 weeks, 0.25 units of insulin twice a week for 20 weeks or 0.5 units of growth hormone twice a week for 10 weeks. Groups of four rats were killed at various times between 7 and 77 weeks after the beginning of the experiment. Hepatocellular carcinomas were induced by 77 weeks in all 17 animals given aflatoxin B_1 alone, in all 12 animals given aflatoxin B_1 and growth hormone, in 6/10 animals given aflatoxin B_1 and insulin and in none of eight untreated controls. The development of hepatocellular carcinomas was preceded by focal and nodular liver lesions (designated as hyperplastic or neoplastic) in all groups, the incidence of which was correlated with the incidence of liver carcinomas in the different groups. Six animals that received aflatoxin B_1 and adrenocorticotropin simultaneously failed to develop hepatocellular carcinomas by 77 weeks, but 3/6 animals had malignant lymphomas at 56 weeks and 8/8 animals had lymphoid hyperplasia at 35 weeks; these lesions were not seen in other groups (Chedid *et al.*, 1977).

Groups of 10–14 male and 7–14 female Wistar rats, 12 weeks of age, were administered 0 and 50 µg/animal aflatoxin B_1 in 0.1 ml DMSO intragastrically twice a week for four weeks, followed by 0 or 70 µg/animal aflatoxin B_1 in 0.15 ml DMSO twice a week for an additional six weeks, with or without thyroidectomy six days before the beginning of aflatoxin B_1 treatment. Further groups received 0 or 0.5 mg/animal 1-methyl-2-mercaptoimidazole in 0.5 ml water intragastrically twice weekly from the beginning of aflatoxin B_1 treatment to the end of the experiment and were given 0 or 0.05 g calcium pantothenate in the drinking-water throughout the experiment. The experiment was terminated 58 weeks after its start. At this time, the incidence of hepatomas [histology not further specified] in animals treated with aflatoxin B_1 alone was 11/12 in males and 10/11 in females, and that of cholangiocellular adenomas 5/12 in males and 3/11 in females. After combined treatment with aflatoxin B_1 and 1-methyl-2-mercaptoimidazole, hepatomas were observed in only 1/11 males and 0/9 females, and no cholangiocellular tumour occurred. There was also a much lower incidence of precancerous liver lesions (designated as focal and nodular hyperplasia) after the combined treatment than after administration of aflatoxin B_1 alone. No hepatoma but discrete focal hyperplastic changes and a single focus of nodular hyperplasia were observed when the combined treatment with aflatoxin B_1 and 1-methyl-2-mercapatoimidazole was preceded by thyroidectomy. No neoplastic or preneoplastic liver lesion was found in other groups, including the untreated controls (Bednarz, 1989).

Groups of 18 and 42 male Wistar rats, nine weeks of age, were each given 0 or 42 µg aflatoxin B_1 in DMSO in drinking-water three times a week for 27 weeks (total dose of aflatoxin B_1, 3.4 mg/animal) and 0, 3 or 6 mg/kg of water sodium selenite up to 79 weeks after the beginning of the experiment. Seven to 10 animals in each group were sacrificed at 18, 30, 38, 51 and 79 weeks. Malignant liver tumours (predominantly hepatocellular carcinomas) developed in 11/18 rats treated with aflatoxin B_1 alone but not in animals that received the combined treatment with aflatoxin B_1 and sodium selenite or in untreated controls. At each interim sacrifice, the number and size of aflatoxin B_1-induced preneoplastic foci of altered hepatocytes (basophilic-, acidophilic- and clear-cell foci) was reduced by additional administration of selenium. The inhibitory effect on the development of preneoplastic liver lesions was more pronounced at the low than at the high dose of selenium (Lei *et al.*, 1990).

Groups of 10–30 male Wistar rats, nine weeks old, were each given 0 or 42 µg aflatoxin B_1 (dissolved in DMSO prior to dilution in water) in drinking-water three times a week for 27 weeks (total dose of aflatoxin B_1 per rat, 3.4 mg at 27 weeks) with and without cultured extracts of *Rhizopus delemar* (an edible yeast). Unspecified numbers of rats were sacrificed 18, 30, 38 and 52 weeks after the beginnng of the experiment [number of surviving animals unspecified]. At 52 weeks, 5/7 rats that had received aflatoxin B_1 alone had liver neoplasms (two hepatocellular carcinomas, three combined hepatocellular–cholangiocellular carcinomas). No liver neoplasm was observed in rats treated with aflatoxin B_1 and extracts of *R. delemar*. Various types of foci of altered hepatocytes (as defined histologically or by histochemical demonstration of adenosinetriphosphatase, glucose-6-phosphatase and GGT) were seen in rats treated with aflatoxin B_1 for more than 18 weeks; the number and size of such foci were significantly lower ($p < 0.01$) in rats that received aflatoxin B_1 with extracts of *R. delemar* (Zhu *et al.*, 1989). [The Working Group noted the incomplete reporting.]

Female Wistar rats, 13 weeks of age, received a single intragastric dose of 5 mg/kg aflatoxin B_1 dissolved in 96.6% olive oil:3.5% DMSO. Three weeks later, 25 treated and 28 untreated rats were fed a diet containing nafenopin, the concentration of which was adjusted to provide a daily dose of 100 mg/kg bw; 24 aflatoxin-B_1 treated and 28 untreated rats were fed normal diet. Subgroups of nine rats killed after 55 weeks had hepatocellular adenomas at incidences of 0 (nafenopin only), 11% (aflatoxin only) and 44% (aflatoxin plus nafenopin); hepatocellular carcinomas occurred in 11% (nafenopin only), 0 (aflatoxin only) and 11% (aflatoxin plus nafenopin). Subgroups of 10–14 rats killed at week 70 had hepatocellular adenomas at incidences of 43% (nafenopin only), 0 (aflatoxin only) and 72% (aflatoxin plus nafenopin); hepatocellular carcinomas occurred at incidences of 29% (nafenopin only), 0 (aflatoxin only) and 45% (aflatoxin plus nafenopin). In a similar experiment, male Wistar rats, weighing 260–300 g, received a single dose of 2000 µg/kg aflatoxin B_1. At termination, after 55–59 weeks, the following liver tumour incidences were obtained: adenoma, 41% (nafenopin only, 17 rats), 0 (aflatoxin only, 22 rats), 80% (aflatoxin plus nafenopin, 10 rats); carcinoma, 29% (nafenopin only), 0 (aflatoxin only), 60% (aflatoxin plus nafenopin). No hepatic adenoma or carcinoma was found in 20 female or 18 male control rats. Foci of altered hepatocytes were checked at various times between 3 and

70 weeks in haematoxylin–eosin-stained sections from animals of each sex. Clear eosinophilic and tigroid foci were found in all aflatoxin-treated groups, the numbers and size increasing with time, but were rarely seen in controls. Nafenopin treatment had little effect on the prevalence of eosinophilic foci but decreased that of tigroid foci; it strongly (approximately 20 fold) enhanced the appearance of a further subpopulation of weakly basophilic foci, which were very rare in rats treated only with aflatoxin and in untreated controls (Kraupp-Grasl *et al.*, 1990).

Trout: Duplicate groups of 150 rainbow trout fingerlings were fed diets containing 0, 4, 8 or 20 µg/kg aflatoxin B_1 or 4 µg/kg of diet aflatoxin B_1 (purified) plus either 250 mg/kg gossypol, 50 µg/kg 3-methylcoumarin, 0.5% polymerized corn oil or 220 mg/kg cyclopropenoid fatty acids. Samples representing 10% of the fish in each group were taken after 3, 6, 9, 12 and 15 months. After administration of aflatoxin B_1 alone, the incidence of hepatomas increased with increasing dose and time (4 µg aflatoxin B_1: 6 months, 0/30; 9 months, 4/20; 12 months, 3/20; 15 months, 4/20; 8 µg aflatoxin B_1; 6 months, 2/30; 9 months, 5/20; 12 months, 8/20; 15 months, 17/20; 20 µg aflatoxin B_1: 6 months, 1/30; 9 months, 11/20; 12 months, 13/20; 15 months, 16/17). Cyclopropenoid fatty acids fed in the diet containing 4 µg/kg aflatoxin increased the incidence and growth of the hepatomas to 27/30 at 6 months, 20/20 at 9 months and 9/10 at 12 months. Gossypol and 3-methylcoumarin did not promote the early development of tumours induced by the diet containing 4 µg/kg aflatoxin B_1 but resulted in a higher incidence of hepatomas after 12 months (gossypol: 12 months, 6/20; 15 months, 12/20; 3-methylcoumarin: 12 months, 8/20; 15 months, 11/20). Polymerized corn oil did not enhance the carcinogenicity of aflatoxin B_1. No hepatoma was found in fish fed control diet or control diet plus 3-methylcoumarin (Sinnhuber *et al.*, 1968b).

Groups of 120 rainbow trout, 10 weeks of age, were fed diets containing 0 or 6 µg/kg aflatoxin B_1 and received 0 or 100 mg/kg of diet of a polychlorinated biphenyl (Aroclor 1254) for 12 months. Samples of fish were taken at different times from one to nine months, and the remaining animals were killed 12 months after the beginning of the experiment. The first liver neoplasms were observed at nine months (untreated controls, 0/10; aflatoxin B_1, 1/8; Aroclor 1254, 0/6; aflatoxin B_1 plus Aroclor 1254, 1/10). After 12 months, the incidence of hepatocellular carcinomas was significantly ($p < 0.001$) reduced in trout treated with aflatoxin plus the Aroclor 1254 compared to aflatoxin B_1 alone (controls, 0/68; aflatoxin B_1, 26/37; Aroclor 1254, 0/39; aflatoxin B_1 plus Aroclor 1254, 14/46). In addition, fewer tumours per liver were produced in trout on the combined treatment, and the tumours were smaller than those of trout treated with aflatoxin B_1 alone (Hendricks *et al.*, 1977).

Duplicate groups of 250 rainbow trout, six weeks old, were fed diets containing 0 or 6 µg/kg aflatoxin B_1 and received 0 or 5 mg/kg of diet dieldrin (purity, 99%) for 12 months. The first hepatocellular carcinomas were observed after nine months (untreated controls, 0/80; aflatoxin B_1, 19/80; dieldrin, 0/51; aflatoxin B_1 plus dieldrin, 31/80). There was no significant difference in the incidence of hepatocellular carcinomas induced by aflatoxin B_1 alone and by the combined treatment with aflatoxin B_1 and dieldrin (untreated controls, 0/147; aflatoxin B_1, 102/146; dieldrin, 0/149; aflatoxin B_1 plus dieldrin, 113/145) (Hendricks *et al.*, 1979).

Duplicate groups of 100 rainbow trout [age unspecified] were fed semi-purified diets containing 0, 1, 4 or 8 µg/kg aflatoxin with or without 50 mg/kg Aroclor 1254. In each group,

51–74 fish were sacrificed at nine months and 118–126 at 12 months. The presence of 50 mg/kg Aroclor 1254 at 9 and 12 months significantly ($p \leq 0.001$) inhibited the dose-dependent incidence of aflatoxin B_1-induced hepatocellular carcinomas (Shelton et al., 1984).

Duplicate groups of 100 rainbow trout fingerlings were fed diets containing 0 or 20 µg/kg aflatoxin B_1 for two weeks and 0, 50 or 500 mg/kg β-naphthoflavone, 1000 mg/kg flavone, 1000 mg/kg tangeretin-nobilitin mixture, 1000 mg/kg indol-3-carbinol, 1000 mg/kg β-ionone or 2000 mg/kg of diet quercetin during and six weeks after the treatment with aflatoxin B_1. The fish were kept on basal diet for another 52 weeks. Samples of 20 fish were taken from each group at 32 and 45 weeks, and the remaining animals were killed 58 weeks after the end of aflatoxin B_1 exposure. At 58 weeks, the incidence of hepatic neoplasms (predominantly hepatocellular carcinomas, some basophilic hepatic foci included) induced by aflatoxin B_1 was significantly reduced by additional feeding with indol-3-carbinol ($p < 0.01$) and β-naphthoflavone (high-dose, $p < 0.01$; low-dose, $p < 0.05$); the other compounds were less effective (Nixon et al., 1984).

Duplicate groups of 100 rainbow trout, 19 weeks old, were fed diets containing 0 or 20 µg/kg aflatoxin B_1 for four weeks and 0, 500 or 2000 mg/kg indol-3-carbinol either eight weeks before and during or 12 weeks after aflatoxin exposure. Samples of fish were taken 30, 40 and 52 weeks after cessation of aflatoxin B_1 exposure. The high dose of indol-3-carbinol given before and during aflatoxin exposure did not affect the incidence of hepatocellular carcinomas, whereas the same dose given after aflatoxin B_1 exposure significantly ($p < 0.0001$) increased the tumour incidence compared to that in fish treated with aflatoxin B_1 alone (Bailey et al., 1987).

Triplicate groups of 150 rainbow trout fingerlings were administered ^3H-aflatoxin B_1 at 0, 10, 20, 40, 80, 160 and 320 µg/l of water and fed a diet containing 0, 1000, 2000, 3000 or 4000 mg/kg indol-3-carbinol four weeks prior to and during aflatoxin B_1 treatment. Groups of 15 trout were selected randomly and killed after 7 and 14 days of treatment with aflatoxin B_1; the remaining 120 fish from each group were observed up to 52 weeks after cessation of treatment. Administration of aflatoxin B_1 alone at doses of 10, 20, 40 and 80 µg/l of water resulted in incidences of liver tumours [histology unspecified] ranging from 10.7 to 68.4%. Tumour incidences in fish treated with aflatoxin B_1 plus indol-3-carbinol showed a dose-related reduction (Dashwood et al., 1989). In another experiment, groups of 100 trout fingerlings were exposed to aflatoxin B_1 at 12.5, 25, 50, 100, 200 or 400 µg/l of water for 30 min and subsequently fed diets containing 0, 750, 1500 or 2000 mg/kg indol-3-carbinol for 24 weeks. The incidences of tumours [histological types unspecified] in fish treated with aflatoxin B_1 followed by indol-3-carbinol were increased compared to those treated with aflatoxin B_1 alone (Dashwood et al., 1990).

Duplicate groups of 100 rainbow trout fry were exposed to aflatoxin B_1 at 12.5 µg/l of water, followed by a diet containing 0 or 2000 µg/kg indol-3-carbinol for three, six and nine months, beginning zero, three and six months after treatment with aflatoxin B_1 or administered on alternating months after aflatoxin B_1 administration. In addition, triplicate groups of 100 rainbow trout fry were exposed to aflatoxin B_1 at 50 µg/l of water, followed by a diet containing 0 or 2000 µg/kg indol-3-carbinol administered on alternating weeks or on two days each week. The experiment was terminated after 36 weeks. Trout exposed to aflatoxin B_1 at 12.5 µg/l of water had a liver tumour incidence of 19.1%; fish exposed to the same dose of aflatoxin B_1 immediately followed by 2000 µg/kg indol-3-carbinol for 12, 24 and 36 weeks had average tumour incidences of 27.3, 41.7 and 53.9%, respectively, the incidence in the latter two groups being significantly increased ($p < 0.05$) compared to that in trout treated with aflatoxin B_1 alone. Trout exposed to aflatoxin B_1 at 12.5 µg/l of water, followed by 2000 µg/kg indol-3-carbinol after a delay of 4, 12 or 24 weeks had average liver tumour indidences, respectively, of 56.8, 53.9 and 33.3%, the incidence being significantly increased ($p < 0.05$) in the two groups with longer treatment. Trout exposed to aflatoxin B_1 at 12.5 µg/l of water, followed by 2000 µg indol-3-carbinol administered for alternating one- or three-month periods had tumour incidences of 35 and 43.6%, only the latter being significantly increased ($p < 0.05$). Exposure to aflatoxin B_1 at 50 µg/l of water followed by 0 or 2000 µg/kg indol-3-carbinol either on alternating weeks or on two days each week induced liver tumour incidences of 40.4, 63.9 and 62.0%, respectively. The liver tumour incidence in trout treated with aflatoxin B_1 alone was significantly ($p < 0.05$) higher than that in fish that received indol-3-carbinol (Dashwood et al., 1991).

(ii) *Intraperitoneal administration*

Rat: Groups of 30 male Fischer 344 rats, four weeks old, were each given 0 or 100 µg aflatoxin B_1 in 0.2 ml saline (after dissolution in dimethylformamide) intraperitoneally and were then administered 0 or 1.5 mg corticosterone or hydrocortisone or 2 units of adrenocorticotropin intramuscularly twice a week for 10 weeks, and were sacrificed 65 weeks after beginning of treatment. The additional treatment with hormones resulted in a significant reduction in the incidence of aflatoxin B_1-induced hepatocellular carcinomas at 65 weeks: aflatoxin B_1 alone, 12/13; plus adrenocorticotropin, 9/27; plus hydrocortisone, 15/24; plus corticosterone, 12/23. In addition to hepatocellular carcinomas, neoplastic hepatic nodules and basophilic lesions were observed in several animals (Chedid et al., 1980).

Groups of 20 male Fischer 344 rats, weighing 150 g, were administered a single intraperitoneal dose of 0, 62.5, 250 or 1000 µg/kg bw aflatoxin B_1 in DMSO; after an interval of two weeks, they were fed 200 mg/kg of diet 2-acetylaminofluorene for two weeks and were then partially hepatectomized at the end of the third week of the experiment. All rats were killed at the end of week four. Administration of aflatoxin B_1 resulted in a dose-dependent increase in the number (and size) of hyperplastic liver lesions (composed of acidophilic and basophilic cells) (Imaida et al., 1981).

Groups of 100 well-fed and 120 malnourished (by food restriction and increased litter size during suckling up to 21 days after birth, resulting in a 60% decrease in body weight) male newborn Holtzmann rats were administered 500, 250 and 250 µg/kg bw of aflatoxin B_1 in DMSO intraperitoneally on days 10, 14 and 18, respectively, and were killed at between 30 and 65 weeks. The effective numbers of animals were 34/100 in the well-fed and 43/120 in the malnourished group. Neoplastic hepatic nodules were observed only in malnourished rats treated with aflatoxin B_1 between 46 and 55 weeks (malnourished, 3/13; well-fed, 0/8) and between 56 and 65 weeks (malnourished 6/17; well-fed, 0/15). The differences in the incidence of neoplastic hepatic nodules were significant ($p = 0.006$; Fisher's exact probability test). Phenotypically altered preneoplastic hepatic lesions (consisting of clear, acidophilic, basophilic or vacuolated cells) were observed in both groups of aflatoxin B_1-treated rats but appeared earlier and progressed more rapidly in the malnourished rats (Rizvi et al., 1987).

Groups of 10–15 male ACI/N rats, six weeks of age, were administered 0 or 1500 µg/kg bw aflatoxin B_1 in 0.2 ml DMSO intraperitoneally twice a week for 10 weeks (total dose of aflatoxin B_1, 30 mg/kg bw), followed, after an interval of one week, by oral administration of ethanol (10% in drinking-water) or plain water for 56 weeks. In order to demonstrate iron-excluding (preneoplastic) liver lesions, all rats were given subcutaneous injections of 125 mg/kg bw elemental iron three times a week for two weeks prior to killing at 67 weeks. The incidence of hepatocellular neoplasms was 3/15 (one hepatocellular carcinoma, two neoplastic nodules) in rats exposed to aflatoxin B_1 followed by ethanol and 0/14 in animals treated with aflatoxin B_1 alone. No hepatocellular tumour was observed in 15 vehicle controls given 10% ethanol in the drinking-water or in 10 vehicle controls. The number of iron-excluding hepatic foci per square centimetre of liver tissue section was 6.98 in rats treated with aflatoxin B_1 alone, 26.39 in rats treated with aflatoxin B_1 followed by ethanol, 0.11 in the vehicle controls treated with ethanol and 0.09 in the vehicle controls. Similar results were obtained by morphometric evaluation of foci of altered hepatocytes as detected in haematoxylin–eosin-stained tissue sections. The increase in the number (and total area) of preneoplastic hepatic foci in rats treated with aflatoxin B_1 followed by ethanol as compared to those treated with aflatoxin B_1 alone was significant ($p < 0.001$) (Tanaka et al., 1989).

Groups of 13 and 18 male Fischer 344 rats, eight weeks old, were administered 150 µg/kg bw aflatoxin B_1 intraperitoneally five times a week for two weeks, submitted to partial hepatectomy three weeks after the last dose of aflatoxin B_1, were given 0 or 75 mg/kg bw phenobarbital by gavage during week 9 of the experiment and were sacrificed one day after the last dose of phenobarbital. The size and area of aflatoxin-induced GGT^+ hyperplastic hepatic nodules were not significantly affected by additional treatment with phenobarbital, which, however, induced the expression of the cytochrome P450 isozyme 2B1/2 in all nodules (Chen & Eaton, 1991).

Groups of 7–16 male Wistar rats, 60 days of age, were administered [route unspecified but assumed to be intraperitoneal] a single dose of 0 or 500 µg/kg bw aflatoxin B_1 in saline, were fed a diet containing 5, 15 or 40% casein 28 days prior to and four days after exposure to aflatoxin B_1, then fed the 15% casein diet for an additional 24 days, followed by 200 mg/kg of diet 2-acetylaminofluorene for a period of two weeks, in the middle of which a two-thirds partial hepatectomy was performed. All animals were killed 10 days after completion of the administration of 2-acetylaminofluorene. The number of foci of altered hepatocytes (as defined by GGT^+ staining or the placental form of glutathione S-transferase and calculated per liver) was increased significantly ($p < 0.05$) with increasing levels of casein in the diet following aflatoxin, while increasing levels of casein without aflatoxin did not increase the number of foci (Blanck et al., 1992).

Trout: Groups of 25 rainbow trout fingerlings were given 0 or 50 µg/kg bw aflatoxin B_1 in propylene glycol and (3 h prior to each injection of aflatoxin B_1) 0 or 5 mg/kg bw of β-diethylaminoethyldiphenyl propylacetate hydrochloride (SKF-525A) intraperitoneally twice weekly for 25 weeks (total doses: aflatoxin B_1, 25 mg/kg bw; SKF-525A, 250 mg/kg bw). The experiment was terminated after 50 weeks. Survival after 25 weeks was 20/25 in the group treated with aflatoxin B_1 alone and 17/25 in the group that received the combined treatment.

Treatment with aflatoxin B_1 alone induced hepatocellular carcinomas in 16/20 fish. When SKF-525A was administered prior to each injection of aflatoxin B_1, the incidence of hepatocellular carcinomas was 1/17. Of the animals that did not develop hepatocellular carcinomas, 3/20 given aflatoxin B_1 alone and 2/17 given the combined treatment had proliferative basophilic hepatic foci (Scarpelli, 1976).

(iii) *Skin application*

Mouse: In a two-stage protocol, groups of eight female CD-1 mice, six weeks old, received a topical application to the skin of aflatoxin B_1 as a single dose of 25, 50 and 100 µg (dissolved in benzene) followed by 1000 µg croton oil (dissolved in acetone) twice weekly for 22 weeks. Two additional groups of eight mice received topical applications of doses of 100 and 500 µg twice weekly for 22 weeks after initiation with a single dose of 50 µg 7,12-dimethylbenz[*a*]anthracene (DMBA) dissolved in acetone. Animals treated with DMBA followed by croton oil served as positive controls and mice treated with the solvents alone as negative controls. Five to six of the eight animals that received aflatoxin B_1 as an initiator, followed by croton oil, developed skin papillomas. No animal receiving DMBA as an initiator, followed by aflatoxin B_1, developed a skin papilloma. All of the positive controls and none of the negative controls developed skin papillomas (Lindenfelser *et al.*, 1974).

(iv) *Exposure of fish eggs*

Trout: Groups of 200 rainbow trout eggs, 21 days after fertilization, were exposed to aflatoxin B_1 (added in 0.5 ml of 95% ethanol) at 0 or 500 µg/l of water for 1 h. After hatching, duplicate groups of 100 fry were fed semi-purified diet with or without 100 mg/kg Aroclor 1254 for one year. Samples of fish were taken at 9 and 12 months. At the end of 12 months, the incidences of aflatoxin B_1-induced hepatocellular carcinomas were essentially the same in fish additionally treated with Aroclor 1254 and those that were not: untreated controls, 0/107; aflatoxin B_1, 79/120; Aroclor 1254, 1/108; aflatoxin B_1 plus Aroclor 1254, 69/108 (Hendricks *et al.*, 1980b).

Duplicate groups of 100 rainbow trout embryos were exposed to a single dose of aflatoxin B_1 at 500 µg/l of water 21 days after fertilization; after hatching, they were fed diets containing 0, 0.03 or 0.3% butylated hydroxyanisole or 0.005 or 0.05% β-naphthoflavone for eight weeks. Fish were kept for up to one year after exposure to aflatoxin B_1. In trout treated with aflatoxin B_1 alone, the incidence of hepatocellular carcinomas was 83/191, and that of cholangiocellular carcinomas was 5/191. There was no significant difference in the incidence of liver tumours in trout treated with aflatoxin B_1 plus butylated hydroxyanisole at either dose level and in trout treated with aflatoxin B_1 alone. Administration of β-naphthoflavone at the high dose level to aflatoxin B_1-treated trout significantly ($p < 0.01$) enhanced the incidence of hepatocellular carcinomas (104/171) (Goeger *et al.*, 1988).

Groups of 200 rainbow trout eggs were exposed 21 days after fertilization to aflatoxin B_1 at 0, 5, 25 or 125 µg/l of water for 30 min. Six weeks later, duplicate groups of 60 trout from each group received 0 or 20 mg/kg of diet 17β-oestradiol for five weeks and, after a treatment-free interval of four weeks, were fed 0 or 10 mg/kg of diet 17β-oestradiol for up to nine months after the beginning of oestradiol treatment. At nine months, a low incidence of hepatocellular and/or mixed hepatocellular/cholangiocellular carcinomas (6/119) and of hepatocellular adenomas (1/119) was observed in trout exposed as embryos to aflatoxin B_1 alone at the highest dose, but no liver neoplasm was found in the medium- and low-dose aflatoxin B_1 group at nine months. 17β-Oestradiol increased the incidence of liver carcinomas in fish exposed to aflatoxin B_1 as embryos at all dose levels (5 μg/l, 1/75; 25 μg/l, 8/90; 125 μg/l, 53/89). This increase was significant ($p < 0.001$) in the groups treated with the medium and high doses of aflatoxin B_1. Additional treatment with 17β-oestradiol of trout exposed as embryos to the highest dose level of aflatoxin B_1 also resulted in an increased incidence of hepatocellular adenomas and/or basophilic hepatic foci (aflatoxin B_1 alone, 1/119; aflatoxin B_1 plus 17β-oestradiol, 9/89). No neoplastic or preneoplastic liver lesion was observed in untreated controls or in fish treated with 17β-oestradiol alone (Núñez et al., 1989).

In another experiment, groups of 200 rainbow trout eggs were exposed 21 days after fertilization to aflatoxin B_1 at 0 or 0.25 μg/l of water for 30 min. Five weeks later, duplicate groups of 70 fish were fed 0, 5, 10 or 15 mg/kg of diet 17β-oestradiol on alternate weeks up to nine months after the beginning of oestradiol treatment. 17β-Oestradiol significantly ($p < 0.05$) increased the incidence of hepatocellular carcinomas induced by aflatoxin B_1 (no oestradiol, 1/130; low-dose oestradiol, 7/136; medium-dose oestradiol, 7/140; high-dose oestradiol, 10/132) (Núñez et al., 1989).

3.6.3 Aflatoxin M_1

Trout: Duplicate groups of 60 rainbow trout, 50 days old, were fed diets containing 0, 4, 16, 32 or 64 μg/kg aflatoxin M_1 or 4 μg/kg aflatoxin B_1 plus 0 or 100 mg/kg of diet cyclopropenoid fatty acid for 12 months, and then placed on basal diet. Random samples of fish were taken at 4, 8, 12 and 16 months. Administration of aflatoxin M_1 increased the incidence of hepatomas after 12 and 16 months at dose levels of 4 μg (12 months, 6/46; 16 months, 8/20), 16 μg (12 months, 30/43; 16 months, 13/20), 32 μg (12 months, 28/46; 16 months, 19/20) and 64 μg/kg of diet aflatoxin M_1 (12 months, 30/50; 16 months, 14/20). Treatment with aflatoxin M_1 at a dose of 4 μg/kg of diet produced a lower incidence of hepatomas than aflatoxin B_1 at 4 μg/kg of diet (12 months, 22/46; 16 months, 12/20). Additional administration of cyclopropenoid fatty acid increased the hepatocarcinogenic effect of both aflatoxin M_1 and aflatoxin B_1 given at a dose level of 4 μg/kg of diet. No neoplasm was found in fish fed control diet (Sinnhuber et al., 1974).

In another experiment, groups of 50 rainbow trout, 50 days old, were fed diets containing 20 μg/kg aflatoxin M_1 and 0 or 100 mg/kg cyclopropenoid fatty acid for 5, 10, 20 or 30 days and then placed on basal diet. Samples of 20 fish were taken from each group at 8 months, and the remaining trout were sacrificed at 12 months. The incidence of hepatomas at 12 months was markedly enhanced by additional administration of cyclopropenoid fatty acid: treatment with aflatoxin M_1 plus cyclopropenoid fatty acid for five days, 3/28; 10 days, 10/27; 20 days, 16/32; 30 days, 20/26 (Sinnhuber et al., 1974).

4. Other Relevant Data

4.1 Absorption, distribution, metabolism and excretion

4.1.1 *Humans*

Aflatoxins can cross the placental barrier in humans. In a study carried out in Songkhla, Thailand, during the dry season, Denning *et al.* (1990) showed that the concentrations of aflatoxins (B_1, G_1 and Q_1) were higher in cord sera obtained at birth than in maternal sera obtained immediately after birth. Seventeen of the 35 cord sera samples contained aflatoxin at concentrations between 0.064 and 13.6 nmol/ml. The sera of two of the mothers contained aflatoxin, the highest value being 1.22 nmol/ml. All of the women stated that they had eaten groundnuts, maize and mouldy food frequently, but the relatively low proportion who had aflatoxin in their sera reflects the fact that the dry season in this area is associated with low levels of aflatoxin contamination.

Aflatoxins M_1, M_2, B_1, B_2, G_1 and G_2 have been detected in cord blood samples from Ghana (63/188, 34%) and aflatoxins M_1, M_2 and B_2 in cord blood from Nigeria (9/78, 12%). Higher levels of aflatoxins M_1 and M_2 were detected in cord blood than in maternal blood samples collected simultaneously in Nigeria, suggesting that aflatoxins may accumulate in fetuses exposed to these toxins *in utero*; aflatoxin B_1 was found only in maternal blood in Nigeria (Lamplugh *et al.*, 1988).

Lactating African women (in the Sudan, Ghana, Kenya and Nigeria) exposed to aflatoxins in the diet secrete principally aflatoxin M_1 in their milk, less frequently aflatoxin M_2 and even less frequently aflatoxins B_1, B_2, G_1 and G_2 (Maxwell *et al.*, 1989). In Ghana, aflatoxin M_1 was detected in mothers' milk most frequently at concentrations ranging from 20 to 1816 ng/l and more often in the wet season (41%) than in the dry season (28%); the mean concentration of aflatoxin M_1 was higher in the wet season (445 ng/l) than in the dry season (293 ng/l). Aflatoxins M_2, B_1 and B_2 were also detected (Lamplugh *et al.*, 1988).

Aflatoxin M_1 is present in the urine of humans exposed to aflatoxin B_1, at levels correlated with that of intake of aflatoxin B_1 (Campbell *et al.*, 1970; Zhu *et al.*, 1987).

Evidence that humans activate aflatoxin B_1 *in vivo* is provided by the detection of bound aflatoxins in serum albumin samples (Gan *et al.*, 1988) and by the excretion of aflatoxin B_1-guanine adducts (Autrup *et al.*, 1983; Groopman *et al.*, 1985; Autrup *et al.*, 1987). In the Gambia, 30 pregnant women were recruited, and sera from umbilical cord and maternal venous blood were assayed for aflatoxin–albumin adducts immediately after delivery. There was a highly significant correlation between adduct levels in maternal venous blood and matched cord sera ($r = 0.52, p = 0.001$) (Wild *et al.*, 1991). In one study, hepatitis B virus carriers had higher levels of aflatoxin–serum albumin adducts (mean, 4.41) than non-carriers (mean, 4.04) or uninfected people in the same village in the Gambia (Allen *et al.*, 1992). Human urine samples from exposed Chinese populations contained aflatoxin B_1-*N*7-guanine at a range of 0.1–10 ng/ml, in addition to aflatoxin M_1 and P_1 (Groopman *et al.*, 1985); the adduct has also been detected in the urine of African populations (Autrup *et al.*, 1983, 1987).

Several studies (Gan *et al.*, 1988; Groopman *et al.*, 1992a; Wild *et al.*, 1992a) have reported dose–response characteristics for aflatoxin–albumin and –guanine adducts. Dose was calculated from individuals' foods, and specific adduct markers were measured (see also section 4.4.1). Although aflatoxin P_1 is a major human urinary metabolite of aflatoxin B_1, its level of excretion correlated poorly with aflatoxin B_1 ingested by 30 men and 12 women in a study of exposed Chinese individuals, whereas aflatoxin B_1–guanine and aflatoxin M_1 excretion showed strong positive correlations (Groopman et al., 1992a). Correlations have been demonstrated between aflatoxin B_1–serum albumin adduct levels, aflatoxin B_1 intake and the urinary excretion of aflatoxin M_1 (Gan et al., 1988).

Aflatoxin G_1 was present in the urine of people in the Gambia who were exposed to this aflatoxin in the diet; no aflatoxin G_1 metabolites were found (Groopman et al., 1992b).

4.1.2 Experimental systems

Aflatoxin B_1 penetrated isolated human epidermis (stratum corneum plus viable epidermis) in vitro. The rate of penetration was low under non-occluded conditions but was approximately 40 times greater under conditions of occlusion (Riley et al., 1985).

Human liver microsomes activate aflatoxin B_1 in vitro to DNA binding species; hydrolysis of the adducts to yield aflatoxin B_1 8,9-dihydrodiol indicates the formation of aflatoxin B_1 8,9-epoxide as an intermediate (Swenson et al., 1974). Metabolism of aflatoxin B_1 in vitro by human liver microsomes produced aflatoxin Q_1 (70–90% of the soluble metabolites) and aflatoxin B_1 8,9-dihydrodiol (10–30%). Aflatoxin M_1 was also detected in traces. Human liver cytosols have a low capacity to catalyse the formation of an aflatoxin B_1–glutathione conjugate (Moss & Neal, 1985). Human liver microsomes activate aflatoxin G_1 at a rate two to three times less than aflatoxin B_1 to form aflatoxin G_1–N7-guanine adducts (Baertschi et al., 1989).

Human liver cytosol fractions from 24 healthy individuals with μ class glutathione S-transferases are more effective in reducing aflatoxin B_1–DNA binding than are corresponding fractions deficient in this class of enzyme, assayed using trans-stilbene oxide as substrate (Liu et al., 1991). Activity in lymphocytes was found to be correlated with that in the liver. In the population studied, about 65% had higher glutathione S-transferase activity (> 100 pmol/min per 10^7 cells) and 40% had lower activity (< 100 pmol/min per 10^7 cells).

The cytochrome P450IIIA4 (nifedipine oxidase) plays a major role in the activation of aflatoxin B_1 (Shimada & Guengerich, 1989); other cytochromes P450 are also active (Forrester et al., 1990).

The metabolic activation of aflatoxin B_1 in human adult and fetal liver to a mutagenic metabolite has been examined using homogenates of fetal liver and microsomes from adult liver. Mutation in the umu gene in a plasmid-containing strain of Salmonella typhimurium was induced by both adult and fetal systems (Kitada et al., 1990). Antibody inhibition studies indicated the involvement of cytochrome P450IIIA4 in the adult system and of a homologous isozyme (P450 HFLa) in the fetal system.

Human bronchus and colon in culture metabolize aflatoxin B_1 to a DNA binding species, bronchus being more active than colon; the adducts formed are aflatoxin B_1–N7-guanine (8,9-dihydro-8-(N7-guanyl)-9-hydroxyaflatoxin B_1) and the open-ring aflatoxin B_1 (8,9-dihydro-8-(N5-formyl-2′,5′,6′-triamino-4′-oxo-N5-pyrimidyl)-9-hydroxyaflatoxin B_1) (Autrup et al., 1979) (see Fig. 2).

Aflatoxin B_1 was absorbed via the skin in rats (Wei et al., 1970). Aflatoxins are absorbed from the gut of sheep (Wilson et al., 1985) and rats (Kumagai, 1989) and are transported via blood, and not by the lymphatic system. Absorption after intratracheal instillation in rats is more rapid than after an oral dose, but the body distribution and excretion patterns are not affected by route of administration (Coulombe & Sharma, 1985). If the tracheally administered dose is adsorbed onto dust, binding to lung and tracheal DNA is increased and retention of aflatoxin B_1 in the trachea is prolonged (Coulombe et al., 1991).

The patterns of aflatoxin B_1 metabolic activation, DNA adduct formation and excretion in liver and kidneys of rats have been reviewed (Essigmann et al., 1982).

Aflatoxin B_1 incubated with rat plasma or administered intraperitoneally binds noncovalently, primarily to albumin which is probably the major transport protein for aflatoxin B_1 (Dirr & Schabort, 1986). A low, non-toxic intraperitoneal dose (0.7 μg/kg bw) of aflatoxin B_1 given to rats is taken up into blood, plasma and liver in a biphasic manner (rapid, 0–2 h, followed by slow, 2–12 h) (Ewaskiewicz et al., 1991).

Multiple binding proteins for aflatoxin B_1 exist in rat liver cytosol (Taggart et al., 1986), some of which may facilitate movement of aflatoxin B_1 within the cell (Mainigi & Sorof, 1977). Rats exposed to ^{14}C-aflatoxin B_1 (2 μCi; 20 μg) intraperitoneally secrete principally ^{14}C-aflatoxin M_1 in the milk. Aflatoxin M_1 accumulates in the liver and lung of offspring, accompanied by macromolecular binding to protein and RNA. Binding to DNA was not detected by radiochemical assay (Allameh et al., 1989).

Aflatoxin M_1 is excreted in the milk of a range of mammalian species (sheep, goats, cows) exposed to aflatoxin B_1 within the order of a 1% conversion level of the toxin (Nabney et al., 1967; Wogan, 1969b; Applebaum et al., 1982; Goto & Hsieh, 1985).

Aflatoxin B_1 as well as aflatoxin M_1 have been shown to be concentrated in the liver of rats 30 min after an intraperitoneal or oral dose of 7 mg/kg bw ^{14}C-aflatoxin B_1; at 24 h, both aflatoxins were detected only as traces (Wogan, 1969b). Whole-body autoradiographic studies in mice have shown that aflatoxin B_1 and/or its metabolites are concentrated in nasal glands (Larsson et al., 1990), in pigmented eye cells (Larsson et al., 1988) and in the pigment of the Harderian glands (Arora et al., 1978). In-vitro studies using bovine melanin have shown that unmetabolized aflatoxin B_1 binds reversibly to the pigment (Larsson et al., 1988).

Aflatoxin B_1 is metabolized to a range of metabolites by microsomal systems (Figs 2 and 3), including aflatoxins P_1, M_1 and Q_1 and the presumed active metabolite aflatoxin B_1 8,9-epoxide, but there is considerable variation in the spectrum of metabolites between species. Aflatoxin P_1 is the major liver microsomal metabolite in mouse but not rat microsomal systems in vitro, whereas aflatoxin Q_1 shows the reverse association (Dahms & Gurtoo, 1976). Aflatoxicol is produced by cytosolic enzymes, whereas aflatoxicol H_1 and M_1 are produced by a combination of cytosolic and microsomal enzymes (Wogan, 1973; Essigmann et al., 1982; Ueno et al., 1985). Species that are more resistant to the carcinogenic effect of aflatoxin B_1 produce more aflatoxin P_1 (which is often excreted in conjugated form (Dalezios et al., 1971)) and have less aflatoxicol in their plasma. Resistant mice excrete ten times more aflatoxin P_1 than sensitive rats. Aflatoxicol was a major aflatoxin B_1 metabolite detected in the serum of rats 50 min after intravenous injection of ^{14}C-aflatoxin B_1, whereas the metabolite was not detected in mouse or monkey serum (Wong & Hsieh, 1980).

Fig. 2. Metabolic activation of aflatoxin B_1 to the 8,9-epoxide, leading to binding to glutathione, DNA and serum albumin

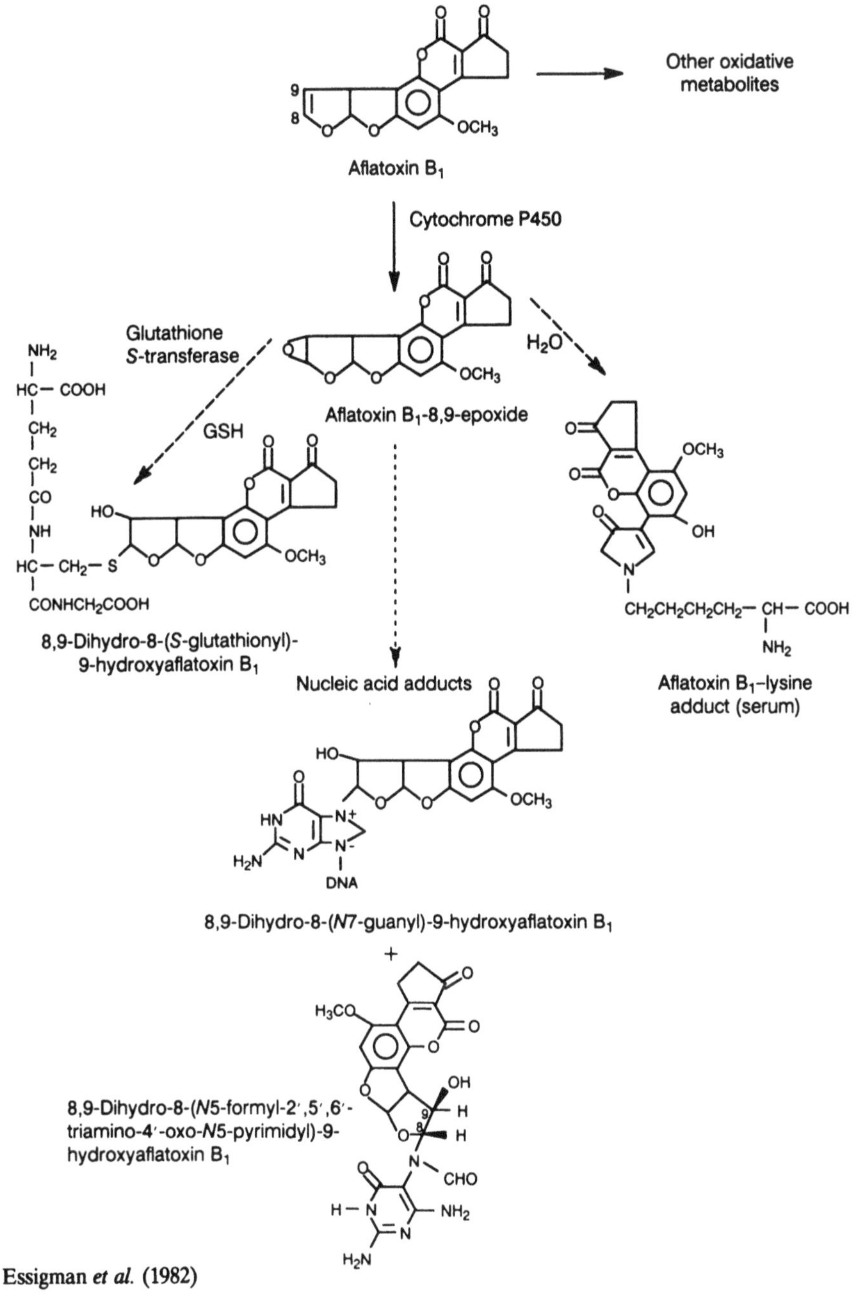

Aflatoxin B_1

→ Other oxidative metabolites

Cytochrome P450

Glutathione S-transferase

Aflatoxin B_1-8,9-epoxide

GSH

H_2O

8,9-Dihydro-8-(S-glutathionyl)-9-hydroxyaflatoxin B_1

Nucleic acid adducts

Aflatoxin B_1–lysine adduct (serum)

8,9-Dihydro-8-(N7-guanyl)-9-hydroxyaflatoxin B_1

8,9-Dihydro-8-(N5-formyl-2′,5′,6′-triamino-4′-oxo-N5-pyrimidyl)-9-hydroxyaflatoxin B_1

From Essigman *et al.* (1982)

Fig. 3. Major metabolites of aflatoxin B_1

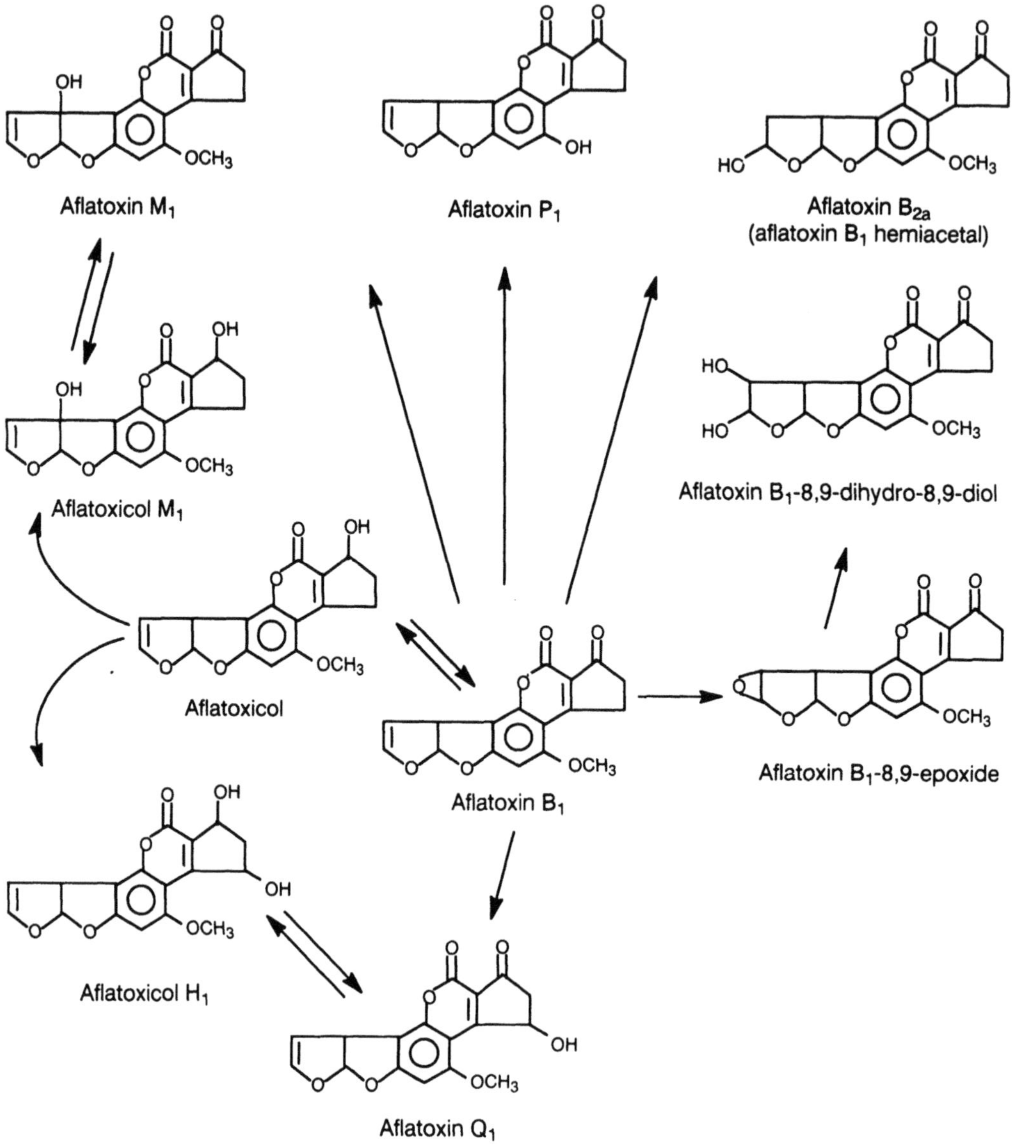

From Essigmann et al. (1982)

Susceptibility to acute toxicity has been suggested to be correlated with glutathione conjugating activity (O'Brien et al., 1983; Lotlikar, 1989), which depends on the presence of appropriate glutathione S-transferases (Neal et al., 1987; Ramsdell & Eaton, 1990; Hayes et al., 1991). The majority of the preneoplastic focal liver lesions (as defined by morphology and immunohistochemical demonstration of glutathione S-transferases) induced by aflatoxin B_1 in rainbow trout showed deficient (67%) or normal (12%) expression of glutathione S-transferases; in the remaining foci of altered hepatocytes (21%), increased expression of glutathione S-transferase was observed. Advanced neoplasms were either deficient or normal in their expression of glutathione S-transferases (Kirby et al., 1990). The relative levels of DNA binding of aflatoxin B_1 in rats in vivo correlate inversely with glutathione–aflatoxin B_1 conjugating activity in the cytosol (Kensler et al., 1986). Primary metabolites can also be conjugated as sulfates (Dalezios et al., 1971) or glucuronides (Rohrig & Yourtee, 1983). The production of primary microsomal metabolites, including activated epoxide, can be influenced by the induction of cytochrome P450 or P448 (Metcalfe et al., 1981). Mitochondrial P450s can also activate aflatoxin B_1 (Niranjan et al., 1984). Aflatoxin B_1 dihydrodiol is formed by the hydration of aflatoxin B_1 epoxide and at neutral pH forms a protein-binding species by Schiff's base reaction (Neal & Colley, 1979). In vivo, aflatoxin B_1 is adducted to the lysine of serum albumin by Schiff's base reaction (Sabbioni et al., 1987). The hemiacetal aflatoxin B_{2a} has been reported as a microsomal metabolite, but this result may have been due in part to misidentification of aflatoxin B_1-8,9-dihydrodiol (Neal et al., 1981a).

In many species—rats (Groopman et al., 1985), sheep (Masri et al., 1967), pigs (Lüthy et al., 1980) and cows (Allcroft et al., 1968)—aflatoxin M_1 is the main unconjugated metabolite of aflatoxin B_1 found in the urine and accounts for 2–9% of the dose. In the urine of rhesus monkeys, aflatoxin M_1 was reported to account for 2.3% of an intraperitoneal dose of aflatoxin B_1 and conjugated aflatoxin P_1 for > 20% of the injected dose (60% of the urinary metabolites: 50% as glucuronide, 10% as sulfite and 3% unconjugated) (Dalezios et al., 1971).

In the DNA of exposed rats, aflatoxin B_1 forms a major adduct, aflatoxin-$N7$-guanine, following metabolic epoxidation. It is released from the DNA by depurination and is excreted in the urine, predominantly over 24 h after exposure, in a dose-dependent manner (Groopman et al., 1992c). Rats administered 1 mg/kg bw aflatoxin B_1 by intraperitoneal injection excreted this adduct in 48 h, which accounted for 30–40% of the initial DNA binding present in the liver (Bennett et al., 1981).

More aflatoxin B_1 metabolites are usually excreted in rat faeces than in urine after intraperitoneal injection of ^{14}C-ring-labelled aflatoxin B_1 (Wogan, 1969b). Excretion of glutathione conjugates of aflatoxin B_1 in rats occurs almost exclusively through the bile; 14% of an intraperitoneally injected dose was disposed of by this route (Emerole, 1981). Degradation of aflatoxin B_1–glutathione conjugate by enzymes of the mercapturic acid pathway has been described in rat kidney preparations in vitro (Moss et al., 1985), and the level of urinary excretion of aflatoxin B_1–mercapturate, together with the sulfate and glucuronide conjugates, correlates with species sensitivity to aflatoxin B_1 (Raj & Lotlikar, 1984).

Aflatoxin B_2 is metabolically converted to aflatoxin B_1 in rats following an intraperitoneal dose of 1 mg/kg bw aflatoxin B_2. The resulting aflatoxin B_1 can then be activated to form aflatoxin B_1-$N7$-guanine adducts in the liver (Groopman et al., 1981a).

In-vitro studies using rat-derived metabolizing systems indicate a decreased capacity to induce DNA binding, overall decreased metabolism but increased aflatoxicol production with age (Jayaraj et al., 1985).

4.2 Toxic effects

4.2.1 Humans

Attempted suicide by ingestion of a mixture of aflatoxins (a total of 1.5 mg/kg over two weeks), containing 15–45% aflatoxin B_1 caused no major symptoms of poisoning (Willis et al., 1980).

Cell-mediated resistance to malaria is increased in individuals exposed to aflatoxins (Hendrickse & Maxwell, 1989).

Aflatoxicosis (jaundice, fever, ascites, oedema of the feet and vomiting) in India was associated with consumption of maize heavily contaminated with *Aspergillus flavus* and containing 6.25–15.6 ppm (mg/kg) aflatoxin. Fatalities occurred in individuals who may have consumed 2–6 mg aflatoxin daily over a period of one month; 106 people died out of a total of 397 patients. Men appeared to be more susceptible than women; no infant was affected. As many as 200 villages were involved in the episode (Krishnamachari et al., 1975). Three- and five-year follow-ups, involving liver biopsies, showed almost complete recovery from acute poisoning (Tandon & Tandon, 1989).

In Kenya, an outbreak of jaundice, accompanied by fatalities, was associated with the consumption of maize which contained up to 12 mg/kg aflatoxin B_1. Livers showed centrolobular necrosis (Ngindu et al., 1982).

A case report of fatal hepatic toxicity in a 15-year-old boy was linked to the consumption of mouldy cassava. Histological changes in the liver were similar to those observed in monkeys treated with aflatoxin (Alpert et al., 1970) and consisted of centrilobular lesions in the liver characterized by loss of retainable cytoplasm, with areas of liver cell necrosis, polymorphonuclear neutrophil infiltration and mild fatty changes in midzonal cells. Two siblings were taken ill at the same time as the affected boy. Analysis of the mouldy cassava in the family's store showed a high aflatoxin content (1.7 mg/kg), and it was concluded that 3.1 kg of cassava was sufficient to produce a lethal dose of aflatoxin (Serck-Hanssen, 1970).

Ingestion of aflatoxin B_1 has been suggested as a possible cause of cirrhosis in Indian children (Amla et al., 1971), but other studies of liver cirrhosis using similar thin-layer chromatographic methods in urine and blood assays failed to detect the fluorescent compounds reported in the earlier studies (Bhandari & Bhandari, 1980). Aflatoxin B_1 was detected in the urine of only 2/35 children with cirrhosis and in 1/35 controls (Dhatt et al., 1982).

Exposure to aflatoxins has been implicated in the etiology of kwashiorkor (Hendrickse & Maxwell, 1989); 5 of 16 liver biopsy samples from children with kwashiorkor contained aflatoxins B_1, B_2 or aflatoxicol; none of 10 samples from children with marasmus did so. The levels of aflatoxins detected were: aflatoxin B_1, 32 and 33 ng/g liver; aflatoxin B_2, 2 ng/g liver; and aflatoxicol, 1 and 4 ng/g liver (Coulter et al., 1986).

Blood and urine samples from 252 Sudanese children were investigated for aflatoxin content. Aflatoxins were detected more frequently (36%) and at higher levels (706 pg/ml) in sera from children with kwashiorkor than in controls (16% and 77 pg/ml); aflatoxicol was frequently found in these samples (14%), but rarely in children who did not have kwashiorkor. A higher proportion of urine samples from kwashiorkor patients contained detectable levels of aflatoxin (33%) than controls (20%), but the mean concentration was lower than in the other children in the study (143 pg/ml, controls 191 pg/ml) (Hendrickse et al., 1982). Children with kwashiorkor placed on aflatoxin-free diet excreted large amounts of aflatoxins in the faeces for up to nine days (Hendrickse & Maxwell, 1989).

The etiology of the development of symptoms of encephalitis and fatty degeneration of the viscera (EFDV), which involves a high level of mortality (similar to Reye's syndrome), was studied in Thai children. Samples were obtained at autopsy from 23 children who had died with symptoms of EFDV and from 15 children who had died from other causes. For the 23 children who had died with EFDV, the presence of aflatoxin B_1 was confirmed in 2/21 liver samples (at levels of 47 and 93 μg/kg), and blue fluorescent spots in chromatograms similar to those produced by aflatoxin B_1 were seen in 15 other samples, suggesting the presence of trace amounts (1–4 μg/kg) of aflatoxin B_1. Of the 15 children who had died from other causes, eight had similar trace amounts of aflatoxin B_1. In a separate study, the urine of eight of 51 children with EFDV contained trace amounts of aflatoxin B_1. A spot similar to that produced by aflatoxin B_2 was seen in one sample, and a spot similar to that of aflatoxin M_1 was detected in two samples. No compound similar to aflatoxins was detected in urine samples from 39 healthy children (Shank et al., 1971).

4.2.2 Experimental systems

The order of potency for both the acute and chronic toxicity of aflatoxins is $B_1 > G_1 > B_2 > G_2$ (Busby & Wogan, 1984).

Species that are more sensitive to the acute toxic effects of aflatoxin B_1 (LD_{50}: monkey, 2.2 mg/kg bw; rat, 5.5–17.9 mg/kg bw) show higher distribution volumes, higher levels of aflatoxins in plasma and liver and longer plasma half-lives than less sensitive species (mouse, 9.0–60.0 mg/kg bw) (Wong & Hsieh, 1980).

Liver lesions occurred in rabbits after percutaneous absorption of aflatoxins B_1 and B_2 at doses of 160–1250 μg/kg bw, and severe glycogen depletion was seen. After topical administration of aflatoxins at > 1400 μg/kg bw, midzonal necrosis was seen, accompanied by fatty changes, in 8/10 treated rabbits, and hyaline-acidophilic changes in the cytoplasm were seen in liver cells. No lesion was observed in the livers of rabbits treated with the vehicle alone or with a dose of aflatoxins < 50 μg/kg bw (Ungar & Joffe, 1969).

The liver is the usual target organ for both acute and chronic toxicity, but lesions in the kidney and glandular stomach have been reported in rats (Butler & Barnes, 1966, 1968). Ethanol ingestion has been found to potentiate aflatoxin B_1 hepatotoxicity (Toskulkao et al., 1986). Ammoniation of aflatoxin B_1-contaminated maize or groundnut meal fed to rats was effective in reducing hepatotoxicity (Norred & Morrissey, 1983; Manson & Neal, 1987). Low-protein diets increased the acute toxic effects of aflatoxin B_1 (Madhavan & Gopalan, 1965). For data on the effects of dietary constituents on the formation of aflatoxin–guanine adducts, see p. 336.

Aflatoxin B_1 has been reported to affect both humoral and cell-mediated immune systems (for reviews, see Pier & McLoughlin, 1985; Richard, 1991). Direct and complex effects of aflatoxins on lymphocytes have been reported in mice, with differing sensitivities in various sub-sets. Inhibitory effects noted on T and natural killer cells, which would compromise cell-mediated immunity (Reddy & Sharma, 1989), were correlated with inhibition of ^{3}H-thymidine uptake in lymphocyte cultures (Reddy *et al.*, 1987) and inhibition of the humoral immune system by aflatoxin B_1 in mice (Reddy *et al.*, 1983) and in a range of other species, such as birds, pigs, horses and rabbits (Ray *et al.*, 1991). Aflatoxins B_1 and M_1 are particularly potent in reducing phagocytosis and superoxide production in rat peritoneal macrophages (Cusumano *et al.*, 1990). Exposure to aflatoxin B_1 resulted in an exaggerated response to mitogens in mouse lymphocytes (Hendrickse & Maxwell, 1989).

Symptoms similar to those of Reyes syndrome were observed in adult macaque monkeys given aflatoxin B_1; fatty degeneration of the liver was the most consistent pathological finding (Bourgeois *et al.*, 1971).

Protein kinase C and phospholipase C in human platelets are activated by aflatoxin B_1 *in vitro* (Van Den Heever & Dirr, 1991).

4.3 Reproductive and developmental toxicity

4.3.1 *Humans*

Aflatoxins cross the placental barrier, and there is some evidence that concentrations in cord blood are higher than in maternal blood (see section 4.1.1).

4.3.2 *Experimental systems*

Sixteeen intraperitoneal injections of approximately 60 µg/kg bw aflatoxin B_1 to rats induced pronounced testicular degeneration and impaired spermatogenesis (Egbunike *et al.*, 1980).

Groups of 8–12 pregnant JCl-ICR mice were treated intraperitoneally with 16 or 32 mg/kg bw aflatoxin B_1 for various two-day period during days 6–13 of gestation. Fetuses were examined on day 18 of gestation. Maternal death, decreased body weight and increased kidney weight were observed in the group given 32 mg/kg bw. Reduced fetal weight and external malformations, such as cleft palate and open eyelids, and skeletal malformations, such as wavy ribs and bent long bones, were also seen in this group (Tanimura *et al.*, 1982).

Groups of 19–36 NMRI mice were treated with doses of 0, 15, 45 or 90 mg/kg bw aflatoxin B_1 intraperitoneally or with 45 mg/kg bw orally on days 12 and 13 of pregnancy. The intraperitoneal treatment with 45 or 90 mg/kg bw aflatoxin B_1 produced retardation in fetal development; incidences of cleft palate were higher than in controls (control, 1.5%; 45 mg/kg, 4.1%; 90 mg/kg, 5.6%). No corresponding effect was seen after oral treatment. At a dose of 45 mg/kg bw, diaphragmatic malformations were observed after either intraperitoneal (18%) or oral (13%) administration. Intraperitoneal injection of aflatoxin G_1 at 90 mg/kg bw induced malformations of the diaphragm (14.7% of fetuses) and kidney (5.5%). No such anomaly was seen in controls (Roll *et al.*, 1990).

A group of eight CBA mice was treated by oral intubation on day 8 of pregnancy and seven mice on day 9 with 4 mg/kg bw aflatoxin B_1. Fetal anomalies—exencephaly (4), open eyes (3) and protrusion of intestines (2)—were observed in 7/61 fetuses exposed on day 8 and in none of the 51 fetuses exposed on day 9 (Arora et al., 1981).

Groups of 10 pregnant rats (*Rattus norvegicus*) were treated subcutaneously with 0.7, 1.4, 3.5 or 7.0 mg/kg bw aflatoxin (75% B_1, 25% B_2) on either day 8 or day 16 of pregnancy. Average fetal weight was decreased, and fetuses showed wrinkled skin and slightly enlarged heads in comparison with controls. Malformations were not observed (Sharma & Sahai, 1987).

A significant increase in liver triglyceride content was found in one- and two-month-old Fischer rat offspring of dams that had been treated intraperitoneally with 2 mg/kg bw aflatoxin B_1 during days 8-10 of pregnancy. Treatment on days 8-10 or on days 15-17 induced a decrease in motor activity in one-month-old offspring. At the age of two or three months, behaviour became normal, but there was persistent neuronal degeneration in the brains of the offspring (Chentanez et al., 1986).

In day-10 rat embryos exposed *in vitro* to aflatoxin B_1 at 15 μM [4.7 mg] or more, neural tube defects occurred. The presence of hepatic fractions from induced adult male rats had no effect on the ability of the compound to produce dysmorphogenesis, but it enhanced the embryolethal effects (Geissler & Faustman, 1988).

4.4 Genetic and related effects (see also Tables 10-19 and Appendices 1 and 2)

4.4.1 *Humans*

(a) *Urinary excretion of guanine adducts*

Aflatoxin B_1-guanine adducts were reported to have been detected in human urine in a previous IARC monograph (IARC, 1987a). Further studies have been published since then.

The diets of 30 men and 12 women (age range, 25-64 years) living in Guangxi Autonomous Region, China, were monitored for one week and aflatoxin intake levels determined daily. Starting on the fourth day, total urine volumes were obtained in consecutive 12-h fractions for three or four days. Each urine sample was analysed by high-performance liquid chromatography (HPLC) and radioimmunoassay, and the relationships between excretion of total aflatoxin metabolites, aflatoxin-$N7$-guanine, aflatoxin M_1, aflatoxin P_1 and aflatoxin B_1, and aflatoxin B_1 intake levels were determined. The average intake of aflatoxin B_1 was 48.4 μg/day by men and 77.4 μg/day by women, giving total mean exposures over seven days of 276.8 and 542.6 μg aflatoxin B_1, respectively. Linear regression analysis of the association between daily aflatoxin B_1 intake and daily total aflatoxin metabolite excretion gave a non-significant correlation coefficient (r) of 0.26. Aflatoxin-$N7$-guanine excretion and aflatoxin B_1 intake from the previous day, however, gave $r = 0.65$ ($p < 0.000001$). Comparison of total aflatoxin-$N7$-guanine excretion throughout the collection period with total dietary aflatoxin B_1 for each individual, smoothing day-to-day variations, gave $r = 0.80$ ($p < 0.0000001$) (Groopman et al., 1992a).

Daily aflatoxin intake levels were measured for one week in the diets of 10 men and 10 women (age range, 15-56 years) living in a village in the Gambia. Starting on the fourth day, total 24-h urines were obtained for four days. The subjects were also tested for hepatitis B virus carrier status. Preparative monoclonal antibody affinity chromatography/HPLC and competitive enzyme-linked immunosorbent assays (ELISA) were performed on each urine sample, and the relationship between aflatoxin intake levels and urinary excretion of all aflatoxin metabolites and of aflatoxin–$N7$-guanine was determined. The average intake of all aflatoxins during the one-week collection period was 8.2 µg for men and 15.7 µg for women [these are almost certainly underestimates of the true intake (Wild et al., 1992a)]. There was considerable day-to-day variation in exposures, from zero to 29.6 µg aflatoxins. Linear regression analysis of the association between urinary excretion of aflatoxins and average daily aflatoxin intake gave $r = 0.65$ ($p < 0.001$). HPLC revealed a preponderance of aflatoxin G_1 in many of the urine samples, in addition to the metabolites aflatoxin P_1 and aflatoxin Q_1 and to aflatoxin–$N7$-guanine adducts. Regression of total urinary excretion of aflatoxin–$N7$-guanine by all subjects over the collection period on total dietary aflatoxin B_1 intake gave $r = 0.82$ ($p < 0.0001$). Separation of the population into hepatitis B virus carriers and non-carriers revealed no difference in urinary aflatoxin–guanine adduct levels for a given dietary exposure (Groopman et al., 1992b).

(b) *Covalent binding to DNA*

Consistent with the presence of aflatoxin B_1–$N7$-guanine in human urine are reports of the presence of aflatoxin–guanine adducts in human liver tissues, as determined by immunoassay on DNA. One study in Taiwan, China, of nine hepatocellular carcinoma patients showed antibody inhibition in a competitive ELISA by tumour DNA from all seven samples tested and from two of eight DNA samples from 'adjacent-normal' tissue (Hsieh et al., 1988). In a study of 27 further hepatocellular carcinoma patients in Taiwan, a positive signal was seen after immunofluorescent staining in eight (30%) of the tumour and seven (26%) of the non-tumour liver tissues. Some of these samples gave positive results in ELISA (Zhang et al., 1991). Reports of aflatoxin–guanine adducts in human tissues have been made from Czechoslovakia (Garner et al., 1988) and California, USA (Lee et al., 1989), on the basis of studies of antibodies; however, the specificity of all the available assay methods is dependent on the specificity of the antibody used, and no other confirmatory analytical approach was employed.

Studies of DNA binding of aflatoxin B_1 *in vitro* and in fish and mammals *in vivo* are summarized in Tables 10 and 11, respectively.

(c) *Modulation of DNA binding* (Table 12)

Many factors have been identified that modulate aflatoxin B_1-induced DNA binding, including several antioxidants and dietary factors. The available data are summarized in Table 12. In studies *in vitro*, retinoid analogues, indol-3-carbinol and allixin inhibited binding of aflatoxin B_1 to DNA.

Antioxidants such as butylated hydroxytoluene (see IARC, 1986), butylated hydroxyanisole (see IARC, 1986) and ethoxyquin and the dithiolthiones, oltipraz and 1,2-dithiol-3-thione, reduced aflatoxin B_1–DNA binding *in vivo*. Pretreatment of trout and salmon with β-naphthoflavone resulted in reduced activation and subsequent binding to DNA. Both pretreatment and concomitant dietary treatment with indol-3-carbinol in trout reduced covalent aflatoxin B_1–DNA binding. Dietary administration of broccoli and brussels sprouts to rats and pretreatment with crocetin or genipiside reduced hepatic aflatoxin B_1–guanine adduct formation, as did concomitant phenobarbital treatment of rats (Holeski et al., 1987). In one study, covalent binding of aflatoxin B_1 to DNA was increased following pretreatment with ethanol 6 h previously (Wang et al., 1990). No such effect was found in another study (Marinovich & Lutz, 1985).

Neonatal treatment of rats with diethylstilboestrol, followed by exposure to aflatoxin B_1 in adulthood resulted in significantly higher levels of DNA adducts in females but not in males (Lamartiniere, 1990). Feeding rats with a lipotrope-deficient diet resulted in significantly lower levels of covalent binding of aflatoxin B_1 to DNA (Campbell et al., 1978). Rats fed a choline-deficient, low-methionine diet had significantly higher levels of total adducts following exposure to multiple doses of aflatoxin B_1 than controls (Schrager et al., 1990).

Treatment of rats with low levels of aflatoxin B_1 followed by exposure to a higher dose resulted in reduced levels of covalent binding to hepatic DNA (Neal et al., 1981b).

Partial reduction of hepatic glutathione levels slightly increased covalent binding of aflatoxin B_1 to DNA; however, when glutathione depletion was almost complete, a 30-fold increase in covalent aflatoxin B_1–DNA binding was found (Monroe & Eaton, 1988). In a large majority of the studies, modulation of aflatoxin B_1–DNA binding was associated with changes in the activity of various enzyme systems involved in the biotransformation of aflatoxin B_1. Protective effects were strongly associated with induction of conjugating enzyme systems, and especially the glutathione S-transferase isozymes.

(d) *Mutations in* p53 *tumour-suppressor gene in human hepatocellular carcinoma*
(Table 13)

Mutations of the *p53* tumour suppressor gene—mainly missense (transitions and transversions)—in those regions of the gene that are highly conserved during evolution (exons 5–8) have been found at high frequency in a variety of common human cancers (Hollstein et al., 1991a,b; Jones et al., 1991; Levine et al., 1991). A high frequency of mutations at a mutational 'hot-spot' (Challen et al., 1992) (the third nucleotide of codon 249 in exon 7) has been found in hepatocellular carcinomas from patients resident in areas where they are considered to be at high risk of exposure to aflatoxins and where there is a high incidence of hepatocellular carcinoma. The data are given in detail in Table 13. Of 101 tumours from patients in high-exposure areas (for example, Qidong in China, Mozambique, Viet Nam and India), 40 contained a G to T transversion (AGG to AGT) in codon 249 in exon 7 of the *p53* gene, and one contained a G to C transversion at that position (Hsu et al., 1991). In contrast, only one of 205 hepatocellular carcinomas from regions where there is low exposure to aflatoxins (including Taiwan, Australia, Japan, southern Africa, Germany, Spain, Italy, Turkey, Israel, Saudi Arabia, the United Kingdom and the USA) contained this mutation. The uncertainties associated with classification of tumours across different studies and the small number of studies in which such data are provided preclude a judgement as to whether tumour size or stage is correlated with the presence or absence of the codon-249 mutation. [The Working Group noted that some of the G to T transversions described by Hsu et al. (1991) may have been included in the study of Scorsone et al. (1992), as the patients were from the same institute. Similarly, the patients studied by Bressac et al. (1991) were included in the study of Ozturk et al. (1991).]

Point mutations of *p53* in hepatocellular carcinomas have been documented at sites other than codon 249. [The Working Group noted that those studies in which restriction analysis alone was used to detect mutations would not have have picked up such mutations.] Murakami *et al.* (1991) detected eight different mutations in 7/22 advanced hepatocellular carcinomas in patients in Tokyo, Japan, where exposure to aflatoxins is low (Table 13). Six were base-pair substitutions in codons other than 249, and two were deletions. There appeared to be an association between tumour size and degree of differentiation, as there were more mutations in larger, poorly differentiated tumours; however, this conclusion was based on small numbers. No mutation was seen in 21 early hepatocellular carcinomas. In a study in another region of low exposure to aflatoxins—the United Kingdom—Challen *et al.* (1992) found *p53* mutations in 2/19 hepatocellular carcinomas, neither of them at codon 249 (Table 13).

Whether there is any relationship between hepatitis B virus (HBV) status and specific *p53* mutations, especially at codon 249, is difficult to establish from the data presently available. Investigators have used various methods of assigning HBV status, and numbers within individual studies are small. What evidence is available does not strongly suggest a direct relationship between codon-249 mutation and HBV status. For example, Ozturk *et al.* (1991) found that of seven patients in Mozambique positive for HBsAg, four had a tumour with a codon-249 *p53* mutation; and four of eight patients negative for HBsAg had a tumour with a codon-249 *p53* mutation and one had a *p53* mutation at codon 157. Of 16 patients in Australia with hepatocellular carcinoma, eight had been exposed to HBV and eight had not; none carried a codon-249 *p53* mutation in their tumour (Hayward *et al.*, 1991). Similarly, 6/19 patients in the United Kingdom had serological evidence of past or present HBV infection, and 14/19 did not; none carried a codon-249 *p53* mutation in their tumour (Challen *et al.*, 1992).

(e) Loss of heterozygosity and hepatitis B virus

Although epidemiological studies show that HBV infection is intimately linked with the development of hepatocellular carcinoma in high-risk populations, the underlying molecular mechanism is unknown. HBV–DNA sequences integrated into host DNA, seemingly at random, are commonly found in hepatocellular carcinomas in patients from areas endemic for HBV infection. The finding that gross chromosomal abnormalities (including deletions) are frequent at sites of viral integration raises the possibility that integration of HBV–DNA and the chromosomal damage that ensues may cause allele loss at tumour suppressor gene loci (Fujimori *et al.*, 1991; Slagle *et al.*, 1991; Walker *et al.*, 1991; Scorsone *et al.*, 1992).

Loss of the second allele of a tumour-suppressor gene in a cell that has already sustained a mutation in the first allele is considered to be a critical step in carcinogenesis (Harris, 1991). Allele loss, as measured by loss of heterozygosity for restriction fragment-length polymorphism markers, is common in hepatocellular carcinomas and occurs in several chromosomes (including 17p, which contains *p53*), suggesting the involvement of several tumour-suppressor genes in the development of liver cancers. No specific association has been found, however, between HBV status and HBV integration (Fujimori *et al.*, 1991; Walker *et al.*, 1991).

Table 10. Studies of aflatoxin B_1 and DNA binding in vitro

Reference	Details of study	DNA binding
DNA binding in vitro		
Puisieux et al. (1991)	Covalent binding of AF-8,9-epoxide to plasmid pC53SN3 containing a full-length human wild-type $p53$ complementary DNA insert in vitro. $p53$ nucleotide residues specifically targeted by AFB_1 were identified by DNA polymerase fingerprinting analysis. Percentage of $p53$ nucleotides in exons 5–8 targeted by AFB_1: G, 62%; C, 6%; T, 2%; A, 3%	0.3–2 adducts/1000 bases
Garner et al. (1979)	Covalent binding to calf thymus DNA in vitro in presence of liver microsomes from phenobarbital-induced male Wistar rats	2.83 μg AFB_1/mg DNA
	Covalent binding to calf thymus DNA in vitro with chloroperoxybenzoic acid oxidation	16.03 μg AFB_1/mg DNA
Groopman et al. (1981b)	Covalent binding to calf thymus DNA in vitro in presence of liver microsomes from phenobarbital-induced rats. Study of chemical stability of AFB_1-modified DNA in vitro	1 adduct/60 or 1500 nucleotides
Pestka et al. (1982)	Covalent binding to DNA in vitro in presence of liver microsomes from male Fischer rats. Study of AFB_{2a} antiserum for detecting AFB_1-modified DNA and related metabolites	[190 μg AFB_1/ml] 8 μg AFB_1/mg DNA at 1 h
Misra et al. (1983)	Covalent binding to ΦX174 and pBR322 DNA in vitro using chloroperoxybenzoic acid oxidation or photoactivation at 350 nm. Study of covalent and noncovalent interactions of AFB_1 with defined DNA sequences: sequence-specific effects abolished in single-stranded DNA, and reactivity of guanine strongly suppressed	
Israel-Kalinsky et al. (1984)	Covalent binding to calf thymus DNA in vitro using photoactivation at 365 nm and 390 nm	2 nmol AFB_1/μmol DNA-P at 20 μM AFB_1
	Covalent binding to DNA of *Salmonella typhimurium* TA100 in presence of liver microsomes from Aroclor 1254-induced male Sprague-Dawley rats	2 μM [0.62 μg/ml/15 min]; 5.3 pmol AFB_1/μmol DNA-P
	Covalent binding to DNA of *S. typhimurium* TA100 photoactivated at 365 nm	5 μM [1.55 μg/ml/120 min]; 0.37 pmol AFB_1/μmol DNA-P
	AFB_1 binding to DNA induced in vitro by mammalian microsomes or by photoactivation proportional to mutagenicity and lethality	
Amstad et al. (1984)	Covalent binding to human lymphocytes in vitro. Clastogenic effects observed at this dose of AFB_1 considered to be caused by indirect effects on cell membrane	0.3 μM [0.09 μg/ml] causing ca. 20% cells with aberrations; 2 adducts/10^8 DNA-P (not certain to be DNA adducts)
Yang et al. (1985)	Covalent binding in vitro to DNA from normal human liver, primary hepatocellular carcinoma cells, Mahlavu hepatocellular carcinoma cells, in presence of liver microsomes from rats or chloroperoxybenzoic acid oxidation. Kinetics of AFB_1 binding to these DNAs similar initially but extents of binding to carcinoma cell. DNA slightly higher	100–400 fmol/μg DNA
Jayaraj et al. (1985)	Covalent binding to calf thymus DNA in vitro, in presence of liver microsomes from male Fischer 344 rats 4, 12 and 26 months old. Formation of AFB_1–DNA adducts with microsomes from 4- and 12-month-old rats was similar but significantly lower [about half] with those from 26-month-old rats	

Table 10 (contd)

Reference	Details of study	DNA binding
Mariën et al. (1987)	Covalent binding of chloroperoxybenzoic acid-oxidized AFB_1 to purified λDNA; 195-bp fragment of φX174 of known sequence; 195-bp fragment of pLV405. Activated AFB_1 reacts primarily with guanyl residues to form N7 adducts; ≤10-fold variation in binding of AFB_1 in 33 different guanyl residues in two different 195-bp DNA fragments	
Benasutti et al. (1988)	Covalent binding to M13mp19 DNA fragments of known sequence in vitro following chloroperoxybenzoic acid oxidation of AFB_1. Study of reaction of binding of AFB_1 to 190 guanyl residues in different pentanucleotide contexts confirmed that N7 of guanine is major adduct. In a pentanucleotide 5'-WXGYZ-3', the influence of neighbouring bases (X and Y) most influential in determining reactivity of G are: X = G > C > A > T and Y = G > T > C > A	
Loechler et al. (1988)	Molecular modelling study of interaction of AFB_1 binding site in double-stranded DNA fragments of different sequences. Two potential binding modes proposed in which AFB_1 moiety is either intercalated or bound externally in the major groove. One particular external binding site could account for the preference that the reactive G has for another G on its 5' side.	
Cole et al. (1988)	Covalent binding to DNA of mouse, rat and human hepatocytes in vitro. Binding in male but not female rat hepatocytes declined over one week	0.2 μm for 24 h: male rat (203 pmol/mg DNA), human (42) > female rat (38); mouse (1.4)
Delcuve et al. (1988)	Covalent binding of 2,3-dichloro-AFB_1 to DNA of mature and immature chicken erythrocytes in vitro	
Loveland et al. (1988)	Covalent binding to Salmo gairdneri (rainbow trout) hepatocyte DNA in vitro. Binding (pmol bound AF/μg DNA)/(μmol dose) versus time fitted a linear function passing close to origin for each AF.	DNA binding at 1 h relative to AFB_1: aflatoxicol, 0.53; AFM_1, 0.81; aflatoxicol M_1, 0.83; significantly less than that of AFB_1
Bailey et al. (1988)	Covalent binding to hepatocyte DNA of S. gairdneri in vitro Covalent binding to hepatocyte DNA of Oncorhynchus kisutch (coho salmon) in vitro	0.2 μg/ml; 667 pmol AFB_1/mg DNA 0.2 μg/ml; 34 pmol AFB_1/mg DNA
Stone et al. (1988)	Non-covalent binding to calf thymus DNA, synthetic polydeoxyribonucleotides and the oligodeoxynucleotide d(ATGCAT)$_2$ in vitro	
Stark et al. (1988)	Covalent binding to calf thymus and supercoiled pBR322 DNA in vitro using photoactivation at 365 nm. Authors concluded that photoactivated aflatoxins bind to guanines, some of which are released, creating apurinic sites which are converted to DNA chain breaks at physiological pH and temperature.	
Baertschi et al. (1989)	Covalent binding of AFB_1 to calf thymus DNA in vitro, in presence of human liver microsomal cytochrome P450$_{NF}$; comparison with AFG_1	Guanyl-N7 adduct formation in pmol/min/nmol cytochrome P protein: 28 (50 μM AFB_1), 12 (50 μM AFG_1)

Table 10 (contd)

Reference	Details of study	DNA binding
Harris et al. (1989)	Covalent binding of AFB_1 epoxide to synthetic oligodeoxynucleotides d(ATCGAT)$_2$ and d(ATG-CAT)$_2$ in vitro. Synthetic AFB_1-epoxide reacts with d(ATCGAT)$_2$ with 1 adduct/duplex and 2 per d(ATGCAT)$_2$ duplex. Both react to give the N7 adduct on guanine. Difference in stoichiometry explained by geometrical constraints, such that two sites of intercalation are allowed in d(ATG-CAT)$_2$ but only one in d(ATCGAT)$_2$.	
Lipsky et al. (1990)	Covalent binding to calf thymus DNA in vitro in presence of microsomes, with and without cytosols, from mouse, rat and human livers. Binding mediated by liver microsomes from 13 humans varied from 0.87 to 14.58 nmol/mg microsomal protein. Binding mediated by mouse and rat liver microsomes was 4.68 (mouse) and 3.14 (rat) nmol/mg microsomal protein. Covalent binding to DNA in human, rat and mouse hepatocytes in vitro	AFB_1 bound (pmol/mg DNA) hepatocyte DNA: human, 53, 69, 25; rat, 203; mouse, 1.4
Ball et al. (1990)	Covalent binding to DNA in tracheal explants from hamsters, rats and rabbits in vitro. Repair rates in hamster and rat constant over time, with removal of N7-guanine adduct accounting for majority of adduct loss. All adduct types removed rapidly during first 12 h after treatment, followed by a slower removal phase	Binding of 0.5 μM AFB_1 (pmol/mg DNA): rabbit, 78; hamster, 28; rat, 3
Stark et al. (1990)	Covalent binding to calf thymus DNA in vitro using photoactivation at 365 nm, under aerobic and anaerobic conditions in H_2O or D_2O. Photoactivated binding decreased markedly under anaerobic conditions and with compounds known to quench or scavenge singlet oxygen ($^1O_2$)	Binding 68–330 pmol/μmol nucleotide in dark; 1–7 nmol/μmol nucleotide after irradiation
Shaulsky et al. (1990)	Covalent binding to nick-translated DNA labelled with ^{14}C in each DNA base in vitro, using liver microsomes from phenobarbital-induced male Sprague-Dawley rats photoactivation at 365 nm The only AFB_1-DNA adducts detected in significant amounts were guanine adducts; no stable adduct with A, C or T detected	2626 pmol/μmol nucleotide 2304 pmol/μmol nucleotide
Crespi et al. (1990)	Covalent binding to DNA of human lymphoblastoid AHH-1 TK$^{+/-}$ cells transfected with and expressing human CYP1A2 gene in vitro. Binding not detectable in cells not carrying or expressing CYP1A2	Exposure to 3 ng/ml: 16.3 dpm/μg DNA
Crespi et al. (1991)	Covalent binding to DNA of human lymphoblastoid AHH-1 TK$^{+/-}$ cells transfected with and expressing either human CYP1A2, human CYP2A3 or human CYP3A4 genes in vitro. Binding not detected in cells carrying and expressing CYP2A3, 3A4; detected in CYP1A2 cells	Exposed to 3 ng/ml: 1.16 pmol/mg DNA
Stark & Liberman (1991)	Covalent binding to calf thymus DNA in vitro using photoactivation at 365 nm, under aerobic and anaerobic conditions in H_2O or D_2O. DNA binding enhanced in D_2O, and a singlet-oxygen scavenger inhibited mutagenesis. DNA photobinding of AFB_1 increased in presence of AFB_2	[10 μM, 30 min irradiation] 128 pmol/μmol nucleotide

[a] AF, aflatoxin

Table 11. Studies of DNA binding in fish and mammals exposed to aflatoxin B_1 *in vivo*

Reference	Details of study	Route and dose	DNA binding
Toledo et al. (1987)	Covalent binding to liver DNA of *Oryzias latipes* (medaka fish) *in vivo*; dose–response linear over the range 70–550 μg AFB_1/kg; maximum binding within the first 24 h after injection, followed by rapid loss of adducts	i.p.; 70, 140, 275, 550 μg/kg bw	1.0–22.0 pmol/μmol DNA at 24 h
Bailey et al. (1988)	Covalent binding to liver DNA of *S. gairdneri* (rainbow trout) *in vivo*	i.p.; 10 μg/kg bw	48 pmol AFB_1–$N7$-guanine/mg DNA at 1 day
		Oral; 80 ppb, 3 weeks	29 pmol/mg DNA at 1 day
		Embryos; 0.5 μg/ml, 1 h	1.4 pmol/mg DNA at 1 day
	Covalent binding to liver DNA of *Oncorhynchus kisutch* (coho salmon) *in vivo*	i.p.; 10 μg/kg bw	4 pmol AFB_1–$N7$-guanine/mg DNA at 1 day
		Oral; 80 ppb, 3 weeks	1.6 pmol/mg DNA at 1 day
		Embryos; 0.5 μg/ml, 1 h	0.07 pmol/mg DNA at 1 day
	DNA binding 7.56 times greater in trout than salmon liver at various times after injection, 20 times greater in embryos, and 18 times greater in trout liver after a three-week dietary (80 ppb) exposure. Major DNA adduct, 8,9-dihydro-8-($N7$-guanyl)-9-hydroxyAFB$_1$, in both species		
Nakatsuru et al. (1989)	Covalent binding to liver DNA of *Salmo mykiss* (a trout species) *in vivo*	i.p.; 0.1, 0.5 mg/kg bw, 24 h	35, 150 fmol AFB_1/μg DNA
	Covalent binding to liver DNA of *O. kisutch in vivo*	i.p.; 0.1, 0.5 mg/kg bw, 24 h	Not detected, 22 fmol AFB_1/μg DNA
	Covalent binding to liver DNA of male Fischer 344 rats *in vivo*	i.p.; 1, 2 mg/kg bw, 2 h	268, 420 fmol AFB_1/μg DNA
Garner et al. (1979)	Covalent binding to kidney and liver DNA of Wistar rats *in vivo*; comparison of AFB_1 and AFG_1 binding to macromolecules; binding to RNA > DNA > protein	i.p.; 0.6 mg/kg bw	Kidney: 10 ng AFB_1/mg DNA at 2 h; Liver: 15 ng AFB_1/mg DNA at 2 h

Table 11 (contd)

Reference	Details of study	Route and dose	DNA binding
Bennett et al. (1981)	Covalent binding to DNA in vivo: detection of urinary AFB$_1$–guanine adducts in male Fischer rats. Administration of AFB$_1$ resulted in urinary excretion of major DNA adduct formed in vivo. Measurement of urinary adducts in rats injected with different doses of AFB$_1$ showed that excretion occurred in a dose-dependent manner. Comparison of dose–response curve for adduct excretion with that previously observed for adduct formation in rat liver DNA in vivo revealed close similarity	i.p.; 0.125, 0.25, 0.5, 1 mg/kg	
Groopman et al. (1982)	Covalent binding to liver DNA of male CDF Fischer rats in vivo. Monoclonal antibodies specific for modified DNA, containing both the 2,3-dihydro-2-(N7-guanyl)-3-hydroxy-AFB$_1$ and putative 2,3-dihydro-2-(N5-formyl-2',5',6'-triamino-4'-oxo-N5-pyrimidyl)-3-hydroxy-AFB$_1$, suggesting that these DNA adducts share a common antigenic determinant	i.p.; 0.01–1 mg/kg bw	1 AFB$_1$ residue per 1.35×10^6, 2.5×10^5, 3×10^4 nucleotides at 2 h
Irvin & Wogan (1984)	Covalent binding to nuclear liver DNA and ribosomal liver DNA sequences of male Fischer rats in vivo. Over a dose range of 0.25–2.0 mg AFB$_1$/kg bw, ribosomal DNA contained 4–5 times more AFB$_1$ residues than nuclear DNA, showing that ribosomal DNA is preferentially accessible to carcinogen modification in vivo.	i.p.; 0.25, 0.5, 1.0, 2.0 mg/kg bw	90–640 pmol AFB$_1$/mg nuclear DNA (30–240 AFB$_1$ adducts/10^6 nucleotides)
Irvin & Wogan (1985)	Covalent binding to liver nuclear DNA and ribosomal DNA of male Fischer rats in vivo: 12 h after injection, ribosomal DNA contained 4–5 times more AFB$_1$ residues than did total nuclear DNA. AFB$_1$–ribosomal DNA residues removed 5.7 times more rapidly than total nuclear DNA residues over 12 h. Levels of the major adduct, N7-guanyl–AFB$_1$ and its stable formamidopyrimidine derivatives also determined: no difference in proportion of these adducts in ribosomal and total nuclear DNA	i.p.; 1 mg/kg bw	288, 227, 158 pmol AFB$_1$/mg nuclear DNA; 1212, 800, 590 pmol AFB$_1$/mg ribosomal DNA, 2, 6, 12 h after dosing
Wild et al. (1986)	Covalent binding to liver DNA of male Wistar rats in vivo given a single dose 24 h before killing; 24 h after single dose, constant ratio found between levels of AFB$_1$ bound to plasma protein and that bound to liver DNA; 0.98–2.15% of dose bound to plasma protein at 24 h	Intra-gastric; 3.5, 10, 100, 200 μg/kg bw	100–2500 pg AFB$_1$/mg DNA
	Covalent binding to liver DNA of male Wistar rats in vivo given multiple doses for 23 days before killing	Intra-gastric, 0.5 μg/rat, twice daily, 5 days/week, 2, 3, 7, 11, 21, 24 days before killing	108.4 pg AFB$_1$/mg at 2 days to peak of 267 pg AFB$_1$/mg DNA at 14 days
	Binding of AFB$_1$ to plasma protein rose three times higher than after single dose; reached plateau after 7–14 days of treatment and remained stable. Binding to DNA accumulated 2.5-fold and in parallel to plasma protein. All detectable plasma protein-bound AFB$_1$ associated with a single peak corresponding to albumin.		

Table 11 (contd)

Reference	Details of study	Route and dose	DNA binding
Yu et al. (1988)	Covalent binding to liver DNA fractions (nuclear, nucleolar and transcriptionally active subnuclear fraction rich in ribosomal RNA DNA) in vivo, in male Sprague-Dawley rats. Binding of AFB_1 to nuclear and nucleolar DNA reached plateau at 3000 μg AFB_1/kg bw; binding to protein linear, although with different slopes, for nuclear and nucleolar fractions up to highest dose used	i.p.; 100, 500, 1000, 3000, 5000 μg/kg bw	Binding (pmol AFB_1/mg) linear to 3000 μg/kg dose. Nuclear DNA, ~ 100 DNA; nucleolar DNA, ~ 400; P3 DNA, ~ 580
Buss et al. (1990)	Covalent binding to liver DNA in vivo in male Fischer F344 rats given AFB_1:		Adducts/10^9 nucleotides:
	One dose level by oral gavage once or 10 times in 10 days	Gavage, 0.1 μg/kg bw	Single dose: ~ 4, falling to ~ 1.2 after 10 days Multiple doses: ~ 20, falling to ~ 10 after 10 days
	Six dose levels by oral gavage once or 10 times in 10 days	Gavage, 1, 10, 100 ng/kg bw; 1, 10, 100 μg/kg bw	Single dose: $>10^{-2}$ to $>10^3$ Multiple doses: $>10^{-1}$ to $>10^3$
	Three dose levels in drinking-water for 4, 6 and 8 weeks	Oral; 2.2, 73, 2110 ng/kg bw per day	At 8 weeks, 0.9, 32, 850
	AFB_1–DNA adduct levels measured after 4, 6 and 8 weeks; did not increase after 4 weeks, indicating steady-state for adduct formation and removal had been reached; at 8 weeks, adduct levels 0.91, 32 and 850 AFB_1 adducts per 10^9 nucleotides, i.e., directly proportional to dose		
Wild et al. (1990b)	Covalent binding to DNA in vivo in various organs of male Sprague-Dawley rats exposed to single and multiple doses of AFB_1. Marked intercellular variation in adduct levels observed in kidney and lung, in contrast to liver, where binding was more homogeneous. No adduct detected in oesophagus, forestomach, colon, spleen or testis (detection limit, ~ 300 pg AFB_1/mg DNA)	Gavage, [^{3}H]AFB_1, 19, 32, 75 μg/kg bw Gavage, AFB_1 daily up to 14 days (165 μg/kg bw on day 1)	Single dose: 83, 146, 444 pg AFB_1/mg DNA at 24 h Multiple doses: rapid accumulation in liver over 3 days to plateau

Table 11 (contd)

Reference	Details of study	Route and dose	DNA binding
Groopman et al. (1992c)	Covalent binding to liver DNA *in vivo* in male Fischer 344 rats. Measurement of AFB_1–$N7$-guanine excretion in rat urine, using combined monoclonal antibody immunoaffinity chromatography/HPLC method, showed dose-dependent excretion in urine. Comparison of dose-dependent residual levels of AFB_1 binding to liver DNA with amount of AFB_1–$N7$-guanine excreted in urine showed a correlation coefficient of 0.98. Urinary AFM_1 and AFP_1 evaluated as molecular dosimeters: AFM_1 found to be excellent marker	i.p.; 1 mg/kg bw p.o.; 0.03, 0.07, 1.0 mg/kg	Not possible to deduce from log–log plot

AF, aflatoxin

Table 12. Modulation of aflatoxin B_1–DNA binding

Reference	Details of study	Route and dose	DNA binding
DNA binding in vitro			
Fukayama & Hsieh (1984)	Covalent binding to calf thymus DNA in vitro in presence of liver microsomes from untreated male Fischer 344 rats		2.11 nmol AFB_1/mg DNA
	Covalent binding to calf thymus DNA in vitro in presence of liver microsomes from rats given dietary BHT (0.5%) for 10 days before killing. Microsomes from BHT-treated rats mediated same level of DNA binding as those from untreated rats, despite significantly lower bacterial mutagenicity in presene of BHT microsomes		2.2 nmol AFB_1/mg DNA
Firozi et al. (1987)	Covalent binding to calf thymus DNA in vitro in presence of liver microsomes from phenobarbital-induced male Wistar rats. Study of effect of vitamin A and derivatives on microsome-mediated AFB_1–DNA binding. Concentrations that induced 50% inhibition of binding were: retinol, 40 nM; retinyl acetate, 50 nM; all-*trans*-retinoic acid, 75 nM; retinyl palmitate, 170 nM		
Jhee et al. (1988, 1989a,b)	Covalent binding to calf thymus DNA or endogenous hepatocyte nuclei in vitro in presence of liver microsomes with or without cytosol from untreated male Fischer 344 rats or rats given dietary BHA (0.75%) for 2 weeks before killing. Cytosol from untreated rats produced more DNA binding than cytosol from BHA-treated rats. GSH conjugation of AFB_1 higher in presence of BHA-treated rat cytosol		
Fong et al. (1990)	Covalent binding to calf thymus DNA in vitro in presence of liver microsomes from *Salmo gairdneri* and rats; I3C under acid conditions caused dose-dependent reduction in DNA binding mediated by rat or trout microsomes	500 pmol AFB_1/ml [0.16 µg/ml] for 1 h	132 pmol AFB_1 bound/mg DNA (trout microsomes)
Yamasaki et al. (1991)	Covalent binding to calf thymus DNA in vitro in presence of liver post-mitochondrial supernatant from Aroclor 1254-induced male Sprague-Dawley rats: allixin (a phenolic from garlic) inhibited DNA binding and adduct formation by AFB_1 (by ~30%)	With 75 µg allixin/ml for 1 h	106 pmol/mg DNA 71 pmol/mg DNA
DNA binding in vivo			
Whitham et al. (1982)	Covalent binding to liver DNA of *S. gairdneri* (rainbow trout) in vivo	i.p.; 5, 25, 100, 300 µg/kg bw	9.2, 56, 315 and 1240 pmol/mg DNA at 24 h
	Covalent binding to liver DNA of *Oncorhynchus kisutch* (coho salmon) in vivo	i.p.; dose not given in a comparative study with trout	10.6 pmol/mg DNA at 24 h in salmon; 243 pmol/mg DNA at 24 h in trout
	In trout, almost linear increase in binding with increasing dose; time-course parameters showed no difference in binding between 4 h and 48 h after ^3H-AFB_1 treatment. Compared with AFB_1-treated controls, binding was two-fold lower in β-naphthoflavone-treated trout and 20-fold lower in salmon. AFB_1 binding not significantly altered by dietary protein or cyclopropenoid fatty acids		

Table 12 (contd)

Reference	Details of study	Route and dose	DNA binding
Dashwood et al. (1988)	Covalent binding to liver DNA of *S. gairdneri in vivo*. Fish exposed to AFB_1 and to I3C by concomitant dietary exposure. Linear increases in DNA binding with increasing dose of AFB_1 and with time of inhibitor/carcinogen co-treatment, at each I3C dose level. Successive increases in I3C dose resulted in corresponding dose-related decreases in DNA binding, suppressed by almost 95% at highest I3C dose tested (4000 ppm)	Dietary AFB_1: 10, 20, 40, 80, 160, 320 ppb Dietary I3C: 1000–4000 ppm	80 ppb AFB_1, ~1800 pmol/mg DNA reduced to 90 pmol/mg DNA by 4000 ppm dose
Goeger et al. (1988)	Covalent binding to liver and red-blood cell DNA of *S. gairdneri in vivo*	i.p.; 180 nmol/kg bw [56 μg/kg]	~150 nmol AFB_1/g liver DNA in control and BHA-treated at 1 day
		i.p.; 64 nmol/kg bw [20 μg/kg]	~45 nmol AFB_1/g liver DNA in control reduced to 25 nmol AFB_1/g liver DNA with β-naphthoflavone at 1 day
Dashwood et al. (1989)	Covalent binding to liver DNA of *S. gairdneri in vivo*. About 10 000 fish pretreated with one of five doses of I3C. After 4 weeks, they received the same dietary level of I3C for a further two weeks, with ^{3}H-AFB_1 in dose range 10–320 ppb. 15 fish randomly selected to control diet for determination of tumour response at 1 year. Linear increases in DNA binding with dose of AFB_1 at each I3C dose level. Successive increases in I3C dose gave dose-related decreases in DNA binding, resulting in a series of curves of decreasing slope. At I3C doses of ≤2000 ppm, inhibitor-altered tumour response predicted precisely by changes in levels of DNA adducts formed in liver	(see Dashwood et al., 1988)	
Campbell et al. (1978)	Covalent binding to liver DNA in male Sprague-Dawley rats *in vivo* given nutritionally complete synthetic diet (1) or marginal lipotrope-deficient diet (2). Value obtained with diet 2 claimed to be significantly lower ($p < 0.05$) than that obtained with diet 1; with both diets, binding to RNA > DNA > protein	i.p.; 1 mg/kg bw	Diet 1: 19.2 ng AFB_1/mg DNA at 6 h Diet 2: 15.6 ng AFB_1/mg DNA at 6 h
Neal et al. (1981a)	Covalent binding to liver DNA of male Fischer 344 rats *in vivo*; controls and rats fed diet with 4 ppm unlabelled AFB_1 10 weeks before injection of ^{3}H-AFB_1	i.p.; 0.5 mg/kg bw	Controls, ~12.5 ng AFB_1/mg DNA Treated, ~2 ng AFB_1/mg DNA
	Study of effect of pre-feeding low level of AFB_1 before measuring DNA binding at high single dose. Increased levels of reduced GSH and GSH S-transferase accompanied reduction in binding in rats fed diet with AFB_1, suggesting that decreased binding is due to increased detoxification of AFB_1 metabolites		

Table 12 (contd)

Reference	Details of study	Route and dose	DNA binding
Appleton et al. (1982)	Covalent binding to liver DNA of male Fischer F344 in vivo; treated rats given single i.p. injection of 2.7 nmol/0.5 ml/kg bw diethyl maleate 1 h before injection of AFB_1, to reduce GSH content	i.p. 10, 25, 65, 160, 390 and 1000 ng/kg bw	Controls: 0.37, 0.48, 1.47, 3.93, 8.54, 16.48 pg AFB_1/mg DNA at 1 h
	In dose range 10–1000 ng/kg radiolabelled AFB_1, produced measurable covalent binding of AFB_1 to DNA, RNA, and protein, which increased linearly over dose range. Adduct formation observed at lowest dose used (10 ng/kg), which is within human exposure range. Although diethyl maleate reduced hepatic GSH from 5 to 2.3 μmol/g of liver and slightly increased binding, the dose–response curve for macromolecular adduct formation remained linear in both pretreated and control groups; binding to RNA > DNA > protein		Treated: 0.35, 0.48, 1.39, 4.70, 11.47, 19.43 pg AFB_1/mg DNA at 1 h
Loury & Hsieh (1984)	Covalent binding to liver DNA in vivo in male Fischer F344 rats fed, for 41 weeks before single dose of radiolabelled AFB_1,	Gavage, 5 μg/kg bw	
	Control semi-synthetic diet		~3.6 pmol/mg DNA at 6 h
	50 ppb AFB_1 in semi-synthetic diet		~1.0 pmol/mg DNA at 6 h
	0.5 ppb AFM_1 in semi-synthetic diet		~2.2 pmol/mg DNA at 6 h
	50 ppm AFM_1 in semi-synthetic diet		~2.1 pmol/mg DNA at 6 h
	In animals pre-exposed to 50 ppb AFB_1, binding to DNA, RNA and protein decreased by 72%, 74% and 61%, respectively. Pre-exposure to AFM_1 resulted in a small reduction in binding to nucleic acids. GSH S-transferase activity increased by 133% in animals fed 50 ppb AFB_1, by 48% in those fed 50 ppb AFM_1, and remained at control values in rats fed 0.5 ppb AFM_1		
Marinovich & Lutz (1985)	Covalent binding to liver DNA in vivo in male Fischer F344 rats given ethanol (0.44–3.4 g/kg bw) in drinking-water for 10 days or by single oral administration. Level of binding not affected by any type of ethanol pretreatment	Oral; 100 ng/kg bw	13 500 μmol AFB_1/mol DNA nucleotide/mmol AFB_1 dose, mg/kg bw (CBI units) at 1 day

Table 12 (contd)

Reference	Details of study	Route and dose	DNA binding
Kensler et al. (1986)	Covalent binding to liver and kidney DNA in male Fischer F344 rats fed semi-purified diet: Alone	I.p.; 1 mg/kg bw	Liver: 859; kidney: 94 pmol AFB_1/mg DNA
	Containing 0.45% BHA for 2 weeks		Liver: 304; kidney: 36 pmol AFB_1/mg DNA
	Containing 0.45% BHT for 2 weeks		Liver: 129; kidney: 61 pmol AFB_1/mg DNA
	Containing 0.5% ethoxyquin for 2 weeks		Liver: 77; kidney: 19 pmol AFB_1/mg DNA
	Containing 0.1% 5-(2-pyrazinyl)-4-methyl-1,2-dithiol-3-thione (oltipraz) for 2 weeks		Liver: 202; kidney: 34 pmol AFB_1/mg DNA
	Several AFB_1 metabolite–DNA adducts formed in both tissues. Principal and related adducts of 8,9-dihydro-8-(N7-guanyl)-9-hydroxy-AFB_1 represented 80–90% of all adducts in both tissues and in all treatment groups. Antioxidants ethoxyquin, BHT, BHA and oltipraz reduced binding of AFB_1 to liver DNA by 91, 85, 65 and 76% and to kidney DNA by 80, 35, 62 and 64%, respectively. Concordantly, the specific activities of three hepatic enzymes involved in AFB_1 detoxification were significantly elevated by all antioxidants. Correlation (r = 0.95) between degree of inhibition of DNA binding and induction of hepatic GSH S-transferase activities by antioxidants		
Fukayama & Hsieh (1985)	Covalent binding to liver DNA of male Fischer F344 rats fed control diet or 0.5% BHT in diet for 10 days	Intra-gastric; 62.5 μg/kg bw	Control: ~7 pmol AFB_1/mg DNA at 6 h Treated: ~13 pmol AFB_1/mg DNA at 6 h
	Radioactivity bound to hepatic nuclear DNA six times less in BHT-pretreated rats than in controls 6 h after administration of AFB_1. Half-lives of DNA binding, 30 and 46 h for control and BHT-pretreated rats, respectively		

Table 12 (contd)

Reference	Details of study	Route and dose	DNA binding
Mandel et al. (1987)	Covalent binding to kidney and liver DNA in male Fischer 344 rats fed: Control diet	i.p.; 1 mg/kg bw	Kidney: 12.2; liver: 126.4 pg/μg DNA at 2 h
	Diet containing 0.5% ethoxyquin 2 days before dosing with AFB$_1$		Kidney, 5.4; liver, 29.5 pg/μg DNA at 2 h
	Diet containing 0.5% ethoxyquin 14 days before dosing with AFB$_1$		Kidney, 3.7; liver, 3.9 pg/μg DNA at 2 h
	Formation of presumed detoxified metabolites AFM$_1$ and AFQ$_1$ enhanced to a greater extent than formation of active metabolite, AFB$_1$-8,9-epoxide. Ethoxyquin reduced binding of AFB$_1$ to DNA of liver and kidney		
Holeski et al. (1987)	Covalent binding to liver DNA in vivo in male Sprague-Dawley, either intact or injected i.p. daily for 4 days with 75 mg/kg bw phenobarbital	i.p.; 0.25 mg/kg bw	Controls: 78 pmol AFB$_1$/mg DNA at 2 h Treated: 35 pmol AFB$_1$/mg DNA at 2 h
	Phenobarbital treatment had no significant effect on amount of AFB$_1$ remaining in liver but decreased amount of binding to liver DNA by 55%. For individual animals from each group, correlation between increase in excretion of AFB$_1$-GSH and decrease in covalent binding was significant ($r = 0.77$)		
Monroe et al. (1986)	Covalent binding to liver DNA in male Sprague-Dawley rats given: Propylene glycol (vehicle) Single oral dose (500 mg/kg bw) of BHA Corn oil (vehicle) Daily s.c. BHA (500 mg/kg bw) for 9 days	i.p.; 0.25 mg/kg bw	97.5 pmol/mg DNA 77.8 pmol/mg DNA 95.8 pmol/mg DNA 16.4 pmol/mg DNA
	In order to differentiate between enzyme induction and direct antioxidant effects, BHA was given for 9 days or as a single dose. Repeated treatment enhanced biliary excretion of both the GSH conjugate of AFB$_1$ and the AFB$_1$ glucuronide to 200% of control values, and reduced covalent binding of AFB$_1$ to liver DNA to 17% of control. A single BHA treatment had no effect on biliary excretion or binding to liver macromolecules, despite high concentrations of BHA in liver during period of AFB$_1$ metabolism		

Table 12 (contd)

Reference	Details of study	Route and dose	DNA binding
Monroe & Eaton (1987)	Covalent binding to liver DNA *in vivo* in Female CD-1 mice fed control diet	0.25 mg/kg bw	~4.5 AFB$_1$ adducts/10^7 nucleotides
	Female CD-1 mice fed diet containing 0.75% BHA		~3.1 AFB$_1$ adducts/10^7 nucleotides
	Male Sprague-Dawley rats fed control diet		~375 AFB$_1$ adducts/10^7 nucleotides
	Male Sprague-Dawley rats fed diet containing 0.75% BHA		~70 AFB$_1$ adducts/10^7 nucleotides
	BHA treatment resulted in a decrease in hepatic AFB$_1$-guanine adduct formation in mice to 68% of control and, in rats, to 18% of control. AFB$_1$-DNA binding in control mice 1.2% of that in control rats		
Monroe & Eaton (1988)	Covalent binding to liver DNA in male Swiss-Webster mice: Controls	i.p.; 0.25 mg/kg bw	6.24 AFB$_1$ adducts/10^7 nucleotides
	Controls fed diet containing 0.75% BHA for 13 days		3.39 AFB$_1$ adducts/10^7 nucleotides
	Given D,L-buthionine-S-sulfoximine (0.6 g/kg) and diethyl maleate (0.75 ml/kg) i.p. 2 and 1.5 h before AFB$_1$ injection		184.9 AFB$_1$ adducts/10^7 nucleotides
	Fed diet containing 0.75% BHA for 13 days and given buthionine-S-sulfoximine and diethyl maleate as above		5.51 AFB$_1$ adducts/10^7 nucleotides
	Depletion of GSH accomplished with buthionine-S-sulfoximine and diethyl maleate before injection of AFB$_1$, giving 97% and 70% in control and BHA-treated mice, respectively. In control mice, GSH depletion associated with 30-fold increase in covalent AFB$_1$-DNA binding. DNA binding in mice treated with dietary BHA alone was reduced to 54% of control. In BHA-treated mice, pre-treatment with buthionine-S-sulfoximine and diethyl maleate increased DNA binding by 62%. Dietary BHA increased hepatic microsome-mediated activation of AFB$_1$ to the AFB$_1$-epoxide by eight-fold in both control and pretreated mice		

Table 12 (contd)

Reference	Details of study	Route and dose	DNA binding
Ramsdell & Eaton (1988)	Covalent binding to liver DNA in male Sprague-Dawley rats fed 21 or 22 days: Purified diet	i.p., 0.25 mg/kg bw	162 pmol AFB_1/mg DNA at 2 h
	Purified diet containing 25% freeze-dried broccoli		72.1 pmol AFB_1/mg DNA at 2 h
	Proprietary rodent diet		142 pmol AFB_1/mg DNA at 2 h
	Binding of AFB_1 to DNA significantly lower in the group given broccoli but not in rats fed rodent diet		
Lotlikar et al. (1989)	Covalent binding to liver DNA in male Sprague-Dawley rats fed: Control diet and water	i.p., 400 µg/kg bw	29.8 pmol AFB_1/mg DNA
	Control diet with 0.1% phenobarbital in drinking-water for 1 week		8.6 pmol AFB_1/mg DNA
	Phenobarbital treatment caused significant reduction in binding *in vivo*		
Salbe & Bjeldanes (1989)	Covalent binding to liver DNA in male Sprague-Dawley rats fed: Basal diet	i.p., 3 µg/kg bw; intra-gastric, 3 µg/kg bw	1.5 pmol AFB_1/mg DNA at 2 h; 0.98 pmol AFB_1/mg DNA at 2 h
	Basal diet containing 25% dry weight freeze-dried brussels sprouts for 2 weeks before dosing with AFB_1	i.p., 3 µg/kg bw; intra-gastric, 3 µg/kg bw	0.59 pmol AFB_1/mg DNA at 2 h; 0.45 pmol AFB_1/mg DNA at 2 h
	Basal diet containing 250 ppm I3C for 2 weeks before dosing with AFB_1	i.p., 3 µg/kg bw; intra-gastric, 3 µg/kg bw	1.11 pmol AFB_1/mg DNA; 0.82 pmol AFB_1/mg DNA at 2 h
	Basal diet with 0.1% phenobarbital in drinking-water for 1 week before dosing with AFB_1	i.p., 3 µg/kg bw; intra-gastric, 3 µg/kg bw	0.83 pmol AFB_1/mg DNA at 2 h; 0.37 pmol AFB_1/mg DNA at 2 h
	Brussels sprouts caused significant 50–60% decrease in AFB_1–DNA binding and increased hepatic and intestinal GSH-S-transferase activities. Route of administration did not alter inhibition of binding compared with control rats in either treatment group, suggesting that small intestine may not play significant role in metabolism of AFB_1. In a second experiment, rats were dosed either i.p. or intragastrically with AFB_1 and killed 2, 6, 12, 24 or 48 h later. Brussels sprouts significantly decreased hepatic AFB_1–DNA binding		

Table 12 (contd)

Reference	Details of study	Route and dose	DNA binding
Jhee et al. (1989b)	Covalent binding to liver DNA in male Fischer F344 rats fed: Control diet	i.p.; 400 µg/kg bw	440 pmol AFB_1/mg DNA at 2 h
	Diet containing 0.75% BHA for 2 weeks		69 pmol AFB_1/mg DNA at 2 h
	BHA treatment reduced DNA binding to 15% of controls with concomitant increase in biliary excretion of AFB_1-reduced GSH conjugate		
Lamartiniere (1990)	Covalent binding to liver DNA in:	i.p.; 1 mg/kg bw	*pmol total AFB_1 adducts/mg DNA*
	Male Sprague-Dawley CD control rats, given AFB_1 at 6 months. Groups of these rats also given corn oil (CO) and 0.1% phenobarbital (PB) in drinking-water for 5 days or 20 mg/kg 3-methylcholanthrene (MC) for 3 days before AFB_1		CO, 255; PB, 28; 3MC, 94
	Male Sprague-Dawley CD rats given 1.45 µmol diethylstilboestrol s.c. 2, 4 and 6 days after birth and AFB_1 at 6 months. Groups of these rats also given CO or 0.1% PB in drinking-water for 5 days or 20 mg/kg MC for 3 days before AFB_1		CO, 291; PB, 46; 3MC, 52
	Female Sprague-Dawley CD control rats, given AFB_1 at 6 months. Groups of these rats also given CO or 0.1% PB in drinking-water for 5 days or 20 mg/kg MC for 3 days before AFB_1		CO, 250; PB, 49; 3MC, 39
	Female Sprague-Dawley CD rats given 1.45 µmol diethylstilboestrol s.c. 2, 4 and 6 days after birth and AFB_1 at 6 months. Groups of these rats also given CO or 0.1% PB in drinking-water for 5 days or 20 mg/kg MC for 3 days before AFB_1		CO, 920; PB, 49; 3MC, 36
	Female rats exposed to diethylstilboestrol had significantly higher DNA adduct levels (3 to 6 fold) than adult female rats treated neonatally with vehicle. DNA adduct levels were not significantly higher in control males than in diethylstilboestrol-treated males. PB and MC treatment followed by AFB_1 injection resulted in significantly decreased AFB_1-guanine adduct levels in all rats.		
Wang et al., 1990	Covalent binding to liver DNA in male Wistar rats: Controls	i.p.; 10 µg/kg bw	1.97 pmol AFB_1/mg DNA
	Given a single oral dose of 100 mmol/kg ethanol 6 h before AFB_1		2.90 pmol AFB_1/mg DNA
	Given a single oral dose of 100 mmol/kg ethanol 18 h before AFB_1		2.22 pmol AFB_1/mg DNA
	Binding increased by 47% and hepatotoxicity potentiated in rats treated with ethanol 6 h (time of maximal GSH depletion) before administration of AFB_1; binding not increased by treatment with ethanol 18 h (time for approximately normal GSH levels) before AFB_1, and no potentiation of hepatotoxicity observed		

Table 12 (contd)

Reference	Details of study	Route and dose	DNA binding
Schrager et al. (1990)	Covalent binding to liver DNA in male rats fed:	i.p.; 25 µg/rat [~ 0.8 mg/kg Fischer 344 rats; ~ 0.6 mg/kg Sprague-Dawley rats]	pmol AFB$_1$/mg DNA:
	Control diet for 21 days then AFB$_1$ 5 days/week for 2 weeks (Fischer 344 rats)		21 at 2 h, day 2
	Control diet for 21 days then AFB$_1$, once (Sprague-Dawley rats)		1.3 at 2 h, 6 at 24 h
	Choline-deficient/methionine-low diet for 21 days then dosed AFB$_1$ 5 days/week for 2 weeks (Fischer 344 rats)		49 at 2 h, day 2
	Choline-deficient/methionine-low diet for 21 days then AFB$_1$, once (Sprague-Dawley rats)		16 at 2h, 4 at 24 h
	Choline-deficient/methionine-low diet increased AFB$_1$ hepatocarcinogenesis and reduced time to first tumours. Total adduct levels in choline-deficient animals increased significantly during multiple-dose schedule. When total adduct levels were integrated over 10-day dose period, 41% increase in adduct burden apparent in choline-deficient animals		
Wang et al. (1991a)	Covalent binding to liver DNA in male Wistar rats:	i.p.; 6 µg/kg bw	
	Controls		142.3 pg AFB$_1$/mg DNA
	Given 6 mg crocetin/kg bw by gastric gavage for 3 days before AFB$_1$		116.6 pg AFB$_1$/mg DNA
	Given 10 mg crocetin/kg bw by gastric gavage for 3 days before AFB$_1$		88.1 pg AFB$_1$/mg DNA
	Crocetin raised hepatic GSH, GSH S-transferase and GSH peroxidase levels and significantly decreased AFB$_1$–guanine adduct formation		
Wang et al. (1991b)	Covalent binding to liver DNA in male Wistar rats:	i.p.; 33.9 µg/kg bw	
	Controls		3.3 pg AFB$_1$/mg DNA
	Given 6 mg geniposide/kg bw by gastric gavage for 3 days before AFB$_1$		1.7 pg AFB$_1$/mg DNA
	Given 10 mg geniposide/kg bw by gastric gavage for 3 days before AFB$_1$		1.8 pg AFB$_1$/mg DNA
	Geniposide raised hepatic GSH, GSH S-transferase and GSH-peroxidase levels and significantly decreased AFB$_1$–guanine adduct formation		
Groopman et al. (1992c)	Covalent binding to liver DNA in male Fischer rats:	Gavage; 250 µg/kg bw	
	Controls		51 pmol AFB-N7-guanine per mg DNA at 2 days
	Given 0.03% 1,2-dithiol-3-thione in the diet		9 pmol AFB-N7-guanine per mg DNA at 2 days
	Reduction in adduct levels in serum and urine of 1,2-dithiol-3-thione-treated rats was similar to that seen in liver		

AF, aflatoxin; BHT, butylated hydroxytoluene; BHA, butylated hydroxyanisole; I3C, indole-3-carbinol; GSH, glutathione; CBI, carcinogen binding index

Table 13. Mutations in the *p53* tumour suppressor gene in human hepatocellular carcinomas

Reference	Details of patients					Details of tumours		Details of mutations	p53 alteration			Remarks
	Sex/age	Ethnic origin/country	Exposure to aflatoxins	HBV status	Cirrhosis	Grade/presentation	Number analysed	Loss of p53 allele	Codon	Base change (nucleotide)	Amino acid change	
Murakami et al. (1991)	No data; no relationship between p53 aberrations, age, sex or chronic liver disease	Tokyo, Japan	Low	No data	See column 1	Early	21	0/21		Wild type 21/21	0/21	9 DNA fragments containing 1 or 2 exons 2–11 examined by PCR and SSCP; mutations detected by SSCP sequenced after asymmetric PCR. Of 22 advanced tumours, 0/3 well-differentiated, 4/11 moderately differentiated, 4/8 poorly differentiated tumours contained p53 aberrations; seen in 0/18 tumours <2 cm, 1/6 2–3 cm and 7/19 >3 cm. No codon-249 mutation in 43 tumours
						Advanced	22	+	246	Wild type 15/22		
								+	244	ATG to GTG	Met to Val	
								+	110	GGC to AGC	Gly to Ser	
								+	381–	CGT to TGT	Arg to Cys	
									382	One base-pair deletion	Frameshift	
								+	168–	31 base-pair deletion	Frameshift	
									178			
								–	175,	CGC to AGC,	Arg to Ser	
									272	GCC to GAC,	Lys to Asp	
								+	276	GTG to ATG	Val to Met	
								7/22		8/22		
Hayward et al. (1991)		*Australia*				Early + advanced	16	7/43			8/43	Study to detect changes in bases 2 or 3 of codon 249 in exon 7 of p53 gene. Assay depends on abolition of HaeIII restriction site. 125 base-pair fragment containing all of exon 7 and part of flanking introns amplified by PCR. All 16 samples cleaved at this site by HaeIII, showing that none of the tumours had a mutation at that position
	F/44	Caucasian	Low	–	–	2ª, earlyᵇ		Not assayed for	249	Wild type		
	F/36	Caucasian	Low	–	–	1ª, early			249	Wild type		
	M/71	Caucasian	Low	–	–	2, late			249	Wild type		
	M/67	Caucasian	Low	–	+	2, late			249	Wild type		
	M/68	Caucasian	Low	–	+	1–3, late			249	Wild type		
	M/76	Caucasian	Low	–	+	3, late			249	Wild type		
	M/79	Caucasian	Low	–	+	2, late			249	Wild type		
	M/68	Caucasian	Low	–	+	2–3, late			249	Wild type		
	M/51	Melanesian	Low	+	+	2, early			249	Wild type		
	M/52	Aboriginal	Low	+	+	1, early			249	Wild type		
	M/52	Caucasian	Low	+	+	1, early			249	Wild type		
	M/29	Melanesian	Low	+	+	3, early			249	Wild type		
	M/57	Vietnamese	Low	+	–	1, late			249	Wild type		
	M/35	Melanesian	Low	+	–	2, late			249	Wild type		
	F/69	Caucasian	Low	+	+	1, late			249	Wild type		
	M/36	Caucasian	Low	+	+	2, late			249	Wild type		
										Total: 0/16		No codon-249 mutation in 16 tumours

Table 13 (contd)

Reference	Details of patients				Details of tumours		Details of mutations				Remarks
	Sex/age	Ethnic origin/ country	Exposure to aflatoxins	HBV status	Cirrhosis	Grade/ presentation	Number analysed	Loss of p53 allele	p53 alteration		
									Codon	Base change (nucleotide)	Amino acid change
Challen et al. (1992)	9 M, 10 F; mean age, 62; range, 21–76	17/19 British	Low	+6/19	+14/19	No data	19	Not assayed	17/19 no mutation	17/19 wild type		Exons 5, 6, 7, 8 of p53 from 19 tumours amplified by PCR and analysed by SSCP and direct sequencing
									1/19 codons 158, 159	6 base-pair deletion		
									1/19 codon 286	GAA to AAA	Glu to Lys	No codon-249 mutation in 19 tumours
Hosono et al. (1991)	No data	Taiwan, China	Low	Most tumours contained integrated HBV DNA	No data	No data	18 2	1/18 0/2		0/9	0/9	Steady-state levels of p53-specific RNA in tumours did not differ from those in normal adult liver. Exons 5, 6, 7, 8 of p53 from 9 tumours amplified by PCR, subcloned and sequenced. No structural alteration seen. Synthesis of p53-specific mRNA or protein in HepG2 human hepatoblastoma cell line not affected by gene expression and replication of human HBV
							Total: 1/20		Total: 0/9			No codon-249 mutation in 9 tumours

Table 13 (contd)

Reference	Details of patients					Details of tumours		Details of mutations				Remarks
	Sex/age	Ethnic origin/ country	Exposure to afla- toxins	HBV status	Cirrhosis	Grade/ presen- tation	Number analysed	Loss of p53 allele	p53 alteration		Amino acid change	
									Codon	Base change (nucleotide)		
Hsu et al. (1991)	M/44	Qidong, China	High	High- risk area for HBV	No data	II, 5 x 4 cm			249	Wild type		Sections of p53 gene containing exons 5, 6, 7, 8 and nearby intron regions amplified by PCR from genomic DNA from tumour samples and sequenced. Each mutation con- firmed by sequencing both a second PCR sample and coding and non-coding strands of products. No other mu- tation found in exons 5, 6 or 8 or remainder of exon 7
	F/33		High			II, 14 x 10 cm			249	Wild type		
	M/49		High			III, 11 x 10 cm		+ (?)	249	AGG to AGT	Arg to Ser	
	M/60		High			II, 4 x 4 cm			249	Wild type		
	F/58		High			III, 10 x 7 cm		+ (?)	249	AGG to AGT	Arg to Ser	
	F/39		High			II, 11 x 8 cm			249	Wild type		
	F/35		High			II, 5 x 5 cm		+ (?)	249	AGG to AGC	Arg to Ser	
	M/65		High			II, 11 x 5 cm			249	Wild type		
	M/54^c		High			II, 13 x 6 cm			249	AGG to AGG/ AGT^d	Arg to Ser	
	M/37^c		High			II, 10 x 15 cm			249	AGG to AGG/ AGT^d	Arg to Ser	
	M/31		High			II, 3 x 3 cm		+ (?)	249	AGG to AGT	Arg to Ser	
	M/52		High			II, 5 x 4 cm			249	Wild type		
	M/35		High			II, 6 x 4 cm		+ (?)	249	AGG to AGT	Arg to Ser	
	M/39		High			III, 5 x 5 cm			249	Wild type		
	M/40		High			III, 9 x 10 cm			249	Wild type		
	M/38		High			III, 6 x 7 cm			249	AGG to AGG/ AGT^d	Arg to Ser	
										Total: 8/16		8 Codon-249 mutations in 16 tumours
Bressac et al. (1991)	No data	Southern Africa	High or low	–	–	Late^f	10	17p LOH^e NI^j				Exons 5, 6, 7, 8 were amplified by PCR from tumour DNA and sequenced
			High	NT^g	–	Late		–				
				+	–	Early		–				
				+	–	Early		+				
				+	–	Late		+				
				+	–	Late		NT	249	AGG to AGT	Arg to Ser	
				+	–	Late		NT	286	8 base-pair deletion	Frameshift	
				+	+	Early		NI	157	GTC to TTC	Val to Phe	
				+	+	Late		NI	249	AGG to AGT	Arg to Ser	
				+	+	Late		+	249	AGG to AGT	Arg to Ser	
								Total: 3/8		Total: 5/10		3 Codon-249 mutations in 10 tumours

Table 13 (contd)

Reference	Details of patients				Details of tumours			Details of mutations				Remarks
	Sex/age	Ethnic origin/ country	Exposure to afla- toxins	HBV status	Cirrhosis	Grade/ presen- tation	Number analysed	Loss of p53 allele	p53 alteration		Amino acid change	
									Codon	Base change (nucleotide)		
Ozturk et al. (1991)	M/23	Mozam- bique	High	HBsAG +	No data	No data	15	No data	249	AGG to AGT	Arg to Ser	167 specimens collected from patients in 14 countries. RNA (re- versed transcribed to cDNA) used for 18 tu- mours, genomic DNA for rest. Exon 7 (includ- ing codon 249) ampli- fied by PCR; separate PCR products analysed by restriction analysis with 2 enzymes. HaeIII digestion of exon 7 of normal p53 results in 2 fragments; mutation at codon 249 abolishes HaeIII site: only one fragment detected after HaeIII digestion. G to T mutation at base 3 of codon 249 creates new PleI recognition site in exon 7; digestion with PleI yields 3 fragments; with no codon-249 mutation, PleI digestion yields only 2 fragments. Restriction analysis with HaeIII used for initial screening. All identified mutations confirmed by additional restriction analysis with HaeIII and PleI
	M/27		High	+					249	AGG to AGT	Arg to Ser	
	M/27#		High	+					249	AGG to AGT	Arg to Ser	
	M/36#		High	+					249	AGG to AGT	Arg to Ser	
	M/40		High	-					249	AGG to AGT	Arg to Ser	
	M/27#		High	-					249	AGG to AGT	Arg to Ser	
	M/32		High	-					249	AGG to AGT	Arg to Ser	
	M/37		High	-					249	AGG to AGT	Arg to Ser	
	M/28		High	+					249	Wild type		
	M/38#		High	+					249	Wild type		
	M/26		High	+					249	Wild type		
	M/41		High	-					249	Wild type		
	M/50		High	-					249	Wild type		
	M/39#		High	-					157	GTC to TTC	Val to Phe	
				Total: 7/15						Total: 8/15		
	M/28		Low	+	No data	No data	12	No data	249	AGG to AGT	Arg to Ser	
	M/29		Low	+					249	Wild type		
	M/35		Low	+					249	Wild type		
	M/46		Low	+					249	Wild type		
	M/49		Low	+					249	Wild type		
	M/24		Low	+					249	Wild type		
	M/48#		Low	+					249	Wild type		
	M/58#		Low	+					249	Wild type		
	M/32#		Low	+					286	8 base-pair deletion	Frameshift	
	M/27		Low	-					249	Wild type		
	M/50		Low	-					249	Wild type		
	M/45#		Low	NT Total: 9/11					249	Wild type Total: 1/12		

Table 13 (contd)

Reference	Details of patients				Details of tumours			Details of mutations				Remarks
	Sex/age	Ethnic origin/country	Exposure to aflatoxins	HBV status	Cirrhosis	Grade/presentation	Number analysed	Loss of p53 allele	p53 alteration		Amino acid change	
									Codon	Base change (nucleotide)		
Ozturk et al. (1991) (contd)	No data	Viet Nam	High	HBsAg 10%			3		249	AGG to AGT 1/3	Arg to Ser	12 Codon-249 mutations in 167 tumours
		China	High	10%			30		249	AGG to AGT 2/30	Arg to Ser	
		S. Africa[b]	Low	10%			24		249	AGG to AGT 1/24	Arg to Ser	
		Korea (Republic of)	Low	5%			5		249	Wild type 5/5		
		Japan	No data	2%			12		249	Wild type 12/12		
		USA	Low	0.3%			27		249	Wild type 27/27		
		Germany	Low	0.5%			20		249	Wild type 20/20		
		Spain	Low	1%			12		249	Wild type 12/12		
		Turkey	Low	1-5%			8		249	Wild type 8/8		
		Saudi Arabia	Low	1-5%			4		249	Wild type 4/4		
		Israel	Low	1-5%			3		249	Wild type 3/3		
		Italy	Low	1-5%			3		249	Wild type 3/3		
		India	High	1-5%			1		249	Wild type 1/1		
Scorsone et al. (1992)		Qidong, China[i]	High	HBV integration sites	No data	No data	36	17p marker loss[j]				DNA extracted from normal and tumour tissue from each patient. Separate PCR products of exon 7 (including codon 249) analysed by restriction with 2 enzymes. HaeIII digestion described above (Ozturk et al., 1991). AG to T transversion at codon 249 produced HinfI site
	M/48			+				–	249	Wild type		
	M/52			+				+i	249	AGG to AGT	Arg to Ser	
	F/49			–				+i	249	Wild type		
	M/42			+				+i	249	AGG to AGT	Arg to Ser	
	M/38			+				+i	249	Wild type		
	M/38			–				+i	249	Wild type		
	M/43			+				–	249	Wild type		
	M/54			+				+i	249	AGG to AGT	Arg to Ser	
	M/37			+				+i	249	AGG to AGT	Arg to Ser	
	M/56			+				+i	249	Wild type		
	F/41			+				+i	249	Wild type		
	M/50			+				–	249	AGG to AGT	Arg to Ser	
	M/51			+				–	249	Wild type		
	m/38			+				+i	249	AGG to AGT	Arg to Ser	
	M/48			+					249	Wild type		

Table 13 (contd)

Reference	Details of patients					Details of tumours			Details of mutations				Remarks
	Sex/age	Ethnic origin/ country	Exposure to aflatoxins	HBV status	Cirrhosis	Grade/ presentation	Number analysed	Loss of p53 allele	p53 alteration		Amino acid change		
									Codon	Base change (nucleotide)			
Scorsone et al. (1992) (contd)	M/40			+				+	249	AGG to AGT	Arg to Ser	All 21 tumour DNAs showing loss of HaeIII site showed gain of HinfI site not present in DNA from normal tissue of same patient; loss of HaeIII site thus due to specific mutation in base 3 of codon 249
	M/37			+				−	249	AGG to AGT	Arg to Ser	
	M/36			+				+	249	Wild type		
	F/31			+				−	249	Wild type		
	M/56			+				−	249	Wild type		
	M/36			+				+i	249	AGG to AGT	Arg to Ser	
	M/26			+				+	249	AGG to AGT	Arg to Ser	
	M/44			+				+	249	AGG to AGT	Arg to Ser	
	M/64			+				+	249	AGG to AGT	Arg to Ser	
	No data			+				NI	249	Wild type		
	No data			+				−	249	Wild type		
	M/48			+				−	249	AGG to AGT	Arg to Ser	
	M/45			+				−	249	AGG to AGT	Arg to Ser	
	M/33			+				−	249	AGG to AGT	Arg to Ser	
	F/50			+				+i	249	AGG to AGT	Arg to Ser	
	M/55			+				+	249	AGG to AGT	Arg to Ser	
	M/41			+				+	249	Wild type		
	F/32			+				NI	249	AGG to AGT	Arg to Ser	
	M/37			+				NI	249	AGG to AGT	Arg to Ser	
	M/52			+				−	249	AGG to AGT	Arg to Ser	
	F/40			+				+	249	AGG to AGT	Arg to Ser	
								Total: 12/20/36[k]		Total: 21/36		21 Codon-249 mutations in 36 tumours

HBV hepatitis B virus; PCR, polymerase chain reaction, SSCP, single-strand conformation polymorphism; NT, not tested; NI, not informative

[a]Grade 1, well differentiated; grade 2, moderately well differentiated; grade 3, poorly differentiated

[b]Early, found incidentally at transplantation or screening for α-fetoprotein and tumour < 2 cm diameter.

[c]See Scorsone et al. (1992)

[d]Heterozygote with both wild-type and mutant bands present for the third base-pair of codon 249.

[e]Loss of heterozygosity on chromosome 17p, determined using pYNZ22.1 probe or by analysis of p53 gene alleles by Southern blotting of BanII- or ScaI-digested genomic DNA using a human p53 cDNA probe

[f]Tumours designated as early or late on basis of size (small, < 5 cm diameter: moderately small, 5–10 cm: large, > 10 cm) and patient prognosis; late tumours large (or massive) and patients died within days or a few weeks after admission to hospital; early tumours small or moderately small and surgically resected

[g]Mentioned in Bressac et al. (1991) who give data on cirrhosis and presentation

[h]Includes Tsanskei, above

[i]From same liver cancer institute as Hsu et al. (1991)

[j]+, i, indirect evidence of loss of p53 allele

[k]No. direct p53 allele losses/all 17p allele losses/total no. of tumours analysed [Authors stated that 22 patients had 17p loss; only 20 found]

Aflatoxin B_1, not only induces point mutations but is also a potent clastogen, inducing chromosomal aberrations in several species *in vivo*. It is possible, therefore, that the allele loss seen in hepatocellular carcinomas is due in part to exposure to aflatoxin.

Some insight into the temporal relationship between codon-249 mutation and loss of the second allele of *p53* in hepatocellular carcinoma was gained in the study of Scorsone *et al.* (1992). Of 21 patients whose tumours had codon-249 mutations, 12 showed evidence of allele loss [numbers different in table and text of paper]. This suggests that the first step in *p53* inactivation in hepatocellular carcinoma is a point mutation in the allele, followed by loss of the wild-type allele.

(f) Point mutations in oncogenes in human hepatocellular carcinoma

Of the limited numbers of human liver tumours examined, only a few have exhibited mutations in proto-oncogenes. Of 34 tissue specimens surgically resected from 30 patients and of five cell lines of human hepatocellular carcinomas in Japan, only two had *ras* point mutations. In one, there was GGT to GTT transversion in codon 12 of c-Ki-*ras*; in the other, there was a CAA to AAA transition in codon 61 of N-*ras* (Tsuda *et al.*, 1989).

In another study conducted in Japan, 23 malignant hepatic tumours were examined. Point mutations at K-*ras* codon 12 or K-*ras* codon 61 were found in six of nine cholangiocarcinomas, but no point mutation around codon 12, 13 or 61 of the *ras* genes was found in 12 hepatocellular carcinomas or two hepatoblastomas (Tada *et al.*, 1990).

Expression of c-N-*ras* proto-oncogene was one to four times greater in tumour tissue from 12 hepatoma patients (in Shanghai, China) than in tissue from two normal livers. Increased expression of c-N-*ras* was also observed in tissue surrounding the tumours. Expression of c-*myc* was highly enhanced to various degrees in all 12 hepatoma patients. No point mutation was seen in c-N-*ras* or c-*myc* in any of the tumours (Zhang *et al.*, 1990).

4.4.2 *Experimental systems*

Aflatoxin B_1

Following its metabolic conversion to the 8,9-epoxide, aflatoxin B_1 reacts almost exclusively at *N*7 of guanine to form 8,9-dihydro-2-(*N*7-guanyl)-9-hydroxyaflatoxin B_1 (Autrup *et al.*, 1979; Shaulsky *et al.*, 1990). No other adduct has been identified.

The genetic effects of aflatoxin B_1 have been reviewed (Stark, 1986; IARC, 1987b). It has been tested extensively for genetic effects in a wide variety of tests *in vivo* and *in vitro*, giving positive results in the majority of assays. It is mutagenic and induces DNA damage in bacteria and binds covalently to isolated DNA. In fungi, aflatoxin B_1 is mutagenic and induces gene conversion and mitotic recombination. It induces sex-linked recessive lethal mutations and somatic mutation and recombination in *Drosophila*. It binds covalently to DNA of fish and chicken cells *in vitro*. It induces cell transformation in several test systems and chromosomal aberrations, sister chromatid exchange, mutation, unscheduled DNA synthesis and DNA strand breaks in rodent cells *in vitro*. In human cells *in vitro*, aflatoxin B_1 induces chromosomal aberrations, micronucleus formation, sister chromatid exchange, mutation and unscheduled DNA synthesis and binds covalently to DNA. It induces DNA damage and mutation in bacteria in host-mediated assays employing mice. It binds covalently to DNA of several species of fish and to DNA of chicken embryos.

Aflatoxin B_1 induces chromosomal aberrations, micronucleus formation, sister chromatid exchange, unscheduled DNA synthesis and DNA strand breaks, and binds covalently to DNA in cells of rodents treated *in vivo*; in one study, it was reported to be weakly active in a dominant lethal mutation assay in mice, but it was inactive in another study. In a single study, a mixture of 75% aflatoxin B_1 and 25% aflatoxin B_2 induced dominant lethal mutations in rats. Aflatoxin B_1 induced chromosomal aberrations in bone marrow of rhesus monkeys treated *in vivo*.

The 8,9-dihydro-2-(N7-guanyl)-9-hydroxyaflatoxin B_1 adduct has been isolated and identified in mammals treated *in vivo*. This adduct has also been identified in human urine (Autrup *et al.*, 1987; Groopman *et al.*, 1992a,b).

(a) *Modulating factors in the genetic activity of aflatoxins*

The genetic activity of aflatoxin B_1 can be inhibited by dietary components, including vitamin A (Qin & Huang, 1986), phenolic compounds (gallic acid, chlorogenic acid, caffeic acid, dopamine, eugenol, *para*-hydroxybenzoic acid) (San & Chan, 1987; Francis *et al.*, 1989b), plant flavonoids (kaempferol, morin, fisetin, biochanin A, rutin) (Francis *et al.*, 1989a), allixin (Yamasaki *et al.*, 1991) and penta-acetyl geniposide (Tseng *et al.*, 1992).

Aflatoxin B_1 induced fewer sister chromatid exchanges in mice fed a diet supplemented with retinyl acetate (vitamin A) than in mice fed a diet deficient in retinyl (Qin & Huang, 1986). The number of chromosomal aberrations induced in bone marrow of Chinese hamsters was reduced when they were given sodium selenite at 2 mg/l in drinking-water for 14 days (Petr *et al.*, 1990), and in mice given ascorbic acid at 10 mg/kg bw for 6 and 12 weeks (Bose & Sinha, 1991).

(b) *Mutations in proto-oncogenes and tumour-suppressor genes in tumours induced by aflatoxins in animals* in vivo (Table 19)

In rainbow trout fed aflatoxin B_1 at 80 ppb (µg/kg) in the diet for two weeks, 8/14 liver tumours of mixed histology contained mutations (seven GGA to GTA transversions, one GGA to AGA transition) in codon 12 of the c-Ki-*ras* proto-oncogene, and 2/14 contained GGT to GTT transversions at codon 13 of that gene (Chang *et al.*, 1991).

Male CF1 mice from the Shell Toxicology Laboratory, seven days old, were given single intraperitoneal injections of aflatoxin B_1 at 6 µg/g bw. Of eight liver tumours that developed 24 weeks or more after injection, one contained a CAA to CTA transversion at codon 61 of c-Ha-*ras* and two contained a CAA to AAA transversion at that codon (Bauer-Hofmann *et al*, 1990).

DNA samples from liver tumours induced in male Fischer 344 rats fed 1 ppm (mg/kg) aflatoxin B_1 and 0.3 ppm (mg/kg) aflatoxin G_1 and from two cell lines derived from the liver tumours were transfected into NIH 3T3 mouse cells, and primary transformants were selected by injection into nude athymic mice; secondary and tertiary transformants were obtained *in vitro* by focus assay. Of four tumours and two cell lines analysed, all contained exogenous *ras* sequences. Activation was found in 1/7 Ha-*ras*, 1/7 Ki-*ras* and 5/7 N-*ras* genes, but only one mutation (a G to A transition at codon 12) was identified in Ki-*ras* (Sinha *et al.*, 1988).

In an assay similar to that used by Sinha *et al* (1988), McMahon *et al.* (1990) showed that 3/8 liver carcinomas that had developed in male Fischer 344 within one to two years after intraperitoneal injection of 25 µg aflatoxin B_1 on five days a week for eight weeks contained c-Ki-*ras* mutations at codon 12. One was a GGT to TGT transversion, and two were GGT to GAT transitions.

In a retrospective study of four rhesus monkeys and four cynomolgus monkeys treated with aflatoxin B_1 by a variety of schedules, no codon 249 mutation was detected in exons 5, 7 or 8 of the *p53* gene in four hepatocellular carcinomas (two from one rhesus monkey), two cholangiocarcinomas, one spindle-cell carcinoma, one haemangioendothelial sarcoma or one osteogenic sarcoma. One hepatocellular carcinoma carried a G to T transversion at codon 175 (Fujimoto *et al.*, 1992).

Aflatoxin B_2

No data were available on the genetic and related effects of aflatoxin B_2 in humans. Aflatoxin B_2 binds to DNA *in vitro* after photoactivation by exposure to radiation at 365 nm. It induces mutation and DNA damage in bacteria but is not mutagenic to fungi in the absence of a metabolic system and does not induce gene conversion or mitotic recombination in yeast. It transforms Syrian hamster embryo cells and induces sister chromatid exchange in Chinese hamster cells and unscheduled DNA synthesis in rat hepatocytes but not in human fibroblasts *in vitro*. It inhibits intercellular communication between Syrian hamster cells *in vitro*. Aflatoxin B_2 binds covalently to DNA in hepatocytes of rats treated *in vivo*.

Aflatoxin G_1

No data were available on the genetic and related effects of aflatoxin G_1 in humans. Aflatoxin G_1 induces mutation and DNA damage in bacteria and binds covalently to isolated DNA. It induces mutation in *Neurospora crassa* but neither mutation nor gene conversion in *Saccharomyces cerevisiae*. It induces unscheduled DNA synthesis in human fibroblasts and rat hepatocytes *in vitro* and causes chromosomal aberrations and sister chromatid exchange in Chinese hamster cells *in vitro*. Aflatoxin G_1 induces chromosomal aberrations in bone-marrow cells of Chinese hamsters and mice treated *in vivo* and binds to DNA in kidney and liver cells of treated rats.

Aflatoxin G_2

No data were available on the genetic and related effects of aflatoxin G_2 in humans. Aflatoxin G_2 gave conflicting results for mutation in bacteria and does not cause DNA damage. It does not induce mutation in cultured rodent cells or in fungi in the absence of a metabolic system. It induces sister chromatid exchange in Chinese hamster cells and unscheduled DNA synthesis in rat and hamster hepatocytes *in vitro*. Aflatoxin G_2 does not induce unscheduled DNA synthesis in human fibroblasts *in vitro*.

Aflatoxin M_1

No data were available on the genetic and related effects of aflatoxin M_1 in humans. Aflatoxin M_1 is mutagenic to bacteria and binds to DNA *in vitro*. It binds to DNA of cultured trout hepatocytes and induces unscheduled DNA synthesis in rat hepatocytes *in vitro*.

Table 14. Genetic and related effects of aflatoxin B_1

Test system	Result		Dose (LED/HID)[a]	Reference
	Without exogenous metabolic system	With exogenous metabolic system		
PRB, Prophage induction/SOS/DNA strand breaks/cross-links	–	+	68.0000	Goze et al. (1975)
PRB, Prophage induction/SOS/DNA strand breaks/cross-links	0	+	25.0000	Sarasin et al. (1977)
PRB, Prophage induction/SOS/DNA strand breaks/cross-links	0	+	0.0500	Elespuru & Yarmolinsky (1979)
PRB, Prophage induction/SOS/DNA strand breaks/cross-links	(+)	+	0.0500	Wheeler et al. (1981)
PRB, Prophage induction/SOS/DNA strand breaks/cross-links	0	+	0.0050	Ho & Ho (1981)
PRB, Prophage induction/SOS/DNA strand breaks/cross-links	+	0	0.0300	Affolter et al. (1983a)
PRB, SOS repair, *Escherichia coli* PQ37	?	+	0.0100	Krivobok et al. (1987)
PRB, SOS repair, *Escherichia coli* PQ37	0	+	0.0100	Auffray & Boutibonnes (1987)
PRB, *umu* expression, *Salmonella typhimurium* TA1535/pSK1002	0	+	1.5000	Shimada et al. (1987)
PRB, *umu* expression, *Salmonella typhimurium* TA1535/pSK1002	0	+	3.1000[b]	Shimada et al. (1989)
PRB, *umu* expression, *Salmonella typhimurium* TA1535/pSK1002	0	+	3.1000[b]	Kitada et al. (1990)
PRB, *umu* expression, *Salmonella typhimurium* TA1535/pSK1002	0	+	3.0000	Baertschi et al. (1989)
ECB, *Escherichia coli*, DNA strand breaks/cross-links/repair	–	0	300.0000	Thielmann & Gersbach (1978)
ECD, *Escherichia coli pol*A, differential toxicity (spot test)	0	+	2.5000	Rosenkranz & Leifer (1980)
ERD, *Escherichia coli rec* strain, differential toxicity	–	+	0.0200	Ichinotsubo et al. (1977)
ERD, *Escherichia coli rec* strain, differential toxicity	0	+	0.8000	Mamber et al. (1983)
BSD, *Bacillus subtilis rec* strains, differential toxicity	+	0	10.0000	Ueno & Kubota (1976)
BSD, *Bacillus subtilis rec* strains, differential toxicity	0	+	13.4000	McCarroll et al. (1981)
BSD, *Bacillus subtilis rec* strains, differential toxicity	0	+	1.0000	Hirano et al. (1982)
SAF, *Salmonella typhimurium*, forward mutation	0	+	0.0300	Stark et al. (1979)
SAF, *Salmonella typhimurium*, forward mutation	0	+	0.0900	Skopek & Thilly (1983)
SA0, *Salmonella typhimurium* TA100, reverse mutation	(+)	+	5.0000	von Engel & von Milczewski (1976)
SA0, *Salmonella typhimurium* TA100, reverse mutation	–	+	0.0250	Coles et al. (1977)
SA0, *Salmonella typhimurium* TA100, reverse mutation	0	+	0.0500	Tang & Friedman (1977)
SA0, *Salmonella typhimurium* TA100, reverse mutation	+	+	0.0500	Ueno et al. (1978)

Table 14 (contd)

Test system	Result		Dose (LED/HID)[a]	Reference
	Without exogenous metabolic system	With exogenous metabolic system		
SA0, *Salmonella typhimurium* TA100, reverse mutation	+	+	0.0500	Wehner *et al.* (1978)
SA0, *Salmonella typhimurium* TA100, reverse mutation	–	+	0.0060	Malaveille *et al.* (1979)
SA0, *Salmonella typhimurium* TA100, reverse mutation	0	+	0.0130	Baker *et al.* (1980)
SA0, *Salmonella typhimurium* TA100, reverse mutation	0	+	0.0050	Booth *et al.* (1980)
SA0, *Salmonella typhimurium* TA100, reverse mutation	0	+	0.0010	Matsushima *et al.* (1980)
SA0, *Salmonella typhimurium* TA100, reverse mutation	0	+	0.0250	Coles *et al.* (1977)
SA0, *Salmonella typhimurium* TA100, reverse mutation	–	+	0.0500	Decloitre & Hamon (1980)
SA0, *Salmonella typhimurium* TA100, reverse mutation	0	+	0.0100	Booth *et al.* (1981)
SA0, *Salmonella typhimurium* TA100, reverse mutation	0	+	0.0100	Nishioka *et al.* (1981)
SA0, *Salmonella typhimurium* TA100, reverse mutation	0	+	0.0250	Wheeler *et al.* (1981)
SA0, *Salmonella typhimurium* TA100, reverse mutation	0	+	0.0300	Dobiáš *et al.* (1982)
SA0, *Salmonella typhimurium* TA100, reverse mutation	–	+	0.0050	Dunn *et al.* (1982)
SA0, *Salmonella typhimurium* TA100, reverse mutation	0	+	0.0060	Sizaret *et al.* (1982)
SA0, *Salmonella typhimurium* TA100, reverse mutation	0	+	0.0300	Stark & Giroux (1982)
SA0, *Salmonella typhimurium* TA100, reverse mutation	0	+	0.0400	Affolter *et al.* (1983b)
SA0, *Salmonella typhimurium* TA100, reverse mutation	0	+	0.0100	Distelrath *et al.* (1983)
SA0, *Salmonella typhimurium* TA100, reverse mutation	0	+	0.0500	Dorange *et al.* (1983)
SA0, *Salmonella typhimurium* TA100, reverse mutation	0	+	0.3000	Malaveille *et al.* (1983)
SA0, *Salmonella typhimurium* TA100, reverse mutation	0	+	0.3700	Södeerkvist *et al.* (1983)
SA0, *Salmonella typhimurium* TA100, reverse mutation	–[c]	+	0.6000	Israel-Kalinsky *et al.* (1984)
SA0, *Salmonella typhimurium* TA100, reverse mutation	–	+	0.2500	Yourtee & Kirk–Yourtee (1986)
SA0, *Salmonella typhimurium* TA100, reverse mutation	0	+	0.1500[b]	Ishii *et al.* (1986)
SA0, *Salmonella typhimurium* TA100, reverse mutation	+	+	0.5000	Yourtee *et al.* (1987a)
SA0, *Salmonella typhimurium* TA100, reverse mutation	0	+	0.2000[b]	Francis *et al.* (1989a)
SA0, *Salmonella typhimurium* TA100, reverse mutation	0	+	0.0250[b]	Yamasaki *et al.* (1991)
SA0, *Salmonella typhimurium* TA100, reverse mutation	–	+	0.0250	Turmo *et al.* (1991)
SA5, *Salmonella typhimurium* TA1535, reverse mutation	–	–	5.0000	Commoner (1976)

Table 14 (contd)

Test system	Result		Dose (LED/HID)[a]	Reference
	Without exogenous metabolic system	With exogenous metabolic system		
SA5, *Salmonella typhimurium* TA1535, reverse mutation	–	–	0.5000	Wehner et al. (1978)
SA5, *Salmonella typhimurium* TA1535, reverse mutation	0	–	0.0500	Wheeler et al. (1981)
SA5, *Salmonella typhimurium* TA1535, reverse mutation	–	–	0.0500	Dunn et al. (1982)
SA5, *Salmonella typhimurium* TA1535, reverse mutation	0	–	0.0500	Dorange et al. (1983)
SA7, *Salmonella typhimurium* TA1537, reverse mutation	–	–	0.5000	Commoner (1976)
SA7, *Salmonella typhimurium* TA1537, reverse mutation	–	+	0.0500	Wehner et al. (1978)
SA7, *Salmonella typhimurium* TA1537, reverse mutation	–	–	0.0500	Dunn et al. (1982)
SA8, *Salmonella typhimurium* TA1538, reverse mutation	–	(+)	5.0000	von Engel & von Milczewski (1976)
SA8, *Salmonella typhimurium* TA1538, reverse mutation	–	+	0.2500	Commoner (1976)
SA8, *Salmonella typhimurium* TA1538, reverse mutation	0	+	0.1300	Stott & Sinnhuber (1978)
SA8, *Salmonella typhimurium* TA1538, reverse mutation	–	–	0.0500	Dunn et al. (1982)
SA8, *Salmonella typhimurium* TA1538, reverse mutation	0	+	0.0500	Dorange et al. (1983)
SA8, *Salmonella typhimurium* TA98, reverse mutation	–	(+)	5.0000	von Engel & von Milczewski (1976)
SA9, *Salmonella typhimurium* TA98, reverse mutation	–	+	0.0250	Coles et al. (1977)
SA9, *Salmonella typhimurium* TA98, reverse mutation	+	+	0.0050	Ueno et al. (1978)
SA9, *Salmonella typhimurium* TA98, reverse mutation	–	+	0.0500	Wehner et al. (1978)
SA9, *Salmonella typhimurium* TA98, reverse mutation	0	+	0.3100	Norpoth et al. (1979)
SA9, *Salmonella typhimurium* TA98, reverse mutation	0	+	0.0500	Booth et al. (1980)
SA9, *Salmonella typhimurium* TA98, reverse mutation	0	+	0.0100	Busk & Ahlborg (1980)
SA9, *Salmonella typhimurium* TA98, reverse mutation	0	+	0.0100	Booth et al. (1981)
SA9, *Salmonella typhimurium* TA98, reverse mutation	0	+	0.0500	Gayda & Pariza (1981)
SA9, *Salmonella typhimurium* TA98, reverse mutation	0	+	0.1000	Jayaraj & Richardson (1981)
SA9, *Salmonella typhimurium* TA98, reverse mutation	0	+	0.0100	Nishioka et al. (1981)
SA9, *Salmonella typhimurium* TA98, reverse mutation	0	+	0.5000	Nix et al. (1981)
SA9, *Salmonella typhimurium* TA98, reverse mutation	0	+	0.0050	Robertson et al. (1981)
SA9, *Salmonella typhimurium* TA98, reverse mutation	0	+	0.0500	Wheeler et al. (1981)
SA9, *Salmonella typhimurium* TA98, reverse mutation	0	+	3.1000	Chan et al. (1982)

Table 14 (contd)

Test system	Result		Dose (LED/HID)[a]	Reference
	Without exogenous metabolic system	With exogenous metabolic system		
SA9, *Salmonella typhimurium* TA98, reverse mutation	0	+	0.0250	Chernesky *et al.* (1982)
SA9, *Salmonella typhimurium* TA98, reverse mutation	–	+	0.0200	Dobiáš *et al.* (1982)
SA9, *Salmonella typhimurium* TA98, reverse mutation	–	–	0.0500	Dunn *et al.* (1982)
SA9, *Salmonella typhimurium* TA98, reverse mutation	0	+	0.1200	Friedman *et al.* (1982)
SA9, *Salmonella typhimurium* TA98, reverse mutation	+	+	2.5000	Higashi *et al.* (1982)
SA9, *Salmonella typhimurium* TA98, reverse mutation	0	+	0.0500	Dorange *et al.* (1983)
SA9, *Salmonella typhimurium* TA98, reverse mutation	0	+	0.0100	Israels *et al.* (1983)
SA9, *Salmonella typhimurium* TA98, reverse mutation	0	+	0.0500	Kawajiri *et al.* (1983)
SA9, *Salmonella typhimurium* TA98, reverse mutation	0	+	0.0050	Robertson *et al.* (1983)
SA9, *Salmonella typhimurium* TA98, reverse mutation	0	+	0.5000	Söderkvist *et al.* (1983)
SA9, *Salmonella typhimurium* TA98, reverse mutation	–	+	0.0050	Walters & Combes (1983)
SA9, *Salmonella typhimurium* TA98, reverse mutation	0	+	0.0120	Coulombe *et al.* (1982)
SA9, *Salmonella typhimurium* TA98, reverse mutation	0	+	0.5000	Yourtee & Kirk-Yourtee (1986)
SA9, *Salmonella typhimurium* TA98, reverse mutation	0	+	0.0500	Qin & Huang (1986)
SA9, *Salmonella typhimurium* TA98, reverse mutation	0	+	9.0000[b]	San & Chan (1987)
SA9, *Salmonella typhimurium* TA98, reverse mutation	0	+	0.2000	Francis *et al.* (1989a)
SA9, *Salmonella typhimurium* TA98, reverse mutation	(+)	+	0.0250	Turmo *et al.* (1991)
SAS, *Salmonella typhimurium* (other strains), reverse mutation	0	+	3.100	Stark & Giroux (1982)
SCG, *Saccharomyces cerevisiae*, gene conversion	+	0	166.0000	Callen & Philpot (1977)
SCG, *Saccharomyces cerevisiae*, gene conversion	–	+	50.0000	Callen *et al.* (1977)
SCG, *Saccharomyces cerevisiae*, gene conversion	+	0	84.2400	Callen *et al.* (1978)
SCH, *Saccharomyces cerevisiae*, homozygosis	+	0	165.0000	Callen & Philpot (1977)
SCH, *Saccharomyces cerevisiae*, homozygosis	–	+	50.0000	Kuczuk *et al.* (1978)
SCH, *Saccharomyces cerevisiae*, homozygosis	–	–	1000.0000	Simmon (1979a)
NCF, *Neurospora crassa*, forward mutation	+	0	40.0000	Ong (1970)
NCF, *Neurospora crassa*, forward mutation	+	0	10.0000	Ong (1971)
NCF, *Neurospora crassa*, forward mutation	+	0	41.0000	Ong & de Serres (1972)

Table 14 (contd)

Test system	Result		Dose (LED/HID)[a]	Reference
	Without exogenous metabolic system	With exogenous metabolic system		
NCF, *Neurospora crassa*, forward mutation	+		137.0000	Matzinger & Ong (1976)
DMG, *Drosophila melanogaster*, genetic crossing over/recombination	+	0	50.0000	Graf et al. (1983)
DMM, *Drosophila melanogaster*, somatic mutation	+	0	12.5000	Fahmy & Fahmy (1983a)
DMM, *Drosophila melanogaster*, somatic mutation	+	0	6.2400	Fahmy & Fahmy (1983b)
DMM, *Drosophila melanogaster*, somatic mutation	+	0	50.0000	Graf et al. (1983)
DMX, *Drosophila melanogaster*, sex-linked recessive lethal mutation	+	0	20.0000	Lamb & Lilly (1971)
DMX, *Drosophila melanogaster*, sex-linked recessive lethal mutation	+	0	15.6000	Fahmy et al. (1978)
DMX, *Drosophila melanogaster*, sex-linked recessive lethal mutation	+	0	0.5000	Nix et al. (1981)
DMX, *Drosophila melanogaster*, sex-linked recessive lethal mutation	+	0	6.2400	Fahmy & Fahmy (1983b)
DIA, DNA strand breaks/cross-links, animal cells *in vitro*	–	0	1.0000	Casto (1983)
DIA, DNA strand breaks/cross-links, rat hepatocytes *in vitro*	+	–	0.9400	Sina et al. (1983)
DIA, DNA strand breaks, Chinese hamster ovary cells *in vitro*	–	0	10.0000	Štětina & Votava (1986)
DIA, DNA strand breaks, AWRF rat fibroblasts *in vitro*	+	0	10.0000	Štětina & Votava (1986)
URP, Unscheduled DNA synthesis, rat primary hepatocytes *in vitro*	+	0	3.1200	Williams (1976)
URP, Unscheduled DNA synthesis, rat primary hepatocytes *in vitro*	+	0	0.0300	Michalopoulos et al. (1978)
URP, Unscheduled DNA synthesis, rat primary hepatocytes *in vitro*	+	0	0.3000	Probst et al. (1981)
URP, Unscheduled DNA synthesis, rat primary hepatocytes *in vitro*	+	0	31.0000	Ito (1982)
URP, Unscheduled DNA synthesis, rat primary hepatocytes *in vitro*	+	0	0.0030	Loury & Byard (1983)

Table 14 (contd)

Test system	Result		Dose (LED/HID)[a]	Reference
	Without exogenous metabolic system	With exogenous metabolic system		
URP, Unscheduled DNA synthesis, rat primary hepatocytes in vitro	+	0	0.0030	McQueen & Way (1991)
UIA, Unscheduled DNA synthesis, other animal cells in vitro	+	0	5.0000	Casto et al. (1976)
UIA, Unscheduled DNA synthesis, other animal cells in vitro	–	0	0.3000	Ide et al. (1981)
UIA, Unscheduled DNA synthesis, other animal cells in vitro	0	+	0.0300	McQueen et al. (1983)
UIA, Unscheduled DNA synthesis, other animal cells in vitro	+	0	10.0000	Tsutsui et al. (1984)
UIA, Unscheduled DNA synthesis, other animal cells in vitro	–	0	80.0000	Sawada et al. (1992)
GCL, Gene mutation, Chinese hamster lung cells hprt locus in vitro	+	0	1.2500	Sawada et al. (1992)
GCO, Gene mutation, Chinese hamster ovary cells in vitro	0	+	0.3000	Bermudez et al. (1982)
GCO, Gene mutation, Chinese hamster ovary cells in vitro	0	+	0.0400	Thompson et al. (1983)
G9H, Gene mutation, Chinese hamster lung V79 cells hprt locus in vitro	–	+	0.1300	Krahn & Heidelberger (1977)
G9H, Gene mutation, Chinese hamster lung V79 cells[d] hprt locus in vitro	0	+	0.1600	Kuroki et al. (1979)
G9H, Gene mutation, Chinese hamster lung V79 cells hprt locus in vitro	–	0	9.4000	Doehmer et al. (1988)
G9H, Gene mutation, Chinese hamster lung V79 cells[e] hprt locus in vitro	+	0	6.2000	Doehmer et al. (1988)
G9O, Gene mutation, Chinese hamster lung V79 cells ouabain[r] in vitro	0	+	0.3000	Langenbach et al. (1978a)
G9O, Gene mutation, Chinese hamster lung V79 cells ouabain[r] in vitro	0	+	0.0620	Langenbach et al. (1978b)
GML, Gene mutation, mouse lymphoma cells exclusive of L5178Y, in vitro	0	+	0.6000	Friedrich & Nass (1983)
GIA, Gene mutation, other animal cells in vitro	+	0	0.3000	Tong & Williams (1978)
GIA, Gene mutation, other animal cells in vitro	+	0	0.1000	Billings et al. (1983)
GIA, Gene mutation, other animal cells in vitro	+	0	0.0200	Link et al. (1983)
GIA, Gene mutation, other animal cells in vitro	+	0	0.3000	Ved Brat et al. (1983)

Table 14 (contd)

Test system	Result		Dose (LED/HID)[a]	Reference
	Without exogenous metabolic system	With exogenous metabolic system		
GIA, Gene mutation, other animal cells *in vitro*	+	0	0.3000	Tong *et al.* (1984)
SIC, Sister chromatid exchange, Chinese hamster cells *in vitro*	(+)	+	0.3100	Wolff & Takehisa (1977)
SIC, Sister chromatid exchange, Chinese hamster cells *in vitro*	+	+	0.3100	Thomson & Evans (1979)
SIC, Sister chromatid exchange, Chinese hamster cells *in vitro*	–	+	0.1000	Batt *et al.* (1980)
SIC, Sister chromatid exchange, Chinese hamster cells *in vitro*	+	+	0.3100	Ray-Chaudhuri *et al.* (1980)
SIC, Sister chromatid exchange, Chinese hamster cells *in vitro*	0	+	0.0250	Huang *et al.* (1982)
SIC, Sister chromatid exchange, Chinese hamster cells *in vitro*	+	+	0.0300	Baker *et al.* (1983)
SIC, Sister chromatid exchange, Chinese hamster cells *in vitro*	+	0	0.3000	Kroeger-Koepke *et al.* (1983)
SIR, Sister chromatid exchange, rat cells *in vitro*	+	0	0.1600	Ved Brat *et al.* (1983)
CIC, Chromosomal aberrations, Chinese hamster cells *in vitro*	0	+	0.5000	Batt *et al.* (1980)
CIC, Chromosomal aberrations, Chinese hamster cells *in vitro*	–	+	1.6000	Stich & Stich (1982)
CIC, Chromosomal aberrations, Chinese hamster cells *in vitro*	–	+	3.1200	Whitehead *et al.* (1983)
CIT, Chromosomal aberrations, transformed cells *in vitro*	+	0	1.0000	Umeda *et al.* (1977)
TBM, Cell transformation, BALB/c 3T3 mouse cells *in vitro*	+	0	0.5000	DiPaolo *et al.* (1972)
TBM, Cell transformation, BALB/3T3 clone A31-1-1 mouse cells *in vitro*	+	0	0.0600	Cortesi *et al.* (1983)
TCM, Cell transformation, C3H 10T½ mouse cells *in vitro*	–	0	15.6000	Boreiko *et al.* (1982)
TCM, Cell transformation, C3H 10T½ mouse cells *in vitro*	+	0	1.0000	Nesnow *et al.* (1982)
TCM, Cell transformation, C3H 10T½ mouse cells *in vitro*	+	0	0.5000	Oshiro & Balwierz (1982)
TCM, Cell transformation, C3H 10T½ mouse cells *in vitro*	+	0	1.0000	Billings *et al.* (1983)
TCS, Cell transformation, Syrian hamster embryo cells, clonal assay *in vitro*	+	0	0.5000	DiPaolo *et al.* (1972)
TCS, Cell transformation, Syrian hamster embryo cells, clonal assay *in vitro*	+	0	0.1000	Pienta *et al.* (1977)
TCS, Cell transformation, Syrian hamster embryo cells, clonal assay *in vitro*	+	0	0.1250	DiPaolo (1980)
TCS, Cell transformation, Syrian hamster embryo cells, clonal assay *in vitro*	+	0	0.0300	Poily *et al.* (1980)

Table 14 (contd)

Test system	Result		Dose (LED/HID)[a]	Reference
	Without exogenous metabolic system	With exogenous metabolic system		
TFS, Cell transformation, Syrian hamster embryo cells, focal assay in vitro	+	0	0.1200	Casto et al. (1977)
TCS, Cell transformation, other cell lines in vitro	+	0	15.6000	Shimada et al. (1983)
T7S, Cell transformation, SA7/Syrian hamster embryo cells in vitro	+	0	0.2500	Casto et al. (1976)
T7S, Cell transformation, SA7/Syrian hamster embryo cells in vitro	+	0	0.0600	Casto (1981)
DIH, DNA strand breaks/cross-links, human cells in vitro	–	0	10.0000	Casto (1983)
RIH, DNA repair exclusive of unscheduled DNA synthesis, human cells in vitro	0	+	0.0800	Leadon et al. (1981)
UHF, Unscheduled DNA synthesis, human fibroblasts in vitro	+	+	3.1200	San & Stich (1975)
UHF, Unscheduled DNA synthesis, human fibroblasts in vitro	(+)	+	18.7000	Stich & Laishes (1975)
UHF, Unscheduled DNA synthesis, human fibroblasts in vitro	+	0	31.2000	Mitchell (1976)
UHL, Unscheduled DNA synthesis, human lymphocytes in vitro	+	0	31.6000	Lake et al. (1980)
UHT, Unscheduled DNA synthesis, transformed human cells in vitro	(+)	+	0.3000	Martin et al. (1977)
UHT, Unscheduled DNA synthesis, transformed human cells in vitro	–	+	0.3000	Yu et al. (1983)
UIH, Unscheduled DNA synthesis, other human cells in vitro	+	0	9.4000	Freeman & San (1980)
UIH, Unscheduled DNA synthesis, other human cells in vitro	+	0	0.3000	Butterworth et al. (1982)
GIH, Gene mutation, human lymphoblastoid AHH-1 tk+/- cells	+	0	5.0000	Crespi et al. (1990)
GIH, Gene mutation, human lymphoblastoid AHH-1 tk+/- cells in vitro	+	0	1.0000	Crespi et al. (1991)
GIH, Gene mutation, human lymphoblastoid AHH-1 tk+/- cells[f] hprt locus in vitro	+	0	0.0030	Crespi et al. (1990)
GIH, Gene mutation, human lymphoblastoid AHH-1 tk+/- cells[g] hprt locus in vitro	+	0	0.0100	Crespi et al. (1991)
SHL, Sister chromatid exchange, human lymphocytes in vitro	+	0	0.0100	El-Zawahri et al. (1977)
SHL, Sister chromatid exchange, human lymphocytes in vitro	+	+	0.3000	Thomson & Evans (1979)

Table 14 (contd)

Test system	Result		Dose (LED/HID)[a]	Reference
	Without exogenous metabolic system	With exogenous metabolic system		
SHL, Sister chromatid exchange, human lymphocytes in vitro	−	0	6.0000	Fabry & Roberfroid (1981)
SHL, Sister chromatid exchange, human lymphocytes in vitro	+	+	0.3000	Inoue et al. (1983)
SHL, Sister chromatid exchange, human lymphocytes in vitro	(+)	0	0.0094	Amstad et al. (1984)
SHL, Sister chromatid exchange, human lymphocytes in vitro	+	0	0.0400	Li et al. (1989)
SHT, Sister chromatid exchange, transformed human cells in vitro	+	0	7.8000	Huh et al. (1982)
SHT, Sister chromatid exchange, transformed human cells in vitro	+	0	0.1600	Abe et al. (1983)
MIH, Micronucleus formation, human cells in vitro	+	+	0.0300	Iskandar & Vijayalaxmi (1981)
CHF, Chromosomal aberrations, human fibroblasts in vitro	−	+	18.7000	Stich & Laishes (1975)
CHL, Chromosomal aberrations, human lymphocytes in vitro	+	0	0.0100	El-Zawahri et al. (1977)
CHL, Chromosomal aberrations, human lymphocytes in vitro	(+)	+	0.0600	Fabry & Roberfroid (1981)
CHL, Chromosomal aberrations, human lymphocytes in vitro	+	0	0.0300	Amstad et al. (1984)
CHL, Chromosomal aberrations, human lymphocytes in vitro	−	+	0.6000	Ferguson et al. (1986)
CHL, Chromosomal aberrations, human lymphocytes in vitro	−	+	3.0000	Suit et al. (1977)
BFA, Body fluids from animals, microbial mutagenicity				
HMM, Host-mediated assay, microbial cells in animal hosts	+		10 × 1 ip	Brennan-Craddock et al. (1990)
HMM, Host-mediated assay, microbial cells in animal hosts	+		5 × 1 ip	Zeilmaker et al. (1991)
HMM, Host-mediated assay, microbial cells in mouse hosts	+		33 × 1 ip	Simmon et al. (1979)
HMM, Host-mediated assay, microbial cells in animal hosts	+		3.1000	Wei & Chang (1982)
DVA, DNA strand breaks/cross-links, rats in vivo	+		1 × 1 ip	Petzold & Swenberg (1978)
UPR, Unscheduled DNA synthesis, rat hepatocytes in vivo	+		2 × 1 po	Mirsalis et al. (1982)
UVA, Unscheduled DNA synthesis, chick embryo liver cells in vivo	+		0.3 (injection into egg)	Hamilton & Bloom (1984)
SVA, Sister chromatid exchange, mouse bone-marrow cells in vivo	+		20 × 1 iv	Nakanishi & Schneider (1979)
SVA, Sister chromatid exchange, chick embryo cells in vivo	+		0.30 (0.01 μg/embryo)	Todd & Bloom (1980)
SVA, Sister chromatid exchange, bone marrow, male and female NMRI mice in vivo	+		100 × 1 po	Madle et al. (1986)

Table 14 (contd)

Test system	Result		Dose (LED/HID)[a]	Reference
	Without exogenous metabolic system	With exogenous metabolic system		
SVA, Sister chromatid exchange, bone marrow, male and female Chinese hamsters *in vivo*	–		25 × 1 po	Madle *et al.* (1986)
SVA, Sister chromatid exchange, bone marrow, male Wistar rats *in vivo*	+		6.25 × 1 po	Madle *et al.* (1986)
SVA, Sister chromatid exchange, bone marrow, female Wistar rats *in vivo*	–		6.25 × 1 po	Madle *et al.* (1986)
SVA, Sister chromatid exchange, bone marrow, male C57/6J mice *in vivo*	+		16.00 × 1 sc	Qin & Huang (1986)
SVA, Sister chromatid exchange, leukocytes, male Wistar rats *in vivo*	+		1.00 × 1 ip	Li & Lin (1990)
MVM, Micronucleus test, mouse bone-marrow cells *in vivo*	+		20.00 × 1 ip	Friedman & Staub (1977)
MVM, Micronucleus test, mice *in vivo*	–		100.00 × 1 ip	Bruce & Heddle (1979)
MVM, Micronucleus test, mice *in vivo*	+		5.00 × 1 ip	Fabry & Roberfroid (1981)
MVM, Micronucleus test, male and female NMRI mice *in vivo*	+		100.00 × 1 po	Madle *et al.* (1986)
MVR, Micronucleus test, rats *in vivo*	+		2.50 × 1 ip	Trzos *et al.* (1978)
MVR, Micronucleus test, male and female Wistar rats *in vivo*	+		3.13 × 1 po	Madle *et al.* (1986)
MVC, Micronucleus test, hamster bone-marrow cells *in vivo*	–		3.00 × 1 ip	Friedman & Staub (1977)
MVC, Micronucleus test, male and female Chinese hamsters *in vivo*	–		25.00 × 1 po	Madle *et al.* (1986)
CBA, Chromosomal aberrations, Chinese hamster bone-marrow cells *in vivo*	+		12.50 × 2 po	Korte & Rückert (1980)
CBA, Chromosomal aberrations, animal bone-marrow cells *in vivo*	+		5.00 × 1 ip	Fabry & Roberfroid (1981)
CBA, Chromosomal aberrations, bone marrow, male and female Swiss albino mice *in vivo*	+		8.00 × 1 ip	Krishnamurthy & Neelaram (1986)
CBA, Chromosomal aberrations, bone marrow, male and female Swiss albino mice *in vivo*	+		0.05 μg/kg bw diet	Kumari & Sinha (1990)
CBA, Chromosomal aberrations, bone marrow, male Wistar rats	+		5.00 × 1 ip	Ito *et al.* (1989)

Table 14 (contd)

Test system	Result		Dose (LED/HID)[a]	Reference
	Without exogenous metabolic system	With exogenous metabolic system		
CBA, Chromosomal aberrations, bone marrow, male Chinese hamsters (*Cricetulus griseus*) *in vivo*	+		0.001 × 1 ip	Bárta *et al.* (1984)
CBA, Chromosomal aberrations, bone marrow, male Chinese hamsters (*Cricetulus griseus*) *in vivo*	+		5.00 × 1 po	Petr *et al.* (1990)
CBA, Chromosomal aberrations, bone marrow, male Chinese hamsters (*Cricetulus griseus*) *in vivo*	+		0.0001 × 1 ip	Bárta *et al.* (1990)
CBA, Chromosomal aberrations, bone marrow, Chinese hamsters (*Cricetulus griseus*) *in vivo*	+		12.50 × 1 po	Roll *et al.* (1990)
CBA, Chromosomal aberrations, bone marrow, male Chinese hamsters (*Cricetulus griseus*) *in vivo*	+		1 × 1 ip	Petr *et al.* (1991)
CBA, Chromosomal aberrations, bone marrow, rhesus (*Macaca mulatta*) monkeys *in vivo*	+		0.1 × 1 ip	Bárta *et al.* (1984)
CLA, Chromosomal aberrations, leukocytes, male Wistar rats *in vivo*	+		1 × 1 ip	Li & Lin (1990)
CGC, Chromosomal aberrations, mouse spermatogonia *in vivo*, spermatocytes	–		5 × 1 ip	Leonard *et al.* (1975)
DLM, Dominant lethal mutation, mice	(+)		75 × 1 ip	Epstein *et al.* (1972)
DLM, Dominant lethal mutation, NMRI mice	–		45 × 1 ip	Roll *et al.* (1990)
DLR, Dominant lethal mutation, rats (strain unspecified)	+[h]		0.7 × 1 sc	Sharma *et al.* (1988)
ICR, Inhibition of intercellular communication, animal cells *in vitro*	–	0	50.0000	Jone *et al.* (1987)

Table 14 (contd)

Test system	Result		Dose (LED/HID)[a]	Reference
	Without exogenous metabolic system	With exogenous metabolic system		
SPM, Sperm morphology, mice in vivo	–		100	Bruce & Heddle (1979)
SPR, Sperm morphology, rats in vivo	–		0.3000	Egbunike (1979)

+, positive; (+), weakly positive; –, negative; 0, not tested

[a] In-vitro tests, µg/ml; in-vivo tests, mg/kg bw

[b] Only dose tested

[c] Photoactivated aflatoxin B_1 positive

[d] Transfected with and stably expressing cynomolgus monkey cytochrome P450IA1 cDNA (not on profile)

[e] Transfected with and stably expressing rat cytochrome P450IIB1 cDNA (not on profile)

[f] Transfected with and expressing human CYP1A2 gene (not on profile)

[g] Transfected with and expressing human CYP3A4 gene (not on profile)

[h] 75% aflatoxin B_1, 25% aflatoxin B_2

Table 15. Genetic and related effects of aflatoxin B$_2$

Test system	Result		Dose (LED/HID)[a]	Reference
	Without exogenous metabolic system	With exogenous metabolic system		
PRB, Prophage induction/SOS/DNA strand breaks/cross-links	–	–	5.0000	Wheeler et al. (1981)
PRB, SOS repair, Escherichia coli PQ37	?	+	3.0000	Krivobok et al. (1987)
ECB, Escherichia coli, DNA strand breaks/cross-links/repair	–	0	24.0000	Thielmann & Gersbach (1978)
ECL, Escherichia coli polA, differential toxicity (liquid)	0	–	250.0000	Rosenkranz & Poirier (1979)
BSD, Bacillus subtilis rec strains, differential toxicity	–	0	100.0000	Ueno & Kubota (1976)
SAF, Salmonella typhimurium, forward mutation	–	+	0.1800	Xu et al. (1984)
SA0, Salmonella typhimurium TA100, reverse mutation	–	–	5.0000	von Engel & von Milczewski (1976)
SA0, Salmonella typhimurium TA100, reverse mutation	0	+	50.0000	McCann et al. (1975)
SA0, Salmonella typhimurium TA100, reverse mutation	0	–	3.0000	Gurtoo et al. (1978)
SA0, Salmonella typhimurium TA100, reverse mutation	0	–	0.2000	Dahl et al. (1980)
SA0, Salmonella typhimurium TA100, reverse mutation	–	–	5.0000	Wheeler et al. (1981)
SA0, Salmonella typhimurium TA100, reverse mutation	+[b]	0	40 μM	Israel-Kalinsky et al. (1984)
SA0, Salmonella typhimurium TA100, reverse mutation	–	–	25.0000	Yourtee et al. (1987b)
SA5, Salmonella typhimurium TA1535, reverse mutation	–	–	250.0000	Rosenkranz & Poirier (1979)
SA5, Salmonella typhimurium TA1535, reverse mutation	–	–	125.0000	Simmon (1979b)
SA5, Salmonella typhimurium TA1535, reverse mutation	–	–	5.0000	Wheeler et al. (1981)
SA7, Salmonella typhimurium TA1537, reverse mutation	–	–	125.0000	Simmon (1979b)
SA8, Salmonella typhimurium TA1538, reverse mutation	–	–	5.0000	von Engel & von Milczewski (1976)
SA8, Salmonella typhimurium TA1538, reverse mutation	–	–	250.0000	Rosenkranz & Poirier (1979)
SA8, Salmonella typhimurium TA1538, reverse mutation	–	–	5.0000	Wheeler et al. (1981)
SA9, Salmonella typhimurium TA98, reverse mutation	–	–	5.0000	von Engel & von Milczewski (1976)
SA9, Salmonella typhimurium TA98, reverse mutation	0	+	50.0000	McCann et al. (1975)
SA9, Salmonella typhimurium TA98, reverse mutation	0	+	0.0250	Hsieh et al. (1977)
SA9, Salmonella typhimurium TA98, reverse mutation	0	+	37.5000	Kleinwächter & Koukalová (1979)
SA9, Salmonella typhimurium TA98, reverse mutation	0	–	0.2000	Dahl et al. (1980)

Table 15 (contd)

Test system	Result		Dose (LED/HID)[a]	Reference
	Without exogenous metabolic system	With exogenous metabolic system		
SA9, *Salmonella typhimurium* TA98, reverse mutation	–	–	5.0000	Wheeler et al. (1981)
SA9, *Salmonella typhimurium* TA98, reverse mutation	–	+	0.5000	Coulombe et al. (1982)
SAS, *Salmonella typhimurium* other strains, reverse mutation	–	–	125.0000	Simmon (1979b)
SCH, *Saccharomyces cerevisiae*, homozygosis	–	–	100.0000	Simmon (1979a)
NCF, *Neurospora crassa*, forward mutation	–	0	38.0000	Ong & de Serres (1972)
URP, Unscheduled DNA synthesis, rat primary hepatocytes *in vitro*	(+)	0	3.1400	Williams (1976)
URP, Unscheduled DNA synthesis, rat primary hepatocytes *in vitro*	+	0	30.0000	Williams (1977)
URP, Unscheduled DNA synthesis, rat primary hepatocytes *in vitro*	+	0	3.1400	Probst et al. (1981)
G9H, Gene mutation, Chinese hamster lung V79 cells *hprt* locus *in vitro*	0	–	14.0000	Krahn & Heidelberger (1977)
SIC, Sister chromatid exchange, Chinese hamster cells *in vitro*	–	+	3.1400	Batt et al. (1980)
TCS, Cell transformation, Syrian hamster embryo cells, clonal assay *in vitro*	–	+	0.5000	Pienta et al. (1977)
UHF, Unscheduled DNA synthesis, human fibroblasts *in vitro*	–	–	314.0000	Stich & Laishes (1975)
HMM, Host-mediated assay, microbial cells in mouse hosts	–		2 × 1 po	Simmon et al. (1979)
DLR, Dominant lethal mutation, rats	+[b]		0.7 × 1 sc	Sharma et al. (1988)
BID, Binding to calf thymus DNA *in vitro*[c]	+	0	30.0000	Israel-Kalinsky et al. (1984)
BID, Binding to calf thymus and supercoiled pBR322 DNA *in vitro*[c]	+	0	90.0000	Stark (1988)
BID, Binding to ^{14}C-labelled nick-translated DNA *in vitro*[c]	+	0	31.2000	Shaulsky et al. (1990)
BID, Binding to calf thymus DNA *in vitro* (aerobic or anaerobic)[c]	+	0	0.6000	Stark et al. (1990)
BID, Binding to calf thymus DNA *in vitro* (aerobic or anaerobic)[c]	+	0	3.2000	Stark & Liberman (1991)
BVD, Binding to liver DNA of rats *in vivo*	(+)		2 × 1 ip	Swenson et al. (1977)
BVD, Binding to liver DNA of male CDF Fischer rats *in vivo*	+		1 × 1 ip	Groopman et al. (1981a)
ICR, Inhibition of gap-junctional intercellular communication, Chinese hamster V79 lung cells *in vitro*	–	0	50.0000	Jone et al. (1987)

+, positive; (+), weakly positive; –, negative; 0, not tested

[a] In-vitro tests, μg/ml; in-vivo tests, mg/kg bw
[b] 75% aflatoxin B$_1$, 25% aflatoxin B$_2$
[c] Photoactivated

Table 16. Genetic and related effects of aflatoxin G_1

Test system	Result		Dose (LED/HID)[a]	Reference
	Without exogenous metabolic system	With exogenous metabolic system		
PRB, Prophage induction/SOS/DNA strand breaks/cross-links	0	+	10.0000	Mamber et al. (1984)
PRB, SOS repair, Escherichia coli PQ37	?	+	1.0000	Krivobok et al. (1987)
PRB, SOS repair, Salmonella typhimurium TA1535/pSK1002	0	+	6.2000	Shimada et al. (1989)
PRB, umu expression, Salmonella typhimurium TA1535/pSK1002	0	+[b]	3.0000	Baertschi et al. (1989)
ECB, Escherichia coli, DNA strand breaks/cross-links/repair	–	0	300.0000	Thielmann & Gersbach (1978)
ERD, Escherichia coli rec strains, differential toxicity	0	+	0.8000	Mamber et al. (1983)
ERD, Escherichia coli rec strains, differential toxicity	0	+	0.8000	Mamber et al. (1984)
BSD, Bacillus subtilis rec strains, differential toxicity	+	0	100.0000	Ueno & Kubota (1976)
SAF, Salmonella typhimurium, forward mutation	–	+	0.0200	Xu et al. (1984)
SA0, Salmonella typhimurium TA100, reverse mutation	–	+	5.0000	von Engel & von Milczewski (1976)
SA0, Salmonella typhimurium TA100, reverse mutation	0	+	1.0000	Gurtoo et al. (1978)
SA0, Salmonella typhimurium TA100, reverse mutation	0	–	0.0250	Booth et al. (1981)
SA0, Salmonella typhimurium TA100, reverse mutation	+	+	2.5000	Yourtee et al. (1987b)
SA8, Salmonella typhimurium TA1538, reverse mutation	–	–	5.0000	von Engel & von Milczewski (1976)
SA9, Salmonella typhimurium TA98, reverse mutation	–	–	5.0000	von Engel & von Milczewski (1976)
SA9, Salmonella typhimurium TA98, reverse mutation	0	+	0.0500	Hsieh et al. (1977)
SA9, Salmonella typhimurium TA98, reverse mutation	–	+	0.5000	Ueno et al. (1978)
SA9, Salmonella typhimurium TA98, reverse mutation	0	+	1.0000	Kleinwächter & Koukalová (1979)
SA9, Salmonella typhimurium TA98, reverse mutation	0	+	1.0000	Booth et al. (1981)
SA9, Salmonella typhimurium TA98, reverse mutation	–	+	0.2500	Coulombe et al. (1982)
SA9, Salmonella typhimurium TA98, reverse mutation	0	+	0.8000	Mamber et al. (1984)
SCG, Saccharomyces cerevisiae, gene conversion	–	–	200.0000	Callen et al. (1977)
SCR, Saccharomyces cerevisiae, reverse mutation	–	–	300.0000	Callen et al. (1977)
NCF, Neurospora crassa, forward mutation	(+)	0	40.0000	Ong (1971)

Table 16 (contd)

Test system	Result		Dose (LED/HID)[a]	Reference
	Without exogenous metabolic system	With exogenous metabolic system		
NCF, *Neurospora crassa*, forward mutation	+	0	39.0000	Ong & de Serres (1972)
NCF, *Neurospora crassa*, forward mutation	−	+	219.7600	Matzinger & Ong (1976)
URP, Unscheduled DNA synthesis, rat primary hepatocytes *in vitro*	+	0	3.2800	Williams (1977)
URP, Unscheduled DNA synthesis, rat primary hepatocytes *in vitro*	+	0	0.2000	Probst *et al.* (1981)
URP, Unscheduled DNA synthesis, rat primary hepatocytes *in vitro*	−	+	32.0000	Ito (1982)
SIC, Sister chromatid exchange, Chinese hamster cells *in vitro*	−	+	2.5000	Batt *et al.* (1980)
CIC, Chromosomal aberrations, Chinese hamster cells *in vitro*	0	+	29.8000	Batt *et al.* (1980)
UHF, Unscheduled DNA synthesis, human fibroblasts *in vitro*	−	+	32.8000	San & Stich (1975)
UHF, Unscheduled DNA synthesis, human fibroblasts *in vitro*	−	+	20.0000	Stich & Laishes (1975)
SHL, Sister chromatid exchange, human lymphocytes *in vitro*	+	0	30.0000	El-Zawahri *et al.* (1990)
CHL, Chromosomal aberrations, human lymphocytes *in vitro*	+	0	1.0000	El-Zawahri *et al.* (1990)
CBA, Chromosomal aberrations, Chinese hamster bone-marrow cells *in vivo*	+		50 × 2 po	Korte & Rückert (1980)
CBA, Chromosomal aberrations, mouse bone-marrow cells *in vivo*	+		50 × 1 po or ip	Roll *et al.* (1990)
BID, Binding to calf thymus DNA, DNA adducts (guanyl-N7)	0	+	15.0000	Baertschi *et al.* (1989)
BID, Binding to DNA *in vitro*	+	0	79.0000	Garner *et al.* (1979)
BVD, Binding to DNA, rats *in vivo*	+		0.6 × 1 ip	Garner *et al.* (1979)
BVD, Binding to DNA, rats *in vivo*	+		0.5 × 1 ip	Wild *et al.* (1990b)

+, positive; (+), weak positive; −, negative; 0, not tested
[a]In-vitro tests, μg/ml; in-vivo tests, mg/kg bw
[b]With rat and human microsomes

Table 17. Genetic and related effects of aflatoxin G$_2$

Test system	Result		Dose (LED/HID)[a]	Reference
	Without exogenous metabolic system	With exogenous metabolic system		
PRB, SOS repair, *Escherichia coli* PQ37	–	–	60.0000	Krivobok *et al.* (1987)
ECB, *Escherichia coli*, DNA strand breaks/cross-links/repair	–	0	25.0000	Thielmann & Gersbach (1978)
BSD, *Bacillus subtilis* rec strains, differential toxicity	–	0	100.0000	Ueno & Kubota (1976)
SA0, *Salmonella typhimurium* TA100, reverse mutation	–	–	5.0000	von Engel & von Milczewski (1976)
SA0, *Salmonella typhimurium* TA100, reverse mutation	0	–	1.0000	Gurtoo *et al.* (1978)
SA8, *Salmonella typhimurium* TA1538, reverse mutation	–	–	5.0000	von Engel & von Milczewski (1976)
SA9, *Salmonella typhimurium* TA98, reverse mutation	–	–	5.0000	von Engel & von Milczewski (1976)
SA9, *Salmonella typhimurium* TA98, reverse mutation	0	+	0.0500	Hsieh *et al.* (1977)
SA9, *Salmonella typhimurium* TA98, reverse mutation	0	–	25.0000	Kleinwächter & Koukalová (1979)
NCF, *Neurospora crassa*, forward mutation	–	0	40.0000	Ong & de Serres (1972)
URP, Unscheduled DNA synthesis, rat primary hepatocytes *in vitro*	+	0	3.8000	Williams (1977)
URP, Unscheduled DNA synthesis, rat primary hepatocytes *in vitro*	+	0	0.0300	Probst *et al.* (1981)
URP, Unscheduled DNA synthesis, female rat primary hepatocytes *in vitro*	–	0	30.0000	McQueen & Way (1991)
UIA, Unscheduled DNA synthesis, Syrian hamster primary hepatocytes *in vitro*	+	0	0.0300	McQueen *et al.* (1983)
G9H, Gene mutation, Chinese hamster lung V79 cells *hprt* locus *in vitro*	–	–	3.3000	Kuroki *et al.* (1979)
G9O, Gene mutation, Chinese hamster lung V79 cells ouabain[r] *in vitro*	0	–	1.1000	Langenbach *et al.* (1978a)
G9O, Gene mutation, Chinese hamster lung V79 cells ouabain[r] *in vitro*	–	–	3.3000	Kuroki *et al.* (1979)
GIA, Gene mutation, other animal cells *in vitro*	–	0	3.3000	Tong & Williams (1978)
GIA, Gene mutation, other animal cells *in vitro*	–	0	165.0000	Ved Brat *et al.* (1983)
GIA, Gene mutation, other animal cells *in vitro*	–	0	3.3000	Tong *et al.* (1984)

Table 17 (contd)

Test system	Result		Dose (LED/HID)[a]	Reference
	Without exogenous metabolic system	With exogenous metabolic system		
SIC, Sister chromatid exchange, Chinese hamster cells *in vitro*	−	+	3.3000	Batt *et al.* (1980)
SIC, Sister chromatid exchange, Chinese hamster cells *in vitro*	−	+	0.3000	Ray-Chaudhuri *et al.* (1980)
SIR, Sister chromatid exchange, rat cells *in vitro*	(+)	0	3.3000	Ved Brat *et al.* (1983)
UHF, Unscheduled DNA synthesis, human fibroblasts *in vitro*	−	−	330.0000	Stich & Laishes (1975)

+, positive; (+), weak positive; −, negative; 0, not tested
[a]In-vitro tests, μg/ml; in-vivo tests, mg/kg bw

Table 18. Genetic and related effects of aflatoxin M_1

Test system	Result		Dose (LED/HID)[a]	Reference
	Without exogenous metabolic system	With exogenous metabolic system		
SA0, *Salmonella typhimurium* TA100, reverse mutation	–	+	5.0000	von Engel & von Milczewski (1976)
SA0, *Salmonella typhimurium* TA100, reverse mutation	0	–	0.5000	Gurtoo et al. (1978)
SA0, *Salmonella typhimurium* TA100, reverse mutation	0	+	0.1000	Uwaifo & Bababunmi (1979)
SA0, *Salmonella typhimurium* TA100, reverse mutation	–	+	0.1000	Uwaifo et al. (1979)
SA5, *Salmonella typhimurium* TA1535, reverse mutation	–	+	0.1500	Uwaifo et al. (1979)
SA7, *Salmonella typhimurium* TA1537, reverse mutation	–	+	0.1500	Uwaifo et al. (1979)
SA8, *Salmonella typhimurium* TA1538, reverse mutation	–	–	5.0000	von Engel & von Milczewski (1976)
SA8, *Salmonella typhimurium* TA1538, reverse mutation	–	+	0.1500	Uwaifo et al. (1979)
SA9, *Salmonella typhimurium* TA98, reverse mutation	–	–	5.0000	von Engel & von Milczewski (1976)
SA9, *Salmonella typhimurium* TA98, reverse mutation	0	+	0.0250	Hsieh et al. (1977)
SA9, *Salmonella typhimurium* TA98, reverse mutation	–	+	0.1500	Uwaifo et al. (1979)
SA9, *Salmonella typhimurium* TA98, reverse mutation	–	+	0.5000	Coulombe et al. (1982)
URP, Unscheduled DNA synthesis, rat primary hepatocytes *in vitro*	+	0	0.0125	Green et al. (1982)
BID, Binding to λDNA *in vitro*	+	0	20.0000	Mariën et al. (1987)
BID, Binding to *Salmo gairdneri* (rainbow trout) hepatocyte DNA *in vitro*	+	0	0.6200	Loveland et al. (1988)

+, positive; (+), weak positive; –, negative; 0, not tested
[a]In-vitro tests, μg/ml; in-vivo tests, mg/kg bw

Table 19. Mutations in oncogenes or tumour-suppressor genes found in animals treated *in vivo* with aflatoxin B_1

Species	Aflatoxin B_1 treatment	Type of tumours examined	Method of analysis	No. mutated genes/ no. tumours analysed	Details of mutations — Gene, codon	Details of mutations — Base change	Reference
Oncorhynchus mykiss (Rainbow trout)	80 ppb in diet for 2 weeks; fish killed 9 months after treatment	25% pure hepatocellular carcinomas, 75% mixed hepatocellular/cholangiocellular carcinomas; not examined histologically and consequently not identified	PCR from tumour DNA of 111 base-pair fragment that included codons 12 and 13 of c-Ki-*ras* gene. PCR fragments analysed by oligonucleotide hybridization with probes carrying different codon 12 and 13 base-pair substitutions. Four tumour DNAs were also cloned and sequenced.	10/14[a]	c-Ki-*ras*, 12	GGA to GTA (7/10), GGA to AGA (1/10), GGT to GTT (2/10)	Chang *et al.* (1991)
					c-Ki-*ras*, 13		
STCF1 male mice (derived from CF1 mice)	7-day-old mice injected i.p. with 6 μg/g bw aflatoxin B_1 and killed sequentially after start of treatment	First liver tumour found in control animals at 78 weeks and in treated mice at 24 weeks. 48% treated mice over 40 weeks had liver tumours; 77% adenoma, 23% adenoma and carcinoma or carcinoma alone	Fragments around codon 61 of c-Ha-*ras* exon 2 and codon 12 amplified by PCR from DNA isolated from formalin-fixed paraffin-embedded tissues. PCR fragments analysed by oligonucleotide hybridization with probes carrying different base-pair substitutions	3/8[b]	c-Ha-*ras*, 61	CAA to CTA, 1/3 CAA to AAA, 2/3[c]	Bauer-Hofmann *et al.* (1990)
Male Fischer 344 rats	Dietary groundnut meal naturally contaminated with aflatoxins to final concentrations of 1 ppm aflatoxin B_1 and 0.3 ppm aflatoxin G_1. Rats, 3 weeks old at start of feeding, given diet for 8 weeks and killed after 1 year	Yield not given; treatment produces liver tumours of mixed types. Tumours analysed not examined histologically	Co-transfection of DNA from each of 4 liver tumours and 2 tumorigenic liver tumour cell-lines with a gene for antibiotic resistance, followed by selection for antibiotic resistance; tumorigenicity testing in nude mice demonstrated DNA-mediated transfer of neoplastic phenotype. DNA extracted from nude mouse tumours used in secondary round of transfection with NIH 3T3 cells, which gave positive results in focus assays. DNA from transfectants analysed by oligonucleotide hybridization with probes carrying different *ras* base-pair substitutions	4/4 tumours	c-Ki-*ras*, codon 12 in 1/4 tumours	GGA to GAA	Sinha *et al.* (1988)
					c-N-*ras* in 3/4 tumours		
				Hepatoma cell lines JB1	c-N-*ras*		
				BL10	c-N-*ras*		
				Immortalized, non-transformed rat liver-cell lines			
				BL8	None		
				BL9	None		
				BL8 exposed to aflatoxin B_1 and transformed			
				L6	c-Ha-*ras*		

Table 19 (contd)

Species	Aflatoxin B$_1$ treatment	Type of tumours examined	Method of analysis	No. mutated genes/ no. tumours analysed	Details of mutations — Gene, codon	Details of mutations — Base change	Reference
Male Fischer 344 rats	Weanling rats given daily i.p. injections of 25 μg aflatoxin B$_1$, 5 days/week for 8 weeks; killed at 12–18 months	9/9 rats developed liver carcinomas within 1–2 years after treatment	DNA from excised tumours transfected into NIH 3T3 mouse cells, which were assayed for ability to form morphologically transformed foci or to induce tumours in nude mice. c-Ki-*ras* and N-*ras* fragments amplified by PCR of DNA from transformed foci or s.c. nude mouse tumours and analysed by oligonucleotide hybridization with probes carrying different *ras* base-pair substitutions. Primary liver tissue from control and treated rats also analysed	3/8[d]	c-Ki-*ras*, 12	GGT to TGT (1/8) GGT to GAT (2/8)	McMahon *et al.* (1990)
Macaca mulatta (rhesus monkey)	Aflatoxin B$_1$ administered by a variety of schedules from 1964 to 1978	4 *rhesus monkeys* 3 hepatocellular carcinomas (2 in 1 animal); 2 cholangiocarcinomas	DNA extracted from paraffin blocks and analysed for *p53* mutations. At least 6 clones from PCR-amplified exons 5, 7, 8 of each of 9 tumours sequenced. Restriction analysis of exon 7 using *Hae*III fragments used to look for codon-249 mutations	1/9	No codon-249 mutation. Exons 5, 7, 8 of *p53* gene at codon 175	CGC to CTC substitution at codon 175 of exon 5	Fujimoto *et al.* (1992)
Macaca fascicularis (cynomolgus monkey)		4 *cynomolgus monkeys* 1 hepatocellular carcinoma; 1 spindle-cell carcinoma; 1 haemangioendothelial sarcoma; 1 osteogenic sarcoma					

PCR, polymerase chain reaction
[a]Similar analysis of exon 1 of c-*ras*-2 gene revealed no mutation
[b]No mutation at codon 12 of c-Ha-*ras* gene
[c]Also seen in 2/8 tumours from phenobarbital-treated mice
[d]A GGC to GTT mutation at codon 13 of N-*ras* seen in livers from controls and from aflatoxin B$_1$-treated rats at equal frequencies (3/3 and 5/5, respectively) [suggesting that Fischer F344 rats carry a germ-line mutation that incurs high sensitivity to liver carcinogenesis by aflatoxin B$_1$ and other chemical carcinogens]

5. Summary of Data Reported and Evaluation

5.1 Exposure data

Aflatoxins are a group of relatively stable toxins produced mainly by two *Aspergillus* species that are ubiquitous in areas of the world with hot, humid climates. Whether exposure is predominantly to aflatoxin B_1 or to mixed B_1 and G_1 depends on the geographical distribution of the *Aspergillus* strains. *Aspergillus flavus*, which produces aflatoxins B_1 and B_2, occurs worldwide; *A. parasiticus*, which produces aflatoxins B_1, B_2, G_1 and G_2, occurs principally in the Americas and in Africa. Exposure occurs primarily through dietary intake of maize and groundnuts. Exposure to aflatoxin M_1 occurs mainly through consumption of milk, including mother's milk. Life-time exposure to aflatoxins in some parts of the world, commencing *in utero*, has been confirmed by biomonitoring.

5.2 Human carcinogenicity data

One cohort study of a small number of Dutch oilpress workers exposed to aflatoxin-containing dusts indicated increased mortality from cancer, but no death from hepatocellular carcinoma was observed. A cohort study in China found significant excess mortality from liver cancer among individuals in villages where foods were heavily contaminated with aflatoxins. A cohort study of Danish workers exposed to aflatoxin from imported feed found an excess of hepatocellular carcinoma among those who had had major exposure to aflatoxin-contaminated feed in the period 10 or more years before diagnosis. In a cohort study in China, a significant elevation in risk for hepatocellular carcinoma was found among people with aflatoxin metabolites in the urine, after adjustment for hepatitis B surface antigen positivity. The elevation in risk was particularly high among those excreting aflatoxin B_1-guanine adducts; however, there was no association between dietary and urinary aflatoxin levels among subjects in whom both were detected.

Of three hospital-based case–control studies in which an attempt was made to evaluate exposure to aflatoxin B_1, one (in the Philippines) found a significantly greater risk for hepatocellular carcinoma among people whose intake of aflatoxin was estimated to be heavy than in those with light aflatoxin intake. The other two studies, one in Hong Kong and one in Thailand, gave negative results. In Thailand, one study on hepatocellular carcinoma and another on cholangiocarcinoma also found no association with the presence of aflatoxin B_1-albumin adducts in sera.

The two cohort studies in China addressed combined exposure to hepatitis B virus and aflatoxins and suggested that each has an independent effect.

Several correlation studies have been performed, the majority showing a strong association between estimated aflatoxin intake and incidence of hepatocellular carcinoma. In only a few was it possible to evaluate simultaneously any correlation with the prevalence of hepatitis infection. Of those that did so, two—one in Swaziland and one in China—showed a stronger correlation with exposure to aflatoxin B_1 than with hepatitis B viral infection. The largest such study, in China, did not show an association with the presence of aflatoxin B_1

metabolites in urine. The study from Swaziland was the only one in which it was shown that subjects had concomitant exposure to aflatoxin B_1 and G_1.

5.3 Carcinogenicity in experimental animals

Mixtures of aflatoxins and aflatoxin B_1 have been tested extensively for carcinogenicity by various routes of administration in several strains of mice and rats, in hamsters, several strains of fish, ducks, tree shrews and monkeys. Following their oral administration, mixtures of aflatoxins and aflatoxin B_1 caused hepatocellular and/or cholangiocellular liver tumours, including carcinomas, in all species tested except mice. In rats, renal-cell tumours and a low incidence of tumours at other sites, including the colon, were also found. In monkeys, liver angiosarcomas, osteogenic sarcomas and adenocarcinomas of the gall-bladder and pancreas developed, in addition to hepatocellular and cholangiocellular carcinomas. In adult mice, aflatoxin B_1 administered intraperitoneally increased the incidence of lung adenomas. Intraperitoneal administration of aflatoxin B_1 to infant mice, adult rats and toads produced high incidences of liver-cell tumours in all of these species. Subcutaneous injection of aflatoxin B_1 resulted in local sarcomas in rats. Exposure of fish embryos to aflatoxin B_1 induced a high incidence of hepatocellular adenomas and carcinomas. Intraperitoneal administration of aflatoxin B_1 to rats during pregnancy and lactation induced benign and malignant tumours in mothers and their progeny in the liver and in various other organs, including those of the digestive tract, the urogenital system and the central and peripheral nervous systems. In several species, aflatoxin B_1 administered by different routes induced foci of altered hepatocytes, the number and size of which was correlated with later development of hepatocellular adenomas and carcinomas.

Aflatoxin B_2 induced foci of altered hepatocytes and hepatocellular adenomas following its oral administration to rats. A low incidence of hepatocellular carcinomas was observed after intraperitoneal administration of aflatoxin B_2 to rats.

Oral administration of aflatoxin G_1 induced foci of altered hepatocytes, hepatocellular adenomas and carcinomas and renal-cell tumours in rats and liver-cell tumours in fish. The hepatocarcinogenic effect of aflatoxin G_1 was weaker than that of aflatoxin B_1. Subcutaneous injection of aflatoxin G_1 in rats resulted in local sarcomas, which developed at a lower incidence and at later times than those induced by aflatoxin B_1 at the same dose level and by the same route. Oral administration of aflatoxin G_2 to trout had no hepatocarcinogenic effect in one experiment.

Aflatoxin M_1, a hydroxy metabolite of aflatoxin B_1, produced fewer hepatocellular carcinomas following its oral administration to rats and fish than aflatoxin B_1 given at the same dose level and by the same route. Aflatoxin Q_1, another metabolite of aflatoxin B_1, produced a high incidence of hepatocellular carcinomas following its oral administration to fish. Administration to rats and fish of aflatoxicol, yet another metabolite of aflatoxin B_1, induced hepatocellular carcinomas in both species; the tumour incidence was lower than that in animals treated with aflatoxin B_1 at the same dose level.

A large number of experiments have been carried out in which aflatoxins were administered in combination (prior to, during and following) with diets, viruses, parasites, known carcinogens and a number of different chemicals in order to study the modulating effects, including chemoprevention, of the agents on aflatoxin-induced carcinogenesis. Enhancing and inhibitory effects on the carcinogenicity of aflatoxins have been observed.

5.4 Other relevant data

Aflatoxin B_1 is consistently genotoxic, producing adducts in humans and animals *in vivo* and chromosomal anomalies in rodents and, in a single study, in rhesus monkeys *in vivo*. In human and animal cells in culture, it produces DNA damage, gene mutation and chromosomal anomalies; in animal cells *in vitro*, it also induces cell transformation. In insects and lower eukaryotes, it induces gene mutation and recombination. In bacteria, it produces DNA damage and gene mutation.

Aflatoxin B_1 is hepatoxic in humans and animals and is nephrotoxic and immunosuppressive in animals.

Aflatoxin B_2 has not been studied extensively, and most data are derived from single reports. Aflatoxin B_2 becomes bound to DNA of rats treated *in vivo*, after metabolic conversion to aflatoxin B_1. In rodent cells, it induces DNA damage, sister chromatid exchange and cell transformation, but not gene mutation. In fungi, it produces neither gene mutation nor recombination, whereas it produced gene mutation in bacteria.

Aflatoxin G_1 binds to DNA and produces chromosomal aberrations in rodents treated *in vivo*. In cultured human and animal cells, it induces DNA damage, and, in single studies, it induced chromosomal anomalies. It induces mutation in fungi and DNA damage and gene mutation in bacteria.

There are few published genetic studies on **aflatoxin G_2** and **aflatoxin M_1**. Aflatoxin G_1 produced DNA damage and sister chromatid exchange in animal cells in culture. Aflatoxin M_1 produced DNA damage in cultured rodent cells and gene mutation in bacteria.

Humans metabolize aflatoxin B_1 to an 8,9-epoxide, forming DNA and albumin adducts by the same activation pathways as susceptible animal species. Humans metabolize aflatoxin B_1 to the major aflatoxin B_1–$N7$-guanine and –serum albumin adduct at levels comparable to those in susceptible animal species (rat).

Glutathione *S*-transferase-mediated conjugation of glutathione to the 8,9-epoxide reduces DNA damage, and this mechanism is important in reducing the tumour burden in experimental animals. Animal species, such as the mouse, that are resistant to aflatoxin carcinogenesis have three to five times more glutathione *S*-transferase activity than susceptible species, such as the rat. Humans have less glutathione *S*-transferase activity for 8,9-epoxide conjugation than rats or mice, suggesting that humans are less capable of detoxifying this important metabolite.

Studies of human microsomal activation of aflatoxin B_1 show that at non-saturating concentrations of aflatoxin B_1 the rate of formation of the 8,9-epoxide is similar to that found in sensitive species (rat and monkey).

The value of aflatoxin B_1–$N7$-guanine as an indicator of risk for developing tumours is demonstrated by experiments with chemoprotective agents that show concordance between reduction of levels of DNA adduct formation and reduced incidence of liver tumours in rats and trout.

The presence of DNA- and protein–aflatoxin adducts in humans, the urinary excretion of aflatoxin B_1-$N7$-guanine adducts by humans, and the ability of human tissues to activate aflatoxin B_1 to form DNA adducts *in vitro* provide evidence that humans have the biochemical pathways required for aflatoxin-induced carcinogenesis. The following evidence is consistent with those biochemical mechanisms.

Studies with bacteria show that activated aflatoxin B_1 specifically induces G to T transversions. On the basis of experiments conducted *in vitro*, aflatoxin B_1 specifically targets the third and not the second nucleotide of codon 249 (AG*G*) of the human *p53* gene, an effect not seen with benzo[*a*]pyrene-7,8-diol-9,10-epoxide when tested at the same level of binding.

A high frequency of mutations at a mutational 'hot-spot' (the third nucleotide of codon 249 in exon 7) has been found in *p53* tumour suppressor genes in hepatocellular carcinomas from patients resident in areas considered to offer a high risk of exposure to aflatoxins and where there is a high incidence of hepatocellular carcinoma. In contrast, this mutation is rare in hepatocellular carcinomas from regions of low exposure to aflatoxins (including Australia, Japan, southern Africa, Germany, Spain, Italy, Turkey, Israel, Saudi Arabia, the United Kingdom and the USA).

5.5 Evaluation[1]

There is *sufficient evidence* in humans for the carcinogenicity of naturally occurring mixtures of aflatoxins.

There is *sufficient evidence* in humans for the carcinogenicity of aflatoxin B_1.

There is *inadequate evidence* in humans for the carcinogenicity of aflatoxin M_1.

There is *sufficient evidence* in experimental animals for the carcinogenicity of naturally occurring mixtures of aflatoxins and aflatoxins B_1, G_1 and M_1.

There is *limited evidence* in experimental animals for the carcinogenicity of aflatoxin B_2.

There is *inadequate evidence* in experimental animals for the carcinogenicity of aflatoxin G_2.

Overall evaluations

Naturally occurring aflatoxins *are carcinogenic to humans (Group 1)*.

Aflatoxin M_1 *is possibly carcinogenic to humans (Group 2B)*.

---

[1]For definitions of the italicized terms, see Preamble, pp. 26-29.

6. References

Abe, S., Nemoto, N. & Sasaki, M. (1983) Sister-chromatid exchange induction by indirect mutagens/carcinogens, aryl hydrocarbon hydroxylase activity and benzo[a]pyrene metabolism in cultured human hepatoma cells. *Mutat. Res.*, **109**, 83–90

Affolter, M., Parent-Vaugeois, C. & Anderson, A. (1983a) Curing and induction of the Fels 1 and Fels 2 prophages in the Ames mutagen tester strains of *Salmonella typhimurium*. *Mutat. Res.*, **110**, 243–262

Affolter, M., Parent-Vaugeois, C. & Anderson, A. (1983b) Mutagenic response of Ames strains cured of their inducible fels 1 and fels 2 prophages. *Cancer Res.*, **43**, 653–659

Allameh, A., Saxena, M. & Raj, H.G. (1989) Interaction of aflatoxin B_1 metabolites with cellular macromolecules in neonatal rats receiving carcinogen through mothers' milk. *Carcinogenesis*, **10**, 2131–2134

Allcroft, R., Roberts, B.A. & Lloyd, M.K. (1968) Excretion of aflatoxin in a lactating cow. *Food Cosmet. Toxicol.*, **6**, 619–625

Allen, S.J., Wild, C.P., Wheeler, J.G., Riley, E.M., Montesano, R., Bennett, S., Whittle, H.C., Hall, A.J. & Greenwood, B.M. (1992) Aflatoxin exposure, malaria and hepatitis B infection in rural Gambian children. *Trans. R. Soc. Trop. Med. Hyg.*, **86**, 426–430

Alpert, E., Serck-Hanssen, A. & Rajagopolan, B. (1970) Aflatoxin-induced hepatic injury in the African monkey. *Arch. environ. Health*, **20**, 723–728

Alpert, M.E., Hutt, M.S.R., Wogan, G.N. & Davidson, C.S. (1971) Association between aflatoxin content of food and hepatoma frequency in Uganda. *Cancer*, **28**, 253–260

Amla, I., Kamala, C.S., Gopalakrishna, G.S., Jayaraj, A.P., Sreenivasamurthy, V. & Parpia, H.A.B. (1971) Cirrhosis in children from peanut meal contaminated by aflatoxin. *Am. J. clin. Nutr.*, **24**, 609–614

Amstad, P., Levey, A., Emerit, I. & Cerutti, P. (1984) Evidence for membrane-mediated chromosomal damage by aflatoxin B_1 in human lymphocytes. *Carcinogenesis*, **5**, 719–723

Angsubhakorn, S., Bhamarapravati, N., Romruen, K. & Sahaphong, S. (1981a) Enhancing effects of dimethylnitrosamine on aflatoxin B_1 hepatocarcinogenesis in rats. *Int. J. Cancer*, **28**, 621–626

Angsubhakorn, S., Bhamarapravati, N., Romruen, K., Sahaphong, S., Thamavit, W. & Miyamoto, M. (1981b) Further study of α-benzene hexachloride inhibition of aflatoxin B_1 hepatocarcinogenesis in rats. *Br. J. Cancer*, **43**, 881–883

Angsubhakorn, S., Bhamarapravati, N., Sahaphong, S. & Sathiropas, P. (1988) Reducing effects of rodent malaria on hepatic carcinogenesis induced by dietary aflatoxin B_1. *Int. J. Cancer*, **41**, 69–73

Angsubhakorn, S., Bhamarapravati, N., Pradermwong, A., Im-Emgamol, N. & Sahaphong, S. (1989) Minimal dose and time protection by lindane (γ-isomer of 1,2,3,4,5,6-hexachlorocyclohexane) against liver tumors induced by aflatoxin B_1. *Int. J. Cancer*, **43**, 531–534

Applebaum, R.S., Brackett, R.E., Wiseman, D.W. & Marth, E.H. (1982) Aflatoxin: toxicity to dairy cattle and occurrence in milk and milk products. A review. *J. Food Prot.*, **45**, 752–777

Appleton, B.S. & Campbell, T.C. (1983a) Effect of high and low dietary protein on the dosing and postdosing periods of aflatoxin B_1-induced hepatic preneoplastic lesion development in the rat. *Cancer Res.*, **43**, 2150–2154

Appleton, B.S. & Campbell, T.C. (1983b) Dietary protein intervention during the postdosing phase of aflatoxin B_1-induced hepatic preneoplastic lesion development. *J. natl Cancer Inst.*, **70**, 547–549

Appleton, B.S., Goetchius, M.P. & Campbell, T.C. (1982) Linear dose–response curve for the hepatic macromolecular binding of aflatoxin B_1 in rats at very low exposures. *Cancer Res.*, **42**, 3659–3662

Armstrong, B. (1980) The epidemiology of cancer in the People's Republic of China. *Int. J. Epidemiol.*, **9**, 305–315

Arora, R.G., Appelgren, L.-E. & Bergman, A. (1978) Distribution of [^{14}C]-labelled aflatoxin B_1 in mice. *Acta pharmacol. toxicol.*, **43**, 273–279

Arora, R.G., Frölén, H. & Nilsson, A. (1981) Interference of mycotoxins with prenatal development of the mouse. I. Influence of aflatoxin B_1, ochratoxin A and zearalenone. *Acta vet. scand.*, **22**, 524–534

Auffray, Y. & Boutibonnes, P. (1987) Genotoxic activity of some mycotoxins using the SOS chromotest. *Mycopathologia*, **100**, 49–53

Autrup, H., Essigmann, J.M., Croy, R.G., Trump, B.F., Wogan, G.N. & Harris, C.C. (1979) Metabolism of aflatoxin B_1 and identification of the major aflatoxin B_1–DNA adducts formed in cultured human bronchus and colon. *Cancer Res.*, **39**, 694–698

Autrup, H., Bradley, K.A., Shamsuddin, A.K.M., Wakhisi, J. & Wasunna, A. (1983) Detection of putative adduct with fluorescence characteristics identical to 2,3-dihydro-2-(7'-guanyl)-3-hydroxyaflatoxin B_1 in human urine collected in Murang'a district, Kenya. *Carcinogenesis*, **4**, 1193–1195

Autrup, H., Seremet, T., Wakhisi, J. & Wasunna, A. (1987) Aflatoxin exposure measured by urinary excretion of aflatoxin B_1–guanine adduct and hepatitis B virus infection in areas with different liver cancer incidence in Kenya. *Cancer Res.*, **47**, 3430–3433

Autrup, J.L., Schmidt, J., Seremet, T. & Autrup, H. (1991) Determination of exposure to aflatoxins among Danish workers in animal-feed production through the analysis of aflatoxin B_1 adducts to serum albumin. *Scand. J. Work Environ. Health*, **17**, 436–440

Ayres, J.L., Lee, D.J., Wales, J.H. & Sinnhuber, R.O. (1971) Aflatoxin structure and hepatocarcinogenicity in rainbow trout (*Salmo gairdneri*). *J. natl Cancer Inst.*, **46**, 561–564

Baertschi, S.W., Raney, K.D., Shimada, T., Harris, T.M. & Guengerich, F.P. (1989) Comparison of rates of enzymatic oxidation of aflatoxin B_1, aflatoxin G_1, and sterigmatocystin and activities of the epoxides in forming guanyl-N^7 adducts and inducing different genetic responses. *Chem. Res. Toxicol.*, **2**, 114–122

Bagshawe, A.F., Gacengi, D.M., Cameron, C.H., Dorman, J. & Dane, D.S. (1975) Hepatitis Bs antigen and liver cancer. A population based study in Kenya. *Br. J. Cancer*, **31**, 581–584

Bailey, G.S., Hendricks, J.D., Shelton, D.W., Nixon, J.E. & Pawlowski, N.E. (1987) Enhancement of carcinogenesis by the natural anticarcinogen indole-3-carbinol. *J. natl Cancer Inst.*, **78**, 931–934

Bailey, G.S., Williams, D.E., Wilcox, J.S., Loveland, P.M., Coulombe, R.A. & Hendricks, J.D. (1988) Aflatoxin B_1 carcinogenesis and its relation to DNA adduct formation and adduct persistence in sensitive and resistant salmonid fish. *Carcinogenesis*, **9**, 1919–1926

Baker, R.S.U., Bonin, A.M., Stupans, I. & Holder, G.M. (1980) Comparison of rat and guinea pig as sources of the S9 fraction in the *Salmonella*/mammalian microsome mutagenicity assay. *Mutat. Res.*, **71**, 43–52

Baker, R.S.U., Mitchell, G.A., Meher-Homji, K.M. & Podobna, E. (1983) Sensitivity of two Chinese hamster cell lines to SCE induction by a variety of chemical mutagens. *Mutat. Res.*, **118**, 103–116

Ball, R.W., Wilson, D.W. & Coulombe, R.A., Jr (1990) Comparative formation and removal of aflatoxin B_1–DNA adducts in cultured mammalian tracheal epithelium. *Cancer Res.*, **50**, 4918–4922

Bannasch, P., Benner, U., Enzmann, H. & Hacker, H.J. (1985) Tigroid cell foci and neoplastic nodules in the liver of rats treated with a single dose of aflatoxin B_1. *Carcinogenesis*, **6**, 1641–1648

Bárta, I., Adámková, M., Markarjan, D., Adžigitov, F. & Prokeš, K. (1984) The mutagenic acitivity of aflatoxin B_1 in the *Cricetulus griseus* hamster and *Macaca mulatta* monkey. *J. Hyg. Epidemiol. Microbiol. Immunol.*, 28, 149–159

Bárta, I., Adámková, M., Petr, T. & Bártová, J. (1990) Dose and time dependence of chromosomal aberration yields of bone marrow cells in male Chinese hamsters after a single i.p. injection of aflatoxin B_1. *Mutat. Res.*, 244, 189–195

Batt, T.R., Hsueh, J.L., Chen, H.H. & Huang, C.C. (1980) Sister chromatid exchanges and chromosome aberrations in V79 cells induced by aflatoxin B_1, B_2, G_1 and G_2 with or without metabolic activation. *Carcinogenesis*, 1, 759–763

Bauer-Hofmann, R., Buchmann, A., Wright, A.S. & Schwarz, M. (1990) Mutations in the Ha-*ras* proto-oncogene in spontaneous and chemically induced liver tumours of the CF1 mouse. *Carcinogenesis*, 11, 1875–1877

Bednarz, W. (1989) Primary and transplantable hepatomas induced by aflatoxin B_1 in hypothyroid rats. *Neoplasma*, 36, 113–126

Benasutti, M., Ejadi, S., Whitlow, M.D. & Loechler, E.L. (1988) Mapping the binding site of aflatoxin B_1 in DNA: systematic analysis of the reactivity of aflatoxin B_1 with guanines in different DNA sequences. *Biochemistry*, 27, 472–481

Bennett, R.A., Essigmann, J.M. & Wogan, G.N. (1981) Excretion of an aflatoxin–guanine adduct in the urine of aflatoxin B_1-treated rats. *Cancer Res.*, 41, 650–654

Bermudez, E., Couch, D.B. & Tillery, D. (1982) The use of primary rat hepatocytes to achieve metabolic activation of promutagens in the Chinese hamster ovary/hypoxanthine-guanine phosphoribosyl transferase mutational assay. *Environ. Mutag.*, 4, 55–64

Bhandari, P.C. & Bhandari, B. (1980) Aflatoxin and Indian childhood cirrhosis. *Indian Pediatr.*, 17, 593–596

Billings, P.C., Uwaifo, A.O. & Heidelberger, C. (1983) Influence of benzoflavone on aflatoxin B_1-induced cytotoxicity, mutation, and transformation of C3H/10T½ cells. *Cancer Res.*, 43, 2659–2663

Black, J.J., Maccubbin, A.E. & Schiffert, M. (1985) A reliable, efficient, microinjection apparatus and methodology for the in vivo exposure of rainbow trout and salmon embryos to chemical carcinogens. *J. natl Cancer Inst.*, 75, 1123–1128

Black, J.J., Maccubbin, A.E., Myers, H.K. & Zeigel, R.F. (1988) Aflatoxin B_1 induced hepatic neoplasia in Great Lakes coho salmon. *Bull. environ. Contam. Toxicol.*, 41, 742–745

Blanck, A., Lindhe, B., Hällström, I.P., Lindeskog, P. & Gustafsson, J.-Å. (1992) Influence of different levels of dietary casein on initiation of male rat liver carcinogenesis with a single dose of aflatoxin B_1. *Carcinogenesis*, 13, 171–176

Booth, S.C., Welch, A.M. & Garner, R.C. (1980) Some factors affecting mutant numbers in the *Salmonella*/microsome assay. *Carcinogenesis*, 1, 911–923

Booth, S.C., Bösenberg, H., Garner, R.C., Hertzog, P.J. & Norpoth, K. (1981) Activation of aflatoxin B_1 in liver slices and in bacterial mutagenicity assays using livers from different species including man. *Carcinogenesis*, 2, 1063–1068

Boreiko, C.J., Ragan, D.L., Abernethy, D.J. & Frazelle, J.H. (1982) Initiation of C3H/10T½ cell transformation by *N*-methyl-*N'*-nitro-*N*-nitrosoguanidine and aflatoxin B_1. *Carcinogenesis*, 3, 391–395

Bosch, F.X. & Muñoz, N. (1988) Prospects for epidemiological studies on hepatocellular cancer as a model for assessing viral and chemical interactions. In: Bartsch, H., Hemminki, K. & O'Neill, I.K., eds, *Methods for Detecting DNA Damaging Agents in Humans: Applications in Cancer Epidemiology and Prevention* (IARC Scientific Publications No. 89), Lyon, IARC, pp. 427–438

Bose, S. & Sinha, S.P. (1991) Aflatoxin-induced structural chromosomal changes and mitotic disruption in mouse bone marrow. *Mutat. Res.*, **261**, 15-19

Bourgeois, C.H., Shank, R.C., Grossman, R.A., Johnsen, D.O., Wooding, W.L. & Chandavimol, P. (1971) Acute aflatoxin B_1 toxicity in the macaque and its similarity to Reye's syndrome. *Lab. Invest.*, **24**, 206-216

Boyd, J.N., Misslbeck, N. & Stoewsand, G.S. (1983) Changes in preneoplastic response to aflatoxin B_1 in rats fed green beans, beets or squash. *Food chem. Toxicol.*, **21**, 37-40

Brennan-Craddock, W.E., Coutts, T.M., Rowland, I.R. & Alldrick, A.J. (1990) Dietary fat modifies the in vivo mutagenicity of some food-borne carcinogens. *Mutat. Res.*, **230**, 49-54

Bressac, B., Kew, M., Wands, J. & Ozturk, M. (1991) Selective G to T mutations of p53 gene in hepatocellular carcinoma from southern Africa. *Nature*, **350**, 429-431

Bruce, R.D. (1990) Risk asessment for aflatoxin: II. Implications of human epidemiology data. *Risk Anal.*, **10**, 561-569

Bruce, W.R. & Heddle, J.A. (1979) The mutagenic activity of 61 agents as determined by the micronucleus, *Salmonella* and sperm abnormality assays. *Can. J. Genet. Cytol.*, **21**, 319-334

Budavari, S., ed. (1989) *The Merck Index*, 11th ed., Rahway, NJ, Merck & Co., pp. 30-31

Bulatao-Jayme, J., Almero, E.M., Castro, M.C.A., Jardeleza, M.T.R. & Salamat, L.A. (1982) A case-control dietary study of primary liver cancer risk from aflatoxin exposure. *Int. J. Epidemiol.*, **11**, 112-119

Busby, W.F., Jr & Wogan, G.N. (1984) Aflatoxins. In: Searle, C.E., ed., *Chemical Carcinogens*, 2nd ed. (ACS Monograph 182), Washington DC, American Chemical Society, pp. 945-1136

Busk, L. & Ahlborg, U.G. (1980) Retinol (vitamin A) as an inhibitor of the mutagenicity of aflatoxin B_1. *Toxicol. Lett.*, **6**, 243-249

Buss, P., Caviezel, M. & Lutz, W.K. (1990) Linear dose-response relationship for DNA adducts in rat liver from chronic exposure to aflatoxin B_1. *Carcinogenesis*, **11**, 2133-2135

Butler, W.H. & Barnes, J.M. (1966) Carcinoma of the glandular stomach in rats given diets containing aflatoxin. *Nature*, **209**, 90

Butler, W.H. & Barnes, J.M. (1968) Carcinogenic action of groundnut meal containing aflatoxin in rats. *Food Cosmet. Toxicol.*, **6**, 135-141

Butler, W.H. & Hempsall, V. (1981) Histochemical studies of hepatocellular carcinomas in the rat induced by aflatoxin. *J. Pathol.*, **134**, 157-170

Butler, W.H., Greenblatt, M. & Lijinsky, W. (1969) Carcinogenesis in rats by aflatoxins B_1, G_1, and B_2. *Cancer Res.*, **29**, 2206-2211

Butler, W.H., Hempsall, V. & Stewart, M.G. (1981) Histochemical studies on the early proliferative lesion induced in the rat liver by aflatoxin. *J. Pathol.*, **133**, 325-340

Butterworth, B.E., Doolittle, D.J., Working, P.K., Strom, S.C., Jirtle, R.L. & Michalopoulos, G. (1982) Chemically-induced DNA repair in rodent and human cells. In: Bridges, V.A., Butterworth, B.E. & Weinstein, I.B., eds, *Indicators of Genotoxic Exposure* (Banbury Report 13), Cold Spring Harbor, NY, CSH Press, pp. 101-114

Cabral, J.R.P. & Neal, G.E. (1983) The inhibitory effects of ethoxyquin on the carcinogenic action of aflatoxin B_1 in rats. *Cancer Lett.*, **19**, 125-132

Callen, D.F. & Philpot, R.M. (1977) Cytochrome P-450 and the activation of promutagens in *Saccharomyces cerevisiae*. *Mutat. Res.*, **45**, 309-324

Callen, D.F., Mohn, G.R. & Ong, T.-M. (1977) Comparison of the genetic activity of aflatoxins B_1 and G_1 in *Escherichia coli* and *Saccharomyces cerevisiae*. *Mutat. Res.*, **45**, 7-11

Callen, D.F., Wolf, C.R. & Philpot, R.M. (1978) Cumene hydroperoxide and yeast cytochrome P-450: spectral interactions and effect on the genetic activity of promutagens. *Biochem. biophys. Res. Commun.*, **83**, 14–20

Campbell, T.C., Caedo, J.P., Jr, Bulatao-Jayme, J., Salamat, L. & Engel, R.W. (1970) Aflatoxin M_1 in human urine. *Nature*, **227**, 403–404

Campbell, T.C., Hayes, J.R. & Newberne, P.M. (1978) Dietary lipotropes, hepatic microsomal mixed-function oxidase activities, and in vivo covalent binding of aflatoxin B_1 in rats. *Cancer Res.*, **38**, 4569–4573

Campbell, T.C., Chen, J., Liu, C., Li, J. & Parpia, B. (1990) Nonassociation of aflatoxin with primary liver cancer in a cross-sectional ecological survey in the People's Republic of China. *Cancer Res.*, **50**, 6882–6893

Canton, J.H., Kroes, R., van Logten, M.J., van Schothorst, M., Stavenuiter, J.F.C. & Verhülsdonk, C.A.H. (1975) The carcinogenicity of aflatoxin M_1 in rainbow trout. *Food Cosmet. Toxicol.*, **13**, 441–443

Carnaghan, R.B.A. (1965) Hepatic tumours in ducks fed a low level of toxic groundnut meal. *Nature*, **208**, 308

Carnaghan, R.B.A. (1967) Hepatic tumours and other chronic liver changes in rats following a single oral administration of aflatoxin. *Br. J. Cancer*, **21**, 811–814

Castegnaro, M., Hunt, D.C., Sansone, E.B., Schuller, P.L., Siriwardana, M.G., Telling, G.M., van Egmond, H.P. & Walker, E.A., eds (1980) *Laboratory Decontamination and Destruction of Aflatoxins B_1, B_2, G_1, G_2 in Laboratory Wastes* (IARC Scientific Publications No. 37), Lyon, IARC

Castegnaro, M., Pleština, R., Dirheimer, G., Chernozemsky, I.N. & Bartsch, H., eds (1991) *Mycotoxins, Endemic Nephropathy and Urinary Tract Tumours* (IARC Scientific Publications No. 115), Lyon, IARC

Casto, B.C. (1981) Detection of chemical carcinogens and mutagens in hamster cells by enhancement of adenovirus transformation. In: Mishra, N., Dunkel, V. & Mehlman, I., eds, *Advances in Modern Environmental Toxicology*, Vol. 1, Princeton, NJ, Senate Press, pp. 241–271

Casto, B.C. (1983) Comparison of the sensitivity of rodent and human cells to chemical carcinogens using viral transformation, DNA damage and cytotoxicity assays. *Basic Life Sci.*, **24**, 429–449

Casto, B.C., Pieczynski, W.J., Janosko, N. & DiPaolo, J.A. (1976) Significance of treatment interval and DNA repair in the enhancement of viral transformation by chemical carcinogens and mutagens. *Chem.-biol. Interactions*, **13**, 105–125

Casto, B.C., Janosko, N. & DiPaolo, J.A. (1977) Development of a focus assay model for transformation of hamster cells *in-vitro* by chemical carcinogens. *Cancer Res.*, **37**, 3508–3515

Challen, C., Lunec, J., Warren, W., Collier, J. & Bassendine, M.F. (1992) Analysis of the p53 tumor-suppressor gene in hepatocellular carcinomas from Britain. *Hepatology*, **16**, 1362–1366

Chan, R.I.M., Stich, H.F., Rosin, M.P. & Powrie, W.D. (1982) Antimutagenic activity of browning reaction products. *Cancer Lett.*, **15**, 27–33

Chang, Y.-J., Mathews, C., Mangold, K., Marien, K., Hendricks, J. & Bailey, G. (1991) Analysis of *ras* gene mutations in rainbow trout liver tumors initiated by aflatoxin B_1. *Mol. Carcinog.*, **4**, 112–119

Chedid, A., Bundeally, A.E. & Mendenhall, C.L. (1977) Inhibition of hepatocarcinogenesis by adrenocorticotropin in aflatoxin B_1-treated rats. *J. natl Cancer Inst.*, **58**, 339–349

Chedid, A., Halfman, C.J. & Greenberg, S.R. (1980) Hormonal influences on chemical carcinogenesis: studies with the aflatoxin B_1 hepatocarcinoma model in the rat. *Dig. Dis. Sci.*, **25**, 869–874

Chen, Z.Y. & Eaton, D.L. (1991) Differential regulation of cytochrome(s) P450 2B1/2 by phenobarbital in hepatic hyperplastic nodules induced by aflatoxin B_1 or diethylnitrosamine plus 2-acetylaminofluorene in male F344 rats. *Toxicol. appl. Pharmacol.*, **111**, 132–144

Chentanez, T., Glinsukon, T., Patchimasiri, V., Klongprakit, C. & Chentanez, V. (1986) The effects of aflatoxin B_1 given to pregnant rats on the liver, brain and the behaviour of their offspring. *Nutr. Rep. int.*, **34**, 379–386

Chernesky, P., Jayaraj, A. & Richardson, A. (1982) Effect of testosterone treatment on the ability of female rat liver S9 to activate aflatoxin B_1 and 2-aminofluorene in the Ames/*Salmonella* system. *Mutat. Res.*, **103**, 267–273

Choke, H.C. (1991) Overview of mycotoxins in the Asian region. In: Merican, Z., Ahamad, N., Lan, Y.Q., Moy, G.G., Othman, N.D.M., Shahabudin, H.A. & Mohamed, A.R., eds, *Proceedings of the First Asian Conference on Food Safety, 2–7 September 1990, Kuala Lumpur, Malaysia*, Selangor Darul Ehsan, Malaysian Institute of Food Technology, pp. 36–45

Cole, R.J. & Cox, R.H. (1981) *Handbook of Toxic Fungal Metabolites*, New York, Academic Press, pp. 1–66

Cole, K.E., Jones, T.W., Lipsky, M.M., Trump, B. & Hsu, I.-C. (1988) In vitro binding of aflatoxin B_1 and 2-acetylaminofluorene to rat, mouse and human hepatocyte DNA: the relationship of DNA binding to carcinogenicity. *Carcinogenesis*, **9**, 711–716

Coles, B.F., Smith, J.R. & Garner, R.C. (1977) Mutagenicity of 3a,8a-dihydrofuro[2,3-*b*]benzofuran, a model of aflatoxin B_1, for *Salmonella* typhimurium TA100. *Biochem. Biophys. Res. Commun.*, **76**, 888–892

Commoner, V. (1976) *Reliability of Bacterial Mutagenesis Techniques to Distinguish Carcinogenic and Noncarcinogenic Chemicals* (US EPA Report No. EPA-600/1-76-022; NTIS No. PB-259 934), Washington DC, US Government Printing Office

Cortesi, E., Saffiotti, U., Donovan, P.J., Rice, J.M. & Kakunaga, T. (1983) Dose–response studies on neoplastic transformation of BALB/3T3 clone A31-1-1 cells by aflatoxin B_1, benzidine, benzo[*a*]pyrene, 3-methyl-cholanthrene, and *N*-methyl-*N*'-nitro-*N*-nitrosoguanidine. *Teratog. Carcinog. Mutag.*, **3**, 101–110

Coulombe, R.A., Jr & Sharma, R.P. (1985) Clearance and excretion of intratracheally and orally administered aflatoxin B_1 in the rat. *Food chem. Toxicol.*, **23**, 827–830

Coulombe, R.A., Shelton, D.W., Sinnhuber, R.O. & Nixon, J.E. (1982) Comparative mutagenicity of aflatoxins using a *Salmonella*/trout hepatic enzyme activation system. *Carcinogenesis*, **3**, 1261–1264

Coulombe, R.A., Jr, Huie, J.M., Ball, R.W., Sharma, R.P. & Wilson, D.W. (1991) Pharmacokinetics of intratracheally administered aflatoxin B_1. *Toxicol. appl. Pharmacol.*, **109**, 196–206

Coulter, J.B.S., Lamplugh, S.M., Suliman, G.I., Omer, M.I.A. & Hendrickse, R.G. (1984) Aflatoxins in human breast milk. *Ann. Trop. Paediatr.*, **4**, 61–66

Coulter, J.B.S., Suliman, G.I., Lamplugh, S.M., Mukhtar, B.I. & Hendrickse, R.G. (1986) Aflatoxins in liver biopsies from Sudanese children. *Am. J. trop. Med. Hyg.*, **35**, 360–365

Cova, L., Wild, C.P., Mehrotra, R., Turusov, V., Shirai, T., Lambert, V., Jacquet, C., Tomatis, L., Trépo, C. & Montesano, R. (1990) Contribution of aflatoxin B_1 and hepatitis B virus infection in the induction of liver tumors in ducks. *Cancer Res.*, **50**, 2156–2163

Crespi, C.L., Steimel, D.T., Aoyama, T., Gelboin, H.V. & Gonzalez, F.J. (1990) Stable expression of human cytochrome P450IA2 cDNA in a human lymphoblastoid cell line: role of the enzyme in the metabolic activation of aflatoxin B_1. *Mol. Carcinog.*, **3**, 5–8

Crespi, C.L., Penman, B.W., Steimel, D.T., Gelboin, H.V. & Gonzalez, F.J. (1991) The development of a human cell line stably expressing human CYP3A4: role in the metabolic activation of aflatoxin B_1 and comparison to CYP1A2 and CYP2A3. *Carcinogenesis*, **12**, 355–359

Cullen, J.M., Ruebner, B.H., Hsieh, L.S., Hyde, D.M. & Hsieh, D.P. (1987) Carcinogenicity of dietary aflatoxin M_1 in male Fischer rats compared to aflatoxin B_1. *Cancer Res.*, **47**, 1913–1917

Cullen, J.M., Marion, P.L., Sherman, G.J., Hong, X. & Newbold, J.E. (1990) Hepatic neoplasms in aflatoxin B_1-treated, congenital duck hepatitis B virus-infected, and virus-free Pekin ducks. *Cancer Res.*, **50**, 4072–4080

Cusumano, V. (1991) Aflatoxins in sera from patients with lung cancer. *Oncology*, **48**, 194–195

Cusumano, V., Costa, G.B. & Seminara, S. (1990) Effect of aflatoxins on rat peritoneal macrophages. *Appl. environ. Microbiol.*, **56**, 3482–3484

Dahl, G.A., Miller, E.C. & Miller, J.A. (1980) Comparative carcinogenicities and mutagenicities of vinyl carbamate, ethyl carbamate, and ethyl *N*-hydroxycarbamate. *Cancer Res.*, **40**, 1194–1203

Dahms, R. & Gurtoo, H.L. (1976) Metabolism of aflatoxin B_1 to aflatoxins Q_1, M_1 and P_1 by mouse and rat. *Res. Commun. chem. Pathol. Pharmacol.*, **15**, 11–20

Dalezios, J., Wogan, G.N. & Weinreb, S.M. (1971) Aflatoxin P_1: a new aflatoxin metabolite in monkeys. *Science*, **171**, 584–585

Dashwood, R.H., Arbogast, D.N., Fong, A.T., Hendricks, J.D. & Bailey, G.S. (1988) Mechanisms of anti-carcinogenesis by indole-3-carbinol: detailed in vivo DNA binding dose–response studies after dietary administration with aflatoxin B_1. *Carcinogenesis*, **9**, 427–432

Dashwood, R.H., Arbogast, D.N., Fong, A.T., Pereira, C., Hendricks, J.D. & Bailey, G.S. (1989) Quantitative inter-relationships between aflatoxin B_1 carcinogen dose, indole-3-carbinol anti-carcinogen dose, target organ DNA adduction and final tumor response. *Carcinogenesis*, **10**, 175–181

Dashwood, R.H., Fong, A.T., Hendricks, J.D. & Bailey, G.S. (1990) Tumor dose–response studies with aflatoxin B_1 and the ambivalent modulator indole-3-carbinol: inhibitory *versus* promotional potency. *Basic Life Sci.*, **52**, 361–365

Dashwood, R.H., Fong, A.T., Williams, D.E., Hendricks, J.D. & Bailey, G.S. (1991) Promotion of aflatoxin B_1 carcinogenesis by the natural tumor modulator indole-3-carbinol: influence of dose, duration, and intermittent exposure on indole-3-carbinol promotional potency. *Cancer Res.*, **51**, 2362–2365

Decloitre, F. & Hamon, G. (1980) Species-dependent effects of dietary lindane and/or zineb on the activation of aflatoxin B_1 into mutagenic derivatives. *Mutat. Res.*, **79**, 185–192

Deger, G.E. (1976) Aflatoxin—human colon carcinogenesis? *Ann. intern. Med.*, **85**, 204–205

Delcuve, G.P., Moyer, R., Bailey, G. & Davie, J.R. (1988) Gene-specific differences in the aflatoxin B_1 adduction of chicken erythrocyte chromatin. *Cancer Res.*, **48**, 7146–7149

Denning, D.W., Onwubalili, J.K., Wilkinson, A.P. & Morgan, M.R.A. (1988) Measurement of aflatoxin in Nigerian sera by enzyme-linked immunosorbent assay. *Trans. R. Soc. trop. Med. Hyg.*, **82**, 169–171

Denning, D.W., Allen, R., Wilkinson, A.P. & Morgan, M.R.A. (1990) Transplacental transfer of aflatoxin in humans. *Carcinogenesis*, **11**, 1033–1035

Dhatt, P.S., Parida, N.K., Das Chaudhury, P.K. & Singh, H. (1982) Aflatoxins and Indian childhood cirrhosis. *Indian Pediatr.*, **19**, 407–408

Dickens, F. & Jones, H.E.H. (1963) The carcinogenic action of aflatoxin after its subcutaneous injection in the rat. *Br. J. Cancer*, **17**, 691–698

Dickens, F. & Jones, H.E.H. (1965) Further studies on the carcinogenic action of certain lactones and related substances in the rat and mouse. *Br. J. Cancer*, **19**, 392–403

Dickens, F., Jones, H.E.H. & Waynforth, H.B. (1966) Oral, subcutaneous and intratracheal administration of carcinogenic lactones and related substances: the intratracheal administration of cigarette tar in the rat. *Br. J. Cancer*, **20**, 134–144

DiPaolo, J.A. (1980) Quantitative in vitro transformation of Syrian golden hamster embryo cells with the use of frozen stored cells. *J. natl Cancer Inst.*, **64**, 1485–1489

DiPaolo, J.A., Nelson, R.L. & Donovan, P.J. (1972) In vitro transformation of Syrian hamster cells by diverse chemical carcinogens. *Nature*, **235**, 278–280

Dirr, H.W. & Schabort, J.C. (1986) Aflatoxin B_1 transport in rat blood plasma. Binding to albumin *in vivo* and *in vitro* and spectrofluorimetric studies into the nature of the interaction. *Biochim. biophys. Acta*, **881**, 383–390

Distlerath, L.M., Loper, J.C. & Tabor, M.W. (1983) Effects of metyrapone on microsomal-dependent *Salmonella* mutagenesis. Studies with chloroallyl ethers and model compounds. *Biochem. Pharmacol.*, **32**, 3739–3748

Dix, K.M. (1984) The development of hepatocellular tumours following aflatoxin B_1 exposure of the partially hepatectomised mouse. *Carcinogenesis*, **5**, 385–390

Dobiáš, L., Paulíková, H., Kůsová, J., Klein, A. & Chatrná, E. (1982) Demonstration of mycotoxins with mutagenic action in foodstuffs by means of bacterial indicator strains of *Salmonella typhimurium* (Czech.). *Česk. Hyg.*, **27**, 275–281

Doehmer, J., Dogra, S., Friedberg, T., Monier, S., Adesnik, M., Glatt, H. & Oesch, F. (1988) Stable expression of rat cytochrome P-450IIB1 cDNA in Chinese hamster cells (V79) and metabolic activation of aflatoxin B_1. *Proc. natl Acad. Sci. USA*, **85**, 5769–5773

Dorange, J.L., Aranda, G., Cornu, A. & Dulieu, H. (1983) Genetic toxicity of methyl methanethiosulfonate on *Salmonella typhimurium*, *Saccharomyces cerevisiae* and *Nicotiana tabacum*. *Mutat. Res.*, **120**, 207–217

Dorner, J.W., Cole, R.J. & Diener, U.L. (1984) The relationship of *Aspergillus flavus* and *Aspergillus parasiticus* with reference to production of aflatoxins and cyclopiazonic acid. *Mycopathologia*, **87**, 13–15

Dunaif, G.E. & Campbell, T.C. (1987a) Dietary protein level and aflatoxin B_1-induced preneoplastic hepatic lesions in the rat. *J. Nutr.*, **117**, 1298–1302

Dunaif, G.E. & Campbell, T.C. (1987b) Relative contribution of dietary protein level and aflatoxin B_1 dose in generation of presumptive preneoplastic foci in rat liver. *J. natl Cancer Inst.*, **78**, 365–369

Dunn, J.J., Lee, L.S. & Ciegler, A. (1982) Mutagenicity and toxicity of aflatoxin precursors. *Environ. Mutag.*, **4**, 19–26

Dvořáčková, I., Stora, C. & Ayraud, N. (1981) Evidence for aflatoxin B_1 in two cases of lung cancer in man. *J. Cancer Res. clin. Oncol.*, **100**, 221–224

Egan, H., Stoloff, L., Scott, P., Castegnaro, M., O'Neill, I.K. & Bartsch, H., eds (1982) *Environmental Carcinogens. Selected Methods of Analysis, Vol. 5, Some Mycotoxins* (IARC Scientific Publications No. 44), Lyon, IARC

Egbunike, G.N. (1979) The effects of micro doses of aflatoxin B_1 on sperm production rates, epididymal sperm abnormality and fertility in the rat. *Zbl. vet. Med. A*, **26**, 66–72

Egbunike, G.N., Emerole, G.O., Aire, T.A. & Ikegwuonu, F.I. (1980) Sperm production rates, sperm physiology and fertility in rats chronically treated with sublethal doses of aflatoxin B_1. *Andrologia*, **12**, 467–475 van Egmond, H.P. (1984) Determination of mycotoxins. In: King, R.D., ed., *Developments in Food Analysis Techniques III*, Barking, Essex, Applied Science Publishers, pp. 99–144

van Egmond, H.P., ed. (1989a) *Mycotoxins in Dairy Products*, Amsterdam, Elsevier

van Egmond, H.P. (1989b) Current situation on regulations for mycotoxins. Overview of tolerances and status of standard methods of sampling and analysis. *Food Addit. Contam.*, **6**, 139–188

van Egmond, H.P. (1992) Worldwide regulations for mycotoxins. In: Mise, K. & Richards, J.L., eds, *Emerging Food Safety Problem Resulting from Microbial Contamination*, Tokyo, Ministry of Health and Welfare

Elespuru, R.K. & Yarmolinsky, M.B. (1979) A colorimetric assay of lysogenic induction designed for screening potential carcinogenic and carcinostatic agents. *Environ. Mutagen.*, **1**, 65–78

El-Mofty, M.M. & Sakr, S.A. (1988) The induction of neoplastic lesions by aflatoxin B_1 in the Egyptian toad (*Bufo regularis*). *Nutr. Cancer*, **11**, 55–59

El-Zawahri, M., Moubasher, A., Morad, M. & El-Kady, I. (1977) Mutagenic effect of aflatoxin B_1. *Ann. Nutr. Aliment.*, **31**, 859–866

El-Zawahri, M.M., Morad, M.M. & Khishin, A.F. (1990) Mutagenic effect of aflatoxin G_1 in comparison with B_1. *J. environ. Pathol. Toxicol. Oncol.*, **10**, 45–51

Emerole, G.O. (1981) Excretion of aflatoxin B_1 as a glutathione conjugate. *Eur. J. Drug Metab. Pharmacokin.*, **6**, 265–268

von Engel, G. & von Milczewski, K.E. (1976) Detection of mycotoxins activated with a set of *Salmonella typhimurium* histidine autotropes (Ger.). *Kieler Milchwirtschaftl. Forsch. ber.*, **28**, 359–365

Epstein, S., Bartus, B. & Farber, E. (1969) Renal epithelial neoplasms induced in male Wistar rats by oral aflatoxin B_1. *Cancer Res.*, **29**, 1045–1050

Epstein, S.S., Arnold, E., Andrea, J., Bass, W. & Bishop, Y. (1972) Detection of chemical mutagens by the dominant lethal assay in the mouse. *Toxicol. appl. Pharmacol.*, **23**, 288–325

Essigmann, J.M., Croy, R.G., Bennett, R.A. & Wogan, G.N. (1982) Metabolic activation of aflatoxin B_1: patterns of DNA adduct formation, removal, and excretion in relation to carcinogenesis. *Drug Metab. Rev.*, **13**, 581–602

Ewaskiewicz, J.I., Devlin, T.M. & Ch'ih, J.J. (1991) The in vivo disposition of aflatoxin B_1 in rat liver. *Biochem. biophys. Res. Commun.*, **179**, 1095–1100

Fabry, L. & Roberfroid, M. (1981) Mutagenicity of aflatoxin B_1: observations *in vivo* and their relation to in vitro activation. *Toxicol. Lett.*, **7**, 245–250

Fahmy, M.J. & Fahmy, O.G. (1983a) Differential induction of altered gene expression by carcinogens at mutant alleles of a *Drosophila* locus with a transposable element. *Cancer Res.*, **43**, 801–807

Fahmy, M.J. & Fahmy, O.G. (1983b) Misregulation versus mutation in the alteration of gene expression by carcinogens through interactions with transposable elements in *Drosophila melanogaster*. *Teratog. Carcinog. Mutag.*, **3**, 27–39

Fahmy, M.J., Fahmy, O.G. & Swenson, D.H. (1978) Aflatoxin-B_1-2,3-dichloride as a model of the active metabolite of aflatoxin B_1 in mutagenesis and carcinogenesis. *Cancer Res.*, **38**, 2608–2616

Ferguson, L.R., Parslow, M.I. & McLarin, J.A. (1986) Chromosome damage by dothistromin in human peripheral blood lymphocyte cultures: a comparison with aflatoxin B_1. *Mutat. Res.*, **170**, 47–53

Firozi, P.F., Aboobaker, V.S. & Bhattacharya, R.K. (1987) Action of vitamin A on DNA adduct formation by aflatoxin B_1 in a microsome catalyzed reaction. *Cancer Lett.*, **34**, 213–220

Fong, L.Y.Y. & Chan, W.C. (1981) Long-term effects of feeding aflatoxin-contaminated market peanut oil to Sprague-Dawley rats. *Food Cosmet. Toxicol.*, **19**, 179–183

Fong, A.T., Swanson, H.I., Dashwood, R.H., Williams, D.E., Hendricks, J.D. & Bailey, G.S. (1990) Mechanisms of anti-carcinogenesis by indole-3-carbinol. Studies of enzyme induction, electrophile-scavenging, and inhibition of aflatoxin B_1 activation. *Biochem. Pharmacol.*, **39**, 19–26

Forrester, L.M., Neal, G.E., Judah, D.J., Glancey, M.J. & Wolf, C.R. (1990) Evidence for involvement of multiple forms of cytochrome P450 in aflatoxin B_1 metabolism in human liver. *Proc. natl Acad. Sci. USA*, **87**, 8306–8310

Francis, A.R., Shetty, T.K. & Bhattacharya, R.K. (1989a) Modifying role of dietary factors on the mutagenicity of aflatoxin B_1: in vitro effect of plant flavonoids. *Mutat. Res.*, **222**, 393–401

Francis, A.R., Shetty, T.K. & Bhattacharya, R.K. (1989b) Modification of the mutagenicity of aflatoxin B_1 and N-methyl-N'-nitro-N-nitrosoguanidine by certain phenolic compounds. *Cancer Lett.*, **45**, 177–182

Fraysinet, C. & Lafarge-Frayssinet, C. (1990) Effect of ammoniation on the carcinogenicity of aflatoxin-contaminated groundnut oil cakes: long-term feeding study in the rat. *Food Addit. Contam.*, **7**, 63–68

Freeman, H.J. & San, R.H.C. (1980) Use of unscheduled DNA synthesis in freshly isolated human intestinal mucosal cells for carcinogen detection. *Cancer Res.*, **40**, 3155–3157

Friedman, M.A. & Staub, J. (1977) Induction of micronuclei in mouse and hamster bone marrow by chemical carcinogens. *Mutat. Res.*, **43**, 255–262

Friedman, M., Wehr, C.M., Schade, J.E. & MacGregor, J.T. (1982) Inactivation of aflatoxin B_1 mutagenicity by thiols. *Food Chem. Toxicol.*, **20**, 887–892

Friedrich, U. & Nass, G. (1983) Evaluation of a mutation test using S49 mouse lymphoma cells and monitoring simultaneously the induction of dexamethasone resistance, 6-thioguanine resistance and ouabain resistance. *Mutat. Res.*, **110**, 147–162

Friesen, M.D. (1989) The International Mycotoxin Check Sample Programme. *J. Toxicol.-Toxin Rev.*, **8**, 363–373

Fujimori, M., Tokino, T., Hino, O., Kitagawa, T., Imamura, T., Okamoto, E., Mitsunobu, M., Ishikawa, T., Nakagama, H., Harada, H., Yagura, M., Matsubara, K. & Nakamura, Y. (1991) Allelotype study of primary hepatocellular carcinoma. *Cancer Res.*, **51**, 89–93

Fujimoto, Y., Hampton, L.L., Luo, L.-D., Wirth, P.J. & Thorgeirsson, S.S. (1992) Low frequency of p53 gene mutation in tumors induced by aflatoxin B_1 in nonhuman primates. *Cancer Res.*, **52**, 1044–1046

Fukayama, M.Y. & Hsieh, D.P.H. (1984) The effects of butylated hydroxytoluene on the in vitro metabolism, DNA-binding and mutagenicity of aflatoxin B_1 in the rat. *Food Chem. Toxicol.*, **22**, 355–360

Fukayama, M.Y. & Hsieh, D.P.H. (1985) Effect of butylated hydroxytoluene pretreatment on the excretion, tissue distribution and DNA binding of [^{14}C]aflatoxin B_1 in the rat. *Food Chem. Toxicol.*, **23**, 567–573

Gan, L.-S., Skipper, P.L., Peng, X., Groopman, J.D., Chen, J.-S., Wogan, G.N. & Tannenbaum, S.R. (1988) Serum albumin adducts in the molecular epidemiology of aflatoxin carcinogenesis: correlation with aflatoxin B_1 intake and urinary excretion of aflatoxin M_1. *Carcinogenesis*, **7**, 1323–1325

Garner, R.C., Martin, C.N., Smith, J.R.L., Coles, B.F. & Tolson, M.R. (1979) Comparison of aflatoxin B_1 and aflatoxin G_1 binding to cellular macromolecules *in vitro*, *in vivo* and after peracid oxidation; characterisation of the major nucleic acid adducts. *Chem.-biol. Interactions*, **26**, 57–73

Garner, R.C., Ryder, R. & Montesano, R. (1985) Monitoring of aflatoxins in human body fluids and application to field studies. *Cancer Res.*, **45**, 922–928

Garner, R.C., Dvořáčková, I. & Tursi, F. (1988) Immunoassay procedures to detect exposure to aflatoxin B_1 and benzo(a)pyrene in animals and man at the DNA level. *Int. Arch. occup. environ. Health*, **60**, 145–150

Gayda, D.P. & Pariza, M.W. (1981) Activation of aflatoxin B_1 by primary cultures of adult rat hepatocytes: effects of hepatocyte density. *Chem.-biol. Interactions*, **35**, 255–265

Geissler, F. & Faustman, E.M. (1988) Developmental toxicity of aflatoxin B_1 in the rodent embryo *in vitro*: contribution of exogenous biotransformation systems to toxicity. *Teratology*, **37**, 101–111

Gil, R., Callaghan, R., Boix, J., Pellin, A. & Llombart-Bosch, A. (1988) Morphometric and cytophotometric nuclear analysis of altered hepatocyte foci induced by *N*-nitrosomorpholine (NNM) and aflatoxin B_1 (AFB_1) in liver of Wistar rats. *Virchows Arch. B Cell Pathol.*, **54**, 341–349

Goeger, D.E., Shelton, D.W., Hendricks, J.D., Pereira, C. & Bailey, G.S. (1988) Comparative effect of dietary butylated hydroxyanisole and β-naphthoflavone on aflatoxin B_1 metabolism, DNA adduct formation, and carcinogenesis in rainbow trout. *Carcinogenesis*, **9**, 1793–1800

Goerttler, K., Löhrke, H., Schweizer, H.-J. & Hesse, B. (1980) Effects of aflatoxin B_1 on pregnant inbred Sprague-Dawley rats and their F_1 generation. A contribution to transplacental carcinogenesis. *J. natl Cancer Inst.*, **64**, 1349–1354

Goldblatt, L.A., ed. (1969) *Aflatoxin. Scientific Background, Control and Implications*, New York, Academic Press

Goodall, C.M. & Butler, W.H. (1969) Aflatoxin carcinogenesis: inhibition of liver cancer induction in hypophysectomized rats. *Int. J. Cancer*, **4**, 422–429

Gopalan, C., Tulpule, P.G. & Krishnamurthi, D. (1972) Induction of hepatic carcinoma with aflatoxin in the rhesus monkey. *Food Cosmet. Toxicol.*, **10**, 519–521

Goto, T. & Hsieh, D.P.H. (1985) Fractionation of radioactivity in the milk of goats administered [^{14}C]aflatoxin B_1. *J. Assoc. off. anal. Chem.*, **68**, 456–458

Goze, A., Sarasin, A., Mouse, Y. & Devoret, R. (1975) Induction and mutagenesis of prophage lambda in *Escherichia coli* K-12 by metabolites of aflatoxin B_1. *Mutat. Res.*, **28**, 1–7

Graf, U., Juon, H., Katz, A.J., Frei, H.J. & Würgler, F.E. (1983) A pilot study on a new *Drosophila* spot test. *Mutat. Res.*, **120**, 233–239

Green, C.E., Rice, D.W., Hsieh, D.P.H. & Byard, J.L. (1982) The comparative metabolism and toxic potency of aflatoxin B_1 and aflatoxin M_1 in primary cultures of adult-rat hepatocytes. *Food Chem. Toxicol.*, **20**, 53–60

Grice, H.C., Moodie, C.A. & Smith, D.C. (1973) The carcinogenic potential of aflatoxin or its metabolites in rats from dams fed aflatoxin pre- and postpartum. *Cancer Res.*, **33**, 262–268

Groopman, J.D. (1993) Molecular dosimetry methods for assessing human aflatoxin exposures. In: Eaton, D.L. & Groopman, J.D., *The Toxicology of Aflatoxins: Human Health, Veterinary and Agricultural Significance*, New York, Academic Press (in press)

Groopman, J.D. & Sabbioni, G. (1991) Detection of aflatoxin and its metabolites in human biological fluids. In: Bray, G.A. & Ryan, D.H., eds, *Mycotoxins, Cancer and Health* (Pennington Center Nutrition Series, Vol. 1), Baton Rouge, LA, Louisiana State University Press, pp. 18–31

Groopman, J.D., Fowler, K.W., Busby, W.F., Jr & Wogan, G.N. (1981a) Interaction of aflatoxin B_2 with rat liver DNA and histones *in vivo*. *Carcinogenesis*, **2**, 1371–1373

Groopman, J.D., Croy, R.G. & Wogan, G.N. (1981b) In vitro reactions of aflatoxin B_1-adducted DNA. *Proc. natl Acad. Sci. USA*, **78**, 5445–5449

Groopman, J.D., Haugen, A., Goodrich, G.R., Wogan, G.N & Harris, C.C. (1982) Quantitation of aflatoxin B_1-modified DNA using monoclonal antibodies. *Cancer Res.*, **42**, 3120–3124

Groopman, J.D., Donahue, P.R., Zhu, J., Chen, J. & Wogan, G.N. (1985) Aflatoxin metabolism in humans: detection of metabolites and nucleic acid adducts in urine by affinity chromatography. *Proc. natl Acad. Sci. USA*, **82**, 6492–6496

Groopman, J.D., Jiaqi, Z., Donahue, P.R., Pikul, A., Lisheng, Z., Chen, J.-S. & Wogan, G.N. (1992a) Molecular dosimetry of urinary aflatoxin–DNA adducts in people living in Guangxi Autonomous Region, People's Republic of China. *Cancer Res.*, **52**, 45–52

Groopman, J.D., Hall, A.J., Whittle, H., Hudson, G.J., Wogan, G.N., Montesano, R. & Wild, C.P. (1992b) Molecular dosimetry of aflatoxin-N7-guanine in human urine obtained in the Gambia, West Africa. *Cancer Epidemiol. Biomarkers Prev.*, **1**, 221–227

Groopman, J.D., Hasler, J.A., Trudel, L.J., Pikul, A., Donahue, P.R. & Wogan, G.N. (1992c) Molecular dosimetry in rat urine of aflatoxin N^7-guanine and other aflatoxin metabolites by multiple monoclonal antibody affinity chromatography and immunoaffinity/high performance liquid chromatography. *Cancer Res.*, **52**, 267–274

Gurtoo, H.L., Dahms, R.P. & Paigen, B. (1978) Metabolic activation of aflatoxins related to their mutagenicity. *Biochem. biophys. Res. Commun.*, **81**, 965–972

Gurtoo, H.L., Koser, P.L., Bansal, S.K., Fox, H.W., Sharma, S.D., Mulhern, A.I. & Pavelic, Z.P. (1985) Inhibition of aflatoxin B_1-hepatocarcinogenesis in rats by β-naphthoflavone. *Carcinogenesis*, **6**, 675–678

Hall, A.J. & Wild, C.P. (1992) Aflatoxin biomarkers (Letter to the Editor). *Lancet*, **339**, 1413–1414

Hall, A.J. & Wild, C.P. (1993) The epidemiology of aflatoxin-related disease. In: Eaton, D.L. & Groopman, J.D., eds, *The Toxicology of Aflatoxins: Human Health, Veterinary and Agricultural Significance*, New York, Academic Press (in press)

Hamilton, J.W. & Bloom, S.E. (1984) Correlation between mixed-function oxidase enzyme induction and aflatoxin B_1-induced unscheduled DNA synthesis in the chick embryo, *in vivo*. *Environ. Mutag.*, **6**, 41–48

Harris, C.C. (1991) Chemical and physical carcinogenesis: advances and perspectives for the 1990s. *Cancer Res.*, **51**, 5023s–5044s

Harris, T.M., Stone, M.P., Gopalakrishnan, S., Baertschi, S.W., Raney, K.D. & Byrd, S. (1989) Aflatoxin B_1 epoxide, the ultimate carcinogenic form of aflatoxin B_1: synthesis and reaction with DNA. *J. Toxicol.-Toxin Rev.*, **8**, 111–120

Hayes, R.B., van Nieuwenhuize, J.P., Raatgever, J.W. & ten Kate, F.J.W. (1984) Aflatoxin exposures in the industrial setting: an epidemiological study of mortality. *Food chem. Toxicol.*, **22**, 39–43

Hayes, J.D., Judah, D.J., McLellan, L.I. & Neal, G.E. (1991) Contribution of the glutathione *S*-transferases to the mechanisms of resistance to aflatoxin B_1. *Pharmacol. Ther.*, **50**, 443–472

Hayward, N.K., Walker, G.J., Graham, W. & Cooksley, E. (1991) Hepatocellular carcinoma mutation (Letter to the Editor). *Nature*, **352**, 764

Hendricks, J.D., Putnam, T.P., Bills, D.D. & Sinnhuber, R.O. (1977) Inhibitory effect of a polychlorinated biphenyl (Aroclor 1254) on aflatoxin B_1 carcinogenesis in rainbow trout (*Salmo gairdneri*). *J. natl Cancer Inst.*, **59**, 1545–1551

Hendricks, J.D., Putnam, T.P. & Sinnhuber, R.O. (1979) Effect of dietary dieldrin on aflatoxin B_1 carcinogenesis in rainbow trout (*Salmo gairdneri*). *J. environ. Pathol. Toxicol.*, **2**, 719–728

Hendricks, J.D., Sinnhuber, R.O., Nixon, J.E., Wales, J.H., Masri, M.S. & Hsieh, D.P.H. (1980a) Carcinogenic response of rainbow trout (*Salmo gairdneri*) to aflatoxin Q_1 and synergistic effect of cyclopropenoid fatty acids. *J. natl Cancer Inst.*, **64**, 523–527

Hendricks, J.D., Putnam, T.P. & Sinnhuber, R.O. (1980b) Null effect of dietary Aroclor 1254 on hepatocellular carcinoma incidence in rainbow trout (*Salmo gairdneri*) exposed to aflatoxin B_1 as embryos. *J. environ. Pathol. Toxicol.*, **4**, 9–16

Hendrickse, R.G. & Maxwell, S.M. (1989) Aflatoxins and child health in the tropics. *J. Toxicol.-Toxin Rev.*, **8**, 31–41

Hendrickse, R.G., Coulter, J.B.S., Lamplugh, S.M., MacFarlane, S.B.J., Williams, T.E., Omer, M.I.A. & Suliman, G.I. (1982) Aflatoxins and kwashiorkor: a study in Sudanese children. *Br. med. J.*, **285**, 843–846

Higashi, K., Ikeuchi, K. & Karasaki, Y. (1982) Use of metabolic activation systems of tulip bulbs in the Ames test for environmental mutagens. *Bull. environ. Contam. Toxicol.*, **29**, 505–510

Hirano, K., Hagiwara, T., Ohta, Y., Matsumoto, H. & Kada, T. (1982) Rec-assay with spores of *Bacillus subtilis* with and without metabolic activation. *Mutat. Res.*, **97**, 339–347

Ho, Y.L. & Ho, S.K. (1981) Screening of carcinogens with the prophage lambda cl*ts*857 induction test. *Cancer Res.*, **41**, 532–536

Holeski, C.J., Eaton, D.L., Monroe, D.H. & Bellamy, G.M. (1987) Effects of phenobarbital on the biliary excretion of aflatoxin P_1-glucuronide and aflatoxin B_1-*S*-glutathione in the rat. *Xenobiotica*, **17**, 139–153

Hollstein, M., Sidransky, D., Vogelstein, B. & Harris, C.C. (1991a) p53 Mutations in human cancers. *Science*, **253**, 49–53

Hollstein, M.C., Peri, L., Mandard, A.M., Welsh, J.A., Montesano, R., Metcalf, R.A., Bak, M. & Harris, C.C. (1991b) Genetic analysis of human esophageal tumors from two high incidence geographic areas: frequent p53 base substitutions and absence of *ras* mutations. *Cancer Res.*, **51**, 4102–4106

Hosono, S., Lee, C.-S., Chou, M.-J., Yang, C.-S. & Shih, C. (1991) Molecular analysis of the p53 alleles in primary hepatocellular carcinomas and cell lines. *Oncogene*, **6**, 237–243

Hsieh, D.P.H., Wong, J.J., Wong, Z.A., Michas, C. & Ruebner, B.H. (1977) Hepatic transformation of aflatoxin and its carcinogenicity. In: Hiatt, H.H., Watson, J.D. & Winsten, J.A., eds, *Origins of Human Cancer*, Book B, Cold Spring Harbor, NY, CSH Press, pp. 697–707

Hsieh, D.P.H., Cullen, J.M. & Ruebner, B.H. (1984) Comparative hepatocarcinogenicity of aflatoxins B_1 and M_1 in the rat. *Food Chem. Toxicol.*, **22**, 1027–1028

Hsieh, L.-L., Hsu, S.-W., Chen, D.-S. & Santella, R.M. (1988) Immunological detection of aflatoxin B_1–DNA adducts formed *in vivo*. *Cancer Res.*, **48**, 6328–6331

Hsu, I.C., Metcalf, R.A., Sun, T., Welsh, J.A., Wang, N.J. & Harris C.C. (1991) Mutational hotspot in the p53 gene in human hepatocellular carcinomas. *Nature*, **350**, 427–428

Huang, C.C., Hsueh, J.L., Chen, H.H. & Batt, T.R. (1982) Retinol (vitamin A) inhibits sister chromatid exchanges and cell cycle delay induced by cyclophosphamide and aflatoxin B_1 in Chinese hamster V79 cells. *Carcinogenesis*, **3**, 1–5

Huh, N., Menoto, N. & Utakoji, T. (1982) Metabolic activation of benzo[*a*]pyrene, aflatoxin B_1, and dimethylnitrosamine by human hepatoma cell line. *Mutat. Res.*, **94**, 339–348

IARC (1972) *IARC Monographs on the Evaluation of Carcinogenic Risk of Chemicals to Man*, Vol. 1, *Some Inorganic Substances, Chlorinated Hydrocarbons, Aromatic Amines, N-Nitroso Compounds, and Natural Products*, Lyon, pp. 145–156

IARC (1976) *IARC Monographs on the Evaluation of Carcinogenic Risk of Chemicals to Man*, Vol. 10, *Some Naturally Occurring Substances*, Lyon, pp. 51–72

IARC (1986) *IARC Monographs on the Evaluation of Carcinogenic Risks to Humans*, Vol. 40, *Some Naturally Occurring and Synthetic Food Components, Furocoumarins and Ultraviolet Radiation*, Lyon, pp. 113–206

IARC (1987a) *IARC Monographs on the Evaluation of Carcinogenic Risks to Humans*, Suppl. 7, *Overall Evaluations of Carcinogenicity: An Updating of* IARC Monographs *Volumes 1 to 42*, Lyon, pp. 83–87

IARC (1987b) *IARC Monographs on the Evaluation of Carcinogenic Risks to Humans*, Suppl. 6, *Genetic and Related Effects: An Updating of Selected* IARC Monographs *from Volumes 1–42*, Lyon, pp. 40–56

Ichinotsubo, D., Mower, H.F., Setliff, J. & Mandel, M. (1977) The use of rec^- bacteria for testing of carcinogenic substances. *Mutat. Res.*, **46**, 53–62

Ide, F., Ishikawa, T. & Takayama, S. (1981) Detection of chemical carcinogens by assay of unscheduled DNA synthesis in rat tracheal epithelium in short-term organ culture. *J. Cancer Res. clin. Oncol.*, **102**, 115–126

Imaida, K., Shirai, T., Tatematsu, M., Takano, T. & Ito, N. (1981) Dose responses of five hepatocarcinogens for the initiation of rat hepatocarcinogenesis. *Cancer Lett.*, **14**, 279–283

Inoue, K., Shibata, T. & Abe, T. (1983) Induction of sister-chromatid exchanges in human lymphocytes by indirect carcinogens with and without metabolic activation. *Mutat. Res.*, **117**, 301–309

Irvin, T.R. & Wogan, G.N. (1984) Quantitation of aflatoxin B_1 adduction within the ribosomal RNA gene sequences of rat liver DNA. *Proc. natl Acad. Sci. USA*, **81**, 664–668

Irvin, T.R. & Wogan, G.N. (1985) Quantitative and qualitative characterization of aflatoxin B_1 adducts formed *in vivo* within the ribosomal RNA genes of rat liver DNA. *Cancer Res.*, **45**, 3497–3502

Ishii, K., Maeda, K., Kamataki, T. & Kato, R. (1986) Mutagenic activation of aflatoxin B_1 by several forms of purified cytochrome P-450. *Mutat. Res.*, **174**, 85–88

Iskandar, O. & Vijayalaxmi, E.T. (1981) The enhancement of the effect of aflatoxin B_1 by metabolic activation with rat-liver microsomes on human lymphocytes assayed with the micronucleus test. *Mutat. Res.*, **91**, 63–66

Israel-Kalinsky, H., Malca-Mor, L. & Stark, A.A. (1984) Comparative aflatoxin B_1 mutagenesis of *Salmonella typhimurium* TA100 in metabolic and photoactivation systems. *Cancer Res.*, **44**, 1831–1839

Israels, L.G., Walls, G.A., Ollmann, D.J., Friesen, E. & Israels, E.D. (1983) Vitamin K as a regulator of benzo[*a*]pyrene metabolism, mutagenesis, and carcinogenesis: studies with rat chromosomes and tumorigenesis in mice. *J. clin. Invest.*, **71**, 1130–1140

Ito, N. (1982) Unscheduled DNA synthesis induced by chemical carcinogens in primary cultures of adult rat hepatocytes. *Mie med. J.*, **32**, 53–60

Ito, Y., Ohnishi, S. & Fujie, K. (1989) Chromosome aberrations induced by aflatoxin B_1 in rat bone marrow cells *in vivo* and their suppression by green tea. Mutat. Res., **222**, 253–261

Iverson, F., Campbell, J., Clayson, D., Hierlihy, S., Labossiere, E. & Hayward S. (1987) Effects of antioxidants on aflatoxin-induced hepatic tumors in rats. Cancer Lett., **34**, 139–144

Jayaraj, A. & Richardson, A. (1981) Metabolic activation of aflatoxin B_1 by liver tissue from male Fischer F344 rats of various ages. *Mech. Ageing Dev.*, **17**, 163–171

Jayaraj, A., Hardwick, J.P., Diller, T.W. & Richardson, A.G. (1985) Metabolism, covalent binding, and mutagenicity of aflatoxin B_1 by liver extracts from rats of various ages. *J. natl Cancer Inst.*, **74**, 95–103

Jelinek, C.F., Pohland, A.E. & Wood, G.E. (1989) Review of mycotoxin contamination. Worldwide occurrence of mycotoxins in foods and feeds—an update. *J. Assoc. off. anal. Chem.*, **72**, 223–230

Jemmali, M. (1990) Decontamination and detoxification of mycotoxins. *J. environ. Pathol. Toxicol. Oncol.*, **10**, 154–159

Jhee, E.-C., Ho, L.L. & Lotlikar, P.D. (1988) Effect of butylated hydroxyanisole pretreatment on in vitro hepatic aflatoxin B_1-binding and aflatoxin B_1-glutathione conjugation in rats. *Cancer Res.*, **48**, 2688–2692

Jhee, E.C., Ho, L.L., Tsuji, K., Gopalan, P. & Lotlikar, P.D. (1989a) Mechanism of inhibition of aflatoxin B_1-DNA binding in the liver by butylated hydroxyanisole pretreatment of rats. *J. Toxicol.-Toxin Rev.*, **8**, 141–153

Jhee, E.C., Ho, L.L., Tsuji, K., Gopalan, P. & Lotlikar, P.D. (1989b) Effect of butylated hydroxyanisole pretreatment on aflatoxin B_1-DNA binding and aflatoxin B_1-glutathione conjugation in isolated hepatocytes from rats. *Cancer Res.*, **49**, 1357–1360

Jone, C., Erickson, L., Trosko, J.E. & Chang, C.C. (1987) Effect of biological toxins on gap-junctional intercellular communication in Chinese hamster V79 cells. *Cell. Biol. Toxicol.*, **3**, 1–15

Jones, P.A., Buckley, J.D., Henderson, B.E., Ross, R.K. & Pike, M.C. (1991) From gene to carcinogen: a rapidly evolving field in molecular epidemiology. *Cancer Res.*, **51**, 3617–3620

Kalengayi, M.M.R., Ronchi, G. & Desmet, V.J. (1975) Histochemistry of gamma-glutamyl transpeptidase in rat liver during aflatoxin B_1-induced carcinogenesis. *J. natl Cancer Inst.*, **55**, 579–588

Kawajiri, K., Yonekawa, H., Gotoh, O., Watanabe, J., Igarashi, S. & Tagashira, Y. (1983) Contributions of two inducible forms of cytochrome P-450 in rat liver microsomes to the metabolic activation of various chemical carcinogens. *Cancer Res.*, **43**, 819–823

Keen, P. & Martin, P. (1971) Is aflatoxin carcinogenic in man? The evidence in Swaziland. *Trop. geogr. Med.*, **23**, 44–53

Kensler, T.W., Egner, P.A., Davidson, N.E., Roebuck, B.D., Pikul, A. & Groopman, J.D. (1986) Modulation of aflatoxin metabolism, aflatoxin-N^7 guanine formation, and hepatic tumorigenesis in rats fed ethoxyquin: role of induction of glutathione S-transferases. *Cancer Res.*, **46**, 3924–3931

Kensler, T.W., Egner, P.A., Dolan, P.M., Groopman, J.D. & Roebuck, B.D. (1987) Mechanism of protection against aflatoxin tumorigenicity in rats fed 5-(2-pyrazinyl)-4-methyl-1,2-dithiol-3-thione (oltipraz) and related 1,2-dithiol-3-thiones and 1,2-dithiol-3-ones. *Cancer Res.*, **47**, 4271–4277

Kirby, G.M., Stalker, M., Metcalfe, C., Kocal, T., Ferguson, H. & Hayes, M.A. (1990) Expression of immunoreactive glutathione S-transferases in hepatic neoplasms induced by aflatoxin B_1 or 1,2-dimethylbenzanthracene in rainbow trout (*Oncorhynchus mykiss*). *Carcinogenesis*, **11**, 2255–2257

Kitada, M., Taneda, M., Ohta, K., Nagashima, K., Itahashi, K. & Kamataki, T. (1990) Metabolic activation of aflatoxin B_1 and 2-amino-3-methylimidazo[4,5-*f*]quinoline by human adult and fetal livers. *Cancer Res.*, **50**, 2641–2645

Kleinwächter, V. & Koukalová, B. (1979) Reduction of mutagenic activity of aflatoxins after UV-irradiation. *Acta biol. med. Ger.*, **38**, 1239–1242

Korte, A. & Rückert, G. (1980) Chromosomal analysis in bone-marrow cells of Chinese hamsters after treatment with mycotoxins. *Mutat. Res.*, **78**, 41–49

Krahn, D.F. & Heidelberger, C. (1977) Liver homogenate-mediated mutagenesis in Chinese hamster V79 cells by polycyclic aromatic hydrocarbons and aflatoxins. *Mutat. Res.*, **46**, 27–44

Kraupp-Grasl, B., Huber, W., Putz, B., Gerbracht, U. & Schulte-Hermann, R. (1990) Tumor promotion by the peroxisome proliferator nafenopin involving a specific subtype of altered foci in rat liver. *Cancer Res.*, **50**, 3701–3708

Krishnamachari, K.A.V.R., Bhat, R.V., Nagarajan, V. & Tilak, T.B.G. (1975) Hepatitis due to aflatoxicosis. An outbreak in western India. *Lancet*, i, 1061–1063

Krishnamurthy, P.B. & Neelaram, G.S. (1986) Effect of dietary fat on aflatoxin B_1-induced chromosomal aberrations in mice. *Toxicol. Lett.*, **31**, 229–234

Krivobok, S., Olivier, P., Marzin, D.R., Seigle-Murandi, F. & Steiman, R. (1987) Study of the genotoxic potential of 17 mycotoxins with the SOS chromotest. *Mutagenesis*, **2**, 433–439

Kroeger-Koepke, M.B., Reuber, M.D., Iype, P.T., Lijinsky, W. & Michejda, C.J. (1983) The effect of substituents in the aromatic ring on carcinogenicity of *N*-nitrosomethylaniline in F344 rats. *Carcinogenesis*, **4**, 157–160

Kuczuk, M.H., Benson, P.M., Heath, H. & Hayes, A.W. (1978) Evaluation of the mutagenic potential of mycotoxins using *Salmonella typhimurium* and *Saccharomyces cerevisiae*. *Mutat. Res.*, **53**, 11–20

Kumagai, S. (1989) Intestinal absorption and excretion of aflatoxin in rats. *Toxicol. appl. Pharmacol.*, **97**, 88–97

Kumari, D. & Sinha, S.P. (1990) Combined effect of aflatoxin and vitamin A on clastogeny in mice chromosomes. *Cytologia*, **55**, 387–390

Kuroki, T., Malaveillė, C., Drevon, C., Piccoli, C., MacLeod, M. & Selkirk, J.K. (1979) Critical importance of microsome concentration in mutagenesis assay with V79 Chinese hamster cells. *Mutat. Res.*, **63**, 259–272

Kurtzman, C.P., Horn, B.W. & Hesseltine, C.W. (1987) *Aspergillus nomius*, a new aflatoxin-producing species related to *Aspergillus flavus* and *Aspergillus tamarii*. *Antonie van Leeuwenhoek*, **53**, 147–158

Lake, R.S., Kropko, M.L., McLachlan, S., Pezzutti, M.R., Shoemaker, R.H. & Igel, H.J. (1980) Chemical carcinogen induction of DNA-repair synthesis in human peripheral blood monocytes. *Mutat. Res.*, **74**, 357–377

Lam, K.C., Yu, M.C., Leung, J.W.C. & Henderson, B.E. (1982) Hepatitis B virus and cigarette smoking: risk factors for hepatocellular carcinoma in Hong Kong. *Cancer Res.*, **42**, 5246–5248

Lamartiniere, C.A. (1990) Neonatal diethylstilbestrol treatment alters aflatoxin B_1–DNA adduct concentrations in adult rats. *J. biochem. Toxicol.*, **5**, 41–46

Lamb, M.J. & Lilly, L.J. (1971) Induction of recessive lethals in *Drosophila melanogaster* by aflatoxin B_1. *Mutat. Res.*, **11**, 430–433

Lamplugh, S.M. & Hendrickse, R.G. (1982) Aflatoxins in the livers of children with kwashiorkor. *Ann. trop. Paediatr.*, **2**, 101–104

Lamplugh, S.M., Hendrickse, R.G., Apeagyei, F. & Mwanmut, D.D. (1988) Aflatoxins in breast milk, neonatal cord blood, and serum of pregnant women (Short report). *Br. med. J.*, **296**, 968

Lancaster, M.C., Jenkins, F.P. & Philp, J.McL. (1961) Toxicity associated with certain samples of groundnuts. *Nature*, **192**, 1095–1096

Langenbach, R., Freed, H.J. & Huberman, E. (1978a) Liver cell-mediated mutagenesis of mammalian cells by liver carcinogens. *Proc. natl Acad. Sci. USA*, **75**, 2864–2867

Langenbach, R., Freed, H.J., Raveh, D. & Huberman, E. (1978b) Cell specificity in metabolic activation of aflatoxin B_1 and benzo[*a*]pyrene to mutagens for mammalian cells. *Nature*, **276**, 277–280

Larsson, P., Larsson, B.S. & Tjälve, H. (1988) Binding of aflatoxin B_1 to melanin. *Food chem. Toxicol.*, **26**, 579–586

Larsson, P., Hoedaya, W.I. & Tjälve, H. (1990) Disposition of ^{3}H-aflatoxin B_1 in mice: formation and retention of tissue bound metabolites in nasal glands. *Pharmacol. Toxicol.*, **67**, 162–171

Leadon, S.A., Tyrrell, R.M. & Cerutti, P.A. (1981) Excision repair of aflatoxin B_1-DNA adducts in human fibroblasts. *Cancer Res.*, **41**, 5125-5129

Lee, D.J., Wales, J.H. & Sinnhuber, R.O. (1969) Hepatoma and renal tubule adenoma in rats fed aflatoxin and cyclopropenoid fatty acids. *J. natl Cancer Inst.*, **43**, 1037-1044

Lee, H.-S., Sarosi, I. & Vyas, G.N. (1989) Aflatoxin B_1 formamidopyrimidine adducts in human hepatocarcinogenesis: a preliminary report. *Gastroenterology*, **97**, 1281-1287

Lei, D.-N., Wang, L.-Q., Ruebner, B.H., Hsieh, D.P.H., Wu, B.-F., Zhu, C.-R. & Du, M.-J. (1990) Effect of selenium on aflatoxin hepatocarcinogenesis in the rat. *Biomed. environ. Sci.*, **3**, 65-80

Leonard, A., Deknudt, G. & Linden, G. (1975) Mutagenicity tests with aflatoxins in the mouse. *Mutat. Res.*, **28**, 137-139

Levine, A.J., Momand, J. & Finlay, C.A. (1991) The p53 tumour suppressor gene. *Nature*, **351**, 453-456

Li, S.-Y. & Lin, J.-K. (1990) Sequential monitoring of cytogenetic damage in rat lymphocytes following in vivo exposure to aflatoxin B_1 and N-nitrosophenacetin. *Mutat. Res.*, **242**, 219-224

Li, G.-Y., Yao, K.-T. & Glaser, R. (1989) Sister chromatid exchange and nasopharyngeal carcinoma. *Int. J. Cancer*, **43**, 613-618

Lin, J.J., Liu, C. & Svoboda, D.J. (1974) Long term effects of aflatoxin B_1 and viral hepatitis on marmoset liver. A preliminary report. *Lab. Invest.*, **30**, 267-278

Lindenfelser, L.A., Lillehoj, E.B. & Burmeister, H.R. (1974) Aflatoxin and trichothecene toxins: skin tumor induction and synergistic acute toxicity in white mice. *J. natl Cancer Inst.*, **52**, 113-116

Link, K.H., Heidelberger, C. & Landolph, J.R. (1983) Induction of ouabain-resistant mutants by chemical carcinogens in rat prostate epithelial cells. *Environ. Mutagen.*, **5**, 33-48

Linsell, C.A. & Peers, F.G. (1977) Aflatoxin and liver cell cancer. *Trans. R. Soc. trop. Med. Hyg.*, **71**, 471-473

Lipsky, M.M., Cole, K.E., Hsu, I.-C., Kahng, M.W., Jones, T.W. & Trump, B.F. (1990) Interspecies comparisons of in vitro hepatocarcinogenesis. *Prog. clin. biol. Res.*, **331**, 395-408

Liu, Y.H., Taylor, J., Linko, P., Lucier, G.W. & Thompson, C.L. (1991) Glutathione S-transferase μ in human lymphocyte and liver: role in modulating formation of carcinogen-derived DNA adducts. *Carcinogenesis*, **12**, 2269-2275

Loechler, E.L., Teeter, M.M. & Whitlow, M.D. (1988) Mapping the binding site of aflatoxin B_1 in DNA: molecular modeling of the binding sites for the N(7)-guanine adduct of aflatoxin B_1 in different DNA sequences. *J. biomol. Struct. Dyn.*, **5**, 1237-1257

Lotlikar, P.D. (1989) Metabolic basis for susceptibility and resistance to aflatoxin B_1 hepatocarcinogenesis in rodents. *J. Toxicol.-Toxin Rev.*, **8**, 97-109

Lotlikar, P.D., Raj, H.G., Bohm, L.S., Ho, L.L., Jhee, E.-C., Tsuji, K. & Gopalan, P. (1989) A mechanism of inhibition of aflatoxin B_1-DNA binding in the liver by phenobarbital pretreatment of rats. *Cancer Res.*, **49**, 951-957

Loury, D.J. & Byard, J.L. (1983) Aroclor 1254 pretreatment enhances the DNA repair response of amino acid pyrolysate mutagens in primary cultures of rat hepatocytes. *Cancer Lett.*, **20**, 283-290

Loury, D.N. & Hsieh, D.P.H. (1984) Effects of chronic exposure to aflatoxin B_1 and aflatoxin M_1 on the in vivo covalent binding of aflatoxin B_1 to hepatic macromolecules. *J. Toxicol. environ. Health*, **13**, 575-587

Loveland, P.M., Wilcox, J.S., Hendricks, J.D. & Bailey, G.S. (1988) Comparative metabolism and DNA binding of aflatoxin B_1, aflatoxin M_1, aflatoxicol and aflatoxicol-M_1 in hepatocytes from rainbow trout (*Salmo gairdneri*). *Carcinogenesis*, **9**, 441-446

Lüthy, J., Zweifel, U. & Schlatter, C. (1980) Metabolism and tissue distribution of [^{14}C]-aflatoxin B_1 in pigs. *Food Cosmet. Toxicol.*, **18**, 253-256

Madhavan, T.V. & Gopalan, C. (1965) Effect of dietary protein on aflatoxin liver injury in weanling rats. *Arch. Pathol.*, **80**, 123–126

Madhavan, T.V. & Gopalan, C. (1968) The effect of dietary protein on carcinogenesis of aflatoxin. *Arch. Pathol.*, **85**, 133–137

Madle, E., Korte, A. & Beek, B. (1986) Species differences in mutagenicity testing: 1. Micronucleus and SCE tests in rats, mice and Chinese hamsters with aflatoxin B_1. *Teratog. Carcinog. Mutag.*, **6**, 1–13

Mainigi, K.D. & Sorof, S. (1977) Carcinogen–protein complexes in liver during hepatocarcinogenesis by aflatoxin B_1. *Cancer Res.*, **37**, 4304–4312

Malaveille, C., Kuroki, T., Brun, G., Hautefeuille, A., Camus, A.-M. & Bartsch, H. (1979) Some factors determining the concentration of liver proteins for optimal mutagenicity of chemicals in the *Salmonella*/microsome assay. *Mutat. Res.*, **63**, 245–258

Malaveille, C., Brun, G. & Bartsch, H. (1983) Studies on the efficiency of the *Salmonella*/rat hepatocyte assay for the detection of carcinogens as mutagens: activation of 1,2-dimethylhydrazine and procarbazine into bacterial mutagens. *Carcinogenesis*, **4**, 449–455

Mamber, S.W., Bryson, V. & Katz, S.E. (1983) The *Escherichia coli* WP2/WP100 rec assay for detection of potential chemical carcinogens. *Mutat. Res.*, **119**, 135–144

Mamber, S.W., Bryson, V. & Katz, S.E. (1984) Evaluation of the *Escherichia coli* K12 inductest for detection of potential chemical carcinogens. *Mutat. Res.*, **130**, 141–151

Mandel, H.G., Manson, M.M., Judah, D.J., Simpson, J.L., Green, J.A., Forrester, L.M., Wolf, C.R. & Neal, G.E. (1987) Metabolic basis for the protective effect of the antioxidant ethoxyquin on aflatoxin B_1 hepatocarcinogenesis in the rat. *Cancer Res.*, **47**, 5218–5223

Manson, M.M. & Neal, G.E. (1987) The effect of feeding aflatoxin-contaminated groundnut meal, with or without ammoniation, on the development of gamma-glutamyl transferase positive lesions in the livers of Fischer 344 rats. *Food Addit. Contam.*, **4**, 141–147

Manson, M.M., Green, J.A. & Driver, H.E. (1987) Ethoxyquin alone induces preneoplastic changes in rat kidney whilst preventing induction of such lesions in liver by aflatoxin B_1. *Carcinogenesis*, **8**, 723–728

Mariën, K., Moyer, R., Loveland, P., Van Holde, K. & Bailey, G. (1987) Comparative binding and sequence interaction specificities of aflatoxin B_1, aflatoxicol, aflatoxin M_1, and aflatoxicol M_1 with purified DNA. *J. biol. Chem.*, **262**, 7455–7462

Marinovich, M. & Lutz, W.K. (1985) Covalent binding of aflatoxin B_1 to liver DNA in rats pretreated with ethanol. *Experientia*, **41**, 1338–1340

Martin, C.N., McDermid, A.C. & Garner, R.C. (1977) Measurement of 'unscheduled' DNA synthesis in HeLa cells by liquid scintillation counting after carcinogen treatment. *Cancer Lett.*, **2**, 355–360

Masri, M.S., Lundin, R.E., Page, J.R. & Garcia, V.C. (1967) Crystalline aflatoxin M_1 from urine and milk. *Nature*, **215**, 753–755

Matsushima, T., Sugimura, T., Nagao, M., Yahagi, T., Shirai, A. & Sawamura, M. (1980) Factors modulating mutagenicity in microbial tests. In: Norpoth, K.H. & Garner, R.C., eds, *Short-term Test Systems for Detecting Carcinogens*, Berlin, Springer, pp. 273–285

Matzinger, P.K. & Ong, T.-M. (1976) Mutation induction in rodent liver microsomal metabolites of aflatoxins B_1 and G_1 in *Neurospora crassa*. *Mutat. Res.*, **37**, 27–32

Maxwell, S.M., Apeagyei, F., de Vries, H.R., Mwanmut, D.D. & Hendrickse, R.G. (1989) Aflatoxins in breast milk, neonatal cord blood and sera of pregnant women. *J. Toxicol.-Toxin Rev.*, **8**, 19–29

McCann, J., Choi, E., Yamasaki, E. & Ames, B.N. (1975) Detection of carcinogens as mutagens in the *Salmonella*/microsome test: assay of 300 chemicals. *Proc. natl Acad. Sci. USA*, **72**, 5135–5139

McCarroll, N.E., Keech, B.H. & Piper, C.E. (1981) A microsuspension adaptation of the *Bacillus subtilis* 'rec' assay. *Environ. Mutag.*, **3**, 607–616

McLean, A.E.M. & Marshall, A. (1971) Reduced carcinogenic effects of aflatoxin in rats given phenobarbitone. *Br. J. exp. Pathol.*, **52**, 322–329

McMahon, G., Davis, E.F., Huber, L.J., Kim, Y. & Wogan, G.N. (1990) Characterization of c-Ki-*ras* and N-*ras* oncogenes in aflatoxin B_1-induced rat liver tumors. *Proc. natl Acad. Sci. USA*, **87**, 1104–1108

McQueen, C.A. & Way, B.M. (1991) Sex and strain differences in the hepatocyte primary culture/DNA repair test. *Environ. mol. Mutag.*, **18**, 107–112

McQueen, C.A., Kreiser, D.M. & Williams, G.M. (1983) The hepatocyte primary culture/DNA repair assay using mouse or hamster hepatocytes. *Environ. Mutag.*, **5**, 1–8

Metcalfe, S.A., Colley, P.J. & Neal, G.E. (1981) A comparison of the effects of pretreatment with phenobarbitone and 3-methylcholanthrene on the metabolism of aflatoxin B_1 by rat liver microsomes and isolated hepatocytes *in vitro*. *Chem.-biol. Interactions*, **35**, 145–157

Michalopoulos, G., Sattler, G.L., O'Connor, L. & Pitot, H.C. (1978) Unscheduled DNA synthesis induced by procarcinogens in suspensions and primary cultures of hepatocytes on collagen membranes. *Cancer Res.*, **38**, 1866–1871

Mirsalis, J.C., Tyson, C.K. & Butterworth, B.E. (1982) Detection of genotoxic carcinogens in the in vivo–in vitro hepatocyte DNA repair assay. *Environ. Mutag.*, **4**, 553–562

Misra, R.P., Muench, K.F. & Humayun, M.Z. (1983) Covalent and noncovalent interactions of aflatoxin with defined deoxyribonucleic acid sequences. *Biochemistry*, **22**, 3351–3359

Misslbeck, N.G., Campbell, T.C. & Roe, D.A. (1984) Effect of ethanol consumed in combination with high or low fat diets on the postinitiation phase of hepatocarcinogenesis in the rat. *J. Nutr.*, **114**, 2311–2323

Mitchell, A.D. (1976) *Potential Prescreens for Chemical Carcinogens: Unscheduled DNA Synthesis. Task 2* (Stanford Research Institute Final Report under Contract No. NO1/CP-33394), Bethesda, MD, National Cancer Institute

Monroe, D.H. & Eaton, D.L (1987) Comparative effects of butylated hydroxyanisole on hepatic in vivo DNA binding and in vitro biotransformation of aflatoxin B_1 in the rat and mouse. *Toxicol. appl. Pharmacol.*, **90**, 401–409

Monroe, D.H. & Eaton, D.L. (1988) Effects of modulation of hepatic glutathione on biotransformation and covalent binding of aflatoxin B_1 to DNA in the mouse. *Toxicol. appl. Pharmacol.*, **94**, 118–127

Monroe, D.H., Holeski, C.J. & Eaton, D.L. (1986) Effects of single-dose and repeated-dose pretreatment with 2(3)-*tert*-butyl-4-hydroxyanisole (BHA) on the hepatobiliary disposition and covalent dinding to DNA of aflatoxin B_1 in the rat. *Food chem. Toxicol.*, **24**, 1273–1281

Moore, M.R., Pitot, H.C., Miller, E.C. & Miller, J.A. (1982) Cholangiocellular carcinomas induced in Syrian golden hamsters administered aflatoxin B_1 in large doses. *J. natl Cancer Inst.*, **68**, 271–278

Moss, E.J. & Neal, G.E. (1985) The metabolism of aflatoxin B_1 by human liver. *Biochem. Pharmacol.*, **34**, 3193–3197

Moss, E.J., Neal, G.E. & Judah, D.J. (1985) The mercapturic acid pathway metabolites of a glutathione conjugate of aflatoxin B_1. *Chem.-biol. Interactions*, **55**, 139–155

Müller, H.-M. (1982) Decontamination of mycotoxins. I. Physical methods (Ger.). *Übersicht. Tierernahr*, **10**, 95–122

Müller, H.-M. (1983) Decontamination of mycotoxins. II. Chemical methods and reaction with components of feedstuffs (Ger.). *Übersicht. Tierernach*, **11**, 47–80

Murakami, Y., Hayashi, K., Hirohashi, S. & Sekiya, T. (1991) Aberrations of the tumor suppressor p53 and retinoblastoma genes in human hepatocellular carcinomas. *Cancer Res.*, **51**, 5520–5525

Nabney, J., Burbage, M.B., Allcroft, R. & Lewis, G. (1967) Metabolism of aflatoxin in sheep: excretion pattern in the lactating ewe. *Food Cosmet. Toxicol.*, **5**, 11–17

Nakanishi, Y. & Schneider, E.L. (1979) In vivo sister-chromatid exchange: a sensitive measure of DNA damage. *Mutat. Res.*, **60**, 329–337

Nakatsuru, Y., Qin, X., Masahito, P. & Ishikawa, T. (1989) Immunological detection of in vivo aflatoxin B_1–DNA adduct formation in rats, rainbow trout and coho salmon. *Carcinogenesis*, **11**, 1523–1526

Neal, G.E. & Cabral, J.R.P. (1980) Effect of partial hepatectomy on the response of rat liver to aflatoxin B_1. *Cancer Res.*, **40**, 4739–4743

Neal, G.E. & Colley, P.J. (1979) The formation of 2,3-dihydro-2,3-dihydroxy aflatoxin B_1 by the metabolism of aflatoxin B_1 *in vitro* by rat liver microsomes. *FEBS Lett.*, **101**, 382–386

Neal, G.E., Judah, D.J., Stirpe, F. & Patterson, D.S.P. (1981a) The formation of 2,3-dihydroxy-2,3-dihydro-aflatoxin B_1 by the metabolism of aflatoxin B_1 by liver microsomes isolated from certain avian and mammalian species and the possible role of this metabolite in the acute toxicity of aflatoxin B_1. *Toxicol. appl. Pharmacol.*, **58**, 431–437

Neal, G.E., Metcalfe, S.A., Legg, R.F., Judah, D.J. & Green, J.A. (1981b) Mechanism of the resistance to cytotoxicity which precedes aflatoxin B_1 hepatocarcinogenesis. *Carcinogenesis*, **2**, 457–461

Neal, G.E., Nielsch, U., Judah, D.J. & Hulbert, P.B. (1987) Conjugation of model substrates of microsomally-activated aflatoxin B_1 with reduced glutathione, catalysed by cytosolic glutathione-S-transferases in livers of rats, mice and guinea pigs. *Biochem. Pharmacol.*, **36**, 4269–4276

Nesnow, S., Garland, H. & Curtis, G. (1982) Improved transformation of C3H10T½CL8 cells by direct- and indirect-acting carcinogens. *Carcinogenesis*, **3**, 377–380

Newberne, P.M. (1965) Carcinogenicity of aflatoxin-contaminated peanut meals. In: Wogan, G.N., ed., *Mycotoxins in Foodstuffs*, Cambridge, MA, MIT Press, pp. 187–208

Newberne, P.M. & Rogers, A.E. (1973) Rat colon carcinomas associated with aflatoxin and marginal vitamin A. *J. natl Cancer Inst.*, **50**, 439–448

Newberne, P.M. & Williams, G. (1969) Inhibition of aflatoxin carcinogenesis by diethylstilbestrol in male rats. *Arch. environ. Health*, **19**, 489–498

Newberne, P.M., Wogan, G.N., Carlton, W.W. & Abdel-Kader, M.M. (1964) Histopathologic lesions in ducklings caused by *Aspergillus flavus* cultures, culture extracts, and crystalline aflatoxins. *Toxicol. appl. Pharmacol.*, **6**, 542–556

Newberne, P.M., Harrington, D.H. & Wogan, G.N. (1966) Effects of cirrhosis and other liver insults on induction of liver tumors by aflatoxin in rats. *Lab. Invest.*, **15**, 962–969

Newberne, P.M., Weigert, J. & Kula, N. (1979) Effects of dietary fat on hepatic mixed-function oxidases and hepatocellular carcinoma induced by aflatoxin B_1 in rats. *Cancer Res.*, **39**, 3986–3991

Ngindu, A., Johnson, B.K., Kenya, P.R., Ngira, J.A., Ocheng, D.M., Nandwa, H., Omondi, T.N., Jansen, A.J., Ngare, W., Kaviti, J.N., Gatei, D. & Siongok, T.A. (1982) Outbreak of acute hepatitis caused by aflatoxin poisoning in Kenya. *Lancet*, **i**, 1346–1348

van Nieuwenhuize, J.P., Herber, R.F.M., de Bruin, A., Meyer, P.B. & Duba, W.C. (1973) Aflatoxins. An epidemiological study of the carcinogenicity of long-term low-level exposure of a factory population (Dutch). *T. soc. Geneesk.*, **51**, 754–760

Niranjan, B.G., Wilson, N.M., Jefcoate, C.R. & Avadhani, N.G. (1984) Hepatic mitochondrial cytochrome P450 system. Distinctive features of cytochrome P450 involved in the activation of aflatoxin B_1 and benzo(a)pyrene. *J. biol. Chem.*, **259**, 12495–12501

Nishioka, H., Nishi, K. & Kyokane, K. (1981) Human saliva inactivates mutagenicity of carcinogens. *Mutat. Res.*, **85**, 323–333

Nishizumi, M., Albert, R.E., Burns, F.J. & Bilger, L. (1977) Hepatic cell loss and proliferation induced by N-2-fluorenylacetamide, diethylnitrosamine, and aflatoxin B_1 in relation to hepatoma induction. *Br. J. Cancer*, **36**, 192–197

Nix, C.E., Brewen, B. & Epler, J.L. (1981) Microsomal activation of selected polycyclic aromatic hydrocarbons and aromatic amines in *Drosophila melanogaster*. *Mutat. Res.*, **88**, 291–299

Nixon, J.E., Sinnhuber, R.O., Lee, D.J., Landers, M.K. & Harr, J.R. (1974) Effect of cyclopropenoid compounds on the carcinogenic activity of diethylnitrosamine and aflatoxin B_1 in rats. *J. natl Cancer Inst.*, **53**, 453–458

Nixon, J.E., Hendricks, J.D., Pawlowski, N.E., Loveland, P.M. & Sinnhuber R.O. (1981) Carcinogenicity of aflatoxicol in Fischer 344 rats. *J. natl Cancer Inst.*, **66**, 1159–1163

Nixon, J.E., Hendricks, J.D., Pawlowski, N.E., Pereira, C.B., Sinnhuber, R.O. & Bailey, G.S. (1984) Inhibition of aflatoxin B_1 carcinogenesis in rainbow trout by flavone and indole compounds. *Carcinogenesis*, **5**, 615–619

Norpoth, K., Grossmeier, R., Bösenberg, H., Themann, H. & Fleischer, M. (1979) Mutagenicity of aflatoxin B_1, activated by S-9 fractions of human, rat, mouse, rabbit, and monkey liver, towards *S. typhimurium* TA 98. *Int. Arch. occup. environ. Health*, **42**, 333–339

Norred, W.P. & Morrissey, R.E. (1983) Effects of long-term feeding of ammoniated, aflatoxin-contaminated corn to Fischer 344 rats. *Toxicol. appl. Pharmacol.*, **70**, 96–104

Núñez, O., Hendricks, J.D., Arbogast, D.N., Fong, A.T., Lee, B.C. & Bailey, G.S. (1989) Promotion of aflatoxin B_1 hepatocarcinogenesis in rainbow trout by 17β-estradiol. *Aquat. Toxicol.*, **15**, 289–302

Nyathi, C.B., Mutiro, C.F., Hasler, J.A. & Chetsanga, C.J. (1987) A survey of urinary aflatoxin in Zimbabwe. *Int. J. Epidemiol.*, **16**, 516–519

O'Brien, K., Moss, E., Judah, D. & Neal, G. (1983) Metabolic basis of the species difference to aflatoxin B_1 induced hepatotoxicity. *Biochem. biophys. Res. Comm.*, **114**, 813–821

Olsen, J.H., Dragsted, L. & Autrup, H. (1988) Cancer risk and occupational exposure to aflatoxins in Denmark. *Br. J. Cancer*, **58**, 392–396

Ong, T.-M. (1970) Mutagenicity of aflatoxins in *Neurospora crassa*. *Mutat. Res.*, **9**, 615–618

Ong, T.-M. (1971) Mutagenic activities of aflatoxin B_1 and G_1 in *Neurospora crassa*. *Mol. gen. Genet.*, **111**, 159–170

Ong, T.-M. & de Serres, F.J. (1972) Mutagenicity of chemical carcinogens in *Neurospora crassa*. *Cancer Res.*, **32**, 1890–1893

Onyemelukwe, G.C., Ogbadu, G.H. & Salifu, A. (1982) Aflatoxins, B_1, B_2, G_1, G_2 in primary liver cell carcinoma. *Toxicol. Lett.*, **10**, 309–312

Oshiro, Y. & Balwierz, P.S. (1982) Morphological transformation of C3H/10T½ CL8 cells by procarcinogens. *Environ. Mutagen.*, **4**, 105–108

Ozturk, M. & 28 others (1991) p53 Mutation in hepatocellular carcinoma after aflatoxin exposure. *Lancet*, **338**, 1356–1359

Park, H.K. (1989) Effect of *Clonorchis sinensis* infection on the histopathology of the liver in rats administered aflatoxin B_1. *Jpn. J. Parasitol.*, **38**, 198–206

Parkin, D.M., Srivatanakul, P., Khlat, M., Chenvidhya, D., Chotiwan, P., Insiripong, S., L'Abbé, K.A. & Wild, C.P. (1991) Liver cancer in Thailand. I. A case–control study of cholangiocarcinoma. *Int. J. Cancer*, **48**, 323–328

Peers, F.G. & Linsell, C.A. (1973) Dietary aflatoxins and liver cancer — a population based study in Kenya. *Br. J. Cancer*, **27**, 473–484

Peers, F.G., Gilman, G.A. & Linsell, C.A. (1976) Dietary aflatoxins and human liver cancer. A study in Swaziland. *Int. J. Cancer*, **17**, 167–176

Peers, F.G., Bosch, X., Kaldor, J., Linsell, C.A. & Pluijmen, M. (1987) Aflatoxin exposure, hepatitis B virus infection and liver cancer in Swaziland. *Int. J. Cancer*, **39**, 545–553

Pestka, J.J., Li, Y.K. & Chu, F.S. (1982) Reactivity of aflatoxin B_2 antibody with aflatoxin B_1-modified DNA and related metabolites. *Appl. environ. Microbiol.*, **44**, 1159–1165

Petr, T., Bárta, I. & Turek, B. (1990) In vivo effect of selenium on the mutagenic activity of aflatoxin B_1. *J. Hyg. Epidemiol. Microbiol. Immunol.*, **34**, 123–128

Petr, T., Bárta, I., Adámková, M., Hrabal, P. & Bártová, J. (1991) The effect of partial hepatectomy on the genotoxicity of aflatoxin B_1. *Neoplasma*, **38**, 77–83

Petzold, G.L. & Swenberg, J.A. (1978) Detection of DNA damage induced *in vivo* following exposure of rats to carcinogens. *Cancer Res.*, **38**, 1589–1594

Pienta, R.J., Poiley, J.A. & Lebherz, W.B., III (1977) Morphological transformation of early passage golden Syrian hamster embryo cells derived from cryopreserved primary cultures as a reliable in vitro bioassay for identifying diverse carcinogens. *Int. J. Cancer*, **19**, 642–655

Pier, A.C. & McLoughlin, M.E. (1985) Mycotoxic suppression of immunity. In: Lacey, J., ed., *Trichothecenes and Other Mycotoxins*, New York, John Wiley & Sons, pp. 507–519

Pitt, J.I., Hocking, A.D., Bhudasamai, K., Miscamble, B.F., Wheeler, K.A. & Tanbook-Ek, P. (1993) The normal mycoflora of commodities from Thailand. I. Nuts and oilseeds. *Int. J. Food Microbiol.* (in press)

Poiley, J.A., Raineri, R., Cavanaugh, D.M., Ernst, M.K. & Pienta, R.J. (1980) Correlation between transformation potential and inducible enzyme levels of hamster embryo cells. *Carcinogenesis*, **1**, 323–328

Pritchard, D.J. & Butler, W.H. (1988) The ultrastructural features of aflatoxin B_1-induced lesions in the rat liver. *Br. J. exp. Pathol.*, **69**, 793–804

Probst, G.S., McMahon, R.E., Hill, L.E., Thompson, C.Z., Epp, J.K. & Neal, S.B. (1981) Chemically-induced unscheduled DNA synthesis in primary rat hepatocyte cultures: a comparison with bacterial mutagenicity using 218 compounds. *Environ. Mutagen.*, **3**, 11–32

Puisieux, A., Lim, S., Groopman, J. & Ozturk, M. (1991) Selective targeting of p53 gene mutational hotspots in human cancers by etiologically defined carcinogens. *Cancer Res.*, **51**, 6185–6189

Qian, G.-S., Yu, M.C., Ross, R.K., Yuan, J.-M., Gao, Y.-T., Henderson, B.E., Wogan, G.N. & Groopman, J.D. (1993) Aflatoxin exposure and hepatocellular carcinoma in Shanghai, People's Republic of China. *Cancer Epidemiol. Biomarkers Prev.* (in press)

Qin, S. & Huang, C.C. (1986) Influence of mouse liver stored vitamin A on the induction of mutations (Ames tests) and SCE of bone marrow cells by aflatoxin B_1, benzo(*a*)pyrene, or cyclophosphamide. *Environ. Mutag.*, **8**, 839–847

Raj, H.G. & Lotlikar, P.D. (1984) Urinary excretion of thiol conjugates of aflatoxin B_1 in rats and hamsters. *Cancer Lett.*, **22**, 125–133

Ramsdell, H.S. & Eaton, D.L. (1988) Modification of aflatoxin B_1 biotransformation *in vitro* and DNA binding *in vivo* by dietary broccoli in rats. *J. Toxicol. environ. Health*, **25**, 269–284

Ramsdell, H.S. & Eaton, D.L. (1990) Mouse liver glutathione *S*-transferase isoenzyme activity towards aflatoxin B_1-8,9-epoxide and benzo[*a*]pyrene-7,8-dihydrodiol-9,10-epoxide. *Toxicol. appl. Pharmacol.*, **105**, 216–225

Ray, P.K., Singh, K.P., Raisuddin, R. & Prasad, A.K. (1991) Immunological responses to aflatoxins and other chemical carcinogens. *J. Toxicol.-Toxin Rev.*, **10**, 63–85

Ray-Chaudhuri, R., Kelley, S. & Iype, P.T. (1980) Induction of sister chromatid exchanges by carcinogens mediated through cultured rat liver epithelial cells. *Carcinogenesis*, 1, 779–786

Reddy, R.V. & Sharma, R.P. (1989) Effects of aflatoxin B_1 on murine lymphocytic functions. *Toxicology*, 54, 31–44

Reddy, J.K., Svoboda, D.J. & Rao, M.S. (1976) Induction of liver tumors by aflatoxin B_1 in the tree shrew (*Tupaia glis*), a nonhuman primate. *Cancer Res.*, 36, 151–160

Reddy, R.V., Sharma, R.P. & Taylor, M.J. (1983) Dose and time related response of immunologic functions to aflatoxin B_1 in mice. In: Hayes, A.W., Schnell, R.C. & Miya, T.S., eds, *Developments in the Science and Practice of Toxicology*, Amsterdam, Elsevier, pp. 431–434

Reddy, R.V., Taylor, M.J. & Sharma, R.P. (1987) Studies of immune function of CD-1 mice exposed to aflatoxin B_1. *Toxicology*, 43, 123–132

van Rensburg, S.J., van der Watt, J.J., Purchase, I.F.H., Coutinho, L.P. & Markham, R. (1974) Primary liver cancer rate and aflatoxin intake in a high cancer area. *S. Afr. med. J.*, 48, 2508a–2508d

van Rensburg, S.J., Cook-Mozaffari, P., van Schalkwyk, D.J., van der Watt, J.J., Vincent, T.J. & Purchase, I.F. (1985) Hepatocellular carcinoma and dietary aflatoxin in Mozambique and Transkei. *Br. J. Cancer*, 51, 713–726

van Rensburg, S.J., van Schalkwyk, G.C. & van Schalkwyk, D.J. (1990) Primary liver cancer and aflatoxin intake in Transkei. *J. environ. Pathol. Toxicol. Oncol.*, 10, 11–16

Richard, J.L. (1991) Mycotoxins as immunomodulators in animal systems. In: Bray, G.A. & Ryan, D., eds, *Mycotoxins, Cancer and Health*, Baton Rouge, LA, Louisiana State University Press, pp. 197–220

Riley, R.T., Kemppainen, B.W. & Norred, W.P. (1985) Penetration of aflatoxins through isolated human epidermis. *J. Toxicol. environ. Health*, 15, 769–777

Rizvi, T.A., Mathur, M. & Nayak, N.C. (1987) Effect of protein calorie malnutrition and cell replication on aflatoxin B_1-induced hepatocarcinogenesis in rats. *J. natl Cancer Inst.*, 79, 817–830

Rizvi, T.A., Mathur, M. & Nayak, N.C. (1989) Enhancement of aflatoxin B1-induced hepatocarcinogenesis in rats by partial hepatectomy. *Virchows Arch. B Cell Pathol.*, 56, 345–350

Robertson, I.G.C., Philpot, R.M., Zeiger, E. & Wolf, C.R. (1981) Specificity of rabbit pulmonary cytochrome P-450 isozymes in the activation of several aromatic amines and aflatoxin B_1. *Mol. Pharmacol.*, 20, 662–668

Robertson, I.G.C., Zeiger, E. & Goldstein, J.A. (1983) Specificity of rat liver cytochrome P-450 isozymes in the mutagenic activation of benzo[*a*]pyrene, aromatic amines and aflatoxin B_1. *Carcinogenesis*, 4, 93–96

Roebuck, B.D., Liu, Y.-L., Rogers, A.E., Groopman, J.D. & Kensler, T.W. (1991) Protection against aflatoxin B_1-induced hepatocarcinogenesis in F344 rats by 5-(2-pyrazinyl)-4-methyl-1,2-dithiole-3-thione (oltipraz): predictive role for short-term molecular dosimetry. *Cancer Res.*, 51, 5501–5506

Rogers, A.E. & Newberne, P.M. (1971) Nutrition and aflatoxin carcinogenesis. *Nature*, 229, 62–63

Rogers, A.E., Lenhart, G. & Morrison, G. (1980) Influence of dietary lipotrope and lipid content on aflatoxin B_1, N-2-fluorenylacetamide, and 1,2-dimethylhydrazine carcinogenesis in rats. *Cancer Res.*, 40, 2802–2807

Rohrig, T.P. & Yourtee, D.M. (1983) In vitro metabolism of aflatoxin Q_1 by rat liver post-mitochondrial homogenates. *Res. Commun. chem. Pathol. Pharmacol.*, 40, 457–464

Roll, R., Matthiaschk, G. & Korte, A. (1990) Embryotoxicity and mutagenicity of mycotoxins. *J. environ. Pathol. Toxicol. Oncol.*, 10, 1–7

Rosenkranz, H.S. & Leifer, Z. (1980) Determining the DNA-modifying activity of chemicals using DNA-polymerase-deficient *Escherichia coli*. In: de Serres, F.J. & Hollaender, A., eds, *Chemical Mutagens: Principles and Methods for Their Detection*, Vol. 6, New York, Plenum, pp. 109–147

Rosenkranz, H.S. & Poirier, L.A. (1979) Evaluation of the mutagenicity and DNA-modifying activity of carcinogens and noncarcinogens in microbial systems. *J. natl Cancer Inst.*, **62**, 873–892

Ross, R.K., Yuan, J.-M., Yu, M.C., Wogan, G.N., Qian, G.-S., Tu, J.-T., Groopman, J.D., Gao, Y.-T. & Henderson, B.E. (1992) Urinary aflatoxin biomarkers and risk of hepatocellular carcinoma. *Lancet*, **339**, 943–946

Sabbioni, G. & Wild, C.P. (1991) Identification of an aflatoxin G_1-serum albumin adduct and its relevance to the measurement of human exposure to aflatoxins. *Carcinogenesis*, **12**, 97–103

Sabbioni, G., Skipper, P.L., Büchi, G. & Tannenbaum, S.R. (1987) Isolation and characterisation of the major serum albumin adduct formed by aflatoxin B_1 *in vivo* in rats. *Carcinogenesis*, **8**, 819–824

Salbe, A.D. & Bjeldanes, L.F. (1989) Effect of diet and route of administration on the DNA binding of aflatoxin B_1 in the rat. *Carcinogenesis*, **10**, 629–634

San, R.H.C. & Chan, R.I.M. (1987) Inhibitory effect of phenolic compounds on aflatoxin B_1 metabolism and induced mutagenesis. *Mutat. Res.*, **177**, 229–239

San, R.H.C. & Stich, H.F. (1975) DNA repair synthesis of cultured human cells as a rapid bioassay for chemical carcinogens. *Int. J. Cancer*, **16**, 284–291

Sarasin, A., Goze, A., Devoret, R. & Moulé, Y. (1977) Induced reactivation of UV-damaged phage lambda in *E. coli* K12 host cells treated with aflatoxin B_1 metabolites. *Mutat. Res.*, **42**, 205–214

Sargeant, K., O'Kelly, J., Carnaghan, R.B.A. & Allcroft, R. (1961) The assay of a toxic principle in certain groundnut meals. *Vet. Rec.*, **73**, 1219–1222

Sato, S., Matsushima, T., Tanaka, N., Sugimura, T. & Takashima, F. (1973) Hepatic tumors in the guppy (*Lebistes reticulatus*) induced by aflatoxin B_1, dimethylnitrosamine, and 2-acetylaminofluorene. *J. natl Cancer Inst.*, **50**, 767–778

Sawada, M., Kitamura, R. & Kamataki, T. (1992) Stable expression of monkey cytochrome P-450IA1 cDNA in Chinese hamster CHL cells and its application for detection of mutagenicity of aflatoxin B_1. *Mutat. Res.*, **265**, 23–29

Scarpelli, D.G. (1976) Drug metabolism and aflatoxin-induced hepatoma in rainbow trout (*Salmo gairdneri*). *Prog. exp. Tumor Res.*, **20**, 339–350

Schoenhard, G.L., Hendricks, J.D., Nixon, J.E., Lee, D.J., Wales, J.H., Sinnhuber, R.O. & Pawlowski, N.E. (1981) Aflatoxicol-induced hepatocellular carcinoma in rainbow trout (*Salmo gairdneri*) and the synergistic effects of cyclopropenoid fatty acids. *Cancer Res.*, **41**, 1011–1014

Schrager, T.F., Newberne, P.M., Pikul, A.H. & Groopman, J.D. (1990) Aflatoxin–DNA adduct formation in chronically dosed rats fed a choline-deficient diet. *Carcinogenesis*, **11**, 177–180

Schulsinger, D.A., Root, M.M. & Campbell, T.C. (1989) Effect of dietary protein quality on development of aflatoxin B_1-induced hepatic preneoplastic lesions. *J. natl Cancer Inst.*, **81**, 1241–1245

Scorsone, K.A., Zhou, Y.-Z., Butel, J.S. & Slagle, B.L. (1992) p53 Mutations cluster at codon 249 in hepatitis B virus-positive hepatocellular carcinomas from China. *Cancer Res.*, **52**, 1635–1638

Scott, P.M. (1989) Methods for determination of aflatoxin M_1 in milk and milk products—a review of performance characteristics. *Food Addit. Contam.*, **6**, 283–305

Scott, P.M. (1990) Natural poisons. In: Helrich, K., ed., *Official Methods of Analysis of the Association of Official Analytical Chemists*, 15th ed., Arlington, VA, Association of Official Analytical Chemists, pp. 1184–1213

Sell, S., Hunt, J.M., Dunsford, H.A. & Chisari, F.V. (1991) Synergy between hepatitis B virus expression and chemical hepatocarcinogens in transgenic mice. *Cancer Res.*, **51**, 1278–1285

Serck-Hanssen, A. (1970) Aflatoxin-induced fatal hepatitis? A case report from Uganda. *Arch. environ. Health*, **20**, 729–731

Shank, R.C., Bourgeois, C.H., Keschamras, N. & Chandavimol, P. (1971) Aflatoxins in autopsy specimens from Thai children with an acute disease of unknown aetiology. *Food Cosmet. Toxicol.*, **9**, 501–507

Shank, R.C., Gordon, J.E., Wogan, G.N., Nondasuta, A. & Subhamani, B. (1972a) Dietary aflatoxins and human liver cancer. III. Field survey of rural Thai families for ingested aflatoxins. *Food Cosmet. Toxicol.*, **10**, 71–84

Shank, R.C., Bhamarapravati, N., Gordon, J.E. & Wogan, G.N. (1972b) Dietary aflatoxins and human liver cancer. IV. Incidence of primary liver cancer in two municipal populations of Thailand. *Food Cosmet. Toxicol.*, **10**, 171–179

Shank, R.C., Wogan, G.N., Gibson, J.B. & Nondasuta, A. (1972c) Dietary aflatoxins and human liver cancer. II. Aflatoxins in market foods and foodstuffs of Thailand and Hong Kong. *Food Cosmet. Toxicol.*, **10**, 61–69

Sharma, A. & Sahai, R. (1987) Teratological effects of aflatoxin on rats (*Rattus norvegicus*). *Indian J. Anim. Res.*, **21**, 35–40

Sharma, A., Sahai, R., Sikka, A.K. & Sarma, H.K. (1988) Induction of dominant lethals in rat (*Rattus norvegicus*) by aflatoxins. *Indian J. anim. Res.*, **22**, 13–19

Shaulsky, G., Johnson, R.L., Shockcor, J.P., Taylor, L.C.E. & Stark, A.-A. (1990) Properties of aflatoxin–DNA adducts formed by photoactivation and characterisation of the major photoadduct as aflatoxin–N7-guanine. *Carcinogenesis*, **11**, 519–527

Shelton, D.W., Hendricks, J.D., Coulombe, R.A. & Bailey, G.S. (1984) Effect of dose on the inhibition of carcinogenesis/mutagenesis by Aroclor 1254 in rainbow trout fed aflatoxin B_1. *J. Toxicol. environ. Health*, **13**, 649–657

Shimada, T. & Guengerich, F.P. (1989) Evidence for cytochrome P-450NF, the nifedipine oxidase, being the principal enzyme involved in the bioactivation of aflatoxins in human liver. *Proc. natl Acad. Sci. USA*, **86**, 462–465

Shimada, T., Furukawa, K., Kreiser, D.M., Cawein, A. & Williams, G.M. (1983) Induction of transformation by six classes of chemical carcinogens in rat liver epithelial cells. *Cancer Res.*, **43**, 5087–5092

Shimada, T., Nakamura, S.-I., Imaoka, S. & Funae, Y. (1987) Genotoxic and mutagenic activation of aflatoxin B_1 by constitutive forms of cytochrome P-450 in rat liver microsomes. *Toxicol. appl. Pharmacol.*, **91**, 13–21

Shimada, T., Iwasaki, M., Martin, M.V. & Guengerich, F.P. (1989) Human liver microsomal cytochrome P-450 enzymes involved in the bioactivation of procarcinogens detected by *umu* gene response in *Salmonella typhimurium* TA1535/pSK1002. *Cancer Res.*, **49**, 3218–3228

Sieber, S.M., Correa, P., Dalgard, D.W. & Adamson, R.H. (1979) Induction of osteogenic sarcomas and tumors of the hepatobiliary system in nonhuman primates with aflatoxin B_1. *Cancer Res.*, **39**, 4545–4554

Silas, J.C., Harrison, M.A., Carpenter, J.A. & Roth, I.L. (1987) Airborne aflatoxin in corn processing facilities in Georgia. *Am. ind. Hyg. Assoc. J.*, **48**, 198–201

Simmon, V.F. (1979a) In vitro assays for recombinogenic activity of chemical carcinogens and related compounds with *Saccharomyces cerevisiae* D3. *J. natl Cancer Inst.*, **62**, 901–909

Simmon, V.F. (1979b) In vitro mutagenicity assays of chemical carcinogens and related compounds with *Salmonella typhimurium*. *J. natl Cancer Inst.*, **62**, 893–899

Simmon, V.F., Rosenkranz, H.S., Zeiger, E. & Poirier, L.A. (1979) Mutagenic activity of chemical carcinogens and related compounds in the intraperitoneal host-mediated assay. *J. natl Cancer Inst.*, **62**, 911–918

Sina, J.F., Bean, C.L., Dysart, G.R., Taylor, V.I. & Bradley, M.O. (1983) Evaluation of the alkaline elution/rat hepatocyte assay as a predictor of carcinogenic/mutagenic potential. *Mutat. Res.*, **113**, 357–391

Sinha, S., Webber, C., Marshall, C.J., Knowles, M.A., Proctor, A., Barrass, N.C. & Neal, G.E. (1988) Activation of *ras* oncogene in aflatoxin-induced rat liver carcinogenesis. *Proc. natl Acad. Sci. USA*, **85**, 3673–3677

Sinnhuber, R.O., Wales, J.H., Ayres J.L., Engebrecht, R.H. & Amend, D.L. (1968a) Dietary factors and hepatoma in rainbow trout (*Salmo gairdneri*). I. Aflatoxins in vegetable protein feedstuffs. *J. natl Cancer Inst.*, **41**, 711–718

Sinnhuber, R.O., Lee, D.J., Wales, J.H. & Ayres, J.L. (1968b) Dietary factors and hepatoma in rainbow trout (*Salmo gairdneri*). II. Cocarcinogenesis by cyclopropenoid fatty acids and the effect of gossypol and altered lipids on aflatoxin-induced liver cancer. *J. natl Cancer Inst.*, **41**, 1293–1301

Sinnhuber, R.O., Lee, D.J., Wales, J.H., Landers, M.K. & Keyl, A.C. (1974) Hepatic carcinogenesis of aflatoxin M_1 in rainbow trout (*Salmo gairdneri*) and its enhancement by cyclopropene fatty acids. *J. natl Cancer Inst.*, **53**, 1285–1288

Sizaret, P., Malaveille, C., Brun, G., Aguelon, A.-M. & Toussaint, G. (1982) Inhibition by specific antibodies of the mutagenicity of aflatoxin B_1 in bacteria. *Oncodev. Biol. Med.*, **3**, 125–134

Skopek, T.R. & Thilly, W.G. (1983) Rate of induced forward mutation at 3 genetic loci in *Salmonella typhimurium*. *Mutat. Res.*, **108**, 45–56

Slagle, B.L., Zhou, Y.-Z. & Butel, J.S. (1991) Hepatitis B virus integration event in human chromosome 17p near the p53 gene identifies the region of the chromosome commonly deleted in virus-positive hepatocellular carcinomas. *Cancer Res.*, **51**, 49–54

Södeerkvist, P., Busk, L., Toftgård, R. & Gustafsson, J.-Å. (1983) Metabolic activation of promutagens, detectable in Ames' *Salmonella* assay, by 5000 × g supernatant of rat ventral prostate. *Chem.-biol. Interactions*, **46**, 151–163

Soffritti, M. & McConnell, E.E. (1988) Liver foci formation during aflatoxin B_1 carcinogenesis in the rat. *Ann. N.Y. Acad. Sci.*, **534**, 531–540

Srivatanakul, P., Parkin, D.M., Jiang, Y.-Z., Khlat, M., Kao-Ian, U.-T., Sontipong, S. & Wild, C.P. (1991a) The role of infection by *Opisthorchis viverrini*, hepatitis B virus, and aflatoxin exposure in the etiology of liver cancer in Thailand. A correlation study. *Cancer*, **68**, 2411–2417

Srivatanakul, P., Parkin, D.M., Khlat, M., Chenvidhya, D., Chotiwan, P., Insiripong, S., L'Abbé, K.A. & Wild, C.P. (1991b) Liver cancer in Thailand. II. A case–control study of hepatocellular carcinoma. *Int. J. Cancer*, **48**, 329–332

Stark, A.-A. (1986) Molecular aspects of aflatoxin B_1 mutagenesis and carcinogenesis. In: Steyn, P.S. & Vleggaar, R., eds, *Mycotoxins and Phycotoxins*, Amsterdam, Elsevier, pp. 435–445

Stark, A.-A. & Giroux, C.N. (1982) Mutagenicity and cytotoxicity of the carcinogen–mutagen aflatoxin B_1 in *Streptococcus pneumoniae* (pneumococcus) and *Salmonella typhimurium*: dependence on DNA repair functions. *Mutat. Res.*, **106**, 195–208

Stark, A.-A. & Liberman, D.F. (1991) Synergism between aflatoxins in covalent binding to DNA and in mutagenesis in the photoactivation system. *Mutat. Res.*, **247**, 77–86

Stark, A.-A., Essigmann, J.M., Demain, A.L., Skopek, T.R. & Wogan, G.N. (1979) Aflatoxin B_1 mutagenesis, DNA binding, and adduct formation in *Salmonella typhimurium*. *Proc. natl Acad. Sci. USA*, **76**, 1343–1347

Stark, A.-A., Malca-Mor, L., Herman, Y. & Liberman, D.F. (1988) DNA strand scission and apurinic sites induced by photoactivated aflatoxins. *Cancer Res.*, **48**, 3070–3076

Stark, A.-A., Gal, Y. & Shaulsky, G. (1990) Involvement of singlet oxygen in photoactivation of aflatoxin B_1 and B_2 to DNA-binding forms *in vitro*. *Carcinogenesis*, **11**, 529–534

Štětina, R. & Votava, M. (1986) Induction of DNA single-strand breaks and DNA synthesis inhibition by patulin, ochratoxin A, citrinin, and aflatoxin B_1 in cell lines CHO and AWRF. *Folia biol.*, **32**, 128–144

Stich, H.F. & Laishes, B.A. (1975) The response of Xeroderma pigmentosum cells and controls to the activated mycotoxins, aflatoxins and sterigmatocystin. *Int. J. Cancer*, **16**, 266–274

Stich, H.F. & Stich, W. (1982) Chromosome-damaging activity of saliva of betel nut and tobacco chewers. *Cancer Lett.*, **15**, 193–202

Stoloff, L. (1982) Mycotoxins as potential environmental carcinogens. In: Stich, H.F., ed., *Carcinogens and Mutagens in the Environment*, Vol. 1, Boca Raton, FL, CRC Press, pp. 97–120

Stoloff, L. (1983) Aflatoxin as a cause of primary liver-cell cancer in the United States: a probability study. *Nutr. Cancer*, **5**, 165–186

Stoloff, L., van Egmond, H.P. & Park, D.L. (1991) Rationales for the establishment of limits and regulations for mycotoxins. *Food Addit. Contam.*, **8**, 213–222

Stone, M.P., Gopalakrishnan, S., Harris, T.M. & Graves, D.E. (1988) Carcinogen–nucleic acid interactions: equilibrium binding studies of aflatoxins B_1 and B_2 with DNA and the oligodeoxynucleotide $d(ATGCAT)_2$. *J. biomol. Struct. Dyn.*, **5**, 1025–1041

Stoner, G.D., Conran, P.B., Greisiger, E.A., Stober, J., Morgan, M. & Pereira, M.A. (1986) Comparison of two routes of chemical administration on the lung adenoma response in strain A/J mice. *Toxicol. appl. Pharmacol.*, **82**, 19–31

Stora, C. & Dvořáčková, I. (1986) Aflatoxin B_1 and viral hepatitis B, relative roles in the genesis of primary liver cancer in man (Fr.). *Bull. Acad. natl Méd.*, **170**, 763–775

Stora, C. & Dvořáčková, I. (1987) Aflatoxin, viral hepatitis and primary liver cancer. *J. Med.*, **18**, 23–41

Stora, C., Dvořáčková, I. & Ayraud, N. (1981) Characterization of aflatoxin B_1 (AFB) in human liver cancer. *Res. Commun. chem. Pathol. Pharmacol.*, **31**, 77–85

Stora, C., Breittmayer, J.P., Donzeau, M. & Grivaux, C. (1987) Influence of a protein and riboflavin deficient diet on the oncogenic expression of aflatoxin B_1 in the Sprague Dawley rat of both sexes. *Tumor Res.*, **22**, 1–13

Stott, W.T. & Sinnhuber, R.O. (1978) Trout hepatic enzyme activation of aflatoxin B_1 in a mutagen assay system and the inhibitory effect of PCBs. *Bull. environ. Contam. Toxicol.*, **19**, 35–41

Suit, J.L., Rogers, A.E., Jetten, M.E.R. & Luria, S.E. (1977) Effects of diet on conversion of aflatoxin B_1 to bacterial mutagen(s) by rats *in vivo* and by rat hepatic microsomes *in-vitro*. *Mutat. Res.*, **46**, 313–323

Sun, T.-T. & Chu, Y.-Y. (1984) Carcinogenesis and prevention strategy of liver cancer in areas of prevalence. *J. cell. Physiol.*, **Suppl. 3**, 39–44

Sun, T.-T. & Wang, N.-J. (1983) Studies on human liver carcinogenesis. In: Harris, C.C. & Autrup, H.N., eds, *Human Carcinogenesis*, New York, Academic Press, pp. 757–780

Swenson, D.H., Miller, E.C. & Miller, J.A. (1974) Aflatoxin B_1-2,3-oxide: evidence for its formation in rat liver *in vivo* and by human liver microsomes *in vitro*. *Biochem. biophys. Res. Commun.*, **60**, 1036–1043

Swenson, D.H., Lin, J.-K., Miller, E.C. & Miller, J.A. (1977) Aflatoxin B_1-2,3-oxide as a probable intermediate in the covalent binding of aflatoxins B_1 and B_2 to rat liver DNA and ribosomal RNA *in vivo*. *Cancer Res.*, **37**, 172–181

Tada, M., Omata, M. & Ohto, M. (1990) Analysis of *ras* gene mutations in human hepatic malignant tumors by polymerase chain reaction and direct sequencing. *Cancer Res.*, **50**, 1121-1124

Taggart, P., Devlin, T.M. & Ch'ih, J.J. (1986) Multiple aflatoxin B_1 binding proteins exist in rat liver cytosol. *Proc. Soc. exp. Biol. Med.*, **182**, 68-72

Tanaka, T., Nishikawa, A., Iwata, H., Mori, Y., Hara, A., Hirono, I. & Mori, H. (1989) Enhancing effect of ethanol on aflatoxin B_1-induced hepatocarcinogenesis in male ACI/N rats. *Jpn. J. Cancer Res.*, **80**, 526-530

Tandon, H.D. & Tandon, B.N. (1989) Pathology of liver in an outbreak of aflatoxicosis in man with a report on the follow up. In: Natori, S., Hashimoto, K. & Ueno, Y., eds, *Mycotoxins and Phycotoxins '88, A Collection of Invited Papers Presented at the Seventh International IUPAC Symposium on Mycotoxins and Phycotoxins, Tokyo, Japan, 16-19 August 1988* (Bioactive Molecules, Vol. 10), Amsterdam, Elsevier, pp. 99-107

Tang, T. & Friedman, M.A. (1977) Carcinogen activation by human liver enzymes in the Ames mutagenicity test. *Mutat. Res.*, **46**, 387-394

Tanimura, T., Kihara, T. & Yamamoto, Y. (1982) Teratogenicity of aflatoxin B_1 in the mouse (Jpn.). *Kankyo Kagaku Kenkyusho Kenkyu Hokoku (Kinki Daigaku)*, **10**, 247-256

Temcharoen, P., Anukarahanonta, T. & Bhamarapravati, N. (1978) Influence of dietary protein and vitamin B_{12} on the toxicity and carcinogenicity of aflatoxins in rat liver. *Cancer Res.*, **38**, 2185-2190

Thielmann, H.W. & Gersbach, H. (1978) The nucleotide-permeable *Escherichia coli* cell, a sensitive DNA repair indicator for carcinogens, mutagens, and antitumor agents binding covalently to DNA. *Z. Krebsforsch.*, **92**, 177-214

Thompson, L.H., Salazar, E.P., Brookman, K.W. & Hoy, C.A. (1983) Hypersensitivity to cell killing and mutation induction by chemical carcinogens in an excision repair-deficient mutant of CHO cells. *Mutat. Res.*, **112**, 329-344

Thomson, V.E. & Evans, H.J. (1979) Induction of sister-chromatid exchanges in human lymphocytes and Chinese hamster cells exposed to aflatoxin B_1 and *N*-methyl-*N*-nitrosourea. *Mutat. Res.*, **67**, 47-53

Tilak, T.B.G. (1974) Induction of cholangiocarcinoma following treatment of a rhesus monkey with aflatoxin. *Food Cosmet. Toxicol.*, **13**, 247-249

Todd, L.A. & Bloom, S.E. (1980) Differential induction of sister chromatid exchanges by indirect-acting mutagen–carcinogens at early and late stages of embryonic development. *Environ. Mutag.*, **2**, 435-445

Toledo, C., Hendricks, J., Loveland, P., Wilcox, J. & Bailey, G. (1987) Metabolism and DNA binding *in vivo* of aflatoxin B_1 in medaka (*Oryzias latipes*). *Comp. Biochem. Physiol.*, **87C**, 275-281

Tong, C. & Williams, G.M. (1978) Induction of purine analog-resistant mutants in adult rat liver epithelial cell lines by metabolic activation-dependent and -independent carcinogens. *Mutat. Res.*, **58**, 339-352

Tong, C., Telang, S. & Williams, G.M. (1984) Differences in responses of 4 adult rat-liver epithelial cell lines to a spectrum of chemical mutagens. *Mutat. Res.*, **130**, 53-61

Toskulkao, C., Yoshida, T., Glinsukon, T. & Kuroiwa, Y. (1986) Potentiation of aflatoxin B_1-induced hepatotoxicity in male Wistar rats with ethanol pretreatment. *J. toxicol. Sci.*, **11**, 41-51

Trzos, R.J., Petzold, G.L., Brunden, M.N. & Swenberg, J.A. (1978) The evaluation of sixteen carcinogens in the rat using the micronucleus test. *Mutat. Res.*, **58**, 79-86

Tseng, T.-H., Chu, C.-Y. & Wang, C.-J. (1992) Inhibition of penta-acetyl geniposide on AFB_1-induced genotoxicity in C3H10T½ cells. *Cancer Lett.*, 62, 233–242

Tsuboi, S., Nakagawa, T., Tomita, M., Seo, T., Ono, H., Kawamura, K. & Iwamura, N. (1984) Detection of aflatoxin B_1 in serum samples of male Japanese subjects by radioimmunoassay and high-performance liquid chromatography. *Cancer Res.*, 44, 1231–1234

Tsuda, H., Hirohashi, S., Shimosato, Y., Ino, Y., Yoshida, T. & Terada, M. (1989) Low incidence of point mutation of c-Ki-*ras* and N-*ras* oncogenes in human hepatocellular carcinoma. *Jpn. J. Cancer Res.*, 80, 196–199

Tsutsui, T., Degen, G.H., Schiffmann, D., Wong, A., Maizumi, H., McLachlan, J.A. & Barrett, J.C. (1984) Dependence on exogenous metabolic activation for induction of unscheduled DNA synthesis in Syrian hamster embryo cells by diethylstilbestrol and related compounds. *Cancer Res.*, 44, 184–189

Turmo, E., Sánchez-Baeza, F., Bujons, J., Camps, F., Casellas, M., Solanas, A.-M. & Messeguer, A. (1991) Synthesis and mutagenicity of the aflatoxin B_1 model 3a,8a-dihydro-4,6-dimethoxyfuro-[2,3-*b*]benzofuran and its 2,3 epoxy derivative. *J. agric. Food Chem.*, 39, 1723–1728

Uchida, T., Suzuki, K., Esumi, M., Arii, M. & Shikata, T. (1988) Influence of aflatoxin B_1 intoxication on duck livers with duck hepatitis B virus infection. *Cancer Res.*, 48, 1559–1565

Ueno, Y. & Kubota, K. (1976) DNA-attacking ability of carcinogenic mycotoxins in recombination-deficient mutant cells of *Bacillus subtilis*. *Cancer Res.*, 36, 445–451

Ueno, Y., Kubota, K., Ito, T. & Nakamura, Y. (1978) Mutagenicity of carcinogenic mycotoxins in *Salmonella typhimurium*. *Cancer Res.*, 38, 536–542

Ueno, Y., Tashiro, F. & Nakaki, H. (1985) Mechanism of metabolic activation of aflatoxin B_1. *Gann Monogr. Cancer Res.*, 30, 111–124

Umeda, M., Tsutsui, T. & Saito, M. (1977) Mutagenicity and inducibility of DNA single-strand breaks and chromosome aberrations by various mycotoxins. *Gann*, 68, 619–625

Ungar, J. & Joffe, A.Z. (1969) Acute liver lesions resulting from percutaneous absorption of aflatoxins. *Pathol. Microbiol.*, 33, 65–76

Uwaifo, A.O. & Bababunmi, E.A. (1979) Reduced mutagenicity of aflatoxin B_1 due to hydroxylation: observations on five *Salmonella typhimurium* tester strains. *Cancer Lett.*, 7, 221–225

Uwaifo, A.O., Emerole, G.O., Bababunmi, E.A. & Bassir, O. (1979) Comparative mutagenicity of palmotoxin Bo and aflatoxins B_1 and M_1. *J. environ. Pathol. Toxicol.*, 2, 1099–1107

Van Den Heever, L.H. & Dirr, H.W. (1991) Effect of aflatoxin B_1 on human platelet protein kinase C. *Int. J. Biochem.*, 23, 839–843

Ved Brat, S., Tong, C., Telang, S. & Williams, G.M. (1983) Comparison of sister chromatid exchange and mammalian cell mutagenesis at the hypoxanthine guanine phosphoribosyl transferase locus in adult rat liver epithelial cells. *Ann. N.Y. Acad. Sci.*, 407, 474–475

Vesselinovitch, S.D., Mihailovich, N., Wogan, G.N., Lombard, L.S. & Rao, K.V.N. (1972) Aflatoxin B_1, a hepatocarcinogen in the infant mouse. *Cancer Res.*, 32, 2289–2291

Wales, J.H., Sinnhuber, R.O., Hendricks, J.D., Nixon, J.E. & Eisele, T.A. (1978) Aflatoxin B_1 induction of hepatocellular carcinoma in the embryos of rainbow trout (*Salmo gairdneri*). *J. natl Cancer Inst.*, 60, 1133–1139

Walker, G.J., Hayward, N.K., Falvey, S. & Cooksley, W.G.E. (1991) Loss of somatic heterozygosity in hepatocellular carcinoma. *Cancer Res.*, 51, 4367–4370

Walters, J.M. & Combes, R.D. (1983) Evaluation of a methodology for the use of preparations from rat small intestine in the *Salmonella*/microsome assay. *Mutat. Res.*, 113, 393–402

Wang, Y., Lan, L., Ye, B., Xu, Y., Liu, Y. & Li, W. (1983) Relation between geographical distribution of liver cancer and climate—aflatoxin B_1 in China. *Sci. sin. (Ser. B)*, **26**, 1166–1175

Wang, C.-J., Wang, S.-W., Shiah, H.-S. & Lin, J.-K. (1990) Effect of ethanol on hepatoxicity and hepatic DNA-binding of aflatoxin B_1 in rats. *Biochem. Pharmacol.*, **40**, 715–721

Wang, C.-J., Shiow, S.-J. & Lin, J.-K. (1991a) Effects of crocetin on the hepatoxicity and hepatic DNA binding of aflatoxin B_1 in rats. *Carcinogenesis*, **12**, 459–462

Wang, C.-J., Wang, S.-W. & Lin, J.-K. (1991b) Suppressive effect of geniposide on the hepatotoxicity and hepatic DNA binding of aflatoxin B_1 in the rat. *Cancer Lett.*, **60**, 95–102

Ward, J.M., Sontag, J.M., Weisburger, E.K. & Brown, C.A. (1975) Effect of lifetime exposure to aflatoxin B_1 in rats. *J. natl Cancer Inst.*, **55**, 107–113

Wehner, F.C., Marasas, W.F.O. & Thiel, P.G. (1978) Lack of mutagenicity to *Salmonella typhimurium* of some Fusarium mycotoxins. *Appl. environ. Microbiol.*, **35**, 659–662

Wei, R.D. & Chang, S.-C. (1982) In vivo mutagenicity of aflatoxin B_1 in mice. *J. Chin. Biochem. Soc.*, **11**, 16–23

Wei, R.D., Liu, G.X. & Lee, S.S. (1970) Uptake of aflatoxin B_1 by the skin of rats. *Experientia*, **26**, 82–83

Wheeler, L., Halula, M. & Demeo, M. (1981) A comparison of aflatoxin B_1-induced cytotoxicity, mutagenicity and prophage induction in *Salmonella typhimurium* mutagen tester strains TA1535, TA1538, TA98 and TA100. *Mutat. Res.*, **83**, 39–48

Whitehead, F.W., San, R.H.C. & Stich, H.F. (1983) An intestinal cell-mediated chromosome aberration test for the detection of genotoxic agents. *Mutat. Res.*, **111**, 209–217

Whitham, M., Nixon, J.E. & Sinnhuber, R.O. (1982) Liver DNA bound *in vivo* with aflatoxin B_1 as a measure of hepatocarcinoma initiation in rainbow trout. *J. natl Cancer Inst.*, **68**, 623–628

WHO (1987) *Evaluation of Certain Food Additives and Contaminants* (Technical Report Series No. 759), Geneva

Wieder, R., Wogan, G.N. & Shimkin, M.B. (1968) Pulmonary tumors in strain A mice given injections of aflatoxin B_1. *J. natl Cancer Inst.*, **40**, 1195–1197

Wild, C.P. (1991) Nonassociation of aflatoxin with primary liver cancer in a cross-sectional ecological survey in the People's Republic of China (Letter to the Editor). *Cancer Res.*, **51**, 3825–3827

Wild, C.P., Garner, R.C., Montesano, R. & Tursi, F. (1986) Aflatoxin B_1 binding to plasma albumin and liver DNA upon chronic administration to rats. *Carcinogenesis*, **6**, 853–858

Wild, C.P., Pionneau, F.A., Montesano, R., Mutiro, C.F. & Chetsanga, C.J. (1987) Aflatoxin detected in human breast milk by immunoassay. *Int. J. Cancer*, **40**, 328–333

Wild, C.P., Chapot, B., Scherer, E., Den Engelse, L. & Montesano, R. (1988) Application of antibody methods to the detection of aflatoxin in human body fluids. In: Bartsch, H., Hemminki, K. & O'Neill, I.K., eds, *Methods for Detecting DNA Damaging Agents in Humans: Applications in Cancer Epidemiology and Prevention* (IARC Scientific Publications No. 89), Lyon, IARC, pp. 67–74

Wild, C.P., Jiang, Y.-Z., Sabbioni, G., Chapot, B. & Montesano, R. (1990a) Evaluation of methods of quantitation of aflatoxin–albumin adducts and their application to human exposure assessment. *Cancer Res.*, **50**, 245–251

Wild, C.P., Montesano, R., Van Benthem, J., Scherer, E. & Den Engelse, L. (1990b) Intercellular variation in levels of adducts of aflatoxin B_1 and G_1 in DNA from rat tissues: a quantitative immunocytochemical study. *J. Cancer Res. clin. Oncol.*, **116**, 134–140

Wild, C.P., Rasheed, F.N., Jawla, M.F.B., Hall, A.J., Jansen, L.A.M. & Montesano, R. (1991) In-utero exposure to aflatoxin in West Africa (Letter to the Editor). *Lancet*, **337**, 1602

Wild, C.P., Hudson, G.J., Sabbioni, G., Chapot, B., Hall, A.J., Wogan, G.N., Whittle, H., Montesano, R. & Groopman, J.D. (1992a) Dietary intake of aflatoxins and the level of albumin-bound aflatoxin in peripheral blood in the Gambia, West Africa. *Cancer Epidemiol. Biomarkers Prev.*, **1**, 229–234

Wild, C.P., Jansen, L.A.M., Cova, L. & Montesano, R. (1992b) Molecular dosimetry of aflatoxin exposure: contribution to understanding the multifactorial aetiopathogenesis of primary hepatocellular carcinoma (PHC) with particular reference to hepatitis B virus (HBV). *Environ. Health Perspectives*, **99**

Wilkinson, A.P., Denning, D.W. & Morgan, M.R.A. (1988) Analysis of UK sera for aflatoxin by enzyme-linked immunosorbent assay. *Hum. Toxicol.*, **7**, 353–356

Williams, G.M. (1976) Carcinogen-induced DNA repair in primary rat liver cell cultures; a possible screen for chemical carcinogens. *Cancer Lett.*, **1**, 231–236

Williams, G.M. (1977) Detection of chemical carcinogens by unscheduled DNA synthesis in rat liver primary cell cultures. *Cancer Res.*, **37**, 1845–1851

Williams, G.M., Tanaka, T. & Maeura, Y. (1986) Dose-related inhibition of aflatoxin B_1-induced hepatocarcinogenesis by the phenolic anti-oxidants butylated hydroxyanisole and butylated hydroxytoluene. *Carcinogenesis*, **7**, 1043–1050

Willis, R.M., Mulvihill, J.J. & Hoofnagle, J.H. (1980) Attempted suicide with purified aflatoxin (Letter to the Editor). *Lancet*, **i**, 1198–1199

Wilson, R., Ziprin, R., Ragsdale, S. & Busbee, D. (1985) Uptake and vascular transport of ingested aflatoxin. *Toxicol. Lett.*, **29**, 169–176

Wogan, G.N. (1969a) Naturally occurring carcinogens in foods. *Prog. exp. Tumor Res.*, **11**, 134–162

Wogan, G.N. (1969b) Metabolism and biochemical effects of aflatoxins. In: Goldblatt, L.A., ed., *Aflatoxin. Scientific Background, Control and Implications*, New York, Academic Press, pp. 151–186

Wogan, G.N. (1973) Aflatoxin carcinogenesis. *Meth. Cancer Res.*, **7**, 309–344

Wogan, G.N. & Newberne, P.M. (1967) Dose–response characteristics of aflatoxin B_1 carcinogenesis in the rat. *Cancer Res.*, **27**, 2370–2376

Wogan, G.N. & Paglialunga, S. (1974) Carcinogenicity of synthetic aflatoxin M_1 in rats. *Food Cosmet. Toxicol.*, **12**, 381–384

Wogan, G.N., Edwards, G.S. & Newberne, P.M. (1971) Structure–activity relationships in toxicity and carcinogenicity of aflatoxins and analogs. *Cancer Res.*, **31**, 1936–1942

Wogan, G.N., Paglialunga, S. & Newberne, P.M. (1974) Carcinogenic effects of low dietary levels of aflatoxin B_1 in rats. *Food Cosmet. Toxicol.*, **12**, 681–685

Wolff, S. & Takehisa, S. (1977) Induction of sister chromatid exchanges in mammalian cells by low concentrations of mutagenic carcinogens that require metabolic activation as well as those that do not. In: Scott, D., Bridges, B.A. & Sobels, F.H., eds, *Progress in Genetic Toxicology, Vol. 2, Developments in Toxicology and Environmental Sciences*, Amsterdam, Elsevier, pp. 193–200

Wong, Z.A. & Hsieh, D.P.H. (1980) The comparative metabolism and toxicokinetics of aflatoxin B_1 in the monkey, rat and mouse. *Toxicol. appl. Pharmacol.*, **55**, 115–125

Wray, B.B. & Hayes, A.W. (1980) Aflatoxin B_1 in the serum of a patient with primary hepatic carcinoma. *Environ. Res.*, **22**, 400–403

Wu, S.-M., Sun, Z.-T., Wu, Y.-Y., Wei, Y.-P., Gu, J.-P. & Lu, Z.-H. (1984) Urinary excretion of aflatoxin M_1 (AFM_1) in Beijing and Qidong inhabitants (Chin.). *Chin. J. Oncol.*, **6**, 163–167

Xu, J., Whong, W.-Z. & Ong, T.-M. (1984) Validation of the *Salmonella* (SV50)/arabinose-resistant forward mutation assay system with 26 compounds. *Mutat. Res.*, **130**, 79–86

Yamasaki, T., Teel, R.W. & Lau, B.H.S. (1991) Effect of allixin, a phytoalexin produced by garlic, on mutagenesis, DNA-binding and metabolism of aflatoxin B_1. *Cancer Lett.*, **59**, 89–94

Yang, S.S., Taub, J.V., Modali, R., Vieira, W., Yasei, P. & Yang, G.C. (1985) Dose dependency of aflatoxin B_1 binding on human high molecular weight DNA in the activation of proto-oncogene. *Environ. Health Perspectives*, **62**, 231–238

Yeh, F.-S., Mo, C.-C. & Yen, R.-C. (1985) Risk factors for hepatocellular carcinoma in Guangxi, People's Republic of China. *Natl Cancer Inst. Monogr.*, **69**, 47–48

Yeh, F.-S., Yu, M.-C., Mo, C.-C., Luo, S., Tong, M.J. & Henderson, B.E. (1989) Hepatitis B virus, aflatoxins, and hepatocellular carcinoma in southern Guangxi, China. *Cancer Res.*, **49**, 2506–2509

Yourtee, D.M. & Kirk-Yourtee, C.L. (1986) The mutagenicity of aflatoxin Q_1 to *Salmonella typhimurium* TA100 with or without rat or human liver microsomal preparations. *Res. Commun. chem. Pathol. Pharmacol.*, **54**, 101–113

Yourtee, D.M., Kirk-Yourtee, C.L. & Searles, S. (1987a) Stereochemical effect in the mutagenicity of the aflatoxicols toward *Salmonella typhimurium*. *Life Sci.*, **41**, 1795–1803

Yourtee, D.M., Kirk-Yourtee, C.L. & Searles, S. (1987b) The direct mutagenicity of aflatoxin B_1 and metabolites to *Salmonella typhimurium*: structure mutagenicity relationships and mechanisms of action. *Res. Commun. chem. Pathol. Pharmacol.*, **57**, 55–76

Yu, Y.-N., Ding, C., Li, Q.-G. & Chen, X.-R. (1983) A modified method of UDS detection in vitro suitable for screening the DNA-damaging effects of chemicals. *Mutat. Res.*, **122**, 377–384

Yu, F.-L., Geronimo, I.H., Bender, W. & Permthamsin, J. (1988) Correlation studies between the binding of aflatoxin B_1 to chromatin components and the inhibition of RNA synthesis. *Carcinogenesis*, **9**, 527–532

Yu, S.-Z., Cheng, Z.-Q., Liu, Y.-K., Huang, Z.-Y. & Zhao, Y.-F. (1989) The aflatoxins and contaminated water in the etiological study of primary liver cancer. In: Natori, S., Hashimoto, K. & Ueno, Y., eds, *Mycotoxins and Phycotoxins '88. A Collection of Invited Papers Presented at the Seventh International IUPAC Symposium on Mycotoxins and Phycotoxins, Tokyo, Japan, 16–19 August 1988* (Bioactive Molecules, Vol. 10), Amsterdam, Elsevier, pp. 37–44

Zarba, A., Wild, C.P., Hall, A.J., Montesano, R., Hudson, G.J. & Groopman, J.D. (1992) Aflatoxin M_1 in human breast milk from the Gambia, West Africa, quantified by combined monoclonal antibody immunoaffinity chromatography and HPLC. *Carcinogenesis*, **13**, 891–894

Zawirska, B. & Bednarz, W. (1981) The particular traits of carcinogenesis induced in Wistar rats by aflatoxin B_1. III. Porphyrins and the activity of gamma-glutamyltranspeptidase in primary hepatomas and in their tissue of origin. *Neoplasma*, **28**, 35–49

Zeilmaker, M.J., van Teylingen, C.M.M., van Helten, J.B.M. & Mohn, G.R. (1991) The use of EDTA-permeabilized *E. coli* cells as indicators of aflatoxin B_1-induced differential lethality in the DNA repair host-mediated assay. *Mutat. Res.*, **263**, 137–142

Zhang, X.-K., Huang, D.-P., Qiu, D.-K. & Chiu, J.-F. (1990) The expression of c-*myc* and c-N-*ras* in human cirrhotic livers, hepatocellular carcinomas and liver tissue surrounding the tumors. *Oncogene*, **5**, 909–914

Zhang, Y.-J., Chen, C.-J., Lee, C.-S., Haghighi, B., Yang, G.-Y., Wang, L.-W., Feitelson, M. & Santella, R. (1991) Aflatoxin B_1–DNA adducts and hepatitis B virus antigens in hepatocellular carcinoma and non-tumorous liver tissue. *Carcinogenesis*, **12**, 2247–2252

Zhu, J.-Q., Zhang, L.-S., Hu, X., Xiao, Y., Chen, J.-S., Xu, Y.-C., Fremy, J. & Chu, F.S. (1987) Correlation of dietary aflatoxin B_1 levels with excretion of aflatoxin M_1 in human urine. *Cancer Res.*, **47**, 1848–1852

Zhu, C.-R., Du, M.-J., Lei, D.-N. & Wan, L.-Q. (1989) A study on the inhibition of aflatoxin B_1 induced hepatocarcinogenesis by the *Rhizopus delemar*. *Mater. med. Pol.*, **21**, 87–91

TOXINS DERIVED FROM *FUSARIUM GRAMINEARUM*, *F. CULMORUM* AND *F. CROOKWELLENSE*: ZEARALENONE, DEOXYNIVALENOL, NIVALENOL AND FUSARENONE X

The most widely distributed toxigenic *Fusarium* species is *Fusarium graminearum*, which causes disease in wheat and maize all over the world, except in dryland wheat and subtropical maize. This fungus produces the type-B triochothecenes deoxynivalenol and nivalenol (Thrane, 1989) and zearalenone, depending on the strain. The closely related species, *F. culmorum* and *F. crookwellense*, produce the same toxins and occur in cooler and slightly warmer areas, respectively. *F. crookwellense* and some strains of *F. graminearum* also produce the type-B trichothecene fusarenone X.

Zearalenone was considered by a previous working group, in October 1982 (IARC, 1983), and fusarenone X was considered (as fusarenon X) by two groups, in February 1976 (IARC, 1976) and October 1982 (IARC, 1983). Since that time, new data have become available, and these have been incorporated into the monograph and taken into consideration in the present evaluations.

1. Exposure Data

1.1 Chemical and physical data

1.1.1 *Synonyms, structural and molecular data*

Zearalenone

   Chem. Abstr. Services Reg. No.: 17924-92-4
   Chem. Abstr. Name: 1H-2-Benzoxacyclotetradecin-1,7(8H)-dione, 3,4,5,6,9,10-hexahydro-14,16-dihydroxy-3-methyl, [S-(E)]-
   IUPAC Systematic Name: (−)-(3S,11E)-3,4,5,6,9,10-Hexahydro-14,16-dihydroxy-3-methyl-1H-2-benzoxacyclotetradecin-1,7(8H)-dione
   Synonyms: F2; compound F-2; fermentation estrogenic substance; FES; (S)-(−)-3,4,5,6,9,10-hexahydro-14,16-dihydroxy-3-methyl-1H-2-benzoxacyclotetradecin-1,7-(8H)-dione; mycotoxin F2; toxin F2; (−)-zearalenone; (S)-zearalenone; *trans*-zearalenone; (10S)-zearalenone; zenone

$C_{18}H_{22}O_5$ Mol. wt: 318.4

Deoxynivalenol

Chem. Abstr. Services Reg. No.: 51481-10-8
Chem. Abstr. Name: Trichothec-9-en-8-one, 12,13-epoxy-3,7,15-trihydroxy-(3α,7α)-
Synonyms: Dehydronivalenol; 4-deoxynivalenol; 12,13-epoxy-3α,7α,15-trihydroxy-9-trichothecen-8-one; Rd toxin; spiro[2,5-methano-1-benzoxepin-10,2'-oxirane], trichothec-9-en-8-one derivative; vomitoxin

$C_{15}H_{20}O_6$ Mol. wt: 296.32

Nivalenol

Chem. Abstr. Services Reg. No.: 23282-20-4
Chem. Abstr. Name: Trichothec-9-en-8-one, 12,13-epoxy-3,4,7,15-tetrahydroxy-, (3α,4β,7α)-
IUPAC Systematic Name: Trichothec-9-en-8-one, 12,13-epoxy-3α,4β,7α,15-tetrahydroxy-
Synonyms: Spiro[2,5-methano-1-benzoxepin-10,2'-oxirane], trichothec-9-en-8-one derivative; 3α,4β,7α,15-tetrahydroxy-12,13-epoxytrichothec-9-en-8-one

$C_{15}H_{20}O_7$ Mol. wt: 312.32

Fusarenone X

Chem. Abstr. Services Reg. No.: 23255-69-8
Chem. Abstr. Name: Trichothec-9-en-8-one, 4-(acetyloxy)-12,13-epoxy-3,7,15-trihydroxy(3α,4β,7β)-
IUPAC Systematic Name: 12,13-Epoxy-3α,4β,7β,15-tetrahydroxytrichothec-9-en-8-one 4-acetate or (2R,3R,4S,5S,5aR,6R,9aR,10S)-2,3,4,5,5a,9a-hexahydro-3,4,6-trihydroxy-5a-(hydroxymethyl)-5,8-dimethylspiro[2,5-methano-1-benzoxepin-10,2'-oxirane]-7-(6H)-one 4-acetate
Synonyms: Fusarenon; fusarenon X; nivalenol 4-O-acetate; nivalenol monoacetate; 3,7,15-trihydroxy-4-acetoxy-8-oxo-12,13-epoxy-Δ^9-trichothecene; 3,7,15-trihydroxy-scirp-4-acetoxy-9-en-8-one

$C_{17}H_{22}O_8$ Mol. wt: 354.1

1.1.2 Chemical and physical properties

Zearalenone

From Urry *et al.* (1966), Pohland *et al.* (1982) and Budavari (1989), unless otherwise specified

(a) *Description*: White crystals
(b) *Melting-point*: 164–165 °C
(c) *Optical rotation*: $[\alpha]_D^{25}$ −170.5° (c = 1.0 in methanol); $[\alpha]_D^{21}$ −189° (c = 3.14 in chloroform)
(d) *Spectroscopy data*: Ultraviolet, infrared, mass spectral and proton nuclear magnetic resonance data have been reported.
(e) *Solubility*: Solubilities at 25 °C in percent by weight are: water, 0.002; *n*-hexane, 0.05; benzene, 1.13; acetonitrile, 8.6; dichloromethane, 17.5; methanol, 18; ethanol, 24; and acetone, 58 (Hidy *et al.*, 1977).
(f) *Stability*: Stable when heated at 120 °C; 29% decomposed when sample heated at 150 °C and 69% when heated at 200 °C for 60 min (Kuiper-Goodman *et al.*, 1987); stable to hydrolysis in neutral or acid buffer solutions (Müller, 1983)

Deoxynivalenol

From Cole and Cox (1981)

(a) *Description*: White needles
(b) *Melting-point*: 151–153 °C
(c) *Optical rotation*: $[\alpha]_D^{25}$ +6.35° (c = 0.07 in ethanol)
(d) *Spectroscopy*: Infrared, ultraviolet, mass spectral and proton nuclear magnetic resonance data have been reported.
(e) *Solubility*: Soluble in ethanol, methanol, ethyl acetate, water and chloroform

Nivalenol

From Cole and Cox (1981), unless otherwise specified

(a) *Description*: White crystals
(b) *Melting-point*: 222–223 °C (decomposition; dried in presence of P_2O_5 at reduced pressure)

(c) *Optical rotation*: $[\alpha]_D^{24}$ +21.54° (c = 1.3 in ethanol)
(d) *Spectroscopy data*: Ultraviolet, infrared, mass spectral (Brumley *et al.*, 1982) and proton nuclear magnetic resonance data have been reported.
(e) *Solubility*: Soluble in methanol, ethanol, ethyl acetate and chloroform; slightly soluble in water; soluble in polar organic solvents (Budavari, 1989)

Fusarenone X

From Ueno *et al.* (1969), Saito and Ohtsubo (1974) and Cole and Cox (1981), unless otherwise specified

(a) *Description*: Transparent bipyramid crystals
(b) *Melting-point*: 91–92 °C
(c) *Optical rotation*: $[\alpha]_D^{25}$ +58° (c = 1.0 in methanol); $[\alpha]_D^{24}$ +56.1° (in ethanol)
(d) *Spectroscopy data*: Ultraviolet, infrared, nuclear magnetic resonance and mass spectra have been reported.
(e) *Solubility*: Soluble in chloroform, ethyl acetate, methanol and water; insoluble in *n*-hexane and *n*-pentane
(f) *Stability*: Generally stable (WHO, 1990); hydrolysed by bases to nivalenol
(g) *Reactivity*: Reacts with acetic anhydride to give tetraacetylnivalenol

1.1.3 *Analysis*

Zearalenone

A detailed review of methods for the analysis of zearalenone is provided by Kuiper-Goodman *et al.* (1987). Analysis can be done by thin-layer chromatography, high-performance liquid chromatography (HPLC)–fluorescence detection and gas chromatography after derivatization. Zearalenone and related alcohols can be analysed by gas chromatography with derivatization (Schwadorf & Müller, 1992). Two methods have undergone trials by the Association of Official Analytical Chemists (AOAC) (USA): a thin-layer chromatographic method and an HPLC–fluorescence detection method (Gilbert, 1991). A number of antibody-based methods also exist (e.g., Warner *et al.*, 1986). Sensitive methods exist for the determination of zearalenone and related compounds in animal tissues (e.g., HPLC with fluorescence) (Kuiper-Goodman *et al.*, 1987).

Deoxynivalenol

Detailed reviews of methods of analysis for deoxynivalenol are provided by Scott (1990) and WHO (1990). Deoxynivalenol can be determined by thin-layer chromatography, HPLC with ultraviolet detection and gas chromatography after derivatization. Two methods have undergone trials at the AOAC: a thin-layer chromatography method and a gas chromatography method involving derivatization (Scott *et al.*, 1981; Trucksess *et al.*, 1986; Gilbert, 1991). Materials contaminated with deoxynivalenol are usually co-contaminated with other trichothecenes, and mass spectrometry confirmation is required, at least for some samples (Scott, 1990). Various antibody-based methods are available for the analysis of deoxynivalenol (Scott, 1990; WHO, 1990; Abouzied *et al.*, 1991), but they suffer from problems of cross-reactivity with other trichothecenes.

Nivalenol

A detailed review of methods for the analysis of nivalenol is given by WHO (1990). There is no AOAC-accepted method for the analysis of nivalenol (Gilbert, 1991). This toxin has been analysed by HPLC–ultraviolet detection and gas chromatography with derivatization, to provide useful data on its occurrence (Lauren & Greenhalgh, 1987; Tanaka *et al.*, 1988). Antibody-based assays with reasonable specificity have been developed by at least two groups (Ikebuchi *et al.*, 1990; Teshima *et al.*, 1990; Wang & Chu, 1991).

Fusarenone X

Most methods for the analysis of fusarenone X were developed for its simultaneous determination with other trichothecenes. They involve thin-layer chromatography, polarographic methods, HPLC with ultraviolet detection and gas chromatography or gas chromatography/mass spectrometry after derivatization (Bata *et al.*, 1983; Bottalico *et al.*, 1983; Visconti & Bottalico, 1983; Karppanen *et al.*, 1985; Visconti *et al.*, 1984).

1.2 Production and use

Zearalenone

Zearalenone was first isolated in 1962 from cultures of the fungus *Fusarium graminearum* (Stob *et al.*, 1962). The structure was determined in 1966 (Urry *et al.*, 1966), and a total synthesis was published two years later (Taub *et al.*, 1968). Zearalenone can be produced by culturing strains of the species that produce it in solid-state fermentations on autoclaved rice or maize or on a nutrient medium absorbed on vermiculite (Hidy *et al.*, 1977; Greenhalgh *et al.*, 1983; Kuiper-Goodman *et al.*, 1987).

Zearalenone is produced commercially as an intermediate in the preparation of zeranol (α-zearalenol) by submerged fermentation (Hidy *et al.*, 1977). Zeranol is used as a growth promoter in beef cattle, feedlot lambs and suckling beef calves (US Food and Drug Administration, 1980).

Zearalenone is produced by *F. graminearum*, *F. crookwellense*, *F. culmorum* and *F. semitectum*. The validity of reports that other species produce it has been questioned (Marasas *et al.*, 1984; Thrane, 1989). Cereals are infected by *F. graminearum* and related species when susceptible cultivars and inbred strains are planted. Disease epidemics occur when wet weather occurs at anthesis or silking (Sutton, 1982). Zearalenone can be produced in maize stored in open cribs if the maize does not dry quickly.

Deoxynivalenol

Deoxynivalenol was first isolated by Japanese workers as 'Rd-toxin', and shortly thereafter as 'vomitoxin' in the USA, from barley infected with *F. graminearum* (Morooka *et al.*, 1972; Vesonder *et al.*, 1973; Yoshizawa & Morooka, 1973; Miller *et al.*, 1983) and *F. culmorum* (Greenhalgh *et al.*, 1986). Deoxynivalenol can be produced by culturing strains of the species that produce it on sterilized rice or maize (e.g., Greenhalgh *et al.*, 1983). It can also be produced as the monoacetate in liquid culture followed by a simple hydrolysis step (Greenhalgh *et al.*, 1986).

Deoxynivalenol is produced by strains of *F. graminearum* and *F. culmorum* (Marasas *et al.*, 1984; Thrane, 1989), which are pathogens of cereals, particularly wheat and maize. The disease is favoured by the planting of susceptible cultivars, and wet weather at anthesis or silking results in epidemic conditions of the disease (Sutton, 1982). *F. graminearum* and *F. culmorum* produce complex mixtures of toxins that vary by region of isolation. Most North American strains of *F. graminearum* produce deoxynivalenol, 15-acetyldeoxynivalenol, zearalenone and many additional metabolites, some of which occur in grains. In contrast, most Japanese, Australian and Italian strains produce either nivalenol, fusarenone X and zearalenone or deoxynivalenol, 3-acetyldeoxynivalenol and zearalenone plus the common metabolites. Strains of *F. graminearum* that produce nivalenol have never been found in North or South America (Miller *et al.*, 1991). An extensive discussion of the major and minor metabolites of *F. graminearum* and related species isolated in different countries is provided by Miller *et al.* (1991).

Nivalenol

Nivalenol was first isolated from '*Fusarium nivale*' Fn2B (Tatsuno *et al.* 1968, 1969; Ueno *et al.*, 1970a); this strain was subsequently shown to be an atypical strain of *F. sporotrichioides* (Marasas *et al.*, 1984). Most naturally occurring nivalenol is produced by strains of *F. graminearum* and *F. crookwellense* (Miller *et al.*, 1991). Nivalenol can be produced by culturing strains of the species that produce it on sterilized rice or maize (Ueno *et al.*, 1970a). It can also be produced in liquid cultures (Ueno *et al.*, 1970a; Lauren *et al.*, 1987).

Nivalenol-producing strains of *F. graminearum* appear to occur primarily in Japan, Australia and New Zealand (Blaney, 1991), although they have also been reported in Italy (Logrieco *et al.*, 1988; Miller *et al.*, 1991). Such strains do not appear to occur in North America. *F. crookwellense* is cosmopolitan, and minor amounts of nivalenol reported in Canadian grain come from this species. *F. graminearum* is a pathogen of cereals, and the disease is favoured by planting susceptible cultivars. Wet weather at anthesis or silking results in epidemic conditions of the disease (Sutton, 1982). *F. crookwellense* is a weak pathogen of cereals and is favoured under warmer conditions than those optimal for *F. graminearum*. Both species produce many other metabolites that occur in contaminated crops (Miller *et al.*, 1991).

Fusarenone X

Fusarenone X was first isolated from '*Fusarium nivale*' Fn2B (Ueno *et al.*, 1969), which was subsequently shown to be an atypical strain of *F. sporotrichioides* (Marasas *et al.*, 1984). In general, only *F. crookwellense* and some strains of *F. graminearum* produce fusarenone X (Ichinoe *et al.*, 1983; Goliński *et al.*, 1988). It has been produced by growing the fungus in liquid culture or on sterilized rice or maize (Ueno *et al.*, 1969; Bottalico *et al.*, 1990).

1.3 Occurrence

Zearalenone

Zearalenone is among the most widely distributed *Fusarium* mycotoxins. It is associated primarily with maize but occurs in modest concentrations in wheat, barley and sorghum, among other commodities (Table 1; Kuiper-Goodman *et al.*, 1987). Concentrations in food in North America, Japan and Europe are generally low; however, in some developing countries, exposures can be high, particularly where maize is grown under north-temperate conditions (including highlands). Zearalenone is not transmitted from feed to milk to any significant extent (Prelusky *et al.*, 1990a). Milling of cereals contaminated with zearalenone concentrates the toxin in the bran fractions (Kuiper-Goodman *et al.*, 1987).

Table 1. Natural occurrence of zearalenone

Country or region	Product	Year	Positive samples/ total no.	Content (mg/kg)	Reference
North America					
Canada	Feed	1972-78	266/2022	0.01-141	Funnell (1979)
Canada	Feed	1975-79	3/493	0.05-0.2	Prior (1981)
Canada	Maize	1978-80	23/81	0.01-0.5	Williams (1985)
USA	Maize	1968-69	5/293	0.45-0.8	Shotwell *et al.* (1971)
USA	Maize	1972	38/223	0.1-5.0	Eppley *et al.* (1974)
USA	Maize	1972	3/3	0.9-7.8	Bennett *et al.* (1976)
USA	Maize	1974	23/372	0.04-10.4	Stoloff *et al.* (1976)
USA	Maize	NR	8/8	0.43-7.62	Shotwell *et al.* (1976)
USA	Maize	1986	17/19	0.2-13.2	Abbas *et al.* (1988)
USA	Feed	1968-70	28/65	0.1-2909	Mirocha & Christensen (1974)
USA	Feed	1981	40/342	0.1-8	Côté *et al.* (1984)
USA	Food	1985	17/19	0.003-0.13	Warner & Pestka (1987)
USA	Feed	NR	9/11	0.01-0.07	Ware & Thorpe (1978)
South America					
Argentina	Maize	NR	16/55	0.2-0.75	López & Tapia (1980)
Argentina	Maize	1981-82	6/94	0.03-0.91	Bean & Echandi (1989)
Argentina	Maize	1987	9/150	0.04-0.35	Chulze *et al.* (1989)
Argentina	Wheat	1983	20/20	Mean, 0.01	Tanaka *et al.* (1988)
Argentina	Barley	1983	13/20	Mean, 0.005	Tanaka *et al.* (1988)
Brazil	Maize	1985-86	16/328	0.26-9.8	Sabino *et al.* (1989)
Europe					
Austria	Maize	1978-79	3/6	0.42-1.0	Bottalico *et al.* (1981)
Austria	Maize	1988-89	39/67	< 40.0	Lew *et al.* (1991)
Bulgaria	Wheat	1983	0/2	ND	Tanaka *et al.* (1988)
France	Maize	1974	62/75	< 170	FAO (1979)
France	Wheat	1984	0/2	ND	Tanaka *et al.* (1988)
Hungary	Maize	1968	NR	70-80	FAO (1979)
Hungary	Wheat	1984	0/2	ND	Tanaka *et al.* (1988)
Norway	Wheat, barley	1984	20/102	0.001-0.023	Sundheim *et al.* (1988)
Poland	Maize	1988	5/5	0.7-350	Visconti *et al.* (1990)
Spain	Maize	1983-86	18/209	0.8-9.6	Muñoz *et al.* (1990)
Yugoslavia	Maize	1972	23/54	0.7-37.5	FAO (1979)

Table 1 (contd)

Country or region	Product	Year	Positive samples /total no.	Content (mg/kg)	Reference
Africa					
Egypt	Cereal, feed	NR	36/64	0.002–0.43	Abdelhamid (1990)
Swaziland	Beer	1976	6/55	0.8–5.3 mg/l	FAO (1979)
Transkei	Maize	1976–77	10/12	0.45–10.0	Marasas et al. (1979)
Transkei	Maize	1979	37/72	< 0.02–5.36	Thiel et al. (1982)
Zambia	Maize	1973–74	1	12.8	Marasas et al. (1977)
Zambia	Beer	1974	23/23	0.01–4.6 mg/l	Lovelace & Nyathi (1977)
Zambia	Maize	1974	17/17	0.1–0.8	Lovelace & Nyathi (1977)
Zambia	Maize	1986	19/33	0.08–6.0	Siame & Lovelace (1989)
Zambia	Feed	1986	16/97	0.05–0.5	Siame & Lovelace (1989)
Australasia					
Australia	Maize	1983	148/174	0.01–0.8	Blaney et al. (1986)
China	Wheat	1984–85	18/38	0.004–0.078	Tanaka et al. (1988)
China	Maize		17/47	0.01–0.17	Luo et al. (1990)
India	Maize	1989	9/86	0.76–1.5	Sinha (1991)
Indonesia	Maize	NR	7/26	0.001–0.01	Widiastuti et al. (1988)
Japan	Barley	1983	13/17	0.004–0.009	Tanaka et al. (1988)
Japan	Wheat	1984	18/18	0.01–0.71	Tanaka et al. (1985a)
Korea, Republic of	Wheat	1983	2/10	0.01–0.04	Lee et al. (1985)
Korea, Republic of	Wheat	1984	5/9	0.003–1.25	Lee et al. (1986)
Korea, Republic of	Wheat	1985	2/2	0.001–2.05	Lee et al. (1987)
Korea, Republic of	Barley	1983	21/28	0.003–1.6	Lee et al. (1985)
Korea, Republic of	Barley	1984	29/31	0.001–0.39	Lee et al. (1986)
Korea, Republic of	Barley	1987, 1989	18/57	0.03–1.13	Park & Lee (1990)
Korea, Republic of	Malt	1983	4/4	0.002–0.04	Lee et al. (1985)
Korea, Republic of	Malt	1984	5/5	0.003–0.05	Lee et al. (1986)
Nepal	Maize	1984	5/9	Mean, 0.82	Tanaka et al. (1988)
Nepal	Barley	1984	4/4	Mean, 0.02	Tanaka et al. (1988)
New Zealand	Maize	1984	15/20	0.1–16	Hussein et al. (1989)

[a]NR, not reported; ND, not detected

Deoxynivalenol

Deoxynivalenol is probably the most widely distributed *Fusarium* mycotoxin. Representative values reported in various crops and processed grains are given in Table 2. Occurrence in foods in North America, Japan and Europe is common, but the concentrations are low (Tanaka et al., 1988; Scott, 1989, 1990; WHO, 1990); its occurrence in cereals in some developing countries, however, particularly in southern China (Luo, 1988) and parts of South America and Africa, is relatively high in some years. Acute mycotoxicoses affecting fairly large numbers of people and caused by ingestion of deoxynivalenol have been reported in

China, India and some other countries (Luo, 1988; Bhat et al., 1989; WHO, 1990; Miller, 1991).

During milling, deoxynivalenol is concentrated in the bran, and cooking flour-based products contaminated with this toxin does not reduce the level appreciably (Scott, 1990; WHO, 1990). It is not transferred into milk, meat or eggs (Prelusky et al., 1984; Trenholm et al., 1989).

Table 2. Natural occurrence of deoxynivalenol

Country or region	Product	Year	Positive samples/ total no.	Content (mg/kg)	Reference
North America					
Canada	Wheat (soft)	1979–88	NR/667	Mean, 0.03–0.74	Scott (1990)
Canada	Wheat (hard)	1979–88	NR/1072	Mean, 0.02–3	Scott (1990)
Canada	Maize	1980–88	NR/203	Mean, 0.2–1.2	Scott (1990)
Canada	Food	1982–89	NR/783	Mean, 0.1–0.2	Scott (1990)
USA	Wheat	1982	54/57	0.2–9.0	Eppley et al. (1984)
USA	Wheat	1982	156/157	0.2–43.0	Shotwell et al. (1985)
USA	Wheat	1984	NR/123	Trace–2.3	Jelinek et al. (1989)
USA	Wheat	1982, 1984–85	163/280	Mean, 0.6–1.4	Scott (1990)
USA	Food	1983/84	NR/132	Trace–0.5	Jelinek et al. (1989)
USA	Food	NR	14/21	< 0.5	Scott (1989)
USA	Food	1989	46/92	1.2–19.0	Abouzied et al. (1991)
USA	Maize	1977, 1984–85	117/250	Mean, 0.4–5.0	Scott (1990)
USA	Maize	NR	24/52	0.5–11	WHO (1990)
USA	Maize	1981	274/342	0.1–42	Côté et al. (1984)
USA	Maize	1970–77	33/66	0.001–28	Scott (1989)
USA	Maize	1972–73	18/20	0.4–65.8	Abbas et al. (1988)
Europe					
Austria	Maize	1988–89	36/67	< 500	Lew et al. (1991)
Austria	Feed	1979–85	1053/1913	< 0.1– > 1.0	Scott (1989)
Austria	Feed	NR	179/389	0.03–22	Scott (1989)
Finland	Feed	1984	11/167	0.001–0.12	Karppanen et al. (1985)
France	Wheat	1982–84	30/43	< 0.27	Snijders (1990)
France	Maize	NR	2/3	0.14, 0.6	Jemmali et al. (1978)
Germany	Wheat	1987	42/44	Mean, 0.13	Scott (1990)
Germany	Oats	1979–80, 1982	35/399	0.01–2.0	Scott (1989)
Germany	Rye	1984	4/22	Mean, 0.40	Tanaka et al. (1988)
Hungary	Maize	NR	2/11	0.2, 1.3	Scott (1989)
Italy	Wheat	1984	1/120	0.12	Tanaka et al. (1988)

Table 2 (contd)

Country or region	Product	Year	Positive samples/ total no.	Content (mg/kg)	Reference
Europe (contd)					
Netherlands	Wheat	1982–84	33/51	< 0.51	Snijders (1990)
Norway	Wheat	1984	32/53	0.008–3.2	Sundheim et al. (1988)
Poland	Wheat	1984	13/48	Mean, 0.1	Tanaka et al. (1988)
Spain	Maize	1984–86	9/209	0.04–0.3	Muñoz et al. (1990)
Sweden	Wheat	1984	8/14	Mean, 0.40	Snijders (1990)
United Kingdom	Wheat	NR	57/148	0.02–> 0.5	WHO (1990)
United Kingdom	Wheat	1984	20/31	Mean, 0.03	Tanaka et al. (1988)
United Kingdom	Barley	1980	34/85	0.01–0.36	Gilbert et al. (1983)
South America					
Argentina	Wheat	1983	3/20	Mean, 0.015	Tanaka et al. (1988)
Argentina	Wheat	1986	7/7	1–20.0	Marpegan et al. (1988)
Argentina	Feed	1986	3/3	1.7–8.0	Marpegan et al. (1988)
Argentina	Maize	1983	2/20	Mean, 0.11	Tanaka et al. (1988)
Argentina	Maize	NR	14/58	0.20–0.40	Chulze et al. (1989)
Africa					
South Africa	Maize	1982–85	50/50	0.007–7.4	Gilbert (1989)
Egypt	Feed	NR	31/64	0.07–4.0	Abdelhamid (1990)
Nigeria	Acha	NR	3/6	0.01–0.06	Scott (1989)
Transkei	Maize	1976–77	8/12	0.07–4.0	Marasas et al. (1979)
Transkei	Maize	NR	43/72	0.01–15.8	Thiel et al. (1982)
Zambia	Maize	1985–86	2/51	Mean, 1.0	Siame & Lovelace (1989)
Zambia	Maize	1985–86	16/33	0.5–16.0	Siame & Lovelace (1989)
Australasia					
Australia	Wheat	1983	11/12	< 6.7	Tobin (1988)
Australia	Triticale	1983	3/3	1.1–11.0	Tobin (1988)
China	Wheat	NR	49/49	Mean, 2.82	Tanaka & Ueno (1989)
China	Wheat	1984	4/4	Mean, 4.28	Tanaka et al. (1988)
China	Wheat	1985	19/19	1.0–40.0	WHO (1990)
China	Wheat	1986	79/150	Mean, 0.03–0.81	Gang et al. (1988)
China	Wheat	1986	77/135	0.01–20.0	Luo (1988)
China	Wheat	1989	14/30	0.01–0.30	Luo et al. (1990)
China	Flour	1984	5/5	0.01–0.69	Ueno et al. (1986)
China	Maize	NR	5/5	0.3–92.8	WHO (1990)
China	Maize	1984–86	29/29	0.36–12.67	Hsia et al. (1988)
China	Maize	1989	34/47	0.01–3.51	Luo et al. (1990)
India	Wheat	NR	1/58	0.31	Ramakrishna et al. (1990)
India	Flour	NR	13/56	0.35–8.38	Ramakrishna et al. (1990)
India	Maize	NR	2/86	0.41, 2.02	Sinha (1991)
Japan	Wheat	1976–80	28/39	0.1–12.4	Yoshizawa (1983)

Table 2 (contd)

Country or region	Product	Year	Positive samples/ total no.	Content (mg/kg)	Reference
Australasia (contd)					
Japan	Wheat	1984	18/18	0.70–6.92	Tanaka et al. (1985a)
Japan	Flour	1982–85	36/36	0.002–0.24	Gilbert (1989)
Japan	Barley	1970–80	73/89	0.05–49.6	Yoshizawa (1983)
Japan	Grain food	NR	14/51	0.02–0.23	Scott (1989)
Korea, Republic of	Wheat	1983	2/10	0.02–0.10	Lee et al. (1985)
Korea, Republic of	Wheat	1984	5/9	0.01–0.17	Lee et al. (1986)
Korea, Republic of	Barley	1983	26/28	0.004–0.51	Lee et al. (1985)
Korea, Republic of	Barley	1984	31/31	0.001–0.90	Lee et al. (1986)
Korea, Republic of	Barley	1987	17/18	0.008–0.50	Park & Lee (1990)
Korea, Republic of	Barley	1989	9/20	0.01–0.16	Park & Lee (1990)
Nepal	Maize	1984	3/9	Mean, 0.54	Tanaka et al. (1988)
Nepal	Wheat	1984	1/10	0.06	Tanaka et al. (1988)
New Zealand	Maize	1984	11/20	0.02–0.3	Hussein et al. (1989)
Taiwan	Wheat	1984–85	12/22	0.03–2.45	Ueno et al. (1986)

NR, not reported

Nivalenol

Nivalenol has been reported extensively in Japanese and Korean grain samples and has been found as a minor contaminant in Europe. It has also been reported in samples from southern Africa (Scott, 1989; WHO, 1990) and Australia (Blaney & Dodman, 1988). It is virtually unknown in grains in North and South America (Table 3).

Little is known about the effects of milling and baking on levels of nivalenol or about its transmission into milk, meat and eggs (WHO, 1990).

Table 3. Natural occurrence of nivalenol

Country or region	Product	Year	Positive sample/ total no.	Content (mg/kg)	Reference
North America					
Canada	Wheat	1980–84	4/10	av. 0.02	Tanaka et al. (1988)
Canada	Maize	1984	1	1.0	Foster et al. (1986)
Europe					
Austria	Maize	1988–89	3/39	< 10.0	Lew et al. (1991)
Austria	Wheat	NR	3/4	0.01–0.04	Scott (1989)
Finland	Feed	1982	1/167	0.01	Karppanen et al. (1985)
France	Wheat	NR	2/2	0.02, 0.06	Ueno et al. (1985)

Table 3 (contd)

Country or region	Product	Year	Positive sample/total no.	Content (mg/kg)	Reference
Europe (contd)					
Germany	Food	NR	27/67	< 0.94	Scott (1989)
Germany	Wheat	NR	16/42	0.01–0.12	Scott (1989)
Hungary	Wheat	NR	1/2	0.004	Ueno et al. (1985)
Italy	Feed	NR	7/7	0.08–0.20	Scott (1989)
Norway	Barley	1984	49/49	0.01–0.25	Sundheim et al. (1988)
Norway	Wheat	1984	53/53	0.02–0.89	Sundheim et al. (1988)
Poland	Wheat	1985	43/48	0.003–0.35	Ueno et al. (1985)
Poland	Maize	1988	2/5	33.2, 42.5	Visconti et al. (1990)
United Kingdom	Wheat	1984	17/31	0.004–0.67	Tanaka et al. (1986)
United Kingdom	Barley	1984	3/8	0.007–1.1	Tanaka et al. (1986)
South America					
Argentina	Cereals	1983	15/60	Mean, 0.03	Tanaka et al. (1988)
Africa					
Transkei	Maize	1985	20/72	0.01–1.41	Thiel et al. (1982)
Transkei	Maize	1985	24/24	0.88–15.2	Sydenham et al. (1990)
Australasia					
China	Wheat	NR	45/49	Mean, 0.04	Tanaka & Ueno (1989)
China	Wheat	1984	1/5	6.66	Ueno et al. (1986)
China	Wheat	1989	7/30	0.01–0.02	Luo et al. (1990)
China	Maize	1985	28/28	0.05–4.05	Hsia et al. (1988)
China	Maize	NR	100%	0.4–12.7	WHO (1990)
India	Flour	NR	2/37	0.03–0.1	Ramakrishna et al. (1990)
Japan	Barley	1970–80	73/89	0.06–22.9	Yoshizawa (1983)
Japan	Barley	1977–82	46/50	< 0.05–11.4	Scott (1989)
Japan	Wheat	1984	7/18	0.05–0.44	Tanaka et al. (1985a)
Japan	Wheat flour	1982–85	12/36	0.004–0.08	Tanaka et al. (1985b)
Korea, Republic of	Wheat	1983	9/10	0.03–0.63	Lee et al. (1985)
Korea, Republic of	Barley	1983	28/28	0.02–3.0	Lee et al. (1985)
Korea, Republic of	Barley	1984	31/31	0.18–1.15	Lee et al. (1986)
Korea, Republic of	Barley	1987	17/18	0.03–1.11	Park & Lee (1990)
Nepal	Maize	1984	6/9	Mean, 0.89	Tanaka et al. (1988)
Taiwan	Barley	1985	4/4	0.29–0.98	Ueno et al. (1986)
Taiwan	Wheat	1984–85	10/22	0.005–0.17	Ueno et al. (1986)

NR, not reported

Fusarenone X

Fusarenone X is the acetylated precursor of nivalenol, and small amounts (10–20%) can occur in nivalenol (Miller *et al.*, 1991).

Reports of the occurrence of fusarenone X are limited to a few samples of maize naturally infected in the field in Europe, with a maximal concentration of 1.8 mg/kg (Bottalico *et al.*, 1983; Scott, 1989; Visconti *et al.*, 1990). It is found together with other *Fusarium* toxins produced by the same fungal species, i.e., nivalenol and zearalenone (Visconti *et al.*, 1990).

1.4 Regulations and guidelines

Zearalenone

An official tolerance level of 1 mg/kg zearalenone in grains, fats and oils was established in the USSR in 1984. Proposed levels in other countries are 0.2 mg/kg in maize in Brazil and 0.03 mg/kg in all foods in Romania (van Egmond, 1989).

Deoxynivalenol

Deoxynivalenol is apparently not subject to regulation in any country; however, guidelines, advisory levels and 'official tolerance levels' exist in some countries. In Canada, a guideline of 2 mg/kg is given for the occurrence of deoxynivalenol in uncleaned soft wheat, except in infant foods for which a guideline of 1 mg/kg is given; a guideline of 1.2 mg/kg is given for uncleaned non-staple foods calculated on the basis of flour or bran. In Romania, there is a tolerance of 0.005 mg/kg in all feeds. In the USA, advisory levels of 2 mg/kg in wheat and wheat products for milling and 1 mg/kg in finished wheat products were established; a level of 4 mg/kg is advised for wheat and wheat products for feed ingredients. Official tolerance levels in the USSR in 1984–85 were 1 mg/kg for durum wheat and 0.5 mg/kg for other wheats (van Egmond, 1989). In China, the suggested tolerance limit in wheat is 1 mg/kg (Luo, 1988).

Nivalenol and fusarenone X

No regulation or guideline exists for these compounds (van Egmond, 1989).

2. Studies of Cancer in Humans

A number of ecological studies have addressed the relationship between exposure to *Fusarium* toxins and oesophageal cancer. Most of the studies refer to mixtures of many toxins from many species of fungi on maize.

Cancer incidence among the Bantu of the Transkei, South Africa, has been reported in a number of surveys covering the period 1955–84 (Rose, 1967, 1973; Rose & Fellingham, 1981; Jaskiewicz *et al.*, 1987). Annual age-standardized incidence (African standard) for all cancer sites combined (1965–69) was 60 and 42 per 100 000 per year for men and women, respectively (Rose & Fellingham, 1981); oesophageal cancer accounted for approximately half of the cases (35 and 17–19 per 100 000) in 1955–69, based on 5095 cases (Rose, 1973). Incidence varied markedly over time and among sub-districts of the 26 districts of the Transkei. The incidence was higher in people of each sex in south-western sub-districts than in north-eastern sub-districts (Jaskiewicz *et al.*, 1987). Similarly high rates for oesophageal cancer are reported from areas in northern China. The age-adjusted mortality rate in the two sexes combined was 100 per 100 000 in the Linxian registry for oesophageal cancer (1959–70), ranging from 140 per 100 000 in high-risk areas to approximately 2 per 100 000 in low-risk areas (Yang, 1980).

Marasas *et al.* (1979) examined the amounts of deoxynivalenol and zearalenone in samples of mouldy maize from randomly selected areas in the high-risk and low-risk oesophageal cancer regions of the Transkei, where maize is the main dietary staple. The level of contamination of maize kernels with each of these two *Fusarium* mycotoxins was apparently higher in the pooled samples from the high-risk than the low-risk region. [The Working Group noted that the actual number of kernels infected with *Fusarium* is not specified; statistical evaluation of the data was not possible.]

In an extension of this study, Marasas *et al.* (1981) included an area of the Transkei with an intermediate rate of oesophageal cancer and collected visibly healthy maize samples at random from each of the three study areas. The proportion of kernels in both mouldy and healthy maize samples infected by *F. graminearum*, which was responsible for contamination of the crops with deoxynivalenol and zearalenone, was not correlated with oesophageal cancer rates.

Marasas *et al.* (1988) studied the prevalence of three *Fusarium* species and other fungi in home-grown maize harvested in 1985 by 12 households situated in a district of high incidence of oesophageal cancer in the Transkei and by 12 households in a low-incidence district. Households in the high-incidence area were identified during a preliminary cytological screening for oesophageal cancer as having one or more adult occupant who showed mild to severe oesophageal abnormalities; households in the other study area was chosen at random. The ears of maize were sorted by the housewife at each domicile into 'good' ears intended for making porridge and 'mouldy' ears intended for brewing beer. No correlation was found between the occurrence of *F. graminearum* in healthy maize and the risk for oesophageal cancer, but an inverse correlation was seen with the occurrence of this fungus in mouldy maize.

Sydenham *et al.* (1990) found high concentrations of various *Fusarium* mycotoxins in the same samples of mouldy home-grown maize collected during 1985 and examined by Marasas *et al.* (1988). The mean levels of nivalenol and zearalenone, produced by *F. graminearum*, were significantly higher in mouldy maize samples from the low-risk than in those from the high risk area.

In a cross-sectional study, Hsia *et al.* (1988) examined the amounts of five *Fusarium* mycotoxins in 109 samples of maize and maize meal stocked by the families of 24 oesophageal cancer patients in 1985–86 and at 68 farms in five villages in Linxian, Henan, China, in 1985, where the death rate from cancer of the oesophagus is excessive and where maize has constituted the main staple food for the past few decades. Nivalenol and deoxynivalenol were found in all of the maize samples from the families of the 24 oesophageal cancer patients, at mean levels of 757 ng/g and 5376 ng/g, respectively. According to the authors, these levels are much higher than those seen in various cereals in foodstuffs in Japan, the United

Kingdom and the USA, where oesophageal cancer rates are low or moderate. The levels were apparently not, however, higher than those observed in all samples of maize from randomly selected farms in the area (1964 ng/g and 4543 ng/g, respectively).

No analytical epidemiological study was available that addressed the carcinogenicity of *Fusarium* toxins.

3. Studies of Cancer in Experimental Animals

No data were available to the Working Group on deoxynivalenol, but they were aware of a study in progress by oral administration to mice (Ghess *et al.*, 1992).

3.1 Oral administration

3.1.1 *Mouse*

Groups of 50 male and 50 female B6C3F$_1$ mice, seven weeks old, were fed diets containing 0, 50 or 100 mg/kg (maximum tolerated dose) *zearalenone* (purity, > 99%) for 103 weeks. All survivors were killed 105–108 weeks after the start of treatment. No significant difference in survival was observed between groups; 64–88% of the mice survived to termination. Hepatocellular adenomas were found in 3/50 (6%) low-dose and 7/49 (14%) high-dose males and in 4/50 (8%) male controls; and in 2/49 (4%) low-dose and 7/49 (14%) high-dose females and 0/50 female controls. The incidence in high-dose females was statistically significantly different ($p < 0.006$) in individual comparisons with the control group. The incidence of hepatocellular adenomas in untreated historical control female B6C3F$_1$ mice at the institute where the study took place was 14/498 (2.8%). A statistically significant positive trend was observed in the incidence of pituitary adenomas in both males (control, 0/40; low-dose, 4/45 (9%); high-dose, 6/44 (14%); $p < 0.022$) and females (control, 3/46 (7%); low-dose, 2/43 (5%); high-dose, 13/42 (31%); $p < 0.001$). The increased incidence of pituitary adenomas was statistically significant in high-dose males ($p < 0.032$) and in high-dose females ($p < 0.003$). The incidence of pituitary adenomas and carcinomas in untreated historical control B6C3F$_1$ mice at the institute where the study was undertaken was 21/428 (4.9%) in females and 0/399 in males (US National Toxicology Program, 1982).

Groups of 42 seven-week-old female C57BL/6CrSlc SPF mice were fed diets containing 0, 6, 12 or 30 mg/kg *nivalenol* (containing less than 100 µg/kg fusarenone X) for two years, to give a daily exposure to nivalenol of 0, 0.66, 1.38 or 3.49 mg/kg bw in the four groups, respectively. Body weights were reduced in all treated groups during the study, but terminal weights were reduced only in the high-dose group (26.4 g *versus* 34.0 g in controls; $p < 0.01$). After two years, the surviving mice were killed, and the liver, thymus, spleen, kidneys and brain examined. The numbers of mice still alive at that time were 22/42 controls, 22/42 fed 6 mg/kg, 20/42 fed 12 mg/kg and 29/42 fed 30 mg/kg. No significant increase in tumour incidence was reported (Ohtsubo *et al.*, 1989). [The Working Group noted the limited number of tissues studied.]

3.1.2 *Rat*

Groups of 50 male and 50 female Fischer 344 rats, five weeks old, were fed diets containing 0, 25 or 50 mg/kg (maximum tolerated dose) *zearalenone* (purity, > 99%) for 103 weeks.

All survivors were killed 104–106 weeks after the start of treatment. Mean body weight gains of treated rats were lower than those of controls, and the depression in mean body weight was dose related. Survival rates of treated and control rats were similar; 74–82% of the animals were still alive at termination of the study. No increase in tumour incidence was observed (US National Toxicology Program, 1982).

A group of 20 male Donryu rats, eight weeks old, were given 0.4 mg/kg bw *fusarenone X* [purity unspecified] weekly by oral intubation for 50 weeks. Of 12 rats that survived 50 weeks, one developed a hepatoma. No tumour occurred in 10 male controls during the experimental period of over 400 days (Saito & Ohtsubo, 1974). [The Working Group noted the incomplete reporting of the experiment.]

Groups of 25 or 49 male Donryu rats, six weeks old, were given diets containing either 3.5 or 7 mg/kg *fusarenone X* [purity unspecified] (isolated from a culture filtrate of *F. nivale*, which is in fact *F. sporotrichioides* (see p. 402)) for two years; a third group of 26 animals was given 7 mg/kg diet for only one year. There were 48 controls. All animals were given a restricted volume of feed (15 g/day, to provide 50 and 105 µg fusarenone X per day per animal). Survivors were killed at 24 months. The mean body weights of the treated animals were in general lower than those of the respective controls, and a treatment-related effect on survival was noted: after 18 months of treatment, 50% of the controls were alive, compared with 15/49 (31%) and 4/25 (16%) in the low- and high-dose groups, respectively; survival at that time in the group receiving the 7 mg/kg diet for one year was 9/52 (17%). The major cause of death was chronic bronchopneumonia. No increase in the incidence of tumours was noted in treated rats (Saito *et al.*, 1980). [The Working Group noted the poor survival of the treated animals.]

3.2 Pre- and postnatal exposure

Rat

Groups of 50 male and 50 female FDRL-Wistar-derived rats, six to eight weeks of age, were fed diets that resulted in daily intakes of 0.1, 1.0 or 3.0 mg/kg bw *zearalenone* (> 96% pure). Groups of 70 rats served as controls. Diets were adjusted weekly according to the previous week's body weights and food consumption. F_0 generation rats were fed the diet for five weeks before mating, during mating (approximately two weeks) and during gestation but not during lactation. After weaning, the F_0 generation was killed. In the F_1 generation, 90 rats per sex were fed zearalenone (0.1, 1.0 and 3.0 mg/kg bw) at 28 days of age; 140 of each sex served as controls. Ten animals per sex per group (selected randomly) were killed at 13, 26, 64 and 104 weeks after initiation of zearalenone feeding. All surviving male and female animals were killed at 108 and 111 weeks, respectively. Male rats fed diets containing 1.0 and 3.0 mg/kg bw of zearalenone had decreased terminal body weights. There was no difference in the incidence of tumours in any of the zearalenone-exposed groups as compared to controls in the F_1 generation (Becci *et al.*, 1982a).

3.3 Subcutaneous administration

3.3.1 Mouse

Two groups of 16 or 18 DDD male mice, eight weeks of age, received 10 or 20 weekly subcutaneous injections of 2.5 mg/kg bw *fusarenone X*. A group of 11 mice served as controls. Most of the animals survived the treatment. No increase in tumour incidence was noted in treated animals when compared with controls; one case of leukaemia was observed (Saito & Ohtsubo, 1974). [The Working Group noted the incomplete reporting of the experiment.]

3.3.2 Rat

A group of 18 male Donryu rats, eight weeks of age, were given weekly subcutaneous injections of 0.4 mg/kg bw *fusarenone X* for 22 weeks; most of the rats survived more than one year, and one developed a lung adenoma. No tumour was seen in 10 controls (Saito & Ohtsubo, 1974). [The Working Group noted the incomplete reporting of the experiment.]

3.4 Skin application

Mouse

Groups of 10 female ICR mice [age unspecified] each received topical applications of 2 or 20 µg (0.4 µg for the first six weeks) *fusarenone X* twice a week for 25 weeks. Other mice each received applications of 51.2 µg 7,8-dimethylbenz[a]anthracene (DMBA) and, two weeks later, fusarenone X at the same doses as the first groups, twice a week for 23 weeks. Groups of 10 positive controls each received DMBA plus 10.5 µg 12-*O*-tetradecanoyl-phorbol 13-acetate (TPA) twice a week for 20 weeks. Eight of 10 mice receiving DMBA plus TPA developed skin papillomas on their backs 4–8 weeks after the beginning of treatment with DMBA [unspecified whether gross or microscopic examinations], but no skin papilloma was found in mice receiving 2 or 20 µg fusarenone X alone or 20 µg fusarenone X in combination with DMBA. One mouse treated with DMBA plus 2 µg fusarenone X developed multiple small skin papillomas at 23 weeks (Ueno, 1984).

3.5 Administration with known carcinogens

Mouse

Groups of 38–56 male and female C57Bl/6 × C3H F_1 mice, one week of age, were given single doses of 6 mg/kg bw aflatoxin B_1 by intraperitoneal injection and seven weeks later were fed diets containing 0, 6 or 12 mg/kg diet *nivalenol* for one year. They were sacrificed at 71 weeks of age. All treated male mice developed liver tumours (mostly hepatocellular carcinomas). In females, the incidences were 0/21 in controls, 8/26 in mice given aflatoxin B_1 alone, 3/15 in mice given aflatoxin B_1 plus 6 mg/kg nivalenol and 0/19 in mice given aflatoxin B_1 plus 12 mg/kg nivalenol (Ueno *et al.*, 1991).

4. Other Relevant Data

4.1 Absorption, distribution, metabolism and excretion

4.1.1 Humans

Zearalenone

An adult man was given a single oral dose of 100 mg zearalenone, and his urine was collected over the following 24 h. Urinary analysis showed the presence of parent compound and α-zearalenol in approximately equal amounts together with lower levels of the isomeric β-zearalenol, all in the form of glucuronic acid conjugates (Mirocha et al., 1981).

No data were available to the Working Group on deoxynivalenol, nivalenol or fusarenone X.

4.1.2 Experimental systems

Zearalenone

Zearalenone is fairly rapidly absorbed in several species following oral administration (Mirocha et al., 1981; Olsen et al., 1985). In mice injected with ^{3}H-zearalenone, specific localization was found in oestrogen target organs such as the uterus, interstitial cells of the testicles and the follicles of the ovary (Appelgren et al., 1982). Some is concentrated in adipose tissue of rats (Ueno et al., 1977), and fat-soluble metabolites accumulate in egg yolk (Dailey et al., 1980).

The primary metabolites of zearalenone are the reduced products, α- and β-zearalenol (Fig. 1), and the glucuronic acid conjugates of both the parent compound and metabolites. The reduction is catalysed by microsomal and cytoplasmic fractions. Species differ in the distribution of the two epimers (Mirocha et al., 1981; Farnworth & Trenholm, 1983). In prepubertal gilt, α-zearalenol was found in plasma at levels exceeding those of zearalenone by three to four times (Olsen et al., 1985). The metabolites were also found in milk produced five days after discontinuation of administration of zearalenone to cows, sheep and pigs (Hagler et al., 1980; Palyusik et al., 1980).

Fig. 1. Metabolism of zearalenone

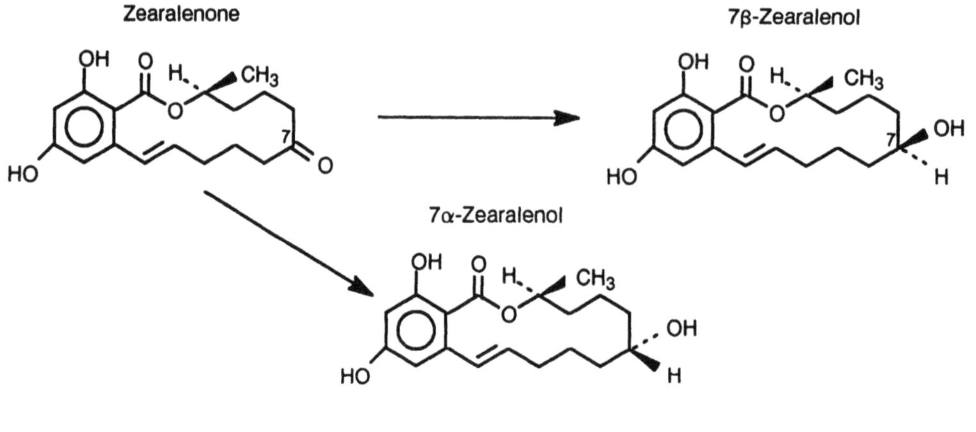

Deoxynivalenol

Following a single intravenous injection (1 mg/kg bw) of deoxynivalenol to swine, the mycotoxin was rapidly distributed to all tissues; the highest concentrations were found in kidney and liver and, correspondingly, in urine and bile. The elimination half-life was estimated at 3.9 h. Trace levels of the toxin were still detectable after 24 h (Prelusky & Trenholm, 1991). In cows, 24 h after an oral dose of 920 mg, trace levels of deoxynivalenol were detectable in blood, and extremely low levels of free and conjugated deoxynivalenol (< 4 ng/ml) were present in the milk (Prelusky et al., 1984).

In rats, 96 h after a single oral dose of 10 mg/kg bw ^{14}C-deoxynivalenol, most of the radiolabel was excreted in the faeces (64%) and urine (25%). Some radiolabel was retained in the liver and very little in other tissues (Lake et al., 1987).

Deoxynivalenol is metabolized by rodents in vivo by an apparently novel reaction involving loss of the epoxide oxygen function (referred to as de-epoxidation) to give de-epoxy deoxynivalenol (Fig. 2). This is the predominant metabolite in faeces after oral administration of deoxynivalenol to rats (Yoshizawa et al., 1983); it was also found in urine, faeces, plasma and milk of lactating cows (Côté et al., 1986; Yoshizawa et al., 1986).

Fig. 2. Metabolism of deoxynivalenol

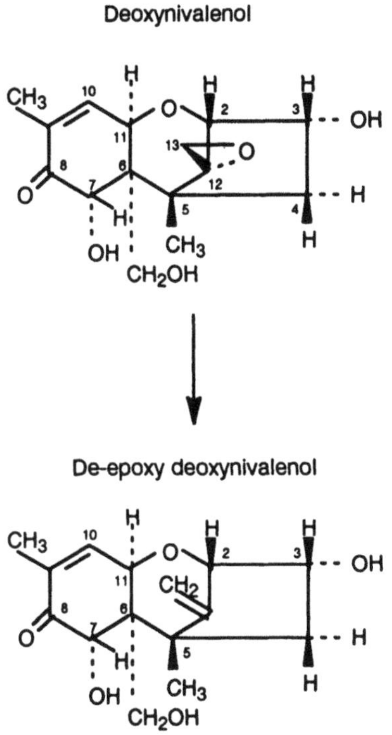

From Yoshizawa et al. (1983)

After oral administration of 5 mg/kg bw deoxynivalenol to two sheep, 54 and 75% of the unmetabolized compound was recovered in the faeces and 7% in the urine of each animal (Prelusky *et al.*, 1986). After intravenous administration to sheep of 4 mg/kg bw, the metabolites were excreted in urine in the order: conjugated deoxynivalenol, conjugated de-epoxy deoxynivalenol, deoxynivalenol and de-epoxy deoxynivalenol. After a single oral administration of deoxynivalenol at 1.32 g for three days or a single intravenous injection at 4 mg/kg to sheep, only trace amounts of deoxynivalenol or its metabolites were found in milk (Prelusky *et al.*, 1987).

Nivalenol

Nivalenol is metabolized to de-epoxy nivalenol. After long-term oral administration of nivalenol to male rats, the dose was recovered as faecal nivalenol (7%), faecal de-epoxy nivalenol (80%), urinary nivalenol (1%) and urinary de-epoxy nivalenol (1%) (Onji *et al.*, 1989).

Fusarenone X

Thirty minutes after subcutaneous administration of uniformly labelled ^{3}H-fusarenone X at 4 mg/kg bw to mice, activity was found in liver, kidneys, intestines, stomach, spleen, bile and plasma; none was detected in heart, brain or testis. The highest activity, corresponding to 3% of the dose, was observed in the liver. Twelve hours after administration, no label was present in the organs, and 25% of the dose was recovered as metabolized forms of fusarenone X in the urine (Ueno *et al.*, 1971).

Fusarenone X was deacetylated by rat and rabbit liver carboxy esterases to nivalenol (Ohta *et al.*, 1978).

4.2 Toxic effects

4.2.1 *Humans*

Two outbreaks of disease related to trichothecenes have been well described—one in China in 1984–85 (Luo, 1988) and one in India in 1987 (Bhat *et al.*, 1989). Each involved several hundred cases.

During the first incident, outbreaks of poisoning from mouldy maize and scabby wheat were reported. After about 600 persons consumed mouldy cereals, 463 cases of poisoning (77% of the total) were reported. The latent period for onset of symptoms was 5–30 min, and these included nausea, vomiting, abdominal pain, diarrhoea, dizziness and headache. No death occurred. Pigs and chicks fed the same mouldy cereals were also affected. *Deoxynivalenol* was detected within a range of 0.34–92.8 mg/kg and *zearalenone* within a range of 0.004–0.587 mg/kg; neither T-2 toxin nor nivalenol was found (WHO, 1990).

The other outbreak occurred in Kashmir, India, in 1987 (Bhat *et al.*, 1989) and was ascribed to the consumption of bread made from mouldy flour. Of 224 randomly selected persons investigated, 97 had symptoms that included abdominal pain (100%), throat irritation (63%), diarrhoea (39%), blood in stools (5%) and vomiting (7%). Symptoms developed 15 min to 1 h after consumption of locally baked bread. Trichothecene toxins were detected in refined and ordinary wheat flour samples at the following ranges of concentrations: *deoxynivalenol*, 0.35–8.38 mg/kg; *nivalenol*, 0.03–0.1 mg/kg; T-2 toxin, 0.55–0.8 mg/kg; and acetyl deoxynivalenol, 0.6–2.4 mg/kg.

4.2.2 *Experimental systems*

Zearalenone

The toxicology of zearalenone has been reviewed (Kuiper-Goodman *et al.*, 1987).

Zearalenone given in the diet of mice for 13 weeks caused atrophy of seminal vesicles and testes, squamous metaplasia of the prostate gland, osteopetrosis, myelofibrosis of the bone marrow, cytoplasmic vacuolization of the adrenal glands, hyperkeratosis of the vagina and endometrial hyperplasia. Rats are about 10 times more sensitive than mice, as osteopetrosis was found in 5/10 female rats given 30 mg/kg in their diet (US National Toxicology Program, 1982). Swine are the most sensitive domestic animals; for example, dietary levels of zearalenone as low as 5 mg/kg induce pseudopregnancy with a failure to cycle (Etienne & Jemmali, 1982).

The visible signs of zearalenone-induced hyperoestrogenism in female swine are swollen vulva and mamma, enlargement of the uterus, ovarian changes and infertility (Mirocha & Christensen, 1974; Bauer *et al.*, 1987). the no-observed-adverse-effect level is less than 5 mg/kg bw (Farnworth & Trenholm, 1983).

Zearalenone binds to oestrogen receptors in human breast cancer cells; however, its relative binding affinity is only 5% of that of 17β-oestradiol (Martin *et al.*, 1978). In oestrogen-sensitive cell lines exposed to zearalenone, oestrogen-specific proteins were expressed. In this system, zearalenone was suggested to activate the oestrogen receptor (Mayr, 1988), inducing expression of oestrogen-controlled genes.

Zearalenone inhibits mitogen-induced blastogenesis in rat and human peripheral blood lymphocytes. The amount of zearalenone necessary to inhibit proliferation by 50% was 13 µg/ml, i.e., 250 times more than the dose required for the action of deoxynivalenol (see below) (Atkinson & Miller, 1984).

The relative binding affinity of α-zearalenol for cytosolic oestrogen receptors was 10–20 times greater than that of zearalenone and some 100 times greater than that of β-zearalenol. Thus, reduction to α-zearalenol is an activation process, while the production of β-zearalenol is a deactivation process: the observed interspecies variations in sensitivity to dietary zearalenone may be due to differences in metabolism and the relative binding activity of these metabolites for oestrogen receptors (Fitzpatrick *et al.*, 1989).

Deoxynivalenol

Deoxynivalenol is one of the least acutely toxic compounds of the trichothecene class of mycotoxins. LD_{50}s for this compound in $B6C3F_1$ mice were reported to be 78 mg/kg bw by gavage and 49 mg/kg bw intraperitoneally (Forsell *et al.*, 1987). In male DDY mice, the LD_{50} was 46 mg/kg bw by oral administration and 70 mg/kg bw by intraperitoneal administration. The subcutaneous LD_{50} in 10-day-old Peking ducklings was 27 mg/kg bw (Yoshizawa & Morooka, 1974), and an oral LD_{50} of about 140 mg/kg bw was found for broiler chickens (Huff *et al.*, 1981).

Intravenous administration of 0.5 mg/kg bw deoxynivalenol to swine induced vomiting, diarrhoea, muscular weakness, tremors and twilight coma. Hypoglycaemia and pancreatic islet cell lesions were observed (Coppock *et al.*, 1985).

Acute intraperitoneal doses of deoxynivalenol (10–1000 mg/kg bw) resulted in extensive necrosis of the gastrointestinal tract, bone marrow and lymphoid tissues and focal lesions in kidney and cardiac tissue in B6C3F$_1$ mice (Forsell *et al.*, 1987). Engorgement of the testes was reported by Yoshizawa and Morooka (1974). In broiler chickens, acute deoxynivalenol toxicosis is characterized by extensive ecchymotic haemorrhaging throughout the carcass, disturbance of the nervous system and irritation of the upper gastrointestinal tract (Huff *et al.*, 1981). Fitzpatrick *et al.* (1988) found elevated concentrations of indoleamines, serotonine and 5-hydroxy-3-indolacetic acid in rat brain 24 h after oral dosing with 2.5 mg/kg bw.

The minimal doses that caused vomiting after subcutaneous administration of deoxynivalenol were 10 mg/kg bw in ducklings and 0.1 mg/kg bw in dogs (Yoshizawa & Morooka, 1974).

Swine fed a diet containing 5 mg/kg deoxynivalenol for nine weeks developed vomiting and depleted hepatic glycogen (Schuh *et al.*, 1982). In young pigs, a dietary level of 20 mg/kg caused vomiting, 12 mg/kg caused almost complete feed refusal, and 1.3 mg/kg induced significant depression in feed intake and rate of weight gain (Forsyth *et al.*, 1977; Young *et al.*, 1983). Feed refusal appears to occur at the level of the central nervous system (for review, see Prelusky *et al.*, 1990b). After intravenous administration (1.0 mg/kg bw), deoxynivalenol was detected in the cerebral spinal fluid of sheep and swine. An analysis of the area under curves for body compartment distribution indicated that about 2.5 times as much toxin eventually reaches the cerebral spinal fluid in pigs as in sheep (Prelusky *et al.*, 1990b).

In rats fed 20 mg/kg deoxynivalenol in the diet for 90 days, no significant effect was seen on serum enzyme levels, haematological parameters or histopathological lesions. There was no sign of feed refusal, but the treated rats were less efficient at converting feed into body mass than controls (Morrissey *et al.*, 1985). Body weight gain was, however, completely inhibited in rats fed diets containing 150 mg/kg deoxynivalenol (Yoshizawa *et al.*, 1978).

Hunder *et al.* (1991) found impairment of intestinal transfer and uptake of nutrients such as glucose and 5-methyltetrahydrofolic acid in mice fed 10 mg/kg deoxynivalenol in the diet.

Feeding studies in poultry demonstrated only minor adverse effects on growth, food consumption and fertility, even at levels up to 38 mg/kg in the diet (Hulan & Proudfoot, 1982; Kubena *et al.*, 1987; Moran *et al.*, 1987). In growing lambs given a wheat diet containing 15 mg/kg deoxynivalenol, no change in body weight gain or food consumption was observed (Harvey *et al.*, 1986). Feeding trials with farm animals showed no serious adverse effect of deoxynivalenol at dietary concentrations of 2 mg/kg in swine, 5 mg/kg in poultry and 6 mg/kg in dairy cattle when fed at a rate of 1% body weight per day (Trenholm *et al.*, 1984).

Deoxynivalenol inhibited protein synthesis at the ribosomal level in an in-vitro system from rabbit reticulocytes and in a suspension of reticulocytes. The inhibition took place at the elongation–termination step of protein synthesis; most other trichothecenes (T-2 toxin and its metabolites, nivalenol and fusarenone X) inhibit the initial step of protein synthesis

(Ueno, 1983). It also inhibits DNA synthesis in murine splenic lymphocytes and human peripheral blood lymphocytes (Mekhancha-Dahel et al., 1990).

At low doses, deoxynivalenol causes immunotoxicity, with a no-observed-effect level in mice of 0.25–0.50 mg/kg bw per day (Tryphonas et al., 1986). Of particular interest is the capacity of dietary deoxynivalenol to induce extremely high levels of immunoglobulin A (IgA) in mice, owing to alteration of IgA production at the mucosal and systemic levels. Dietary deoxynivalenol enhances terminal differentiation of IgA-secreting cells in Peyer's patches. This and resultant migration of IgA-secreting cells into the systemic compartment favour a shift from IgG to IgA as the primary serum isotype (Bondy & Pestka, 1991).

Deoxynivalenol inhibits phytohaemaglutinin-induced lymphocyte proliferation, reducing ^{3}H-thymidine incorporation into DNA by 50% at a concentration of 90 ng/ml in rat lymphocytes and at a concentration of 220 ng/ml in human lymphocytes (Atkinson & Miller, 1984).

Nivalenol

LD_{50}s for nivalenol in six-week-old male DDY mice were found to be 38.9 (oral), 7.4 (intraperitoneal), 7.2 (subcutaneous) and 7.3 (intravenous) mg/kg bw. Post-mortem examination revealed marked congestion and haemorrhage in the intestine (Ryu et al., 1988). Tatsuno (1968) determined the intraperitoneal LD_{50} for nivalenol in ddS mice to be 4.0 mg/kg bw; pathological changes included cell degeneration of bone marrow, lymph nodes, intestines, testes and thymus. Nivalenol induced radiomimetic damage in animal cells, and newborn mice were found to be much more sensitive than adult mice, having a subcutaneous LD_{50} of 0.16 mg/kg bw (Ueno, 1987). In Fischer 344 rats, the oral LD_{50} was 19.5 mg/kg bw. Sedation, eyelid closure, staggering gait, diarrhoea and congestion of the lungs and digestive tract were observed (Kawasaki et al., 1990). The subcutaneous LD_{50} in rats was found to be 0.9 mg/kg bw (Ueno, 1983).

Subacute toxicity studies were performed for 24 days, during which female mice were given 30 mg/kg nivalenol in the diet. Significant erythropenia and slight leukopenia were observed, but no marked change was seen in other haematological parameters, feed consumption, body weight gain or weights of liver, spleen or thymus. Ultrastructural studies revealed polyribosomal breakdown of bone-marrow cells. No effect was seen in groups receiving 10 mg/kg nivalenol or less in the feed (Ryu et al., 1987).

In another subacute toxicity test, nivalenol was given orally at daily doses of 0.4 and 2.0 mg/kg bw to rats for 30 days. No significant change was observed in biological or haematological parameters. Liver and spleen weights were slightly increased with the dose of 2.0 mg/kg bw, but no histological change was seen (Kawasaki et al., 1990).

The long-term toxicity of nivalenol was studied for one (Ryu et al., 1988) and two (Ohtsubo et al., 1989) years in female mice fed diets containing 0, 6, 12 or 30 mg/kg nivalenol. Body weight gain and feed consumption showed dose-dependent decreases throughout the study period, indicating that nivalenol retards growth. The absolute weight of the liver in the group given 30 mg/kg and that of the kidneys in the groups given 12 and 30 mg/kg were significantly reduced compared with those of controls. When kidney weight was expressed relative to brain weight, a reduction was seen only in the group given 12 mg/kg. Some leukopenia was seen in nivalenol-treated animals, and dose-dependent increases in the serum concentrations of alkaline phosphatase and non-esterified fatty acids were observed. No ultrastructural change in the bone marrow was noted after one year.

Mice given nivalenol developed changes in proliferating cells of the small intestine and germ centres of lymph follicles in spleen, lymph node, thymus and bone marrow; they also developed testicular lesions: the spermatogenic cells were reduced in number, some of the blastic cells were necrotic, and multinucleated spermatic giant cells were present in the seminiferous tubules (Saito et al., 1969).

The lowest subcutaneous dose that caused vomiting in ducklings was 1.0 mg/kg bw (Ueno, 1987).

Skin necrotization occurred in guinea-pigs and mice painted with 100 μg nivalenol, showing that this compound is much less active than diacetoxyscirpenol or fusarenone X (Ueno et al., 1970b).

Nivalenol inhibited multiplication of HeLa cells at every phase of the growth cycle. By means of autoradiography, nivalenol at concentrations higher than 0.5 μg/ml was shown to block G1 about 2 h before the beginning of S phase and to block G2 just before mitosis (Ohtsubo et al., 1968). It did not inhibit RNA synthesis in HeLa cells, but at a dose of 15 μg/ml it caused complete breakdown of polyribosomes in HeLa cells after only 1 min (Cundliffe et al., 1974).

Nivalenol was also highly toxic to other cells of human origin—uterine carcinoma, embryonic kidney and lymphocytes—at $ID_{50}s$ between 0.3 and 1.0 μg/ml (Tanaka et al., 1978).

Nivalenol inhibited protein synthesis in rabbit reticulocytes at an ID_{50} of 2.5 μg/ml. It also inhibited poly U-directed synthesis of polyphenylalanine in a rabbit reticulocyte cell-free system at the ribosomal level, at an ID_{50} of 0.5 μg/ml (Ueno et al., 1968). It inhibited protein synthesis (ID_{50}, 6 μg/ml) and DNA synthesis (ID_{50}, > 10 μg/ml) in Ehrlich ascites tumour cells (Ueno et al., 1973).

Fusarenone X

The LD_{50} of fusarenone X in mice was 3–5 mg/kg bw, irrespective of the route of administration or the sex of the animal used. The survival time of mice injected intraperitoneally with the toxin was usually three to four days; when a dose several times higher than the LD_{50} was given, mice survived for less than 24 h, regardless of the route of administration. Newborn mice were highly sensitive, succumbing to lethal intoxication at doses of less than 0.1 mg/kg bw. Of the species tested, guinea-pigs are the most sensitive. In cats and ducklings, vomiting is a major symptom (Ueno et al., 1971).

In mice administered fusarenone X, severe cellular destruction and karyorrhexis were induced in actively dividing cells of the gastrointestinal tract, thymus, lymph nodes, spleen, bone marrow, ovary and testes. These effects were independent of the route of administration (Ueno et al., 1971).

Administration of fusarenone X to mice, rats and guinea-pigs induced multiple symptoms, including diarrhoea, food refusal, vomiting and hyperaemia in the intestine (Matsuoka & Kubota, 1981, 1985). Fusarenone X was highly irritating to the skin of mice, rabbits and guinea-pigs, causing haemorrhage and necrosis of the epidermis and degeneration and necrosis of hair follicles and dermis. Medium to severe toxicity (depending on the animal species) was observed after application of 10–100 µg of the toxin for two days (Ueno et al., 1970b). Foot oedema was induced in rats after subplantar injection of 10–100 µg (Matsuoka et al., 1979).

Fusarenone X caused diarrhoea in mice and rats by increasing the permeability of intestinal epithelial cells, resulting in exudation of plasma contents and an eventual increase in the volume of intestinal fluid (Matsuoka & Kubota, 1981, 1987a,b). Fusarenone X (intraperitoneally injected; 1, 2, 4 mg/kg bw) caused a dose-related increase in capillary permeability in mice (Matsuoka & Kubota, 1987b).

Fusarenone X is a potent inhibitor of protein and DNA syntheses in rat hepatocytes, rabbit and guinea-pig reticulocytes, Ehrlich ascites tumour cells, HeLa cells, guinea-pig splenic cells, mouse fibroblasts and *Escherichia coli* (Ohtsubo & Saito, 1970; Ohtsubo et al., 1972; Ueno et al., 1973; Ito et al., 1982). The ID_{50} for protein synthesis was 0.25 µg/ml in whole rabbit reticulocytes, 0.2 in a cell-free rabbit reticulocyte system and 8 µg/ml in a cell-free rat liver system. The toxin binds to the 80 S eukaryotic ribosomes and polysomes (Ueno, 1977). No inhibition of RNA synthesis was found. The ID_{50}s for incorporation of ^{3}H-thymidine into DNA and of ^{3}H-leucine into protein of HeLa S3 cells were 0.1 and 0.13 µg/ml (2.8 and 3.6 × 10^{-7} M), respectively (Ohtsubo & Saito, 1970).

Fusarenone X binds *in vitro* to active SH groups of creatine phosphokinase, lactate dehydrogenase and alcohol dehydrogenase, inhibiting their activities (Ueno & Matsumoto, 1975).

Daily intraperitoneal administration of fusarenone X from 25 µg for seven days had an immunosuppressive effect in BALB/c mice. IgE and IgG1 antibody formation *in vivo*, as well as in-vitro antibody formation by splenic lymphocytes raised by T-dependent and independent mitogens, was suppressed. Strongly adherent, phagocytic, suppressive cellular elements—possibly activated macrophages—were found in the spleens of treated animals (Obara et al., 1984).

Fusarenone X injected intraperitoneally to guinea-pigs several times at a dose of 0.75 mg/kg bw inhibited the responses of splenic cells towards mitogenic stimulation with lipopolysaccharide of *E. coli* and concanavalin A *in vitro*. *In vivo*, it did not appreciably affect the antibody response to 2,4-dinitrophenyl-bovine serum albumin. The levels of IgG_1 and IgG_2 were not significantly reduced after four weeks (Ito et al., 1982).

4.3 Reproductive and developmental toxicity

4.3.1 *Humans*

No data were available to the Working Group.

4.3.2 *Experimental systems*

Zearalenone

Groups of 10 Wistar rats received 1, 5 or 10 mg/kg bw zearalenone by oral intubation on days 6–15 of gestation. Mean fetal weight was significantly reduced in the high-dose group, and there was a higher prevalence of minor skeletal anomalies in the fetuses (Ruddick et al., 1976).

Groups of 50 male and 50 female Wistar rats received 0.1, 1.0 or 10.0 mg/kg bw zearalenone daily in the diet. Fertility was greatly impaired in females receiving the high dose: only 26 pregnancies resulted from 50 matings in the F_1 generation, and the number of resorptions and stillbirths was greatly increased; none of the 12 females in the F_2 generation that were mated became pregnant. No teratogenic response was seen at any dose (Becci *et al.*, 1982b).

Newborn female mice were injected daily for five days with 1 μg zearalenone. At eight months of age, corpora lutea were absent from 25/34 treated mice, indicating ovarian dysfunction; 56% of the exposed mice had dense collagen deposition in the uterine stroma and lacked uterine glands. Altered vaginal epithelium was found in 32% of the zearalenone-treated mice (Williams *et al.*, 1989). Similar effects were induced in rats after a single subcutaneous injection of 1 mg zearalenone to three- or five-day-old animals (Kumagai & Shimizu, 1982).

Deoxynivalenol

Groups of 15–19 pregnant Swiss-Webster mice were treated daily on days 8–11 of pregnancy with deoxynivalenol (purity, 96%; containing 4% 4,7-*d*-dideoxynivalenol as an impurity) at 0.5–15 mg/kg bw by oral intubation. Doses of 2.5 mg/kg bw and higher significantly increased the resorption rate; doses of 10 mg/kg bw or more induced complete resorption of all embryos. Low incidences of several anomalies (lumbar vertebrae, ribs, sternebrae) were observed in fetuses from the 1.0-, 2.5- and 5.0-mg/kg bw groups. No increase in the incidence of adverse effects was observed in the 0.5-mg/kg bw dose group (Khera *et al.*, 1982).

Male and female Swiss-Webster mice were fed diets resulting in daily doses of 0.375–2.0 mg/kg bw deoxynivalenol before mating and during pregnancy, and progeny (F_{1a}) were examined up to 21 days of age. Mice were then rebred to produce F_{1b} litters, which were evaluated on day 19 of gestation for gross, visceral and skeletal malformations. The highest dose decreased food consumption and reduced body weight in animals of each sex in the F_0 generation; transient body weight reduction during the last week of pregnancy was also observed in female mice exposed to 1.5 mg/kg bw. Fertility was not impaired by the treatment. Pronounced postnatal mortality was observed in the highest-dose group only. No major malformation was found (Khera *et al.*, 1984).

Male and female Sprague-Dawley rats were fed diets containing deoxynivalenol at 0.25, 0.5 or 1.0 mg/kg bw for six weeks prior to mating; females were treated throughout pregnancy. Dilatation of the renal pelvis (15, 9 and 29% compared to 0 in controls) and urinary bladder (24, 39 and 34% compared to 11% in controls) was observed in exposed fetuses; no other adverse effect was noted (Khera *et al.*, 1984).

Fischer 344 rats were fed a diet containing deoxynivalenol at 0, 0.5, 2.0 or 5.0 mg/kg during pregnancy. At the end of the experiment, the body weights of females (without fetuses and uterus) in the two highest-dose groups were significantly lower than those in controls. Fetal weight was unaffected by treatment; and no significant adverse effect on the incidence of gross, skeletal or visceral abnormalities was noted (Morrissey, 1984).

Administration of a diet containing 20 mg/kg deoxynivalenol to male and female Sprague-Dawley rats before mating and throughout pregnancy induced a slight decrease in fertility of females, but there was no significant difference from the control group with respect to postnatal survival of pups, number of pups born, sex ratio, mean pup weight or percentage of animals alive at four days that survived the 21-day lactation period (Morrissey & Vesonder, 1985).

New Zealand White rabbits were fed diets containing deoxynivalenol from day 0 to day 30 of gestation to give daily intakes of 0, 0.3, 0.6, 1.0, 1.6, 1.8 and 2.0 mg/kg bw. A decrease in mean fetal body weight was observed after daily intakes of 1.0 and 1.6 mg/kg bw. Complete resorption of fetuses was noticed in the two highest-dose groups. Doses that had no maternal toxic effect (0.3 and 0.6 mg/kg bw) had no adverse effect on fetuses at term. No teratogenic effect was observed (Khera *et al.*, 1986).

Nivalenol

Nivalenol was injected intraperitoneally into pregnant ICR mice at doses of 0, 0.1, 0.5 or 1.5 mg/kg bw daily from day 7 to day 15 of gestation. The highest dose caused stillbirths after vaginal haemorrhage in 6 of 10 animals. High percentages of embryolethality (48.4 and 87.8%) were recorded in the two highest-dose groups. No fetal malformation was observed. A single administration of 3 mg/kg bw on day 7 affected embryos within 10 h, damaged the placenta within 24 h and induced stillbirths by 48 h (Ito *et al.*, 1986).

Fusarenone X

All of a group of four DDD female mice treated with a single subcutaneous injection of 2.6 mg fusarenone X [purity unspecified] on day 10 of gestation aborted the following day. Abortion occurred less frequently (16–20%) at doses of 0.63–1.6 mg/kg bw and at longer intervals after injection. The weight and length of surviving fetuses from dams given 1.6 mg/kg bw on day 6 or 8 of gestation were significantly reduced as compared with those of controls. All mice fed diets containing 10 or 20 mg/kg fusarenone X (approximately 50 or 100 µg/animal per day) aborted in early pregnancy or throughout pregnancy. Feeding of diets containing 10 and 20 mg/kg fusarenone X for seven days during the middle of pregnancy caused 40 and 100% of dams to abort, respectively. No teratogenic effect was observed (Ito *et al.*, 1980).

4.4 Genetic and related effects

4.4.1 *Humans*

No data were available to the Working Group.

4.4.2 *Experimental systems* (see also Tables 4–7 and Appendices 1 and 2)

Zearalenone

The genotoxic effects of zearalenone and some of its derivatives have been reviewed (Kuiper-Goodman *et al.*, 1987).

Zearalenone did not induce SOS error-prone DNA repair in *Escherichia coli*, although the *rec* assay indicated that the compound induced differential toxicity in repair-deficient and -proficient *Bacillus* strains. Zearalenone did not induce mutation in *Salmonella typhimurium* or mitotic crossing over in *Saccharomyces cerevisiae*. In Chinese hamster cells *in vitro*, zearalenone induced sister chromatid exchange, chromosomal aberrations and polyploidy; sister chromatid exchange was weakly induced in cultured human lymphocytes.

β-Zearalenol, but not α-zearalenol, induced differential toxicity in *Bacillus subtilis* strains.

Deoxynivalenol

Deoxynivalenol was not mutagenic to *Salmonella typhimurium*. In Chinese hamster V79 cells *in vitro*, it inhibited gap-junctional intercellular communication and induced chromosomal aberrations, but it did not produce gene mutation at the *hprt* locus. A sample of maize from Linxian County, China, was extracted with acetonitrile and water and fractionated by HPLC. The fraction that co-eluted with deoxynivalenol induced chromosomal aberrations in V79 cells (Hsia *et al.*, 1988). Deoxynivalenol did not induce unscheduled DNA synthesis in rat primary hepatocyte cultures, but it enhanced cell transformation in mouse embryo cells *in vitro*. A precursor, 3-acetyldeoxynivalenol, also induced chromosomal aberrations in mammalian cells *in vitro* (Hsia *et al.*, 1988).

Nivalenol

In Chinese hamster V79 cells *in vitro*, nivalenol slightly increased the frequencies of chromosomal aberrations and sister chromatid exchange. The fraction of maize described above that co-eluted with nivalenol induced chromosomal aberrations in V79 cells (Hsia *et al.*, 1988).

Fusarenone X

Fusarenone X did not induce DNA damage and was not mutagenic to *Salmonella typhimurium* in one study; another study, which was considered to provide positive results by a previous working group (IARC, 1983), was considered to be inadequate. Fusarenone X increased the frequency of petite mutations in yeast. It did not induce gene mutation in cultured mouse carcinoma cells, but it induced chromosomal aberrations and (weakly) sister chromatid exchange in Chinese hamster cells *in vitro*. Weak induction of DNA single-strand breaks was described in cultured human cells.

5. Summary of Data Reported and Evaluation

5.1 Exposure data

The mycotoxins considered are produced by *Fusarium* species that occur primarily on wheat, barley and maize. The toxins occur whenever these cereals are grown under humid conditions. Exposure occurs through dietary consumption of contaminated cereals. Deoxynivalenol has been held responsible for large-scale human poisonings this century in China and India. Chronic exposures to deoxynivalenol, zearalenone and nivalenol occur in several parts of the world; humans are rarely exposed to fusarenone X.

5.2 Human carcinogenicity data

A few ecological studies that considered *F. graminearum* suggested no correlation with the incidence of oesophageal cancer.

Table 4. Genetic and related effects of zearalenone and α– and β–zearalenol

Test system	Result		Dose (LED/HID)[a]	Reference
	Without exogenous metabolic system	With exogenous metabolic system		
PRB, SOS spot test, *Escherichia coli*	–	–	0.0000	Auffray & Boutibonnes (1986)
PRB, SOS chromotest test, *Escherichia coli* PQ37	–	–	30.0000	Krivobok et al. (1987)
BSD, *Bacillus subtilis* rec strains, differential toxicity	+	0	100 µg/plate	Ueno & Kubota (1976)
BSD, *Bacillus subtilis* rec strains, differential toxicity	+	0	500.0000	Boutibonnes et al. (1984)
SA0, *Salmonella typhimurium* TA100, reverse mutation	–	–	200.0000	Wehner et al. (1978)
SA0, *Salmonella typhimurium* TA100, reverse mutation	–	–	25.0000	Boutibonnes & Loquet (1979)
SA0, *Salmonella typhimurium* TA100, reverse mutation	–	–	100.0000	Bartholomew & Ryan (1980)
SA0, *Salmonella typhimurium* TA100, reverse mutation	–	–	250.0000	Ingerowski et al. (1981)
SA0, *Salmonella typhimurium* TA100, reverse mutation	–	–	167.0000	Mortelmans et al. (1986)
SA5, *Salmonella typhimurium* TA1535, reverse mutation	–	–	50.0000	Kuczuk et al. (1978)
SA5, *Salmonella typhimurium* TA1535, reverse mutation	–	–	200.0000	Wehner et al. (1978)
SA5, *Salmonella typhimurium* TA1535, reverse mutation	–	–	25.0000	Ingerowski et al. (1981)
SA5, *Salmonella typhimurium* TA1535, reverse mutation	–	–	167.0000	Mortelmans et al. (1986)
SA7, *Salmonella typhimurium* TA1537, reverse mutation	–	–	50.0000	Kuczuk et al. (1978)
SA7, *Salmonella typhimurium* TA1537, reverse mutation	–	–	200.0000	Wehner et al. (1978)
SA7, *Salmonella typhimurium* TA1537, reverse mutation	–	–	25.0000	Ingerowski et al. (1981)
SA7, *Salmonella typhimurium* TA1537, reverse mutation	–	–	167.0000	Mortelmans et al. (1986)
SA8, *Salmonella typhimurium* TA1538, reverse mutation	–	–	50.0000	Kuczuk et al. (1978)
SA8, *Salmonella typhimurium* TA1538, reverse mutation	–	–	25.0000	Bartholomew & Ryan (1980)
SA8, *Salmonella typhimurium* TA1538, reverse mutation	–	–	250.0000	Ingerowski et al. (1981)
SA9, *Salmonella typhimurium* TA98, reverse mutation	–	–	200.0000	Wehner et al. (1978)
SA9, *Salmonella typhimurium* TA98, reverse mutation	–	–	25.0000	Boutibonnes & Loquet (1979)
SA9, *Salmonella typhimurium* TA98, reverse mutation	–	–	100.0000	Bartholomew & Ryan (1980)
SA9, *Salmonella typhimurium* TA98, reverse mutation	–	–	250.0000	Ingerowski et al. (1981)
SA9, *Salmonella typhimurium* TA98, reverse mutation	–	–	167.0000	Mortelmans et al. (1986)

Table 4 (contd)

Test system	Result		Dose (LED/HID)a	Reference
	Without exogenous metabolic system	With exogenous metabolic system		
SCH, *Saccharomyces cerevisiae* D-3, mitotic crossing over, ade2 locus	–	–	100 µg/plate	Kuczuk et al. (1978)
GML, Gene mutation, mouse lymphoma L5178Y tk+/tk– cells *in vitro*	–	–	60.0000	McGregor et al. (1988)
SIC, Sister chromatid exchange, Chinese hamster V79 lung cells *in vitro*	–	–	32.0000	Thust et al. (1983)
SIC, Sister chromatid exchange, Chinese hamster ovary cells *in vitro*	+	–	10.0000	Galloway et al. (1987)
CIC, Chromosomal aberrations, Chinese hamster V79 lung cells *in vitro*	–	–	32.0000 (no concurrent control)	Thust et al. (1983)
CIC, Chromosomal aberrations, Chinese hamster ovary cells *in vitro*	+	(+)	15.0000	Galloway et al. (1987)
PIA, Polyploidy, Chinese hamster ovary cells *in vitro*	+	–	10.0000	Galloway et al. (1987)
SHL, Sister chromatid exchange, human lymphocytes *in vitro*	(+)	0	3.0000	Cooray (1984)

Zearalenol (α and β-mixture or unidentified)

Test system	Result		Dose (LED/HID)a	Reference
PRB, SOS chromotest test, *Escherichia coli* PQ37	–	0	60.0000	Krivobok et al. (1987)
BSD, *Bacillus subtilis* rec strains, differential toxicity	–	0	500.0000 (single dose)	Boutibonnes et al. (1984)

α-Zearalenol

Test system	Result		Dose (LED/HID)a	Reference
BSD, *Bacillus subtilis* rec strains, differential toxicity	–	0	100 µg/plate	Ueno & Kubota (1976)
SA0, *Salmonella typhimurium* TA100, reverse mutation	–	–	50.0000	Bartholomew & Ryan (1980)
SA0, *Salmonella typhimurium* TA100, reverse mutation	–	–	250.0000	Ingerowski et al. (1981)
SA5, *Salmonella typhimurium* TA1535, reverse mutation	–	–	250.0000	Ingerowski et al. (1981)
SA7, *Salmonella typhimurium* TA1537, reverse mutation	–	–	250.0000	Ingerowski et al. (1981)
SA8, *Salmonella typhimurium* TA1538, reverse mutation	–	–	250.0000	Ingerowski et al. (1981)
SA8, *Salmonella typhimurium* TA1538, reverse mutation	–	–	250.0000	Bartholomew & Ryan (1980)

Table 4 (contd)

Test system	Result		Dose (LED/HID)a	Reference
	Without exogenous metabolic system	With exogenous metabolic system		
α-Zearalenol (contd)				
SA9, *Salmonella typhimurium* TA98, reverse mutation	–	–	250.0000	Bartholomew & Ryan (1980)
SA9, *Salmonella typhimurium* TA98, reverse mutation	–	–	250.0000	Ingerowski *et al.* (1981)
β-Zearalenol				
BSD, *Bacillus subtilis rec* strains, differential toxicity	+	0	100 μg/plate	Ueno & Kubota (1976)
SA9, *Salmonella typhimurium* TA98, reverse mutation	–	–	250.0000	Ueno *et al.* (1978)
SA0, *Salmonella typhimurium* TA100, reverse mutation	–	–	50.0000	Bartholomew & Ryan (1980)
SA8, *Salmonella typhimurium* TA1538, reverse mutation	–	–	250.0000	Bartholomew & Ryan (1980)
SA9, *Salmonella typhimurium* TA98, reverse mutation	–	–	250.0000	Bartholomew & Ryan (1980)

+, positive; (+), weakly positive; –, negative; 0, not tested; ?, inconclusive (e.g., variable response in several experiments within an adequate study)
[a]In-vitro tests, μg/ml; 0.0000, not given

Table 5. Genetic and related effects of deoxynivalenol

Test system	Result		Dose (LED/HID)[a]	Reference
	Without exogenous metabolic system	With exogenous metabolic system		
SA0, *Salmonella typhimurium* TA100, reverse mutation	–	–	200.0000	Wehner et al. (1978)
SA5, *Salmonella typhimurium* TA1535, reverse mutation	–	–	200.0000	Wehner et al. (1978)
SA7, *Salmonella typhimurium* TA1537, reverse mutation	–	–	200.0000	Wehner et al. (1978)
SA9, *Salmonella typhimurium* TA98, reverse mutation	–	–	200.0000	Wehner et al. (1978)
URP, Unscheduled DNA synthesis, rat primary hepatocytes *in vitro*	–	0	1000.0000	Bradlaw et al. (1985)
G9H, Gene mutation, Chinese hamster V79 cells, thioguanine[r] *in vitro*	–	–[b]	3.0000	Rogers & Héroux-Metcalf (1983)
CIC, Chromosomal aberrations, Chinese hamster V79 cells *in vitro*	+	0	0.1 μg/l[c]	Hsia et al. (1988)
CIC, Chromosomal aberrations, Chinese hamster V79 cells *in vitro*	+	0	0.1 μg/l	Hsia et al. (1988)
TBM, Cell transformation, BALB/c 3T3 A31-1-1 mouse embryo cells *in vitro*	+	0	0.2000	Sheu et al. (1988)
ICR, Inhibition of gap-junctional intercellular communication, Chinese hamster V79 cells *in vitro*	+	0	0.3000	Jone et al. (1987)

+, positive; –, negative; 0, not tested

[a]In-vitro tests, μg/ml

[b]Hepatocyte-mediated activation

[c]HPLC fraction of maize sample that co-eluted with deoxynivalenol

Table 6. Genetic and related effects of nivalenol

Test system	Result		Dose (LED/HID)[a]	Reference
	Without exogenous metabolic system	With exogenous metabolic system		
SIC, Sister chromatid exchange, Chinese hamster V79 cells *in vitro*	(+)[b]	(+)	3.0000	Thust et al. (1983)
CIC, Chromosomal aberrations, Chinese hamster V79 cells *in vitro*	–	(+)[c]	15.0000 (5×10^{-5}M) (no concurrent control)	Thust et al. (1983)
CIC, Chromosomal aberrations, Chinese hamster V79 cells *in vitro*	+	0	0.001[d]	Hsia et al. (1988)

+, positive; (+), weakly positive; –, negative; 0, not tested; ?, inconclusive (e.g., variable response in several experiments within an adequate study)
[a]In-vitro tests, μg/ml
[b]Toxic doses not reached
[c]All are chromatid exchanges
[d]HPLC fraction of maize sample that co-eluted with nivalenol

Table 7. Genetic and related effects of fusarenone X

Test system	Result		Dose (LED/HID)[a]	Reference
	Without exogenous metabolic system	With exogenous metabolic system		
PRB, Prophage induction, *Escherichia coli* K-12	–	0	100.0000	Ueno *et al.* (1971)
BSD, *Bacillus subtilis* rec strains, differential toxicity	–	0	100 µg/plate	Ueno & Kubota (1976)
SA0, *Salmonella typhimurium* TA100, reverse mutation	–	–	250.0000	Ueno *et al.* (1978)
SA9, *Salmonella typhimurium* TA98, reverse mutation	–	–	250.0000	Ueno *et al.* (1978)
SCF, *Saccharomyces cerevisiae*, petite mutation	+	0	250.0000	Ueno *et al.* (1971)
GIA, Gene mutation, mouse mammary carcinoma FM3A cells, 8-azaguanine[r] *in vitro*	–	0	1.0000	Umeda *et al.* (1977)
SIC, Sister chromatid exchange, Chinese hamster V79 cells *in vitro*	(+)	(+)	3.0000	Thust *et al.* (1983)
CIC, Chromosomal aberrations, Chinese hamster V79 cells *in vitro*	+	+	3.0000[b]	Thust *et al.* (1983)
DIH, DNA single-strand breaks, human HeLa cells *in vitro*	(+)	0	32.0000	Umeda *et al.* (1972)

+, positive; (+), weakly positive; –, negative; 0, not tested
[a]In-vitro tests, µg/ml
[b]No concurrent control

5.3 Animal carcinogenicity data

Zearalenone was tested for carcinogenicity by administration in the diet in one experiment in mice and in two experiments in rats. An increased incidence of hepatocellular adenomas was observed in female mice and of pituitary adenomas in mice each sex. No increase in the incidence of tumours was observed in rats.

No data were available to the Working Group on the carcinogenicity in experimental animals of deoxynivalenol.

Nivalenol was tested for carcinogenicity in one experiment in female mice by oral administration in the diet. No increase in tumour incidence was observed.

Fusarenone X was tested for carcinogenicity in two studies in male rats by oral administration and in male mice and male rats by subcutaneous injection. The studies were inadequate for evaluation.

5.4 Other relevant data

In episodes of food poisoning in humans caused by deoxynivalenol, severe gastrointestinal involvement was the primary sign.

Zearalenone has oestrogenic effects in domestic pigs and experimental animals. Deoxynivalenol causes outbreaks of feed refusal and vomiting in domestic pigs. Deoxynivalenol and fusarenone X cause immunosuppression in mice. Nivalenol causes bone-marrow toxicity in experimental animals.

No data were available on the genetic and related effects of zearalenone, deoxynivalenol, nivalenol or fusarenone X in humans.

Zearalenone induces chromosomal anomalies in cultured rodent cells. It does not induce recombination in yeast or gene mutation or DNA damage in bacteria.

Deoxynivalenol induces cell transformation, chromosomal aberrations and inhibition of gap-junctional intercellular communication in cultured mammalian cells. It does not induce unscheduled DNA synthesis or mutation in cultured mammalian cells and does not induce mutation in bacteria.

Nivalenol and fusarenone X have not been studied adequately for genetic effects.

5.5 Evaluation[1]

There is *inadequate evidence* in humans for the carcinogenicity of toxins derived from *Fusarium graminearum*.

No data were available on the carcinogenicity to humans of toxins derived from *F. crookwellense* and *F. culmorum*.

There is *limited evidence* in experimental animals for the carcinogenicity of zearalenone.

There is *inadequate evidence* in experimental animals for the carcinogenicity of deoxynivalenol.

---

[1]For definition of the italicized terms, see Preamble, pp. 26–29.

There is *inadequate evidence* in experimental animals for the carcinogenicity of nivalenol.

There is *inadequate evidence* in experimental animals for the carcinogenicity of fusarenone X.

Overall evaluation

Toxins derived from *Fusarium graminearum*, *F. culmorum* and *F. crookwellense* are not classifiable as to their carcinogenicity to humans (Group 3).

6. References

Abbas, H.K., Mirocha, C.J., Meronuck, R.A., Pokorny, J.D., Gould, S.L. & Kommedahl, T. (1988) Mycotoxins and *Fusarium* spp. associated with infected ears of corn in Minnesota. *Appl. environ. Microbiol.*, 54, 1930–1933

Abdelhamid, A.M. (1990) Occurrence of some mycotoxins (aflatoxins, ochratoxin A, citrinin, zearalenone and vomitoxin) in various Egyptian feeds. *Arch. anim. Nutr.*, 40, 647–664

Abouzied, M.M., Azcona, J.I., Braselton, W.E. & Pestka, J.J. (1991) Immunochemical assessment of mycotoxins in 1989 grain foods: evidence for deoxynivalenol (vomitoxin) contamination. *Appl. environ. Microbiol.*, 57, 672–677

Appelgren, L.-E., Arora, R.G. & Larsson, P. (1982) Autoradiographic studies of [^{3}H]zearalenone in mice. *Toxicology*, 25, 243–253

Atkinson, H.A.C. & Miller, K. (1984) Inhibitory effect of deoxynivalenol, 3-acetyldeoxynivalenol and zearalenone on induction of rat and human lymphocyte proliferation. *Toxicol. Lett.*, 23, 215–221

Auffray, Y. & Boutibonnes, P. (1986) Evaluation of the genotoxic activity of some mycotoxins using *Escherichia coli* in the SOS spot test. *Mutat. Res.*, 171, 79–82

Bartholomew, R.M. & Ryan, D.S. (1980) Lack of mutagenicity of some phytoestrogens in the *Salmonella*/mammalian microsome assay. *Mutat. Res.*, 78, 317–321

Bata, A., Ványi, A. & Lásztity, R. (1983) Simultaneous detection of some fusariotoxins by gas–liquid chromatography. *J. Assoc. off. anal. Chem.*, 66, 577–581

Bauer, J., Heinritzi, K., Gareis, M. & Gedek, B. (1987) Changes in the genital tract of female pigs after feeding practice-relevant amounts of zearalenone (Ger.). *Tierärztl. Prax.*, 15, 33–36

Bean, G.A. & Echandi, R. (1989) Maize mycotoxins in Latin America. *Plant Dis.*, 73, 597–600

Becci, P.J., Voss, K.A., Hess, F.G., Gallo, M.A., Parent, R.A., Stevens, K.R. & Taylor, J.M. (1982a) Long-term carcinogenicity and toxicity study of zearalenone in the rat. *J. appl. Toxicol.*, 2, 247–254

Becci, P.J., Johnson, W.D., Hess, F.G., Gallo, M.A., Parent, R.A. & Taylor, J.M. (1982b) Combined two-generation reproduction–teratogenesis study of zearalenone in the rat. *J. appl. Toxicol.*, 2, 201–206

Bennett, G.A., Peplinski, A.J., Brekke, O.L., Jackson, L.K. & Wichser, W.R. (1976) Zearalenone: distribution in dry-milled fractions of contaminated corn. *Cereal Chem.*, 53, 299–307

Bhat, R.V., Beedu, S.R., Ramakrishna, Y. & Munshi, K.L. (1989) Outbreak of trichothecene mycotoxicosis associated with consumption of mould-damaged wheat products in Kashmir Valley, India. *Lancet*, i, 35–37

Blaney, B.J. (1991) *Fusarium* and *Alternaria* toxins. In: Champ, B.R., Highley, E., Hocking, A.D. & Pitt, J.I., eds, *Fungi and Mycotoxins in Stored Products* (ACIAR Proceedings 36), Canberra, Australian Centre for International Agricultural Research, pp. 86–98

Blaney, B.J. & Dodman, R.L (1988) Production of the mycotoxins zearalenone, 4-deoxynivalenol and nivalenol by isolates of *Fusarium graminearum* groups 1 and 2 from cereals in Queensland. *Austr. J. agric. Res.*, 39, 21–29

Blaney, B.J., Ramsey, M.D. & Tyler, A.L. (1986) Mycotoxins and toxigenic fungi in insect-damaged maize harvested during 1983 in far north Queensland. *Austr. J. agric. Res.*, 37, 235–244

Bondy, G.S. & Pestka, J.J. (1991) Dietary exposure to the trichothecene vomitoxin (deoxynivalenol) stimulates terminal differentiation of Peyer's patch B cells to IgA secreting plasma cells. *Toxicol. appl. Pharmacol.*, 108, 520–530

Bottalico, A., Lerario, P. & Visconti, A. (1981) Occurrence of trichothecenes and zearalenone in pre-harvest *Fusarium*-infected ears of maize from some Austrian localities. *Phytopathol. mediterr.*, 20, 1–6

Bottalico, A., Lerario, P. & Visconti, A. (1983) Mycotoxins occurring in *Fusarium*-infected maize ears in the field, in some European countries. In: *Proceedings of a Symposium on Mycotoxins, Cairo, 6–8 September 1981*, Cairo, National Research Centre, pp. 375–382

Bottalico, A., Logrieco, A. & Visconti, A. (1990) Mycotoxins produced by *Fusarium crookwellense*. *Phytopath. mediterr.*, 29, 124–127

Boutibonnes, P. & Loquet, C. (1979) Antibacterial activity, DNA-attacking ability and mutagenic ability of the mycotoxin zearalenone. *Int. Res. Commun. System med. Sci.*, 7, 204

Boutibonnes, P., Auffray, Y., Malherbe, C., Kogbo, W. & Marais, C. (1984) Antibacterial and genotoxic properties of 33 mycotoxins (Fr.). *Mycopathologia*, 87, 43–49

Bradlaw, J.A., Swentzel, K.C., Alterman, E. & Hauswirth, J.W. (1985) Evaluation of purified 4-deoxynivalenol (vomitoxin) for unscheduled DNA synthesis in the primary rat hepatocyte–DNA repair assay. *Food chem. Toxicol.*, 23, 1063–1067

Brumley, W.C., Andrzejewski, D., Trucksess, E.W., Dreifuss, P.A., Roach, J.A.G., Eppley, R.M., Thomas, F.S., Thorpe, C.W. & Sphon, J.A. (1982) Negative ion chemical ionization mass spectrometry of trichothecenes. Novel fragmentation under OH$^-$ conditions. *Biomed. Mass Spectrom.*, 9, 451–457

Budavari, S., ed. (1989) *Merck Index*, 11th ed., Rahway, NJ, Merck & Co., pp. 1052, 1596

Chulze, S., Bertinetti, C., Dalcero, A., Etcheverry, M., Farnochi, C., Torres, A., Rizzo, I. & Varsavsky, E. (1989) Incidence of aflatoxin, zearalenone and deoxynivalenol on corn in Argentina. *Mycotoxin Res.*, 5, 9–12

Cole, R.J. & Cox, R.H. (1981) *Handbook of Toxic Fungal Metabolites*, New York, Academic Press, pp. 202–205, 206–208, 213–216

Cooray, R. (1984) Effects of some mycotoxins on mitogen-induced blastogenesis and SCE frequency in human lymphocytes. *Food chem. Toxicol.*, 22, 529–534

Coppock, R.W., Swanson, S.P., Gelberg, H.B., Koritz, G.D., Hoffmann, W.E., Buck, W.B. & Vesonder, R.F. (1985) Preliminary study of the pharmacokinetics and toxicopathy of deoxynivalenol (vomitoxin) in swine. *Am. J. vet. Res.*, 46, 169–174

Côté, L.-M., Reynolds, J.D., Vesonder, R.F., Buck, W.B., Swanson, S.P., Coffey, R.T. & Brown, D.C. (1984) Survey of vomitoxin-contaminated feed grains in midwestern United States, and associated health problems in swine. *J. Am. vet. Med. Assoc.*, 184, 189–192

Côté, L.-M., Dahlem, A.M., Yoshizawa, T., Swanson, S.P. & Buck, W.B. (1986) Excretion of deoxynivalenol and its metabolite in milk, urine, and feces of lactating dairy cows. *J. Dairy Sci.*, 69, 2416–2423

Cundliffe, E., Cannon, M. & Davies, J. (1974) Mechanism of inhibition of eukaryotic protein synthesis by trichothecene fungal toxins. *Proc. natl Acad. Sci. USA*, 71, 30–34

Dailey, R.E., Reese, R.E. & Brouwer, E.A. (1980) Metabolism of [^{14}C]zearalenone in laying hens. *J. agric. Food Chem.*, **28**, 286–291

van Egmond, H.P. (1989) Current situation on regulations for mycotoxins. Overview of tolerances and status of standard methods of sampling and analysis. *Food Addit. Contam.*, **6**, 139–188

Eppley, R.M., Stoloff, L., Trucksess, M.W. & Chung, C.W. (1974) Survey of corn for Fusarium toxins. *J. Assoc. off. anal. Chem.*, **57**, 632–635

Eppley, R.M., Trucksess, M.W., Nesheim, S., Thorpe, C.W., Wood, G.E. & Pohland, A.E. (1984) Deoxynivalenol in winter wheat: thin layer chromatographic method and survey. *J. Assoc. off. anal. Chem.*, **67**, 43–45

Etienne, M. & Jemmali, M. (1982) Effects of zearalenone (F_2) on estrous activity and reproduction in gilts. *J. Anim. Sci.*, **55**, 1–10

FAO (1979) *Perspective on Mycotoxins. Proceedings of the Joint FAO/WHO/UNEP Conference on Mycotoxins, Nairobi, Kenya, 1977* (Food and Nutrition Paper 13, MYC 4a), Rome

Farnworth, E.R. & Trenholm, H.L. (1983) The metabolism of the mycotoxin zearalenone and its effects on the reproductive tracts of young male and female pigs. *Can. J. Anim. Sci.*, **63**, 967–975

Fitzpatrick, D.W., Boyd, K.E. & Watts, B.M. (1988) Comparison of the trichothecenes deoxynivalenol and T-2 toxin for their effects on brain biogenic monoamines in the rat. *Toxicol. Lett.*, **40**, 241–245

Fitzpatrick, D.W., Picken, C.A., Murphy, L.C. & Buhr, M.M. (1989) Measurement of the relative binding affinity of zearalenone, α-zearalenol and β-zearalenol for uterine and oviduct estrogen receptors in swine, rats and chickens: an indicator of estrogenic potencies. *Comp. Biochem. Physiol.*, **94C**, 691–694

Forsell, J.H., Jensen, R., Tai, J.-H., Witt, M., Lin, W.S. & Pestka, J.J. (1987) Comparison of acute toxicities of deoxynivalenol (vomitoxin) and 15-acetyldeoxynivalenol in the B6C3F$_1$ mouse. *Food chem. Toxicol.*, **25**, 155–162

Forsyth, D.M., Yoshizawa, T., Morooka, N. & Tuite, J. (1977) Emetic and refusal activity of deoxynivalenol to swine. *Appl. environ. Microbiol.*, **34**, 547–552

Foster, B.C., Neish, G.A., Lauren, D.R., Trenholm, H.L., Prelusky, D.B. & Hamilton, R.M.G. (1986) Fungal and mycotoxin content of slashed corn. *Microbiol. Alim. Nutr.*, **4**, 199–203

Funnel, H.S. (1979) Mycotoxins in animal feedstuffs in Ontario 1972 to 1977. *Can. J. comp. Med.*, **43**, 243–246

Galloway, S.M., Armstrong, M.J., Reuben, C., Colman, S., Brown, B., Cannon, C., Bloom, A.D., Nakamura, F., Ahmed, M., Duk, S., Rimpo, J., Margolin, B.H., Resnick, M.A., Anderson, B. & Zeiger, E. (1987) Chromosome aberrations and sister chromatid exchanges in Chinese hamster ovary cells: evaluations of 108 chemicals. *Environ. mol. Mutag.*, **10** (Suppl. 10), 1–175

Gang, L., Ying, X., Zhang, Z.H., Fang, C.M. & Li, L. (1988) Investigation of deoxynivalenol and zearalenone in wheat from Anhui Province. In: Aibara, K., Kumagai, S., Ohtsubo, K. & Yoshizawa, T., eds, *Proceedings of the 7th International IUPAC Symposium on Mycotoxins and Phycotoxins, Tokyo, 16–19 August 1988*, Tokyo, Japanese Association of Mycotoxicology, pp. 67–68

Ghess, M.-J., Wilbourn, J.D. & Vainio, H., eds (1992) *Directory of Agents Being Tested for Carcinogenicity*, No. 15, Lyon, IARC, p. 3

Gilbert, J. (1989) Current views on the occurrence and significance of *Fusarium* toxins. *J. appl. Bacteriol.*, **18** (Symp. Suppl.), 89S–98S

Gilbert, J. (1991) Accepted and collaboratively tested methods of sampling, detection, and analysis of mycotoxins. In: Champ, B.R., Highley, E., Hocking, A.D. & Pitt, J.I., eds, *Fungi and Mycotoxins in Stored Products* (ACIAR Proceedings 36), Canberra, Australian Centre for International Agricultural Research, pp. 108–114

Gilbert, J., Shepherd, M.J. & Startin, J.R. (1983) A survey of the occurrence of the trichothecene mycotoxin deoxynivalenol (vomitoxin) in UK grown barley and in imported maize by combined gas chromatography–mass spectrometry. *J. Sci. Food Agric.*, 34, 86–92

Goliński, P., Vesonder, R.F., Latus-Zietkiewicz, D. & Perkowski, J. (1988) Formation of fusarenone X, nivalenol, zearalenone, α-*trans*-zearalenol, β-*trans*-zearalenol, and fusarin C by *Fusarium crookwellense*. *Appl. environ. Microbiol.*, 54, 2147–2148

Greenhalgh, R., Neish, G.A. & Miller, J.D. (1983) Deoxynivalenol, acetyl deoxynivalenol and zearalenone formation by Canadian isolates of *Fusarium graminearum* on solid substrates. *Appl. environ. Microbiol.*, 46, 625–629

Greenhalgh, R., Levandier, D., Adams, W., Miller, J.D., Blackwell, B.A., McAlees, A.J. & Taylor, A. (1986) Production and characterization of deoxynivalenol and other secondary metabolites of *Fusarium culmorum* (CMI 14764; HLX 1503). *J. agric. Food Chem.*, 34, 98–102

Hagler, W.M., Dankó, G., Horváth, L., Palyusik, M. & Mirocha, C.J. (1980) Transmission of zearalenone and its metabolite into ruminant milk. *Acta vet. acad. sci. hung.*, 28, 209–216

Harvey, R.B., Kubena, L.F., Corrier, D.E., Witzel, D.A., Phillips, T.D. & Heidelbaugh, N.D. (1986) Effects of deoxynivalenol in a wheat ration fed to growing lambs. *Am. J. vet. Res.*, 47, 1630–1632

Hidy, P.H., Baldwin, R.S., Greasham, R.L., Keith, C.L. & McMullen, J.R. (1977) Zearalenone and some derivatives: production and biological activities. *Adv. appl. Microbiol.*, 22, 59–82

Hsia, C.C., Wu, J.L., Lu, X.Q. & Li, Y.S. (1988) Natural occurrence and clastogenic effects of nivalenol, deoxynivalenol, 3-acetyl-deoxynivalenol, 15-acetyl-deoxynivalenol, and zearalenone in corn from a high risk area of esophageal cancer. *Cancer Detect. Prev.*, 13, 79–86

Huff, W.E., Doerr, J.A., Hamilton, P.B. & Vesonder, R.F. (1981) Acute toxicity of vomitoxin (deoxynivalenol) in broiler chickens. *Poultry Sci.*, 60, 1412–1414

Hulan, H.W. & Proudfoot, F.G. (1982) Effects of feeding vomitoxin contaminated wheat on the performance of broiler chickens. *Poultry Sci.*, 61, 1653–1659

Hunder, G., Schümann, K., Strugala, G., Gropp, J., Fichtl, B. & Forth, W. (1991) Influence of subchronic exposure to low dietary deoxynivalenol, a trichothecene mycotoxin, on intestinal absorption of nutrients in mice. *Food chem. Toxicol.*, 29, 809–814

Hussein, H.M., Franich, R.A., Baxter, M. & Andrew, I.G. (1989) Naturally occurring *Fusarium* toxins in New Zealand maize. *Food Addit. Contam.*, 6, 49–58

IARC (1976) *IARC Monographs on the Evaluation of Carcinogenic Risk of Chemicals to Man*, Vol. 11, *Cadmium, Nickel, Some Epoxides, Miscellaneous Industrial Chemicals and General Considerations on Volatile Anaesthetics*, Lyon, pp. 169–173

IARC (1983) *IARC Monographs on the Evaluation of the Carcinogenic Risk of Chemicals to Humans*, Vol. 31, *Some Food Additives, Feed Additives and Naturally Occurring Substances*, Lyon, pp. 153–161, 279–291

Ichinoe, M., Kurata, H., Sugiura, Y. & Ueno, Y. (1983) Chemotaxonomy of *Gibberella zeae* with special reference to production of trichothecenes and zearalenone. *Appl. environ. Microbiol.*, 46, 1364–1369

Ikebuchi, H., Teshima, R., Hirai, K., Sato, M., Ichinoe, M. & Terao, T. (1990) Production and characterization of monoclonal antibodies to nivalenol tetraacetate and their application to enzyme-linked immunoassay of nivalenol. *Biol. Chem. Hoppe-Seyler*, 371, 31–36

Ingerowski, G.H., Scheutwinkel-Reich, M. & Stan, H.-J. (1981) Mutagenicity studies on veterinary anabolic drugs with the *Salmonella*/microsome test. *Mutat. Res.*, **91**, 93–98

Ito, Y., Ohtsubo, K. & Saito, M. (1980) Effects of fusarenon-X, a trichothecene produced by *Fusarium nivale*, on pregnant mice and their fetuses. *Jpn. J. exp. Med.*, **50**, 167–172

Ito, H., Watanabe, K. & Koyama, J. (1982) The immunosuppressive effects of trichothecenes and cyclochlorotine on the antibody responses of guinea pigs. *J. Pharm. Dyn.*, **5**, 403–409

Ito, Y., Ohtsubo, K., Ishii, K. & Ueno, Y. (1986) Effects of nivalenol on pregnancy and fetal development of mice. *Mycotoxin Res.*, **2**, 71–77

Jaskiewicz, K., Marasas, W.F.O. & van der Walt, F.E. (1987) Oesophageal and other main cancer patterns in four districts of Transkei, 1981-1984. *South Afr. med. J.*, **72**, 27–30

Jelinek, C.F., Pohland, A.E. & Wood, G.E. (1989) Review of mycotoxin contamination. Worldwide occurrence of mycotoxins in foods and feeds. An update. *J. Assoc. off. anal. Chem.*, **72**, 223–230

Jemmali, M., Ueno, Y., Ishii, K., Frayssinet, C. & Etienne, M. (1978) Natural occurrence of trichothecenes (nivalenol, deoxynivalenol, T_2) and zearalenone in corn. *Experientia*, **34**, 1333–1334

Jone, C., Erickson, L., Trosko, J.E. & Chang, C.C. (1987) Effect of biological toxins on gap-junctional intercellular communication in Chinese hamster V79 cells. *Cell Biol. Toxicol.*, **3**, 1–15

Karppanen, E., Rizzo, A., Berg, S., Lindfors, E. & Aho, R. (1985) Fusarium mycotoxins as a problem in Finnish feeds and cereals. *J. agric. Sci. Finl.*, **57**, 195–206

Kawasaki, Y., Uchida, O., Sekita, K., Matsumoto, K., Ochiai, T., Usui, A., Nakaji, Y., Furuya, T., Kurokawa, Y. & Tobe, M. (1990) Single and repeated oral administration toxicity studies of nivalenol in F344 rats. *J. Food. Hyg. Soc. Jpn*, **31**, 144–154

Khera, K.S., Whalen, C., Angers, G., Vesonder, R.F. & Kuiper-Goodman, T. (1982) Embryotoxicity of 4-deoxynivalenol (vomitoxin) in mice. *Bull. environ. Contam. Toxicol.*, **29**, 487–491

Khera, K.S., Arnold, D.L., Whalen, C., Angers, G. & Scott, P.M. (1984) Vomitoxin (4-deoxynivalenol): effects on reproduction of mice and rats. *Toxicol. appl. Pharmacol.*, **74**, 345–356

Khera, K.S., Whalen, C. & Angers, G. (1986) A teratology study on vomitoxin (4-deoxynivalenol) in rabbits. *Food chem. Toxicol.*, **24**, 421–424

Krivobok, S., Olivier, P., Marzin, D.R., Seigle-Murandi, F. & Steiman, R. (1987) Study of the genotoxic potential of 17 mycotoxins with the SOS chromotest. *Mutagenesis*, **2**, 433–439

Kubena, L.F., Harvey, R.B., Corrier, D.E., Huff, W.E. & Phillips, T.D. (1987) Effects of feeding deoxynivalenol (DON, vomitoxin)-contaminated wheat to female white leghorn chickens from 1 day old through egg production. *Poultry Sci.*, **66**, 1612–1618

Kuczuk, M.H., Benson, P.M., Heath, H. & Wallace Hayes, A.W. (1978) Evaluation of the mutagenic potential of mycotoxins using *Salmonella typhimurium* and *Saccharomyces cerevisiae*. *Mutat. Res.*, **53**, 11–20

Kuiper-Goodman, T., Scott, P.M. & Watanabe, H. (1987) Risk assessment of the mycotoxin zearalenone. *Regul. Toxicol. Pharmacol.*, **7**, 253–306

Kumagai, S. & Shimizu, T. (1982) Neonatal exposure to zearalenone causes persistent anovulatory estrus in the rat. *Arch. Toxicol.*, **50**, 279–286

Lake, B.G., Phillips, J.C., Walters, D.G., Bayley, D.L., Cook, M.W., Thomas, L.V., Gilbert, J., Startin, J.R., Baldwin, N.C.P., Bycroft, B.W. & Dewick, P.M. (1987) Studies on the metabolism of deoxynivalenol in the rat. *Food chem. Toxicol.*, **25**, 589–592

Lauren, D.R. & Greenhalgh, R. (1987) Simultaneous analysis of nivalenol and deoxynivalenol in cereals by liquid chromatography. *J. Assoc. off. anal. Chem.*, **70**, 479–483

Lauren, D.R., Ashley, A., Blackwell, B.A., Greenhalgh, R., Miller, J.D. & Neish, G.A. (1987) Trichothecenes produced by *Fusarium crookwellense* DAOM 193611. *J. agric. Food Chem.*, **35**, 884–889

Lee, U.-S., Jang, H.-S., Tanaka, T., Hasegawa, A., Oh, Y.-J. & Ueno, Y. (1985) The coexistence of the *Fusarium* mycotoxins nivalenol, deoxynivalenol and zearalenone in Korean cereals harvested in 1983. *Food Addit. Contam.*, 2, 185–192

Lee, U.-S., Jang, H.-S., Tanaka, T., Hasegawa, A., Oh, Y.-J., Cho, C.-M., Sugiura, Y. & Ueno, Y. (1986) Further survey on the *Fusarium* mycotoxins in Korean cereals. *Food Addit. Contam.*, 3, 253–261

Lee, U.-S., Jang, H.-S., Tanaka, T., Oh, Y.-J., Cho, C.-M. & Ueno, Y. (1987) Effect of milling on decontamination of *Fusarium* mycotoxins nivalenol, deoxynivalenol and zearalenone in Korean wheat. *J. agric. Food Chem.*, 35, 126–129

Lew, H., Adler, A. & Edinger, W. (1991) Moniliformin and the European corn borer (*Ostrinia nubilalis*). *Mycotoxin Res.*, 7, 71–76

Logrieco, A., Bottalico, A. & Altomare, C. (1988) Chemotaxonomic observations on zearalenone and trichothecenes production by *Gibberella zeae* from cereals in southern Italy. *Mycologia*, 80, 892–895

López, T.A. & Tapia, M.O. (1980) Identification of the mycotoxin zearalenone in Argentina (Span.). *Rev. Argent. Microbiol.*, 12, 29–33

Lovelace, C.E.A. & Nyathi, C.B. (1977) Estimation of the fungal toxins, zearalenone and aflatoxin, contaminating opaque maize beer in Zambia. *J. Sci. Food Agric.*, 28, 288–292

Luo, Y. (1988) Fusarium toxins contamination of cereals in China. In: Aibara, K., Kumagai, S., Ohtsubo, K. & Yoshizawa, T., eds, *Proceedings of the 7th International IUPAC Symposium on Mycotoxins and Phycotoxins, Tokyo, 16–19 August 1988*, Tokyo, Japanese Association of Mycotoxicology, pp. 97–98

Luo, Y., Yoshizawa, T. & Katayama, T. (1990) Comparative study on the natural occurrence of *Fusarium* mycotoxins (trichothecenes and zearalenone) in corn and wheat from high- and low-risk areas for human esophageal cancer in China. *Appl. environ. Microbiol.*, 56, 3723–3726

Marasas, W.F.O., Kriek, N.P.J., van Rensburg, S.J. & van Schalkwyk, G.C. (1977) Occurrence of zearalenone and deoxynivalenol, mycotoxins produced by *Fusarium graminearum* Schwabe, in maize produced in southern Africa. *S. Afr. J. Sci.*, 73, 346–349

Marasas, W.F.O., van Rensburg, S.J. & Mirocha, C.J. (1979) Incidence of *Fusarium* species and the mycotoxins, deoxynivalenol and zearalenone, in corn produced in esophageal cancer areas in Transkei. *J. agric. Food Chem.*, 27, 1108–1112

Marasas, W.F.O., Wehner, F.C., van Rensburg, S.J. & van Schalkwyk, D.J. (1981) Mycoflora of corn produced in human esophageal cancer areas in Transkei, southern Africa. *Phytopathology*, 71, 792–796

Marasas, W.F.O., Nelson, P.E. & Toussoun, T.A. (1984) *Toxigenic Fusarium Species*, University Park, PA, Pennsylvania State University Press

Marasas, W.F.O., Jaskiewicz, K., Venter, F.S. & van Schalkwyk, D.J. (1988) *Fusarium moniliforme* contamination of maize in oesophageal cancer areas in Transkei. *S. Afr. med. J.*, 74, 110–114

Marpegan, M.R., Perfumo, C.J., Godoy, H.M., Sala de Miguel, M., Diaz, E. & Risso, M.A. (1988) Feed refusal of pigs caused by Fusarium mycotoxins in Argentina. *J. vet. Med.*, A35, 610–616

Martin, P.M., Horwitz, K.B., Ryan, D.S. & McGuire, W.L. (1978) Phytoestrogen interaction with estrogen receptors in human breast cancer cells. *Endocrinology*, 103, 1860–1867

Matsuoka, Y. & Kubota, K. (1981) Studies on mechanisms of diarrhea induced by fusarenon-X, a trichothecene mycotoxin from *Fusarium* species. *Toxicol. appl. Pharmacol.*, 57, 293–301

Matsuoka, Y. & Kubota, K. (1985) Studies on mechanisms of diarrhea induced by fusarenon-X, a trichothecene mycotoxin from *Fusarium* species: the effects of fusarenon-X and various cathartics on the digestion and absorption in the mouse intestine (Jpn.). *Yakugaku Zasshi*, **105**, 77–82

Matsuoka, Y. & Kubota, K. (1987a) Studies on mechanisms of diarrhea induced by fusarenon-X, a trichothecene mycotoxin from *Fusarium* species: fusarenon-X induced diarrhea is not mediated by cyclic nucleotides. *Toxicol. appl. Pharmacol.*, **91**, 326–332

Matsuoka, Y. & Kubota, K. (1987b) Characteristics of inflammation induced by fusarenon-X, a trichothecene mycotoxin from *Fusarium* species. *Toxicol. appl. Pharmacol.*, **91**, 333–340

Matsuoka, Y., Kubota, K. & Ueno, Y. (1979) General pharmacological studies of fusarenon-X, a trichothecene mycotoxin from *Fusarium* species. *Toxicol. appl. Pharmacol.*, **50**, 87–94

Mayr, U.E. (1988) Estrogen-controlled gene expression in tissue culture cells by zearalenone. *FEBS Lett.*, **239**, 223–226

McGregor, D.B., Brown, A., Cattanach, P., Edwards, I., McBride, D., Riach, C. & Caspary, W.J. (1988) Responses of the L5178Y tk^+/tk^- mouse lymphoma cell forward mutation assay: III. 72 coded chemicals. *Environ. mol. Mutag.*, **12**, 85–154

Mekhancha-Dahel, C., Lafarge-Frayssinet, C. & Frayssinet, C. (1990) Immunosuppressive effects of four trichothecene mycotoxins. *Food Addit. Contam.*, **7**, S94–S96

Miller, J.D. (1991) Significance of grain mycotoxins for health and nutrition. In: Champ, B.R., Highley, E., Hocking, A.D. & Pitt, J.I., eds, *Fungi and Mycotoxins in Stored Products* (ACIAR Proceedings 36), Canberra, Australian Centre for International Agricultural Research, pp. 126–135

Miller, J.D., Taylor, A. & Greenhalgh, R. (1983) Production of deoxynivalenol and related compounds in liquid culture by *Fusarium graminearum*. *Can. J. Microbiol.*, **29**, 1171–1178

Miller, J.D., Greenhalgh, R., Wang, Y.-Z. & Lu, M. (1991) Trichothecene chemotypes of three *Fusarium* species. *Mycologia*, **83**, 121–130

Mirocha, C.J. & Christensen, C.M. (1974) Oestrogenic mycotoxins synthesized by *Fusarium*. In: Purchase, I.F.H., ed., *Mycotoxins*, Amsterdam, Elsevier, pp. 129–148

Mirocha, C.J., Pathre, S.V. & Robison, T.S. (1981) Comparative metabolism of zearalenone and transmission into bovine milk. *Food Cosmet. Toxicol.*, **19**, 25–30

Moran, E.T., Ferket, P.R. & Lun, A.K. (1987) Impact of high dietary vomitoxin on yolk yield and embryonic mortality. *Poultry Sci.*, **66**, 977–982

Morooka, N., Uratsuji, N., Yoshizawa, T. & Yamamoto, H. (1972) Studies on the toxic substances in barley infected with *Fusarium* spp. *J. Food Hyg. Soc. Jpn*, **13**, 368–375

Morrissey, R.E. (1984) Teratological study of Fischer rats fed diet containing added vomitoxin. *Food chem. Toxicol.*, **22**, 453–457

Morrissey, R.E. & Vesonder, R.F. (1985) Effect of deoxynivalenol (vomitoxin) on fertility, pregnancy, and postnatal development of Sprague-Dawley rats. *Appl. environ. Microbiol.*, **49**, 1062–1066

Morrissey, R.E., Norred, W.P. & Vesonder, R.F. (1985) Subchronic toxicity of vomitoxin in Sprague-Dawley rats. *Food chem. Toxicol.*, **23**, 995–999

Mortelmans, K., Haworth, S., Lawlor, T., Speck, W., Tainer, B. & Zeiger, E. (1986) *Salmonella* mutagenicity test: II. Results from testing of 270 chemicals. *Environ. Mutag.*, **8** (Suppl. 7), 1–119

Müller, H.-M. (1983) A survey of methods of decontaminating mycotoxins. II. Chemical methods and reaction with components of feedstuffs. *Übersicht. Tierernach*, **11**, 7–37

Muñoz, L., Cardelle, M., Pereiro, M. & Riguera, R. (1990) Occurrence of corn mycotoxins in Galicia (northwest Spain). *J. agric. Food Chem.*, **38**, 1004–1006

Obara, T., Masuda, E., Takemoto, T. & Tatsuno, T. (1984) Immunosuppressive effect of a trichothecene mycotoxin, fusarenon-X, FSN. In: Kurata, H. & Ueno, Y., eds, *Toxigenic Fungi. Their Toxins and Health Hazard* (Developments in Food Science 7), Amsterdam, Elsevier, pp. 301–311

Ohta, M., Matsumoto, H., Ishii, K. & Ueno, Y. (1978) Metabolism of trichothecene mycotoxins. II. Substrate specificity of microsomal deacetylation of trichothecenes. *J. Biochem.*, **84**, 697–706

Ohtsubo, K. & Saito, M. (1970) Cytotoxic effects of scirpene compounds, fusarenon-X produced by *Fusarium nivale*, dihydronivalenol and dihydrofusarenon-X, on HeLa cells. *Jpn. J. med. Sci. Biol.*, **23**, 217–225

Ohtsubo, K., Yamada, M.-A. & Saito, M. (1968) Inhibitory effect of nivalenol, a toxic metabolite of *Fusarium nivale* on the growth cycle and biopolymer synthesis of HeLa cells. *Jpn. J. med. Sci. Biol.*, **21**, 185–194

Ohtsubo, K., Kaden, P. & Mittermayer, C. (1972) Polyribosomal breakdown in mouse fibroblasts (L-cells) by fusarenon-X, a toxic principle isolated from *Fusarium nivale*. *Biochim. biophys. Acta*, **287**, 520–525

Ohtsubo, K., Ryu, J.-C., Nakamura, K., Izumiyama, N., Tanaka, T., Yamamura, H., Kobayashi, T. & Ueno, Y. (1989) Chronic toxicity of nivalenol in female mice: a 2-year feeding study with *Fusarium nivale* Fn 2b-moulded rice. *Food Chem. Toxicol.*, **27**, 591–598

Olsen, M., Malmlöf, H., Pettersson, H., Sandholm, K. & Kiessling, K.-H. (1985) Plasma and urinary levels of zearalenone and α-zearalenol in a prepubertal gilt fed zearalenone. *Acta pharmacol. toxicol.*, **56**, 239–243

Onji, Y., Dohi, Y., Aoki, Y., Moriyama, T., Nagami, H., Uno, M., Tanaka, T. & Yamazoe, Y. (1989) Deepoxynivalenol: a new metabolite of nivalenol found in the excreta of orally administered rats. *J. agric. Food Chem.*, **37**, 478–481

Palyusik, M., Harrach, B., Mirocha, C.J. & Pathre, S.V. (1980) Transmission of zearalenone and zearalenol into porcine milk. *Acta vet. acad. sci. hung.*, **28**, 217–222

Park, K.-J. & Lee, Y.-W. (1990) Natural occurrence of *Fusarium* mycotoxins in Korean barley samples harvested in 1987 and 1989. *Proc. Jpn. Assoc. Mycotoxicol.*, **31**, 37–41

Pohland, A.E., Schuller, P.L., Steyn, P.S. & van Egmond, H.P. (1982) Physicochemical data for some selected mycotoxins. *Pure appl. Chem.*, **54**, 2219–2284

Prelusky, D.B. & Trenholm, H.L. (1991) Tissue distribution of deoxynivalenol in swine dosed intravenously. *J. agric. Food Chem.*, **39**, 748–751

Prelusky, D.B., Trenholm, H.L., Lawrence, G.A. & Scott, P.M. (1984) Nontransmission of deoxynivalenol (vomitoxin) to milk following oral administration to dairy cows. *J. environ. Sci. Health*, **B19**, 593–609

Prelusky, D.B., Veira, D.M., Trenholm, H.L. & Hartin, K.E. (1986) Excretion profiles of the mycotoxin deoxynivalenol, following oral and intravenous administration to sheep. *Fundam. appl. Toxicol.*, **6**, 356–363

Prelusky, D.B., Veira, D.M., Trenholm, H.L. & Foster, B.C. (1987) Metabolic fate and elimination in milk, urine and bile of deoxynivalenol following administration to lactating sheep. *J. environ. Sci. Health*, **B22**, 125–148

Prelusky, D.B., Scott, P.M., Trenholm, H.L. & Lawrence, G.A. (1990a) Minimal transmission of zearalenone to milk of dairy cows. *J. environ. Sci. Health*, **B25**, 87–103

Prelusky, D.B., Hartin, K.E. & Trenholm, H.L. (1990b) Distribution of deoxynivalenol in cerebral spinal fluid following administration to swine and sheep. *J. environ. Sci. Health*, **B25**, 395–413

Prior, M.G. (1981) Mycotoxins in animal feedstuffs and tissues in western Canada 1975 to 1979. *Can. J. comp. Med.*, **45**, 116–119

Ramakrishna, Y., Bhat, R.V. & Vasanthi, S. (1990) Natural occurrence of mycotoxins in staple foods in India. *J. agric. Food Chem.*, **38**, 1857–1859

Rogers, C.G. & Héroux-Metcalf, C. (1983) Cytotoxicity and absence of mutagenic activity of vomitoxin (4-deoxynivalenol) in a hepatocyte-mediated mutation assay with V79 Chinese hamster lung cells. *Cancer Lett.*, **20**, 29–35

Rose, E.F. (1967) A study of esophageal cancer in the Transkei. *Natl Cancer Inst. Monogr.*, **25**, 83–96

Rose, E.F. (1973) Esophageal cancer in the Transkei: 1955–69. *J. natl Cancer Inst.*, **51**, 7–16

Rose, E.F. & Fellingham, S.A. (1981) Cancer patterns in Transkei. *South Afr. J. Sci.*, **77**, 555–561

Ruddick, J.A., Scott, P.M. & Harwig, J. (1976) Teratological evaluation of zearalenone administered orally to the rat. *Bull. environ. Contam. Toxicol.*, **15**, 678–681

Ryu, J.-C., Ohtsubo, K., Izumiyama, N., Mori, M., Tanaka, T. & Ueno, Y. (1987) Effects of nivalenol on the bone marrow in mice. *J. toxicol. Sci.*, **12**, 11–21

Ryu, J.-C., Ohtsubo, K., Izumiyama, N., Nakamura, K., Tanaka, T., Yamamura, H. & Ueno, Y. (1988) The acute and chronic toxicities of nivalenol in mice. *Fundam. appl. Toxicol.*, **11**, 38–47

Sabino, M., Prado, G., Inomata, E.I., de Oliveira Pedroso, M. & Valeiro Garcia, R. (1989) Natural occurrence of aflatoxins and zearalenone in maize in Brazil. Part II. *Food Addit. Contam.*, **6**, 327–331

Saito, M. & Ohtsubo, K. (1974) Trichothecene toxins of *Fusarium* species. In: Purchase, I.F.H., ed., *Mycotoxins*, Amsterdam, Elsevier, pp. 263–281

Saito, M., Enomoto, M. & Tatsuno, T. (1969) Radiomimetic biological properties of the new scirpene metabolites of *Fusarium nivale*. *Gann*, **60**, 599–603

Saito, M., Horiuchi, T., Ohtsubo, K., Hatanaka, J. & Ueno, Y. (1980) Low tumor incidence in rats with long-term feeding of fusarenon-X, a cytotoxic trichothecene produced by *Fusarium nivale*. *Jpn. J. exp. Med.*, **50**, 293–302

Schuh, M., Leibetseder, J. & Glawischnig, E. (1982) Chronic effects of different levels of deoxynivalenol (vomitoxin) on weight gain, feed consumption, blood parameters, pathological as well as histopathological changes in fattening pigs. In: Pfannhauser, W. & Czedik-Eysenberg, P.B., eds, *Proceedings of the 5th International IUPAC Symposium on Mycotoxins and Phycotoxins, Vienna, 1–3 September 1982*, Vienna, Austrian Chemical Society, pp. 273–276

Schwadorf, K. & Müller, H.-M. (1992) Determination of α- and β-zearalenol and zearalenone in cereals by gas chromatography with ion-trap detection. *J. Chromatogr.*, **595**, 259–267

Scott, P.M. (1989) The natural occurrence of trichothecenes. In: Beasley, V.R., ed., *Trichothecene Mycotoxicosis: Pathophysiological Effects*, Vol. 1, Boca Raton, FL, CRC Press, pp. 1–26

Scott, P.M. (1990) Trichothecenes in grains. *Cereal Foods World*, **35**, 661–669

Scott, P.M., Lau, P.-Y. & Kanhere, S.R. (1981) Gas chromatography with electron capture and mass spectrometric detection of deoxynivalenol in wheat and other grains. *J. Assoc. off. anal. Chem.*, **64**, 1364–1371

Sheu, C.W., Moreland, F.M., Lee, J.K. & Dunkel, V.C. (1988) Morphological transformation of BALB/3T3 mouse embryo cells *in vitro* by vomitoxin. *Food chem. Toxicol.*, **26**, 243–245

Shotwell, O.L., Hesseltine, C.W., Vandegraft, E.E. & Goulden, M.L. (1971) Survey of corn from different regions for aflatoxin, ochratoxin and zearalenone. *Cereal Sci. Today*, **16**, 266–273

Shotwell, O.L., Goulden, M.L. & Bennett, G.A. (1976) Determination of zearalenone in corn: collaborative study. *J. Assoc. off. anal. Chem.*, **59**, 666–670

Shotwell, O.L., Bennett, G.A., Stubblefield, R.D., Shannon, G.M., Kwolek, W.F. & Plattner, R.D. (1985) Deoxynivalenol in hard red winter wheat: relationship between toxin levels and factors that could be used in grading. *J. Assoc. off. anal. Chem.*, **68**, 954–957

Siame, B.A. & Lovelace, C.E.A. (1989) Natural occurrence of zearalenone and trichothecene toxins in maize-based animal feeds in Zambia. *J. Sci. Food Agric.*, **49**, 25–35

Sinha, K.K. (1991) Contamination of freshly harvested maize kernels with *Fusarium* mycotoxins in Bihar State, India. *Mycotoxin Res.*, **7**, 178–183

Snijders, C.H.A. (1990) Fusarium head blight and mycotoxin contamination in wheat, a review. *Neth. J. Plant Pathol.*, **96**, 187–197

Stob, M., Baldwin, R.S., Tuite, J., Andrews, F.N. & Gillette, K.G. (1962) Isolation of an anabolic, uterotrophic compound from corn infected with *Gibberella zeae*. *Nature*, **196**, 1318

Stoloff, L., Henry, S. & Francis, O.J., Jr (1976) Survey for aflatoxins and zearalenone in 1973 crop corn stored on farms and in country elevators. *J. Assoc. off. anal. Chem.*, **59**, 118–121

Sundheim, L., Nagayama, S., Kawamura, O., Tanaka, T., Brodal, G. & Ueno, O. (1988) Trichothecenes and zearalenone in Norwegian barley and wheat. *Norw. J. agric. Sci.*, **2**, 49–59

Sutton, J.C. (1982) Epidemiology of wheat head blight and maize ear rot caused by *Fusarium graminearum*. *Can. J. Plant Pathol.*, **4**, 195–209

Sydenham, E.W., Thiel, P.G., Marasas, W.F.O., Shephard, G.S., Van Schalkwyk, D.J. & Koch, K.R. (1990) Natural occurrence of some *Fusarium* mycotoxins in corn from low and high esophageal cancer prevalence areas of the Transkei, southern Africa. *J. agric. Food Chem.*, **38**, 1900–1903

Tanaka, T. & Ueno, Y. (1989) Worldwide natural occurrence of *Fusarium* mycotoxins, nivalenol, deoxynivalenol and zearalenone. In: Natori, S., Hashimoto, K. & Ueno, Y., eds, *Mycotoxins and Phycotoxins '88*, Amsterdam, Elsevier, pp. 51–56

Tanaka, T., Matsuda, Y., Toyasaki, N., Ogawa, K., Matsuki, Y. & Ueno, Y. (1978) Screening of trichothecene-producing *Fusarium* species from river sediments by mammalian cell culture techniques. *Proc. Jpn. Assoc. Mycotoxicol.*, **5/6**, 50–53

Tanaka, T., Hasegawa, A., Matsuki, Y., Matsui, Y., Lee, U.-S. & Ueno, Y. (1985a) Co-contamination of the *Fusarium* mycotoxins nivalenol, deoxynivalenol and zearalenone, in scabby wheat grains harvested in Hokkaido, Japan. *J. Food Hyg. Soc. Jpn*, **26**, 519–522

Tanaka, T., Hasegawa, A., Matsuki, Y. & Ueno, Y. (1985b) A survey of the occurrence of nivalenol, deoxynivalenol and zearalenone in foodstuffs and health foods in Japan. *Food Addit. Contam.*, **2**, 259–265

Tanaka, T., Hasegawa, A., Matsuki, Y., Lee, U.-S. & Ueno, Y. (1986) A limited survey of *Fusarium* mycotoxins nivalenol, deoxynivalenol and zearalenone in 1984 UK harvested wheat and barley. *Food Addit. Contam.*, **3**, 247–252

Tanaka, T., Hasegawa A., Yamamoto, S., Lee, U.-S., Sugiura, Y. & Ueno, Y. (1988) Worldwide contamination of cereals by the *Fusarium* mycotoxins nivalenol, deoxynivalenol and zearalenone. I. Survey of 19 countries. *J. agric. Food Chem.*, **36**, 979–983

Tatsuno, T. (1968) Toxicologic research on substances from *Fusarium nivale*. *Cancer Res.*, **28**, 2393–2396

Tatsuno, T., Saito, M., Enomoto, M. & Tsunoda, H. (1968) Nivalenol, a toxic principle of *Fusarium nivale*. *Chem. pharm. Bull.*, **16**, 2519–2520

Tatsuno, T., Fujimoto, Y. & Morita, Y. (1969) Toxicological research on substances from *Fusarium nivale*. III. The structure of nivalenol and its monoacetate. *Tetrahedron Lett.*, **33**, 2823–2826

Taub, D., Girotra, N.N., Hoffsommer, R.D., Kuo, C.H., Slates, H.L., Weber, S. & Wendler, N.L. (1968) Total synthesis of the macrolide, zearalenone. *Tetrahedron*, **24**, 2443–2461

Teshima, R., Hirai, K., Sato, M., Ikebuchi, H., Ichinoe, M. & Terao, T. (1990) Radioimmunoassay of nivalenol in barley. *Appl. environ. Microbiol.*, **56**, 764–768

Thiel, P.G., Marasas, W.F.O. & Meyer, C.J. (1982) Natural occurrence of *Fusarium* toxins in maize from Transkei. In: Pfannhauser, W. & Czedik-Eysenberg, P.B., eds, *Proceedings of the Vth International IUPAC Symposium on Mycotoxins and Phycotoxins, 1-3 September 1982, Vienna*, Vienna, Austrian Chemical Society, pp. 126-129

Thrane, U. (1989) *Fusarium* species and their specific profiles of secondary metabolites. In: Chelkowski, J., ed., Fusarium *Mycotoxins, Taxonomy and Pathogenicity*, Amsterdam, Elsevier, pp. 199-225

Thust, R., Kneist, S. & Hühne, V. (1983) Genotoxicity of Fusarium mycotoxins (nivalenol, fusarenon-X, T-2 toxin, and zearalenone) in Chinese hamster V79-E cells *in vitro*. *Arch. Geschwulstforsch.*, **53**, 9-15

Tobin, N.F. (1988) Presence of deoxynivalenol in Australian wheat and triticale—New South Wales Northern Rivers region, 1983. *Austr. J. exp. Agric.*, **28**, 107-110

Trenholm, H.L., Hamilton, R.M.G., Friend, D.W., Thompson, B.K. & Hartin, K.E. (1984) Feeding trials with vomitoxin (deoxynivalenol)-contaminated wheat: effects on swine, poultry and dairy cattle. *J. Am. vet. Med. Assoc.*, **185**, 527-531

Trenholm, H.L., Prelusky, D.B., Young, J.C. & Miller, J.D. (1989) A practical guide to the prevention of *Fusarium* mycotoxins in grain and animal feedstuffs. *Arch. environ. Contam. Toxicol.*, **18**, 443-451

Trucksess, M.W., Flood, M.T. & Page, S.W. (1986) Thin layer chromatographic determination of deoxynivalenol in processed grain products. *J. Assoc. off. anal. Chem.*, **69**, 35-36

Tryphonas, H., Iverson, F., So., Y., Nera, E.A., McGuire, P.F., O'Grady, L., Clayson, D.B. & Scott, P.M. (1986) Effects of deoxynivalenol (vomitoxin) on the humoral and cellular immunity of mice. *Toxicol. Lett.*, **30**, 137-150

Ueno, Y. (1977) Mode of action of trichothecenes. *Pure appl. Chem.*, **49**, 1737-1745

Ueno, Y. (1983) General toxicology. In: Ueno, Y., ed., *Developments in Food Science. IV. Trichothecenes. Chemical, Biological and Toxicological Aspects*, Amsterdam, Elsevier, pp. 135-146

Ueno, Y. (1984) Toxicological features of T-2 toxin and related trichothecenes. *Fundam. appl. Toxicol.*, **4**, S124-S132

Ueno, Y. (1987) Trichothecenes in food. In: Krogh, P., ed., *Mycotoxins in Food*, London, Academic Press, pp. 123-147

Ueno, Y. & Kubota, K. (1976) DNA-attacking ability of carcinogenic mycotoxins in recombination-deficient mutant cells of *Bacillus subtilis*. *Cancer Res.*, **36**, 445-451

Ueno, Y. & Matsumoto, H. (1975) Inactivation of some thiol-enzymes by trichothecene mycotoxins from *Fusarium* species. *Chem. pharm. Bull.*, **23**, 2439-2442

Ueno, Y., Hosoya, M., Morita, Y., Ueno, I. & Tatsuno, T. (1968) Inhibition of the protein synthesis in rabbit reticulocyte by nivalenol, a toxic principle isolated from *Fusarium nivale*-growing rice. *J. Biochem.*, **64**, 479-485

Ueno, Y., Ueno, I., Tatsuno, T., Ohokubo, K. & Tsunoda, H. (1969) Fusarenon-X, a toxic principle of *Fusarium nivale*-culture filtrate. *Experientia*, **25**, 1062

Ueno, Y., Ishikawa, Y., Saito-Amakai, K. & Tsunoda, H. (1970a) Environmental factors influencing the production of fusarenon-X, a cytotoxic mycotoxin of *Fusarium nivale* Fn2B. *Chem. pharm. Bull.*, **18**, 304-312

Ueno, Y., Ishikawa, Y., Amakai, K., Nakajima, M., Saito, M., Enomoto, M. & Ohtsubo, K. (1970b) Comparative study on skin-necrotizing effect of scirpene metabolites of *Fusaria*. *Jpn J. exp. Med.*, **40**, 33-38

Ueno, Y., Ueno, I., Iitoi, Y., Tsunoda, H., Enomoto, M. & Ohtsubo, K. (1971) Toxicological approaches to the metabolites of *Fusaria*. III. Acute toxicity of fusarenon-X. *Jpn. J. exp. Med.*, **41**, 521–539

Ueno, Y., Nakajima, M., Sakai, K., Ishii, K., Sato, N. & Shimada, N. (1973) Comparative toxicology of trichothec mycotoxins: inhibition of protein synthesis in animal cells. *J. Biochem.*, **74**, 285–296

Ueno, Y., Ayaki, S., Sato, N. & Ito, T. (1977) Fate and mode of action of zearalenone. *Ann. Nutr. Aliment.*, **31**, 935–948

Ueno, Y., Kubota, K., Ito, T. & Nakamura, Y. (1978) Mutagenicity of carcinogenic mycotoxins in *Salmonella typhimurium*. *Cancer Res.*, **38**, 536–542

Ueno, Y., Lee, U.-S., Tanaka, T., Hasagawa, A. & Strzelecki, E. (1985) Natural co-occurrence of nivalenol and deoxynivalenol in Polish cereals. *Microbiol. Aliments Nutr.*, **3**, 321–326

Ueno, Y., Lee, U.S., Tanaka, T., Hasagawa, A. & Matsuki, Y. (1986) Examination of Chinese and USSR cereals for the *Fusarium* mycotoxins nivalenol, deoxynivalenol, and zearalenone. *Toxicon*, **24**, 618–621

Ueno, Y., Kobayashi, T., Yamamura, H., Kato, T., Tashiro, F., Nakamura, K. & Ohtsubo, K. (1991) Effect of long-term feeding of nivalenol on aflatoxin-B_1-initiated hepatocarcinogenesis in mice. In: O'Neill, I.K., Chen, J. & Bartsch, H., eds, *Relevance to Human Cancer of N-Nitroso Compounds, Tobacco Smoke and Mycotoxins* (IARC Scientific Publications No. 105), Lyon, IARC, pp. 420–423

Umeda, M., Yamamoto, T. & Saito, M. (1972) DNA-strand breakage of HeLa cells induced by several mycotoxins. *Jpn. J. exp. Med.*, **42**, 527–535

Umeda, M., Tsutsui, T. & Saito, M. (1977) Mutagenicity and inducibility of DNA single-strand breaks and chromosome aberrations by various mycotoxins. *Gann*, **68**, 619–625

Urry, W.H., Wehrmeister, H.L., Hodge, E.B. & Hidy, P.H. (1966) The structure of zearalenone. *Tetrahedron Lett.*, **27**, 3109–3114

US Food and Drug Administration (1980) Foods and drugs. *US Code fed. Regul.*, **Title 21**, part 522, pp. 192, 239

US National Toxicology Program (1982) *Carcinogenesis Bioassay of Zearalenone (CAS No. 17924-92-4) in F344/N rats and B6C3F₁ Mice (Feed Study)* (Technical Report Series No. 235, NIH Publ. No. 83-1791), Research Triangle Park, NC

Vesonder, R.F., Ciegler, A. & Jensen, A.H. (1973) Isolation of the emetic principle from *Fusarium*-infected corn. *Appl. Microbiol.*, **16**, 1008–1010

Visconti, A. & Bottalico, A. (1983) Detection of *Fusarium* trichothecenes (nivalenol, deoxynivalenol, fusarenone and 3-acetyldeoxynivalenol) by high-performance chromatography. *Chromatographia*, **17**, 97–100

Visconti, A., Bottalico, A., Palmisano, F. & Zambonin, P.G. (1984) Differential-pulse polarography of trichothecene mycotoxins. Determination of deoxynivalenol, nivalenol and fusarenone-X in maize. *Anal. chim. Acta*, **159**, 111–118

Visconti, A., Chelkowski, J., Solfrizzo, M. & Bottalico, A. (1990) Mycotoxins in corn ears naturally infected with *Fusarium graminearum* and *F. crookwellense*. *Can. J. Plant Pathol.*, **12**, 187–189

Wang, C.-R. & Chu, F.S. (1991) Production and characterization of antibodies against nivalenol tetraacetate. *Appl. environ. Microbiol.*, **57**, 1026–1030

Ware, G.M. & Thorpe, C.W. (1978) Determination of zearalenone in corn by high pressure liquid chromatography and fluorescence detection. *J. Assoc. off. anal. Chem.*, **61**, 1058–1062

Warner, R.L. & Pestka, J.J. (1987) ELISA (Enzyme linked immunosorbent assays) survey of retail grain-based food products for zearalenone and aflatoxin B_1. *J. Food Prot.*, **50**, 502–503

Warner, R.L., Ram, B.P., Hart, L.P. & Pestka, J.J. (1986) Screening for zearalenone in corn by competitive direct enzyme-linked immunosorbent assay. *J. agric. Food Chem.*, **34**, 714–717

Wehner, F.C., Marasas, W.F.O. & Thiel, P.G. (1978) Lack of mutagenicity to *Salmonella typhimurium* of some *Fusarium* mycotoxins. *Appl. environ. Microbiol.*, **35**, 659–662

WHO (1990) *Selected Mycotoxins: Ochratoxins, Trichothecenes, Ergot* (Environmental Health Criteria 105), Geneva

Widiastuti, R., Maryam, R., Blaney, B.J., Stoltz, S. & Stoltz, D.R. (1988) Cyclopiazonic acid in combination with aflatoxins, zearalenone and ochratoxin A in Indonesian corn. *Mycopathologia*, **104**, 153–156

Williams, B.C. (1985) Mycotoxins in foods and foodstuffs. In: Scott, P.M., Trenholm, H.L. & Sutton, M.D., eds, *Mycotoxins, a Canadian Perspective*, Ottawa, National Research Council of Canada, pp. 49–53

Williams, B.A., Mills, K.T., Burroughs, C.D. & Bern, H.A. (1989) Reproductive alterations in female C57Bl/Crgl mice exposed neonatally to zearalenone, an estrogenic mycotoxin. *Cancer Lett.*, **46**, 225–230

Yang, C.S. (1980) Research on esophageal cancer in China: a review. *Cancer Res.*, **40**, 2633–2644

Yoshizawa, T. (1983) Red-mold diseases and natural occurrence in Japan. In: Ueno, Y., ed., *Trichothecenes, Chemical, Biological and Toxicological Aspects*, Tokyo, Kodansha, pp. 195–209

Yoshizawa, T. & Morooka, N. (1973) Deoxynivalenol and its monoacetate: new mycotoxins from *Fusarium roseum* and moldy barley. *Agric. Biol. Chem.*, **37**, 2933–2934

Yoshizawa, T. & Morooka, N. (1974) Studies on the toxic substances in the infected cereals. III. Acute toxicities of new trichothecene mycotoxins: deoxynivalenol and its monoacetate (Jpn.). *J. Food Hyg. Soc. Jpn*, **15**, 261–269

Yoshizawa, T., Shirota, T. & Morooka, N. (1978) Deoxynivalenol and its acetate as feed refusal principles in rice cultures of *Fusarium roseum* No. 117 (ATCC 28114). *J. Food Hyg. Soc. Jpn*, **19**, 178–184

Yoshizawa, T., Takeda, H. & Ohi, T. (1983) Structure of a novel metabolite from deoxynivalenol, a trichothecene mycotoxin, in animals. *Agric. Biol. Chem.*, **47**, 2133–2135

Yoshizawa, T., Coté, L.-M., Swanson, S.P. & Buck, W.B. (1986) Confirmation of DOM-1, a deepoxidation metabolite of deoxynivalenol, in biological fluids of lactating cows. *Agric. Biol. Chem.*, **50**, 227–229

Young, L.C., McGirr, L., Valli, V.E., Lumsden, J.H. & Sun, A. (1983) Vomitoxin in corn fed to young pigs. *J. Anim. Sci.*, **57**, 655–664

TOXINS DERIVED FROM *FUSARIUM MONILIFORME*: FUMONISINS B_1 AND B_2 AND FUSARIN C

Fusarium moniliforme and a number of related species are ubiquitous on maize. These fungi produce fumonisins and fusarins. Six fumonisins have so far been isolated from *F. moniliforme* and characterized. Fumonisin B_1, fumonisin B_2 and fumonisin B_3 are the major ones produced in nature, while fumonisin B_4 is produced in relatively minor quantities (Thiel *et al.*, 1993). Only fumonisins B_1 and B_2 are considered in this monograph. A review on fumonisins is available (Riley & Richard, 1992). Fusarin C is a member of a family of unstable compounds which includes fusarins A, B, C, D, E and F (Savard & Miller, 1992).

1. Exposure Data

1.1 Chemical and physical data

1.1.1 *Synonyms, structural and molecular data*

Fumonisin B_1

Chem. Abstr. Services Reg. No.: 116355-83-0
Chem. Abstr. Name: 1,2,3-Propanetricarboxylic acid, 1,1'-[1-(12-amino-4,9,11-trihydroxy-2-methyltridecyl)-2-(1-methylpentyl)-1,2-ethanediyl] ester
Synonym: Macrofusine

$C_{34}H_{59}NO_{15}$ Mol. wt: 721

Fumonisin B_2

Chem. Abstr. Services Reg. No.: 116355-84-1
Chem. Abstr. Name: 1,2,3-Propanetricarboxylic acid, 1,1'-[1-(12-amino-9,11-dihydroxy-2-methyltridecyl)-2-(1-methylpentyl)-1,2-ethanediyl] ester

$C_{34}H_{59}NO_{14}$ Mol. wt: 705

Fusarin C

Chem. Abstr. Services Reg. No.: 79748-81-5
Chem. Abstr. Name: 3,5,7,9-Undecatetraenoic acid, 2-ethylidene-11-[4-hydroxy-4-(2-hydroxyethyl)-2-oxo-6-oxa-3-azabicyclo-[3.1.0]hex-1-yl]-4,6,10-trimethyl-11-oxo, methyl ester, [1R-[1α(2E,3E,5E,7E,9E),4α,5α]]-
Synonym: 6-Oxa-3-azabicyclo[3.1.0]hexane, 3,5,7,9-undecatetraenoic acid derivative

$C_{23}H_{29}NO_7$ Mol. wt: 431.5

1.1.2 *Chemical and physical properties*

Fumonisins B_1 and B_2

(a) *Description*: Powder, very hygroscopic
(b) *Melting-point*: Not known (has not been crystallized)
(c) *Spectroscopy data*: Mass spectral and nuclear magnetic resonance data have been reported (Bezuidenhout *et al.*, 1988; Laurent *et al.*, 1989; Plattner *et al.*, 1990).
(d) *Solubility*: Soluble in methanol (Sydenham *et al.*, 1992a)
(e) *Stability*: Stable in acetonitrile:water (1:1) and to light

Fusarin C

From Farber and Scott (1989), unless otherwise specified
(a) *Description*: Yellow oil
(b) *Melting-point*: Not known
(c) *Optical rotation*: $[\alpha]_D^{23}$ +47.04° (2% in methanol)

(d) *Spectroscopy data*: Ultraviolet, infrared, nuclear magnetic resonance and mass spectral data have been reported (Wiebe & Bjeldanes, 1981).
(e) *Solubility*: Soluble in ethanol and methanol
(f) *Stability*: Unstable on exposure to light and heat; decomposes rapidly as pH increases (Zhu & Jeffrey, 1992)

1.1.3 Analysis

Fumonisins B_1 and B_2

The analysis of fumonisins in maize presents major difficulties. A water/methanol extraction of grain followed by ion-exchange chromatography is used to extract fumonisins (Shephard *et al.*, 1990; Sydenham *et al.*, 1992a). Analytical, thin-layer chromatographic methods exist, but these are not useful for quantification (Sydenham *et al.*, 1992a). There are two principal approaches for quantification: (i) hydrolysis followed by derivatization and detection by gas chromatography/mass spectroscopy of the esterified tricarballylic acid (propane-1,2,3-tricarboxylic acid) or the derivatized aminopentol backbone (Plattner *et al.*, 1990; Sydenham *et al.*, 1990a; Scott, 1992; Shephard *et al.*, 1992; Thiel *et al.*, 1993), which has the disadvantage that all fumonisins are determined together; and (ii) preparation of a fluorescent derivative followed by high-performance liquid chromatography (HPLC)–fluorescence detection (Shephard *et al.*, 1990; Plattner *et al.*, 1991). Two derivatives have been tested: The first was a fluorescamine derivative, which unfortunately results in two peaks (Sydenham *et al.*, 1990a); the other was *ortho*-phthaldialdehyde (Shephard *et al.*, 1990). Use of the method has proven satisfactory for the analysis of fumonisins in maize and mixed horse-feed samples, with a limit of detection of 50 ng/g for fumonisin B_1 and 100 ng/g for fumonisin B_2 (Shephard *et al.*, 1990; Thiel *et al.*, 1993). As the *ortho*-phthaldialdehyde derivative has been reported to be unstable, use of a 4-fluoro-7-nitrobenzofurazan derivative was investigated and has also proven useful (Scott & Lawrence, 1991).

A method has been reported for the analysis of fumonisin B_1 in plasma and urine involving solid-phase anion-exchange clean-up, precolumn derivatization with *ortho*-phthaldialdehyde and reverse-phase HPLC with fluorescence detection, at a detection limit of 50 ng/ml (Shephard *et al.*, 1992).

Fusarin C

A detailed review of the few available analytical methods is provided by Farber and Scott (1989). Thin-layer chromatographic methods have been developed, but reliable quantification requires the use of HPLC with ultraviolet detection. Serious questions have been raised, however, as to whether these methods are adequate, in view of the instability of fusarins under extraction conditions: The mutagenicity of extracts containing fusarin C is greater than can be explained on the basis of measured fusarin C concentrations. Rearrangements of the co-occurring fusarins (E and F) under analytical conditions may result in underestimates of the amount of fusarins present in *Fusarium*-infected maize (Savard & Miller, 1992).

1.2 Production and use

1.2.1 Production

Fumonisins B_1 and B_2

Fumonisins B_1 and B_2 were first isolated in 1988 by Bezuidenhout *et al.* (1988); shortly thereafter, fumonisin B_1 was isolated as 'macrofusine' by Laurent *et al.* (1989) from cultures of *Fusarium moniliforme* (*F. verticillioides*) (Marasas *et al.*, 1981). Fumonisins can be produced at concentrations of several grams per kilogram by culturing strains of the fungi that produce this toxin on sterilized maize (Vesonder *et al.*, 1990; Cawood *et al.* 1991): *F. moniliforme* MRC 826 produced 7100 mg/kg fumonisin B_1 and 3000 mg/kg fumonisin B_2 (Thiel *et al.*, 1991a). They can also be produced in liquid cultures (Jackson & Bennett, 1990). In Southeast Asian strains, 0.66–192 mg/l fumonisin B_1 plus fumonisin B_2 have been obtained in liquid fermentations, and high recoveries of the toxins are possible (Miller *et al.*, 1993).

Of 90 strains of *F. moniliforme* that have been grown on autoclaved maize, 61 were found to contain fumonisin B_1, at levels ranging from 48 to 6400 mg/kg (Nelson *et al.*, 1991). Fumonisins B_1 and B_2 are also produced by some related species, including *F. anthophilum, F. dlamini, F. proliferatum, F. napiforme* and *F. nygamai*. These strains have been isolated in Africa, Australia, Nepal, New Caledonia and the USA (Nelson *et al.*, 1991; Thiel *et al.*, 1991a; Nelson *et al.*, 1992) and in Indonesia, Italy, the Philippines, Poland and Thailand (Miller *et al.*, 1993; Nelson *et al.*, 1992; Visconti, 1992). Fumonisins B_1 and B_2 invariably occur together; however, because B_2 always represents 15–35% of B_1, low levels of B_1 may appear to contain 'no detectable' B_2.

Fusarin C

Fusarin C was first isolated from a culture of *F. moniliforme* grown on sterilized corn (Wiebe & Bjeldanes, 1981). The absolute configuration was determined in 1984 (Gelderblom *et al.*, 1984a), and it was reported as a natural product of maize in 1984 (Gelderblom *et al.*, 1984b). Fusarin C can be produced by culturing strains of the species that produce this toxin on sterilized maize (Gelderblom *et al.*, 1983). It can also be produced in liquid culture (Farber & Scott, 1989).

Fusarin C is produced by several species of *Fusarium*, including *F. moniliforme, F. poae, F. avenaceum, F. crookwellense, F. culmorum, F. graminearum* and *F. sambucinum* (Marasas *et al.*, 1984a; Thrane, 1989).

1.3 Occurrence

Fumonisins B_1 and B_2

Fumonisins are widely distributed in maize products, including those from Europe, although limited numbers of analyses have been published outside South Africa and the USA (Thiel *et al.*, 1992) (Table 1).

No data are available about the effects of milling and baking on levels of fumonisins nor on their transmission into milk, meat and eggs. Ammonia treatment was ineffective in reducing fumonisin concentrations in maize (Norred et al., 1991b). After fermentation of maize containing fumonisins, most of the toxin was recovered in the spent grains (Bothast et al., 1992).

Table 1. Natural occurrence of fumonisins B_1 and B_2

Country	Product	Year	Fumonisin B_1		Fumonisin B_2		Reference
			Positive samples/ total no.	Content (mg/kg)	Positive samples/ total no.	Content (mg/kg)	
North America							
USA	Maize feed	1989	3/3	37–122	3/3	2–23	Wilson et al.
USA	Damaged maize kernels		2/2	144, 148	2/2	31, 41	(1990)
USA	Maize feed	1983–86	14/14	1.3–27.0	14/14	0.1–12.6	Thiel et al. (1991b)
USA	Maize feed	1989–90	177/232	1–330			Ross et al. (1991)
USA	Maize feed	NR	2/2	20, 150			Plattner et al. (1990)
USA	Maize meal	1990–91	15/16	av. 1.0	13/16	0.3	Sydenham
USA	Maize grits		10/10	av. 0.6	5/10	0.4	et al. (1991)
USA	Cornflakes		0/2		0/2		
Canada	Maize meal		1/2	0.05	0/2		
South America							
Brazil	Maize	1985, 1990	20/21	0.2–38.5	18/21	0.1–12.0	Sydenham et al. (1992b)
Peru	Maize meal	1990–91	1/2	0.66	1/2	0.13	Sydenham et al. (1991)
Europe							
Austria	Maize	1988–89	3/9	< 15			Lew et al. (1991)
Italy	Maize feed	NR	23/25	0.01–8.4	13/25	0.01–1.33	Minervini et al. (1993)
Africa							
Transkei	Maize	NR	3/3	< 10–83			Sydenham et al. (1990a)
Transkei	Maize	1985	39/48	0.2–46.9	37/48	0.15–16.3	Sydenham et al. (1990b)
Transkei	Maize	1985, 1989	61/74	0.05–117.5	55/74	0.05–22.9	Rheeder et al. (1992)
South Africa	Maize meal	1990–91	46/52	Mean, 0.14	11/52	Mean, 0.08	Sydenham
South Africa	Maize grits		10/18	Mean, 0.13	4/18	Mean, 0.09	et al. (1991)

NR, not reported

Fusarin C

Fusarin C was determined in two samples of maize from southern Africa at concentrations of 0.02 and 0.28 mg/kg (Gelderblom *et al.*, 1984b). One maize sample from the USA was reported to contain 0.39 mg/kg (Thiel *et al.*, 1986), but analysis of 12 maize samples from Canada revealed no fusarin C (Farber & Scott, 1989). Natural occurrence in maize has also been reported from China (Cheng *et al.*, 1985).

1.4 Regulations and guidelines

No regulation or guidelines exists for these compounds (van Egmond, 1989).

2. Studies of Cancer in Humans

Ecological studies of the relationship between exposure to *Fusarium* toxins and oesophageal cancer are summarized in the monograph on toxins derived from *F. graminearum*, *F. culmorum* and *F. crookwellense*, p. 409. Most of the studies refer to mixtures of many toxins from many species of fungi on maize.

In the study of Marasas *et al.* (1981), the proportion of kernels in both mouldy and healthy maize samples infected by *F. moniliforme*, one of the most prevalent fungi in maize in the Transkei, was significantly correlated with oesophageal cancer rates.

In the study of Marasas *et al.* (1988a), described in detail on p. 410, the mean proportions of maize kernels infected with *F. moniliforme* in both healthy and mouldy maize samples from households in the high-incidence oesophageal cancer area were significantly higher (42% and 68%, respectively) than those in the low-incidence area (8% and 35%, respectively). A similar survey was conducted one year later, with the same criteria for high and low incidence but adding 24 households from a study area with an intermediate incidence of oesophageal cancer. Although the proportion of kernels infected with *F. moniliforme* in healthy maize from the latter area was in between those from the high- and low-incidence areas, further subdivision of households in the intermediate-incidence area into 12 situated in a low-risk zone and 12 in a high-risk zone (with an estimated six-fold difference in oesophageal cancer rates) did not reveal a difference in the proportion of infected kernels in the corresponding samples (26 and 24%, respectively). Furthermore, there was no difference in the prevalence of cytological abnormalities of the oesophagus in adult occupants of the low- and high-risk zones of the intermediate-incidence area. [The Working Group noted that the sampling strategy was different in the high- and low-risk areas.]

In the study of Sydenham *et al.* (1990b), significantly higher mean numbers of kernels infected with *F. moniliforme* and correspondingly higher levels of the mycotoxins fumonisin B_1 and fumonisin B_2 were found in mouldy maize samples in the high-risk oesophageal cancer area than in the low-risk area ($p < 0.01$). Fumonisin B_1 and B_2 levels in healthy maize samples from the low-risk area were approximately 20 times lower than those in healthy samples from the high-risk area, and only three out of 12 samples contained these toxins. [The Working Group noted that the sampling strategy was different in the high- and low-risk areas.]

Rheeder *et al.* (1992) reanalysed the samples of healthy and mouldy home-grown maize collected from high-risk and low-risk oesophageal cancer areas during 1976–86 which had been examined by Marasas *et al.* (1979, 1981, 1988a) and by Sydenham *et al.* (1990b). The material was supplemented by samples obtained during 1989 from the same study areas. Samples collected from high-risk areas in 1985 and 1986 were taken from preselected households (Marasas *et al.*, 1988a); all other samples were taken at random. During the entire period, the percentages of kernels infected by *F. moniliforme* in healthy as well as mouldy samples were significantly higher in the high-risk oesophageal cancer area than in the low-risk area. The samples collected in 1985 and 1989 were analysed for the presence of fumonisin B_1 and fumonisin B_2: Both toxins occurred in more samples and at significantly higher levels in healthy and mouldy maize obtained in 1985 and at significantly higher levels in mouldy maize obtained from high-risk areas than from low-risk areas in 1986.

Zhen *et al.* (1984) cultured and isolated fungal strains from samples of wheat, maize, dried sweet potato, rice and soya beans in five counties with a high incidence of oesophageal cancer and three with a low incidence, in Henan Province, northern China. Mortality rates for males ranged between 26.5 and 37.0/100 000 in the low-risk counties and between 76.6 and 161.3 in the high-risk counties. The frequency of contamination by *F. moniliforme* was higher in samples from high-risk counties (6.8% out of 2009 measurements) than in samples from low-risk areas (5.4% out of 830 measurements) ($p < 0.001$). The frequency of contamination by all other fungi analysed was also significantly higher in samples from high-risk counties.

3. Studies of Cancer in Experimental Animals

3.1 Oral administration

3.1.1 *Mouse*

A group of 29 female DBA mice, 8–10 weeks of age, were given 0.5 mg fusarin C [purity unspecified] by gavage twice a week; when toxic effects became apparent, the dose was decreased to 0.05 mg twice a week. A control group of 20 mice was available. Animals were evaluated for development of forestomach and oesophageal tumours and were observed to a maximum of 655 days after initiation of dosing. Dysplasia in the forestomach and oesophagus was observed in 2/28 treated animals, papillomas of the forestomach and oesophagus in 3/28 and carcinomas of the forestomach and oesophagus in 3/28. There was no evidence of such lesions in the control group (Li *et al.*, 1992).

3.1.2 *Rat*

A group of 31 female Wistar rats [age unspecified] were fed a diet containing maize bread inoculated with *F. moniliforme*. After 554–701 days of feeding, four papillomas and two early carcinomas had developed in forestomachs. No epithelial lesion of the forestomach was seen in a control group of 10 female rats given conventional maize bread for 330–700 days (Li *et al.*, 1982). [The Working Group noted the inadequate reporting of the study.]

Groups of 20 male inbred BDIX rats [age unspecified] were fed commercial rat feed containing 0 (8% uninoculated maize) or 4% freeze-dried or 4% oven-dried culture material that had been inoculated with *F. moniliforme* MRC 826 for 286 days and then 2% until termination of the study at day 763. The incidence of liver tumours (hepatocellular and cholangiocellular carcinomas combined) was increased: control, 0/20; 4% freeze-dried, 13/20; and 4% oven-dried, 16/20 (Marasas *et al.*, 1984b).

Two groups of 12 male Fischer rats, weighing approximately 125 g, were fed either commercial rodent chow or maize contaminated with *F. moniliforme* that had caused an outbreak of leukoencephalomalacia in horses (see p. 454). Individually treated rats were necropsied on days 123–145 and the remaining 8 treated and 12 control rats on day 176 after the start of the experiment. All treated rats had hepatic nodules, cholangiofibrosis or cholangiocarcinomas [numbers not given], while no such lesion was found in the controls (Wilson *et al.*, 1985).

Groups of 30 male inbred BDIX rats, weighing approximately 110 g, were fed semipurified diets containing 0 (5% maize meal), 0.5% *F. moniliforme* MRC 826 culture material (containing 364 mg/kg fusarin C and later found to produce fumonisins B_1 and B_2) or 5% *F. moniliforme* MRC 1069 culture material (containing 104 mg/kg fusarin C). In rats treated with 0.5% MRC 826 and examined at 23–27 months, neoplastic hepatic nodules occurred in all 21 surviving animals and in none of 22 controls; in addition, two hepatocellular and eight cholangiocellular carcinomas were observed, and there were increased incidences of forestomach papillomas (13/21 *versus* 5/22) and carcinomas (4/21 *versus* 0/22). In rats treated with 5% MRC 1069, the incidence of neither liver nor forestomach tumours was significantly increased (Jaskiewicz *et al.*, 1987).

Groups of 25 male inbred BDIX rats, weighing 70–80 g, were fed a modified cereal-based diet containing 0 or 50 mg/kg fumonisin B_1 (purified from culture material of *F. moniliforme* MRC 826). Five rats from each group were killed at 6, 12 and 20 months to assess the progression of liver lesions; the remaining rats were killed at 26 months. No hepatocellular carcinoma was observed in treated or control rats at 6 or 12 months; at 18–26 months, however, 10/15 treated rats and 0/15 controls had hepatocellular carcinomas (Gelderblom *et al.*, 1991).

A group of 20 female Wistar rats, weighing 80–120 g, were given 2 mg fusarin C [purity unspecified] by gavage twice a week; as body weights increased, the dose of fusarin C was increased to 3 mg twice a week. A control group of 25 rats was available. Animals were observed to a maximum of 742–814 days. Dysplasia of the forestomach and oesophagus was observed in 1/20 treated animals, papillomas of the forestomach and oesophagus in 5/20 and carcinomas of the forestomach and oesophagus in 5/20. There was no evidence of such lesions in the control group (Li *et al.*, 1992).

3.2 Administration with known carcinogens

3.2.1 *Mouse*

Groups of 10 female ICR/Ha mice, seven weeks of age, were each treated with a single application of 220 or 500 µg fusarin C [purity unspecified] in 0.1 ml acetone on the shaved back, followed one week later by twice weekly skin applications of 2 µg 12-*O*-tetradecanoylphorbol 13-acetate (TPA) in 0.1 ml acetone for 16 weeks. A group of nine positive controls each received a single application of 50 µg 7,12-dimethylbenz[a]anthracene (DMBA) on the dorsal skin, followed one week later by twice weekly applications of 2 µg TPA for 16 weeks. A group of eight mice each received an application of 50 µg DMBA followed by acetone, and 10 mice each received an application of 0.1 ml acetone followed by TPA. Nine of nine mice treated with DMBA followed by TPA had multiple skin papillomas (34), while no skin papilloma was seen in mice painted with DMBA plus acetone or acetone plus TPA. One mouse that received 220 µg fusarin C followed by TPA had two skin papillomas, but there was no skin tumour in mice that received 500 µg fusarin C (Gelderblom et al., 1986).

3.2.2 Rat

Groups of five to seven male BDIX and four to five female Wistar rats, seven weeks old, were fed a synthetic diet and subjected to a two-thirds partial hepatectomy. On the day after surgery, the rats were given a single intraperitoneal injection of 0, 50 or 100 mg/kg bw fusarin C [purity unspecified] in dimethyl sulfoxide and one week later were maintained on 0.05% phenobarbital in drinking-water for 14 weeks. At that time, their livers were examined for the presence of foci of altered hepatocytes, as revealed by γ-glutamyltranspeptidase-positive (GGT^+) staining. Fusarin C did not increase the number of foci of altered hepatocytes (Gelderblom et al., 1986).

Groups of four to five male inbred BDIX rats, weighing approximately 150 g, were given an intraperitoneal injection of 0 or 200 mg/kg bw N-nitrosodiethylamine (NDEA), were held for one week and were then fed diets containing 0 or 0.1% (1 g/kg diet) fumonisin B_1 (purity, 92%) for four weeks. Rats fed the control diet following exposure to NDEA or no treatment had no detectable foci of altered hepatocytes, determined by GGT^+ staining. Rats fed only fumonisin B_1 had an average of 20 ± 8 foci/cm^2, while those fed fumonisin B_1 following NDEA treatment had an average of 55 ± 10 foci/cm^2 (Gelderblom et al., 1988a).

Groups of four male Fischer rats, weighing 100–120 g, were fed a diet containing 0.1% fumonisin B_1 (purity, 90–95%) for 26 days and then subjected to partial hepatectomy. The resected liver lobes contained 2.9 ± 0.7 foci/cm^2 of altered hepatocytes (identified by GGT^+ staining) and a total area of foci of $0.42 \pm 0.34\%$ of liver section area; untreated controls had 1.0 ± 0.2 foci/cm^2 and a total area of $0.01 \pm 0.00\%$. These differences were significant. Two weeks after partial hepatectomy, rats received a daily dose of 2-acetylaminofluorene (20 mg/kg bw by gavage) for three consecutive days, followed by an oral dose of 2 ml/kg bw carbon tetrachloride on the fourth day. Ten days later they were sacrificed and found to have 7.1 ± 0.8 foci/cm^2 comprising $15.6 \pm 8.2\%$ of section area in the liver. Controls subjected to the same treatment without fumonisin B_1 had 1.2 ± 0.3 foci/cm^2 and a total area of $0.07 \pm 0.05\%$. These differences were significant (Gelderblom et al., 1992a).

4. Other Relevant Data

4.1 Absorption, distribution, metabolism and excretion

4.1.1 Humans

No data were available to the Working Group.

4.1.2 *Experimental systems*

Lu *et al.* (1990) studied the distribution and elimination of ^{3}H-fusarin C given by gavage to rats. The highest levels of radiolabel were found in the intestines, stomach and liver; lower levels were found in the kidney, bladder, oesophagus and spleen. Levels of radiolabel in the lungs and brain were low. Those in the blood reached a peak at 3 h after administration, but about 50% remained in the blood even after 24 h. Total urinary excretion of radiolabel was found to be about 31% within 48 h, and about 28% was excreted in the faeces. Only 5.4% unchanged fusarin C was excreted in the urine, and metabolites accounted for 94.6% of the total urinary radiolabel.

Two rat liver microsomal enzymes, carboxylesterase and a monooxygenase, are involved in the metabolism of fusarin C. The carboxyesterase catalyses the conversion of fusarin C to the water-soluble fusarin PM_1 (see Fig. 1) (Gelderblom *et al.*, 1988b). The conversion of fusarin C to an active mutagenic metabolite(s) is catalysed by a monooxygenase (Gelderblom *et al.*, 1984c, 1988b). It has been suggested that esterases metabolize fusarin C to a less mutagenic form: the mutagenicity of fusarin C to *Salmonella typhimurium* TA100 could be doubled by pretreating the microsomes with an esterase inhibitor (1 µM diisopropyl fluorophosphate) (Lu *et al.*, 1989).

Glutathione interacts *in vitro* both chemically and enzymatically with fusarin C, resulting in the formation of fusarin A (see Fig. 1), which does not have the C13–C14 epoxide group of fusarin C, and a compound that lacks the 2-pyrrolidone moiety, suggesting an interaction at the C13–C14 epoxide (Gelderblom *et al.*, 1988c).

4.2 Toxic effects

4.2.1 *Humans*

No data were available to the Working Group.

4.2.2 *Experimental systems*

Equine leukoencephalomalacia is a neurotoxic disease of horses, donkeys and mules and is characterized by multifocal liquefactive necrosis predominantly of the white matter in the cerebral hemispheres. This disease was reproduced experimentally by feeding pure cultures of *F. moniliforme* MRC 826 containing fumonisin B_1 to horses. A horse injected intravenously seven times from day 0 to day 9 with 0.125 mg/kg bw fumonisin B_1 per day showed clinical signs of neurotoxicosis on day 8. On day 10, the horse was killed while in tetanic convulsion. The principal lesions were severe oedema of the brain and early, bilaterally symmetrical leukoencephalomalacia in the brain stem (Marasas *et al.*, 1988b). The characteristic lesions of leukoencephalomalacia were also induced in two horses by oral administration of fumonisin B_1 (Kellerman *et al.*, 1990).

Three pigs were treated by daily intravenous injections of 0.17 mg/kg bw fumonisin B_1 for seven days, 0.4 mg/kg bw fumonisin B_1 for four days or 0.3 mg/kg fumonisin B_2 for five days. The highest dose of fumonisin B_1 was lethal on day 5; necropsy showed marked pulmonary oedema, hydrothorax and pancreatic lesions (focal to massive necrosis, acinar-cell

Fig. 1. Metabolites of fusarin C

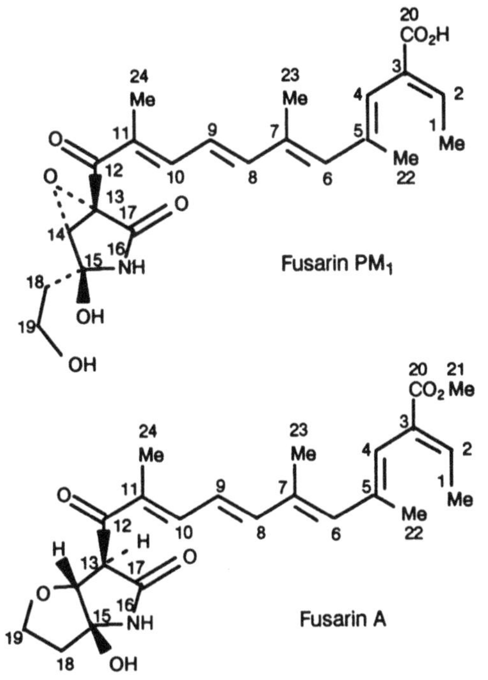

Fusarin PM$_1$

Fusarin A

From Gelderblom *et al.* (1988b)

dissociation and rounded individual acinar cells). No liver damage was detected. No such alteration was found after seven injections of the lower dose of fumonisin B$_1$ or with fumonisin B$_2$ (Harrison *et al.*, 1990).

Fumonisin B$_1$ was administered intravenously to two female SPF cross-bred swine, weighing 6–13 kg (total dose, 4.6 and 7.9 mg/kg or 67 and 72 mg per pig); and corn screenings naturally contaminated with 166 mg/kg fumonisin B$_1$ and 48 mg/kg fumonisin B$_2$ were given orally to three swine (total doses of fumonisins B$_1$ and B$_2$ ranged from 176 to 645 mg per pig). All treated pigs had pulmonary and hepatic changes. At the lower doses, slowly progressive hepatic disease was the most prominent feature, while at the higher doses, acute pulmonary oedema was superimposed on hepatic injury. Pancreatic acinar cell degeneration and many other types of ultrastructural change were noted. The target organs in the pig were concluded to be the lung, liver and pancreas (Haschek *et al.*, 1992).

Rats fed diets containing 1 g/kg fumonisin B$_1$ for four weeks had a significant reduction in weight gain and developed an insidious, progressive toxic hepatitis similar to that induced by culture material of *F. moniliforme* MRC 826. Fatty changes and scant necrosis were present in the proximal convoluted tubules of the kidney. Prominent lymphoid necrosis in Peyer's patches and scattered focal, superficial and mid-zonal epithelial necrosis occurred in the mucosa of the stomach. Severe, disseminated acute myocardial necrosis and severe pulmonary oedema were observed in two of the four rats (Gelderblom et al., 1988a). Rats and mice were fed diets containing seed maize inoculated with *F. moniliforme* (containing 99 mg/kg fumonisin B_1) or ammoniated seed maize inoculated in the same way (containing 75 mg/kg fumonisin B_1) for four weeks. Hepatotoxicity, renal toxicity and adrenal cortical vacuolation were noted in rats and hepatotoxicity in mice, irrespective of ammoniation (Voss et al., 1992). Fumonisin B_2, isolated and characterized along with fumonisin B_1 from *F. moniliforme* MRC 826, given at dietary levels of 0.05–0.1% induced early liver lesions in rats, similar to those induced by fumonisin B_1 (Gelderblom et al., 1992b).

In a long-term experiment, rats were fed a diet containing 50 mg/kg fumonisin B_1 for 26 months. Treatment significantly increased the levels of the serum enzymes alanine aminotransferase, GGT and alkaline phosphatase, as well as those of creatinine and bilirubin (conjugated and non-conjugated). The results were considered to corroborate the histopathological observation that the liver is the main target organ of fumonisin B_1 (Gelderblom et al., 1991).

The molecular mechanism of action of fumonisins is not known; however, these compounds bear a remarkable structural similarity to sphingosine, the long-chain (sphingoid) base backbone of sphingomyelin, cerebrosides, sulfatides, gangliosides and other sphingolipids. Sphingolipids are thought to be involved in the regulation of cell growth, differentiation and neoplastic transformation through participation in cell–cell communication and cell–substratum interactions and possibly through interactions with cell receptors and signalling systems. Incubation of rat hepatocytes with fumonisins B_1 and B_2 inhibited incorporation of ^{14}C-serine into the sphingosine moiety of cellular sphingolipids, at an IC_{50} of 0.1 μM. In contrast, fumonisin B_1 increased the amount of the biosynthetic intermediate, sphinganine, which suggests that fumonisins inhibit the conversion of ^{14}C-sphinganine to N-acyl-^{14}C-sphinganines, a step that is thought to precede introduction of the 4,5-*trans* double bond of sphingosine. In agreement with this mechanism, fumonisin B_1 inhibited the activity of sphingosine N-acyltransferase (ceramide synthase) in rat liver microsomes, with 50% inhibition at approximately 0.1 μM, and reduced the conversion of ^3H-sphingosine to ^3H-ceramide by intact hepatocytes. Fumonisin B_1 (1 μM) almost completely inhibited ^{14}C-sphingosine formation by hepatocytes (Wang et al., 1991).

Fumonisins B_1 and B_2 were toxic *in vitro* to rat hepatoma and dog kidney MDCK cell lines, at IC_{50} values ranging from 2 to 10 μg/ml (Mirocha et al., 1992).

Using a pig kidney cell line, LLC-PK_1, Yoo et al. (1992) demonstrated that both fumonisin B_1 and B_2 inhibit cell proliferation at concentrations between 10 and 35 μM, whereas concentrations > 35 μM caused cell death. Sphingolipid biosynthesis using ^3H-serine as the precursor was reduced at an EC_{50} of 10–15 μM; this effect was accompanied by a decrease in the ratio of ^3H-sphingosine: ^3H-sphinganine, which preceded the effect on cell proliferation. The two processes had a similar dependence on fumonisin concentration. The level of free sphinganine was elevated by 128 fold after exposure to 35 μM fumonisin B_1 for 24 h. These results support the hypothesis that inhibition of sphingolipid biosynthesis *de novo* is an early event in the toxic action of fumonisins on the pig kidney cell line LLC-PK_1, which is less sensitive than hepatocytes.

Shier et al. (1991) examined the effects of fumonisins B_1 and B_2 in a series of cultured mammalian cells lines. Approximate IC_{50} values for the most sensitive hepatoma line,

H4TG, were 4 and 2 µg/ml for fumonisins B_1 and B_2, respectively. An increase in the amount of free sphinganine and a reduction in complex sphingolipids were seen in serum samples from ponies given feed contaminated with 15–44 µg/g fumonisin B_1 (Wang et al., 1992).

Treatment of macrophages *in vitro* with fusarin C (6 µg/ml) inhibited their activation by macrophage activating factor and muramyl dipeptide. Fusarin C also inhibited the cytotoxic activity of already activated macrophages. These effects were dose-dependent; they disappeared partially after 24 h in the absence of fusarin C, and completely after 72 h, and could be overcome by high concentrations of macrophage activating factor and antiserum, suggesting that fusarin C is not generally toxic to cells (Dong & Zhang, 1987). Fusarin C at a concentration of 2.5 µg/ml inhibited the growth of cultured lymphoma cells (Chen & Zhang, 1987). It inhibited valine incorporation into proteins of rat hepatocytes at 10^{-4}M and cell death at 10^{-3}M (Norred et al., 1991a).

4.3 Reproductive and developmental toxicity

No data were available to the Working Group.

4.4 Genetic and related effects

4.4.1 *Humans*

No data were available to the Working Group.

4.4.2 *Experimental systems* (see also Tables 2–4 and Appendices 1 and 2)

Fumonisins B_1 and B_2 were not mutagenic to *S. typhimurium* and did not induce unscheduled DNA synthesis in rat hepatocytes, either *in vitro* or *in vivo*.

Fusarin C induced DNA strand breakage in *S. typhimurium* TA100, but attempts to detect DNA adducts of fusarin C in the same organism by ^{32}P-postlabelling were unsuccessful. HPLC-purified fusarin C did not produce DNA adducts in calf thymus that could be measured by ^{32}P-postlabelling. HPLC-purified fusarin C induced asynchronous replication of polyoma virus in rat fibroblasts and was reported to bind *in vitro* to the DNA of rat oesophageal explants.

Fusarin C was mutagenic to *S. typhimurium* in the presence of an exogenous metabolic activation system. Crude extracts of *F. moniliforme* cultures were also reported to have direct mutagenic activity in *S. typhimurium* TA100; this activity decreased with increasing purity (Lu et al., 1988). Fusarin C induced gene mutation, sister chromatid exchange, chromosomal aberrations and micronucleus formation in cultured Chinese hamster V79 cells.

Table 2. Genetic and related effects of fumonisin B_1

Test system	Result		Dose (LED/HID)[a]	Reference
	Without exogenous metabolic system	With exogenous metabolic system		
SA0, *Salmonella typhimurium* TA100, reverse mutation	–	–	2500.0000	Gelderblom & Snyman (1991)
SA2, *Salmonella typhimurium* TA102, reverse mutation	–	–	2500.0000	Gelderblom & Snyman (1991)
SA9, *Salmonella typhimurium* TA98, reverse mutation	–	–	5000.0000	Gelderblom & Snyman (1991)
SAS, *Salmonella typhimurium* TA97a, reverse mutation	–	–	2500.0000	Gelderblom & Snyman (1991)
URP, Unscheduled DNA synthesis, rat primary hepatocytes *in vitro*	–	0	60.0000	Gelderblom *et al.* (1992a)
UPR, Unscheduled DNA synthesis, rat hepatocytes *in vivo*	–		100 × 1 po	Gelderblom *et al.* (1992a)

–, negative; 0, not tested
[a]In-vitro tests, µg/ml; in-vivo tests, mg/kg bw

Table 3. Genetic and related effects of fumonisin B_2

Test system	Result		Dose (LED/HID)[a]	Reference
	Without exogenous metabolic system	With exogenous metabolic system		
SA0, *Salmonella typhimurium* TA100, reverse mutation	–	–	5000.0000	Gelderblom & Snyman (1991)
SA2, *Salmonella typhimurium* TA102, reverse mutation	–	–	2500.0000	Gelderblom & Snyman (1991)
SA9, *Salmonella typhimurium* TA98, reverse mutation	–	–	5000.0000	Gelderblom & Snyman (1991)
SAS, *Salmonella typhimurium* TA97a, reverse mutation	–	–	5000.0000	Gelderblom & Snyman (1991)
URP, Unscheduled DNA synthesis, rat primary hepatocytes *in vitro*	–	0	30.0000	Gelderblom *et al.* (1992a)
UPR, Unscheduled DNA synthesis, rat hepatocytes *in vivo*	–		100 × 1 po	Gelderblom *et al.* (1992a)

–, negative; 0, not tested
[a]In-vitro tests, µg/ml; in-vivo tests, mg/kg bw

Table 4. Genetic and related effects of fusarin C

Test system	Result without exogenous metabolic system	Result with exogenous metabolic system	Dose (LED/HID)[a]	Reference
PRB, DNA strand breaks, *Salmonella typhimurium* TA100	0		10.0000 (TLC)	Lu et al. (1988)
SA0, *Salmonella typhimurium* TA100, reverse mutation	–	+	0.1000 (TLC)	Wiebe & Bjeldanes (1981)
SA0, *Salmonella typhimurium* TA100, reverse mutation	0	+	0.1000	Gelderblom et al. (1984c)
SA0, *Salmonella typhimurium* TA100, reverse mutation	–	+	2.5000	Cheng et al. (1985)
SA0, *Salmonella typhimurium* TA100, reverse mutation	+[b]	+	2.5000 (TLC)	Lu et al. (1988)
SA0, *Salmonella typhimurium* TA100, reverse mutation	–	+	0.2500	Gelderblom et al. (1988c)
SA0, *Salmonella typhimurium* TA100, reverse mutation	0	+	0.0500 (HPLC)	Lu et al. (1989)
SA2, *Salmonella typhimurium* TA102, reverse mutation	–	+	0.5000	Gelderblom & Snyman (1991)
SA2, *Salmonella typhimurium* TA102, reverse mutation	–	–	5.0000	Gelderblom & Snyman (1991)
SA5, *Salmonella typhimurium* TA1535, reverse mutation	–	–	0.0000 (TLC)	Wiebe & Bjeldanes (1981)
SA5, *Salmonella typhimurium* TA1535, reverse mutation	0	–	25.0000 (TLC)	Lu et al. (1988)
SA8, *Salmonella typhimurium* TA1538, reverse mutation	–	(+)	0.0000 (TLC)	Wiebe & Bjeldanes (1981)
SA9, *Salmonella typhimurium* TA98, reverse mutation	–	(+)	0.0000 (TLC)	Wiebe & Bjeldanes (1981)
SA9, *Salmonella typhimurium* TA98, reverse mutation	–	+	200.0000	Cheng et al. (1985)
SA9, *Salmonella typhimurium* TA98, reverse mutation	–	+	2.5000	Gelderblom & Snyman (1991)
SAS, *Salmonella typhimurium* TA97a, reverse mutation	–	+	2.5000	Gelderblom & Snyman (1991)
DIA, DNA replication, polyoma-transformed rat embryo fibroblast H3	+	0	0.0200 (HPLC)	Lu et al. (1988)
G9H, Gene mutation, Chinese hamster V79 lung cells *hprt* locus *in vitro*	–	+	50.0000	Cheng et al. (1985)
SIC, Sister chromatid exchange, Chinese hamster V79 lung cells *in vitro*	–	+	25.0000	Cheng et al. (1985)
CIC, Chromosomal aberrations, Chinese hamster V79 lung cells *in vitro*	–	+	50.0000	Cheng et al. (1985)

Table 4 (contd)

Test system	Result		Dose (LED/HID)[a]	Reference
	Without exogenous metabolic system	With exogenous metabolic system		
MIA, Micronucleus test, Chinese hamster V79 lung cells *in vitro*	–	+	50.0000	Cheng et al. (1985)
BID, DNA adducts, isolated calf thymus DNA *in vitro*	–	0	50.0000 (HPLC)	Lu et al. (1988)
BID, Binding to DNA of rat oesophageal explants *in vitro*	+	0	0.0000	Lu et al. (1990)
BID, DNA adducts, *Salmonella typhimurium* TA100	–	–	10.0000 (TLC)	Lu et al. (1988)

+, positive; (+), weakly positive; –, negative; 0, not tested
[a]In-vitro tests, μg/ml; purified by TLC (thin-layer chromatography) or HPLC (high-performance liquid chromatography)
[b]With preincubation only; one dose (9.9 μg/ml) tested

5. Summary of Data Reported and Evaluation

5.1 Exposure data

Fumonisin B_1, fumonisin B_2 and fusarin C are produced by *Fusarium* species that occur primarily on maize. These toxins may occur particularly when maize is grown under warm, dry conditions. Exposure occurs through dietary consumption of contaminated maize. Populations that eat milled or ground maize as a dietary staple can therefore be exposed to significant amounts of fumonisins and to lesser amounts of fusarin C.

5.2 Human carcinogenicity data

The only studies available were correlation studies, most of which indicated some relationship between oesophageal cancer rates and the occurrence of *F. moniliforme* or its toxins in maize.

5.3 Animal carcinogenicity data

Cultures of *F. moniliforme* were tested by oral administration in two experiments in male rats of one strain. A culture of *F. moniliforme* known to produce significant amounts of fumonisins B_1 and B_2 induced neoplastic nodules, hepatocellular carcinomas and cholangiocellular carcinomas; in addition, forestomach papillomas and carcinomas were observed. A culture of *F. moniliforme*, known to contain mainly fusarin C, did not induce such tumours.

Two studies in which male rats were fed maize naturally contaminated with *F. moniliforme* were inadequate for evaluation.

Fumonisin B_1 was tested for carcinogenicity by oral administration in the diet in one experiment in male rats, producing hepatocellular carcinomas. It induced the formation of foci of altered (γ-glutamyltranspeptidase-positive) hepatocytes.

No data were available to the Working Group on the carcinogenicity of fumonisin B_2.

Fusarin C was tested in one study in female mice and female rats by oral gavage. It induced papillomas and carcinomas of the oesophagus and forestomach in mice and rats.

5.4 Other relevant data

Fumonisin B_1 causes outbreaks of leukoencephalomalacia in horses and pulmonary oedema in pigs. It is toxic to the central nervous system, liver, pancreas, kidney and lung in a number of animal species. Fumonisin B_2 is hepatotoxic in rats.

Fumonisins B_1 and B_2 do not induce unscheduled DNA synthesis in rat hepatocytes *in vivo* or *in vitro* or mutation in bacteria.

In single studies, fusarin C induces chromosomal anomalies, gene mutation and DNA damage in cultured rodent cells. It induces mutations in bacteria.

5.5 Evaluation[1]

There is *inadequate evidence* in humans for the carcinogenicity of toxins derived from *F. moniliforme*.

There is *sufficient evidence* in experimental animals for the carcinogenicity of cultures of *F. moniliforme* that contain significant amounts of fumonisins.

There is *limited evidence* in experimental animals for the carcinogenicity of fumonisin B_1.

There is *inadequate evidence* in experimental animals for the carcinogenicity of fumonisin B_2.

There is *limited evidence* in experimental animals for the carcinogenicity of fusarin C.

Overall evaluation

Toxins derived from *Fusarium moniliforme are possibly carcinogenic to humans (Group 2B).*

6. References

Bezuidenhout, S.C., Gelderblom, W.C.A., Gorst-Allman, C.P., Horak, R.M., Marasas, W.F.O., Spiteller, G. & Vleggaar, R. (1988) Structure elucidation of the fumonisins, mycotoxins from *Fusarium moniliforme. J. chem. Soc. chem. Commun.*, **11**, 743–745

Bothast, R.J., Bennett, G.A., Vancauwenberge, J.E. & Richard, J.L. (1992) Fate of fumonisin B_1 in naturally contaminated corn during ethanol fermentation. *Appl. environ. Microbiol.*, **58**, 233–236

Cawood, M.E., Gelderblom, W.C.A., Vleggaar, R., Behrend, Y., Thiel, P.G. & Marasas, W.F.O. (1991) Isolation of the fumonisin mycotoxins: a quantitative approach. *J. agric. Food Chem.*, **39**, 1958–1962

Chen, L.-P. & Zhang, Y.-H. (1987) Suppression of the in vitro lymphoproliferative response to syngeneic L5178Y tumor cells by fusarin C in mice. *J. exp. clin. Cancer Res.*, **6**, 25–29

Cheng, S.J., Jiang, Y.Z., Li, M.H. & Lo, H.Z. (1985) A mutagenic metabolite produced by *Fusarium moniliforme* isolated from Linxian County, China. *Carcinogenesis*, **6**, 903–905

Dong, Z.-Y. & Zhang, Y.-H. (1987) Inhibitory effect of a mycotoxin, fusarin C, on macrophage activation and macrophage mediated cytotoxicity to tumor cells in mice. *J. exp. Cancer Res.*, **6**, 31–38

van Egmond, H.P. (1989) Current situation on regulations for mycotoxins. Overview of tolerances and status of standard methods of sampling and analysis. *Food Addit. Contam.*, **6**, 139–188

Farber, J.M. & Scott, P.M. (1989) Fusarin C. In: Chelkowski, J., ed., *Fusarium Mycotoxins, Taxonomy and Pathogenicity*, Amsterdam, Elsevier, pp. 41–52

Gelderblom, W.C.A. & Snyman, S.D. (1991) Mutagenicity of potentially carcinogenic mycotoxins produced by *Fusarium moniliforme. Mycotoxin Res.*, **7**, 46–52

Gelderblom, W.C.A., Thiel, P.G., van der Merwe, K.J., Marasas, W.F.O. & Spies, H.S.C. (1983) A mutagen produced by *Fusarium moniliforme. Toxicon*, **21**, 467–473

Gelderblom, W.C.A., Marasas, W.F.O., Steyn, P.S., Thiel, P.G., van der Merwe, K.J., van Rooyen, P.H., Vleggaar, R. & Wessels, P.L. (1984a) Structure elucidation of fusarin C, a mutagen produced by *Fusarium moniliforme. J. chem. Soc. chem. Commun.*, **7**, 122–124

Gelderblom, W.C.A., Thiel, P.G., Marasas, W.F.O. & van der Merwe, K.J. (1984b) Natural occurrence of fusarin C, a mutagen produced by *Fusarium moniliforme* in corn. *J. agric. Food Chem.*, **32**, 1064–1067

---

[1]For definition of the italicized terms, see Preamble, pp. 26–29.

Gelderblom, W.C.A., Thiel, P.G. & van der Merwe, K.J. (1984c) Metabolic activation and deactivation of fusarin C, a mutagen produced by *Fusarium moniliforme*. *Biochem. Pharmacol.*, 33, 1601–1603

Gelderblom, W.C.A., Thiel, P.G., Jaskiewicz, K. & Marasas, W.F.O. (1986) Investigations on the carcinogenicity of fusarin C—a mutagenic metabolite of *Fusarium moniliforme*. *Carcinogenesis*, 7, 1899–1901

Gelderblom, W.C.A., Jaskiewicz, K., Marasas, W.F.O., Thiel, P.G., Horak, R.M., Vleggaar, R. & Kriek, N.P.J (1988a) Fumonisins—novel mycotoxins with cancer-promoting activity produced by *Fusarium moniliforme*. *Appl. environ. Microbiol.*, 54, 1806–1811

Gelderblom, W.C.A., Thiel, P.G. & van der Merwe, K.J. (1988b) The role of rat liver microsomal enzymes in the metabolism of the fungal metabolite fusarin C. *Food chem. Toxicol.*, 26, 31–36

Gelderblom, W.C.A., Thiel, P.G. & van der Merwe, K.J. (1988c) The chemical and enzymatic interaction of glutathione with the fungal metabolite, fusarin C. *Mutat. Res.*, 199, 207–214

Gelderblom, W.C.A., Kriek, N.P.J., Marasas, W.F.O. & Thiel, P.G. (1991) Toxicity and carcinogenicity of the *Fusarium moniliforme* metabolite, fumonisin B_1, in rats. *Carcinogenesis*, 12, 1247–1251

Gelderblom, W.C.A., Semple, E., Marasas, W.F.O. & Farber, E. (1992a) The cancer-initiating potential of the fumonisin B mycotoxins. *Carcinogenesis*, 13, 433–437

Gelderblom, W.C.A., Marasas, W.F.O., Vleggaar, R., Thiel, P.G. & Cawood, M.E. (1992b) Fumonisins: isolation, chemical characterization and biological effects. *Mycopathologia*, 117, 11–16

Harrison, L.R., Colvin, B.M., Green, J.T., Newman, L.E. & Cole, J.R., Jr (1990) Pulmonary edema and hydrothorax in swine produced by fumonisin B_1, a toxic metabolite of *Fusarium moniliforme*. *J. Vet. Diagn. Invest.*, 2, 217–221

Haschek, W.M., Motelin, G., Ness, D.K., Harlin, K.S., Hall, W.F., Vesonder, R.F., Peterson, R.E. & Beasley, V.R. (1992) Characterization of fumonisin toxicity in orally and intravenously dosed swine. *Mycopathologia*, 117, 83–96

Jackson, M.A. & Bennett, G.A. (1990) Production of fumonisin B_1 by *Fusarium moniliforme* NRRL 13616 in submerged culture. *Appl. environ. Microbiol.*, 56, 2296–2298

Jaskiewicz, K., van Rensburg, S.J., Marasas, W.F. & Gelderblom, W.C. (1987) Carcinogenicity of *Fusarium moniliforme* culture material in rats. *J. natl Cancer Inst.*, 78, 321–325

Kellerman, T.S., Marasas, W.F.O., Thiel, P.G., Gelderblom, W.C.A., Cawood, M. & Coetzer, J.A.W. (1990) Leukoencephalomalacia in two horses induced by oral dosing of fumonisin B_1. *Onderstepoort J. vet. Res.*, 57, 269–275

Laurent, D., Platzer, N., Kohler, F., Sauviat, M.P. & Pellegrin, F. (1989) Macrofusine and microlmoniline: two new mycotoxins isolated from maize infested with *Fusarium moniliforme* Sheld (Fr.). *Microbiol. Aliments Nutr.*, 7, 9–16

Lew, H., Adler, A. & Edinger, W. (1991) Moniliformin and the European corn borer (*Ostrinia nubilaris*). *Mycotoxin Res.*, 7, 71–76

Li, M.-X., Tian, G.-Z., Lu, S.-X., Kuo, S.-P., Ji, C.-L. & Wang, Y.-L. (1982) Forestomach carcinoma induced in rats by cornbread inoculated with *Fusarium moniliforme* (Chin.). *Chin. J. Oncol.*, 4, 241–244

Li, M.-X., Jian, Y.-Z., Han, N.-J., Fan, W.-G., Ma, J.-L. & Bjeldanes, L.E. (1992) Fusarin C induced esophageal and forestomach carcinoma in mice and rats (Chin.). *Chin. J. Oncol.*, 14, 27–29

Lu, S.-J., Ronai, Z.A., Li, M.H. & Jeffrey, A.M. (1988) *Fusarium moniliforme* metabolites: genotoxicity of culture extracts. *Carcinogenesis*, 9, 1523–1527

Lu, S.-J., Li, M.-H., Park, S.S., Gelboin, H.V. & Jeffrey, A.M. (1989) Metabolism of fusarin C by rat liver microsomes. Role of esterase and cytochrome P-450 enzymes with respect to the mutagenicity of fusarin C in *Salmonella typhimurium*. *Biochem. Pharmacol.*, 38, 3811–3817

Lu, F.-X., Lee, M.-X. & Shen, D.-C. (1990) The metabolism and DNA binding of ^{3}H-fusarin C in rats (Chin.). *Acta acad. med. sin.*, **12**, 231–235

Marasas, W.F.O., van Rensburg, S.J. & Mirocha, C.J. (1979) Incidence of *Fusarium* species and the mycotoxins deoxynivalenol and zearalenone in corn produced in esophageal cancer areas in Transkei. *J. agric. Food Chem.*, **27**, 1108–1112

Marasas, W.F.O., Wehner, F.C., van Rensburg, S.J. & van Schalkwyk, D.J. (1981) Mycoflora of corn produced in human esophageal cancer areas in Transkei, southern Africa. *Phytopathology*, **71**, 792–796

Marasas, W.F.O., Nelson, P.E. & Toussoun, T.A. (1984a) *Toxigenic* Fusarium *Species. Identity and Mycotoxicology*, University Park, PA, Pennsylvania State University Press

Marasas, W.F.O., Kriek, N.P.J., Fincham, J.E. & van Rensburg, S.J. (1984b) Primary liver cancer and oesophageal basal cell hyperplasia in rats caused by *Fusarium moniliforme*. *Int. J. Cancer*, **34**, 383–387

Marasas, W.F.O., Jaskiewicz, K., Venter, F.S. & van Schalkwyk, D.J. (1988a) *Fusarium moniliforme* contamination of maize in oesophageal cancer areas in Transkei. *S. Afr. med. J.*, **74**, 110–114

Marasas, W.F.O., Kellerman, T.S., Gelderblom, W.C.A., Coetzer, J.A.W., Thiel, P.G. & van der Lugt, J.J. (1988b) Leukoencephalomalacia in a horse induced by fumonisin B$_1$ isolated from *Fusarium moniliforme*. *Onderstepoort J. vet. Res.*, **55**, 197–203

Miller, J.D., Savard, M.E., Sibilia, A. & Rapior, S. (1993) Production of fumonisins and fusarins by *Fusarium moniliforme* from Southeast Asia. *Mycologia* (in press)

Minervini, F., Bottalico, C., Pestka, J.J. & Visconti, A. (1993) On the occurrence of fumonisins in feed in Italy (Ital.). In: Scimone, G., ed., *Proceedings of the 46th National Congress of the Società Italiana delle Scienze veterinarie, 30 September–3 October 1992, Venice*, Messina, Grafiche Scuderi (in press)

Mirocha, C.J., Gilchrist, D.G., Shier, W.T., Abbas, H.K., Wen, Y. & Vesonder, R.F. (1992) AAL toxins, fumonisins (biology and chemistry) and host-specificity concepts. *Mycopathologia*, **117**, 47–56

Nelson, P.E., Plattner, R.D., Shackelford, D.D. & Desjardins, A.E. (1991) Production of fumonisins by *Fusarium moniliforme* strains from various substrates and geographic areas. *Appl. environ. Microbiol.*, **57**, 2410–2412

Nelson, P.E., Plattner, R.D., Shackelford, D.D. & Desjardins, A.E. (1992) Fumonisin B$_1$ production by *Fusarium* species other than *F. moniliforme* in section *Liseola* and by some related species. *Appl. environ. Microbiol.*, **58**, 984–989

Norred, W.P., Bacon, C.W., Plattner, R.D. & Vesonder, R.F. (1991a) Differential cytotoxicity and mycotoxin content among isolates of *Fusarium moniliforme*. *Mycopathologia*, **115**, 37–43

Norred, W.P., Voss, K.A., Bacon, C.W. & Riley, R.T. (1991b) Effectiveness of ammonia treatment in detoxification of fumonisin-contaminated corn. *Food chem. Toxicol.*, **29**, 815–819

Plattner, R.D., Norred, W.P., Bacon, C.W., Voss, K.W., Peterson, R., Shackelford, D.D. & Weisleder, D. (1990) A method of detection of fumonisins in corn samples associated with field cases of equine leucoencephalomalacia. *Mycologia*, **82**, 698–702

Plattner, R.D., Ross, P.F., Reagor, J., Stedelin, J. & Rice, L.G. (1991) Analysis of corn and cultured corn for fumonisin B$_1$ by HPLC and GC/MS by four laboratories. *J. vet. Diagn. Invest.*, **3**, 357–358

Rheeder, J.P., Marasas, W.F.O., Thiel, P.G., Sydenham, E.W., Shephard, G.S. & van Schalkwyk, D.J. (1992) *Fusarium moniliforme* and fumonisins in corn in relation to human esophageal cancer in Transkei. *Phytopathology*, **82**, 353–357

Riley, R.T. & Richard, J.L., eds (1992) Fumonisins. A current perspective and view to the future. *Mycopathologia*, **117**, 1–126

Ross, P.F., Rice, L.G., Plattner, R.D., Osweiler, G.D., Wilson, T.M., Owens, D.L., Nelson, H.A. & Richard, J.L. (1991) Concentrations of fumonisin B_1 in feeds associated with animal health problems. *Mycopathologia*, **114**, 129–135

Savard, M.E. & Miller, J.D. (1992) Characterization of fusarin F, a new fusarin from *Fusarium moniliforme*. *J. nat. Prod.*, **55**, 64–70

Scott, P.M. (1992) Mycotoxins. *J. Assoc. off. anal. Chem.*, **75**, 95–102

Scott, P.M. & Lawrence, G.A. (1991) Liquid determination of fumonisins with 4-fluoro-7-nitrobenzofurazan (NBD-F) (Abstract No. 251). In: *Proceedings of the 105th AOAC Annual International Meeting, 12–15 August, Phoenix, Arizona*, Arlington, VA, Association of Official Analytical Chemists, p. 147

Shephard, G.S., Sydenham, E.W., Thiel, P.G. & Gelderblom, W.C.A. (1990) Quantitative determination of fumonisins B_1 and B_2 by high-performance liquid chromatography with fluorescence detection. *J. liq. Chromatogr.*, **13**, 2077–2087

Shephard, G.S., Thiel, P.G. & Sydenham, E.W. (1992) Determination of fumonisin B_1 in plasma and urine by high-performance liquid chromatography. *J. Chromatogr.*, **574**, 299–304

Shier, W.T., Abbas, H.K. & Mirocha, C.J. (1991) Toxicity of the mycotoxins fumonisins B_1 and B_2 and *Alternaria alternata* f. sp. *lycopersici* toxin (AAL) in cultured mammalian cells. *Mycopathologia*, **116**, 97–104

Sydenham, E.W., Gelderblom, W.C.A., Thiel, P.G. & Marasas, W.F.O. (1990a) Evidence for the natural occurrence of fumonisin B_1, a mycotoxin produced by *Fusarium moniliforme*, in corn. *J. agric. Food Chem.*, **38**, 285–290

Sydenham, E.W., Thiel, P.G., Marasas, W.F.O., Shephard, G.S., Van Schalkwyk, D.J. & Koch, K.R. (1990b) Natural occurrence of some *Fusarium* mycotoxins in corn from low and high esophageal cancer prevalence areas of the Transkei, southern Africa. *J. agric. Food Chem.*, **38**, 1900–1903

Sydenham, E.W., Shephard, G.S., Thiel, P.G., Marasas, W.F.O. & Stockenström, S. (1991) Fumonisin contamination of commercial corn-based human foodstuffs. *J. agric. Food Chem.*, **39**, 2014–2018

Sydenham, E.W., Shephard, G.S. & Thiel, P.G. (1992a) Liquid chromatographic determination of fumonisins B_1, B_2 and B_3 in foods and feeds. *J. Assoc. off. anal. Chem.*, **75**, 313–318

Sydenham, E.W., Marasas, W.F.O., Shephard, G.S., Thiel, P.G. & Hirooka, E.Y. (1992b) Fumonisin concentrations in Brazilian feeds associated with field outbreaks of confirmed and suspected animal mycotoxicoses. *J. agric. Food Chem.*, **40**, 994–997

Thiel, P.G., Gelderblom, W.C.A., Marasas, W.F.O., Nelson, P.E. & Wilson, T.M. (1986) Natural occurrence of moniliformin and fusarin C in corn screenings known to be hepatocarcinogenic to rats. *J. agric. Food Chem.*, **34**, 773–775

Thiel, P.G., Marasas, W.F.O., Sydenham, E.W., Shephard, G.S., Gelderblom, W.C.A. & Nieuwenhuis, J.J. (1991a) Survey of fumonisin production by *Fusarium* species. *Appl. environ. Microbiol.*, **57**, 1089–1093

Thiel, P.G., Shephard, G.S., Sydenham, E.W., Marasas, W.F.O., Nelson, P.E. & Wilson, T.M. (1991b) Levels of fumonisin B_1 and B_2 associated with confirmed cases of equine leukoencephalomalacia. *J. agric. Food Chem.*, **39**, 109–111

Thiel, P.G., Marasas, W.F.O., Sydenham, E.W., Shephard, G.S. & Gelderblom, W.C.A. (1992) The implications of naturally occurring levels of fumonisins in corn for human and animal health. *Myctopathologia*, **117**, 3–9

Thiel, P.G., Sydenham, E.W., Shephard, G.S. & van Schalkwyk, D.J. (1993) Liquid chromatographic method for the determination of fumonisins B_1 and B_2 in corn: collaborative study. *J. Assoc. off. anal. Chem.* (in press)

Thrane, U. (1989) *Fusarium* species and their specific profiles of secondary metabolites. In: Chelkowski, J., ed., Fusarium *Mycotoxins, Taxonomy and Pathogenicity*, Amsterdam, Elsevier, pp. 199–225

Vesonder, R., Peterson, R., Plattner, R. & Weisleder, D. (1990) Fumonisin B_1: isolation from corn culture, and purification by high performance liquid chromatography. *Mycotoxin Res.*, 8, 85–87

Visconti, A. (1992) Examination of European isolates of *Fusarium* for production of fumonisins (Abstract). In: *Proceedings of the 106th AOAC International Meeting, Cincinnati, OH, 31 August–3 September 1992*, Arlington, VA, Association of Official Analytical Chemists

Voss, K.A., Norred, W.P. & Bacon, C.W. (1992) Subchronic toxicological investigations of *Fusarium moniliforme*-contaminated corn, culture material, and ammoniated culture material. *Mycopathologia*, 117, 97–104

Wang, E., Norred, W.P., Bacon, C.W., Riley, R.T. & Merrill, A.H., Jr (1991) Inhibition of sphingolipid biosynthesis by fumonisins. Implications for diseases associated with *Fusarium moniliforme*. *J. biol. Chem.*, 266, 14486–14490

Wang, E., Ross, P.F., Wilson, T.M., Riley, R.T. & Merrill, A.H., Jr (1992) Increases in serum sphingosine and sphinganine and decreases in complex sphingolipids in ponies given feed containing fumonisins, mycotoxins produced by *Fusarium moniliforme*. *J. Nutr.*, 122, 1706–1716

Wiebe, L.A. & Bjeldanes, L.F. (1981) Fusarin C, a mutagen from *Fusarium moniliforme* grown on corn. *J. Food Sci.*, 46, 1424–1426

Wilson, T.M., Nelson, P.E. & Knepp, C.R. (1985) Hepatic neoplastic nodules, adenofibrosis, and cholangiocarcinomas in male Fischer 344 rats fed corn naturally contaminated with *Fusarium moniliforme*. *Carcinogenesis*, 6, 1155–1160

Wilson, T.M., Ross, P.F., Rice, L.G., Osweiler, G.D., Nelson, H.A., Owens, D.L., Plattner, R.D., Reggiardo, C., Noon, T.H. & Pickrell, J.W. (1990) Fumonisin B_1 levels associated with an epizootic of equine leukoencephalomalacia. *J. vet. Diagn. Invest.*, 2, 213–216

Yoo, H.S., Norred, W.P., Wang, E., Merrill, A.H., Jr & Riley, R.T. (1992) Fumonisin inhibition of *de novo* sphingolipid biosynthesis and cytotoxicity are correlated in LLC-PK_1 cells. *Toxicol. appl. Pharmacol.*, 114, 9–15

Zhen, Y., Yang, S., Ding, L., Han, F., Yang, W. & Liu, Q. (1984) The culture and isolation of fungi from the cereals in five high and three low incidence counties of oesophageal cancer in Henan province (Chin.). *Chin. J. Oncol.*, 6, 27–29

Zhu, B. & Jeffrey, A.M. (1992) Stability of fusarin C: effects of the normal cooking procedure used in China and pH. *Nutr. Cancer*, 18, 53–58

TOXINS DERIVED FROM *FUSARIUM SPOROTRICHIOIDES*: T-2 TOXIN

Fusarium sporotrichioides may occur in cereals, particularly in north temperate climates. This species produces T-2 toxin and other metabolites, such as diacetoxyscirpenol, which also occurs naturally but is not considered in this monograph. T-2 toxin is a type-A trichothecene (Thrane, 1989).

T-2 Toxin was considered by a previous Working Group (as T_2-trichothecene), in October 1982 (IARC, 1983). Since that time, new data have become available, and these have been incorporated into the monograph and taken into consideration in the present evaluation.

1. Exposure Data

1.1 Chemical and physical data

1.1.1 Synonyms, structural and molecular data

Chem. Abstr. Services Reg. No.: 21259-20-1

Chem. Abstr. Name: Trichothec-9-ene-3,4,8,15-tetrol, 12,13-epoxy, 4,15-diacetate 8-(3-methylbutanoate), (3α,4β,8α)-

IUPAC Systematic Name: 12,13-Epoxytrichothec-9-ene-3α,4β,8α,15-tetrol, 4,15-diacetate 8-isovalerate or (2R,3R,4S,5S,5aR,7S,9aR,10S)-2,3,4,5,7,9a-hexahydro-3,4,7-trihydroxy-5,8-dimethyl spiro[2,5-methano-1-benzoxepin-10,2′-oxirane]-5a(6H)-methanol, 4,5a-diacetate 7-isovalerate

Synonyms: Fusariotoxin T2; insariotoxin; 8α(3-methylbutyryloxy)4β,15-diacetoxyscirp-9-en-3α-ol; mycotoxin T2; NSC 138780; T-2 mycotoxin; toxin T2; T_2-toxin; T_2-trichothecene; 3α-hydroxy-4β,15-diacetoxy-8α-(3-methylbutyryloxy)-12,13-epoxy-δ9-tricothecene

$C_{24}H_{34}O_9$ Mol. wt: 466.5

1.1.2 Chemical and physical properties

From Bamburg et al. (1968) and Wei et al. (1971), unless otherwise specified

(a) *Description*: White needles
(b) *Melting-point*: 151–152 °C
(c) *Optical rotation*: $[\alpha]_D^{26}$ +15° (c = 2.58 in 95% ethanol; $[\alpha]_D^{21}$ −15.5° (c = 2.14 in chloroform) (Pohland et al., 1982)
(d) *Spectroscopy data*: Ultraviolet, infrared, nuclear magnetic resonance and mass spectral data have been reported (Pohland et al., 1982).
(e) *Solubility*: Soluble in acetone, acetonitrile, chloroform, diethyl ether, ethyl acetate, methanol, ethanol and dichloromethane (Yates et al., 1968; Lauren & Agnew, 1991)
(f) *Stability*: Stable in the solid state; ester groups are saponified by alkalis, and the epoxide is opened by strong mineral acids (Wei et al., 1971)

1.1.3 Analysis

Available methods for the analysis of T-2 toxin include thin-layer chromatography and, after derivatization, gas chromatography, gas chromatography/mass spectrometry, and high-performance liquid chromatography with fluorescence detection (Scott, 1990; Cohen & Boutin-Muma, 1992). Methods involving hydrolysis followed by derivatization and gas chromatography have been developed for screening feedstuffs and biological fluids (Rood et al., 1988; Lauren & Agnew, 1991).

Materials contaminated with T-2 toxin are usually co-contaminated with other trichothecenes, and mass spectrometric confirmation is normally required (Scott, 1990). Methods based on the use of polyclonal antibodies suffer from the same difficulty as those for deoxynivalenol in terms of cross-reactivity to T-2 analogues (Chu et al., 1979; Kawamura et al., 1990; see also Fan et al., 1988).

1.2 Production and use

T-2 Toxin was first isolated by Bamburg et al. (1968) from a culture of *Fusarium sporotrichioides*, wrongly identified as *F. tricinctum* (Marasas et al., 1984). This toxin is readily produced in liquid fermentations at yields approaching 1 g/l (Greenhalgh et al., 1988). It can also be produced in large quantities in solid fermentations (Cullen et al., 1982). This species produces a range of toxic metabolites (Marasas et al., 1984; Greenhalgh et al., 1990) and is notorious as the probable cause of the 'alimentary toxic aleukia' (see p. 475) that occurred among men in Russia during the Second World War. The disease was caused by eating over-wintered grain now known to have contained high concentrations of T-2 toxin (Mirocha, 1984).

T-2 Toxin is produced by *F. sporotrichioides*, *F. poae*, *F. equiseti* and *F. acuminatum*. The validity of reports that other species of *Fusarium* produce it has been questioned (Marasas et al., 1984; Thrane, 1989). T-2 Toxin is not produced by any other fungal genus. *F. sporotrichioides* is a saprophyte found primarily on cereals left in the field under wet conditions at harvest.

1.3 Occurrence

T-2 Toxin has been reported in cereals in many parts of the world (Table 1; Scott, 1989; WHO, 1990). It is formed in large quantities under the unusual circumstance of prolonged wet weather at harvest.

Table 1. Natural occurrence of T-2 toxin

Country or region	Product	Year	Positive samples/ total no.	Content (mg/kg)	Reference
North America					
Canada	Barley	1973	1/1	25	Puls & Greenway (1976)
Canada	Feed	1979–82	1/51	2.5	Abramson et al. (1983)
Canada	Maize	1984	1/1	3.0	Foster et al. (1986)
USA	Maize	1971	1/1	2.0	Hsu et al. (1972)
			9/118	0.08–0.7	Scott (1989)
South America					
Argentina	Maize	NR	22/100	ND	Scott (1989)
Europe					
Czechoslovakia	Grain	1980	3/100	0.1–1.0	Scott (1989)
Finland	Feed	1976–77	3/230	0.01–0.05	Scott (1989)
Finland	Feed	1984	8/167	0.02–0.71	Karppanen et al. (1985)
France	Maize	NR	1/3	0.02	Jemmali et al. (1978)
France	Maize	1976–78	19/301	< 0.2–1.4	Scott (1989)
France	Feed	NR	4/8	< 1–2.0	Scott (1989)
France	Maize flour	NR	2/38	Trace, 0.03	Scott (1989)
Germany	Feed	NR	1/188	0.065	Scott (1989)
Germany	Oats	NR	2/19	0.08, 0.09	Scott (1989)
Germany	Wheat	NR	1/26	0.1	Scott (1989)
Hungary	Maize	1976–78	5/491	ND	Scott (1989)
Hungary	Maize	1980–81	8/23	0.1–4.4	Bata et al. (1983)
Hungary	Wheat	1980–81	2/23	0.2, 1.9	Bata et al. (1983)
Hungary	Feed	1980–81	2/23	4.1, 5.8	Bata et al. (1983)
Italy	Feed	NR	9/9	0.05–0.3	Scott (1989)
Norway	Barley	1984	2/49	0.02, 0.04	Sundheim et al. (1988)
United Kingdom	Feed	NR	4/ > 2000	Trace	Scott (1989)
Australasia					
India	Wheat	NR	3/58	0.55–4.0	Ramakrishna et al. (1990)
India	Wheat flour	NR	1/37	0.8	Ramakrishna et al. (1990)
India	Maize	NR	1/1	2.0	Ueno (1986)
India	Rice	1981	1/32	0.03	Reddy et al. (1983)
India	Groundnuts	1984–85	6/56	0.17–38.9	Bhavanishankar & Shantha (1987)

Table 1 (contd)

Country or region	Product	Year	Positive samples/ total no.	Content (mg/kg)	Reference
India	Sorghum	1981	2/20	0.01–0.05	Reddy et al. (1983)
India	Sorghum	1984–85	4/67	1.67–15.0	Bhavanishankar & Shantha (1987)
New Zealand	Maize	1984	13/20	0.01–0.2	Hussein et al. (1989)

[a]NR, not reported; ND, not detected

1.4 Regulations and guidelines

An official tolerance level of 0.1 mg/kg was established for T-2 toxin in grains in the USSR in 1984 (van Egmond, 1989).

2. Studies of Cancer in Humans

Ecological studies of the relationship between exposure to *Fusarium* toxins and oesophageal cancer are summarized in the monographs on toxins derived from *F. graminearum*, *F. culmorum* and *F. crookwellense* and from *F. moniliforme*, pp. 409 and 450.

3. Studies of Cancer in Experimental Animals

3.1 Oral administration

3.1.1 *Mouse*

Groups of 50 male and 50 female CD-1 mice, six weeks of age, were fed a semi-synthetic diet containing 0, 1.5 or 3.0 mg/kg T-2 toxin (> 99% pure) for 71 weeks. Mice found moribund (up to 70 weeks) and those at termination of the study (71 weeks) were killed and tissues examined. There was no difference in food consumption or weight gain among the groups. More than 50% of male mice and 75% of females in each group survived to the end of the study; controls had the poorest survival. The incidence of pulmonary adenomas was increased only in males: controls, 4 (10%); low-dose, 7 (15%); and high-dose, 11 (23%); two control males and three high-dose males also had pulmonary adenocarcinomas. Male mice also had an increased incidence of hepatocellular adenomas: controls, 3 (7%); low-dose, 3 (6%); and high-dose, 10 (21%) [actual effective numbers not given]; one mouse in each group also had a hepatocellular carcinoma. The incidences of both hepatocellular and pulmonary adenomas in male mice fed the high dose were reported to be significantly greater than that in the control group ($p < 0.05$). Hyperplasia of the squamous mucosa of the forestomach was increased in a dose-related fashion in both males and females (Schiefer et al., 1987).

A group of 50 male Kunming mice, six to eight weeks of age, received oral administrations of 100 μg/kg bw T-2 toxin in ethanol:saline solution three times a week for 25 weeks. A control group of 30 mice received 10 ml/kg bw ethanol solution. Small numbers of mice were killed at 1, 2, 3, 4, 5, 6, 12, 20 and 25 weeks. Forestomach papillomas occurred in 5/35 treated animals that were available for analysis: one killed at week 6, one killed at week 20 and three killed at week 25. No papilloma of the forestomach was seen in controls (Yang & Xia, 1988a). [The Working Group noted the small number of animals.]

3.1.2 Rat

Approximately 40 weanling male and female Wistar-Porton rats [sex distribution unspecified] were administered one to eight doses of 0.2–4 mg/kg bw T-2 toxin [purity unspecified] intragastrically at approximately monthly intervals [duration unspecified]. Another 30 rats of the same strain received the same treatment but were also given intraperitoneal injections of 200–250 mg/kg bw nicotinamide 10 min before and 2 h after each dose of T-2 toxin. Ten rats were given nicotinamide only, and another 10 rats served as untreated controls. About 65% of the rats treated with T-2 toxin alone or with nicotinamide died within a few days after the T-2 toxin treatment; 25 rats given 1–3 mg/kg bw T-2 toxin alone or with nicotinamide survived for 12–27.5 months, and the authors reported increased incidences of pancreatic and other tumours in these rats (Schoental et al., 1979). [The Working Group noted the lack of detail concerning the experimental protocol.]

Rats were reported to have developed papillomas and carcinomas of the forestomach after prolonged gavage with T-2 toxin. Tumours were also seen in other organs (Li et al., 1988). [The Working Group noted the lack of detail given.]

3.1.3 Trout

Groups of 1000 (reduced to 400 after nine months) rainbow trout, seven to eight months of age, were given 0.2 or 0.4 mg/kg T-2 toxin [purity unspecified] in the diet. Five fish from each group were killed every three months and the livers examined. The experiment was terminated after 12 months of treatment. No evidence of neoplasia was found (Marasas et al., 1969). [The Working Group noted the short duration of the experiment.]

3.2 Administration with known carcinogens

Mouse

Groups of 20 white mice [strain, sex and age unspecified] each received single applications of 10 or 20 μg T-2 toxin [purity unspecified] on the dorsal skin. Two weeks later, the mice received topical applications of croton oil (two drops of a 0.5% solution) twice a week for 10 weeks. The treatment induced no skin papilloma. In a second set of experiments, groups of 20 white mice received an application of 25 μg 7,12-dimethylbenz[a]anthracene (DMBA), followed two weeks later by applications of 10 μg T-2 toxin once a week for 10 weeks. Skin papillomas were observed in 2/20 mice (Marasas et al., 1969). [The Working Group noted the incomplete reporting.]

Groups of eight female CD-1 mice, six weeks old, received single applications of 50 μg DMBA on the shaved back, followed four days later by topical skin applications of T-2 toxin at 10 μg weekly or 25 μg every three weeks for 22 weeks. Positive (DMBA and croton oil), T-2 toxin-treated and solvent controls were also available. One of eight mice administered DMBA followed by 25 μg T-2 toxin developed a skin papilloma. All the positive controls developed papillomas. No such tumour was observed in the solvent controls or in the T-2 toxin-treated mice (Lindenfelser *et al.*, 1974).

Groups of 15–35 male strain 615 mice, 8–10 weeks of age, were given a single skin application of 100 μg DMBA, followed one week later by applications of acetone (15 mice), 2 μg 12-*O*-tetradecanoylphorbol 13-acetate (TPA) (20 mice) or 0.5 μg T-2 toxin (45 mice) three times a week for 26 weeks. One skin papilloma was observed among the mice treated with DMBA alone; in the mice treated with DMBA plus T-2 toxin, skin papillomas were observed in 8/45 and one skin carcinoma was observed. No skin tumour was observed in mice given DMBA and TBA. In a second part of the experiment, mice were treated with six daily doses of 5 μg T-2 toxin followed one week later by applications of 2 μg TPA (35 mice) or acetone (30 mice) three times a week for 26 weeks. No skin tumour occurred in either group. Similar experiments were carried out for 20 weeks with groups of 6–23 male and female BALB/c mice (about twice as many females as males), 8–12 weeks of age. No skin papilloma occurred in five mice treated with DMBA or in 21 mice treated with T-2 toxin alone; however, skin papillomas were found in 2/22 mice treated with DMBA and T-2 toxin, in 9/9 mice treated with DMBA and TPA and in 4/21 mice treated with T-2 toxin followed by TPA. One skin carcinoma was observed in the group given DMBA and T-2 toxin (Yang & Xia, 1988b).

4. Other Relevant Data

4.1 Absorption, distribution, metabolism and excretion

4.1.1 *Humans*

No data were available to the Working Group.

4.1.2 *Experimental systems*

^{3}H-T-2 Toxin given orally to mice and rats was distributed rapidly to tissues and eliminated in faeces and urine. Maximal levels of radiolabel were found after 30 min in plasma of mice after oral administration (Matsumoto *et al.*, 1978) and of guinea-pigs after intramuscular injection (Pace *et al.*, 1985). In chicks administered ^{3}H-T-2 toxin in the diet, maximal levels were reached by 4 h in blood, plasma, abdominal fat, heart, kidneys, gizzard, liver and the remainder of the carcass and by 12 h in muscle, skin, bile and gall-bladder (Chi *et al.*, 1978a). The distribution of T-2 toxin in tissues of swine was similar to that in chickens (Robison *et al.*, 1979a).

Following intravascular administration, the plasma elimination half-time of T-2 toxin and its metabolites (total radiolabel) in swine was approximately 90 min (Corley *et al.*, 1986). After intravenous administration to swine and calves, T-2 toxin was rapidly metabolized (mean elimination half-times of 13.8 and 17.4 min, respectively). A negligible fraction of the dose was recovered unmetabolized in urine. Detectable amounts were present in the spleen and mesenteric lymph nodes 3 h after administration. No T-2 toxin was detected in the liver (Beasley *et al.*, 1986).

In dogs, T-2 toxin administered intravenously was biotransformed rapidly to HT-2 toxin (4-deacetyl-T-2 toxin; see Fig. 1), with a mean plasma elimination half-time of 5.3 min. The mean half-time of HT-2 toxin was four times longer than that of T-2 toxin (19.6 min). Both toxins had high total body clearance (Sintov *et al.*, 1986). The urinary metabolites of T-2 toxin were HT-2 toxin, T-2-triol and T-2-tetraol (free and conjugated forms) (Sintov *et al.*, 1987).

The hepatobiliary system is the major route for the metabolism, detoxification and elimination of T-2 toxin; the metabolized compounds are found mainly in bile (Matsumoto *et al.*, 1978). During a single pass through perfused rat liver, 93% of the delivered ^{3}H-T-2 toxin label was extracted and metabolized by the liver and 55% appeared in bile (Pace, 1986). Oral administration of T-2 toxin to rabbits over 10 days at a subtoxic dose of 1 mg/kg bw per day resulted in a gradual decrease in the toxin-metabolizing capacity of the liver (Ványi *et al.*, 1988). Quantitative and qualitative differences exist between species in the hepatic microsomal metabolism of T-2 toxin (Kobayashi *et al.*, 1987). Human liver enzymes deacetylate T-2 toxin to HT-2 toxin *in vitro* (Ellison & Kotsonis, 1974). Other tissues, in particular intestinal tissues, can also metabolize T-2 toxin (Conrady-Lorck *et al.*, 1989).

After intravascular administration of T-2 toxin to swine, 21 metabolites were detected by reverse-phase high-performance liquid chromatography and radiochromatography, but the structures of some are not known (Corley *et al.*, 1986).

The metabolism of T-2 toxin *in vivo* has been studied in chickens, rats, swine and cows and found to involve (i) deacylation, (ii) hydroxylation, (iii) glucuronide conjugation, (iv) acetylation and (v) de-epoxidation (for a review, see Sintov *et al.*, 1987). The C4 acetyl residue of T-2 toxin is removed rapidly to give HT-2 toxin, which is then deacylated to T-2-tetraol *via* 4-deacetylneosolaniol (15-acetyl T-2-tetraol) (Yoshizawa *et al.*, 1980a,b). Additional deacetylation of HT-2 toxin at C15 gives T-2-triol (Ványi *et al.*, 1988). Another pathway is hydroxylation of the C8 isovaleroxy residue of T-2 toxin and HT-2 toxin to 3'-hydroxy T-2 and 3'-hydroxy-HT-2 toxins (Yoshizawa *et al.*, 1982; Visconti & Mirocha, 1985). Deepoxidation of 3'-hydroxy-HT-2 toxin and T-2-tetraol has been shown to occur *in vivo*, leading to the formation of the deepoxy derivatives 3'-hydroxydeepoxy-HT-2 toxin, 3'-hydroxydeepoxy-T-2-triol, 15-acetyldeepoxy-T-2-tetraol (deepoxy-4-diacetylneosolaniol) and deepoxy-T-2-tetraol (the only one of these structures shown in Fig. 1), which were identified in rat excreta (Yoshizawa *et al.*, 1985). Deepoxy T-2-tetraol was also found in cow blood and urine as a metabolite of T-2 toxin (Chatterjee *et al.*, 1986). Neosolaniol (4,15-diacetyl-T-2-tetraol) is produced when isovaleric acid (3-methylbutanoic acid) is removed from the ester group at the C8 position of T-2 toxin (Chi *et al.*, 1978a). In addition, acetylation of T-2 toxin followed by deacetylation to HT-2 toxin may occur *via iso*-T-2 toxin (in which the hydroxy in C3 and the acetyl in C4 of T-2 toxin are reversed) (Visconti *et al.*, 1985; Sintov *et al.*, 1986). 4-Acetyl-, 8-acetyl- and 15-acetyl-T-2-tetraols have also been identified (Visconti & Mirocha, 1985). Glucuronic acid conjugates represented 63% of total metabolites in urine and 77% of those in bile of swine (Corley *et al.*, 1985).

Robison *et al.* (1979b) observed T-2 toxin at 10–160 µg/l in the milk of a pregnant cow which had been intubated with 182 mg of the toxin on 15 consecutive days. In a lactating cow given a single oral dose of 157 mg ^{3}H-T-2 toxin, maximal levels (37 µg/l) of radiolabel were

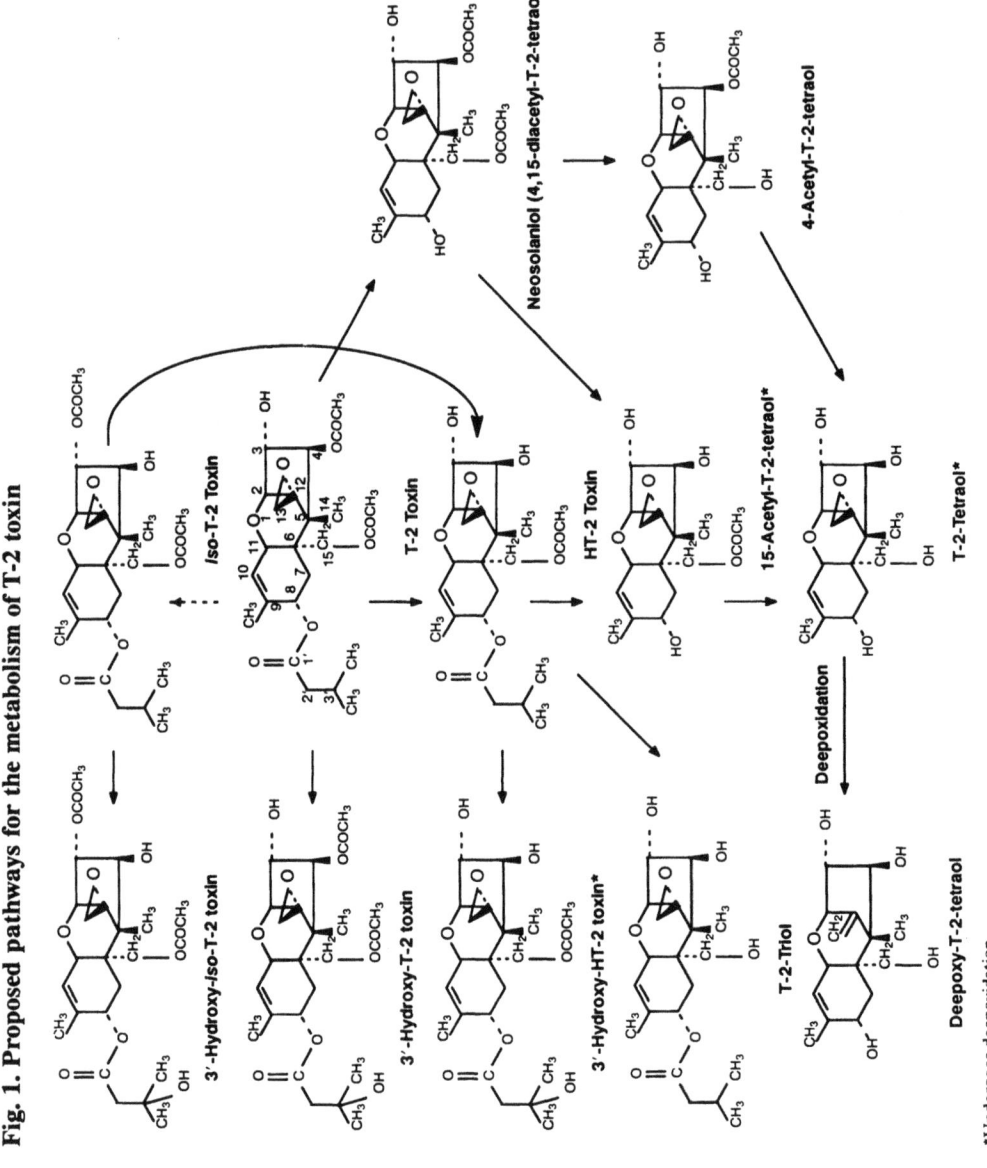

Fig. 1. Proposed pathways for the metabolism of T-2 toxin

*Undergoes deepoxidation

reached in milk after 16 h; 3'-hydroxy-T-2 toxin and 3'-hydroxy-HT-2 toxin were present at concentrations of 2.5–8.5 μg/l after 12–24 h (Yoshizawa et al., 1981, 1982). *iso*-3'-Hydroxy-T-2 toxin, an isomer in which the hydroxy and acetyl groups at the C3 and C4 positions are reversed, was found as the main metabolite of T-2 toxin in cow's urine 13.5 h after oral administration of 200 mg (Visconti et al., 1985). An unknown metabolite was present at a level of 2 mg/l 48 h after administration of the toxin (Ueno, 1987).

The toxicity of T-2 toxin in one-day-old broiler chick diminished gradually as a result of deacetylations: T-2 toxin (oral LD_{50}, 5 mg/kg bw) was 1.5 times more toxic than HT-2 toxin (LD_{50}, 7 mg/kg bw), which was 3.4 times more toxic than neosolaniol (LD_{50}, 25 mg/kg bw), which in turn was 1.4 times more toxic than T-2-tetraol (LD_{50}, 34 mg/kg bw) (Chi et al., 1978b). After intraperitoneal injection to mice, 3'-hydroxy-T-2 toxin (LD_{50}, 4.6 mg/kg bw) was slightly more toxic than the parent compound (LD_{50}, 5.3 mg/kg bw), whereas 3'-hydroxy-HT-2 toxin (LD_{50}, 22.8 mg/kg bw) was approximately 3.5 times less toxic than HT-2 toxin (LD_{50}, 6.5 mg/kg bw) (Yoshizawa et al., 1982).

T-2 Toxin and its metabolites were eliminated in faeces and urine at ratios of 3:1 in mice, 5:1 in rats (Matsumoto et al., 1978) and 1:4 in guinea-pigs (Pace et al., 1985).

4.2 Toxic effects

The toxicology and pharmacology of trichothecenes have been reviewed (Ueno, 1983a,b, 1987).

4.2.1 Humans

Little definitive information is available on the toxic effects in humans of specific trichothecenes. A disease known as 'alimentary toxic aleukia' first identified in Siberia in the former USSR, is, however, believed to be due to the consumption of grain contaminated with T-2 toxin. The aleukia usually occurs in four stages: In the first stage, hyperaemia of the oral mucosa occurs, accompanied by weakness, fever, nausea and vomiting. In more severe cases, the high fever continues and is accompanied by acute oesophagitis, gastritis and gastroenteritis. In rare cases, circulatory failure and convulsions occur. In a second stage, leukopenia, granulopenia and progressive lymphocytosis occur. A third stage is characterized by severe haemorrhagic diathesis, severe necrotic pharyngitis and laryngitis, which causes death in some cases by total closure of the larynx, described by the authors as strangulation. At this stage, exposed individuals have marked leukopenia, the leukocyte count being as low as 100–200/mm^3 or less, platelet diminution and anaemia, resulting in anoxia. The fourth stage is recovery, during which exposed individuals are susceptible to secondary infections; convalescence lasts several weeks. In severe outbreaks of poisoning, fatality rates have been as high as 50% (Joffe, 1974). Mirocha and Pathre (1973) identified T-2 toxin in a grain sample from the USSR that was associated with an episode of alimentary toxic aleukia.

In the outbreak of poisoning in Kashmir, India, in 1987 (Bhat et al., 1989), described on p. 416, T-2 toxin was detected in flour samples at 0.55–0.8 mg/kg.

Bamburg and Strong (1971) reported that accidental contact of laboratory workers with crude extracts containing T-2 toxin (about 200 mg/l) caused severe irritation, loss of sensitivity and desquamation of the skin of the hands. Normal sensitivity was restored 18 days after contact.

4.2.2 Experimental systems

Signs somewhat similar to those in human alimentary toxic aleukia were observed in rhesus monkeys fed T-2 toxin at 0.5 mg/kg bw for 15 days (Rukmini *et al.*, 1980).

The oral LD_{50}s for T-2 toxin are (mg/kg bw): chickens, 4; mice, 10.5; rats, 5.2; guinea-pigs, 3.1 (WHO, 1990); and swine, 4.0 (Cole & Cox, 1981). The intraperitoneal LD_{50} in mice is 5.2 mg/kg bw. After subcutaneous injection, newborn animals (LD_{50}, 0.15 mg/kg bw) were more sensitive than adults (LD_{50}, 1.6 mg/kg bw) to the toxic effects of T-2 toxin (Ueno, 1984).

Once it enters the systemic circulation, by any route of exposure, T-2 toxin rapidly affects proliferating cells in the thymus, lymph nodes, testes and ovaries, spleen, bone marrow and crypts of the small intestine (Ueno, 1977a,b). Oral, parenteral and cutaneous exposures produced gastric and intestinal lesions, haematopoietic and immunosuppressive effects (Paucod *et al.*, 1990) described as radiomimetic (Saito *et al.*, 1969), central nervous system toxicity resulting in anorexia, reduced food intake, lassitude and nausea, suppression of reproductive organ function and acute vascular effects leading to hypotension and shock (Smalley, 1973; Parker *et al.*, 1984, Abstract).

Gross morphological lesions were observed in pigs given a single intravenous injection of T-2 toxin at 1.2 mg/kg bw or more, resulting in oedema, congestion and haemorrhage in lymph nodes and pancreas; congestion and haemorrhage in the gastrointestinal mucosa, subendocardium, adrenal glands and meninges; and oedema in the gall-bladder. Degeneration and necrosis of the lymphoid tissues and gastrointestinal mucosa were seen. Scattered foci of necrosis were present in the pancreas, myocardium, bone marrow, adrenal cortex and the tubular epithelium of the renal medulla (Pang *et al.*, 1987a). Similar effects were observed after exposure by inhalation (Pang *et al.*, 1987b).

Fatty degeneration and enlargement of the liver were observed after administration of T-2 toxin to rats (Suneja *et al.*, 1983) and cattle (Kosuri *et al.*, 1970). No effect was detected in swine fed 8 mg/kg of feed for eight weeks (Weaver *et al.*, 1978).

Vomiting is one of the commonest symptoms of T-2 toxin toxicosis. It induced vomiting at doses of 0.1–10 mg/kg bw in cats, dogs, pigs and ducklings (WHO, 1990).

In rats, the minimal effective dose of T-2 toxin for dermal toxicity, which is characterized by red spots and inflammation on painted skin, was 10 ng (Ueno, 1987). Rabbits are more sensitive to the dermal toxicity of this compound than rats (Hayes & Schiefer, 1979). The no-observed-effect level for weight gain and oral lesions in broiler chicks was 0.2 mg/kg in the diet for nine weeks (WHO, 1990).

Transient leukocytosis due to increased numbers of both neutrophils and lymphocytes shortly after a single injection of T-2 toxin in mice, rats and cats is believed to be caused by a sudden release of leukocytes from lymph nodes (Sato *et al.*, 1975, 1978). The effect seems to involve the action of inflammatory mediators, since it can be prevented by some anti-inflammatory steroids. Repeated administration of T-2 toxin to chickens, mice, guinea-pigs, cats and monkeys, however, severely decreased the number of leukocytes (reviewed by Ueno, 1983a).

Detailed studies of the effects of T-2 toxin on coagulation in several species have revealed multiple mechanisms for haemorrhage, including tissue necrosis, thrombocytopenia, platelet dysfunction, decreased activity of coagulation factors and (possibly) altered vascular integrity (for review, see WHO, 1990).

T-2 Toxin affects the immune system and thereby modifies the immune response in experimental animals. The impairment comprises the following functions: antibody formation, allograft rejection, delayed hypersensitivity and blastogenic response to lectins; it results in decreased resistance to microbial infection. The impairment of the immune system is thought to be linked to the inhibitory effect of T-2 toxin on macromolecule synthesis (reviewed in WHO, 1990).

Increased blood pressure and severe vascular damage were reported in a study of rats administered four doses of 1–3 mg/kg bw T-2 toxin intragastrically over 12 months (Wilson et al., 1982).

Sirkka et al. (1992) observed acute behavioural effects, such as decreased motor activity and performance in the passive avoidance test, and reduced body weight gain in rats given a single oral dose of 2 mg/kg bw T-2 toxin, but not in those given 0.4 mg/kg bw. Fitzpatrick et al. (1988) found elevated concentrations of indoleamines in rat brain 24 h after an oral dose of 2.5 mg/kg bw T-2 toxin. MacDonald et al. (1988) found that T-2 toxin increased the concentrations of tryptophan, serotonin and dopamine in the brain of rats, but decreased those of 3,4-dihydroxyphenylacetic acid.

Feeding of 1.5 mg/kg bw T-2 toxin to young male albino rats daily for four days significantly decreased liver protein and DNA and intestinal mucosal protein content (Suneja et al., 1983). The compound inhibits DNA and protein synthesis in a variety of cell types (reviewed by Ueno, 1983a,b). After HeLa cells were cultured with 30 ng/ml for three days, they underwent complete cytolysis (Saito & Ohtsubo, 1974). DNA synthesis in spleen cells in vitro was inhibited by 73% with a dose of 0.25 ng/ml, and protein synthesis was inhibited by 55% at a dose of 0.5 ng/ml (Rosenstein & Lafarge-Frayssinet, 1983).

The IC_{50} for T-2 toxin in rat hepatoma cells and dog kidney cells was 1–5 ng/ml (Mirocha et al., 1992). The ID_{50} for protein synthesis was 7 ng/ml in reticulocytes from guinea-pigs and 30 ng/ml in reticulocytes from rabbits. T-2 Toxin induced rapid disaggregation of the polysomes, indicating inhibition at the initiation step of protein synthesis. In rabbit reticulocytes in vitro, using poly-U as messenger RNA, the ID_{50} for polyphenylalanine synthesis was 0.15 µg/ml (Ueno et al., 1973). The molecular mechanism of inhibition of protein synthesis may be the high affinity of T-2 toxin for the 60S ribosomal subunit (Hobden & Cundliffe, 1980). The binding affinity of T-2 toxin for yeast ribosomes was 0.03 µM [14 µg]; the binding was reversible at 37 °C and specific (Middlebrook & Leatherman, 1989).

T-2 Toxin at 0.2–1.2 ng/ml induced proliferation in cultured human fetal oesophagus, including focal basal-cell hyperplasia, dysplasia and an increased number of mitoses (Hsia et al., 1983). The changes were reported to be similar to the premalignant lesions seen in epithelium adjacent to human oesophageal carcinomas. At higher doses of T-2 toxin (2–4 ng/ml for six days), the cultured epithelium became necrotic.

T-2 toxin at low concentrations (0.4 pg/ml to 4 ng/ml) appears to have multiple effects on cell membrane function, which are independent of inhibition of protein synthesis. It may act on amino acid, nucleotide and glucose transporters or calcium and potassium channel activities (Bunner & Morris, 1988).

4.3 Reproductive and developmental toxicity

4.3.1 *Humans*

No data were available to the Working Group.

4.3.2 *Experimental systems*

Pregnant mice [strain unspecified] were injected intraperitoneally with T-2 toxin at 0.5, 1.0 or 1.5 mg/kg bw on one of days 7–11 of gestation. With the two higher doses, there was significant maternal mortality and decreased prenatal survival. In eight litters evaluated from the group that received 1.0 mg/kg bw and in four litters from the group that received 1.5 mg/kg bw on day 10, fetal weight was significantly decreased, and 38 and 29% of the fetuses, respectively, had gross malformations, including missing tails, limb malformations, exencephaly, open eyes and retarded jaw development (Stanford *et al.*, 1975).

A dose of 0.5 mg/kg bw T_2-toxin given intraperitoneally to pregnant CD-1 mice on day 10 of gestation induced tail and limb anomalies in 12.5% of the offspring. Additional treatment with 4 mg/kg bw ochratoxin A induced a reduction in fetal weight and a higher incidence of malformations (Hood *et al.*, 1976, 1978). In a similar experiment, T-2 toxin induced tail and limb anomalies in 8.4% of offspring, but additional treatment with 0.4 mg/kg bw rubratoxin B did not significantly increase the incidence of malformations (Hood, 1986).

CD-1 mice were treated orally on day 9 of gestation with 0, 0.5, 1.0, 2.0, 3.0, 3.5 or 4.0 mg/kg bw T-2 toxin. The two highest doses caused fetal and maternal deaths; skeletal defects occurred at low incidences with the 3.0-mg/kg bw dose. In a second experiment, treatment with 3.0 mg/kg bw on days 6, 7, 8, 10, 11 or 12 of gestation caused less fetal mortality; treatment on day 7 significantly reduced litter size (Rousseaux & Schiefer, 1987). Continuous feeding of 1.5 or 3.0 mg/kg T-2 toxin to CD-1 mice in the diet for two generations was neither embryo- nor fetotoxic and had only minimal effects on growth rates of mice (Rousseaux *et al.*, 1986).

Intraperitoneal injection of T-2 toxin to pregnant WAG rats at doses of 0.1, 0.2 or 0.4 mg/kg bw daily or feeding of the toxin at doses of 0.1 or 0.4 mg/kg bw on days 14–20 of gestation produced a decrease in thymus weight in newborn rats, which lasted for about one week (Bertin *et al.*, 1978).

4.4 Genetic and related effects

4.4.1 *Humans*

No data were available to the Working Group.

4.4.2 *Experimental systems* (see also Table 2 and Appendices 1 and 2)

The genotoxicity of T-2 toxin has been reviewed (Haschek, 1989).

Table 2. Genetic and related effects of T-2 toxin

Test system	Result - Without exogenous metabolic system	Result - With exogenous metabolic system	Dose (LED/HID)[a]	Reference
PRB, SOS spot test, *Escherichia coli* PQ37	–	–	0.0000	Auffray & Boutibonnes (1986)
PRB, SOS chromotest test, *Escherichia coli* PQ37	–	–	1.0000	Krivobok *et al.* (1987)
BSD, *Bacillus subtilis* rec strains, differential toxicity	–	0	100 µg/plate	Ueno & Kubota (1976)
SA0, *Salmonella typhimurium* TA100, reverse mutation	–	–	50.0000	Wehner *et al.* (1978)
SA5, *Salmonella typhimurium* TA1535, reverse mutation	–	–	50.0000	Kuczuk *et al.* (1978)
SA5, *Salmonella typhimurium* TA1535, reverse mutation	–	–	50.0000	Wehner *et al.* (1978)
SA7, *Salmonella typhimurium* TA1537, reverse mutation	–	–	50.0000	Kuczuk *et al.* (1978)
SA7, *Salmonella typhimurium* TA1537, reverse mutation	–	–	50.0000	Wehner *et al.* (1978)
SA8, *Salmonella typhimurium* TA1538, reverse mutation	–	–	50.0000	Kuczuk *et al.* (1978)
SA9, *Salmonella typhimurium* TA98, reverse mutation	–	–	50.0000	Wehner *et al.* (1978)
SCH, *Saccharomyces cerevisiae* D-3, mitotic crossing over, *ade2* locus	–	–	100 µg/plate	Kuczuk *et al.* (1978)
SCF, *Saccharomyces cerevisiae*, petite forward mutations	–	0	50.0000	Schappert & Khachatourians (1986)
ACC, *Allium cepa*, polyploidy induction	+	0	20.0000	Linnainmaa *et al.* (1979)
DMX, *Drosophila melanogaster*, sex-linked recessive lethal mutations	(+)	0	63.0000	Sorsa *et al.* (1980)
DMX, *Drosophila melanogaster*, sex-linked recessive lethal mutations	–	–	100–1000 ppm (2–3 days in feed)	Sorsa *et al.* (1980)
DMN, *Drosophila melanogaster*, sex chromosome loss, adult feeding	+	0	20 ppm, 48 h	Sorsa *et al.* (1980)
DIA, DNA single-strand breaks, BALB/c mouse primary hepatocytes *in vitro*	(+)	0	0.0005 (single dose)	Lafarge-Frayssinet *et al.* (1981)
DIA, DNA single-strand breaks, BALB/c mouse spleen lymphocytes *in vitro*	+	0	0.0005	Lafarge-Frayssinet *et al.* (1981)
DIA, DNA single-strand breaks, BALB/c mouse thymic lymphocytes *in vitro*	+	0	0.0005 (single dose)	Lafarge-Frayssinet *et al.* (1981)
G9H, Gene mutation, Chinese hamster V79 fibroblasts, thioguanine[r] *in vitro*	–	+[b]	0.1000	Zhu *et al.* (1987)
SIC, Sister chromatid exchange, Chinese hamster V79 fibroblasts *in vitro*	(+)	(+)	2.3000	Thust *et al.* (1983)
SIC, Sister chromatid exchange, Chinese hamster V79 fibroblasts *in vitro*	–	(+)[c]	0.1000	Zhu *et al.* (1987)

Table 2 (contd)

Test system	Result		Dose (LED/HID)[a]	Reference
	Without exogenous metabolic system	With exogenous metabolic system		
CIC, Chromosomal aberrations, Chinese hamster V79 fibroblasts *in vitro*	+	(+)	0.5000	Thust et al. (1983)
CIC, Chromosomal aberrations, Chinese hamster V79 fibroblasts *in vitro*	+	0	0.0005	Hsia et al. (1986)
CIC, Chromosomal aberrations, Chinese hamster V79 fibroblasts *in vitro*	(+)	(+)[b,c]	0.0500	Zhu et al. (1987)
CIC, Chromosomal aberrations, Chinese hamster V79 fibroblasts *in vitro*	+	0	0.0010	Hsia et al. (1988)
MIA, Micronucleus formation, Chinese hamster V79 fibroblasts *in vitro*	+	+[b,c]	0.0500	Zhu et al. (1987)
UHF, Unscheduled DNA synthesis, human fibroblasts *in vitro*	+	0	0.0050	Oldham et al. (1980)
SHL, Sister chromatid exchange, human lymphocytes *in vitro*	−	−[b]	0.0030	Cooray (1984)
CHL, Chromosomal aberrations, human lymphocytes *in vitro*	+[d]	0	0.0001	Hsia et al. (1986)
DVA, DNA single-strand breaks, BALB/c mouse liver *in vivo*	−		3.00 × 1 ip	Lafarge-Frayssinet et al. (1981)
DVA, DNA single-strand breaks, BALB/c mouse spleen *in vivo*	+		3.00 × 1 ip	Lafarge-Frayssinet et al. (1981)
DVA, DNA single-strand breaks, BALB/c mouse thymus *in vivo*	(+)		3.00 × 1 ip	Lafarge-Frayssinet et al. (1981)
MVC, Micronucleus test, Chinese hamster bone marrow *in vivo*	−		3.00 × 1 ip	Norppa et al. (1980)
CBA, Chromosomal aberrations, Chinese hamster bone marrow *in vivo*	(+)		1.70 × 1 ip	Norppa et al. (1980)
ICR, Inhibition of intercellular communication, Chinese hamster V79 cells *in vitro*	+	0	0.0030	Jone et al. (1987)
T-2-Tetraol				
BSD, *Bacillus subtilis rec* strains, differential toxicity	−	0	5000.0000	Boutibonnes et al. (1984)
UHF, Unscheduled DNA synthesis, human fibroblasts *in vitro*	+	0	0.1000	Oldham et al. (1980)

+, positive; (+), weak positive; −, negative; 0, not tested

[a] In-vitro tests, μg/ml; in-vivo tests, mg/kg bw; 0.0000, dose not given
[b] Activation by rat hepatocytes
[c] Activation by rat oesophageal epithelium
[d] Control aberration frequency extremely low (6/3000)

T-2 Toxin did not induce DNA damage in bacteria, mutation in *Salmonella typhimurium* or mitotic crossing over or mitochondrial petite mutations in yeast. In *Allium cepa*, T-2 toxin increased the number of polyploid root-tip cells. In *Drosophila melanogaster*, it induced sex chromosome loss, but the results for sex-linked recessive lethal mutations were inconclusive. T-2 Toxin inhibited gap-junctional intercellular communication in Chinese hamster V79 cells *in vitro* and induced DNA single-strand breaks in mouse spleen and thymic lymphocytes and [marginally] hepatocytes *in vitro*. In cultured Chinese hamster V79 cells, T-2 toxin induced gene mutation, sister chromatid exchange and chromosomal aberrations. After treatment with T-2 toxin, unscheduled DNA synthesis was induced in cultured human fibroblasts, and chromosomal aberrations, but not sister chromatid exchange, were induced in cultured human lymphocytes. Administration of T-2 toxin to mice *in vivo* resulted in single-strand breaks in spleen and thymus, but not in liver. In Chinese hamsters treated *in vivo*, a slight increase in the frequency of chromosomal aberrations, but not [one dose only] of micronuclei, was observed in bone marrow.

The genotoxic effects of T-2 toxin were observed only at very low doses, since toxic effects interfered at higher doses.

T-2-Tetraol, a metabolic hydrolysis product of T-2 toxin, induced unscheduled DNA synthesis in cultured human fibroblasts.

5. Summary of Data Reported and Evaluation

5.1 Exposure data

T-2 Toxin is produced primarily by *Fusarium sporotrichioides*, which occurs rarely on cereals such as wheat and maize. The toxin is considered to have played a role in large-scale human poisonings in Siberia during this century.

5.2 Human carcinogenicity data

No data were available to the Working Group.

5.3 Animal carcinogenicity data

T-2 Toxin was tested for carcinogenicity in mice and in trout by oral administration in the diet and in rats by intragastric administration. In mice, it increased the incidences of pulmonary and hepatic adenomas in males. The studies in trout and rats were inadequate for evaluation.

5.4 Other relevant data

T-2 Toxin causes outbreaks of haemorrhagic disease in animals and has been associated with alimentary toxic aleukia in humans.

No data were available on the genetic and related effects of T-2 toxin in humans.

Experimental data were drawn mainly from single studies. T-2 Toxin induced DNA damage and chromosomal aberrations in rodents *in vivo*, in cultured human cells and in cultured rodent cells. It inhibited protein synthesis in various mammalian and human cell types *in vitro*. Chromosomal aberrations were also induced in insects. It induced gene mutation in cultured rodent cells but not in bacteria. It did not induce DNA damage in bacteria.

5.5 Evaluation[1]

No data were available on the carcinogenicity to humans of toxins derived from *Fusarium sporotrichioides*.

There is *limited evidence* in experimental animals for the carcinogenicity of T-2 toxin.

Overall evaluation

Toxins derived from *Fusarium sporotrichioides are not classifiable as to their carcinogenicity to humans (Group 3)*.

6. References

Abramson, D., Mills, J.T. & Boycott, B.R. (1983) Mycotoxins and mycoflora in animal feedstuffs in western Canada. *Can. J. comp. Med.*, **47**, 23–26

Auffray, Y. & Boutibonnes, P. (1986) Evaluation of the genotoxic activity of some mycotoxins using *Escherichia coli* in the SOS spot test. *Mutat. Res.*, **171**, 79–82

Bamburg, J.R. & Strong, F.M. (1971) 12,13-Epoxytrichothecenes. In: Kadis, S., Ciegler, A. & Ajl, S.J., eds, *Microbiological Toxins*, Vol. 7, *Algal and Fungal Toxins*, New York, Academic Press, pp. 207–292

Bamburg, J.R., Riggs, N.V. & Strong, F.M. (1968) The structures of toxins from two strains of *Fusarium tricinctum*. *Tetrahedron*, **24**, 3329–3336

Bata, A., Ványi, A. & Lásztity, R. (1983) Simultaneous detection of some fusariotoxins by gas–liquid chromatography. *J. Assoc. off. anal. Chem.*, **66**, 577–581

Beasley, V.R., Swanson, S.P., Corley, R.A., Buck, W.B., Koritz, G.D. & Burmeister, H.R. (1986) Pharmacokinetics of the trichothecene mycotoxin, T-2 toxin, in swine and cattle. *Toxicon*, **24**, 13–23

Bertin, G., Chakor, K., Lafont, P. & Frayssinet, C. (1978) Transmission to the progeny of contamination of maternal diet by mycotoxins (Fr.). *Coll. Med. leg. Toxicol. med.*, **107**, 95–100

Bhavanishankar, T.N. & Shantha, T. (1987) Natural occurrence of *Fusarium* toxins in peanut, sorghum and maize from Mysore (India). *J. Sci. Food Agric.*, **40**, 327–332

Boutibonnes, P., Auffray, Y., Malherbe, C., Kogbo, W. & Marais, C. (1984) Antibacterial and genotoxic properties of 33 mycotoxins (Fr.). *Mycopathologia*, **87**, 43–49

---

[1]For definition of the italicized terms, see Preamble, pp. 26–29.

Bunner, D.L. & Morris, E.R. (1988) Alteration of multiple cell membrane functions in L-6 myoblasts by T-2 toxin: an important mechanism of action. *Toxicol. appl. Pharmacol.*, **92**, 113–121

Chatterjee, K., Visconti, A. & Mirocha, C.J. (1986) Deepoxy T-2 tetraol: A metabolite of T-2 toxin found in cow urine. *J. agric. Food Chem.*, **34**, 695–697

Chi, M.S., Robison, T.S., Mirocha, C.J., Swanson, S.P. & Shimoda, W. (1978a) Excretion and tissue distribution of radioactivity from tritium-labeled T-2 toxin in chicks. *Toxicol. appl. Pharmacol.*, **45**, 391–402

Chi, M.S., Robison, T.S., Mirocha, C.J. & Reddy, K.R. (1978b) Acute toxicity of 12,13-epoxy-trichothecenes in one-day-old broiler chicks. *Appl. environ. Microbiol.*, **35**, 636–640

Chu, F.S., Grossman, S., Wei, R.-D. & Mirocha, C.J. (1979) Production of antibody against T-2 toxin. *Appl. environ. Microbiol.*, **37**, 104–108

Cohen, H. & Boutin-Muma, B. (1992) Fluorescence detection of trichothecene mycotoxins as coumarin-3-carbonyl chloride derivatives by high-performance liquid chromatography. *J. Chromatogr.*, **595**, 143–148

Cole, R.J. & Cox, R.H. (1981) *Handbook of Toxic Fungal Metabolites*, New York, Academic Press, pp. 185–188

Conrady-Lorck, S., Strugala, G.J., Feng, X.-C. & Fichtl, B. (1989) Intestinal metabolism of T-2 toxin, a major trichothecene mycotoxin. *Prog. Pharmacol. clin. Pharmacol.*, **7**, 319–326

Cooray, R. (1984) Effects of some mycotoxins on mitogen-induced blastogenesis and SCE frequency in human lymphocytes. *Food chem. Toxicol.*, **22**, 529–534

Corley, R.A., Swanson, S.P. & Buck, W.B. (1985) Glucuronide conjugates of T-2 toxin and metabolites in swine bile and urine. *J. agric. Food Chem.*, **33**, 1085–1089

Corley, R.A., Swanson, S.P., Gullo, G.J., Johnson, L., Beasley, V.R. & Buck, W.B. (1986) Disposition of T-2 toxin, a trichothecene mycotoxin, in intravascularly dosed swine. *J. agric. Food Chem.*, **34**, 868–875

Cullen, D., Smalley, E.B. & Caldwell, R.W. (1982) New process for T-2 production. *Appl. environ. Microbiol.*, **44**, 371–375

van Egmond, H.P. (1989) Current situation on regulations for mycotoxins. Overview of tolerances and status of standard methods of sampling and analysis. *Food Addit. Contam.*, **6**, 139–188

Ellison, R.A. & Kotsonis, F.N. (1974) In vitro metabolism of T-2 toxin. *Appl. Microbiol.*, **27**, 423–424

Fan, T.S.L., Schubring, S.L., Wei, R.D. & Chu, F.S. (1988) Production and characterization of a monoclonal antibody cross-reactive with most group A trichothecenes. *Appl. environ. Microbiol.*, **54**, 2959–2963

Fitzpatrick, D.W., Boyd, K.E. & Watts, B.M. (1988) Comparison of the trichothecenes deoxynivalenol and T-2 toxin for their effects on brain biogenic monoamines in the rat. *Toxicol. Lett.*, **40**, 241–245

Foster, B.C., Neish, G.A., Lauren, D.R., Trenholm, H.L., Prelusky, D.B. & Hamilton, R.M.G. (1986) Fungal and mycotoxin content of slashed corn. *Microbiol. Aliments Nutr.*, **4**, 199–203

Greenhalgh, R., Blackwell, B.A., Savard, M., Miller, J.D. & Taylor, A. (1988) Secondary metabolites produced by *Fusarium sporotrichioides* DAOM 165006 in liquid culture. *J. agric. Food Chem.*, **36**, 216–219

Greenhalgh, R., Fielder, D.A., Blackwell, B.A., Miller, J.D., Charland, J.-P. & Simon, J.W. (1990). Some minor secondary metabolites of *Fusarium sporotrichioides* DAOM 165006. *J. agric. Food Chem.*, **38**, 1978–1984

Haschek, W.M. (1989) Mutagenicity and carcinogenicity of T-2 toxin. In: Beasley, V.R., ed., *Trichothecene Mycotoxicosis, Pathophysiological Effects*, Vol. 1, Boca Raton, FL, CRC Press, pp. 63–72

Hayes, M.A. & Schiefer, H.B. (1979) Quantitative and morphological aspects of cutaneous irritation by trichothecene mycotoxins. *Food Cosmet. Toxicol.*, **17**, 611–621

Hobden, A.N. & Cundliffe, E. (1980) Ribosomal resistance to the 12,13-epoxy trichothecene antibiotics in the producing organism *Myrothecium verrucaria*. *Biochem. J.*, **190**, 765–770

Hood, R.D. (1986) Effects of concurrent prenatal exposure to rubratoxin B and T-2 toxin in the mouse. *Drug chem. Toxicol.*, **9**, 185–190

Hood, R.D., Kuczuk, M.H. & Szczech, G.M. (1976) Prenatal effects in mice of mycotoxins in combination: ochratoxin A and T-2 toxin (Abstract). *Teratology*, **13**, 25A

Hood, R.D., Kuczuk, M.H. & Szczech, G.M. (1978) Effects in mice of simultaneous prenatal exposure to ochratoxin A and T-2 toxin. *Teratology*, **17**, 25–30

Hsia, C.-C., Tzian, B.-L. & Harris, C.C. (1983) Proliferative and cytotoxic effects of *Fusarium* T_2 toxin on cultured human fetal esophagus. *Carcinogenesis*, **4**, 1101–1107

Hsia, C.-C., Gao, Y., Wu, J.-L. & Tzian, B.-L. (1986) Induction of chromosome aberrations by *Fusarium* T-2 toxin in cultured human peripheral blood lymphocytes and Chinese hamster fibroblasts. *J. cell. Physiol.*, **Suppl. 4**, 65–72

Hsia, C.-C., Wu, J.-L., Lu, X.-Q. & Li, Y.-S. (1988) Natural occurrence and clastogenic effects of nivalenol, deoxynivalenol, 3-acetyldeoxynivalenol, 15-acetyldeoxynivalenol, and zearalenone in corn from a high risk area of esophageal cancer. *Cancer Detect. Prev.*, **13**, 79–86

Hsu, I.-C., Smalley, E.B., Strong, F.M. & Ribelin, W.E. (1972) Identification of T-2 toxin in moldy corn associated with a lethal toxicosis in dairy cattle. *Appl. Microbiol.*, **24**, 684–690

Hussein, H.M., Franich, R.A., Baxter, M. & Andrew, I.G. (1989) Naturally occurring *Fusarium* toxins in New Zealand maize. *Food Addit. Contam.*, **6**, 49–58

IARC (1983) *IARC Monographs on the Evaluation of the Carcinogenic Risk of Chemicals to Humans*, Vol. 31, *Some Food Additives, Feed Additives and Naturally Occurring Substances*, Lyon, pp. 265–278

Jemmali, M., Ueno, Y., Ishii, K., Frayssinet, C. & Etienne, M. (1978) Natural occurrence of trichothecenes (nivalenol, deoxynivalenol, T-2) and zearalenone in corn. *Experientia*, **34**, 1333–1334

Joffe, A.Z. (1974) Toxicity of *Fusarium poae* and *F. sporotrichioides* and its relation to alimentary toxic aleukia. In: Purchase, I.F.H., ed., *Mycotoxins*, Amsterdam, Elsevier, pp. 229–262

Jone, C., Erickson, L., Trosko, J.E. & Chang, C.C. (1987) Effect of biological toxins on gap-junctional intercellular communication in Chinese hamster V79 cells. *Cell Biol. Toxicol.*, **3**, 1–15

Karppanen, E., Rizzo, A., Berg, S., Lindfors, E. & Aho, R. (1985) *Fusarium* mycotoxins as a problem in Finnish feeds and cereals. *J. agric. Sci. Finl.*, **57**, 195–206

Kawamura, O., Nagayama, S., Sato, S., Ohtani, K., Sugimura, Y., Tanaka, T. & Ueno, Y. (1990) Survey of T-2 toxin in cereals by an indirect enzyme-linked immunosorbent assay. *Food agric. Immunol.*, **2**, 173–180

Kobayashi, J., Horikoshi, T., Ryu, J.-C., Tashiro, F., Ishii, K. & Ueno, Y. (1987) The cytochrome P-450-dependent hydroxylation of T-2 toxin in various animal species. *Food chem. Toxicol.*, **25**, 539–544

Kosuri, N.R., Grove, M.D., Yates, S.G., Tallent, W.H., Ellis, J.J., Wolff, I.A. & Nichols, R.E. (1970) Response of cattle to mycotoxins of *Fusarium tricinctum* isolated from corn and fescue. *J. Am. vet. Med. Assoc.*, **157**, 938–940

Krivobok, S., Olivier, P., Marzin, D.R., Seiglemurandi, F. & Steinman, R. (1987) Study of the genotoxic potential of 17 mycotoxins with the SOS chromostest. *Mutagenesis*, **2**, 433–439

Kuczuk, M.H., Benson, P.M., Heath, H. & Hayes, A.W. (1978) Evaluation of the mutagenic potential of mycotoxins using *Salmonella typhimurium* and *Saccharomyces cerevisiae*. *Mutat. Res.*, **53**, 11–20

Lafarge-Frayssinet, C., Decloitre, F., Mousset, S., Martin, M. & Frayssinet, C. (1981) Induction of DNA single-strand breaks by T_2 toxin, a trichothecene metabolite of *Fusarium*. Effect on lymphoid organs and liver. *Mutat. Res.*, **88**, 115–123

Lauren, D.R. & Agnew, M.P. (1991) Multitoxin screening method for *Fusarium* mycotoxins in grains. *J. agric. Food Chem.*, **39**, 502–507

Li, M.-X., Zhu, G.-F., Cheng, S.-J., Jian, Y.-Z. & Fan, W.-G. (1988) Mutagenicity and carcinogenicity of T-2 toxin, a trichothecene produced by *Fusarium* fungi (Chin.). *Chin. J. Oncol.*, **10**, 326–329

Lindenfelser, L.A., Lillehoj, E.B. & Burmeister, H.R. (1974) Aflatoxin and trichothecene toxins: skin tumor induction and synergistic acute toxicity in white mice. *J. natl Cancer Inst.*, **52**, 113–116

Linnainmaa, K., Sorsa, M. & Ilus, T. (1979) Epoxytrichothecene mycotoxins as c-mitotic agents in Allium. *Hereditas*, **90**, 151–156

MacDonald, E.J., Cavan, K.R. & Smith, T.K. (1988) Effect of acute oral doses of T-2 toxin on tissue concentrations of biogenic amines in the rat. *J. Anim. Sci.*, **66**, 434–441

Marasas, W.F.O., Bamburg, J.R., Smalley, E.B., Strong, F.M., Ragland, W.L. & Degurse, P.E. (1969) Toxic effects on trout, rats and mice of T-2 toxin produced by the fungus *Fusarium tricinctum* (Cd.) Snyd. et Hans. *Toxicol. appl. Pharmacol.*, **15**, 471–482

Marasas, W.F.O., Nelson, P.E. & Toussoun, T.A. (1984) *Toxigenic Fusarium Species. Identity and Mycotoxicology*, University Park, PA, Pennsylvania State University Press

Matsumoto, H., Ito, T. & Ueno, Y. (1978) Toxicological approaches to the metabolites of fusaria. XII. Fate and distribution of T-2 toxin in mice. *Jpn. J. exp. Med.*, **48**, 393–399

Middlebrook, J.L. & Leatherman, D.L. (1989) Binding of T-2 toxin to eukaryotic cell ribosomes. *Biochem. Pharmacol.*, **38**, 3103–3110

Mirocha, C.J. (1984) Mycotoxicoses associated with *Fusarium*. In: Moss, M.O. & Smith, J.E., eds, *The Applied Mycology of* Fusarium, Cambridge, Cambridge University Press, pp. 141–155

Mirocha, C.J. & Pathre, S. (1973) Identification of the toxic principle in a sample of poaefusarin. *Appl. Microbiol.*, **26**, 719–724

Mirocha, C.J., Gilchrist, D.G., Shier, W.T., Abbas, H.K., Wen, Y. & Vesonder, R.F. (1992) AAL toxins, fumonisins (biology and chemistry) and host-specificity concepts. *Mycopathologia*, **117**, 47–56

Norppa, H., Penttilä, M., Sorsa, M., Hintikka, E.-L. & Ilus, T. (1980) Mycotoxin T-2 of *Fusarium tricinctum* and chromosome changes in Chinese hamster bone marrow. *Hereditas*, **93**, 329–332

Oldham, J.W., Allred, L.E., Milo, G.E., Kindig, O. & Capen, C.C. (1980) The toxicological evaluation of the mycotoxins T-2 and T-2 tetraol using normal human fibroblasts *in vitro*. *Toxicol. appl. Pharmacol.*, **52**, 159–168

Pace, J.G. (1986) Metabolism and clearance of T-2 mycotoxin in perfused rat livers. *Fundam. appl. Toxicol.*, **7**, 424–433

Pace, J.G., Watts, M.R., Burrows, E.P., Dinterman, R.E., Matson, C., Hauer, E.C. & Wannemacher, R.W., Jr (1985) Fate and distribution of ^{3}H-labeled T-2 mycotoxin in guinea pigs. *Toxicol. appl. Pharmacol.*, **80**, 377–385

Pang, V.F., Lorenzana, R.M., Beasley, V.R., Buck, W.B. & Haschek, W.M. (1987a) Experimental T-2 toxicosis in swine. III. Morphologic changes following intravascular administration of T-2 toxin. *Fundam. appl. Toxicol.*, **8**, 298–309

Pang, V.F., Lambert, R.J., Felsburg, P.J., Beasley, V.R., Buck, W.B. & Haschek, W.M. (1987b) Experimental T-2 toxicosis in swine following inhalation exposure: effects on pulmonary and systemic immunity and morphologic changes. *Toxicol. Pathol.*, **15**, 308–319

Parker, G.W., Wannemacher, R.W., Jr and Gilman, F.J. (1984) The effect of T-2 mycotoxin on the cardiovascular system in the guinea pig (Abstract No. 1710). *Fed. Proc.*, **43**, 578

Paucod, J.-C., Krivobok, S. & Vidal, D. (1990) Immunotoxicity testing of mycotoxins T-2 and patulin on BALB/c mice. *Acta microbiol. Hung.*, **37**, 331–339

Pohland, A.E., Schuller, P.L., Steyn, P.S. & van Egmond, H.P. (1982) Physicochemical data for some selected mycotoxins. *Pure appl. Chem.*, **54**, 2219–2284

Puls, R. & Greenway, J.A. (1976) Fusariotoxicosis from barley in British Columbia. II. Analysis and toxicity of suspected barley. *Can. J. comp. Med.*, **40**, 16–19

Ramakrishna, Y., Bhat, R.V. & Vasanthi, S. (1990) Natural occurrence of mycotoxins in staple foods in India. *J. agric. Food Chem.*, **38**, 1857–1859

Reddy, B.N., Nusrath, M., Kumari, C.K. & Nahdi, S. (1983) Mycotoxin contamination in some food commodities from tribal areas of Medak district, Andhra Pradesh. *Indian Phytopathol.*, **36**, 683–686

Robison, T.S., Mirocha, C.J., Kurtz, H.J., Behrens, J.C., Weaver, G.A. & Chi, M.S. (1979a) Distribution of tritium-labeled T-2 toxin in swine. *J. agric. Food Chem.*, **27**, 1411–1413

Robison, T.S., Mirocha, C.J., Kurtz, H.J., Behrens, J.C., Chi, M.S., Weaver, G.A. & Nystrom, S.D. (1979b) Transmission of T-2 toxin into bovine and porcine milk. *J. Dairy Sci.*, **62**, 637–641

Rood, H.D., Jr, Buck, W.B. & Swanson, S.P. (1988) Diagnostic screening method for the determination of trichothecenes exposure in animals. *J. agric. Food Chem.*, **36**, 74–79

Rosenstein, Y. & Lafarge-Frayssinet, C. (1983) Inhibitory effect of *Fusarium* T2-toxin on lymphoid DNA and protein synthesis. *Toxicol. appl. Pharmacol.*, **70**, 283–288

Rousseaux, C.G. & Schiefer, H.B. (1987) Maternal toxicity, embryolethality and abnormal fetal development in CD-1 mice following one oral dose of T-2 toxin. *J. appl. Toxicol.*, **7**, 281–288

Rousseaux, C.G., Schiefer, H.B. & Hancock, D.S. (1986) Reproductive and teratological effects of continuous low-level dietary T-2 toxin in female CD-1 mice for two generations. *J. appl. Toxicol.*, **6**, 179–184

Rukmini, C., Prasad, J.S. & Rao, K. (1980) Effects of feeding T-2 toxin to rats and monkeys. *Food Cosmet. Toxicol.*, **18**, 267–269

Saito, M. & Ohtsubo, K. (1974) Trichothecene toxins of *Fusarium* species. In: Purchase, I.F.H., ed., *Mycotoxins*, Amsterdam, Elsevier, pp. 263–281

Saito, M., Enomoto, M. & Tatsuno, T. (1969) Radiomimetic biological properties of the new scirpene metabolites of *Fusarium nivale*. *Gann*, **60**, 599–603

Sato, N., Ueno, Y. & Enomoto, M. (1975) Toxicological approaches to the toxic metabolites of fusaria. III. Acute and subacute toxicities of T-2 toxin in cats. *Jpn. J. Pharmacol.*, **25**, 263–270

Sato, N., Ito, T., Kumada, H., Ueno, Y., Asano, K., Saito, M., Ohtsubo, K., Ueno, I. & Hatanaka, Y. (1978) Toxicological approaches to the metabolites of fusaria. XIII. Hematological changes in mice by a single and repeated administrations of trichothecenes. *J. toxicol. Sci.*, **3**, 335–356

Schappert, K.T. & Khachatourians, G.C. (1986) Effects of T-2 toxin on induction of petite mutants and mitochondrial function in *Saccharomyces cerevisiae*. *Curr. Genet.*, **10**, 671–679

Schiefer, H.B., Rousseaux, C.G., Handcock, D.S. & Blakley, B.R. (1987) Effects of low-level long-term oral exposure to T-2 toxin in CD-1 mice. *Food chem. Toxicol.*, **25**, 593–601

Schoental, R., Joffe, A.Z. & Yagen, B. (1979) Cardiovascular lesions and various tumors found in rats given T-2 toxin, a trichothecene metabolite of *Fusarium*. *Cancer Res.*, **39**, 2179–2189

Scott, P.M. (1989) The natural occurrence of trichothecenes. In: Beasley, V.R., ed., *Trichothecene Mycotoxicosis: Pathophysiologic Effects*, Vol. 1, Boca Raton, FL, CRC Press, pp. 1–26

Scott, P.M. (1990) Trichothecenes in grains. *Cereal Foods World*, **35**, 661–666

Sintov, A., Bialer, M. & Yagen, B. (1986) Pharmacokinetics of T-2 toxin and its metabolite HT-2 toxin, after intravenous administration in dogs. *Drug Metab. Disposition*, **14**, 250–254

Sintov, A., Bialer, M. & Yagen, B. (1987) Pharmacokinetics of T-2 tetraol, a urinary metabolite of the trichothecene mycotoxin, T-2 toxin, in dog. *Xenobiotica*, **17**, 941–950

Sirkka, U., Nieminen, S.A. & Ylitalo, P. (1992) Acute neurobehavioural toxicity of trichothecene T-2 toxin in the rat. *Pharmacol. Toxicol.*, **70**, 111–114

Smalley, E.B. (1973) T-2 Toxin. *J. Am. vet. Med. Assoc.*, **163**, 1278–1281

Sorsa, M., Linnainmaa, K., Penttilä, M. & Ilus, T. (1980) Evaluation of the mutagenicity of epoxytrichothecene mycotoxins in *Drosophila melanogaster*. *Hereditas*, **92**, 163–165

Stanford, G.K., Hood, R.D. & Hayes, A.W. (1975) Effect of prenatal administration of T-2 toxin to mice. *Res. Commun. chem. Pathol. Pharmacol.*, **10**, 743–746

Sundheim, L., Nagayama, S., Kawamura, O., Tanaka, T., Brodal, G. & Ueno, O. (1988) Trichothecenes and zearalenone in Norwegian barley and wheat. *Norw. J. agric. Sci.*, **2**, 49–59

Suneja, S.K., Ram, G.C. & Wagle, D.S. (1983) Effect of feeding T-2 toxin on RNA, DNA and protein contents of liver and intestinal mucosa of rats. *Toxicol. Lett.*, **18**, 73–76

Thrane, U. (1989) *Fusarium* species and their specific profiles of secondary metabolites. In: Chelkowski, J., ed., Fusarium *Mycotoxins, Taxonomy and Pathogenicity*, Amsterdam, Elsevier, pp. 199–225

Thust, R., Kneist, S. & Hühne, V. (1983) Genotoxicity of *Fusarium* mycotoxins (nivalenol, fusarenon-X, T-2 toxin, and zearalenone) in Chinese hamster V79 cells *in vitro*. *Arch. Geschwulstforsch.*, **53**, 9–15

Ueno, Y. (1977a) Trichothecenes: overview address. In: Rodrick, J.V., Hesseltine, C.W. & Mehlman, M.A., eds, *Mycotoxins in Human and Animal Health*, Park Forest South, IL, Pathotox, pp. 189–207

Ueno, Y. (1977b) Mode of action of trichothecenes. *Pure appl. Chem.*, **49**, 1737–1745

Ueno, Y. (1983a) General toxicology. In: Ueno, Y., ed., *Developments in Food Science. IV. Trichothecenes*, Amsterdam, Elsevier, pp. 135–146

Ueno, Y., ed. (1983b) *Trichothecenes: Chemical, Biological and Toxicological Aspects*, Amsterdam, Elsevier

Ueno, Y. (1984) Toxicological features of T-2 toxin and related trichothecenes. *Fundam. appl. Toxicol.*, **4**, S124–S132

Ueno, Y. (1986) Trichothecenes as environmental toxicants. *Rev. environ. Toxicol.*, **2**, 303–341

Ueno, Y. (1987) Trichothecenes in food. In: Krogh, P., ed., *Mycotoxins in Food*, London, Academic Press, pp. 123–147

Ueno, Y. & Kubota, K. (1976) DNA-attacking ability of carcinogenic mycotoxins in recombination-deficient mutant cells of *Bacillus subtilis*. *Cancer Res.*, **36**, 445–451

Ueno, Y., Nakajima, M., Sakai, K., Ishii, K., Sato, N. & Shimada, N. (1973) Comparative toxicology of trichothec mycotoxins: inhibition of protein synthesis in animal cells. *J. Biochem.*, **74**, 285–296

Ványi, A., Bata, Á., Fekete, S. & Tamás, J. (1988) Study of the metabolism and excretion of T-2 toxin, a trichothecene fusariotoxin, in rabbits. *Acta vet. hung.*, **36**, 213–220

Visconti, A. & Mirocha, C.J. (1985) Identification of various T-2 toxin metabolites in chicken excreta and tissues. *Appl. environ. Microbiol.*, **49**, 1246–1250

Visconti, A., Treeful, L.M. & Mirocha, C.J. (1985) Identification of *iso*-TC-1 as a new T-2 toxin metabolite in cow urine. *Biomed. Mass Spectrom.*, **12**, 689–694

Weaver, G.A., Kurtz, H.J., Bates, F.Y., Chi, M.S., Mirocha, C.J., Behrens, J.C. & Robison, T.S. (1978) Acute and chronic toxicity of T-2 mycotoxin in swine. *Vet. Rec.*, **103**, 531–535

Wehner, F.C., Marasas, W.F.O. & Thiel, P.G. (1978) Lack of mutagenicity to *Salmonella typhimurium* of some *Fusarium* mycotoxins. *Appl. environ. Microbiol.*, **35**, 659–662

Wei, R.-D., Strong, F.M., Smalley, E.B. & Schnoes, H.K. (1971) Chemical interconversion of T-2 and HT-2 toxins and related compounds. *Biochem. biophys. Res. Commun.*, **45**, 396–401

WHO (1990) *Selected Mycotoxins: Ochratoxins, Trichothecenes, Ergot* (Environmental Health Criteria 105), Geneva

Wilson, C.A., Everard, D.M. & Schoental, R. (1982) Blood pressure changes and cardiovascular lesions found in rats given T-2 toxin, a trichothecene secondary metabolite of certain *Fusarium* microfungi. *Toxicol. Lett.*, **10**, 35–40

Yang, S. & Xia, Q.J. (1988a) Papilloma of forestomach induced by *Fusarium* T-2 toxin in mice (Chin.). *Chin. J. Oncol.*, **10**, 339–341

Yang, S. & Xia, Q.-J. (1988b) Studies on the promoting and initiating effects of *Fusarium* T-2 toxin (Chin.). *Chin. J. Oncol.*, **17**, 107–110

Yates, S.G., Tookey, H.L., Ellis, J.J. & Burkhardt, H.J. (1968) Mycotoxins produced by *Fusarium nivale* isolated from tall fescue (*Festuca arundinacea* Schreb.). *Phytochemistry*, **7**, 139–146

Yoshizawa, T., Swanson, S.P. & Mirocha, C.J. (1980a) In vitro metabolism of T-2 toxin in rats. *Appl. environ. Microbiol.*, **40**, 901–906

Yoshizawa, T., Swanson, S.P. & Mirocha, C.J. (1980b) T-2 Metabolites in the excreta of broiler chickens administered ^{3}H-labeled T-2 toxin. *Appl. environ. Microbiol.*, **39**, 1172–1177

Yoshizawa, T., Mirocha, C.J., Behrens, J.C. & Swanson, S.P. (1981) Metabolic fate of T-2 toxin in a lactating cow. *Food Cosmet. Toxicol.*, **19**, 31–39

Yoshizawa, T., Sakamoto, T., Ayano, Y. & Mirocha, C.J. (1982) 3'-Hydroxy T-2 and 3'-hydroxy HT-2 toxins: new metabolites of T-2 toxin, a trichothecene mycotoxin, in animals. *Agric. Biol. Chem.*, **46**, 2613–2615

Yoshizawa, T., Sakamoto, T. & Kuwamura, K. (1985) Structures of deepoxytrichothecene metabolites from 3'-hydroxy HT-2 toxin and T-2 tetraol in rats. *Appl. environ. Microbiol.*, **50**, 676–679

Zhu, G.-F., Cheng, S.-J. & Li, M.-H. (1987) The genotoxic effects of T-2 toxin, a trichothecene produced by *Fusarium* fungi (Chin.). *Acta biol. exp. sin.*, **20**, 129–134

OCHRATOXIN A

This substance was considered by previous working groups, in October 1975 (IARC, 1976), October 1982 (IARC, 1983) and March 1987 (IARC, 1987). Since that time, new data have become available, and these have been incorporated into the monograph and taken into consideration in the present evaluation.

1. Exposure Data

1.1 Chemical and physical data

1.1.1 *Synonyms, structural and molecular data*

Chem. Abstr. Services Reg. No.: 303-47-9
Chem. Abstr. Name: L-Phenylalanine, *N*-[(5-chloro-3,4-dihydro-8-hydroxy-3-methyl-1-oxo-1*H*-2-benzopyran-7-yl)-carbonyl]-, (*R*)-
IUPAC Systematic Name: *N*-[[(3*R*)-5-Chloro-8-hydroxy-3-methyl-1-oxo-7-isochromanyl]carbonyl]-3-phenyl-L-alanine
Synonym: (–)-*N*-[(5-Chloro-8-hydroxy-3-methyl-1-oxo-7-isochromanyl)carbonyl]-3-phenylalanine

$C_{20}H_{18}ClNO_6$ Mol. wt: 403.8

1.1.2 *Chemical and physical properties*

(a) *Description*: Crystals (recrystallized from xylene); intensely fluorescent in ultraviolet light, emitting green and blue fluorescence in acid and alkaline solutions, respectively (IARC, 1983; Budavari, 1989)

(b) *Melting-point*: 169 °C (recrystallized from xylene) (van der Merwe *et al.*, 1965a,b; Kuiper-Goodman & Scott, 1989)

(c) *Optical rotation*: $[\alpha]_D^{21}$ –46.8° (c = 2.65 mmol/l [1.07 g/l] in chloroform) (Pohland *et al.*, 1982)

(d) *Spectroscopy data*: Ultraviolet, infrared, mass spectral and proton nuclear magnetic resonance data have been reported (van der Merwe *et al.*, 1965a,b; Pohland *et al.*, 1982).

(e) *Solubility*: The free acid is moderately soluble in organic solvents (e.g., chloroform, ethanol, methanol, xylene) (WHO, 1990).

(f) *Stability*: Partially degraded under normal cooking conditions (Müller, 1982). Solutions of ochratoxin A are completely degraded by treatment with an excess of sodium hypochlorite solution (Castegnaro *et al.*, 1991a).

1.1.3 *Analysis*

Methods of analysis for ochratoxins in various matrices have been the subject of three recent reviews (Kuiper-Goodman & Scott, 1989; WHO, 1990; van Egmond, 1991a). Only one method (AOAC 973.37, Nesheim *et al.*, 1992) has been subject to a trial by the US Association of Official Analytical Chemists: a thin-layer chromatography method for grain, dating from 1973 (Gilbert, 1991). This and similar methods are widely used (van Egmond, 1991a). A liquid chromatographic method for determining ochratoxin A in maize, barley and kidney was tested in a IUPAC collaborative study (Nesheim *et al.*, 1992). Several high-performance liquid chromatographic methods for various commodities have been proposed and appear to be useful, and a number of enzyme-linked immunosorbent assay methods are valuable for screening and providing semiquantitative data, e.g., for cereals and porcine kidney (van Egmond, 1991a). Methods also have been reported for determining ochratoxins in blood, e.g., by fluorescence with high-performance liquid chromatographic confirmation (Hult *et al.*, 1982; Bauer & Gareis, 1987).

1.2 Production and use

Ochratoxin A was first isolated in 1965 from a culture of *Aspergillus ochraceus* Wilh. grown on sterile maize meal (van der Merwe *et al.*, 1965a). The structure was established by synthesis (Steyn & Holzapfel, 1967; Roberts & Woollven, 1970). Ochratoxin A is produced by inoculating strains of the fungi that produce this compound on autoclaved grains and oilseed (Peterson & Ciegler, 1978; Madhyastha *et al.*, 1990).

The taxonomy of *Aspergillus* and *Penicillium* has been the subject of disagreement since the 1920s. Numerous erroneous reports have been made of ochratoxin A production by more than a dozen species of these two genera (e.g., WHO, 1990; Frank, 1991). An international commission on *Aspergillus* and *Penicillium* was convened to harmonize the taxonomy (Samson & Pitt, 1990); in addition, in extensive studies of the metabolites of *Penicillium* species, their taxonomy has been made consonant with internationally accepted norms. There is now general agreement that ochratoxin A is produced by only one species of *Penicillium*, *P. verrucosum* (Frisvad & Filtenborg, 1989). Among the aspergilli, *A. ochraceus* is the most important ochratoxin-producing species; however, rare species in the *ochraceus* group, including *A. sclerotiorium*, *A. melleus*, *A. alliaceus* and *A. sulphureus*, also produce ochratoxin A (Ciegler, 1972).

Authentic strains of *P. verrucosum* produce ochratoxin A, verrucosin and often high concentrations of citrinin (Frisvad & Filtenborg, 1989). Authentic strains of *A. ochraceus* produce ochratoxin A and penicillic acid (Ciegler, 1972). Other species of *Penicillium* have been reported in grains containing ochratoxin A, including *P. aurantiogriseum* and *P. commune* (Mantle *et al.*, 1991). *P. aurantiogriseum* produces an array of toxic metabolites, including brevianamide A and B, and *P. commune* produces penicillic acid (Frisvad & Filtenborg, 1989; Frank *et al.*, 1990). Grain containing ochratoxin A probably contains additional fungal metabolites.

1.3 Occurrence

Although ochratoxin A occurs in many commodities—from grains to coffee beans—all over the world, it has been found primarily in north-temperate barley- and wheat-growing areas (Kuiper-Goodman & Scott, 1989; WHO, 1990). Representative values reported in various plant products are given in Table 1. Ochratoxin A concentrations in food-grade wheat are generally low, the highest values and frequencies of occurrence being found in some parts of Europe. One sample from Australia contained relatively high ochratoxin A concentrations, indicating that more study of this toxin outside Europe is warranted. A few studies have reported ochratoxin A in retail pork products in Europe (Table 2), suggesting that feed-grade cereals contain more ochratoxin A than is implied by Table 1. Pork products can be a significant human dietary source of ochratoxin A: ochratoxin A occurs at high frequency in the blood of swine produced in several countries (Table 3).

Table 1. Natural occurrence of ochratoxin A in plant products

Country	Product	Year	Positive samples/ total no.	Content (μg/kg)	Reference
North America					
Canada	Wheat, hay	1971-75	7/95	30-6000	Prior (1976)
Canada	Grain, forage	1975-79	5/474	30-4000	Prior (1981)
Canada	Cereals	1976-78	6/315	3-8	Williams (1985)
Canada	Cereals	1981-83	5/440	10-50	Sinha et al. (1986)
Canada	Peas, beans	1979	1/84	20	Williams (1985)
USA	Maize	1968-69	3/293	80-170	Shotwell et al. (1971)
USA	Wheat	1970-73	11/577	Trace-120	Shotwell et al. (1976)
USA	Barley	NR	23/182	10-29	Nesheim (1971), cited in Krogh & Nesheim (1983)
USA	Barley	NR	18/127	10-40	Nesheim (1971), cited in WHO (1990)
South America					
Brazil	Cassava flour	1985-86	2/33	30, 70	Valente Soares & Rodriguez-Amaya (1989)
Brazil	Beans	1985-86	2/13	90, 160	Rodriguez-Amaya (1989)
Brazil	Dried white corn	1985-86	1/12	30	
Chile	Maize	NR	1/28	55	Vega et al. (1988)
Europe					
Austria	Feed	1986	30/170	100-1000	Böhm & Leibetseder (1987)
Bulgaria	Maize	1984-89	103/264	0.2-1418	Petkova-Bocharova et al. (1991)
Bulgaria	Beans	1984-89	86/260	0.05-285	Petkova-Bocharova et al. (1991)
France	Maize	1973	18/924	5-200	Galtier et al. (1977)
Germany	Barley	1982-87	10/68	0.1-206	Bauer & Gareis (1987)
Germany	Wheat	1982-87	94/719	0.1-12.5	Bauer & Gareis (1987)

Table 1 (contd)

Country	Product	Year	Positive samples/ total no.	Content (µg/kg)	Reference
Europe (contd)					
Germany	Cereals	1973–88	24/765	Mean, 11.8	Frank (1991)
Germany	Bran	1973–88	9/84	6.8	Frank (1991)
Germany	Flour	1973–88	17/93	2.2	Frank (1991)
Italy	Bread	1976–79	1/1	80 000	Visconti & Bottalico (1983)
Norway	Cereal	1973–88	11/538	2–180	Olberg & Yndestad (1982); Kuiper-Goodman & Scott (1989)
Poland	Feed (mixed)	1966, 87	18/1240	10–200	Goliński et al. (1991)
Poland	Cereals	1984–85	158/1353	5–2400	Goliński et al. (1991)
Poland	Bread	1984–85	63/368	Mean, 1360	Goliński et al. (1991)
Poland	Flour	1984–85	48/215	Mean, 4370	Goliński et al. (1991)
Sweden	Beans	1976–79	6/91	10–442	Åkerstrand & Josefsson (1979)
Sweden	Cereals	1972	7/84	16–410	Krogh et al. (1974)
United Kingdom	Bread	NR	1/50	210	Osborne (1980)
United Kingdom	Barley	1976–79	51/376	< 25–5000	Buckle (1983)
United Kingdom	Wheat	1976–79	15/101	< 25–2700	Buckle (1983)
United Kingdom	Feeds	1976–79	27/812	< 25–250	Buckle (1983)
United Kingdom	Breakfast cereals	1976–79	12/243	5–108	Lindsay (1981), cited by Kuiper-Goodman & Scott (1989)
Yugoslavia	Maize	1972–76	45/542	19–140	Pavlović et al. (1979)
Yugoslavia	Wheat	1972–76	11/130	19–> 100	Pavlović et al. (1979)
Africa					
Egypt	Wheat	NR	1/3	10	Abdelhamid (1990)
Egypt	Maize	NR	1/3	12	Abdelhamid (1990)
Egypt	Mixed feed	NR	2/3	Mean, 19	Abdelhamid (1990)
Senegal	Cowpea	1984–88	5/31	Mean, 34	Kane et al. (1991)
Tunisia	Wheat	1982–83	8/28	34–360	Bacha et al. (1988)
Tunisia	Feed	1982–83	3/10	140–360	Bacha et al. (1988)
Australasia					
Australia	Feed	1971–80	1/25	70 000	Connole et al. (1981)
India	Rice	1981	2/32	8, 25	Reddy et al. (1983)
India	Copra	1982–83	1/384	50	Kumari & Nusrath (1987)
Indonesia	Maize	1985–86	1/26	3	Widiastuti et al. (1988)
Japan	Flour	1977–82	11/11	< 2.5–20	Nishijima (1984)

NR, not reported

Table 2. Occurrence of ochratoxin A in retail animal products

Country	Product	Year	Positive samples/ total no.	Content (μg/kg)	Reference
Germany	Kidneys (pork)	1983	41/300	0.5–10	Scheuer & Leistner (1986)
Germany	Sausage (pork)	1984	58/325	0.1–3.4	Scheuer & Leistner (1986)
Switzerland	Sausage	NR	1/12	0.8	Baumann & Zimmerli (1988)
Yugoslavia	Smoked meat products	NR	206	0.01–9	Pepeljnjak & Blažević (1982)
	Bacon		18.9%	37–200	
	Ham		28.9%	40–70	
	Sausages		12%	10–920	
	'Kulen' (specially prepared sausages)		13.3%	10–460	

Table 3. Occurrence of ochratoxin A in swine blood at slaughter

Country	Year	Positive samples/ total no.	Content (ng/ml)	Reference
Canada	1986	813/1006	< 10–229	Frohlich et al. (1991)
Germany	1982–83	93/191	0.1–67	Bauer & Gareis (1987)
Poland	1983–84	335/894	Mean, 2.03	Goliński et al. (1991)
Sweden	1982–83	26/122	2–62	Hult et al. (1984)
Yugoslavia	1979	16.6%	36–77	Pepeljnjak et al. (1982)

Ochratoxin A has been found in blood from individuals in several European countries (Creppy et al., 1991; Hald, 1991), at levels ranging, e.g., in Sweden from 0.3 to 6 ng/ml (Breitholz et al., 1991). In inhabitants of the Balkans, concentrations of up to 100 ng/ml have been found (Fuchs et al., 1991; Petkova-Bocharova & Castegnaro, 1991). Concentrations found in human blood and milk are presented in Table 4. The limits of detection in these studies vary, so that the proportion of positive samples may not be a good indication of relative exposures.

1.4 Regulations and guidelines

In 1990, existing or proposed regulations for ochratoxin A were available in Brazil, Czechoslovakia, Denmark, France, Greece, Hungary, Israel, the Netherlands, Romania, Sweden and the United Kingdom, the levels ranging from 1 to 50 μg/kg in food and from 100 to 1000 μg/kg in animal feed. In Denmark, the levels of ochratoxin A in pork kidney are used to determine if the animal (25 μg/kg) or certain organs (10 μg/kg) can be used as food (van Egmond, 1991b).

Table 4. Occurrence of ochratoxin A in human blood and milk

Country	Year	Positive samples/ total no.	Content (ng/ml)[a]	Reference
Bulgaria	1984–90	82/576	1–35	Petkova-Bocharova & Castegnaro (1991)
Canada	1988	63/159	0.27–35.3	Frohlich et al. (1991)
Denmark	1986–88	78/144	0.1–13.2	Hald (1991)
France	NR	~ 18%	0.1–6	Creppy et al. (1991)
Germany	1977–85	173/306	0.1–14.4	Bauer & Gareis (1987)
Germany	NR	4/36	(0.003) 0.017–0.03 (milk)	Gareis et al. (1988)
Italy	1989–90	9/50	(1.2) 1.7–6.6 (milk)	Micco et al. (1991)
Poland		9/216	1.3–4.8	Goliński & Grabarkiewicz-Szczęsna (1985)
Poland	1983–84	77/1065	Mean, 0.10	Goliński et al. (1991)
Sweden	1989	38/297	0.3–6.7	Breitholtz et al. (1991)
Yugoslavia	1980	42/639	1–40	Hult et al. (1982)
Yugoslavia	1981–89	240/17175	5–100	Fuchs et al. (1991)

NR, not reported
[a]Lower end of range or value in parentheses is the detection limit.

In 1990, a WHO/FAO Joint Expert Committee on Food Additives reviewed the literature on ochratoxin A and recommended a provisional tolerable weekly intake of 112 ng/kg bw (WHO, 1991).

2. Studies of Cancer in Humans

During the 1950s, a fatal chronic renal disease was identified in certain geographically limited areas of Bulgaria, Yugoslavia and Romania (Tanchev & Dorossiev, 1991). In 1964, the disease was recognized as a new nosological entity and was referred to as Balkan endemic nephropathy (BEN). Studies in experimental animals have shown a relationship between exposure to ochratoxin A and a porcine nephropathy that has striking similarity to BEN (Krogh, 1974). The geographical distribution of BEN, in turn, is linked with areas of high incidence of urinary tract tumours. The epidemiology of BEN is therefore considered below.

2.1 Case reports

In Bulgaria, Petrinska-Venkovska (1960) reported 16 cases of urinary tract tumours among 33 autopsied patients with BEN (48%); Tanchev et al. (1970) reported such tumours among 6–7% of BEN patients; in 1968, urinary tract tumours represented 16% of tumours in the endemic area. Other reports appeared subsequently (Petković et al., 1974; Čeović et al., 1976).

2.2 Descriptive studies

The occurrence of BEN and of cancer in the populations of 27 villages in Vratza district, Bulgaria, an endemic area for BEN, during the period 1965–74 has been considered in several publications (Chernozemsky et al., 1977; Stojanov et al., 1977, 1978). The villages were divided into two groups: 15 villages with a high incidence of BEN (population, 147 321) and 12 villages with a low incidence (population, 120 687). Age-adjusted incidence rates for BEN in the high-incidence villages were reported to be 506 and 315 per 100 000 in women and men, respectively; and the rates for urinary system (kidney and urinary tract) tumours, 104 and 89 per 100 000, respectively. In the same villages, BEN accounted for about 30% of total mortality in women and for over 40% in men, and urinary tract tumours were the most frequent neoplasms recorded, with incidences far higher than those in other parts of Bulgaria and in Europe. A ratio of 28.3 (95% confidence interval [CI], 16.5–47.4) was computed by comparing the urinary tract tumour incidence among the patients with BEN with that in the population of the low-incidence villages. The rate ratio was particularly high for tumours of the renal pelvis and ureter (88.9; 50.0–143.2) among patients with BEN. People in the high-incidence villages not affected by BEN also had increased rates of urinary system tumours in general (4.3; 1.6–11.3) and of tumours of the renal pelvis and ureter in particular (6.7; 2.3–16.4). [The Working Group noted that the statistical analysis was not adequately described and that bias due to better tumour ascertainment among cases of BEN could not be ruled out.]

A survey of the geographical distribution of BEN and of the incidence of tumours of the urinary system was carried out in central Serbia by Radovanović and Krajinović (1979) during the period 1970–74. The population of the region was divided into two groups, each of approximately 2.6 million people, on the basis of the topographical distribution of BEN. The incidence of urinary system tumours was determined in each of the study groups from cancer registry data. The annual incidence of cancer of the renal pelvis and ureter overall was 39 per 100 000 in counties affected by BEN and 15 per 100 000 in the non-endemic counties, on the basis of 52 and 20 tumour cases, respectively. The incidence of tumours of the renal parenchyma was slightly higher in the endemic than the non-endemic area, but no difference in the incidence of urinary bladder cancer was seen between the two. It was indicated, however, that distinction of the two subpopulations according to endemic area was very crude. This may have led to an underestimate of the true differences in the incidences of urinary system tumours.

Using records of cases of urinary tract tumour collected over a period of 16 years (1974–89) at the medical centre of the County of Slavonski Brod in Croatia, Šoštarić and Vukelić (1991) studied the distribution of these tumours in the areas of the County that were endemic for BEN (population, 10 094) and in the rest of the County (96 306). A total of 67 tumours were recorded in the endemic areas (estimated cumulative incidence, 0.664%) and 126 in the non-endemic area (0.131%). The difference in the recorded relative number of cases was highest for tumours of the renal pelvis (0.287 *versus* 0.021%) and ureter (0.089 *versus* 0.013%) and only moderate for bladder cancer (0.228 *versus* 0.089%). In general, more cases were seen among women than among men in the endemic area. Tumours of the renal parenchyma were not mentioned in this survey.

In a study carried out in Bulgaria, Petkova-Bocharova and Castegnaro (1985) examined contamination by ochratoxin A of 65 samples of beans, maize or wheat flour from households in an area endemic for BEN and with a high incidence of urinary system tumours and in 65 samples from households in non-endemic areas. None of the samples was visibly mouldy. Samples from high-risk areas were collected from families in which cases of BEN and/or urinary tract tumours had been diagnosed ('affected' households), while samples from non-endemic areas were taken at random. All samples of home-produced beans and maize were taken during February and March 1982 from the 1981 harvest; wheat flour samples were purchased from local shops, which were supplied from central state stocks. Although there was no significant difference between endemic and non-endemic areas in the mean values of ochratoxin A in contaminated samples of beans (range, 25–27 µg/kg in endemic area, 25–50 µg/kg in non-endemic area) and maize (25–35 and 10–25 µg/kg, respectively), a larger proportion of samples from the endemic area were contaminated: (16.7% of beans (95% CI, 4.8–33.9) and 27.3% of maize (11.2–47.4) in endemic areas, and 7.1% of beans and 9.0% of maize in non-endemic areas). None of the samples of wheat flour analysed contained measurable amounts of ochratoxin A.

In an extension of this survey, Petkova-Bocharova *et al.* (1991) collected 524 samples of home-produced, home-stored beans and maize from the harvests of 1984–86 and 1989–90. Of these, 298 were taken from 'affected' and 'non-affected' households in the endemic areas of Bulgaria and 226 from non-affected households in non-endemic areas. Overall, significantly more samples from endemic areas than from control areas were contaminated with ochratoxin A [54 *versus* 12%].

Petkova-Bocharova *et al.* (1988) reported a study in Bulgaria of the association between BEN and/or urinary system tumours and ochratoxin A content in blood samples taken from 187 subjects living in endemic villages and 125 individuals in non-endemic villages. Among 61 patients with BEN and/or urinary system tumours, 14.8% had levels of 1–2 ng/ml and 11.5% had more than 2 ng/ml ochratoxin A in their blood. This proportion was significantly higher than that in a control group of 63 healthy individuals from unaffected families in the endemic villages (7.9 and 3.2%, respectively). The percentage of positive blood samples in the control group was similar to that measured among a random sample of healthy individuals from non-endemic villages (6.2 and 1.5% respectively). Intermediate proportions of positive blood samples were found among healthy individuals from families of patients with either endemic nephropathy and/or a urinary system tumour (9.5 and 6.3%, respectively). [The Working Group noted that no attempt was made by the authors to present separate results for the subgroup of urinary system tumours].

In an extension of this study, Petkova-Bocharova and Castegnaro (1991) collected blood samples from 576 people living inside and outside the endemic areas in Bulgaria during 1984, 1986, 1989 and 1990. Overall, a significantly larger proportion of the blood samples from 105 patients with urinary tract tumours and/or BEN [26.7%] contained ochratoxin A than those from 116 healthy people living in villages in the endemic area [12.1%], or from 119 healthy people from non-affected villages in the endemic area [10.9%] or from 125 healthy people living in non-endemic areas of Bulgaria [7.2%]; however, the proportion among the patients was not significantly higher than that among 111 healthy relatives of patients with urinary tract tumours and/or BEN living in affected villages [16.2%].

No analytical epidemiological study was available to the Working Group.

3. Studies of Cancer in Experimental Animals

3.1 Oral administration

3.1.1 *Mouse*

A group of 10 male ddY mice (average weight, 26 g) received a diet containing 40 mg/kg diet ochratoxin A [purity unspecified] for 44 weeks. A group of 10 untreated controls were fed the basal diet. All survivors were killed 49 weeks after the start of treatment. A significant increase in the incidence of hepatocellular tumours (designated as well-differentiated trabecular adenomas) was seen in 5/9 treated mice and in 0/10 controls; hyperplastic liver nodules were found in 1/9 treated mice and in 2/10 controls. A significant increase in the incidence of renal-cell tumours (nine renal cystadenomas and two solid renal-cell tumours) was found in 9/9 treated mice, with 0/10 controls (Kanisawa & Suzuki, 1978). [The Working Group noted the small size of the experimental groups.]

Groups of 16 male ddY mice [age unspecified] received a diet containing 50 mg/kg diet ochratoxin A for 0, 5, 10, 15, 20, 25 and 30 weeks, followed by a basal diet for 40 weeks. The experiment was terminated at 70 weeks. The estimated mean cumulative doses of ochratoxin A were 0, 11, 21, 30, 33, 40 and 50 mg/mouse, respectively. Hepatomas were seen in animals fed ochratoxin A in the diet for 20 weeks or more: control 0/15; 20 weeks, 2/14; 25 weeks, 5/15; and 30 weeks, 6/17. Renal-cell tumours [not further specified] were seen in mice fed ochratoxin A in the diet for 15 weeks or more: control, 0/15; 15 weeks, 3/15; 20 weeks, 1/14; 25 weeks, 2/15; and 30 weeks, 4/17. Lung tumours [unspecified] were found in all groups, but the incidence was not related to exposure level (Kanisawa, 1984).

Two groups of 20 male DDD mice, six weeks of age, received diets containing ochratoxin A at 0 or 25 mg/kg for 70 weeks. Twenty mice exposed to ochratoxin A had renal-cell tumours, designated as cystadenomas (20 tumours) and solid renal-cell tumours (six tumours); no such tumour occurred in controls. The incidence of hyperplastic hepatic nodules was significantly increased: 16/20 in treated mice *versus* 4/17 in controls; as was that of hepatomas (exhibiting the trabecular structure): 8/20 in treated animals *versus* 1/17 in controls (Kanisawa, 1984).

Groups of 50 male and 50 female $B6C3F_1$ mice, three weeks old, received diets containing crude ochratoxin A at 0, 1 or 40 mg/kg for two years. The crude ochratoxin preparation contained 84% ochratoxin A, 7% ochratoxin B and 9% benzene. Both male and female mice in the high-dose group had terminal body weights that were about 40% lower than those of controls, but survival was not decreased after exposure to ochratoxin. The incidence of renal-cell adenomas was significantly increased (26/49) in high-dose male mice that survived 21 months, as none were found in the low-dose and control groups. A significant increase in the incidence of renal-cell carcinomas (14/49) was also found in high-dose male mice that survived at least 20 months, with none in the low-dose or control groups. Nine of the 26 mice with renal-cell adenomas also had renal-cell carcinomas. No renal tumour was observed in female mice. The incidence of hepatocellular tumours was increased in animals of each sex: males—adenomas: 1/50, 5/47 and 6/50 [$p = 0.03$, Cochran-Armitage test]; carcinomas: 0/50, 3/47 and 4/50 [$p = 0.03$, Cochran-Armitage test]; females—adenomas: 0/47, 1/45 and 2/49; carcinomas: 0/47, 1/45 and 5/49 [$p = 0.007$, Cochran-Armitage test] in control, low-dose and high-dose animals, respectively. The incidences of neoplasms in other tissues were not significantly increased in treated animals compared to controls (Bendele et al., 1985a).

3.1.2 *Rat*

Groups of 80 male and 80 female Fischer 344 rats, 8–10 weeks old, received 0, 21, 70 or 210 µg/kg bw (maximum tolerated dose) ochratoxin A (98% pure) in 5 ml/kg maize oil by gavage on five days per week for 103 weeks. All survivors were killed 104–105 weeks after initiation of exposure. Mean body weights of high-dose rats were generally 4–7% lower than those of vehicle controls. The survival rates of treated and vehicle control female rats were comparable, but the survival of all groups of dosed male rats was decreased. Renal-cell tumours were increased in a dose-related manner in animals of each sex: males—renal-cell adenomas: 1/50, 1/51, 6/51 and 10/50; renal-cell adenocarcinomas: 0/50, 0/51, 16/51 and 30/50; females—renal-cell adenomas: 0/50, 0/51, 1/50 and 5/50; renal-cell adenocarcinomas 0/50, 0/51, 1/50 and 3/50, in control, low-dose, mid-dose and high-dose animals, respectively. Metastasis of the renal-cell tumours occurred in 17 males and one female (US National Toxicology Program, 1989).

3.2 Administration with known carcinogens

Rat

In a medium-term carcinogenicity study, groups of 20 male Fischer 344 rats weighing about 150 g, were kept for one week on a basal diet and were then given ochratoxin A [purity unspecified] at 0 or 50 mg/kg diet for six weeks; one week after initiation of the study, the rats were given a partial hepatectomy, and during weeks 7–9 the animals received 200 mg/kg diet *N*-2-fluorenylacetamide [purity unspecified]. The rats also received carbon tetrachloride (1 ml/kg bw) at the end of week eight. The animals were killed 10 weeks after the start of the experiment. In 14 rats, 0.63 ± 0.69 hepatic hyperplastic nodules/cm^2 were observed ($p < 0.02$), compared to an average of 0.11 ± 0.22/cm^2 in the 20 rats that received no ochratoxin A. In the same study, groups of 20 male rats were fed the ochratoxin A diets during weeks 3–9 and received the *N*-2-fluorenylacetamide during the first two weeks and the carbon tetrachloride at week 1; the partial hepatectomy was done at week 4. Sixteen rats had an average of 0.82 ± 0.54 hepatic hyperplastic nodules/cm^2 ($p < 0.05$) compared with an average of 0.36 ± 0.43/cm^2 in 20 control rats that received similar treatment without exposure to ochratoxin A (Imaida et al., 1982).

4. Other Relevant Data

4.1 Absorption, distribution, metabolism and excretion

4.1.1 *Humans*

No data were available to the Working Group.

4.1.2 *Experimental systems*

Ochratoxin A was injected into the lumina at various sites of the gastrointestinal tract in rats; the highest concentration in portal blood was observed after injection into the proximal jejunum (Kumagai & Aibara, 1982). Ruminants do not absorb much ochratoxin A when the concentration in the feed is low because it is hydrolysed rapidly by their ruminal flora (Galtier & Alvinerie, 1976; Hult *et al.*, 1976); however, ochratoxin A has been detected in kidney, milk and urine of cows given high doses of ochratoxin A (Ribelin *et al.*, 1978; Shreeve *et al.*, 1979).

In human plasma, ochratoxin A binds *in vitro* to certain, as yet unidentified macromolecules (relative molecular mass, 20 000) with extremely high affinity, the association constant being 2.3×10^{10}/mol (Stojković *et al.*, 1984). Saturation of these macromolecules occurs at low levels of ochratoxin A: 10–20 ng/ml of serum (Hult & Fuchs, 1986). Ochratoxin A binds to serum albumin in plasma of different animal species (Chang & Chu, 1977; Galtier, 1974), but at relatively low affinity (up to 10^6/mol) *in vivo* and *in vitro* (Chu, 1971). Studies of the binding of ochratoxin A to human plasma proteins *in vitro* showed that only 0.02% of the total concentration of 10^{-9}–10^{-6}M was unbound; 2% of ochratoxin B, the dechlorinated form of ochratoxin A, was unbound. In monkeys, 0.08% of the toxin remained in the free form (Hagelberg *et al.*, 1989).

In rats given ochratoxin A by gavage at levels corresponding to the parts per million range found in contaminated foodstuffs (~ 4 ppm), the compound was found after 24 h in the following tissues (in decreasing order of concentration): fat, small intestine, testis, kidney, liver, lung, heart, spleen, stomach, muscle and brain. After 48 h, the concentrations had decreased in all tissues except fat, where increased concentrations were found (Kane *et al.*, 1986a). Intravenous administration of 2.5 mg/kg bw to rats resulted after 48 h in the following distribution (in decreasing order of concentration): thyroid, skin, parotid gland, lung, submaxillary gland, heart, seminal vesicle, kidney, large intestine, testis, liver and lachrymal gland (Galtier *et al.*, 1979a). A similar distribution pattern of labelled ochratoxin A was found in mice and rats (Fuchs *et al.*, 1988b; Breitholtz-Emanuelsson *et al.*, 1992).

Secondary distribution peaks of radiolabel in the intestinal contents and serum of rats and mice after oral or intramuscular administration of ^{3}H-ochratoxin A are a consequence of enterohepatic circulation, since the biliary excretion of ochratoxin A is very efficient (Fuchs *et al.*, 1988a; Roth *et al.*, 1988).

The apparent plasma elimination half-time of ochratoxin A after oral administration at 50 μg/kg bw varied from 0.68 h in fish to 120 h in rats and 510 h in monkeys (Hagelberg *et al.*, 1989).

When ochratoxin A was incubated with rat, rabbit, pig and human liver microsomes, (4*R*)- and (4*S*)-4-hydroxyochratoxin A were formed (see Fig. 1), the ratio between the two epimers depending on the animal species (Størmer *et al.*, 1981, 1983). The transformations are cytochrome P450-dependent (Størmer *et al.*, 1981, 1983; Oster *et al.*, 1991). Thus, liver and kidney microsomes from female dark Agouti rats (poor metabolizers of debrisoquine) had three to four times less ochratoxin A 4-hydroxylase activity than liver and kidney microsomes from female Lewis rats (extensive metabolizers of debrisoquine). The ochratoxin A hydroxylase was highly inducible by phenobarbital and 3-methylcholanthrene (Hietanen

Fig. 1. Ochratoxin A and its metabolites

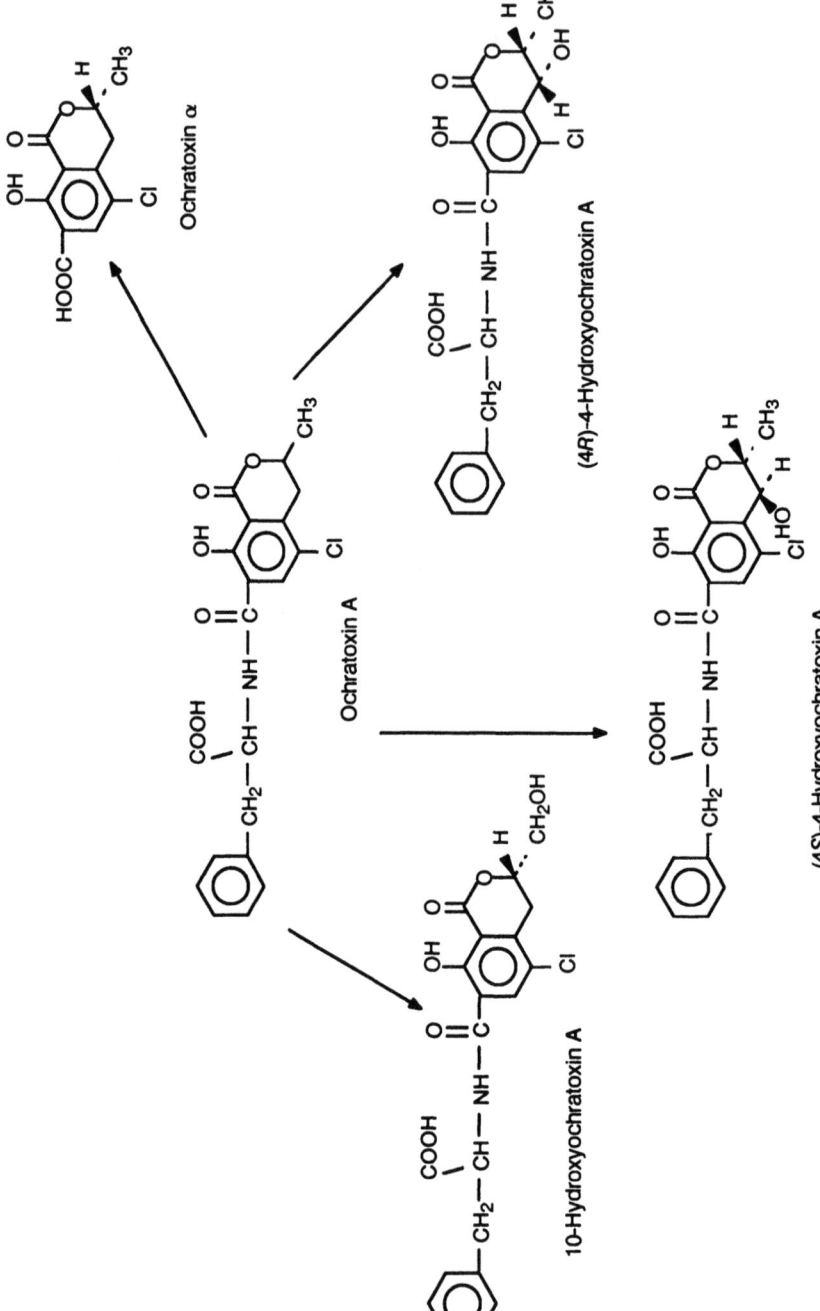

From Størmer et al. (1982); Størmer et al. (1983)

et al., 1986). When primary rat hepatocytes were incubated at 37 °C with ochratoxin A, (4*R*)-4-hydroxyochratoxin A was detected in the incubation medium and in the cells, but only small amounts of (4*S*)-4-hydroxyochratoxin A were found (Hansen *et al.*, 1982). 10-Hydroxyochratoxin A was formed from ochratoxin A with a rabbit liver microsomal system (Størmer *et al.*, 1983). The presence of 4-hydroxyochratoxin A was observed in intestine, liver and bile of mice given ochratoxin A by gavage. Up to 68% of ochratoxin A in bile was in a conjugated form after 24 h (Roth *et al.*, 1988). Ochratoxin A is also hydrolysed to the non-toxic ochratoxin α (Chakor *et al.*, 1988) at various sites, particularly by homogenates of duodenal, ileal, pancreatic (Suzuki *et al.*, 1977) and caecal tissues (Galtier, 1978).

Both biliary excretion and glomerular filtration are important in the plasma clearance of ochratoxin A, but tubular secretion might also be active (Stein *et al.*, 1985); the relative contribution of each excretory route varies with species (Kuiper-Goodman & Scott, 1989). In albino rats, regardless of the route of administration, 6% of a given dose was excreted in urine as ochratoxin A, 1–1.5% as (4*R*)-4-hydroxyochratoxin A and 25–27% as ochratoxin α (Støren *et al.*, 1982).

The amount of 4-hydroxyochratoxin A found in urine depends on the rat strain: the ratio of ochratoxin A:4-hydroxyochratoxin A is two to five times higher in dark Agouti female rats than Lewis rats (Castegnaro *et al.*, 1989).

4.2 Toxic effects

4.2.1 *Humans*

Krogh (1974) suggested that excessive exposure to ochratoxin A plays a role in the development of BEN; however, the causal role of other nephrotoxic agents cannot be excluded (for review, see Castegnaro *et al.*, 1991b). Evidence for a causal relationship between exposure to ochratoxin A and BEN derives from (i) similarities in the morphological and functional renal impairments induced by BEN and those induced by ochratoxin A in animals, and (ii) the finding that food from endemic areas is more heavily contaminated with ochratoxin A than food from disease-free areas (Pavlović *et al.*, 1979; Pleština *et al.*, 1990; Petkova-Bocherova *et al.*, 1991). BEN is a bilateral, non-inflammatory, chronic nephropathy, in which the kidneys are extremely reduced in size and weight and show diffuse cortical fibrosis extending into the corticomedullary junction, hyalinized glomeruli and severely degenerated tubules (Vukelić *et al.*, 1991). The clinical picture is characterized by progressive hypercreatininaemia, hyperuraemia and normo- or slightly hypochromic anaemia, but blood pressure is generally not elevated (Radovanović, 1991; Tanchev & Dorossiev, 1991).

4.2.2 *Experimental systems*

The toxicokinetic parameters of ochratoxin A vary considerably with dose and between species (for review, see Galtier, 1991). Species also vary in their susceptibility to acute poisoning by ochratoxin A, with oral LD_{50} values ranging from 0.2 to 50 mg/kg bw. Dogs and pigs are the most sensitive species (0.2 and 1 mg/kg bw, respectively) (for review, see Frank *et al.*, 1990). Simultaneous oral administration of phenylalanine either prevents death

(Creppy *et al.*, 1980) or raises the oral LD_{50} in mice (Moroi *et al.*, 1985). Pretreatment of mice with phenobarbital or 3-methylcholanthrene also raises the LD_{50} (Moroi *et al.*, 1985; Chakor *et al.*, 1988), while treatment with piperonyl butoxide lowers it (Chakor *et al.*, 1988). Synergistic effects of ochratoxin A with citrinin and with penicillic acid on the LD_{50} were seen in mice following intraperitoneal injection (Sansing *et al.*, 1976).

Ochratoxin A has a nephrotoxic effect on all mammalian species tested. The nephropathy is characterized by polyuria, glucosuria, proteinuria, enzymuria, decreased osmolarity of the urine, impaired tubular function and histopathological changes in the proximal convoluted tubule (for review, see Delacruz & Bach, 1990). Dogs and pigs are very sensitive to the nephrotoxic effects of ochratoxin A following oral exposure (Kuiper-Goodman & Scott, 1989). In pigs, dietary levels as low as 0.2 mg/kg (equivalent to approximately 8 μg/kg bw per day) for 90 days caused a reduction in the activity of several renal enzymes and decreased kidney function (Krogh & Elling, 1977; Elling, 1979, 1983; Elling *et al.*, 1985; Meisner & Krogh, 1986). The presence of ochratoxin A in feed is believed to be the most important cause of spontaneous mycotoxic porcine (Krogh, 1978) and poultry nephropathy (Elling *et al.*, 1975).

In a 90-day study, rats given 2 mg/kg ochratoxin A in the diet (equivalent to 145 μg/kg bw per day) showed increased levels of urinary enzymes originating from the brush border of the proximal convoluted tubules (Kane *et al.*, 1986b).

In chronic toxicity studies, progressive nephropathy but no renal failure was seen in female pigs fed diets containing 1 mg/kg ochratoxin A (equivalent to approximately 41 μg/kg bw per day) for two years (Elling, 1979). In rats, the no-observed-effect level for ability to concentrate urine in a 6–12-month gavage study was 70 μg/kg bw per day for males and 21 μg/kg bw for females (Kuiper-Goodman & Scott, 1989; US National Toxicology Program, 1989).

Ochratoxin A induced haematological changes in rats and mice (Galtier *et al.*, 1979b; Gupta *et al.*, 1983), but the doses used were relatively high (4–5 mg/kg bw over 4–10 days in rats).

Several studies have shown that ochratoxin A affects structural components of the immune system in several species (for reviews, see Kuiper-Goodman & Scott, 1989; Frank *et al.*, 1990). Chickens fed 2 mg/kg ochratoxin A in the diet for 20 days showed depression of immunoglobulin (Ig) G, IgA and IgM in lymphoid tissues and serum (Dwivedi & Burns, 1984); complement activity was slightly affected when ochratoxin A was fed at 2 mg/kg in the diet for five to six weeks (Campbell *et al.*, 1983). Very low levels of ochratoxin A (one intraperitoneal injection of 1 μg/kg bw) suppressed both the IgG and the IgM response in BALB/c mice to a single injection of sheep red blood cells in the standard plaque counting assay for estimation of antibody-producing spleen lymphocytes. (4R)-4-Hydroxyochratoxin A is as effective as ochratoxin A in immunosuppression, whereas ochratoxin α has no effect. Phenylalanine can prevent the effect (Haubeck *et al.*, 1981; Creppy *et al.*, 1983a).

The biochemistry and molecular aspects of the action of ochratoxin A in prokaryotes and eukaryotes have been reviewed (Dirheimer & Creppy, 1991). On the basis of work in prokaryotes (Konrad & Röschenthaler, 1977; Bunge *et al.*, 1978), eukaryotic microorganisms (Creppy *et al.*, 1979a), mammalian cell cultures (Creppy *et al.*, 1979b, 1983b) and experimental animals *in vivo* (Creppy *et al.*, 1984), the initial action of ochratoxin A has been shown to be inhibition of protein synthesis by competition with phenylalanine in the phenylalanyl-tRNA synthetase-catalysed reaction. This inhibition can be reversed by an increased phenylalanine concentration both *in vitro* and *in vivo*. The degree of inhibition of protein synthesis is greater in Madin Darby canine kidney (MDCK) cells than in hepatoma cells (Creppy *et al.*, 1986). It also varies *in vivo* within different organs, with 26% inhibition in the liver, 68% in the kidney and 75% in the spleen 5 h after intraperitoneal injection of 1 mg/kg bw ochratoxin A to mice (Creppy *et al.*, 1984). (4R)-4-Hydroxyochratoxin A is as effective as ochratoxin A in inhibiting protein synthesis, whereas ochratoxin α has no effect (Creppy *et al.*, 1983b). Ochratoxin A also inhibits phenylalanine hydroxylase both *in vitro* and *in vivo* (Creppy *et al.*, 1990).

Synthetic analogues of ochratoxin A in which the phenylalanine moiety was replaced by other amino acids also inhibited protein synthesis *in vitro* and *in vivo* (Creppy *et al.*, 1983c).

Ochratoxin A has been reported to decrease mitochondrial oxidative phosphorylation and respiration by impairing mitochondrial transport carriers located in the inner membrane (Moore & Truelove, 1970; Meisner & Chan, 1974; Wei *et al.*, 1985). Moreover, Elling (1979) suggested that ochratoxin A induces a reduction in the activity of enzymes in the tricarboxylic acid cycle. These suggestions were corroborated by Jung and Endou (1989), who showed in isolated S_2 segments of the nephron of rats that ATP-dependent reactions were significantly inhibited by incubation with ochratoxin A at a concentration of 10^{-8}M, the ID_{50} being reached with a concentration of 10^{-6}M.

Oral administration of ochratoxin A (2 mg/kg bw for two days) decreased the level of cytosolic phosphoenolpyruvate carboxykinase in rat kidneys. The decrease correlated well with that in renal gluconeogenesis (Meisner & Selanik, 1979; Meisner & Meisner, 1981). Krogh *et al.* (1988) showed that the level of cytosolic phosphoenolpyruvate carboxykinase was also decreased in renal cytosol of pigs fed low concentrations of ochratoxin A in capsules (corresponding to 0.2 and 1.00 mg/kg diet) for one to five weeks.

Rahimtula *et al.* (1988) found that addition of ochratoxin A to rat liver and kidney microsomes greatly enhanced the rate of NADPH-dependent lipid peroxidation.

4.3 Reproductive and developmental toxicity

4.3.1 *Humans*

No data were available to the Working Group.

4.3.2 *Experimental systems*

Ochratoxin A causes embryofetal lethality and induces gross structural malformations in mice, rats and hamsters.

After a single intraperitoneal injection of 5 mg/kg bw ochratoxin A on day 11 or 13 of pregnancy to ICR mice, concentrations of the toxin reached highest levels in maternal serum and tissues within 2 h. The half-times in serum were 29 h on day 11 and 24 h on day 13. Concentrations in the embryos increased up to 48 h and 30 h after injection on days 11 and 13, respectively, and reached maximal levels of 360 and 492 ng/g wet weight, respectively.

The peak incidence of pyknotic cells in the embryonic telencephalon coincided with the peak concentration of the toxin in the embryo (Fukui et al., 1987).

A single intraperitoneal injection of 5 mg/kg bw ochratoxin A to pregnant SAF/ICR mice on one of days 7–12 of gestation increased the frequencies of prenatal mortality and fetal malformations and decreased fetal weight. The largest number of malformations was seen in fetuses of animals treated on day 8 of gestation (35% dead or resorbed; 58% grossly malformed; 84% with skeletal malformations) (Hayes et al., 1974).

A single intraperitoneal injection of 2, 3 or 4 mg/kg bw ochratoxin A to CD-1 or Jcl:ICR mice induced craniofacial malformations when given on day 7 or 8, but not when given on day 10 of gestation. Typical findings were exencephaly, median facial clefts, anophthalmia, microphthalmia and other ocular abnormalities (Hood et al., 1978; Shirai et al., 1984; Ohshika et al., 1988).

Hoshino et al. (1988) injected mice intraperitoneally with either 2 or 3 mg/kg bw ochratoxin A on day 10 of gestation and found significant decreases in brain weight, brain size and cerebral cortical thickness in offspring six weeks after birth.

Cerebral necrosis occurred in most fetuses of CD-1 and ICR mice treated orally or intraperitoneally with 3–5 mg/kg bw on days 15–17 of gestation (Szczech & Hood, 1981). Doses of 1.25 and 2.25 mg/kg bw ochratoxin A given by intraperitoneal injection on days 15, 16 and 17 to ICR mice decreased body weight and induced developmental delay in offspring, as indicated by the results of behavioural tests. The effects were more pronounced in female offspring. No compound-related brain alteration was found by histological examination under these conditions (Poppe et al., 1983).

The teratogenic effect of ochratoxin A in CBA mice treated orally was decreased when zearalenone was given concomitantly on day 9 of gestation (Arora & Frölén, 1981; Arora et al., 1981, 1983). Maternal protein deprivation was associated with an increased incidence of malformed fetuses in ochratoxin A-treated CD-1 mice (Singh & Hood, 1985).

In several studies, pregnant Wistar and Sprague-Dawley rats were treated with ochratoxin A on various days during the period of organogenesis. Single oral doses of 6.25, 12.5 or 25 mg/kg bw on day 10 of gestation induced high resorption rates (25, 82 and 78%, respectively) in Sprague-Dawley rats (Still et al., 1971). An abnormal proportion of haemorrhagic fetuses, significant decreases in fetal body weight, an increased resorption rate and decreased litter size were noted after intraperitoneal or oral doses of 4 or 5 mg/kg bw to Wistar rats (Moré & Galtier, 1974, 1975; Moré et al. 1978). Furthermore, multiple oral doses of 0.75 mg/kg bw on days 6–15 were embryotoxic, inducing various gross, visceral and skeletal anomalies (Brown et al., 1976).

The effects of single subcutaneous injections of 0.5–5.0 mg/kg bw on one of days 4–10 of gestation were studied in rats. The minimal teratogenic dose was 1.75 mg/kg bw, which caused decreased fetal weight and various fetal malformations; higher doses caused fetal resorption. Ochratoxin A had the greatest effect when given on day 5, 6 or 7 of gestation (Mayura et al., 1982). In a rat whole-embryo culture system, concentrations of $\geq$ 75 mg/l ochratoxin A in the medium impaired the development of 10-day-old embryos (Mayura et al., 1989).

Several combination experiments have been performed to study the teratogenic action of ochratoxin A in rats more closely. An increased prevalence of gross and skeletal malformations was seen in Sprague-Dawley rats maintained on a 10% protein diet (normal, 27%) after treatment with 1.75 mg/kg bw ochratoxin A on day 6 of gestation (Mayura *et al.*, 1983). Simultaneous treatment with a sub-threshold teratogenic dose of 1 mg/kg bw ochratoxin A and 30 mg/kg bw of the mycotoxin citrinin on days 5, 6, 7, 8, 10, 11 or 14 of gestation was lethal to 22–40% of treated females and resulted in a significant increase in the frequency of malformations (Mayura *et al.*, 1984a). The teratogenic action of a single subcutaneous injection of 1.75 mg/kg bw ochratoxin A on day 7 of gestation was enhanced in rats with surgically impaired renal function in comparison with sham-operated rats (Mayura *et al.*, 1984b). In contrast, phenylalanine given in combination with a single subcutaneous injection of 1.75 mg/kg bw ochratoxin A on day 7 of gestation partially inhibited its prenatal toxicity (Mayura *et al.*, 1984c).

Ochratoxin A also induced prenatal toxicity in golden hamsters after single intraperitoneal injections of 2.5–20 mg/kg bw on one of days 7–10 of gestation; days 8 and 9 were the most critical with regard to malformations. There was no skeletal malformation, but micrognathia, hydrocephalus, short tail, oligodactyly, syndactyly, cleft lip, micromalia and cardiac defects were observed. The fetal mortality rate was significantly increased in litters exposed to the highest dose on day 7, 8 or 9 (Hood *et al.*, 1976).

4.4 Genetic and related effects

4.4.1 *Humans*

No adequate data were available to the Working Group.

4.4.2 *Experimental systems* (see also Table 5 and Appendices 1 and 2)

The genotoxicity of ochratoxin A has been reviewed (Kuiper-Goodman & Scott, 1989; Dirheimer & Creppy, 1991). Ochratoxin A induced mixed responses in tests for DNA repair activity in *Escherichia coli*: it gave a positive result in one study but negative results in three others. It did not induce DNA damage in *Bacillus subtilis* and did not generally induce mutation in *Salmonella typhimurium*. Positive results were obtained in several strains of *S. typhimurium* incubated for 2 h with cell-free medium obtained from 24-h incubations of primary rat hepatocytes with 100 μM ochratoxin A. The frequency of mitotic crossing-over was not increased in yeast.

Ochratoxin A induced DNA single-strand breaks in cultured mouse and Chinese hamster cells, but not in rat fibroblasts. It slightly increased the frequency of unscheduled DNA synthesis in one study in rat and mouse primary hepatocytes *in vitro* but not in another study with rat hepatocytes. [The Working Group noted that the concentration range was lower in the latter study.] Gene mutations were not induced in mouse cells *in vitro*. Ochratoxin A weakly increased the frequency of sister chromatid exchange in the presence of an exogenous metabolic system in Chinese hamster ovary cells; no increase was seen in cultured human lymphocytes in the absence of exogenous metabolic activation.

DNA single-strand breaks were induced after treatment with ochratoxin A *in vivo* in mouse kidney, liver and spleen and in rat kidney and liver. Ochratoxin A also formed DNA

Table 5. Genetic and related effects of ochratoxins

Test system	Result		Dose (LED/HID)[a]	Reference
	Without exogenous metabolic system	With exogenous metabolic system		
Ochratoxin A				
PRB, SOS spot test, *Escherichia coli* PQ37	–	–	0.0000	Auffray & Boutibonnes (1986)
PRB, SOS chromotest test, *Escherichia coli*	–	–	0.0000	Reiss (1986)
PRB, SOS chromotest test, *Escherichia coli* PQ37	–	+[b]	100.0000	Krivobok et al. (1987)
PRB, SOS chromotest test, *Escherichia coli* PQ37	+	0	800.0000	Malaveille et al. (1991)
BSD, *Bacillus subtilis* rec strains, differential toxicity	–	–	100 µg/plate	Ueno & Kubota (1976)
SA0, *Salmonella typhimurium* TA100, reverse mutation	–	–	5.0000	von Engel & von Milczewski (1976)
SA0, *Salmonella typhimurium* TA100, reverse mutation	–	–	0.0000	Nagao et al. (1976)
SA0, *Salmonella typhimurium* TA100, reverse mutation	–	–	150.0000	Wehner et al. (1978)
SA0, *Salmonella typhimurium* TA100, reverse mutation	–	–	100.0000	Bartsch et al. (1980)
SA0, *Salmonella typhimurium* TA100, reverse mutation	–	–	300.0000	Bendele et al. (1985b)
SA0, *Salmonella typhimurium* TA100, reverse mutation	–	–	50.0000	Zeiger et al. (1988)
SA0, *Salmonella typhimurium* TA100, reverse mutation	–	–	50.0000	US National Toxicology Program (1989)
SA0, *Salmonella typhimurium* TA100, reverse mutation	–	+[c]	40.0000	Hennig et al. (1991)
SA2, *Salmonella typhimurium* TA102, reverse mutation	–	–	500.0000	Würgler et al. (1991)
SA5, *Salmonella typhimurium* TA1535, reverse mutation	–	–	50.0000	Kuczuk et al. (1978)
SA5, *Salmonella typhimurium* TA1535, reverse mutation	–	–	150.0000	Wehner et al. (1978)
SA5, *Salmonella typhimurium* TA1535, reverse mutation	–	–	300.0000	Bendele et al. (1985b)
SA5, *Salmonella typhimurium* TA1535, reverse mutation	–	–	50.0000	Zeiger et al. (1988)
SA5, *Salmonella typhimurium* TA1535, reverse mutation	–	–	50.0000	US National Toxicology Program (1989)
SA5, *Salmonella typhimurium* TA1535, reverse mutation	–	+[c]	40.0000	Hennig et al. (1991)
SA7, *Salmonella typhimurium* TA1537, reverse mutation	–	–	150.0000	Wehner et al. (1978)
SA7, *Salmonella typhimurium* TA1537, reverse mutation	–	–	300.0000	Bendele et al. (1985b)
SA7, *Salmonella typhimurium* TA1537, reverse mutation	–	–[c]	40.0000	Hennig et al. (1991)
SA8, *Salmonella typhimurium* TA1538, reverse mutation	–	–	5.0000	von Engel & von Milczewski (1976)
SA8, *Salmonella typhimurium* TA1538, reverse mutation	–	–	50.0000	Kuczuk et al. (1978)

Table 5 (contd)

Test system	Result		Dose (LED/HID)[a]	Reference
	Without exogenous metabolic system	With exogenous metabolic system		
SA8, Salmonella typhimurium TA1538, reverse mutation	–	–	100.0000	Bartsch et al. (1980)
SA8, Salmonella typhimurium TA1538, reverse mutation	–	–	300.0000	Bendele et al. (1985b)
SA8, Salmonella typhimurium TA1538, reverse mutation	–	+[c]	40.0000	Hennig et al. (1991)
SA9, Salmonella typhimurium TA98, reverse mutation	–	–	5.0000	von Engel & von Milczewski (1976)
SA9, Salmonella typhimurium TA98, reverse mutation	–	–	0.0000	Nagao et al. (1976)
SA9, Salmonella typhimurium TA98, reverse mutation	–	–	150.0000	Wehner et al. (1978)
SA9, Salmonella typhimurium TA98, reverse mutation	–	–	300.0000	Bendele et al. (1985b)
SA9, Salmonella typhimurium TA98, reverse mutation	–	–	50.0000	Zeiger et al. (1988)
SA9, Salmonella typhimurium TA98, reverse mutation	–	–	50.0000	US National Toxicology Program (1989)
SA9, Salmonella typhimurium TA98, reverse mutation	–	–[c]	40.0000	Hennig et al. (1991)
SAS, Salmonella typhimurium G46, reverse mutation	–	–	500.0000	Bendele et al. (1985b)
SAS, Salmonella typhimurium C3076, reverse mutation	–	–	500.0000	Bendele et al. (1985b)
SAS, Salmonella typhimurium D3052, reverse mutation	–	–	500.0000	Bendele et al. (1985b)
SAS, Salmonella typhimurium TA97, reverse mutation	–	–	50.0000	Zeiger et al. (1988)
SAS, Salmonella typhimurium TA97, reverse mutation	–	–	50.0000	US National Toxicology Program (1989)
ECW, Escherichia coli WP2 uvrA⁻, reverse mutation	–	–	1000 μg/plate	Bendele et al. (1985b)
EC2, Escherichia coli WP2, reverse mutation	–	–	1000 μg/plate	Bendele et al. (1985b)
SCH, Saccharomyces cerevisiae D-3, mitotic crossing over, ade2 locus	–	–	100.0000	Kuczuk et al. (1978)
DIA, DNA single-strand breaks, BALB/c mouse spleen cells in vitro	+	0	10.0000	Creppy et al. (1985)
DIA, DNA single-strand breaks, Chinese hamster ovary cells in vitro	+	0	200.0000	Štětina & Votava (1986)
DIA, DNA single-strand breaks, AWRF rat fibroblasts in vitro	+	0	200.0000	Štětina & Votava (1986)
URP, Unscheduled DNA synthesis, ACI rat primary hepatocytes in vitro	+	0	0.4000	Mori et al. (1984)
URP, Unscheduled DNA synthesis, rat primary hepatocytes in vitro	–	0	0.0250	Bendele et al. (1985b)
UIA, Unscheduled DNA synthesis, C3H/HeN mouse primary hepatocytes in vitro	+	0	4.0000	Mori et al. (1984)
G5T, Gene mutation, mouse lymphoma L5178Y cells, tk locus in vitro	–	–	12.5000	Bendele et al. (1985b)

Table 5 (contd)

Test system	Result		Dose (LED/HID)[a]	Reference
	Without exogenous metabolic system	With exogenous metabolic system		
GIA, Gene mutation, C3H mammary carcinoma cells, 8-aza[r] in vitro	–	0	10.0000	Umeda et al. (1977)
SIC, Sister chromatid exchange, Chinese hamster ovary cells in vitro	–	(+)	16.0000	US National Toxicology Program (1989)
SHL, Sister chromatid exchange, human lymphocytes in vitro	–	0	10.0000	Cooray (1984)
SHL, Sister chromatide exchange, human lymphocytes in vitro	0	+[c]	0.0400	Hennig et al. (1991)
BVD, Binding (covalent) to DNA, Swiss mouse kidney cells[d] in vivo	+		0.6 × 1 mg/kg p.o.	Pfohl-Leszkowicz et al. (1991)
BVD, Binding (covalent) to DNA, Swiss mouse liver cells[d] in vivo	+		0.6 × 1 mg/kg p.o	Pfohl-Leszkowicz et al. (1991)
BVD, Binding (covalent) to DNA, Swiss mouse spleen cells[d] in vivo	+		0.6 × 1 mg/kg p.o.	Pfohl-Leszkowicz et al. (1991)
DVA, DNA single-strand breaks, BALB/c mouse kidney cells in vivo	+		2.5 × 1 i.p.	Creppy et al. (1985)
DVA, DNA single-strand breaks, BALB/c mouse liver cells in vivo	+		2.5 × 1 i.p.	Creppy et al. (1985)
DVA, DNA single-strand breaks, BALB/c mouse spleen cells in vivo	+		2.5 × 1 i.p.	Creppy et al. (1985)
DVA, DNA single-strand breaks, Wistar rat kidney cells in vivo	+		0.2 × 1 p.o.	Kane et al. (1986a)
DVA, DNA single-strand breaks, Wistar rat liver cells in vivo	+		0.2 × 1 p.o.	Kane et al. (1986a)
SVA, Sister chromatid exchange, Chinese hamster bone marrow in vivo	–		400 × 1 p.o.	Bendele et al. (1985b)
Ochratoxin α				
PRB, SOS chromotest, Escherichia coli PQ37	–	–	400.0000	Malaveille et al. (1991)

+, positive; (+), weak positive; –, negative; 0, not tested
[a]In-vitro tests, μg/ml; in-vivo tests, mg/kg bw; 0.0000, dose not given
[b]Rat liver or kidney S9 mix decreased genotoxicity
[c]Conditioned medium of rat hepatocyte cultures treated with ochratoxin A
[d]32P-Post-labelling

adducts in mouse kidney and to a lesser extent in liver and spleen. The incidence of sister chromatid exchange was not increased in the bone marrow of Chinese hamsters treated *in vivo*.

In a single study, ochratoxin α did not damage DNA.

5. Summary of Data Reported and Evaluation

5.1 Exposure data

Ochratoxin A is produced by *Aspergillus ochraceus* and *Penicillium verrucosum*. Human exposure occurs mainly through consumption of contaminated grain and pork products, as confirmed by detection of ochratoxin A in human blood and milk.

5.2 Human carcinogenicity data

A number of descriptive studies have suggested a correlation between exposure to ochratoxin A and Balkan endemic nephropathy and have found a correlation between the geographical distribution of Balkan endemic nephropathy and a high incidence of and mortality from urothelial urinary tract tumours. In the only study in which ochratoxin A was measured, levels were higher in the blood of patients with Balkan endemic nephropathy and/or urothelial urinary tract tumours than in unaffected people; no distinction was made between the two diseases.

5.3 Animal carcinogenicity data

Ochratoxin A was tested for carcinogenicity by oral administration in mice and rats. It increased the incidence of hepatocellular tumours in mice of each sex and produced renal-cell adenomas and carcinomas in male mice and in rats of each sex.

5.4 Other relevant data

Ochratoxin A caused renal toxicity, nephropathy and immunosuppression in several animal species.

No adequate data were available on the genetic and related effects of ochratoxin A in humans. It induces DNA damage in rodents *in vivo* and in rodent cells *in vitro*.

5.5 Evaluation[1]

There is *inadequate evidence* in humans for the carcinogenicity of ochratoxin A.

There is *sufficient evidence* in experimental animals for the carcinogenicity of ochratoxin A.

---

[1]For definition of the italicized terms, see Preamble, pp. 26–29.

Overall evaluation

Ochratoxin A *is possibly carcinogenic to humans (Group 2B).*

6. References

Abdelhamid, A.M. (1990) Occurrence of some mycotoxins (aflatoxin, ochratoxin A, citrinin, zearalenone and vomitoxin) in various Egyptian feeds. *Arch. anim. Nutr.*, **40**, 647–664

Åkerstrand, A.K. & Josefsson, E. (1979) Fungi and mycotoxins in beans and peas (Swed.). *Vår Föda*, **31**, 405–414

Arora, R.G. & Frölén, H. (1981) Interference of mycotoxins with prenatal development of the mouse. II. Ochratoxin A induced teratogenic effects in relation to the dose and stage of gestation. *Acta vet. scand.*, **22**, 535–552

Arora, R.G., Frölén, H. & Nilsson, A. (1981) Interference of mycotoxins with prenatal development of the mouse. I. Influence of aflatoxin B_1, ochratoxin A and zearalenone. *Acta vet. scand.*, **22**, 524–534

Arora, R.G., Frölén, H. & Fellner-Feldegg, H. (1983) Inhibition of ochratoxin A teratogenesis by zearalenone and diethylstilboestrol. *Food chem. Toxicol.*, **21**, 779–783

Auffray, Y. & Boutibonnes, P. (1986) Evaluation of the genotoxic activity of some mycotoxins using *Escherichia coli* in the SOS spot test. *Mutat. Res.*, **171**, 79–82

Bacha, H., Hadidane, R., Creppy, E.E., Regnault, C., Ellouze, F. & Dirheimer, G. (1988) Monitoring and identification of fungal toxins in food products, animal feed and cereals in Tunisia. *J. Storage Prod. Res.*, **24**, 199–206

Bartsch, H., Malaveille, C., Camus, A.-M., Martel-Planche, G., Brun, G., Hautefeuille, A., Sabadie, N., Barbin, A., Kuroki, T., Drevon, C., Piccoli, C. & Montesano, R. (1980) Validation and comparative studies on 180 chemicals with *S. typhimurium* strains and V79 Chinese hamster cells in the presence of various metabolizing systems. *Mutat. Res.*, **76**, 1–50

Bauer, J. & Gareis, M. (1987) Ochratoxin A in the food chain (Ger.). *J. Vet. Med.*, **B34**, 613–627

Baumann, U. & Zimmerli, B. (1988) Simple determination of ochratoxin A in foods (Ger.). *Mitt. Geb. Lebensmittel. Hyg.*, **79**, 151–158

Bendele, A.M., Carlton, W.W., Krogh, P. & Lillehoj, E.B. (1985a) Ochratoxin A carcinogenesis in the (C57Bl/6J × C3H)F_1 mouse. *J. natl Cancer Inst.*, **75**, 733–742

Bendele, A.M., Neal, S.B., Oberly, T.J., Thompson, C.Z., Bewsey, B.J., Hill, L.E., Rexroat, M.A., Carlton, W.W. & Probst, G.S. (1985b) Evaluation of ochratoxin A for mutagenicity in a battery of bacterial and mammalian cell assays. *Food chem. Toxicol.*, **23**, 911–918

Böhm, J. & Leibetseder, J. (1987) Ochratoxin A—occurrence in kidneys of Austrian slaughter pigs (Ger.). *Z. Tierphysiol. Tierernähr.*, **58**, 41–42

Breitholz, A., Olsen, M., Dahlbäck, Å. & Hult, K. (1991) Plasma ochratoxin A levels in three Swedish populations surveyed using an ion-pair HPLC technique. *Food Addit. Contam.*, **8**, 183–192

Breitholz-Emanuelsson, A., Fuchs, R., Hult, K. & Appelgren, L.E. (1992) Syntheses of ^{14}C-ochratoxin A and ^{14}C-ochratoxin B and a comparative study of their distribution in rats using whole body autoradiography. *Pharmacol. Toxicol.*, **70**, 255–261

Brown, M.H., Szczech, G.M. & Purmalis, B.P. (1976) Teratogenic and toxic effects of ochratoxin A in rats. *Toxicol. appl. Pharmacol.*, **37**, 331–338

Buckle, A.E. (1983) The occurrence of mycotoxins in cereals and animal feedstuffs. *Vet. Res. Commun.*, **7**, 171–186

Budavari, S., ed. (1989) *The Merck Index*, 11th ed., Rahway, NJ, Merck & Co., p. 1068

Bunge, I., Dirheimer, G. & Röschenthaler, R. (1978) In vivo and in vitro inhibition of protein synthesis in *Bacillus stearothermophilus* by ochratoxin A. *Biochem. biophys. Res. Commun.*, 83, 398–405

Campbell, M.L., Jr, May, J.D., Huff, W.E. & Doerr, J.A. (1983) Evaluation of immunity of young broiler chickens during simultaneous aflatoxicosis and ochratoxicosis. *Poult. Sci.*, 62, 2138–2144

Castegnaro, M., Bartsch, H., Bereziat, J.C., Arvela, P., Michelon, J. & Broussolle, L. (1989) Polymorphic ochratoxin A hydroxylation in rat strains phenotyped as poor and extensive metabolizers of debrisoquine. *Xenobiotica*, 19, 225–230

Castegnaro, M., Barek, J., Frémy, J.-M., Lafontaine, M., Miraglia, M., Sansone, E.B. & Telling, G.M., eds (1991a) *Laboratory Decontamination and Destruction of Carcinogens in Laboratory Wastes: Some Mycotoxins* (IARC Scientific Publications No. 113), Lyon, IARC, pp. 9–16

Castegnaro, M., Pleština, R., Dirheimer, G., Chernozemsky, I.N. & Bartsch, H., eds (1991b) *Mycotoxins, Endemic Nephropathy and Urinary Tract Tumours* (IARC Scientific Publications No. 115), Lyon, IARC

Čeović, S., Grims, P. & Mitar, J. (1976) The incidence of tumours of the urinary organs in the region of endemic nephropathy and in control region (Yugosl.). *Lijec. Vjesn.*, 98, 301–304

Chakor, K., Creppy, E.E. & Dirheimer, G. (1988) In vivo studies on the relationship between hepatic metabolism and toxicity of ochratoxin A. *Arch. Toxicol.*, 12, 201–204

Chang, F.C. & Chu, F.S. (1977) The fate of ochratoxin A in rats. *Food Cosmet. Toxicol.*, 15, 199–204

Chernozemsky, I.N., Stoyanov, I.S., Petkova-Bocharova, T.K., Nicolov, I.G, Draganov, I.V., Stoichev, I.I., Tanchev, Y., Naidenov, D. & Kalcheva, N.D. (1977) Geographic correlation between the occurrence of endemic nephropathy and urinary tract tumours in Vratza district, Bulgaria. *Int. J. Cancer*, 19, 1–11

Chu, F.S. (1971) Interaction of ochratoxin A with bovine serum albumin. *Arch. Biochem. Biophys.*, 147, 359–366

Ciegler, A. (1972) Bioproduction of ochratoxin A and penicillic acid by members of the *Aspergillus ochraceus* group. *Can. J. Microbiol.*, 18, 631–636

Connole, M.D., Blaney, B.J. & McEwan, T. (1981) Mycotoxins in animal feeds and toxic fungi in Queensland, 1971-80. *Austr. vet. J.*, 57, 314–318

Cooray, R. (1984) Effects of some mycotoxins on mitogen-induced blastogenesis and SCE frequency in human lymphocytes. *Food chem. Toxicol.*, 22, 529–534

Creppy, E.E., Lugnier, A.A.J., Fasiolo, F., Heller, K., Röschenthaler, R. & Dirheimer, G. (1979a) In vitro inhibition of yeast phenylalanyl-tRNA synthetase by ochratoxin A. *Chem.-biol. Interactions*, 24, 257–261

Creppy, E.E., Lugnier, A.A.J., Beck, G., Röschenthaler, R. & Dirheimer, G. (1979b) Action of ochratoxin A on cultured hepatoma cells—reversion of inhibition by phenylalanine. *FEBS Lett.*, 104, 287–290

Creppy, E.E., Schlegel, M., Röschenthaler, R. & Dirheimer, G. (1980) Phenylalanine prevents acute poisoning by ochratoxin A in mice. *Toxicol. Lett.*, 6, 77–80

Creppy, E.E., Størmer, F.C., Röschenthaler, R. & Dirheimer, G. (1983a) Effects of two metabolites of ochratoxin A, (4R)-4-hydroxyochratoxin A and ochratoxin α, on immune response in mice. *Infect. Immunol.*, 39, 1015–1018

Creppy, E.E., Størmer, F.C., Kern, D., Röschenthaler, R. & Dirheimer, G. (1983b) Effect of ochratoxin A metabolites on yeast phenylalanyl-tRNA synthetase and on the growth and in vivo protein synthesis of hepatoma cells. *Chem.-biol. Interactions*, 47, 239–247

Creppy, E.E., Kern, D., Steyn, P.S., Vleggaar, R., Röschenthaler, R. & Dirheimer, G. (1983c) Comparative study of the effect of ochratoxin A analogues on yeast aminoacyl-tRNA synthetases and on the growth and protein synthesis of hepatoma cells. *Toxicol. Lett.*, **19**, 217–224

Creppy, E.E., Röschenthaler, R. & Dirheimer, G. (1984) Inhibition of protein synthesis in mice by ochratoxin A and its prevention by phenylalanine. *Food chem. Toxicol.*, **22**, 883–886

Creppy, E.E., Kane, A., Dirheimer, G., Lafarge-Frayssinet, C., Mousset, S. & Frayssinet, C. (1985) Genotoxicity of ochratoxin A in mice: DNA single-strand break evaluation in spleen, liver and kidney. *Toxicol. Lett.*, **28**, 29–35

Creppy, E.E., Kane, A., Giessen-Crouse, E., Roth, A., Röschenthaler, R. & Dirheimer, G. (1986) Effect of ochratoxin A on enzyme activities and macromolecules synthesis in MDCK cells. *Arch. Toxicol.*, **Suppl. 9**, 310–314

Creppy, E.E., Chakor, K., Fisher, M.J. & Dirheimer, G. (1990) The mycotoxin ochratoxin A is a substrate for phenylalanine hydroxylase in isolated rat hepatocytes and *in vivo*. *Arch. Toxicol.*, **64**, 279–284

Creppy, E.E., Betbeder, A.M., Gharbi, A., Counord, J., Castegnaro, M., Bartsch, H., Moncharmont, P., Fouillet, B., Chambon, P. & Dirheimer, G. (1991) Human ochratoxicosis in France. In: Castegnaro, M., Pleština, R., Dirheimer, G., Chernozemsky, I.N. & Bartsch, H., eds, *Mycotoxins, Endemic Nephropathy and Urinary Tract Tumours* (IARC Scientific Publications No. 115), Lyon, IARC, pp. 145–151

Delacruz, L. & Bach, P.H. (1990) The role of ochratoxin A metabolism and biochemistry in animal and human nephrotoxicity. *J. biopharmacol. Sci.*, **1**, 277–304

Dirheimer, G. & Creppy, E.E. (1991) Mechanism of action of ochratoxin A. In: Castegnaro, M., Pleština, R., Dirheimer, G., Chernozemsky, I.N. & Bartsch, H., eds, *Mycotoxins, Endemic Nephropathy and Urinary Tract Tumours* (IARC Scientific Publications No. 115), Lyon, IARC, pp. 171–186

Dwivedi, P. & Burns, R.B. (1984) Effect of ochratoxin A on immunoglobulins in broiler chicks. *Res. vet. Sci.*, **36**, 117–121

van Egmond, H.P. (1991a) Methods for determining ochratoxin A and other nephrotoxic mycotoxins. In: Castegnaro, M., Pleština, R., Dirheimer, G., Chernozemsky, I.N. & Bartsch, H., eds, *Mycotoxins, Endemic Nephropathy and Urinary Tract Tumours* (IARC Scientific Publications No. 115), Lyon, IARC, pp. 57–70

van Egmond, H.P. (1991b) Worldwide regulations for ochratoxin A. In: Castegnaro, M., Pleština, R., Dirheimer, G., Chernozemsky, I.N. & Bartsch, H., eds, *Mycotoxins, Endemic Nephropathy and Urinary Tract Tumours* (IARC Scientific Publications No. 115), Lyon, IARC, pp. 331–336

Elling, F. (1979) Ochratoxin A-induced mycotoxic porcine nephropathy: alterations in enzyme activity in tubular cells. *Acta pathol. microbiol. scand. Sect. A*, **87**, 237–243

Elling, F. (1983) Feeding experiments with ochratoxin A-contaminated barley to bacon pigs. IV. Renal lesions. *Acta agric. scand.*, **33**, 153–159

Elling, F., Hald, B., Jacobsen, C. & Krogh, P. (1975) Spontaneous toxic nephropathy in poultry associated with ochratoxin A. *Acta pathol. microbiol. scand. Sect. A*, **83**, 739–741

Elling, F., Nielsen, J.P., Lillehøj, E.B., Thomassen, M.S. & Størmer, F.C. (1985) Ochratoxin A-induced porcine nephropathy: enzyme and ultrastructure changes after short-term exposure. *Toxicon*, **23**, 247–254

von Engel, G. & von Milczewski, K.E. (1976) Detection of mycotoxins activated with a set of *Salmonella typhimurium* histidine autotropes (Ger.). *Kieler Milchwirtsch. Forschungsber.*, **28**, 359–365

Frank, H.K. (1991) Food contamination by ochratoxin A in Germany. In: Castegnaro, M., Pleština, R., Dirheimer, G., Chernozemsky, I.N. & Bartsch, H., eds, *Mycotoxins, Endemic Nephropathy and Urinary Tract Tumours* (IARC Scientific Publications No. 115), Lyon, IARC, pp. 77–81

Frank, H.K., Dirheimer, G., Grunow, W., Netter, K.J., Osswald, H. & Schlatter, J. (1990) *Ochratoxin A. Vorkommen und Toxikologische Bewertung* [Ochratoxin A. Occurrence and toxicological evaluation], Weinheim, Verlagsgesellschaft mbH

Frisvad, J.C. & Filtenborg, O. (1989) Terverticillate penicillia: chemotaxonomy and mycotoxin production. *Mycologia*, **81**, 837–861

Frohlich, A.A., Marquardt, R.R. & Ominski, K.H. (1991) Ochratoxin A as a contaminant in the human food chain: a Canadian perspective. In: Castegnaro, M., Pleština, R., Dirheimer, G., Chernozemsky, I.N. & Bartsch, H., eds, *Mycotoxins, Endemic Nephropathy and Urinary Tract Tumours* (IARC Scientific Publications No. 115), Lyon, IARC, pp. 139–143

Fuchs, R., Radić, B., Peraica, M., Hult, K. & Pleština, R. (1988a) Enterohepatic circulation of ochratoxin A in rats. *Period. Biol.*, **90**, 39–42

Fuchs, R., Appelgren, L.-E. & Hult, K. (1988b) Distribution of ^{14}C-ochratoxin A in the mouse monitored by whole-body autoradiography. *Pharmacol. Toxicol.*, **63**, 355–360

Fuchs, R., Radić, B., Čeović, S., Šoštarić, B. & Hult, K. (1991) Human exposure to ochratoxin A. In: Castegnaro, M., Pleština, R., Dirheimer, G., Chernozemsky, I.N. & Bartsch, H., eds, *Mycotoxins, Endemic Nephropathy and Urinary Tract Tumours* (IARC Scientific Publications No. 115), Lyon, IARC, pp. 131–134

Fukui, Y., Hoshino, K., Kameyama, Y., Yasui, T., Toda, C. & Nagano, H. (1987) Placental transfer of ochratoxin A and its cytotoxic effect on the mouse embryonic brain. *Food chem. Toxicol.*, **25**, 17–24

Galtier, P. (1974) Fate of ochratoxin A in the animal organism. I. Transport of the toxin in the blood of the rat (Fr.). *Ann. Rech. vét.*, **5**, 311–318

Galtier, P. (1978) Contribution of pharmacokinetic studies to mycotoxicology—ochratoxin A. *Vet. Sci. Commun.*, **1**, 349-358

Galtier, P. (1991) Pharmacokinetics of ochratoxin A in animals. In: Castegnaro, M., Pleština, R., Dirheimer, G., Chernozemsky, I.N. & Bartsch, H., eds, *Mycotoxins, Endemic Nephropathy and Urinary Tract Tumours* (IARC Scientific Publications No. 115), Lyon, IARC, pp. 187–200

Galtier, P. & Alvinerie, M. (1976) In vitro transformation of ochratoxin A by animal microbial floras. *Ann. Rech. veter.*, **7**, 91–98

Galtier, P., Jemmali, M. & Larrieu, G. (1977) Survey of the possible occurrence of aflatoxin and ochratoxin A in maize harvested in France in 1973 and 1974 (Fr.). *Ann. Nutr. Aliment.*, **31**, 381–389

Galtier, P., Charpenteau, J.-L., Alvinerie, M. & Labouche, C. (1979a) The pharmacokinetic profile of ochratoxin A in the rat after oral and intravenous administration. *Drug Metab. Disposition*, **7**, 429–434

Galtier, P., Boneu, B., Charpenteau, J.-L., Bodin, G., Alvinerie, M. & Moré, J. (1979b) Physiopathology of haemorrhagic syndrome related to ochratoxin A intoxication in rats. *Food Cosmet. Toxicol.*, **17**, 49–53

Gareis, M., Maertlbauer, E., Bauer, J. & Gedek, B. (1988) Determination of ochratoxin A in human milk. In: Aibara, K., Kumagai, S., Ohtsubo, K. & Yoshizawa, T., eds, *Proceedings of the 7th International IUPAC Symposium on Mycotoxins and Phycotoxins, Tokyo, 16–19 August 1988*, Tokyo, Japanese Association of Mycotoxicology, pp. 61–62

Gilbert, J. (1991) Accepted and collaboratively tested methods of sampling, detection and analysis of mycotoxins. In: Champ, B.R., Highley, E., Hocking, A.D. & Pitt, J.I., eds, *Fungi and Mycotoxins in Stored Products* (ACIAR Proceedings 36), Canberra, Australian Centre for International Agricultural Research, pp. 108–114

Goliński, P. & Grabarkiewicz-Szczęsna, J. (1985) The first Polish cases of the detection of ochratoxin A residues in human blood (Pol.). *Rocz. Panstw. Zakl. Hig.*, **36**, 378–381

Goliński, P., Grabarkiewicz-Szczęsna, J., Chelkowski, J., Hult, K. & Kostecki, M. (1991) Possible sources of ochratoxin A in human blood in Poland. In: Castegnaro, M., Pleština, R., Dirheimer, G., Chernozemsky, I.N. & Bartsch, H., eds, *Mycotoxins, Endemic Nephropathy and Urinary Tract Tumours* (IARC Scientific Publications No. 115), Lyon, IARC, pp. 153–157

Gupta, M., Sasmal, D., Bandyopadhyay, S., Bagchi, G., Chatterjee, T. & Dey, S. (1983) Hematological changes produced in mice by ochratoxin A and citrinin. *Toxicology*, **26**, 55–62

Hagelberg, S., Hult, K. & Fuchs, R. (1989) Toxicokinetics of ochratoxin A in several species and its plasma-binding properties. *J. appl. Toxicol.*, **9**, 91–96

Hald, B. (1991) Ochratoxin A in human blood in European countries. In: Castegnaro, M., Pleština, R., Dirheimer, G., Chernozemsky, I.N. & Bartsch, H., eds, *Mycotoxins, Endemic Nephropathy and Urinary Tract Tumours* (IARC Scientific Publications No. 115), Lyon, IARC, pp. 159–164

Hansen, C.E., Dueland, S., Drevon, C.A. & Størmer, F.C. (1982) Metabolism of ochratoxin A by primary cultures of rat hepatocytes. *Appl. environ. Microbiol.*, **43**, 1267–1271

Haubeck, H.-D., Lorkowski, G., Kölsch, E. & Röschenthaler, R. (1981) Immunosuppression by ochratoxin A and its prevention by phenylalanine. *Appl. environ. Microbiol.*, **41**, 1040–1042

Hayes, A.W., Hood, R.D. & Lee, H.L. (1974) Teratogenic effects of ochratoxin A in mice. *Teratology*, **9**, 93–98

Hennig, A., Fink-Gremmels, J. & Leistner, L. (1991) Mutagenicity and effects of ochratoxin A on the frequency of sister chromatid exchange after metabolic activation. In: Castegnaro, M., Pleština, R., Dirheimer, G., Chernozemsky, I.N. & Bartsch, H., eds, *Mycotoxins, Endemic Nephropathy and Urinary Tract Tumours* (IARC Scientific Publications No. 115), Lyon, IARC, pp. 255–260

Hietanen, E., Malaveille, C., Camus, A.-M., Béréziat, J.-C., Brun, G., Castegnaro, M., Michelon, J., Idle, J.R. & Bartsch, H. (1986) Interstrain comparison of hepatic and renal microsomal carcinogen metabolism and liver S9-mediated mutagenicity in DA and Lewis rats phenotyped as poor and extensive metabolizers of debrisoquine. *Drug Metab. Disposition*, **14**, 118–126

Hood, R.D., Naughton, M.J. & Hayes, A.W. (1976) Prenatal effects of ochratoxin A in hamsters. *Teratology*, **13**, 11–14

Hood, R.D., Kuczuk, M.H. & Szczech, G.M. (1978) Effects in mice of simultaneous prenatal exposure to ochratoxin A and T-2 toxin. *Teratology*, **17**, 25–30

Hoshino, K., Fukui, Y., Hayasaka, I. & Kameyama, Y. (1988) Developmental disturbance of the cerebral cortex of mouse offspring from dams treated with ochratoxin A during pregnancy. *Congenital Anomalies*, **28**, 287–294

Hult, K. & Fuchs, R. (1986) Analysis and dynamics of ochratoxin A in biological systems. In: Steyn, P.S. & Vleggaar, R., eds, *Mycotoxins and Phycotoxins*, Amsterdam, Elsevier, pp. 365–376

Hult, K., Teiling, A. & Gatenbeck, S. (1976) Degradation of ochratoxin A by a ruminant. *Appl. environ. Microbiol.*, **32**, 443–444

Hult, K., Pleština, R., Habazin-Kovak, V., Radić, B. & Čeović, S. (1982) Ochratoxin A in human blood and Balkan endemic nephropathy. *Arch. Toxicol.*, **51**, 313–321

Hult, K., Rutqvist, L., Holmberg, T., Thafvelin, B. & Gatenbeck, S. (1984) Ochratoxin A in blood of slaughter pigs. *Nord vet. Med.*, **36**, 314–316

IARC (1976) *IARC Monographs on the Evaluation of Carcinogenic Risk of Chemicals to Man*, Vol. 10, *Some Naturally Occurring Substances*, Lyon, pp. 191-197

IARC (1983) *IARC Monographs on the Evaluation of the Carcinogenic Risk of Chemicals to Humans*, Vol. 31, *Some Food Additives, Feed Additives and Naturally Occurring Substances*, Lyon, pp. 191-206

IARC (1987) *IARC Monographs on the Evaluation of Carcinogenic Risks to Humans*, Suppl. 7, *Overall Evaluations of Carcinogenicity: An Updating of* IARC Monographs *Volumes 1 to 42*, Lyon, pp. 271-272

Imaida, K., Hirose, M., Ogiso, T., Kurata, Y. & Ito, N. (1982) Quantitative analysis of initiating and promoting activities of five mycotoxins in liver carcinogenesis in rats. *Cancer Lett.*, **16**, 137-143

Jung, K.Y. & Endou, H. (1989) Nephrotoxicity assessment by measuring cellular ATP content. II. Intranephron site of ochratoxin A nephrotoxicity. *Toxicol. appl. Pharmacol.*, **100**, 383-390

Kane, A., Creppy, E.E., Roth, A., Röschenthaler, R. & Dirheimer, G. (1986a) Distribution of the [^{3}H]-label from low doses of radioactive ochratoxin A ingested by rats, and evidence for DNA single-strand breaks caused in liver and kidneys. *Arch. Toxicol.*, **58**, 219-224

Kane, A., Creppy, E.E., Röschenthaler, R. & Dirheimer, G. (1986b) Changes in urinary and renal tubular enzymes caused by subchronic administration of ochratoxin A in rats. *Toxicology*, **42**, 233-243

Kane, A., Diop, N. & Diack, T.S. (1991) Natural occurrence of ochratoxin A in food and feed in Senegal. In: Castegnaro, M., Pleština, R., Dirheimer, G., Chernozemsky, I.N. & Bartsch, H., eds, *Mycotoxins, Endemic Nephropathy and Urinary Tract Tumours* (IARC Scientific Publications No. 115), Lyon, IARC, pp. 93-96

Kanisawa, M. (1984) Synergistic effect of citrinin on hepatorenal carcinogenesis of ochratoxin A in mice. In: Kurata, H. & Ueno, Y., eds, *Toxigenic Fungi, Their Toxins and Health Hazard*, Tokyo, Japanese Association of Mycotoxicology, pp. 245-254

Kanisawa, M. & Suzuki, S. (1978) Induction of renal and hepatic tumors in mice by ochratoxin A, a mycotoxin. *Gann*, **69**, 599-600

Konrad, I. & Röschenthaler, R. (1977) Inhibition of phenylalanine tRNA synthetase from *Bacillus subtilis* by ochratoxin A. *FEBS Lett.*, **83**, 341-347

Krivobok, S., Olivier, P., Marzin, D.R., Seiglemurandi, F. & Steiman, R. (1987) Study of the genotoxic potential of 17 mycotoxins with the SOS chromostest. *Mutagenesis*, **2**, 433-439

Krogh, P. (1974) Mycotoxin porcine nephropathy: a possible model for Balkan endemic nephropathy. In: Puchlev, A., Dinev, I.V., Milev, B. & Doichinov, D., eds, *Endemic Nephropathy*, Sofia, Bulgarian Academy of Sciences, pp. 266-270

Krogh, P. (1978) Causal associations of mycotoxic nephropathy. *Acta pathol. microbiol. scand. Sect. A*, **Suppl. 269**

Krogh, P. & Elling, F. (1977) Mycotoxic nephropathy. *Vet. Sci. Commun.*, **1**, 51-63

Krogh, P. & Nesheim, S. (1983) Ochratoxin A. In: Egan, H., Stoloff, L., Castegnaro, M., Scott, P., O'Neill, I.K. & Bartsch, H., eds, *Environmental Carcinogens: Selected Methods of Analysis*, Vol. 5, *Some Mycotoxins* (IARC Scientific Publications No. 44), Lyon, IARC, pp. 247-253

Krogh, P., Hald, B., Englund, P., Rutqvist, L. & Swahn, O. (1974) Contamination of Swedish cereals with ochratoxin A. *Acta pathol. microbiol. Scand. Sect. B*, **82**, 301-302

Krogh, P., Gyrd-Hansen, N., Hald, B., Larsen, S., Nielsen, J.P., Smith, M., Ivanoff, C. & Meisner, H. (1988) Renal enzyme activities in experimental ochratoxin A-induced porcine nephropathy: diagnostic potential of phosphoenolpyruvate carboxykinase and γ-glutamyl transpeptidase activity. *J. Toxicol. environ. Health*, **23**, 1-14

Kuczuk, M.H., Benson, P.M., Heath, H. & Hayes, A.W. (1978) Evaluation of the mutagenic potential of mycotoxins using *Salmonella typhimurium* and *Saccharomyces cerevisiae*. *Mutat. Res.*, **53**, 11–20

Kuiper-Goodman, T. & Scott, P.M. (1989) Risk assessment of the mycotoxin ochratoxin A. *Biomed. environ. Sci.*, **2**, 179–248

Kumagai, S. & Aibara, K. (1982) Intestinal absorption and secretion of ochratoxin A in the rat. *Toxicol. appl. Pharmacol.*, **64**, 94–102

Kumari, C.K. & Nusrath, M. (1987) Natural occurrence of citrinin and ochratoxin A in coconut products. *Natl Acad. Sci. Lett. India*, **10**, 303–305

Madhyastha, S.M., Marquardt, R.R., Frohlich, A.A., Platford, G. & Abramson, D. (1990) Effect of different cereal and oilseed substrates on the growth and production of toxins by *Aspergillus alutaceus* and *Penicillium verrucosum*. *J. agric. Food Chem.*, **38**, 1506–1510

Malaveille, C., Brun, G. & Bartsch, H. (1991) Genotoxicity of ochratoxin A and structurally related compounds in *Escherichia coli* strains: studies on their mode of action. In: Castegnaro, M., Pleština, R., Dirheimer, G., Chernozemsky, I.N. & Bartsch, H., eds, *Mycotoxins, Endemic Nephropathy and Urinary Tract Tumours* (IARC Scientific Publications No. 115), Lyon, IARC, pp. 261–266

Mantle, P.G., McHugh, K.M., Adatia, R., Heaton, J.M., Gray, T. & Turner, D.R. (1991) *Penicillium aurantiogriseum*-induced, persistent renal histopathological changes in rats; an experimental model for Balkan endemic nephropathy competitive with ochratoxin A. In: Castegnaro, M., Pleština, R., Dirheimer, G., Chernozemsky, I.N. & Bartsch, H., eds, *Mycotoxins, Endemic Nephropathy and Urinary Tract Tumours* (IARC Scientific Publications No. 115), Lyon, IARC, pp. 119–127

Mayura, K., Reddy, R.V., Hayes, A.W. & Berndt, W.O. (1982) Embryocidal, fetotoxic and teratogenic effects of ochratoxin A in rats. *Toxicology*, **25**, 175–185

Mayura, K., Hayes, A.W. & Berndt, W.O. (1983) Effects of dietary protein on teratogenicity of ochratoxin A in rats. *Toxicology*, **27**, 147–157

Mayura, K., Parker, R., Berndt, W.O. & Phillips, T.D. (1984a) Effect of simultaneous prenatal exposure to ochratoxin A and citrinin in the rat. *J. Toxicol. environ. Health*, **13**, 553–561

Mayura, K., Stein, A.F., Berndt, W.O. & Phillips, T.D. (1984b) Teratogenic effects of ochratoxin A in rats with impaired renal function. *Toxicology*, **32**, 277–285

Mayura, K., Parker, R., Berndt, W.O. & Phillips, T.D. (1984c) Ochratoxin A-induced teratogenesis in rats: partial protection by phenylalanine. *Appl. environ. Microbiol.*, **48**, 1186–1188

Mayura, K., Edwards, J.F., Maull, E.A. & Phillips, T.D. (1989) The effects of ochratoxin A on postimplantation rat embryos in culture. *Arch. environ. Contam. Toxicol.*, **18**, 411–415

Meisner, H. & Chan, S. (1974) Ochratoxin A, an inhibitor of mitochondrial transport system. *Biochemistry*, **13**, 2795–2800

Meisner, H. & Krogh, P. (1986) Phosphoenolpyruvate carboxykinase as a selective indicator of ochratoxin A induced nephropathy. In: Tanabe, T., Hook, J.B. & Endou, N., eds, *Nephrotoxicity of Antibiotics and Immunosuppressants*, Amsterdam, Elsevier, pp. 199–206

Meisner, H. & Meisner, P. (1981) Ochratoxin A, an in vivo inhibitor of renal phosphoenolpyruvate carboxykinase. *Arch. Biochem. Biophys.*, **208**, 146–153

Meisner, H. & Selanik, P. (1979) Inhibition of renal gluconeogenesis in rats by ochratoxin. *Biochem. J.*, **180**, 681–684

van der Merwe, K.J., Steyn, P.S., Fourie, L., De Scott, B. & Theron, J.J. (1965a) Ochratoxin A, a toxic metabolite produced by *Aspergillus ochraceus* Wilh. *Nature*, **205**, 1112–1113 van der Merwe, K.J., Steyn, P.S. & Fourie, L. (1965b) Mycotoxins. Part II. The constitution of ochratoxins A, B and C metabolites of *Aspergillus ochraceus* Wilh. *J. chem. Soc.*, 7083–7088

Micco, C., Ambruzzi, M.A., Miraglia, M., Brera, C., Onori, R. & Benelli, L. (1991) Contamination of human milk with ochratoxin A. In: Castegnaro, M., Pleština, R., Dirheimer, G., Chernozemsky, I.N. & Bartsch, H., eds, *Mycotoxins, Endemic Nephropathy and Urinary Tract Tumours* (IARC Scientific Publications No. 115), Lyon, IARC, pp. 105–108

Moore, J.H. & Truelove, B. (1970) Ochratoxin A: inhibition of mitochondrial respiration. *Science*, **168**, 1102–1103

Moré, J. & Galtier, P. (1974) Toxicity of ochratoxin A. I.—Embryotoxic and teratogenic effects in the rat (Fr.). *Ann. Rech. vét.*, **5**, 167–178

Moré, J. & Galtier, P. (1975) Toxicity of ochratoxin A. II.—Effects of the treatment on the progeny (F_1 and F_2) of intoxicated rats (Fr.). *Ann. Rech. vét.*, **6**, 379–389

Moré, J., Galtier, P. & Alvinerie, M. (1978) Toxicity of ochratoxin A. 3. Effects during early gestation in the rat (Fr.). *Ann. Rech. vét.*, **9**, 169–173

Mori, H., Kawai, K., Ohbayashi, F., Kuniyasu, T., Yamazaki, M., Hamasaki, T. & Williams, G.M. (1984) Genotoxicity of a variety of mycotoxins in the hepatocyte primary culture/DNA repair test using rat and mouse hepatocytes. *Cancer Res.*, **44**, 2918–2923

Moroi, K., Suzuki, S., Kuga, T., Yamazaki, M. & Kanisawa, M. (1985) Reduction of ochratoxin A toxicity in mice treated with phenylalanine and phenobarbital. *Toxicol. Lett.*, **25**, 1–5

Müller, H.M. (1982) Decontamination of mycotoxins. I. Physical process (Ger.). *Übersicht. Tierernähr.*, **10**, 95–122

Nagao, M., Honda, M., Hamasaki, T., Natori, S., Ueno, Y., Yamasaki, M., Seino, Y., Yahagi, T. & Sugimura, T. (1976) Mutagenicity of mycotoxins on *Salmonella* (Jpn.). *Proc. Jpn. Assoc. Mycotoxicol.*, **3–4**, 41–43

Nesheim, S., Stack, M.E., Trucksess, M.W., Eppley, R.M. & Krogh, P. (1992) Rapid solvent-efficient method for liquid chromatographic determination of ochratoxin A in corn, barley and kidney: collaborative study. *J. Assoc. off. anal. Chem.*, **75**, 481–487

Nishijima, M. (1984) Survey for mycotoxins in commercial foods. *Dev. Food Sci.*, **7**, 172–181

Ohshika, S., Shirai, S. & Majima, A. (1988) Developmental abnormalities of the optic disc induced in mouse fetuses by ochratoxin A (Jpn.). *Acta soc. ophthalmol. jpn.*, **92**, 414–422

Olberg, I.H. & Yndestad, M. (1982) A Norwegian survey of ochratoxin A in cereals and animal tissues. *Nord. Jordbruksforsk.*, **64**, 296

Osborne, B.G. (1980) The occurrence of ochratoxin A in mouldy bread and flour. *Food Cosmet. Toxicol.*, **18**, 615–617

Oster, T., Jayyosi, Z., Creppy, E.E., El Amri, H.S. & Batt, A.-M. (1991) Characterization of pig liver purified cytochrome P-450 isoenzymes for ochratoxin A metabolism studies. *Toxicol. Lett.*, **57**, 203–214

Pavlović, M., Pleština, R. & Krogh, P. (1979) Ochratoxin A contamination of foodstuffs in an area with Balkan (endemic) nephropathy. *Acta pathol. microbiol. scand. B*, **87**, 243–246

Pepeljnjak, S. & Blažević, N. (1982) Contamination with moulds and occurrence of ochratoxin A in smoked meat products from endemic nephropathy region of Yugoslavia. In: Pfannhauser, W. & Czedik-Eysenberg, P.B., eds, *Proceedings of the Vth IUPAC Symposium on Mycotoxins and Phycotoxins, 1–3 September, Vienna, Austria*, Vienna, Austrian Chemical Society, pp. 102–105

Pepeljnjak, S., Blažević, N. & Čuljak, K. (1982) Histopathological changes and findings of ochratoxin A in organs of pigs, in the area of endemic nephropathy in Yugoslavia. In: Pfannhauser, W. & Czedik-Eysenberg, P.B., eds, *Proceedings of the Vth IUPAC Symposium on Mycotoxins and Phycotoxins, 1–3 September, Vienna, Austria*, Vienna, Austrian Chemical Society, pp. 102–105

Peterson, R.E. & Ciegler, A. (1978) Ochratoxin A: isolation and subsequent purification by high-pressure liquid chromatography. *Appl. environ. Microbiol.*, 36, 613–614

Petkova-Bocharova, T. & Castegnaro, M. (1985) Ochratoxin A contamination of cereals in an area of high incidence of Balkan endemic nephropathy in Bulgaria. *Food Addit. Contam.*, 2, 267–270

Petkova-Bocharova, T. & Castegnaro, M. (1991) Ochratoxin A in human blood in relation to Balkan endemic nephropathy and urinary tract tumours in Bulgaria. In: Castegnaro, M., Pleština, R., Dirheimer, G., Chernozemsky, I.N. & Bartsch, H., eds, *Mycotoxins, Endemic Nephropathy and Urinary Tract Tumours* (IARC Scientific Publications No. 115), Lyon, IARC, pp. 135–137

Petkova-Bocharova, T., Chernozemsky, I.N. & Castegnaro, M. (1988) Ochratoxin A in human blood in relation to Balkan endemic nephropathy and urinary system tumours in Bulgaria. *Food Addit. Contam.*, 5, 299–301

Petkova-Bocharova, T., Castegnaro, M., Michelon, J. & Maru, V. (1991) Ochratoxin A and other mycotoxins in cereals from an area of Balkan endemic nephropathy and urinary tract tumours in Bulgaria. In: Castegnaro, M., Plestina, R., Dirheimer, G., Chernozemsky, I.N. & Bartsch, H., eds, *Mycotoxins, Endemic Nephropathy and Urinary Tract Tumours* (IARC Scientific Publications No. 115), Lyon, IARC, pp. 83–87

Petković, S., Mutavdžić, M., Petronić, N. & Marković, M. (1974) Comparative incidence pattern of urothelium tumours in and beyond endemic nephropathy regions. In: Puchlev, A., Dinev, I.V., Milev, B. & Doichinov, D., eds, *Endemic Nephropathy*, Sofia, Bulgarian Academy of Sciences, pp. 114–116

Petrinska-Venkovska, S. (1960) Morphological studies on endemic nephritis (Bulg.). In: Puchlev, A., ed., *Endemic Nephritis in Bulgaria*, Sofia, Medizina i Fizkultura, pp. 72–90

Pfohl-Leszkowicz, A., Chakor, K., Creppy, E.E. & Dirheimer, G. (1991) DNA adduct formation in mice treated with ochratoxin A. In: Castegnaro, M., Pleština, R., Dirheimer, G., Chernozemsky, I.N. & Bartsch, H., eds, *Mycotoxins, Endemic Nephropathy and Urinary Tract Tumours* (IARC Scientific Publications No. 115), Lyon, IARC, pp. 245–253

Pleština, R., Čeović, S., Gatenbeck, S., Habazin-Novak, V., Hult, K., Hökby, E., Krogh, P. & Radić, B. (1990) Human exposure to ochratoxin A in areas of Yugoslavia with endemic nephropathy. *J. environ. Pathol. Toxicol. Oncol.*, 10, 145–148

Pohland, A.E., Schuller, P.L., Steyn, P.S. & van Egmond, H.P. (1982) Physicochemical data for some selected mycotoxins. *Pure appl. Chem.*, 54, 2219–2284

Poppe, S.M., Stuckhardt, J.L. & Szczech, G.M. (1983) Postnatal behavioral effects of ochratoxin A in offspring of treated mice. *Teratology*, 27, 293–300

Prior, M.G. (1976) Mycotoxin determinations on animal feedstuffs and tissues in western Canada. *Can. J. comp. Med.*, 40, 75–79

Prior, M.G. (1981) Mycotoxins in animal feedstuffs and tissues in western Canada 1975 to 1979. *Can. J. comp. Med.*, 45, 116–119

Radovanović, Z. (1991) Epidemiological characteristics of Balkan endemic nephropathy in eastern regions of Yugoslavia. In Castegnaro, M., Pleština, R., Dirheimer, G., Chernozemsky, I.N. & Bartsch, H., eds, *Mycotoxins, Endemic Nephropathy and Urinary Tract Tumours* (IARC Scientific Publications No. 115), Lyon, IARC, pp. 11–20

Radovanović, Z. & Krajinović, S. (1979) The Balkan endemic nephropathy and urinary tract tumors. *Arch. Geschwulstforsch.*, **49**, 444–447

Rahimtula, A.D., Béréziat, J.-C., Bussacchini-Griot, V. & Bartsch, H. (1988) Lipid peroxidation as a possible cause of ochratoxin A toxicity. *Biochem. Pharmacol.*, **37**, 4469–4477

Reddy, B.N., Nusrath, M., Krishna Kumari, C. & Nahdi, S. (1983) Mycotoxin contamination in some food commodities from tribal areas of Medak District, Andhra Pradesh. *Indian Phytopathol.*, **36**, 683–686

Reiss, J. (1986) Detection of genotoxic properties of mycotoxins with the SOS chromotest. *Naturwissenschaften*, **73**, 677–678

Ribelin, W.E., Fukushima, K. & Still, P.E. (1978) The toxicity of ochratoxin to ruminants. *Can. J. Comp. Med.*, **42**, 172–176

Roberts, J.C. & Woollven, P. (1970) Studies in mycological chemistry. Part XXIV. Synthesis of ochratoxin A, a metabolite of *Aspergillus ochraceus* Wilh. *J. chem. Soc. (C)*, 278–281

Roth, A., Chakor, K., Creppy, E.E., Kane, A., Roschenthaler, R. & Dirheimer, G. (1988) Evidence for an enterohepatic circulation of ochratoxin A in mice. *Toxicology*, **48**, 293–308

Samson, R.A. & Pitt, J.I., eds (1990) *Modern Concepts in* Penicillium *and* Aspergillus *Classification*, New York, Plenum Press

Sansing, G.A., Lillehoj, E.B., Detroy, R.W. & Miller, M.A. (1976) Synergistic toxic effects of citrinin, ochratoxin A and penicillic acid in mice. *Toxicon*, **14**, 2130–220

Scheuer, R. & Leistner, L. (1986) Occurrence of ochratoxin A in pork and pork products. In: *Proceedings of the 32nd European Meeting of Meat Research Workers, 24–29 August 1986, Ghent, Belgium*, Vol. I, Madrid, European Meat Research Workers, p. 191

Shirai, S., Ohshika, S., Yuguchi, S. & Majima, A. (1984) Developmental eye abnormalities in mouse fetuses induced by ochratoxin A (Jpn.). *Acta soc. ophthalmol. jpn.*, **88**, 627–634

Shotwell, O.L., Hesseltine, C.W., Vandegraft, E.E. & Goulden, M.L. (1971) Survey of corn from different regions for aflatoxin, ochratoxin and zearalenone. *Cereal Sci. Today*, **16**, 266–273

Shotwell, O.L., Goulden, M.L. & Hesseltine, C.W. (1976) Survey of US wheat for ochratoxin and aflatoxin. *J. Assoc. off. anal. Chem.*, **59**, 122–124

Shreeve, B.J., Patterson, D.S.P. & Roberts, B.A. (1979) The 'carry-over' of aflatoxin, ochratoxin and zearalenone from naturally contaminated feed to tissues, urine and milk of dairy cows. *Food Cosmet. Toxicol.*, **17**, 151–152

Singh, J. & Hood, R.D. (1985) Maternal protein deprivation enhances the teratogenicity of ochratoxin A in mice. *Teratology*, **32**, 381–388

Sinha, R.N., Abramson, D. & Mills, J.T. (1986) Interactions among ecological variables in stored cereals and associations with mycotoxin production in the climatic zones of western Canada. *J. Food Prot.*, **49**, 608–614

Šoštarić, B. & Vukelić, M. (1991) Characteristics of urinary tract tumours in the area of Balkan endemic nephropathy in Croatia. In: Castegnaro, M., Pleština, R., Dirheimer, G., Chernozemsky, I.N. & Bartsch, H., eds, *Mycotoxins, Endemic Nephropathy and Urinary Tract Tumours* (IARC Scientific Publications No. 115), Lyon, IARC, p. 29–35

Stein, A.F., Phillips, T.D., Kubena, L.F. & Harvey, R.B. (1985) Renal tubular secretion and reabsorption as factors in ochratoxicosis. Effects of probenecid on nephrotoxicity. *J. Toxicol. environ. Health*, **16**, 593–605

Štětina, R. & Votava, M. (1986) Induction of DNA single-strand breaks and DNA synthesis induction by patulin, ochratoxin A, citrinin, and aflatoxin B_1 in cell lines CHO and AWRF. *Folia biol.*, **32**, 128–144

Steyn, P.S. & Holzapfel, C.W. (1967) The synthesis of ochratoxins A and B, metabolites of *Aspergillus ochraceus* Wilh. *Tetrahedron*, **23**, 4449–4461

Still, P.E., Macklin, A.W., Ribelin, W.E. & Smalley, E.B. (1971) Relationship of ochratoxin A to foetal death in laboratory and domestic animals. *Nature*, **234**, 563–564

Stojanov, I.S., Stojchev, I.I., Nicolov, I.G., Draganov, I.V., Petkova-Bocharova, T.K. & Chernozemsky, I.N. (1977) Cancer mortality in a region with endemic nephropathy. *Neoplasma*, **24**, 625–632

Stojanov, I.S., Chernozemsky, I.N., Nicolov, I.G., Stoichev, I.I. & Petkova-Bocharova, T.K. (1978) Epidemiologic association between endemic nephropathy and urinary system tumors in an endemic region. *J. chronic Dis.*, **31**, 721–724

Stojković, R., Hult, K., Gamulin, S. & Pleština, R. (1984) High affinity binding of ochratoxin A to plasma constituents. *Biochem. Int.*, **9**, 33–38

Støren, O., Holm, M. & Størmer, F.C. (1982) Metabolism of ochratoxin A by rats. *Appl. environ. Microbiol.*, **44**, 785–789

Størmer, F.C., Hansen, C.E., Pedersen, J.I., Hvistendahl, G. & Aasen, A.J. (1981) Formation of (4R)- and (4S)-4-hydroxyochratoxin A from ochratoxin A by liver microsomes from various species. *Appl. environ. Microbiol.*, **42**, 1051–1056

Størmer, F.C., Støren, O., Hansen, C.E., Pedersen, J.I. & Aasen, A.J. (1983) Formation of (4R)- and (4S)-4-hydroxyochratoxin A and 10-hydroxyochratoxin A from ochratoxin A by rabbit liver microsomes. *Appl. environ. Microbiol.*, **45**, 1183–1187

Suzuki, S., Satoh, T. & Yamazaki, M. (1977) The pharmacokinetics of ochratoxin A in rats. *Jpn. J. Pharmacol.*, **27**, 735–744

Szczech, G.M. & Hood, R.D. (1981) Brain necrosis in mouse fetuses transplacentally exposed to the mycotoxin ochratoxin A. *Toxicol. appl. Pharmacol.*, **57**, 127–137

Tanchev, Y. & Dorossiev, D. (1991) The first clinical description of Balkan endemic nephropathy (1956) and its validity 35 years later. In: Castegnaro, M., Pleština, R., Dirheimer, G., Chernozemsky, I.N. & Bartsch, H., eds, *Mycotoxins, Endemic Nephropathy and Urinary Tract Tumours* (IARC Scientific Publications No. 115), Lyon, IARC, pp. 21–28

Tanchev, Y., Naidenov, D., Dimitrov, C. & Karlova, E. (1970) Neoplastic diseases and endemic nephritis (Bulg.). *Vŭtr. Bod.*, **9**, 21–29

Ueno, Y. & Kubota, K. (1976) DNA-Attacking ability of carcinogenic mycotoxins in recombination-deficient mutant cells of *Bacillus subtilis*. *Cancer Res.*, **36**, 445–451

Umeda, M., Tsutsui, T. & Saito, M. (1977) Mutagenicity and inducibility of DNA single-strand breaks and chromosome aberrations by various mycotoxins. *Gann*, **68**, 619–625

US National Toxicology Program (1989) *Toxicology and Carcinogenesis Studies of Ochratoxin A (CAS No. 303-47-9) in F344/N Rats (Gavage Studies)* (NTP TR 358; DHHS Publ. No. (NIH) 89-2813), Research Triangle Park, NC

Valente Soares, L.M. & Rodriguez-Amaya, D.B. (1989) Survey of aflatoxins, ochratoxin A, zearalenone, and sterigmatocystin in some Brazilian foods by using multi-toxin thin-layer chromatographic method. *J. Assoc. off. anal. Chem.*, **72**, 22–26

Vega, M., Saelzer, R., Rubeffel, R. & Sepulveda, M.J. (1988) Screening of mycotoxins in Chilean maize. A preliminary report. In: Aibara, K., Kumagai, S., Ohtsubo, K. & Yoshizawa, T., eds, *Proceedings of the 7th International IUPAC Symposium on Mycotoxins and Phycotoxins, Tokyo, 16–19 August 1988*, Tokyo, Japanese Association of Mycotoxicology, pp. 51–52

Visconti, A. & Bottalico, A. (1983) High levels of ochratoxins A and B in moldy bread responsible for mycotoxicosis in farm animals. *J. agric. Food Chem.*, **31**, 1122–1123

Vukelić, M., Šoštarić, B. & Fuchs, R. (1991) Some pathomorphological features of Balkan endemic nephropathy in Croatia. In: Castegnaro, M., Pleština, R., Dirheimer, G., Chernozemsky, I.N. & Bartsch, H., eds, *Mycotoxins, Endemic Nephropathy and Urinary Tract Tumours* (IARC Scientific Publications No. 115), Lyon, IARC, pp. 37–42

Wehner, F.C., Thiel, P.G., van Rensburg, S.J. & Demasius, I.P.C. (1978) Mutagenicity to *Salmonella typhimurium* of some *Aspergillus* and *Penicillium* mycotoxins. *Mutat. Res.*, **58**, 193–203

Wei, Y.-H., Lu, C.-Y., Lin, T.-N. & Wei, R.-D. (1985) Effect of ochratoxin A on rat liver mitochondrial respiration and oxidative phosphorylation. *Toxicology*, **36**, 119–130

WHO (1990) *Selected Mycotoxins: Ochratoxins, Trichothecenes, Ergot* (Environmental Health Criteria 105), Geneva

WHO (1991) *Evaluation of Certain Food Additives and Contaminants, 37th Report of the Joint FAO/WHO Expert Committee on Food Additives* (WHO Tech. Rep. Ser. No. 806), Geneva, pp. 28–31

Widiastuti, R., Maryam, R., Blaney, B.J., Stolz, S. & Stoltz, D.R. (1988) Cyclopiazonic acid in combination with aflatoxins, zearalenone and ochratoxin A in Indonesian corn. *Mycopathologia*, **104**, 153–156

Williams, B.C. (1985) Mycotoxins in foods and foodstuffs (Abstr.). In: Scott, P.M., Trenholm, H.L. & Sutton, M.D., eds, *Mycotoxins, A Canadian Perspective*, Ottawa, National Research Council of Canada, p. 49

Würgler, F.E., Friederich, U. & Schlatter, J. (1991) Lack of mutagenicity of ochratoxin A and B, citrinin, patulin and cnestine in *Salmonella typhimurium* TA102. *Mutat. Res.*, **261**, 209–216

Zeiger, E., Anderson, B., Haworth, S., Lawlor, T. & Mortelmans, K. (1988) *Salmonella* mutagenicity tests: IV. Results from the testing of 300 chemicals. *Environ. mol. Mutag.*, **11** (Suppl. 12), 1–158

SUMMARY OF FINAL EVALUATIONS

Agent	Degree of evidence of carcinogenicity		Overall evaluation of carcinogenicity to humans
	Human	Animal	
Aflatoxins, naturally occurring mixtures of	S	S	1
Aflatoxin B_1	S	S	
Aflatoxin B_2		L	
Aflatoxin G_1		S	
Aflatoxin G_2		I	
Aflatoxin M_1	I	S	2B
Caffeic acid	I^a	S	2B
IQ (2-Amino-3-methylimidazo[4,5-f]-quinoline)	I	S	$2A^b$
d-Limonene	I^a	L	3
MeIQ (2-Amino-3,4-dimethylimidazo-[4,5-f]quinoline)	I	S	2B
MeIQx (2-Amino-3,8-dimethylimidazo-[4,5-f]quinoxaline)	I	S	2B
Ochratoxin A	I	S	2B
PhIP (2-Amino-1-methyl-6-phenylimidazo-[4,5-b]pyridine)	I	S	2B
Pickled vegetables, traditional Asian	L	I	2B
Salted fish, Chinese-style	S	L	1
Toxins derived from *Fusarium graminearum*, *F. culmorum* and *F. crookwellense*	I		3
Zearalenone		L	
Deoxynivalenol		I	
Nivalenol		I	
Fusarenone X		I	
Toxins derived from *Fusarium moniliforme*	I	S	2B
Fumonisin B_1		L	
Fumonisin B_2		I	
Fusarin C		L	

SUMMARY OF FINAL EVALUATIONS
(contd)

Agent	Degree of evidence of carcinogenicity		Overall evaluation of carcinogenicity to humans
	Human	Animal	
Toxins derived from *Fusarium sporotrichioides*	I[a]		3
T-2 Toxin		L	

S, sufficient evidence; L, limited evidence; I, inadequate evidence; for definitions of degrees of evidence and groupings of evaluations, see Preamble, pp. 26–29.
[a]No data available
[b]Other relevant data taken into account in making the overall evaluation

APPENDIX 1

SUMMARY TABLES OF
GENETIC AND RELATED EFFECTS

APPENDIX 1

Summary table of genetic and related effects of caffeic acid

Non-mammalian systems												Mammalian systems																			
Prokaryotes		Lower eukaryotes				Plants			Insects			In vitro													In vivo						
												Animal cells							Human cells						Animals				Humans		
D	G	D	R	G	A	D	G	C	R	G	C	A	D	G	S	M	C	A	T	I	D	G	S	M	C	A	T	I	D	G	S	M	C	DL	A	D	S	M	C	A
–		–[1]											+[1]					+															–							

A, aneuploidy; C, chromosomal aberrations; D, DNA damage; DL, dominant lethal mutation; G, gene mutation; I, inhibition of intercellular communication; M, micronuclei; R, mitotic recombination and gene conversion; S, sister chromatid exchange; T, cell transformation

In completing the tables, the following symbols indicate the consensus of the Working Group with regard to the results for each endpoint:

+   considered to be positive for the specific endpoint and level of biological complexity
+[1] considered to be positive, but only one valid study was available to the Working Group
–   considered to be negative
–[1] considered to be negative, but only one valid study was available to the Working Group
?   considered to be equivocal or inconclusive (e.g., there were contradictory results from different laboratories; there were confounding exposures; the results were equivocal)

Summary table of genetic and related effects of *d*-limonene and related compounds

d-Limonene

Nonmammalian systems					Mammalian systems			
Proka-ryotes	Lower eukaryotes	Plants	Insects		In vitro			In vivo
					Animal cells	Human cells		Animals	Humans
D G	D R G A	A D G C	R G C A		A D G S M C A T	I D G S M C A T		I D G S M C DL A D S M C A
–					– – – –	–		

d-Limonene-1,2-oxide

Nonmammalian systems					Mammalian systems			
Proka-ryotes	Lower eukaryotes	Plants	Insects		In vitro			In vivo
					Animal cells	Human cells		Animals	Humans
D G	D R G A	A D G C	R G C A		A D G S M C A T	I D G S M C A T		I D G S M C DL A D S M C A
					–			

Essential oils containing *d*-limonene

Nonmammalian systems					Mammalian systems			
Proka-ryotes	Lower eukaryotes	Plants	Insects		In vitro			In vivo
					Animal cells	Human cells		Animals	Humans
D G	D R G A	A D G C	R G C A		A D G S M C A T	I D G S M C A T		I D G S M C DL A D S M C A
– –								

A, aneuploidy; C, chromosomal aberrations; D, DNA damage; DL, dominant lethal mutation; G, gene mutation; I, inhibition of intercellular communication; M, micronuclei; R, mitotic recombination and gene conversion; S, sister chromatid exchange; T, cell transformation

In completing the tables, the following symbols indicate the consensus of the Working Group with regard to the results for each end-point:

– considered to be negative
–¹ considered to be negative, but only one valid study was available to the Working Group

APPENDIX 1

Summary table of genetic and related effects of 2-amino-3-methylimidazo[4,5-f]quinoline (IQ)

Nonmammalian systems				Mammalian systems		
Prokaryotes	Lower eukaryotes	Plants	Insects	In vitro		In vivo
				Animal cells	Human cells	Animals	Humans

Prokaryotes		Lower eukaryotes				Plants		Insects			Animal cells							Human cells								Animals								Humans				
D	G	D	R	G	A	D	G	C	R	G	C	A	D	G	S	M	C	A	T	I	D	G	S	M	C	A	T	I	D	G	S	M	C	DL	A	D	S	M	C	A
+	+									+			+	+	+		?				−¹	+¹			?				+¹	+¹	+		−¹	+¹						

A, aneuploidy; C, chromosomal aberrations; D, DNA damage; DL, dominant lethal mutation; G, gene mutation; I, inhibition of intercellular communication; M, micronuclei; R, mitotic recombination and gene conversion; S, sister chromatid exchange; T, cell transformation

In completing the tables, the following symbols indicate the consensus of the Working Group with regard to the results for each end-point:

+   considered to be positive for the specific endpoint and level of biological complexity
+¹ considered to be positive, but only one valid study was available to the Working Group
−¹ considered to be negative, but only one valid study was available to the Working Group
?   considered to be equivocal or inconclusive (e.g., there were contradictory results from different laboratories; there were confounding exposures; the results were equivocal)

Summary table of genetic and related effects of 2-amino-3,4-dimethylimidazo[4,5-*f*]quinoline (MeIQ)

Nonmammalian systems												Mammalian systems																				
Prokaryotes		Lower eukaryotes				Plants				Insects		In vitro														In vivo						
												Animal cells							Human cells							Animals					Humans		
D	G	D	R	G	A	D	G	C	R	G	C	A	D	G	S	M	C	A	T	I	D	G	S	M	C	A	T	I	D	G	S	M	C	DL	A	D	S	M	C	A
+	+										+¹		+	+	?	+¹													+¹		+¹								

A, aneuploidy; C, chromosomal aberrations; D, DNA damage; DL, dominant lethal mutation; G, gene mutation; I, inhibition of intercellular communication; M, micronuclei; R, mitotic recombination and gene conversion; S, sister chromatid exchange; T, cell transformation

In completing the tables, the following symbols indicate the consensus of the Working Group with regard to the results for each end-point:

+ considered to be positive for the specific endpoint and level of biological complexity
+¹ considered to be positive, but only one valid study was available to the Working Group

Summary table of genetic and related effects of 2-amino-3,8-dimethylimidazo[4,5-f]quinoxaline (MeIQx)

Nonmammalian systems											Mammalian systems																				
Proka-ryotes		Lower eukaryotes				Plants				Insects		In vitro														In vivo					
												Animal cells							Human cells							Animals				Humans			
D	G	D	R	G	A	D	G	C	R	G	C	A	D	G	S	M	C	A	T	I	D	G	S	M	C	A	T	I	D	G	S	M	C	DL	A	D	S	M	C	A
+	+										+¹		+	+	+¹	+¹				–¹			+¹						+	–¹	+¹ᵃ	–¹	+¹							

A, aneuploidy; C, chromosomal aberrations; D, DNA damage; DL, dominant lethal mutation; G, gene mutation; I, inhibition of intercellular communication; M, micronuclei; R, mitotic recombination and gene conversion; S, sister chromatid exchange; T, cell transformation

In completing the tables, the following symbols indicate the consensus of the Working Group with regard to the results for each end-point:

+   considered to be positive for the specific endpoint and level of biological complexity
+¹   considered to be positive, but only one valid study was available to the Working Group
–   considered to be negative
–¹   considered to be negative, but only one valid study was available to the Working Group

ᵃPositive in rat liver; negative in mouse bone marrow

Summary table of genetic and related effects of 2-amino-1-methyl-6-phenylimidazo[4,5-b]pyridine (PhIP)

Nonmammalian systems														Mammalian systems																				
Proka-ryotes		Lower eukaryotes				Plants			Insects					In vitro														In vivo						
														Animal cells							Human cells							Animals						Humans
D	G	D	R	G	A	D	G	C	R	G	C	A		D	G	S	M	C	A	T	I	D	G	S	M	C	A	T	I	D	G	S	M	C	DL	A	D	S	M	C	A
+¹	+													+	+¹	+		+¹												+		?¹	?	?	?						

A, aneuploidy; C, chromosomal aberrations; D, DNA damage; DL, dominant lethal mutation; G, gene mutation; I, inhibition of intercellular communication; M, micronuclei; R, mitotic recombination and gene conversion; S, sister chromatid exchange; T, cell transformation

In completing the tables, the following symbols indicate the consensus of the Working Group with regard to the results for each end-point:

+ considered to be positive for the specific endpoint and level of biological complexity
+¹ considered to be positive, but only one valid study was available to the Working Group
? considered to be equivocal or inconclusive (e.g., there were contradictory results from different laboratories; there were confounding exposures; the results were equivocal)

Summary table of genetic and related effects of aflatoxins

Aflatoxin B₁

Nonmammalian systems																			Mammalian systems																						
Proka-ryotes					Lower eukaryotes					Plants				Insects					In vitro													In vivo								
																			Animal cells									Human cells							Animals					Humans			
D	G	D	R	G	A	D	G	C	A	D	G	C	R	G	C	A			D	G	S	M	C	A	T	I	D	G	S	M	C	A	T	I	D	G	S	M	C	DL	A	D	S	M	C	A
+	+			+	+								+¹	+					+	−¹	+	+	+	+		+¹	+	+	+	+	+¹	+			+		+	+	+	+					+*

Sperm morphology, mice, −
*See Table 13 (pp. 329–334) for listing and references

Aflatoxin B₂

Nonmammalian systems																			Mammalian systems																						
Proka-ryotes					Lower eukaryotes					Plants				Insects					In vitro													In vivo								
																			Animal cells									Human cells							Animals					Humans			
D	G	D	R	G	A	D	G	C	A	D	G	C	R	G	C	A			D	G	S	M	C	A	T	I	D	G	S	M	C	A	T	I	D	G	S	M	C	DL	A	D	S	M	C	A
?	+	−¹	−¹																+	−¹	+¹					+¹						−¹			+											

Aflatoxin G₁

Nonmammalian systems																			Mammalian systems																						
Proka-ryotes					Lower eukaryotes					Plants				Insects					In vitro													In vivo								
																			Animal cells									Human cells							Animals					Humans			
D	G	D	R	G	A	D	G	C	A	D	G	C	R	G	C	A			D	G	S	M	C	A	T	I	D	G	S	M	C	A	T	I	D	G	S	M	C	DL	A	D	S	M	C	A
+	+	−¹	−¹																+		+¹	+¹				+¹		+¹				+¹			+				+							

Summary table of genetic and related effects of aflatoxins (contd)

Aflatoxin G_2

Nonmammalian systems																	Mammalian systems																					
Prokaryotes		Lower eukaryotes				Plants				Insects							In vitro												In vivo							
																	Animal cells						Human cells						Animals					Humans			
D	G	D	R	G	A	D	G	C	R	G	C	A					A	D	G	S	M	C	A	T	I	D	G	S	M	C	A	T	I	D	G	S	M	C	DL	A	D	S	M	C	A
-	?	-[1]															+	-	-	+								-[1]																	

Aflatoxin M_1

Nonmammalian systems																	Mammalian systems																					
Prokaryotes		Lower eukaryotes				Plants				Insects							In vitro												In vivo							
																	Animal cells						Human cells						Animals					Humans			
D	G	D	R	G	A	D	G	C	R	G	C	A					A	D	G	S	M	C	A	T	I	D	G	S	M	C	A	T	I	D	G	S	M	C	DL	A	D	S	M	C	A
+																				+																									

A, aneuploidy; C, chromosomal aberrations; D, DNA damage; DL, dominant lethal mutation; G, gene mutation; I, inhibition of intercellular communication; M, micronuclei; R, mitotic recombination and gene conversion; S, sister chromatid exchange; T, cell transformation

In completing the tables, the following symbols indicate the consensus of the Working Group with regard to the results for each end-point:

+ considered to be positive for the specific endpoint and level of biological complexity
+[1] considered to be positive, but only one valid study was available to the Working Group
− considered to be negative
−[1] considered to be negative, but only one valid study was available to the Working Group
? considered to be equivocal or inconclusive (e.g., there were contradictory results from different laboratories; there were confounding exposures; the results were equivocal)

APPENDIX 1

Summary table of genetic and related effects of toxins derived from *Fusarium graminearum*, *F. culmorum* and *F. crookwellense*

Zearalenone

Nonmammalian systems				Mammalian systems			
Prokaryotes	Lower eukaryotes	Plants	Insects	In vitro		In vivo	
				Animal cells	Human cells	Animals	Humans
D G R A D G C	D R G A D G C	A D G C	R G C A	A D G S M C A T	I D G S M C A T	I D G S M C DL	A D S M C A
– –	–			+? –	?		

Deoxynivalenol

Nonmammalian systems				Mammalian systems			
Prokaryotes	Lower eukaryotes	Plants	Insects	In vitro		In vivo	
				Animal cells	Human cells	Animals	Humans
D G R A D G C	D R G A D G C	A D G C	R G C A	A D G S M C A T	I D G S M C A T	I D G S M C DL	A D S M C A
–				– –	+? +?		

Nivalenol

Nonmammalian systems				Mammalian systems			
Prokaryotes	Lower eukaryotes	Plants	Insects	In vitro		In vivo	
				Animal cells	Human cells	Animals	Humans
D G R A D G C	D R G A D G C	A D G C	R G C A	A D G S M C A T	I D G S M C A T	I D G S M C DL	A D S M C A
				?	+		

Summary table of genetic and related effects of toxins derived from *Fusarium graminearum*, *F. culmorum* and *F. crookwellense* (contd)

Fusarenone X

Nonmammalian systems													Mammalian systems																				
Proka-ryotes	Lower eukaryotes				Plants				Insects				In vitro															In vivo					
													Animal cells							Human cells								Animals					Humans	
D	G	D	R	G	A	D	G	C	R	G	C	A	D	G	S	M	C	A	T	I	D	G	S	M	C	A	T	I	D	G	S	M	C	DL	A	D	S	M	C	A
–		–¹			+¹								–¹	?		+¹									?															

A, aneuploidy; C, chromosomal aberrations; D, DNA damage; DL, dominant lethal mutation; G, gene mutation; I, inhibition of intercellular communication; M, micronuclei; R, mitotic recombination and gene conversion; S, sister chromatid exchange; T, cell transformation

In completing the tables, the following symbols indicate the consensus of the Working Group with regard to the results for each end-point:

+   considered to be positive for the specific endpoint and level of biological complexity
+¹ considered to be positive, but only one valid study was available to the Working Group
–   considered to be negative
–¹ considered to be negative, but only one valid study was available to the Working Group
?   considered to be equivocal or inconclusive (e.g., there were contradictory results from different laboratories; there were confounding exposures; the results were equivocal)

APPENDIX 1

Summary table of genetic and related effects of toxins derived from *Fusarium moniliforme*

Fumonisin B$_1$

Nonmammalian systems															Mammalian systems																						
Proka-ryotes		Lower eukaryotes						Plants		Insects			In vitro																In vivo							
													Animal cells									Human cells								Animals						Humans	
D	G	D	R	G	A	D	G	C	R	G	C	A	A	D	G	S	M	C	A	T	I	D	G	S	M	C	A	T	I	D	G	S	M	C	DL	A	D	S	M	C	A

$-^1$

Fumonisin B$_2$

Nonmammalian systems															Mammalian systems																						
Proka-ryotes		Lower eukaryotes						Plants		Insects			In vitro																In vivo							
													Animal cells									Human cells								Animals						Humans	
D	G	D	R	G	A	D	G	C	R	G	C	A	A	D	G	S	M	C	A	T	I	D	G	S	M	C	A	T	I	D	G	S	M	C	DL	A	D	S	M	C	A

$-^1$

Summary table of genetic and related effects of toxins derived from *Fusarium moniliforme* (contd)

Fusarin C

Nonmammalian systems																Mammalian systems																			
Proka-ryotes		Lower eukaryotes				Plants				Insects						In vitro														In vivo					
																Animal cells							Human cells							Animals					Humans	
D	G	D	R	G	A	D	G	C	R	G	C	A	D	G	C	A	D	G	S	M	C	A	T	I	D	G	S	M	C	A	T	I	D	G	S	M	C	DL	A	D	S	M	C	A
?	+																	+¹	+¹	+¹	+¹																							

A, aneuploidy; C, chromosomal aberrations; D, DNA damage; DL, dominant lethal mutation; G, gene mutation; I, inhibition of intercellular communication; M, micronuclei; R, mitotic recombination and gene conversion; S, sister chromatid exchange; T, cell transformation

In completing the tables, the following symbols indicate the consensus of the Working Group with regard to the results for each end-point:

+ considered to be positive for the specific endpoint and level of biological complexity
+¹ considered to be positive, but only one valid study was available to the Working Group
− considered to be negative
−¹ considered to be negative, but only one valid study was available to the Working Group
? considered to be equivocal or inconclusive (e.g., there were contradictory results from different laboratories; there were confounding exposures; the results were equivocal)

APPENDIX 1

Summary table of genetic and related effects of toxins derived from *Fusarium sporotrichioides*: T-2 toxin

Nonmammalian systems				Mammalian systems																					
Proka-ryotes		Lower eukaryotes		Plants		Insects		In vitro													In vivo				
								Animal cells							Human cells							Animals			Humans
D	G	D	R	G	A	D	A	D	G	C	R	G	C	A	D	G	S	M	C	A	T	I	D	G	S	M	C	A	T	D	G	S	M	C	DL	A	D	S	M	C	A
-	-		-¹	-¹							?		+¹		+	+¹	?	+¹	+			+¹	+¹	-¹						+¹				-¹						(+)?	

A, aneuploidy; C, chromosomal aberrations; D, DNA damage; DL, dominant lethal mutation; G, gene mutation; I, inhibition of intercellular communication; M, micronuclei; R, mitotic recombination and gene conversion; S, sister chromatid exchange; T, cell transformation

In completing the tables, the following symbols indicate the consensus of the Working Group with regard to the results for each end-point:

+   considered to be positive for the specific endpoint and level of biological complexity
+¹   considered to be positive, but only one valid study was available to the Working Group
−   considered to be negative
−¹   considered to be negative, but only one valid study was available to the Working Group
?   considered to be equivocal or inconclusive (e.g., there were contradictory results from different laboratories; there were confounding exposures; the results were equivocal)

Summary table of genetic and related effects of ochratoxin A

Nonmammalian systems													Mammalian systems																					
Proka-ryotes	Lower eukaryotes				Plants			Insects					In vitro													In vivo								
													Animal cells							Human cells						Animals						Humans		
D	G	D	R	G	A	D	G	C	R	G	C	A	D	G	S	M	C	A	T	I	D	G	S	M	C	A	T	I	D	G	S	M	C	DL	A	D	S	M	C	A
?	?	–¹											+	–	(+)¹								–¹							+		–¹									

A, aneuploidy; C, chromosomal aberrations; D, DNA damage; DL, dominant lethal mutation; G, gene mutation; I, inhibition of intercellular communication; M, micronuclei; R, mitotic recombination and gene conversion; S, sister chromatid exchange; T, cell transformation

In completing the tables, the following symbols indicate the consensus of the Working Group with regard to the results for each end-point:

+   considered to be positive for the specific endpoint and level of biological complexity
+¹ considered to be positive, but only one valid study was available to the Working Group
–   considered to be negative
–¹ considered to be negative, but only one valid study was available to the Working Group
?   considered to be equivocal or inconclusive (e.g., there were contradictory results from different laboratories; there were confounding exposures; the results were equivocal)

APPENDIX 2

ACTIVITY PROFILES FOR
GENETIC AND RELATED EFFECTS

APPENDIX 2

ACTIVITY PROFILES FOR GENETIC AND RELATED EFFECTS

Methods

The x-axis of the activity profile (Waters *et al.*, 1987, 1988) represents the bioassays in phylogenetic sequence by endpoint, and the values on the y-axis represent the logarithmically transformed lowest effective doses (LED) and highest ineffective doses (HID) tested. The term 'dose', as used in this report, does not take into consideration length of treatment or exposure and may therefore be considered synonymous with concentration. In practice, the concentrations used in all the in-vitro tests were converted to µg/ml, and those for in-vivo tests were expressed as mg/kg bw. Because dose units are plotted on a log scale, differences in molecular weights of compounds do not, in most cases, greatly influence comparisons of their activity profiles. Conventions for dose conversions are given below.

Profile-line height (the magnitude of each bar) is a function of the LED or HID, which is associated with the characteristics of each individual test system—such as population size, cell-cycle kinetics and metabolic competence. Thus, the detection limit of each test system is different, and, across a given activity profile, responses will vary substantially. No attempt is made to adjust or relate responses in one test system to those of another.

Line heights are derived as follows: for negative test results, the highest dose tested without appreciable toxicity is defined as the HID. If there was evidence of extreme toxicity, the next highest dose is used. A single dose tested with a negative result is considered to be equivalent to the HID. Similarly, for positive results, the LED is recorded. If the original data were analysed statistically by the author, the dose recorded is that at which the response was significant ($p < 0.05$). If the available data were not analysed statistically, the dose required to produce an effect is estimated as follows: when a dose-related positive response is observed with two or more doses, the lower of the doses is taken as the LED; a single dose resulting in a positive response is considered to be equivalent to the LED.

In order to accommodate both the wide range of doses encountered and positive and negative responses on a continuous scale, doses are transformed logarithmically, so that effective (LED) and ineffective (HID) doses are represented by positive and negative numbers, respectively. The response, or logarithmic dose unit (LDU_{ij}), for a given test system i and chemical j is represented by the expressions

$LDU_{ij} = -\log_{10}$ (dose), for HID values; $LDU \leq 0$
and (1)
$LDU_{ij} = -\log_{10}$ (dose × 10^{-5}), for LED values; $LDU \geq 0$.

These simple relationships define a dose range of 0 to −5 logarithmic units for ineffective doses (1–100 000 µg/ml or mg/kg bw) and 0 to +8 logarithmic units for effective doses (100 000–0.001 µg/ml or mg/kg bw). A scale illustrating the LDU values is shown in Figure 1. Negative responses at doses less than 1 µg/ml (mg/kg bw) are set equal to 1. Effectively, an LED value ≥100 000 or an HID value ≤1 produces an LDU = 0; no quantitative information is gained from such extreme values. The dotted lines at the levels of log dose units 1 and −1 define a 'zone of uncertainty' in which positive results are reported at such high doses (between 10 000 and 100 000 µg/ml or mg/kg bw) or negative results are reported at such low dose levels (1 to 10 µg/ml or mg/kg bw) as to call into question the adequacy of the test.

Fig. 1. Scale of log dose units used on the y-axis of activity profiles

Positive (µg/ml or mg/kg bw)		Log dose units	
0.001		8	—
0.01		7	—
0.1		6	—
1.0		5	—
10		4	—
100		3	—
1000		2	—
10 000		1	—
100 000	1	0	—
	10	−1	—
	100	−2	—
	1000	−3	—
	10 000	−4	—
	100 000	−5	—
	Negative (µg/ml or mg/kg bw)		

LED and HID are expressed as µg/ml or mg/kg bw.

In practice, an activity profile is computer generated. A data entry programme is used to store abstracted data from published reports. A sequential file (in ASCII) is created for each compound, and a record within that file consists of the name and Chemical Abstracts Service number of the compound, a three-letter code for the test system (see below), the qualitative test result (with and without an exogenous metabolic system), dose (LED or HID), citation number and additional source information. An abbreviated citation for each publication is stored in a segment of a record accessing both the test data file and the citation file. During processing of the data file, an average of the logarithmic values of the data subset is calculated, and the length of the profile line represents this average value. All dose values are plotted for each profile line, regardless of whether results are positive or negative. Results obtained in the absence of an exogenous metabolic system are indicated by a bar (–), and results obtained in the presence of an exogenous metabolic system are indicated by an upward-directed arrow (↑). When all results for a given assay are either positive or negative, the mean of the LDU values is plotted as a solid line; when conflicting data are reported for the same assay (i.e., both positive and negative results), the majority data are shown by a solid line and the minority data by a dashed line (drawn to the extreme conflicting response). In the few cases in which the numbers of positive and negative results are equal, the solid line is drawn in the positive direction and the maximal negative response is indicated with a dashed line.

Profile lines are identified by three-letter code words representing the commonly used tests. Code words for most of the test systems in current use in genetic toxicology were defined for the US Environmental Protection Agency's GENE-TOX Program (Waters, 1979; Waters & Auletta, 1981). For *IARC Monographs* Supplement 6, Volume 44 and subsequent volumes, including this publication, codes were redefined in a manner that should facilitate inclusion of additional tests. Naming conventions are described below.

Data listings are presented in the text and include endpoint and test codes, a short test code definition, results [either with (M) or without (NM) an exogenous activation system], the associated LED or HID value and a short citation. Test codes are organized phylogenetically and by endpoint from left to right across each activity profile and from top to bottom of the corresponding data listing. Endpoints are defined as follows: A, aneuploidy; C, chromosomal aberrations; D, DNA damage; F, assays of body fluids; G, gene mutation; H, host-mediated assays; I, inhibition of intercellular communication; M, micronuclei; P, sperm morphology; R, mitotic recombination or gene conversion; S, sister chromatid exchange; and T, cell transformation.

Dose conversions for activity profiles

Doses are converted to μg/ml for in-vitro tests and to mg/kg bw per day for in-vivo experiments.

1. In-vitro test systems
   (a) Weight/volume converts directly to μg/ml.
   (b) Molar (M) concentration × molecular weight = mg/ml = 10^3 μg/ml; mM concentration × molecular weight = μg/ml.
   (c) Soluble solids expressed as % concentration are assumed to be in units of mass per volume (i.e., 1% = 0.01 g/ml = 10 000 μg/ml; also, 1 ppm = 1 μg/ml).
   (d) Liquids and gases expressed as % concentration are assumed to be given in units of volume per volume. Liquids are converted to weight per volume using the density (D) of the solution (D = g/ml). Gases are converted from volume to mass using the ideal gas law, PV = nRT. For exposure at 20–37°C at standard atmospheric pressure, 1% (v/v) = 0.4 μg/ml × molecular weight of the gas. Also, 1 ppm (v/v) = 4×10^{-5} μg/ml × molecular weight.

(e) In microbial plate tests, it is usual for the doses to be reported as weight/plate, whereas concentrations are required to enter data on the activity profile chart. While remaining cognisant of the errors involved in the process, it is assumed that a 2-ml volume of top agar is delivered to each plate and that the test substance remains in solution within it; concentrations are derived from the reported weight/plate values by dividing by this arbitrary volume. For spot tests, a 1-ml volume is used in the calculation.

(f) Conversion of particulate concentrations given in $\mu g/cm^2$ are based on the area (A) of the dish and the volume of medium per dish; i.e., for a 100-mm dish: $A = \pi R^2 = \pi \times (5\ cm)^2 = 78.5\ cm^2$. If the volume of medium is 10 ml, then $78.5\ cm^2 = 10\ ml$ and $1\ cm^2 = 0.13\ ml$.

2. In-vitro systems using in-vivo activation

For the body fluid–urine (BF–) test, the concentration used is the dose (in mg/kg bw) of the compound administered to test animals or patients.

3. In-vivo test systems

(a) Doses are converted to mg/kg bw per day of exposure, assuming 100% absorption. Standard values are used for each sex and species of rodent, including body weight and average intake per day, as reported by Gold *et al.* (1984). For example, in a test using male mice fed 50 ppm of the agent in the diet, the standard food intake per day is 12% of body weight, and the conversion is dose = 50 ppm × 12% = 6 mg/kg bw per day.

Standard values used for humans are: weight—males, 70 kg; females, 55 kg; surface area, 1.7 m^2; inhalation rate, 20 l/min for light work, 30 l/min for mild exercise.

(b) When reported, the dose at the target site is used. For example, doses given in studies of lymphocytes of humans exposed *in vivo* are the measured blood concentrations in $\mu g/ml$.

Codes for test systems

For specific nonmammalian test systems, the first two letters of the three-symbol code word define the test organism (e.g., SA– for *Salmonella typhimurium*, EC– for *Escherichia coli*). If the species is not known, the convention used is –S–. The third symbol may be used to define the tester strain (e.g., SA8 for *S. typhimurium* TA1538, ECW for *E. coli* WP2*uvrA*). When strain designation is not indicated, the third letter is used to define the specific genetic endpoint under investigation (e.g., ––D for differential toxicity, ––F for forward mutation, ––G for gene conversion or genetic crossing-over, ––N for aneuploidy, ––R for reverse mutation, ––U for unscheduled DNA synthesis). The third letter may also be used to define the general endpoint under investigation when a more complete definition is not possible or relevant (e.g., ––M for mutation, ––C for chromosomal aberration).

For mammalian test systems, the first letter of the three-letter code word defines the genetic endpoint under investigation: A–– for aneuploidy, B–– for binding, C–– for chromosomal aberration, D–– for DNA strand breaks, G–– for gene mutation, I–– for inhibition of intercellular communication, M–– for micronucleus formation, R–– for DNA repair, S— for sister chromatid exchange, T— for cell transformation and U— for unscheduled DNA synthesis.

For animal (i.e., non-human) test systems *in vitro*, when the cell type is not specified, the code letters –IA are used. For such assays *in vivo*, when the animal species is not specified, the code letters –VA are used. Commonly used animal species are identified by the third letter (e.g., —C for Chinese hamster, —M for mouse, —R for rat, —S for Syrian hamster).

For test systems using human cells *in vitro*, when the cell type is not specified, the code letters –IH are used. For assays on humans *in vivo*, when the cell type is not specified, the code letters –VH are used. Otherwise, the second letter specifies the cell type under investigation (e.g., -BH for bone marrow, -LH for lymphocytes).

Some other specific coding conventions used for mammalian systems are as follows: BF– for body fluids, HM– for host-mediated, —L for leukocytes or lymphocytes *in vitro* (-AL, animals; -HL, humans), -L- for leukocytes *in vivo* (-LA, animals; -LH, humans), —T for transformed cells.

Note that these are examples of major conventions used to define the assay code words. The alphabetized listing of codes must be examined to confirm a specific code word. As might be expected from the limitation to three symbols, some codes do not fit the naming conventions precisely. In a few cases, test systems are defined by first-letter code words, for example: MST, mouse spot test; SLP, mouse specific locus test, postspermatogonia; SLO, mouse specific locus test, other stages; DLM, dominant lethal test in mice; DLR, dominant lethal test in rats; MHT, mouse heritable translocation test.

The genetic activity profiles and listings were prepared in collaboration with Environmental Health Research and Testing Inc. (EHRT) under contract to the US Environmental Protection Agency; EHRT also determined the doses used. The references cited in each genetic activity profile listing can be found in the list of references in the appropriate monograph.

References

Garrett, N.E., Stack, H.F., Gross, M.R. & Waters, M.D. (1984) An analysis of the spectra of genetic activity produced by known or suspected human carcinogens. *Mutat. Res.*, *134*, 89–111

Gold, L.S., Sawyer, C.B., Magaw, R., Backman, G.M., de Veciana, M., Levinson, R., Hooper, N.K., Havender, W.R., Bernstein, L., Peto, R., Pike, M.C. & Ames, B.N. (1984) A carcinogenic potency database of the standardized results of animal bioassays. *Environ. Health Perspect.*, *58*, 9–319

Waters, M.D. (1979) *The GENE-TOX program*. In: Hsie, A.W., O'Neill, J.P. & McElheny, V.K., eds, *Mammalian Cell Mutagenesis: The Maturation of Test Systems* (Banbury Report 2), Cold Spring Harbor, NY, CSH Press, pp. 449–467

Waters, M.D. & Auletta, A. (1981) The GENE-TOX program: genetic activity evaluation. *J. chem. Inf. comput. Sci.*, *21*, 35–38

Waters, M.D., Stack, H.F., Brady, A.L., Lohman, P.H.M., Haroun, L. & Vainio, H. (1987) Appendix 1: Activity profiles for genetic and related tests. In: *IARC Monographs on the Evaluation of the Carcinogenic Risk of Chemicals to Humans*, Suppl. 6, *Genetic and Related Effects: An Updating of Selected IARC Monographs from Volumes 1 to 42*, Lyon, IARC, pp. 687–696

Waters, M.D., Stack, H.F., Brady, A.L., Lohman, P.H.M., Haroun, L. & Vainio, H. (1988) Use of computerized data listings and activity profiles of genetic and related effects in the review of 195 compounds. *Mutat. Res.*, *205*, 295–312

APPENDIX 2

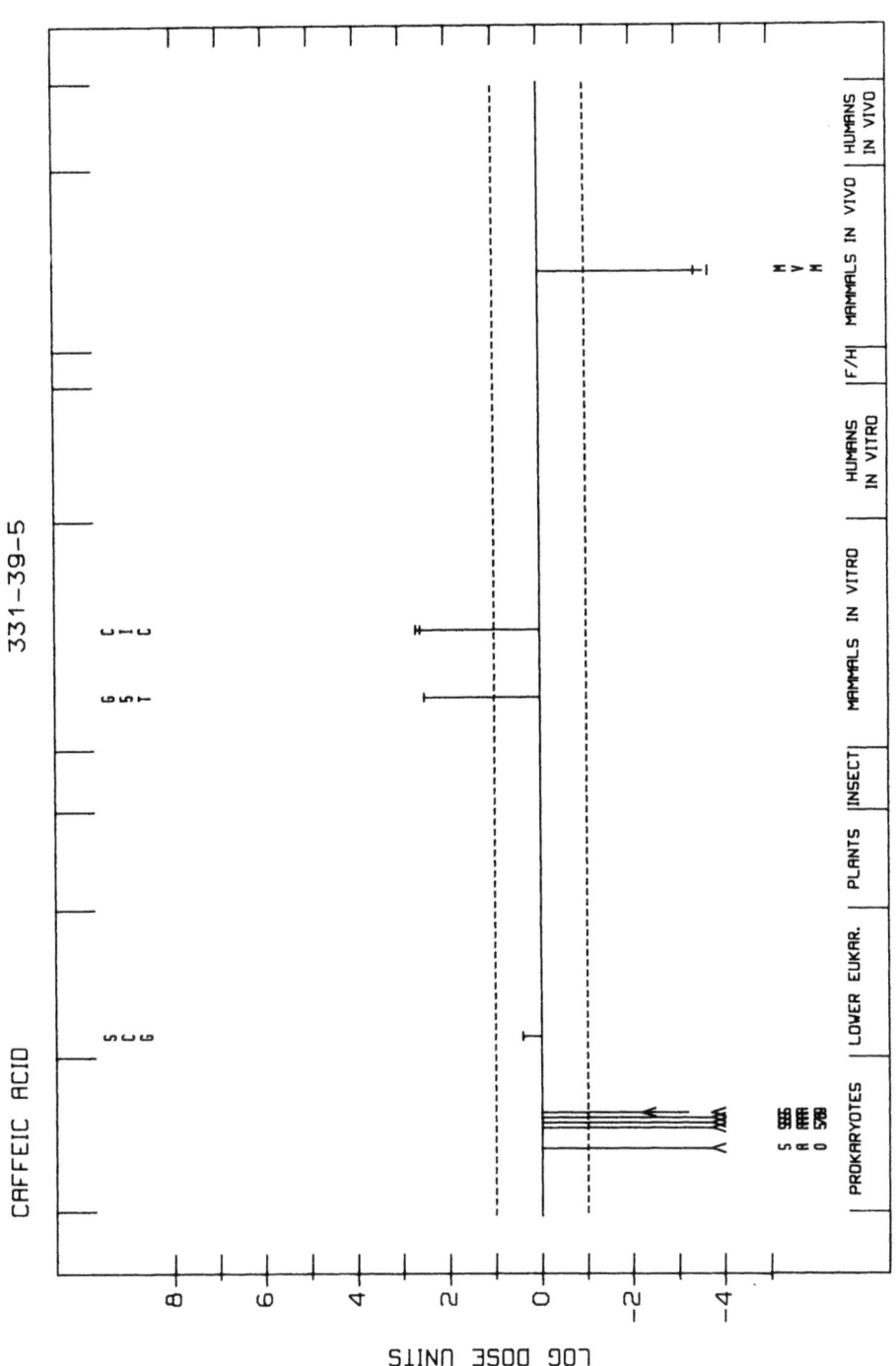

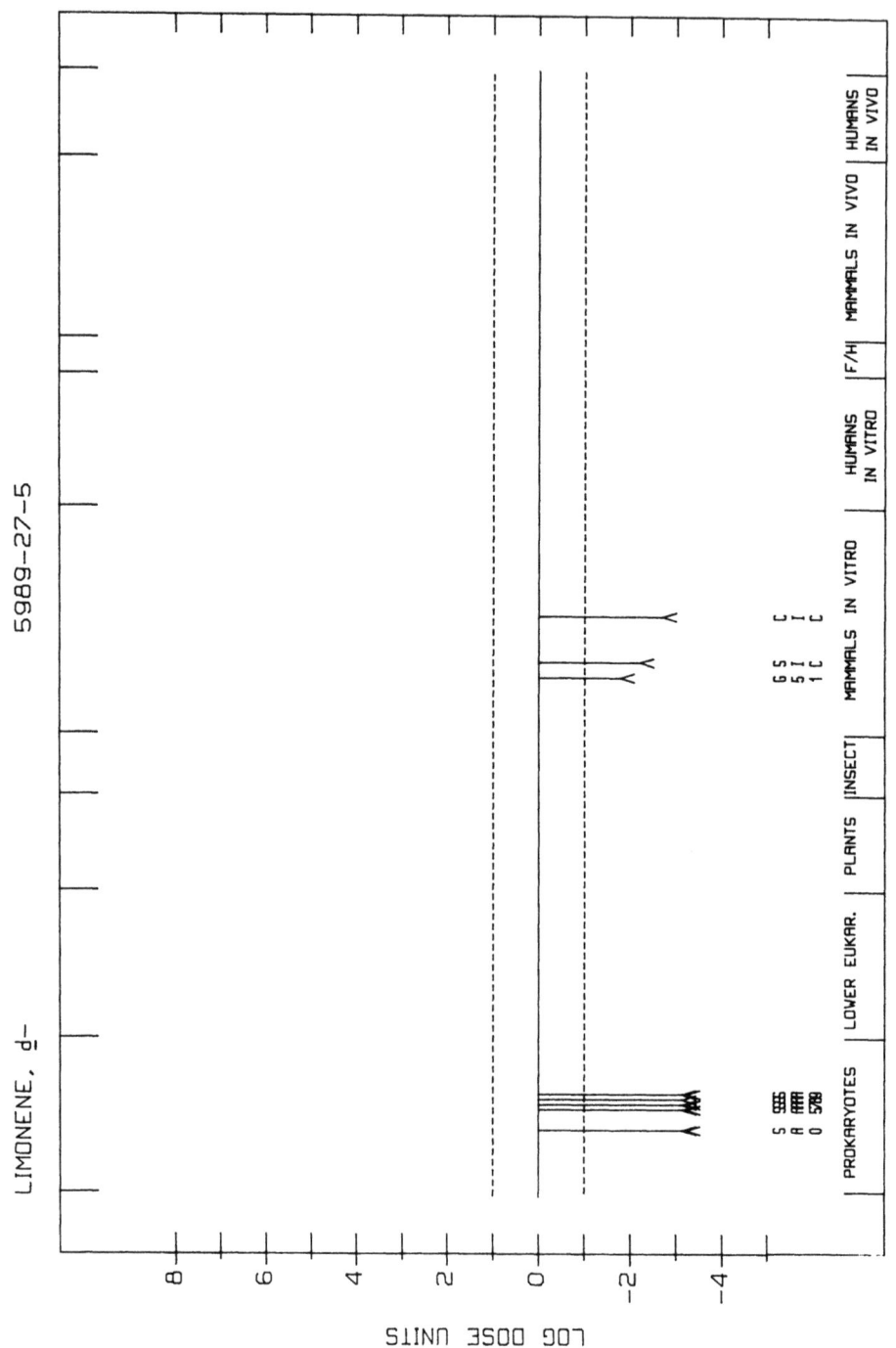

APPENDIX 2

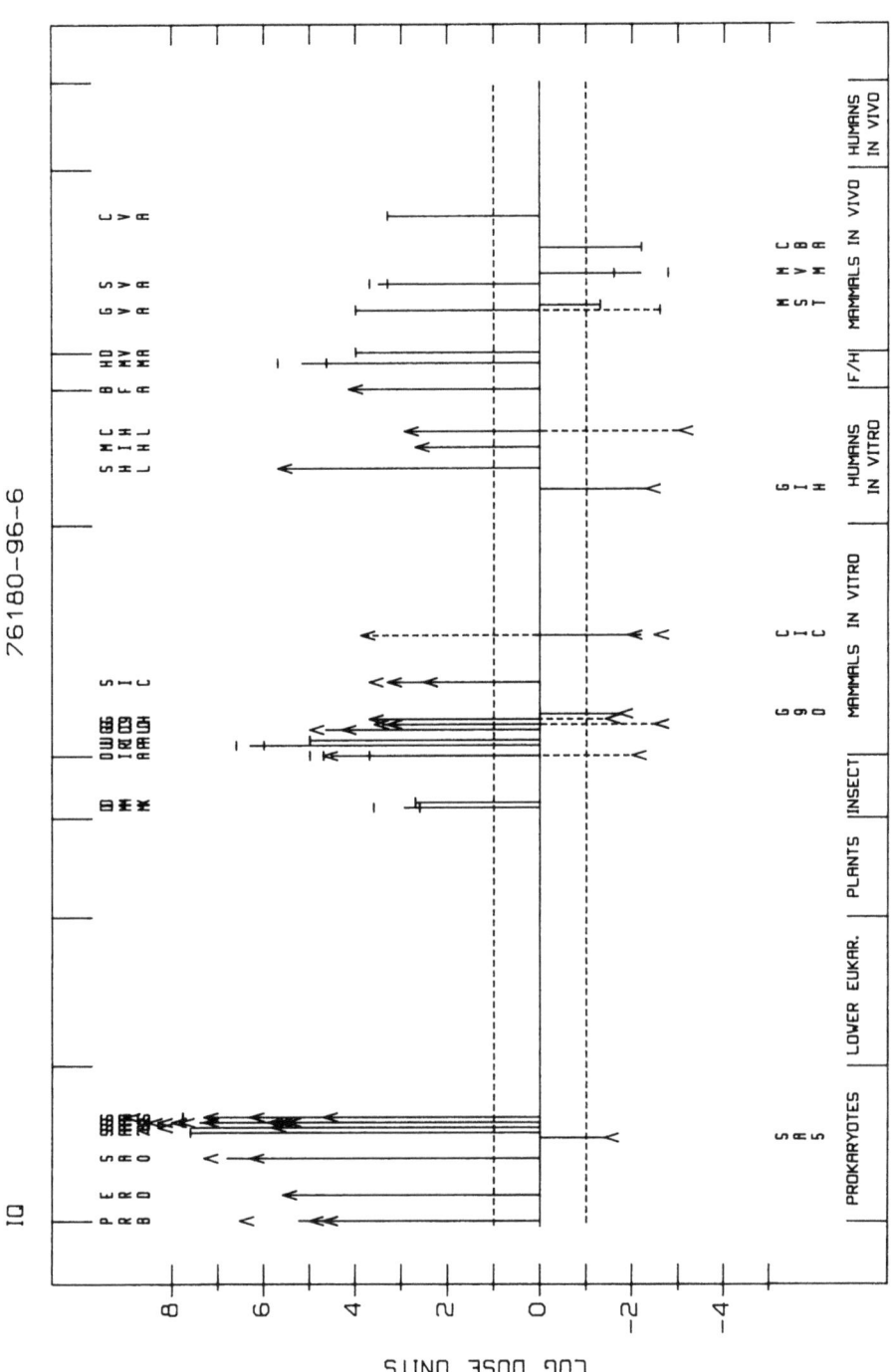

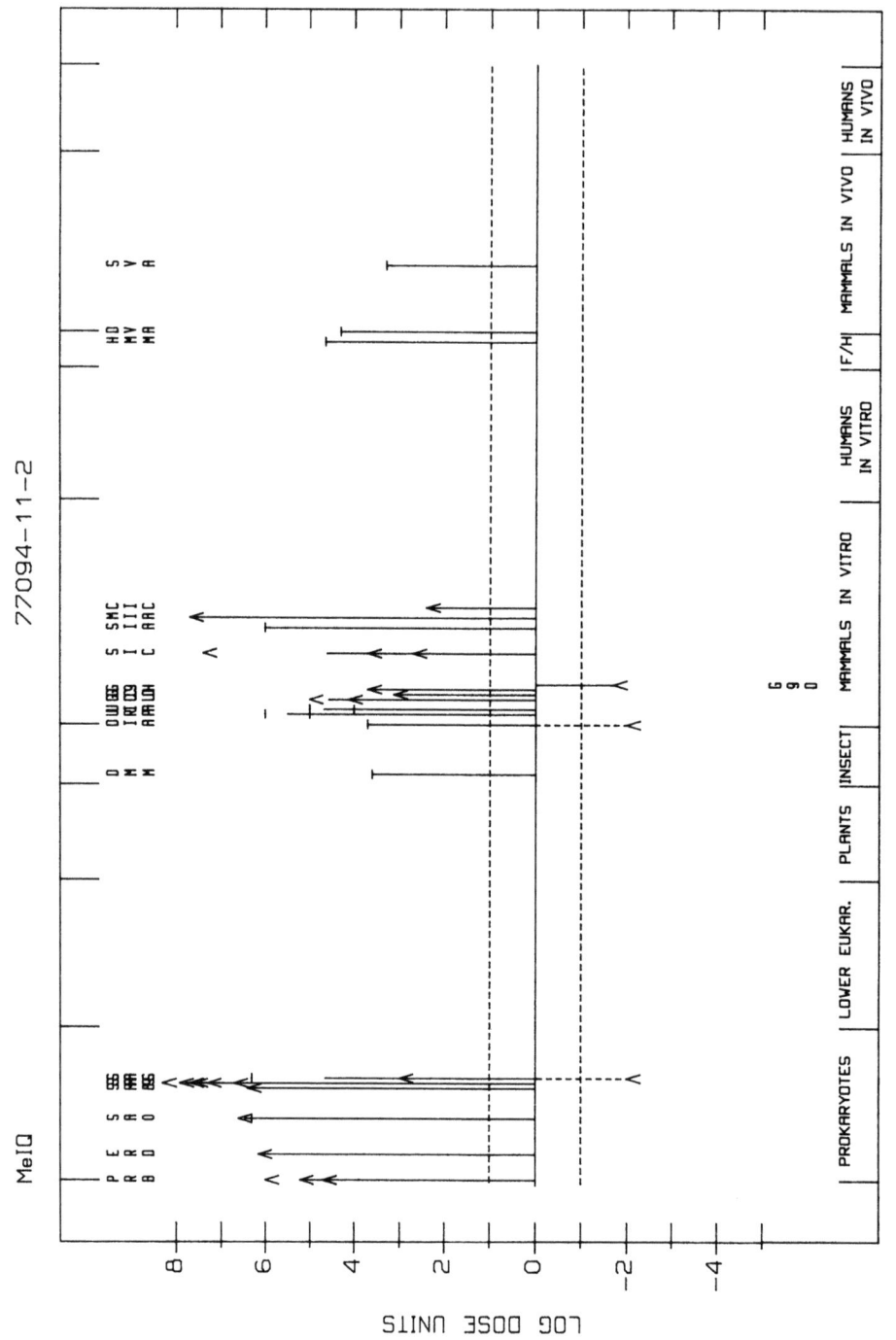

APPENDIX 2

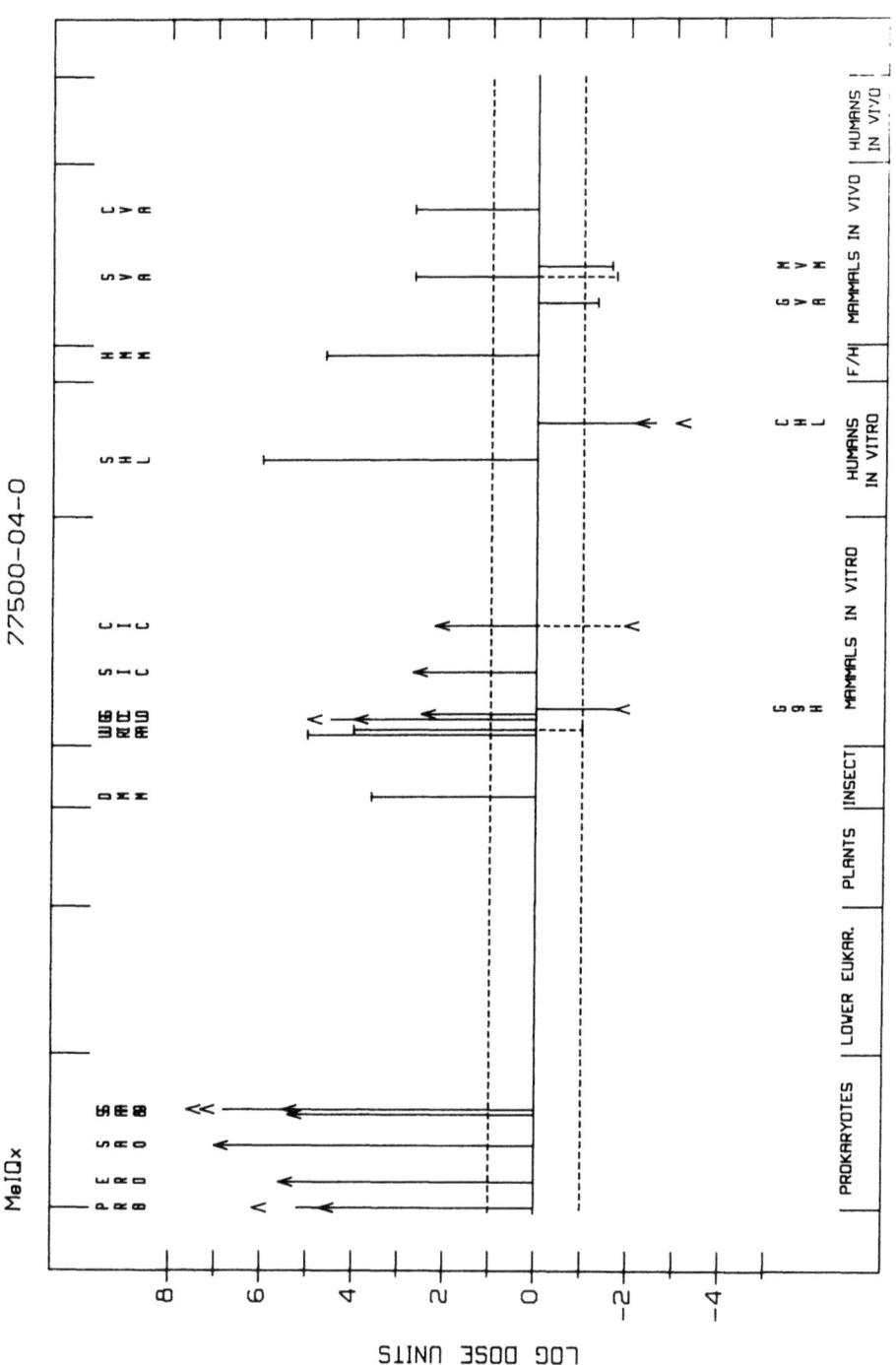

554 IARC MONOGRAPHS VOLUME 56

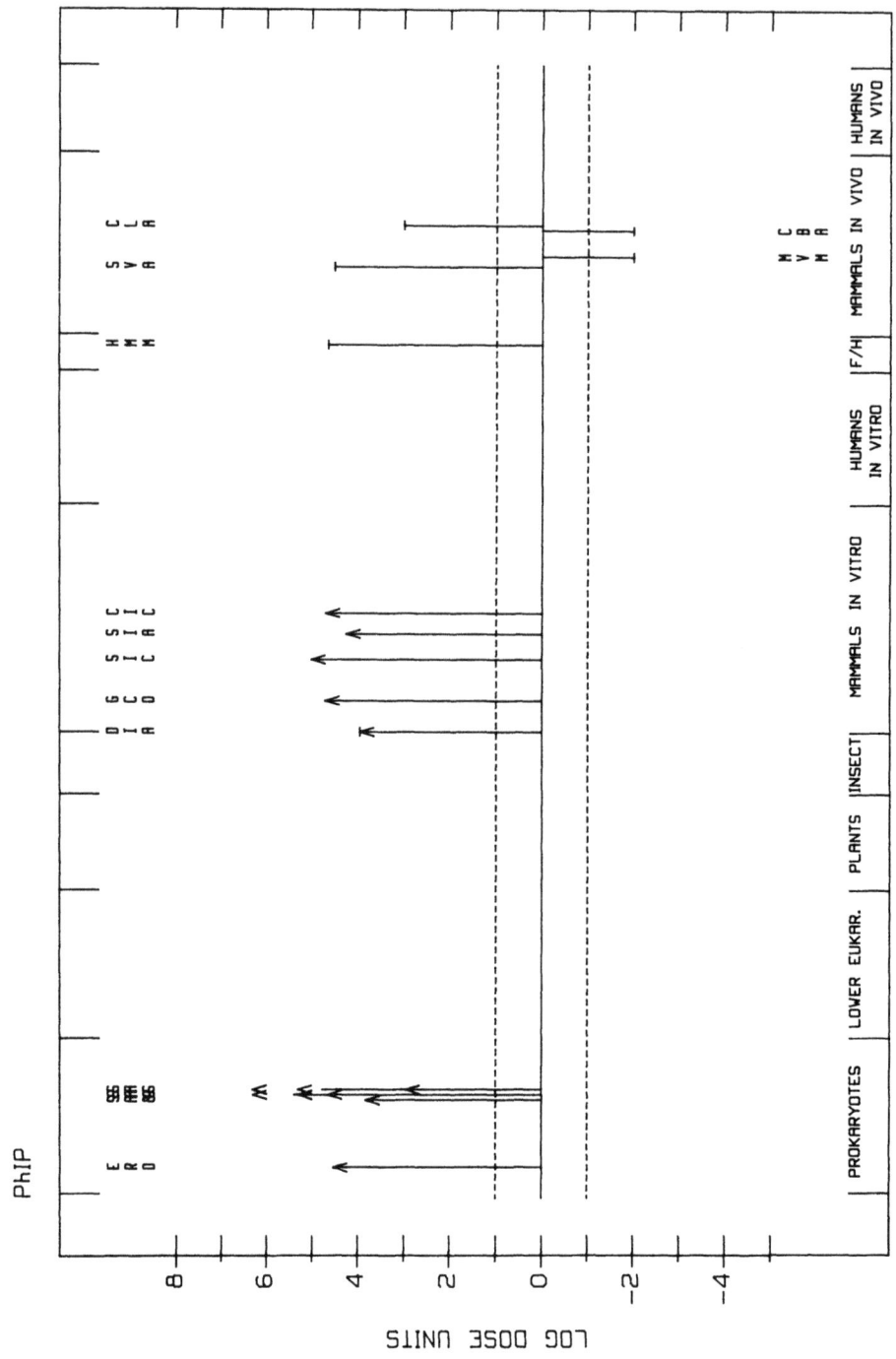

APPENDIX 2

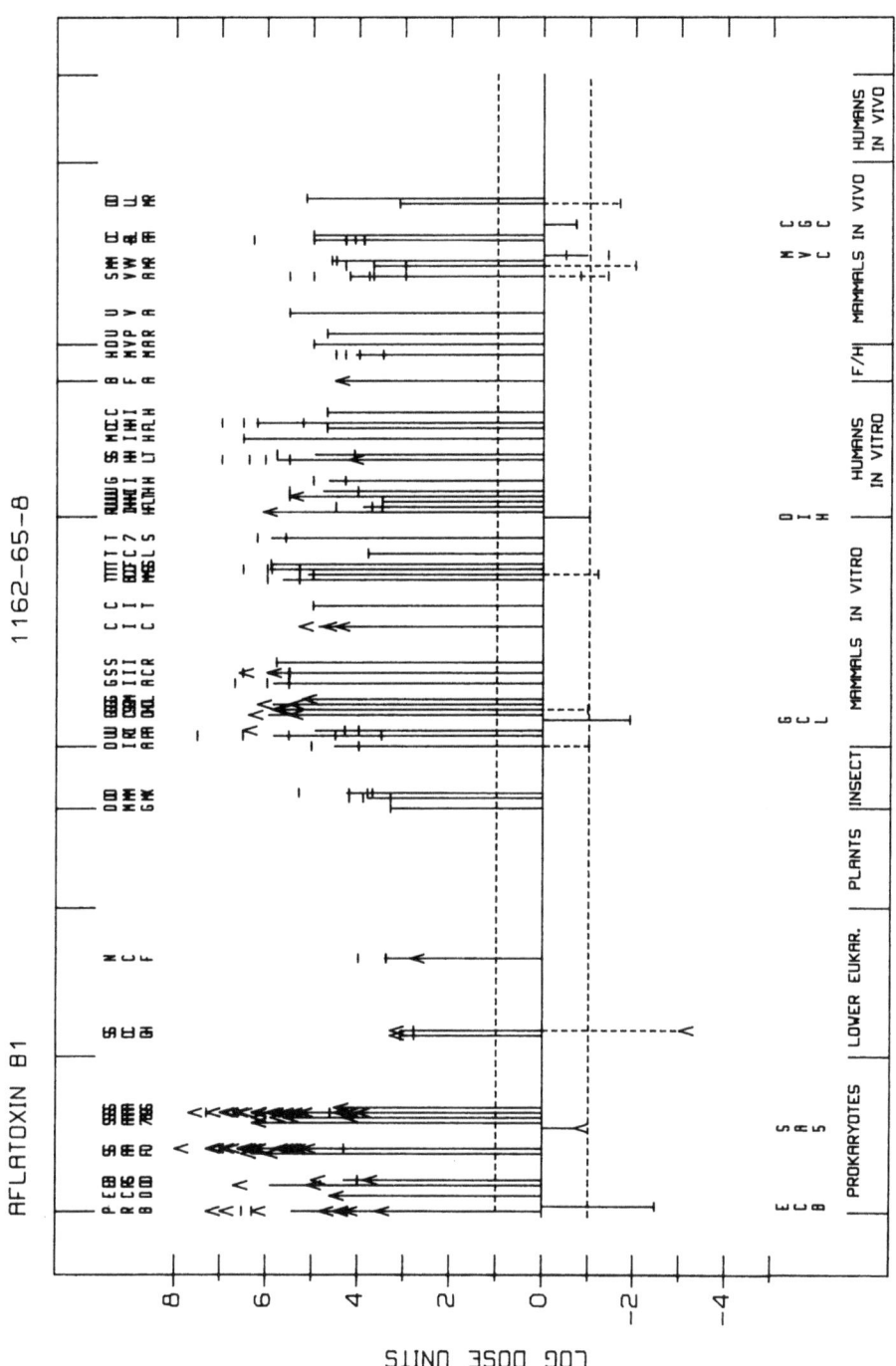

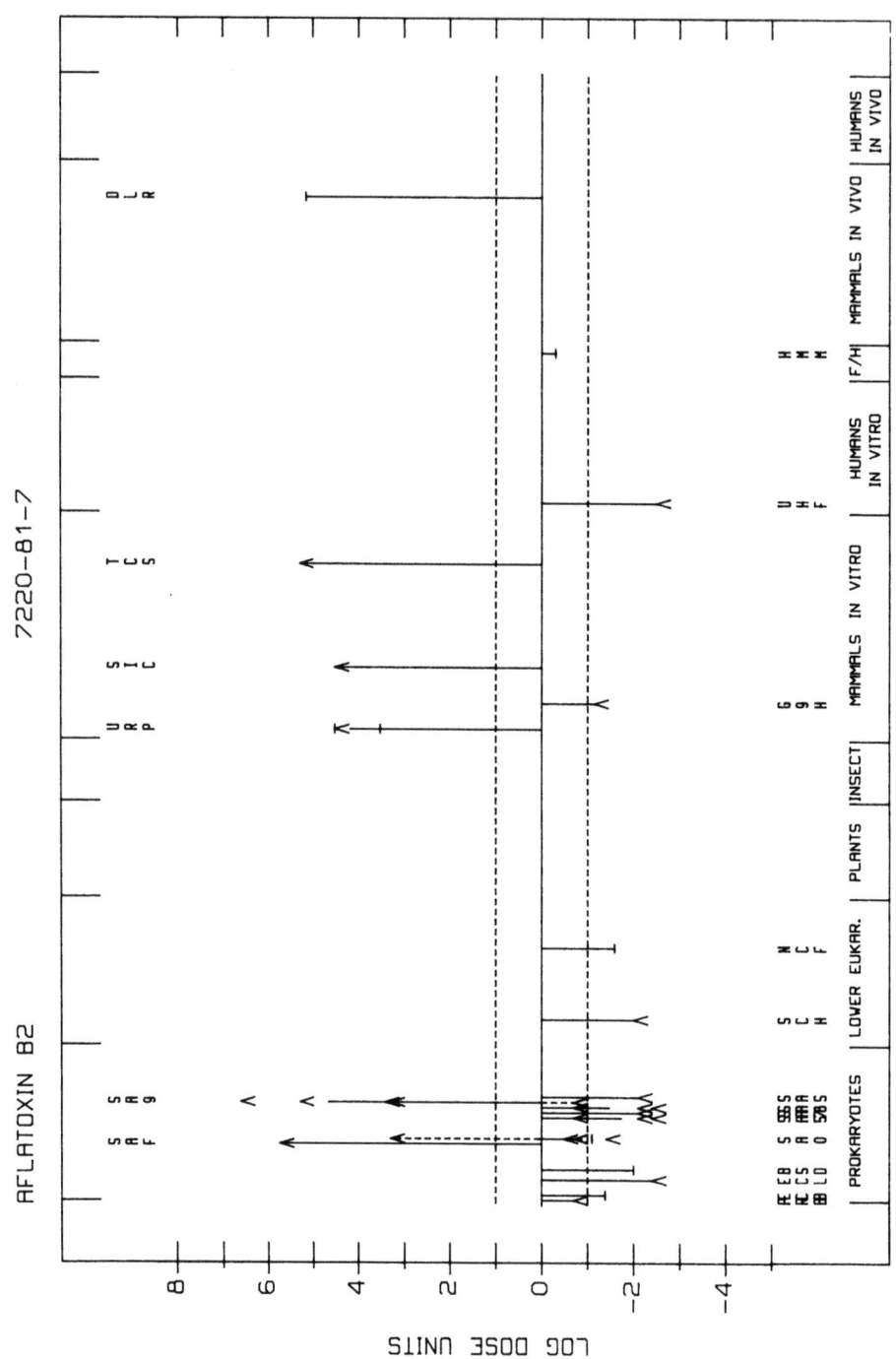

APPENDIX 2

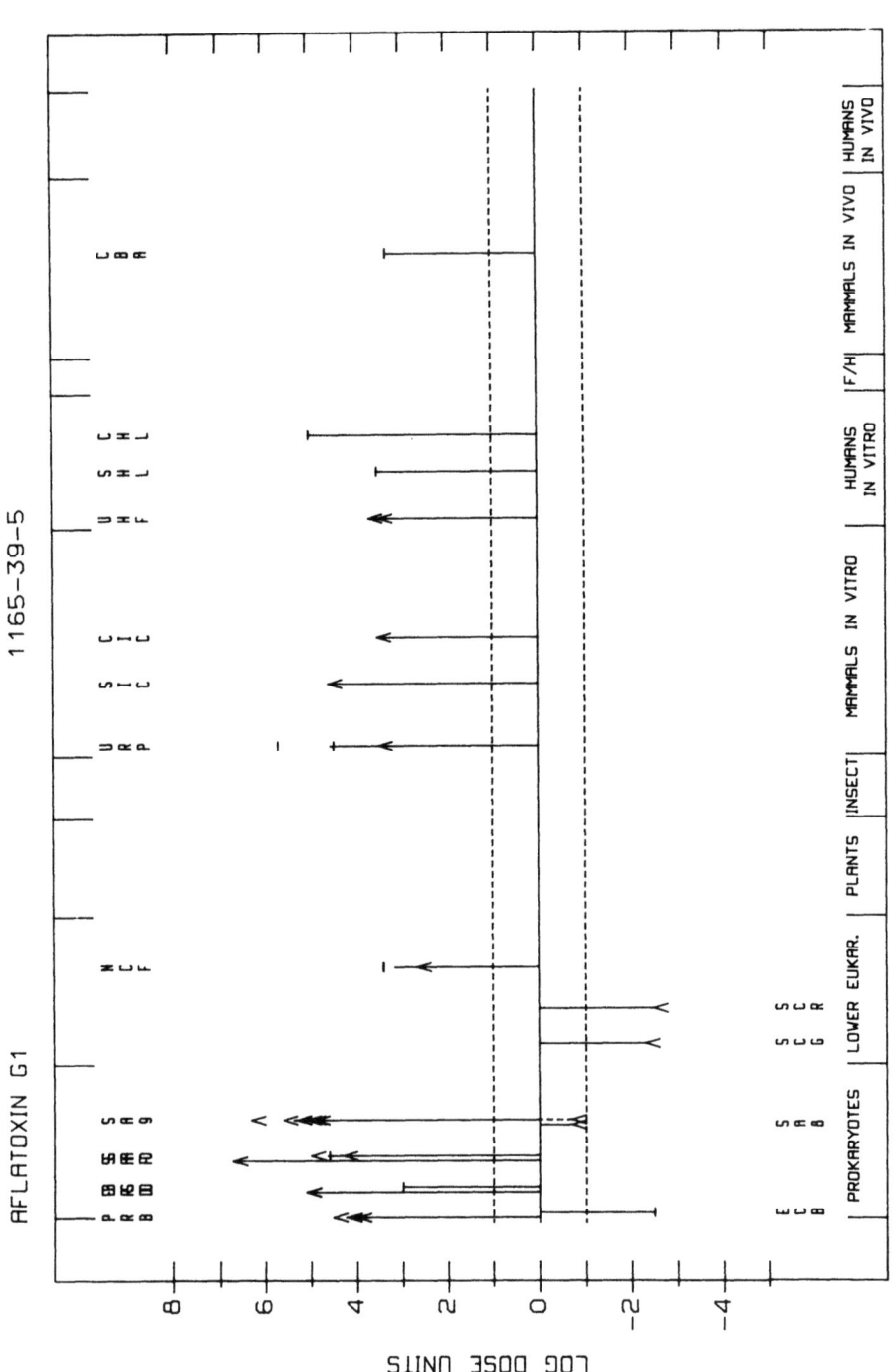

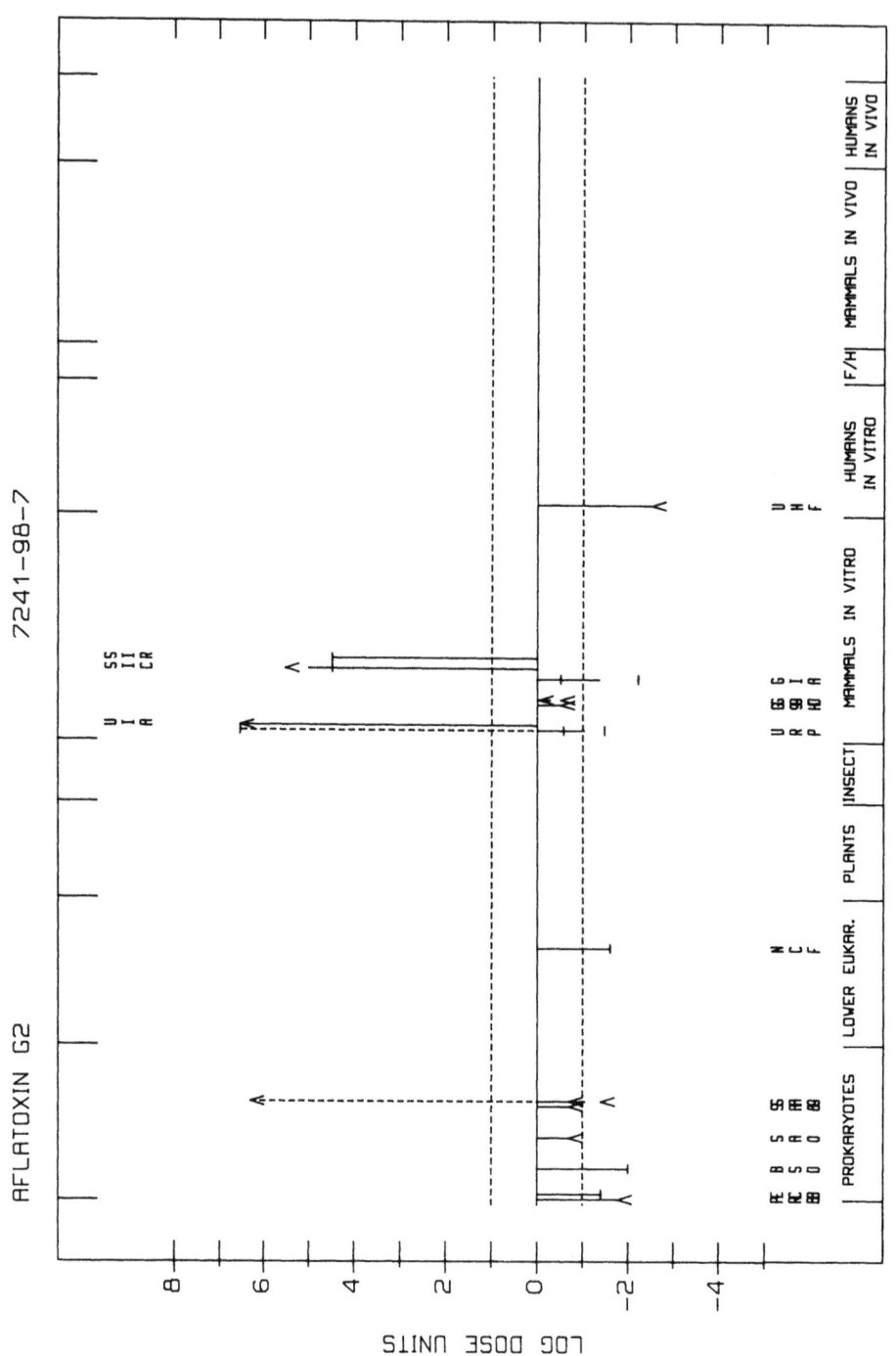

APPENDIX 2

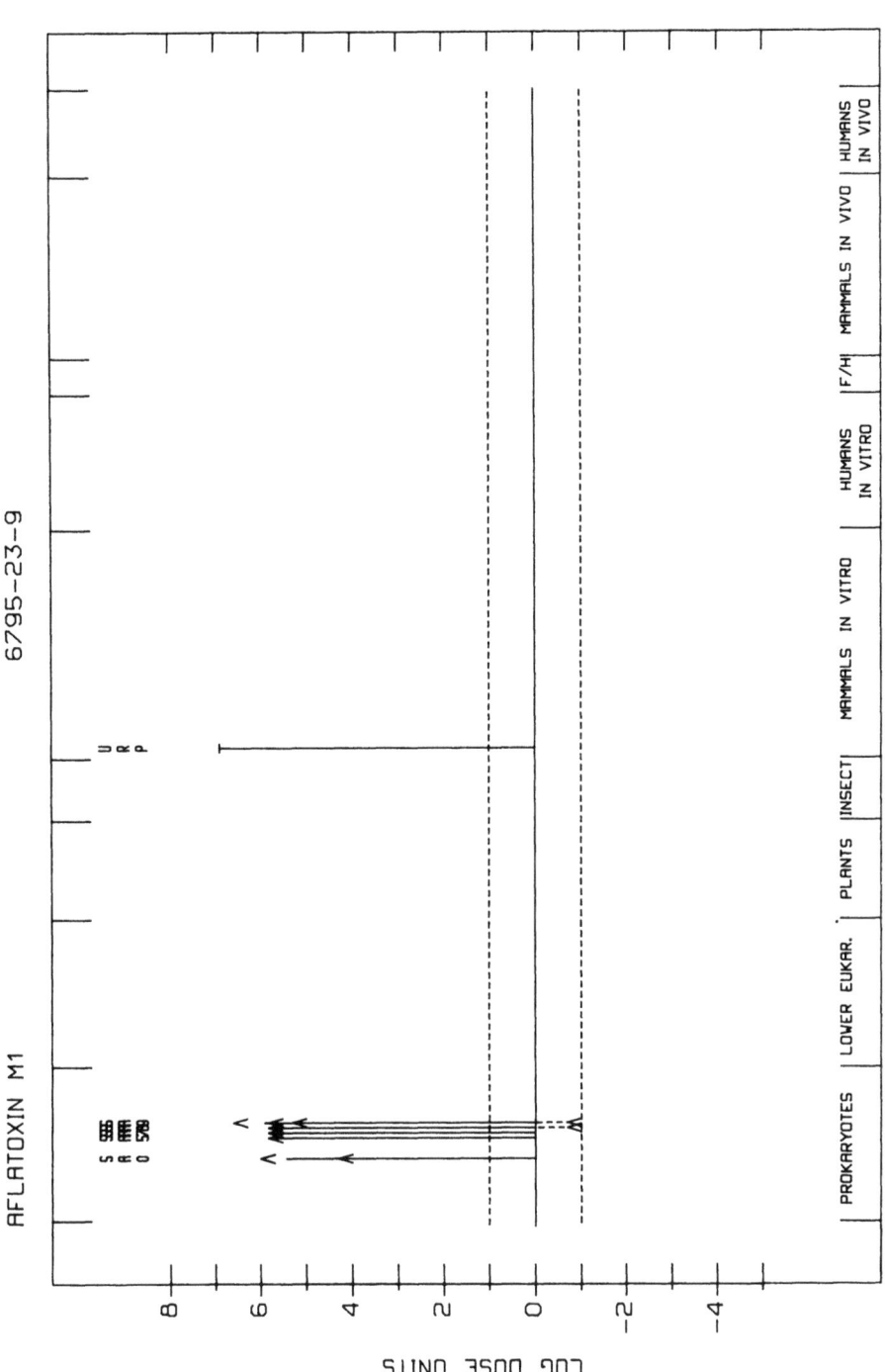

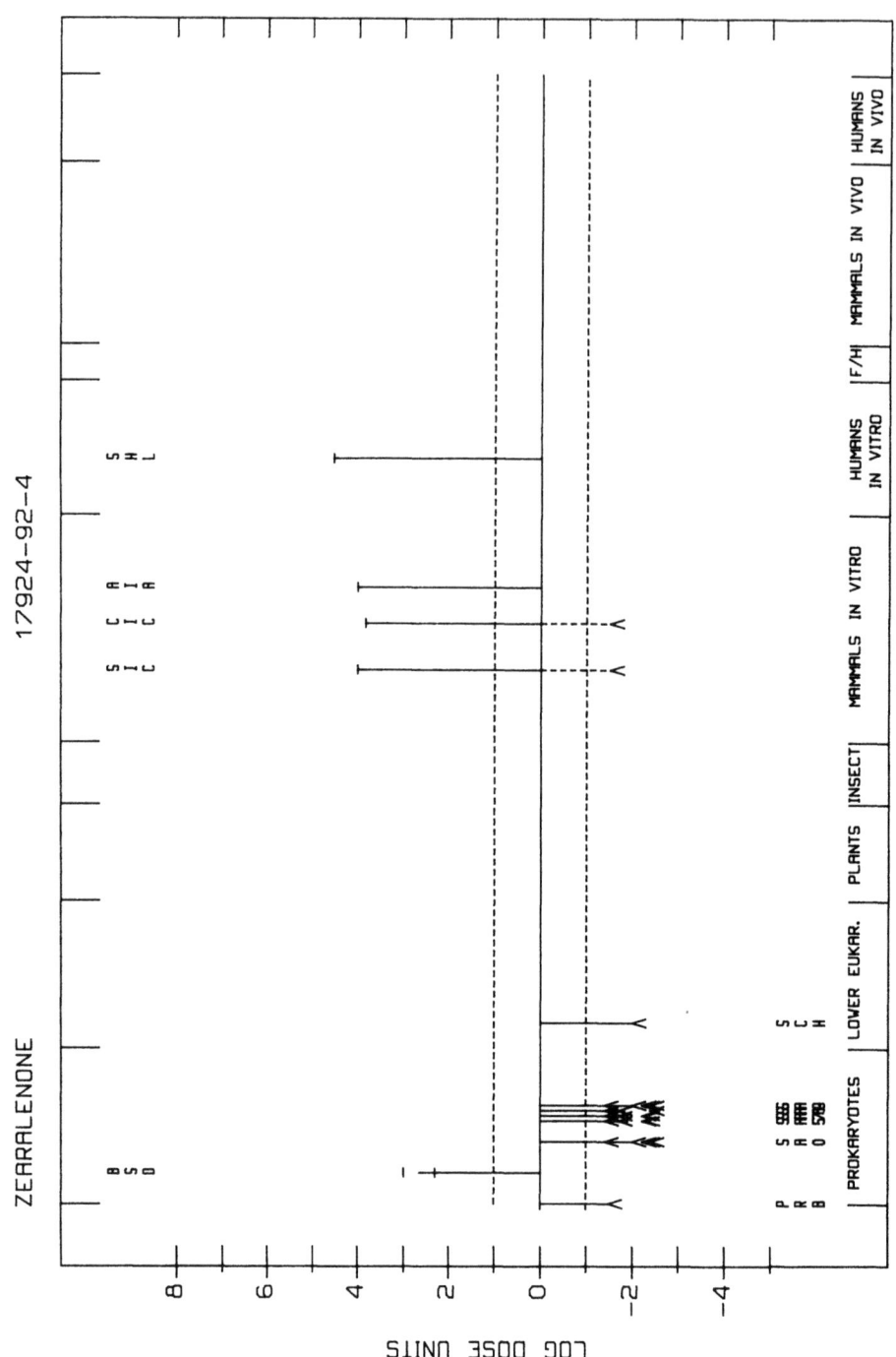

APPENDIX 2

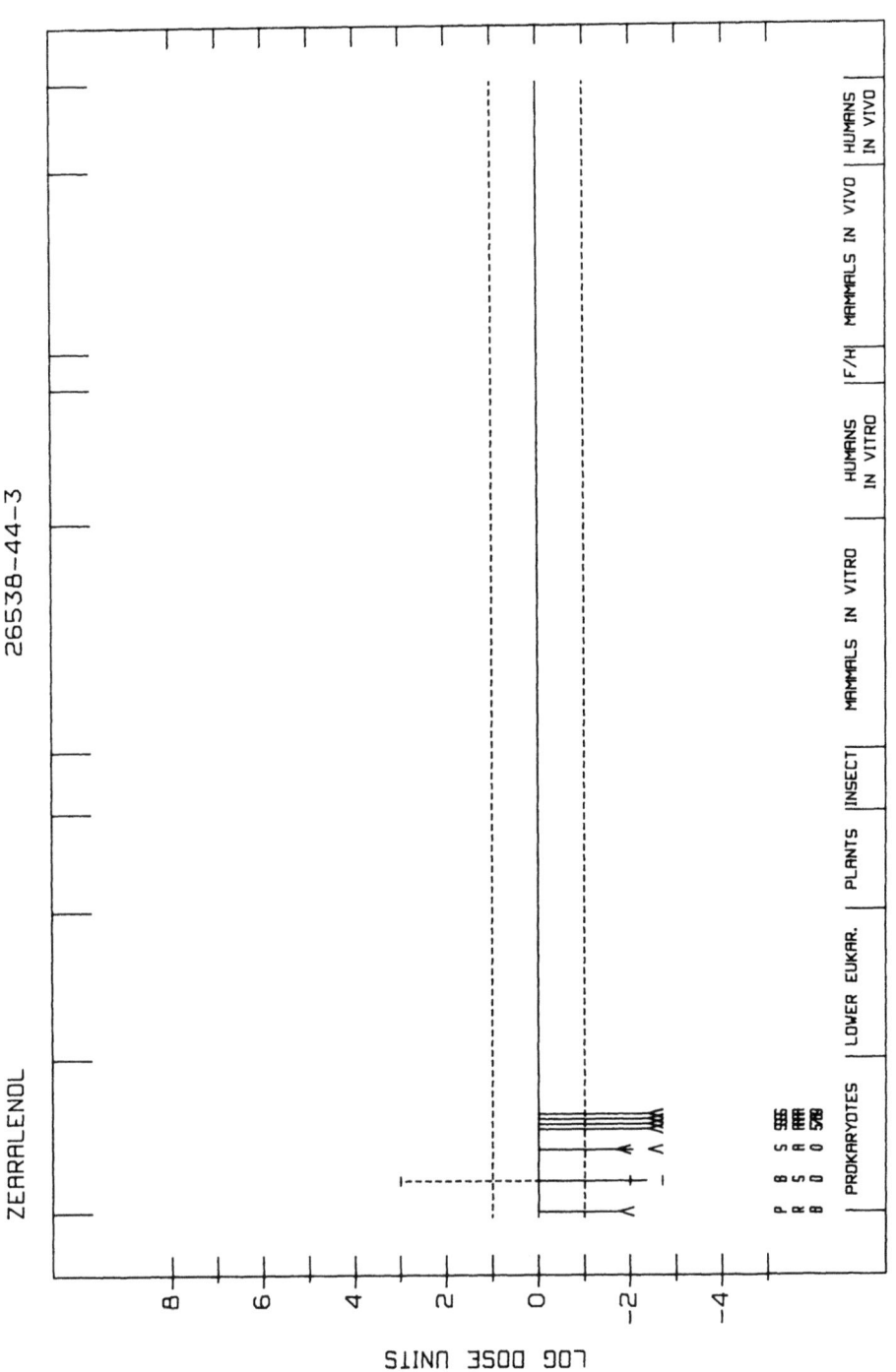

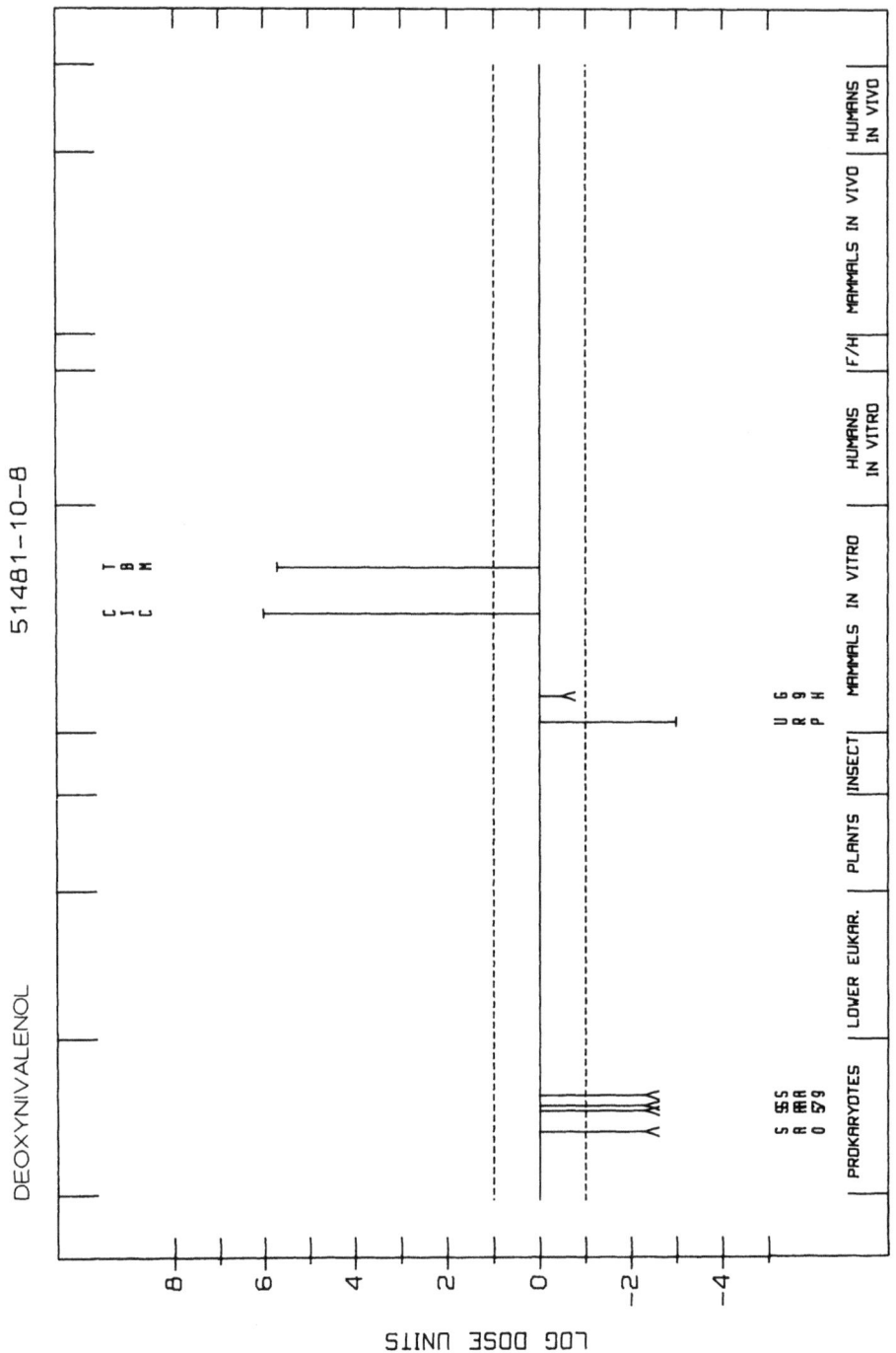

APPENDIX 2

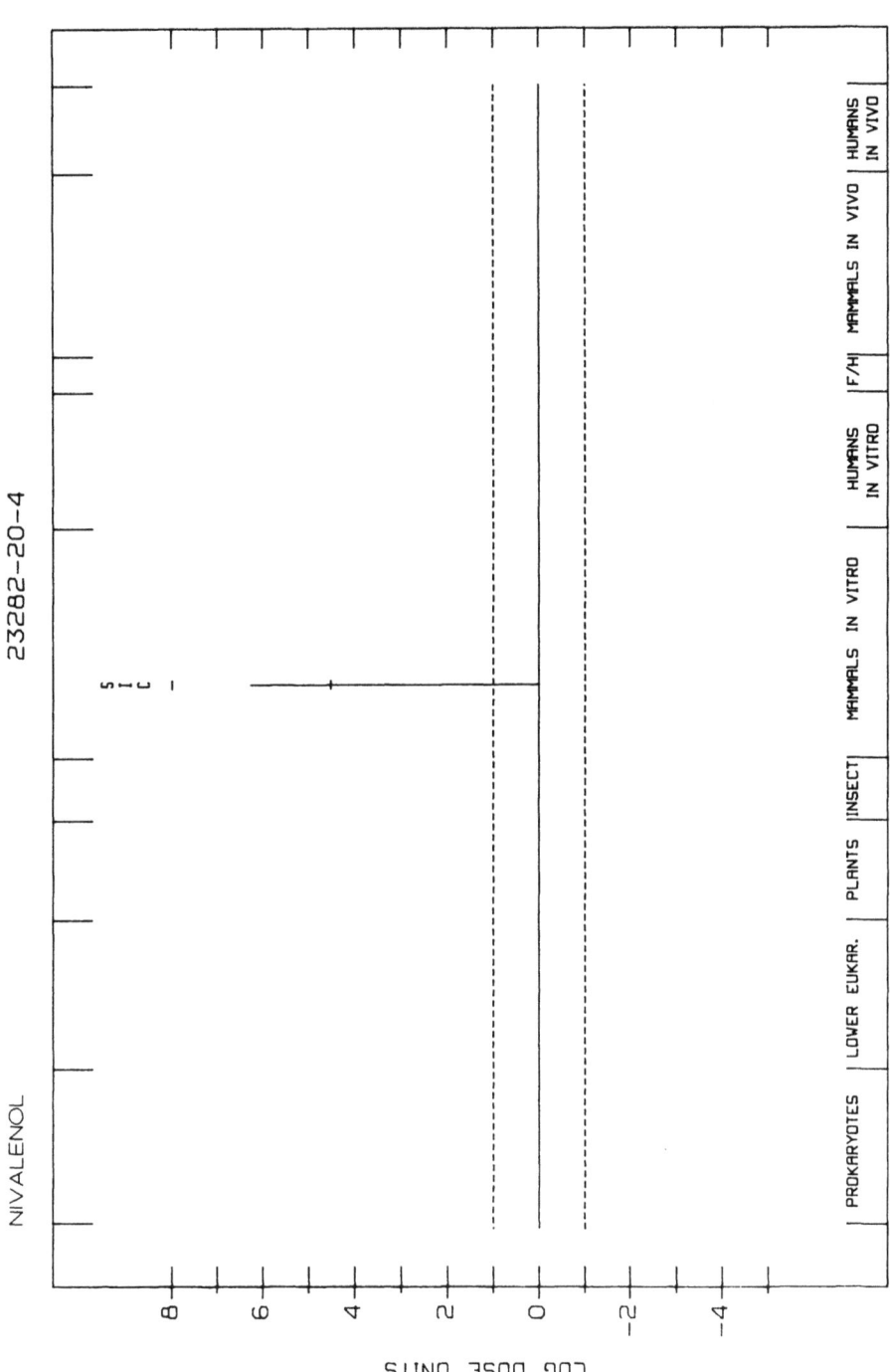

564   IARC MONOGRAPHS VOLUME 56

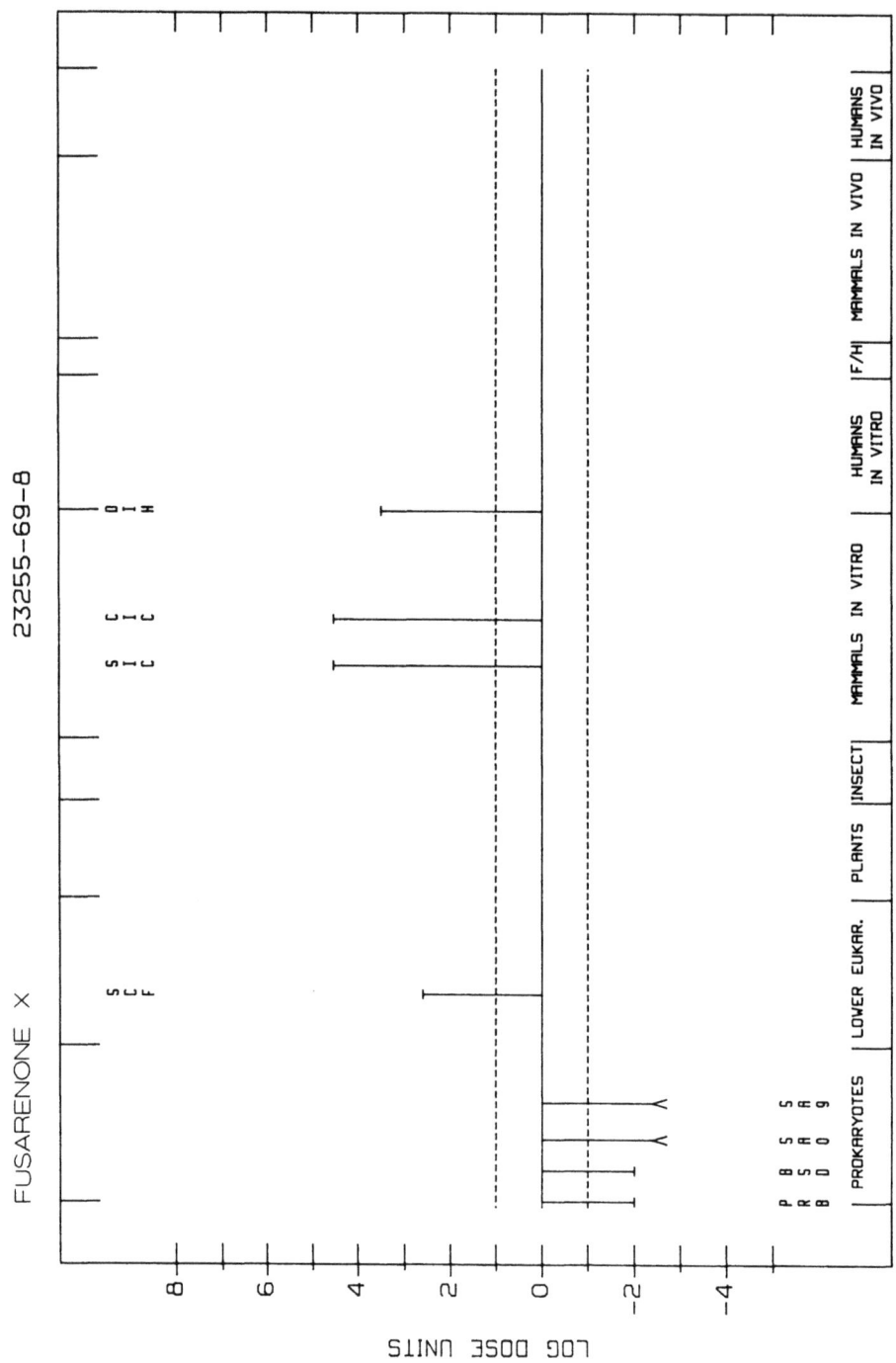

APPENDIX 2

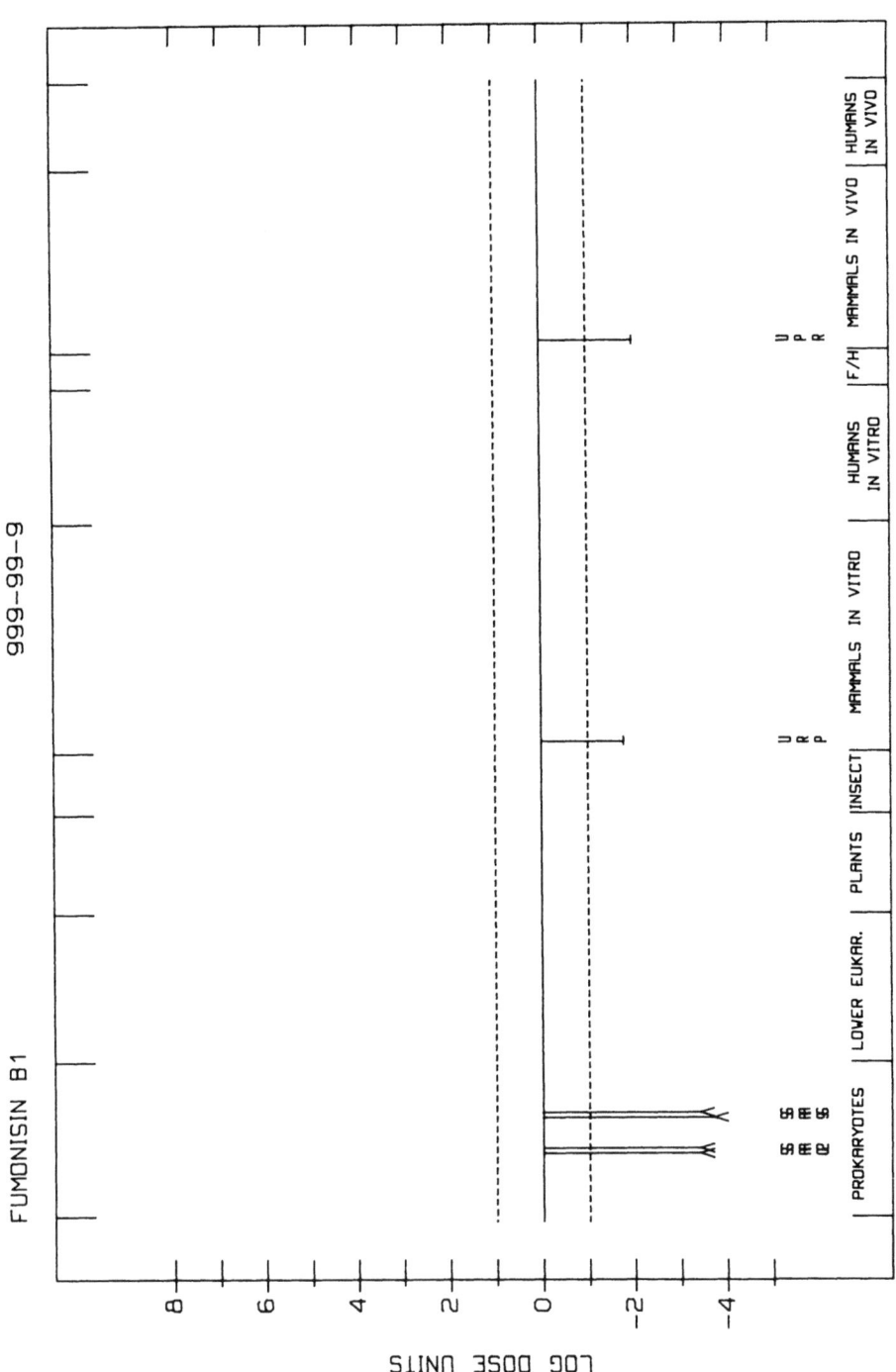

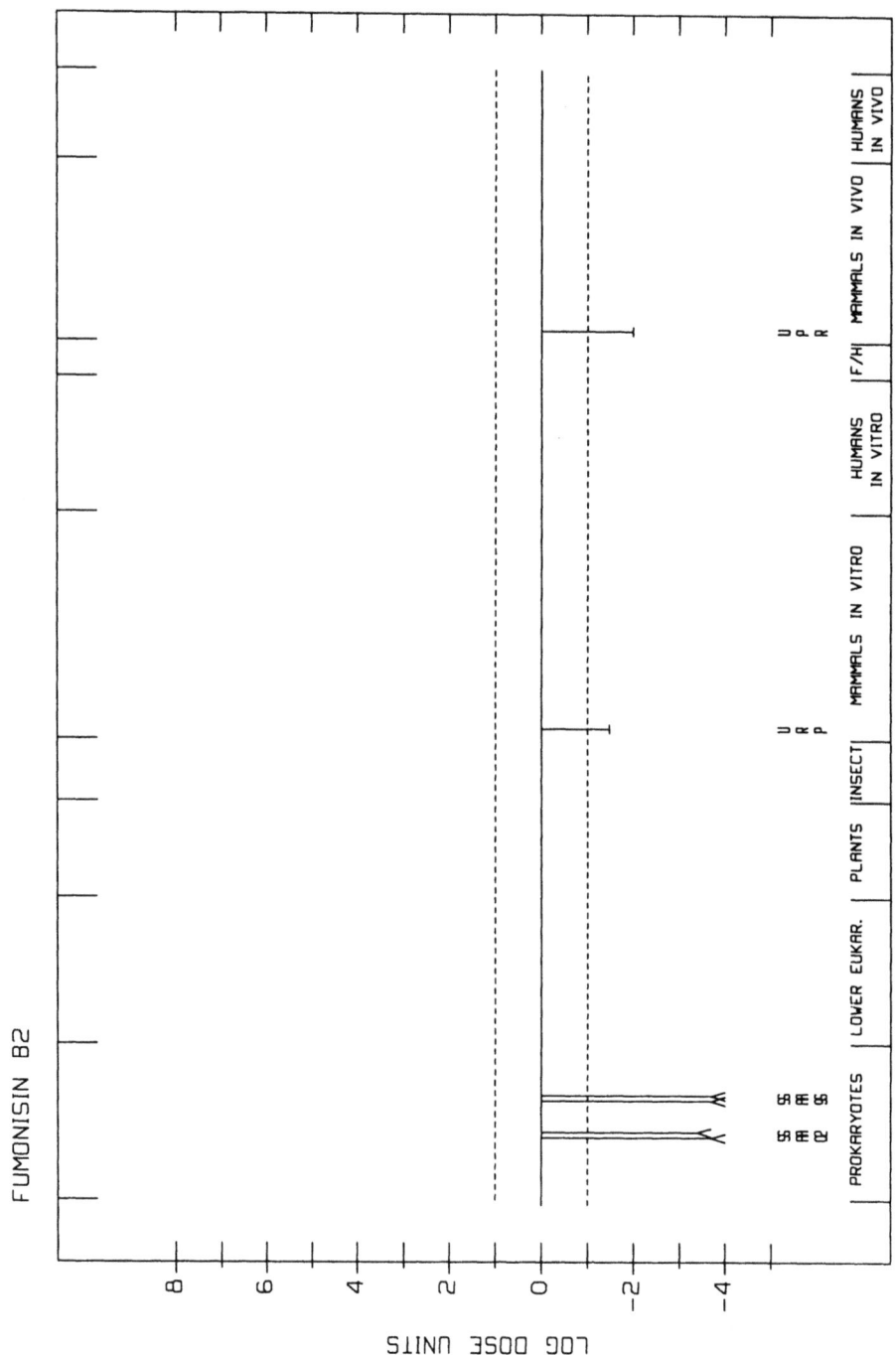

APPENDIX 2

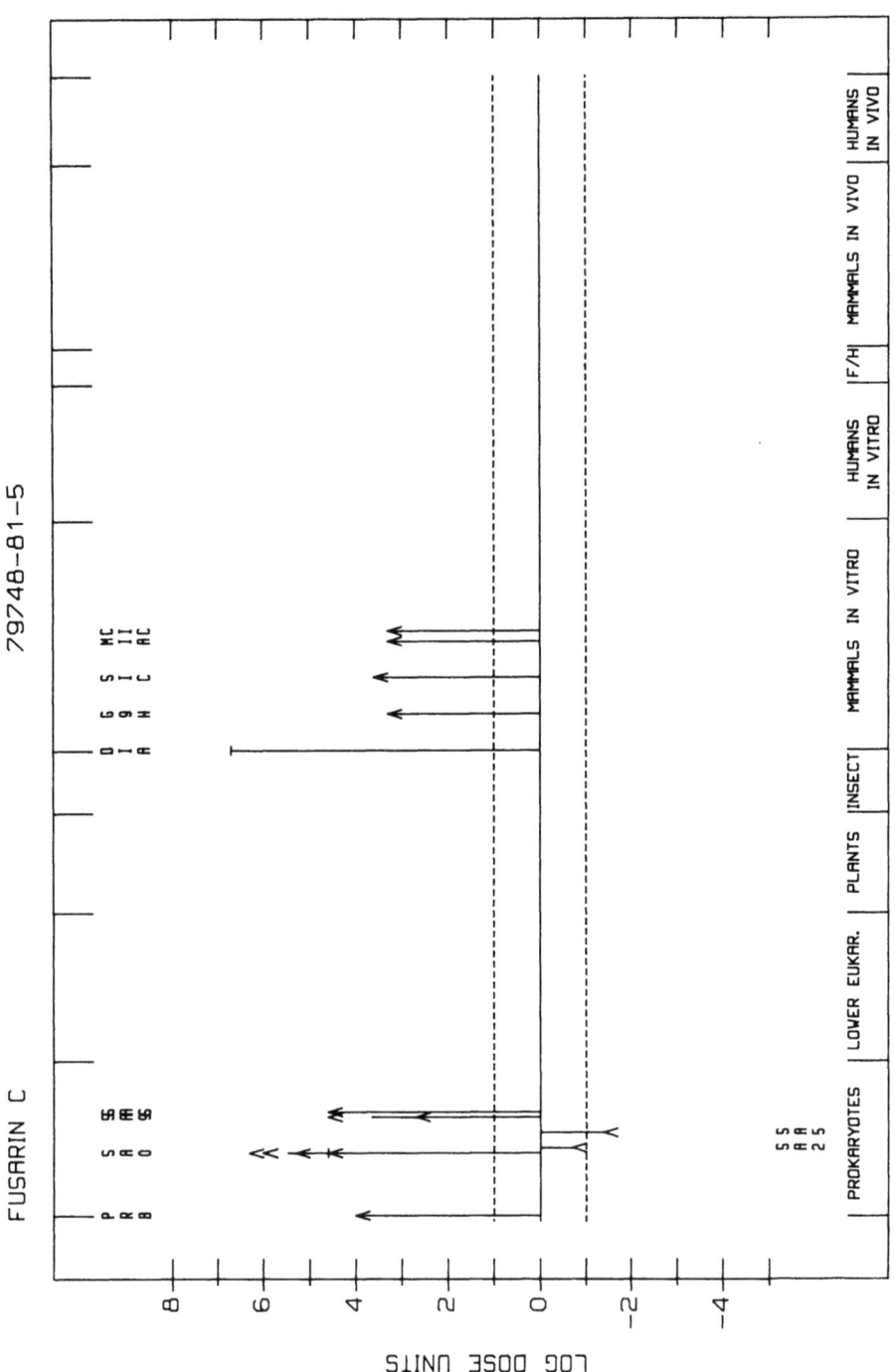

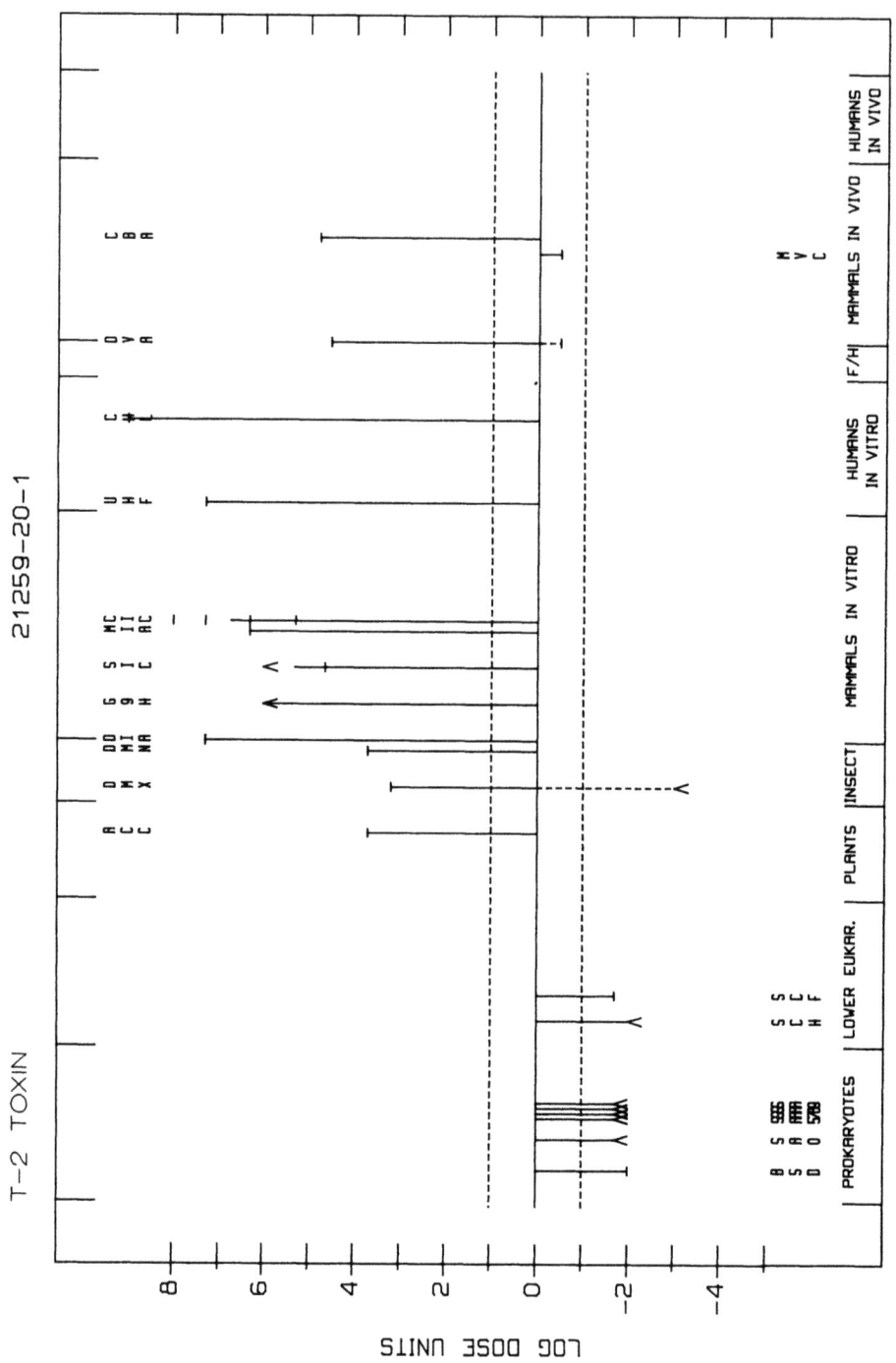

APPENDIX 2

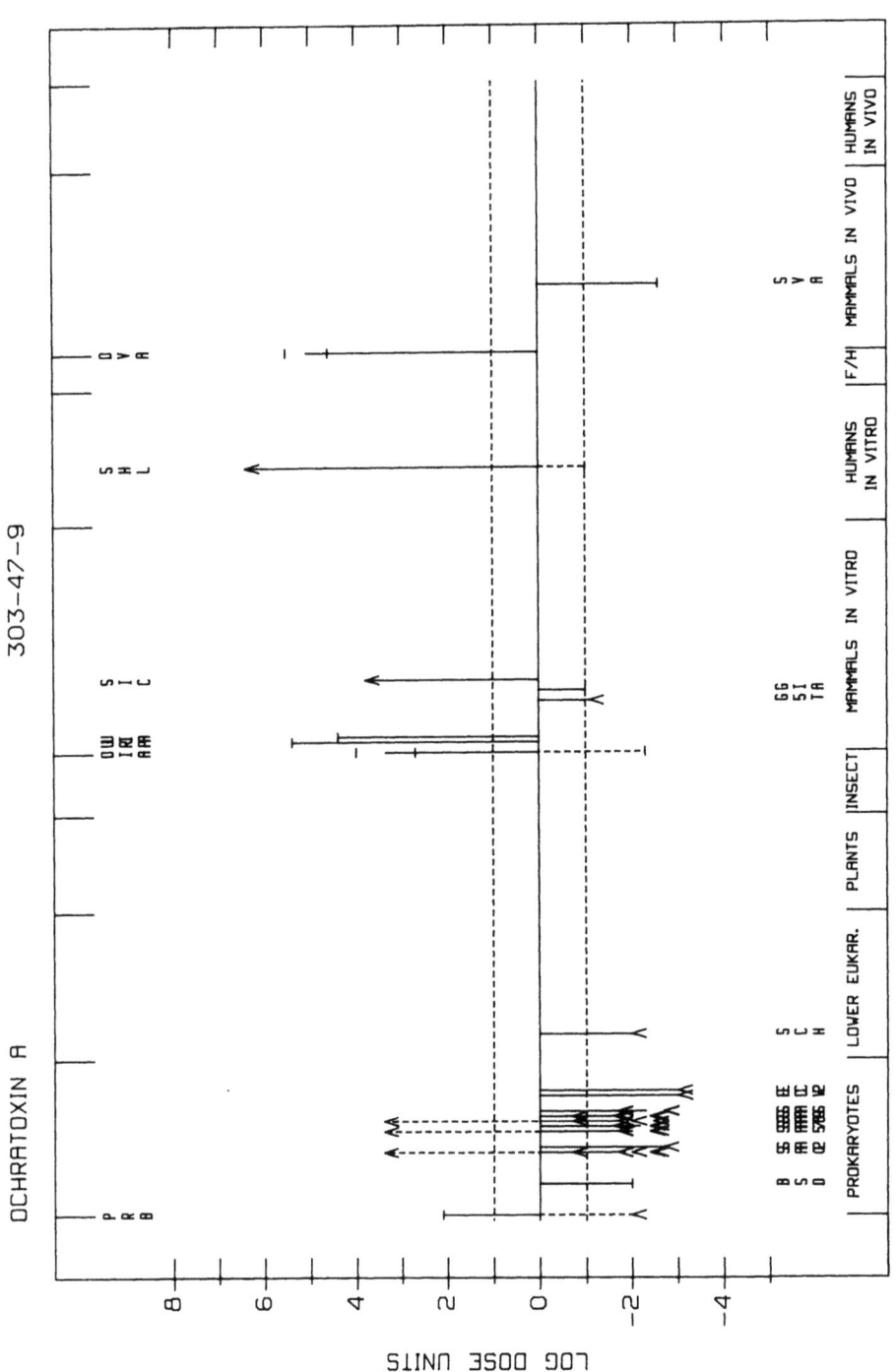

570

IARC MONOGRAPHS VOLUME 56

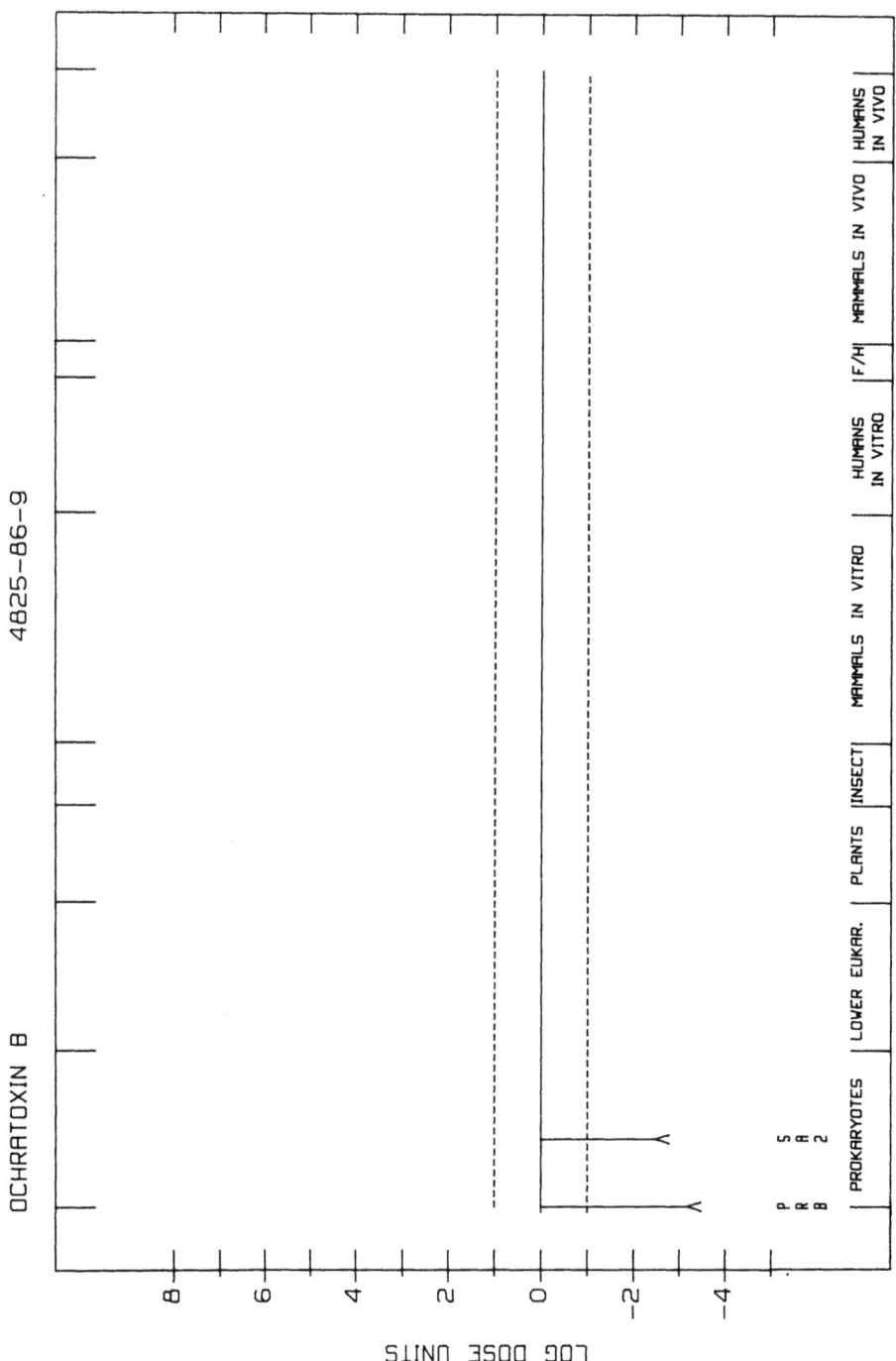

APPENDIX 2

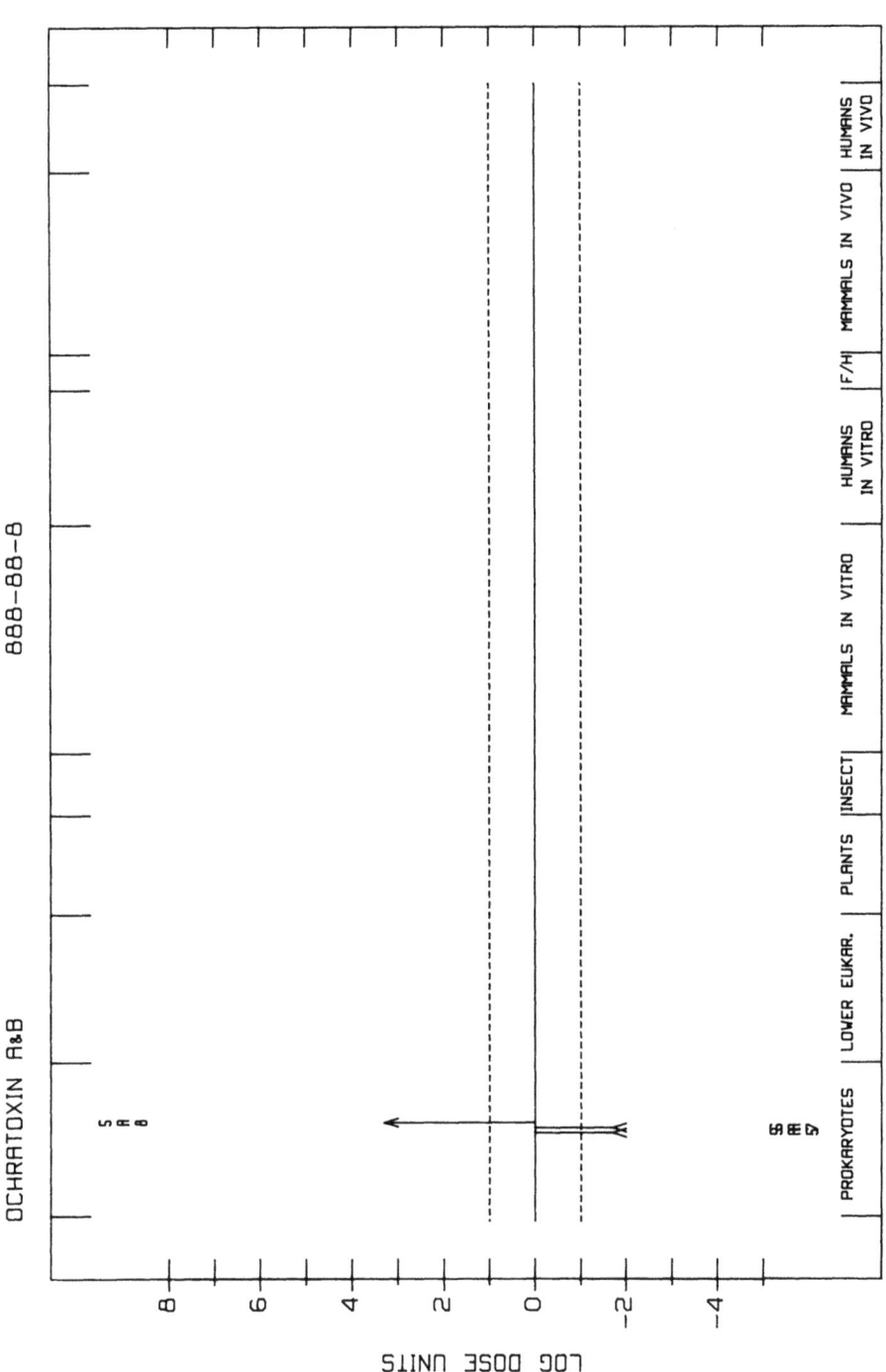

CUMULATIVE CROSS INDEX TO *IARC MONOGRAPHS ON THE EVALUATION OF CARCINOGENIC RISKS TO HUMANS*

The volume, page and year are given. References to corrigenda are given in parentheses.

A

A-α-C	40, 245 (1986); *Suppl.* 7, 56 (1987)
Acetaldehyde	36, 101 (1985) (*corr.* 42, 263); *Suppl.* 7, 77 (1987)
Acetaldehyde formylmethylhydrazone (*see* Gyromitrin)	
Acetamide	7, 197 (1974); *Suppl.* 7, 389 (1987)
Acetaminophen (*see* Paracetamol)	
Acridine orange	16, 145 (1978); *Suppl.* 7, 56 (1987)
Acriflavinium chloride	13, 31 (1977); *Suppl.* 7, 56 (1987)
Acrolein	19, 479 (1979); 36, 133 (1985); *Suppl.* 7, 78 (1987)
Acrylamide	39, 41 (1986); *Suppl.* 7, 56 (1987)
Acrylic acid	19, 47 (1979); *Suppl.* 7, 56 (1987)
Acrylic fibres	19, 86 (1979); *Suppl.* 7, 56 (1987)
Acrylonitrile	19, 73 (1979); *Suppl.* 7, 79 (1987)
Acrylonitrile–butadiene–styrene copolymers	19, 91 (1979); *Suppl.* 7, 56 (1987)
Actinolite (*see* Asbestos)	
Actinomycins	10, 29 (1976) (*corr.* 42, 255); *Suppl.* 7, 80 (1987)
Adriamycin	10, 43 (1976); *Suppl.* 7, 82 (1987)
AF-2	31, 47 (1983); *Suppl.* 7, 56 (1987)
Aflatoxins	1, 145 (1972) (*corr.* 42, 251); 10, 51 (1976); *Suppl.* 7, 83 (1987); 56, 245 (1993)
Aflatoxin B_1 (*see* Aflatoxins)	
Aflatoxin B_2 (*see* Aflatoxins)	
Aflatoxin G_1 (*see* Aflatoxins)	
Aflatoxin G_2 (*see* Aflatoxins)	
Aflatoxin M_1 (*see* Aflatoxins)	
Agaritine	31, 63 (1983); *Suppl.* 7, 56 (1987)
Alcohol drinking	44 (1988)
Aldicarb	53, 93 (1991)
Aldrin	5, 25 (1974); *Suppl.* 7, 88 (1987)
Allyl chloride	36, 39 (1985); *Suppl.* 7, 56 (1987)
Allyl isothiocyanate	36, 55 (1985); *Suppl.* 7, 56 (1987)
Allyl isovalerate	36, 69 (1985); *Suppl.* 7, 56 (1987)
Aluminium production	34, 37 (1984); *Suppl.* 7, 89 (1987)
Amaranth	8, 41 (1975); *Suppl.* 7, 56 (1987)

5-Aminoacenaphthene	*16*, 243 (1978); *Suppl. 7*, 56 (1987)
2-Aminoanthraquinone	*27*, 191 (1982); *Suppl. 7*, 56 (1987)
para-Aminoazobenzene	*8*, 53 (1975); *Suppl. 7*, 390 (1987)
ortho-Aminoazotoluene	*8*, 61 (1975) (*corr. 42*, 254); *Suppl. 7*, 56 (1987)
para-Aminobenzoic acid	*16*, 249 (1978); *Suppl. 7*, 56 (1987)
4-Aminobiphenyl	*1*, 74 (1972) (*corr. 42*, 251); *Suppl. 7*, 91 (1987)
2-Amino-3,4-dimethylimidazo[4,5-*f*]quinoline (*see* MeIQ)	
2-Amino-3,8-dimethylimidazo[4,5-*f*]quinoxaline (*see* MeIQx)	
3-Amino-1,4-dimethyl-5*H*-pyrido[4,3-*b*]indole (*see* Trp-P-1)	
2-Aminodipyrido[1,2-*a*:3',2'-*d*]imidazole (*see* Glu-P-2)	
1-Amino-2-methylanthraquinone	*27*, 199 (1982); *Suppl. 7*, 57 (1987)
2-Amino-3-methylimidazo[4,5-*f*]quinoline (*see* IQ)	
2-Amino-6-methyldipyrido[1,2-*a*:3',2'-*d*]imidazole (*see* Glu-P-1)	
2-Amino-1-methyl-6-phenylimidazo[4,5-*b*]pyridine (*see* PhIP)	
2-Amino-3-methyl-9*H*-pyrido[2,3-*b*]indole (*see* MeA-α-C)	
3-Amino-1-methyl-5*H*-pyrido[4,3-*b*]indole (*see* Trp-P-2)	
2-Amino-5-(5-nitro-2-furyl)-1,3,4-thiadiazole	*7*, 143 (1974); *Suppl. 7*, 57 (1987)
4-Amino-2-nitrophenol	*16*, 43 (1978); *Suppl. 7*, 57 (1987)
2-Amino-5-nitrothiazole	*31*, 71 (1983); *Suppl. 7*, 57 (1987)
2-Amino-9*H*-pyrido[2,3-*b*]indole (*see* A-α-C)	
11-Aminoundecanoic acid	*39*, 239 (1986); *Suppl. 7*, 57 (1987)
Amitrole	*7*, 31 (1974); *41*, 293 (1986) (*corr. 52*, 513; *Suppl. 7*, 92 (1987)
Ammonium potassium selenide (*see* Selenium and selenium compounds)	
Amorphous silica (*see also* Silica)	*42*, 39 (1987); *Suppl. 7*, 341 (1987)
Amosite (*see* Asbestos)	
Ampicillin	*50*, 153 (1990)
Anabolic steroids (*see* Androgenic (anabolic) steroids)	
Anaesthetics, volatile	*11*, 285 (1976); *Suppl. 7*, 93 (1987)
Analgesic mixtures containing phenacetin (*see also* Phenacetin)	*Suppl. 7*, 310 (1987)
Androgenic (anabolic) steroids	*Suppl. 7*, 96 (1987)
Angelicin and some synthetic derivatives (*see also* Angelicins)	*40*, 291 (1986)
Angelicin plus ultraviolet radiation (*see also* Angelicin and some synthetic derivatives)	*Suppl. 7*, 57 (1987)
Angelicins	*Suppl. 7*, 57 (1987)
Aniline	*4*, 27 (1974) (*corr. 42*, 252); *27*, 39 (1982); *Suppl. 7*, 99 (1987)
ortho-Anisidine	*27*, 63 (1982); *Suppl. 7*, 57 (1987)
para-Anisidine	*27*, 65 (1982); *Suppl. 7*, 57 (1987)
Anthanthrene	*32*, 95 (1983); *Suppl. 7*, 57 (1987)
Anthophyllite (*see* Asbestos)	
Anthracene	*32*, 105 (1983); *Suppl. 7*, 57 (1987)
Anthranilic acid	*16*, 265 (1978); *Suppl. 7*, 57 (1987)
Antimony trioxide	*47*, 291 (1989)
Antimony trisulfide	*47*, 291 (1989)
ANTU (*see* 1-Naphthylthiourea)	
Apholate	*9*, 31 (1975); *Suppl. 7*, 57 (1987)
Aramite®	*5*, 39 (1974); *Suppl. 7*, 57 (1987)
Areca nut (*see* Betel quid)	
Arsanilic acid (*see* Arsenic and arsenic compounds)	

Arsenic and arsenic compounds	*1*, 41 (1972); *2*, 48 (1973); *23*, 39 (1980); Suppl. 7, 100 (1987)
Arsenic pentoxide (*see* Arsenic and arsenic compounds)	
Arsenic sulfide (*see* Arsenic and arsenic compounds)	
Arsenic trioxide (*see* Arsenic and arsenic compounds)	
Arsine (*see* Arsenic and arsenic compounds)	
Asbestos	*2*, 17 (1973) (*corr.* 42, 252); *14* (1977) (*corr.* 42, 256); Suppl. 7, 106 (1987) (*corr.* 45, 283)
Atrazine	*53*, 441 (1991)
Attapulgite	*42*, 159 (1987); Suppl. 7, 117 (1987)
Auramine (technical-grade)	*1*, 69 (1972) (*corr.* 42, 251); Suppl. 7, 118 (1987)
Auramine, manufacture of (*see also* Auramine, technical-grade)	Suppl. 7, 118 (1987)
Aurothioglucose	*13*, 39 (1977); Suppl. 7, 57 (1987)
Azacitidine	*26*, 37 (1981); Suppl. 7, 57 (1987); *50*, 47 (1990)
5-Azacytidine (*see* Azacitidine)	
Azaserine	*10*, 73 (1976) (*corr.* 42, 255); Suppl. 7, 57 (1987)
Azathioprine	*26*, 47 (1981); Suppl. 7, 119 (1987)
Aziridine	*9*, 37 (1975); Suppl. 7, 58 (1987)
2-(1-Aziridinyl)ethanol	*9*, 47 (1975); Suppl. 7, 58 (1987)
Aziridyl benzoquinone	*9*, 51 (1975); Suppl. 7, 58 (1987)
Azobenzene	*8*, 75 (1975); Suppl. 7, 58 (1987)

B

Barium chromate (*see* Chromium and chromium compounds)	
Basic chromic sulfate (*see* Chromium and chromium compounds)	
BCNU (*see* Bischloroethyl nitrosourea)	
Benz[*a*]acridine	*32*, 123 (1983); Suppl. 7, 58 (1987)
Benz[*c*]acridine	*3*, 241 (1973); *32*, 129 (1983); Suppl. 7, 58 (1987)
Benzal chloride (*see also* α-Chlorinated toluenes)	*29*, 65 (1982); Suppl. 7, 148 (1987)
Benz[*a*]anthracene	*3*, 45 (1973); *32*, 135 (1983); Suppl. 7, 58 (1987)
Benzene	*7*, 203 (1974) (*corr.* 42, 254); *29*, 93, 391 (1982); Suppl. 7, 120 (1987)
Benzidine	*1*, 80 (1972); *29*, 149, 391 (1982); Suppl. 7, 123 (1987)
Benzidine-based dyes	Suppl. 7, 125 (1987)
Benzo[*b*]fluoranthene	*3*, 69 (1973); *32*, 147 (1983); Suppl. 7, 58 (1987)
Benzo[*j*]fluoranthene	*3*, 82 (1973); *32*, 155 (1983); Suppl. 7, 58 (1987)
Benzo[*k*]fluoranthene	*32*, 163 (1983); Suppl. 7, 58 (1987)
Benzo[*ghi*]fluoranthene	*32*, 171 (1983); Suppl. 7, 58 (1987)
Benzo[*a*]fluorene	*32*, 177 (1983); Suppl. 7, 58 (1987)
Benzo[*b*]fluorene	*32*, 183 (1983); Suppl. 7, 58 (1987)
Benzo[*c*]fluorene	*32*, 189 (1983); Suppl. 7, 58 (1987)
Benzo[*ghi*]perylene	*32*, 195 (1983); Suppl. 7, 58 (1987)

Benzo[c]phenanthrene	32, 205 (1983); *Suppl. 7*, 58 (1987)
Benzo[a]pyrene	3, 91 (1973); 32, 211 (1983); *Suppl. 7*, 58 (1987)
Benzo[e]pyrene	3, 137 (1973); 32, 225 (1983); *Suppl. 7*, 58 (1987)
para-Benzoquinone dioxime	29, 185 (1982); *Suppl. 7*, 58 (1987)
Benzotrichloride (*see also* α-Chlorinated toluenes)	29, 73 (1982); *Suppl. 7*, 148 (1987)
Benzoyl chloride	29, 83 (1982) (*corr.* 42, 261); *Suppl. 7*, 126 (1987)
Benzoyl peroxide	36, 267 (1985); *Suppl. 7*, 58 (1987)
Benzyl acetate	40, 109 (1986); *Suppl. 7*, 58 (1987)
Benzyl chloride (*see also* α-Chlorinated toluenes)	11, 217 (1976) (*corr.* 42, 256); 29, 49 (1982); *Suppl. 7*, 148 (1987)
Benzyl violet 4B	16, 153 (1978); *Suppl. 7*, 58 (1987)
Bertrandite (*see* Beryllium and beryllium compounds)	
Beryllium and beryllium compounds	1, 17 (1972); 23, 143 (1980) (*corr.* 42, 260); *Suppl. 7*, 127 (1987)
Beryllium acetate (*see* Beryllium and beryllium compounds)	
Beryllium acetate, basic (*see* Beryllium and beryllium compounds)	
Beryllium–aluminium alloy (*see* Beryllium and beryllium compounds)	
Beryllium carbonate (*see* Beryllium and beryllium compounds)	
Beryllium chloride (*see* Beryllium and beryllium compounds)	
Beryllium–copper alloy (*see* Beryllium and beryllium compounds)	
Beryllium–copper–cobalt alloy (*see* Beryllium and beryllium compounds)	
Beryllium fluoride (*see* Beryllium and beryllium compounds)	
Beryllium hydroxide (*see* Beryllium and beryllium compounds)	
Beryllium–nickel alloy (*see* Beryllium and beryllium compounds)	
Beryllium oxide (*see* Beryllium and beryllium compounds)	
Beryllium phosphate (*see* Beryllium and beryllium compounds)	
Beryllium silicate (*see* Beryllium and beryllium compounds)	
Beryllium sulfate (*see* Beryllium and beryllium compounds)	
Beryl ore (*see* Beryllium and beryllium compounds)	
Betel quid	37, 141 (1985); *Suppl. 7*, 128 (1987)
Betel-quid chewing (*see* Betel quid)	
BHA (*see* Butylated hydroxyanisole)	
BHT (*see* Butylated hydroxytoluene)	
Bis(1-aziridinyl)morpholinophosphine sulfide	9, 55 (1975); *Suppl. 7*, 58 (1987)
Bis(2-chloroethyl)ether	9, 117 (1975); *Suppl. 7*, 58 (1987)
N,N-Bis(2-chloroethyl)-2-naphthylamine	4, 119 (1974) (*corr.* 42, 253); *Suppl. 7*, 130 (1987)
Bischloroethyl nitrosourea (*see also* Chloroethyl nitrosoureas)	26, 79 (1981); *Suppl. 7*, 150 (1987)
1,2-Bis(chloromethoxy)ethane	15, 31 (1977); *Suppl. 7*, 58 (1987)
1,4-Bis(chloromethoxymethyl)benzene	15, 37 (1977); *Suppl. 7*, 58 (1987)
Bis(chloromethyl)ether	4, 231 (1974) (*corr.* 42, 253); *Suppl. 7*, 131 (1987)
Bis(2-chloro-1-methylethyl)ether	41, 149 (1986); *Suppl. 7*, 59 (1987)
Bis(2,3-epoxycyclopentyl)ether	47, 231 (1989)
Bisphenol A diglycidyl ether (*see* Glycidyl ethers)	
Bisulfites (*see* Sulfur dioxide and some sulfites, bisulfites and metabisulfites)	
Bitumens	35, 39 (1985); *Suppl. 7*, 133 (1987)
Bleomycins	26, 97 (1981); *Suppl. 7*, 134 (1987)
Blue VRS	16, 163 (1978); *Suppl. 7*, 59 (1987)

Boot and shoe manufacture and repair	25, 249 (1981); *Suppl. 7*, 232 (1987)
Bracken fern	40, 47 (1986); *Suppl. 7*, 135 (1987)
Brilliant Blue FCF, disodium salt	16, 171 (1978) *(corr. 42, 257)*; *Suppl. 7*, 59 (1987)
Bromochloroacetonitrile *(see* Halogenated acetonitriles*)*	
Bromodichloromethane	52, 179 (1991)
Bromoethane	52, 299 (1991)
Bromoform	52, 213 (1991)
1,3-Butadiene	39, 155 (1986) *(corr. 42, 264)*; *Suppl. 7*, 136 (1987); 54, 237 (1992)
1,4-Butanediol dimethanesulfonate	4, 247 (1974); *Suppl. 7*, 137 (1987)
n-Butyl acrylate	39, 67 (1986); *Suppl. 7*, 59 (1987)
Butylated hydroxyanisole	40, 123 (1986); *Suppl. 7*, 59 (1987)
Butylated hydroxytoluene	40, 161 (1986); *Suppl. 7*, 59 (1987)
Butyl benzyl phthalate	29, 193 (1982) *(corr. 42, 261)*; *Suppl. 7*, 59 (1987)
β-Butyrolactone	11, 225 (1976); *Suppl. 7*, 59 (1987)
γ-Butyrolactone	11, 231 (1976); *Suppl. 7*, 59 (1987)

C

Cabinet-making *(see* Furniture and cabinet-making*)*	
Cadmium acetate *(see* Cadmium and cadmium compounds*)*	
Cadmium and cadmium compounds	2, 74 (1973); 11, 39 (1976) *(corr. 42, 255)*; *Suppl. 7*, 139 (1987)
Cadmium chloride *(see* Cadmium and cadmium compounds*)*	
Cadmium oxide *(see* Cadmium and cadmium compounds*)*	
Cadmium sulfate *(see* Cadmium and cadmium compounds*)*	
Cadmium sulfide *(see* Cadmium and cadmium compounds*)*	
Caffeic acid	56, 115 (1993)
Caffeine	51, 291 (1991)
Calcium arsenate *(see* Arsenic and arsenic compounds*)*	
Calcium chromate *(see* Chromium and chromium compounds*)*	
Calcium cyclamate *(see* Cyclamates*)*	
Calcium saccharin *(see* Saccharin*)*	
Cantharidin	10, 79 (1976); *Suppl. 7*, 59 (1987)
Caprolactam	19, 115 (1979) *(corr. 42, 258)*; 39, 247 (1986) *(corr. 42, 264)*; *Suppl. 7*, 390 (1987)
Captafol	53, 353 (1991)
Captan	30, 295 (1983); *Suppl. 7*, 59 (1987)
Carbaryl	12, 37 (1976); *Suppl. 7*, 59 (1987)
Carbazole	32, 239 (1983); *Suppl. 7*, 59 (1987)
3-Carbethoxypsoralen	40, 317 (1986); *Suppl. 7*, 59 (1987)
Carbon blacks	3, 22 (1973); 33, 35 (1984); *Suppl. 7*, 142 (1987)
Carbon tetrachloride	1, 53 (1972); 20, 371 (1979); *Suppl. 7*, 143 (1987)
Carmoisine	8, 83 (1975); *Suppl. 7*, 59 (1987)
Carpentry and joinery	25, 139 (1981); *Suppl. 7*, 378 (1987)
Carrageenan	10, 181 (1976) *(corr. 42, 255)*; 31, 79 (1983); *Suppl. 7*, 59 (1987)

Catechol	15, 155 (1977); Suppl. 7, 59 (1987)
CCNU (see 1-(2-Chloroethyl)-3-cyclohexyl-1-nitrosourea)	
Ceramic fibres (see Man-made mineral fibres)	
Chemotherapy, combined, including alkylating agents (see MOPP and other combined chemotherapy including alkylating agents)	
Chlorambucil	9, 125 (1975); 26, 115 (1981); Suppl. 7, 144 (1987)
Chloramphenicol	10, 85 (1976); Suppl. 7, 145 (1987); 50, 169 (1990)
Chlordane (see also Chlordane/Heptachlor)	20, 45 (1979) (corr. 42, 258)
Chlordane/Heptachlor	Suppl. 7, 146 (1987); 53, 115 (1991)
Chlordecone	20, 67 (1979); Suppl. 7, 59 (1987)
Chlordimeform	30, 61 (1983); Suppl. 7, 59 (1987)
Chlorendic acid	48, 45 (1990)
Chlorinated dibenzodioxins (other than TCDD)	15, 41 (1977); Suppl. 7, 59 (1987)
Chlorinated drinking-water	52, 45 (1991)
Chlorinated paraffins	48, 55 (1990)
α-Chlorinated toluenes	Suppl. 7, 148 (1987)
Chlormadinone acetate (see also Progestins; Combined oral contraceptives)	6, 149 (1974); 21, 365 (1979)
Chlornaphazine (see N,N-Bis(2-chloroethyl)-2-naphthylamine)	
Chloroacetonitrile (see Halogenated acetonitriles)	
Chlorobenzilate	5, 75 (1974); 30, 73 (1983); Suppl. 7, 60 (1987)
Chlorodibromomethane	52, 243 (1991)
Chlorodifluoromethane	41, 237 (1986) (corr. 51, 483); Suppl. 7, 149 (1987)
Chloroethane	52, 315 (1991)
1-(2-Chloroethyl)-3-cyclohexyl-1-nitrosourea (see also Chloroethyl nitrosoureas)	26, 137 (1981) (corr. 42, 260); Suppl. 7, 150 (1987)
1-(2-Chloroethyl)-3-(4-methylcyclohexyl)-1-nitrosourea (see also Chloroethyl nitrosoureas)	Suppl. 7, 150 (1987)
Chloroethyl nitrosoureas	Suppl. 7, 150 (1987)
Chlorofluoromethane	41, 229 (1986); Suppl. 7, 60 (1987)
Chloroform	1, 61 (1972); 20, 401 (1979); Suppl. 7, 152 (1987)
Chloromethyl methyl ether (technical-grade) (see also Bis(chloromethyl)ether)	4, 239 (1974); Suppl. 7, 131 (1987)
(4-Chloro-2-methylphenoxy)acetic acid (see MCPA)	
Chlorophenols	Suppl. 7, 154 (1987)
Chlorophenols (occupational exposures to)	41, 319 (1986)
Chlorophenoxy herbicides	Suppl. 7, 156 (1987)
Chlorophenoxy herbicides (occupational exposures to)	41, 357 (1986)
4-Chloro-*ortho*-phenylenediamine	27, 81 (1982); Suppl. 7, 60 (1987)
4-Chloro-*meta*-phenylenediamine	27, 82 (1982); Suppl. 7, 60 (1987)
Chloroprene	19, 131 (1979); Suppl. 7, 160 (1987)
Chloropropham	12, 55 (1976); Suppl. 7, 60 (1987)
Chloroquine	13, 47 (1977); Suppl. 7, 60 (1987)
Chlorothalonil	30, 319 (1983); Suppl. 7, 60 (1987)
para-Chloro-*ortho*-toluidine and its strong acid salts (see also Chlordimeform)	16, 277 (1978); 30, 65 (1983); Suppl. 7, 60 (1987); 48, 123 (1990)
Chlorotrianisene (see also Nonsteroidal oestrogens)	21, 139 (1979)

2-Chloro-1,1,1-trifluoroethane	*41*, 253 (1986); *Suppl. 7*, 60 (1987)
Chlorozotocin	*50*, 65 (1990)
Cholesterol	*10*, 99 (1976); *31*, 95 (1983); *Suppl. 7*, 161 (1987)
Chromic acetate (*see* Chromium and chromium compounds)	
Chromic chloride (*see* Chromium and chromium compounds)	
Chromic oxide (*see* Chromium and chromium compounds)	
Chromic phosphate (*see* Chromium and chromium compounds)	
Chromite ore (*see* Chromium and chromium compounds)	
Chromium and chromium compounds	*2*, 100 (1973); *23*, 205 (1980); *Suppl. 7*, 165 (1987); *49*, 49 (1990) (*corr. 51*, 483)
Chromium carbonyl (*see* Chromium and chromium compounds)	
Chromium potassium sulfate (*see* Chromium and chromium compounds)	
Chromium sulfate (*see* Chromium and chromium compounds)	
Chromium trioxide (*see* Chromium and chromium compounds)	
Chrysazin (*see* Dantron)	
Chrysene	*3*, 159 (1973); *32*, 247 (1983); *Suppl. 7*, 60 (1987)
Chrysoidine	*8*, 91 (1975); *Suppl. 7*, 169 (1987)
Chrysotile (*see* Asbestos)	
Ciclosporin	*50*, 77 (1990)
CI Disperse Yellow 3 (*see* Disperse Yellow 3)	
Cimetidine	*50*, 235 (1990)
Cinnamyl anthranilate	*16*, 287 (1978); *31*, 133 (1983); *Suppl. 7*, 60 (1987)
Cisplatin	*26*, 151 (1981); *Suppl. 7*, 170 (1987)
Citrinin	*40*, 67 (1986); *Suppl. 7*, 60 (1987)
Citrus Red No. 2	*8*, 101 (1975) (*corr. 42*, 254); *Suppl. 7*, 60 (1987)
Clofibrate	*24*, 39 (1980); *Suppl. 7*, 171 (1987)
Clomiphene citrate	*21*, 551 (1979); *Suppl. 7*, 172 (1987)
Coal gasification	*34*, 65 (1984); *Suppl. 7*, 173 (1987)
Coal-tar pitches (*see also* Coal-tars)	*35*, 83 (1985); *Suppl. 7*, 174 (1987)
Coal-tars	*35*, 83 (1985); *Suppl. 7*, 175 (1987)
Cobalt[III] acetate (*see* Cobalt and cobalt compounds)	
Cobalt–aluminium–chromium spinel (*see* Cobalt and cobalt compounds)	
Cobalt and cobalt compounds	*52*, 363 (1991)
Cobalt[II] chloride (*see* Cobalt and cobalt compounds)	
Cobalt–chromium alloy (*see* Chromium and chromium compounds)	
Cobalt–chromium–molybdenum alloys (*see* Cobalt and cobalt compounds)	
Cobalt metal powder (*see* Cobalt and cobalt compounds)	
Cobalt naphthenate (*see* Cobalt and cobalt compounds)	
Cobalt[II] oxide (*see* Cobalt and cobalt compounds)	
Cobalt[II,III] oxide (*see* Cobalt and cobalt compounds)	
Cobalt[II] sulfide (*see* Cobalt and cobalt compounds)	
Coffee	*51*, 41 (1991) (*corr. 52*, 513)
Coke production	*34*, 101 (1984); *Suppl. 7*, 176 (1987)
Combined oral contraceptives (*see also* Oestrogens, progestins and combinations)	*Suppl. 7*, 297 (1987)

Conjugated oestrogens (*see also* Steroidal oestrogens)	*21*, 147 (1979)
Contraceptives, oral (*see* Combined oral contraceptives; Sequential oral contraceptives)	
Copper 8-hydroxyquinoline	*15*, 103 (1977); *Suppl. 7*, 61 (1987)
Coronene	*32*, 263 (1983); *Suppl. 7*, 61 (1987)
Coumarin	*10*, 113 (1976); *Suppl. 7*, 61 (1987)
Creosotes (*see also* Coal-tars)	*35*, 83 (1985); *Suppl. 7*, 177 (1987)
meta-Cresidine	*27*, 91 (1982); *Suppl. 7*, 61 (1987)
para-Cresidine	*27*, 92 (1982); *Suppl. 7*, 61 (1987)
Crocidolite (*see* Asbestos)	
Crude oil	*45*, 119 (1989)
Crystalline silica (*see also* Silica)	*42*, 39 (1987); *Suppl. 7*, 341 (1987)
Cycasin	*1*, 157 (1972) (*corr.* 42, 251); *10*, 121 (1976); *Suppl. 7*, 61 (1987)
Cyclamates	*22*, 55 (1980); *Suppl. 7*, 178 (1987)
Cyclamic acid (*see* Cyclamates)	
Cyclochlorotine	*10*, 139 (1976); *Suppl. 7*, 61 (1987)
Cyclohexanone	*47*, 157 (1989)
Cyclohexylamine (*see* Cyclamates)	
Cyclopenta[*cd*]pyrene	*32*, 269 (1983); *Suppl. 7*, 61 (1987)
Cyclopropane (*see* Anaesthetics, volatile)	
Cyclophosphamide	*9*, 135 (1975); *26*, 165 (1981); *Suppl. 7*, 182 (1987)

D

2,4-D (*see also* Chlorophenoxy herbicides; Chlorophenoxy herbicides, occupational exposures to)	*15*, 111 (1977)
Dacarbazine	*26*, 203 (1981); *Suppl. 7*, 184 (1987)
Dantron	*50*, 265 (1990)
D & C Red No. 9	*8*, 107 (1975); *Suppl. 7*, 61 (1987)
Dapsone	*24*, 59 (1980); *Suppl. 7*, 185 (1987)
Daunomycin	*10*, 145 (1976); *Suppl. 7*, 61 (1987)
DDD (*see* DDT)	
DDE (*see* DDT)	
DDT	*5*, 83 (1974) (*corr.* 42, 253); *Suppl. 7*, 186 (1987); *53*, 179 (1991)
Decabromodiphenyl oxide	*48*, 73 (1990)
Deltamethrin	*53*, 251 (1991)
Deoxynivalenol (*see* Toxins derived from *Fusarium graminearum, F. culmorum* and *F. crookwellense*)	
Diacetylaminoazotoluene	*8*, 113 (1975); *Suppl. 7*, 61 (1987)
N,N'-Diacetylbenzidine	*16*, 293 (1978); *Suppl. 7*, 61 (1987)
Diallate	*12*, 69 (1976); *30*, 235 (1983); *Suppl. 7*, 61 (1987)
2,4-Diaminoanisole	*16*, 51 (1978); *27*, 103 (1982); *Suppl. 7*, 61 (1987)
4,4'-Diaminodiphenyl ether	*16*, 301 (1978); *29*, 203 (1982); *Suppl. 7*, 61 (1987)
1,2-Diamino-4-nitrobenzene	*16*, 63 (1978); *Suppl. 7*, 61 (1987)
1,4-Diamino-2-nitrobenzene	*16*, 73 (1978); *Suppl. 7*, 61 (1987)

2,6-Diamino-3-(phenylazo)pyridine (*see* Phenazopyridine hydrochloride)
2,4-Diaminotoluene (*see also* Toluene diisocyanates)   *16*, 83 (1978); *Suppl. 7*, 61 (1987)
2,5-Diaminotoluene (*see also* Toluene diisocyanates)   *16*, 97 (1978); *Suppl. 7*, 61 (1987)
ortho-Dianisidine (*see* 3,3'-Dimethoxybenzidine)
Diazepam   *13*, 57 (1977); *Suppl. 7*, 189 (1987)
Diazomethane   *7*, 223 (1974); *Suppl. 7*, 61 (1987)
Dibenz[*a,h*]acridine   *3*, 247 (1973); *32*, 277 (1983); *Suppl. 7*, 61 (1987)
Dibenz[*a,j*]acridine   *3*, 254 (1973); *32*, 283 (1983); *Suppl. 7*, 61 (1987)
Dibenz[*a,c*]anthracene   *32*, 289 (1983) (*corr. 42*, 262); *Suppl. 7*, 61 (1987)
Dibenz[*a,h*]anthracene   *3*, 178 (1973) (*corr. 43*, 261); *32*, 299 (1983); *Suppl. 7*, 61 (1987)
Dibenz[*a,j*]anthracene   *32*, 309 (1983); *Suppl. 7*, 61 (1987)
7*H*-Dibenzo[*c,g*]carbazole   *3*, 260 (1973); *32*, 315 (1983); *Suppl. 7*, 61 (1987)
Dibenzodioxins, chlorinated (other than TCDD) [*see* Chlorinated dibenzodioxins (other than TCDD)]
Dibenzo[*a,e*]fluoranthene   *32*, 321 (1983); *Suppl. 7*, 61 (1987)
Dibenzo[*h,rst*]pentaphene   *3*, 197 (1973); *Suppl. 7*, 62 (1987)
Dibenzo[*a,e*]pyrene   *3*, 201 (1973); *32*, 327 (1983); *Suppl. 7*, 62 (1987)
Dibenzo[*a,h*]pyrene   *3*, 207 (1973); *32*, 331 (1983); *Suppl. 7*, 62 (1987)
Dibenzo[*a,i*]pyrene   *3*, 215 (1973); *32*, 337 (1983); *Suppl. 7*, 62 (1987)
Dibenzo[*a,l*]pyrene   *3*, 224 (1973); *32*, 343 (1983); *Suppl. 7*, 62 (1987)
Dibromoacetonitrile (*see* Halogenated acetonitriles)
1,2-Dibromo-3-chloropropane   *15*, 139 (1977); *20*, 83 (1979); *Suppl. 7*, 191 (1987)
Dichloroacetonitrile (*see* Halogenated acetonitriles)
Dichloroacetylene   *39*, 369 (1986); *Suppl. 7*, 62 (1987)
ortho-Dichlorobenzene   *7*, 231 (1974); *29*, 213 (1982); *Suppl. 7*, 192 (1987)
para-Dichlorobenzene   *7*, 231 (1974); *29*, 215 (1982); *Suppl. 7*, 192 (1987)
3,3'-Dichlorobenzidine   *4*, 49 (1974); *29*, 239 (1982); *Suppl. 7*, 193 (1987)
trans-1,4-Dichlorobutene   *15*, 149 (1977); *Suppl. 7*, 62 (1987)
3,3'-Dichloro-4,4'-diaminodiphenyl ether   *16*, 309 (1978); *Suppl. 7*, 62 (1987)
1,2-Dichloroethane   *20*, 429 (1979); *Suppl. 7*, 62 (1987)
Dichloromethane   *20*, 449 (1979); *41*, 43 (1986); *Suppl. 7*, 194 (1987)
2,4-Dichlorophenol (*see* Chlorophenols; Chlorophenols, occupational exposures to)
(2,4-Dichlorophenoxy)acetic acid (*see* 2,4-D)
2,6-Dichloro-*para*-phenylenediamine   *39*, 325 (1986); *Suppl. 7*, 62 (1987)
1,2-Dichloropropane   *41*, 131 (1986); *Suppl. 7*, 62 (1987)
1,3-Dichloropropene (technical-grade)   *41*, 113 (1986); *Suppl. 7*, 195 (1987)

Dichlorvos	20, 97 (1979); *Suppl. 7*, 62 (1987); 53, 267 (1991)
Dicofol	30, 87 (1983); *Suppl. 7*, 62 (1987)
Dicyclohexylamine (*see* Cyclamates)	
Dieldrin	5, 125 (1974); *Suppl. 7*, 196 (1987)
Dienoestrol (*see also* Nonsteroidal oestrogens)	21, 161 (1979)
Diepoxybutane	11, 115 (1976) (*corr.* 42, 255); *Suppl. 7*, 62 (1987)
Diesel and gasoline engine exhausts	46, 41 (1989)
Diesel fuels	45, 219 (1989) (*corr.* 47, 505)
Diethyl ether (*see* Anaesthetics, volatile)	
Di(2-ethylhexyl)adipate	29, 257 (1982); *Suppl. 7*, 62 (1987)
Di(2-ethylhexyl)phthalate	29, 269 (1982) (*corr.* 42, 261); *Suppl. 7*, 62 (1987)
1,2-Diethylhydrazine	4, 153 (1974); *Suppl. 7*, 62 (1987)
Diethylstilboestrol	6, 55 (1974); 21, 173 (1979) (*corr.* 42, 259); *Suppl. 7*, 273 (1987)
Diethylstilboestrol dipropionate (*see* Diethylstilboestrol)	
Diethyl sulfate	4, 277 (1974); *Suppl. 7*, 198 (1987); 54, 213 (1992)
Diglycidyl resorcinol ether	11, 125 (1976); 36, 181 (1985); *Suppl. 7*, 62 (1987)
Dihydrosafrole	1, 170 (1972); 10, 233 (1976); *Suppl. 7*, 62 (1987)
1,8-Dihydroxyanthraquinone (*see* Dantron)	
Dihydroxybenzenes (*see* Catechol; Hydroquinone; Resorcinol)	
Dihydroxymethylfuratrizine	24, 77 (1980); *Suppl. 7*, 62 (1987)
Diisopropyl sulfate	54, 229 (1992)
Dimethisterone (*see also* Progestins; Sequential oral contraceptives)	6, 167 (1974); 21, 377 (1979)
Dimethoxane	15, 177 (1977); *Suppl. 7*, 62 (1987)
3,3'-Dimethoxybenzidine	4, 41 (1974); *Suppl. 7*, 198 (1987)
3,3'-Dimethoxybenzidine-4,4'-diisocyanate	39, 279 (1986); *Suppl. 7*, 62 (1987)
para-Dimethylaminoazobenzene	8, 125 (1975); *Suppl. 7*, 62 (1987)
para-Dimethylaminoazobenzenediazo sodium sulfonate	8, 147 (1975); *Suppl. 7*, 62 (1987)
trans-2-[(Dimethylamino)methylimino]-5-[2-(5-nitro-2-furyl)-vinyl]-1,3,4-oxadiazole	7, 147 (1974) (*corr.* 42, 253); *Suppl. 7*, 62 (1987)
4,4'-Dimethylangelicin plus ultraviolet radiation (*see also* Angelicin and some synthetic derivatives)	*Suppl. 7*, 57 (1987)
4,5'-Dimethylangelicin plus ultraviolet radiation (*see also* Angelicin and some synthetic derivatives)	*Suppl. 7*, 57 (1987)
Dimethylarsinic acid (*see* Arsenic and arsenic compounds)	
3,3'-Dimethylbenzidine	1, 87 (1972); *Suppl. 7*, 62 (1987)
Dimethylcarbamoyl chloride	12, 77 (1976); *Suppl. 7*, 199 (1987)
Dimethylformamide	47, 171 (1989)
1,1-Dimethylhydrazine	4, 137 (1974); *Suppl. 7*, 62 (1987)
1,2-Dimethylhydrazine	4, 145 (1974) (*corr.* 42, 253); *Suppl. 7*, 62 (1987)
Dimethyl hydrogen phosphite	48, 85 (1990)
1,4-Dimethylphenanthrene	32, 349 (1983); *Suppl. 7*, 62 (1987)
Dimethyl sulfate	4, 271 (1974); *Suppl. 7*, 200 (1987)
3,7-Dinitrofluoranthene	46, 189 (1989)

3,9-Dinitrofluoranthene	*46*, 195 (1989)
1,3-Dinitropyrene	*46*, 201 (1989)
1,6-Dinitropyrene	*46*, 215 (1989)
1,8-Dinitropyrene	*33*, 171 (1984); *Suppl. 7*, 63 (1987); *46*, 231 (1989)
Dinitrosopentamethylenetetramine	*11*, 241 (1976); *Suppl. 7*, 63 (1987)
1,4-Dioxane	*11*, 247 (1976); *Suppl. 7*, 201 (1987)
2,4'-Diphenyldiamine	*16*, 313 (1978); *Suppl. 7*, 63 (1987)
Direct Black 38 (*see also* Benzidine-based dyes)	*29*, 295 (1982) (*corr. 42*, 261)
Direct Blue 6 (*see also* Benzidine-based dyes)	*29*, 311 (1982)
Direct Brown 95 (*see also* Benzidine-based dyes)	*29*, 321 (1982)
Disperse Blue 1	*48*, 139 (1990)
Disperse Yellow 3	*8*, 97 (1975); *Suppl. 7*, 60 (1987); *48*, 149 (1990)
Disulfiram	*12*, 85 (1976); *Suppl. 7*, 63 (1987)
Dithranol	*13*, 75 (1977); *Suppl. 7*, 63 (1987)
Divinyl ether (*see* Anaesthetics, volatile)	
Dulcin	*12*, 97 (1976); *Suppl. 7*, 63 (1987)

E

Endrin	*5*, 157 (1974); *Suppl. 7*, 63 (1987)
Enflurane (*see* Anaesthetics, volatile)	
Eosin	*15*, 183 (1977); *Suppl. 7*, 63 (1987)
Epichlorohydrin	*11*, 131 (1976) (*corr. 42*, 256); *Suppl. 7*, 202 (1987)
1,2-Epoxybutane	*47*, 217 (1989)
1-Epoxyethyl-3,4-epoxycyclohexane	*11*, 141 (1976); *Suppl. 7*, 63 (1987)
3,4-Epoxy-6-methylcyclohexylmethyl-3,4-epoxy-6-methyl-cyclohexane carboxylate	*11*, 147 (1976); *Suppl. 7*, 63 (1987)
cis-9,10-Epoxystearic acid	*11*, 153 (1976); *Suppl. 7*, 63 (1987)
Erionite	*42*, 225 (1987); *Suppl. 7*, 203 (1987)
Ethinyloestradiol (*see also* Steroidal oestrogens)	*6*, 77 (1974); *21*, 233 (1979)
Ethionamide	*13*, 83 (1977); *Suppl. 7*, 63 (1987)
Ethyl acrylate	*19*, 57 (1979); *39*, 81 (1986); *Suppl. 7*, 63 (1987)
Ethylene	*19*, 157 (1979); *Suppl. 7*, 63 (1987)
Ethylene dibromide	*15*, 195 (1977); *Suppl. 7*, 204 (1987)
Ethylene oxide	*11*, 157 (1976); *36*, 189 (1985) (*corr. 42*, 263); *Suppl. 7*, 205 (1987)
Ethylene sulfide	*11*, 257 (1976); *Suppl. 7*, 63 (1987)
Ethylene thiourea	*7*, 45 (1974); *Suppl. 7*, 207 (1987)
Ethyl methanesulfonate	*7*, 245 (1974); *Suppl. 7*, 63 (1987)
N-Ethyl-*N*-nitrosourea	*1*, 135 (1972); *17*, 191 (1978); *Suppl. 7*, 63 (1987)
Ethyl selenac (*see also* Selenium and selenium compounds)	*12*, 107 (1976); *Suppl. 7*, 63 (1987)
Ethyl tellurac	*12*, 115 (1976); *Suppl. 7*, 63 (1987)
Ethynodiol diacetate (*see also* Progestins; Combined oral contraceptives)	*6*, 173 (1974); *21*, 387 (1979)
Eugenol	*36*, 75 (1985); *Suppl. 7*, 63 (1987)
Evans blue	*8*, 151 (1975); *Suppl. 7*, 63 (1987)

F

Fast Green FCF	16, 187 (1978); Suppl. 7, 63 (1987)
Fenvalerate	53, 309 (1991)
Ferbam	12, 121 (1976) (corr. 42, 256); Suppl. 7, 63 (1987)
Ferric oxide	1, 29 (1972); Suppl. 7, 216 (1987)
Ferrochromium (see Chromium and chromium compounds)	
Fluometuron	30, 245 (1983); Suppl. 7, 63 (1987)
Fluoranthene	32, 355 (1983); Suppl. 7, 63 (1987)
Fluorene	32, 365 (1983); Suppl. 7, 63 (1987)
Fluorescent lighting (exposure to) (see Ultraviolet radiation)	
Fluorides (inorganic, used in drinking-water)	27, 237 (1982); Suppl. 7, 208 (1987)
5-Fluorouracil	26, 217 (1981); Suppl. 7, 210 (1987)
Fluorspar (see Fluorides)	
Fluosilicic acid (see Fluorides)	
Fluroxene (see Anaesthetics, volatile)	
Formaldehyde	29, 345 (1982); Suppl. 7, 211 (1987)
2-(2-Formylhydrazino)-4-(5-nitro-2-furyl)thiazole	7, 151 (1974) (corr. 42, 253); Suppl. 7, 63 (1987)
Frusemide (see Furosemide)	
Fuel oils (heating oils)	45, 239 (1989) (corr. 47, 505)
Fumonisin B_1 (see Toxins derived from *Fusarium moniliforme*)	
Fumonisin B_2 (see Toxins derived from *Fusarium moniliforme*)	
Furazolidone	31, 141 (1983); Suppl. 7, 63 (1987)
Furniture and cabinet-making	25, 99 (1981); Suppl. 7, 380 (1987)
Furosemide	50, 277 (1990)
2-(2-Furyl)-3-(5-nitro-2-furyl)acrylamide (see AF-2)	
Fusarenon-X (see Toxins derived from *Fusarium graminearum, F. culmorum* and *F. crookwellense*)	
Fusarenone-X (see Toxins derived from *Fusarium graminearum, F. culmorum* and *F. crookwellense*)	
Fusarin C (see Toxins derived from *Fusarium moniliforme*)	

G

Gasoline	45, 159 (1989) (corr. 47, 505)
Gasoline engine exhaust (see Diesel and gasoline engine exhausts)	
Glass fibres (see Man-made mineral fibres)	
Glasswool (see Man-made mineral fibres)	
Glass filaments (see Man-made mineral fibres)	
Glu-P-1	40, 223 (1986); Suppl. 7, 64 (1987)
Glu-P-2	40, 235 (1986); Suppl. 7, 64 (1987)
L-Glutamic acid, 5-[2-(4-hydroxymethyl)phenylhydrazide] (see Agaritine)	
Glycidaldehyde	11, 175 (1976); Suppl. 7, 64 (1987)
Glycidyl ethers	47, 237 (1989)
Glycidyl oleate	11, 183 (1976); Suppl. 7, 64 (1987)
Glycidyl stearate	11, 187 (1976); Suppl. 7, 64 (1987)
Griseofulvin	10, 153 (1976); Suppl. 7, 391 (1987)
Guinea Green B	16, 199 (1978); Suppl. 7, 64 (1987)
Gyromitrin	31, 163 (1983); Suppl. 7, 391 (1987)

H

Haematite	1, 29 (1972); *Suppl. 7*, 216 (1987)
Haematite and ferric oxide	*Suppl. 7*, 216 (1987)
Haematite mining, underground, with exposure to radon	1, 29 (1972); *Suppl. 7*, 216 (1987)
Hair dyes, epidemiology of	16, 29 (1978); 27, 307 (1982)
Halogenated acetonitriles	52, 269 (1991)
Halothane (*see* Anaesthetics, volatile)	
α-HCH (*see* Hexachlorocyclohexanes)	
β-HCH (*see* Hexachlorocyclohexanes)	
γ-HCH (*see* Hexachlorocyclohexanes)	
Heating oils (*see* Fuel oils)	
Heptachlor (*see also* Chlordane/Heptachlor)	5, 173 (1974); 20, 129 (1979)
Hexachlorobenzene	20, 155 (1979); *Suppl. 7*, 219 (1987)
Hexachlorobutadiene	20, 179 (1979); *Suppl. 7*, 64 (1987)
Hexachlorocyclohexanes	5, 47 (1974); 20, 195 (1979) (*corr.* 42, 258); *Suppl. 7*, 220 (1987)
Hexachlorocyclohexane, technical-grade (*see* Hexachlorocyclohexanes)	
Hexachloroethane	20, 467 (1979); *Suppl. 7*, 64 (1987)
Hexachlorophene	20, 241 (1979); *Suppl. 7*, 64 (1987)
Hexamethylphosphoramide	15, 211 (1977); *Suppl. 7*, 64 (1987)
Hexoestrol (*see* Nonsteroidal oestrogens)	
Hycanthone mesylate	13, 91 (1977); *Suppl. 7*, 64 (1987)
Hydralazine	24, 85 (1980); *Suppl. 7*, 222 (1987)
Hydrazine	4, 127 (1974); *Suppl. 7*, 223 (1987)
Hydrochloric acid	54, 189 (1992)
Hydrochlorothiazide	50, 293 (1990)
Hydrogen peroxide	36, 285 (1985); *Suppl. 7*, 64 (1987)
Hydroquinone	15, 155 (1977); *Suppl. 7*, 64 (1987)
4-Hydroxyazobenzene	8, 157 (1975); *Suppl. 7*, 64 (1987)
17α-Hydroxyprogesterone caproate (*see also* Progestins)	21, 399 (1979) (*corr.* 42, 259)
8-Hydroxyquinoline	13, 101 (1977); *Suppl. 7*, 64 (1987)
8-Hydroxysenkirkine	10, 265 (1976); *Suppl. 7*, 64 (1987)
Hypochlorite salts	52, 159 (1991)

I

Indeno[1,2,3-*cd*]pyrene	3, 229 (1973); 32, 373 (1983); *Suppl. 7*, 64 (1987)
Inorganic acids (*see* Sulfuric acid and other strong inorganic acids, occupational exposures to mists and vapours from)	
Insecticides, occupational exposures in spraying and application of	53, 45 (1991)
IQ	40, 261 (1986); *Suppl. 7*, 64 (1987); 56, 165 (1993)
Iron and steel founding	34, 133 (1984); *Suppl. 7*, 224 (1987)
Iron-dextran complex	2, 161 (1973); *Suppl. 7*, 226 (1987)
Iron-dextrin complex	2, 161 (1973) (*corr.* 42, 252); *Suppl. 7*, 64 (1987)
Iron oxide (*see* Ferric oxide)	
Iron oxide, saccharated (*see* Saccharated iron oxide)	
Iron sorbitol–citric acid complex	2, 161 (1973); *Suppl. 7*, 64 (1987)

Isatidine	10, 269 (1976); *Suppl. 7*, 65 (1987)
Isoflurane (*see* Anaesthetics, volatile)	
Isoniazid (*see* Isonicotinic acid hydrazide)	
Isonicotinic acid hydrazide	4, 159 (1974); *Suppl. 7*, 227 (1987)
Isophosphamide	26, 237 (1981); *Suppl. 7*, 65 (1987)
Isopropyl alcohol	15, 223 (1977); *Suppl. 7*, 229 (1987)
Isopropyl alcohol manufacture (strong-acid process) (*see also* Isopropyl alcohol; Sulfuric acid and other strong inorganic acids, occupational exposures to mists and vapours from)	*Suppl. 7*, 229 (1987)
Isopropyl oils	15, 223 (1977); *Suppl. 7*, 229 (1987)
Isosafrole	1, 169 (1972); *10*, 232 (1976); *Suppl. 7*, 65 (1987)

J

Jacobine	10, 275 (1976); *Suppl. 7*, 65 (1987)
Jet fuel	45, 203 (1989)
Joinery (*see* Carpentry and joinery)	

K

Kaempferol	31, 171 (1983); *Suppl. 7*, 65 (1987)
Kepone (*see* Chlordecone)	

L

Lasiocarpine	10, 281 (1976); *Suppl. 7*, 65 (1987)
Lauroyl peroxide	36, 315 (1985); Suppl. 7, 65 (1987)
Lead acetate (*see* Lead and lead compounds)	
Lead and lead compounds	*1*, 40 (1972) (*corr. 42*, 251); 2, 52, 150 (1973); *12*, 131 (1976); *23*, 40, 208, 209, 325 (1980); *Suppl. 7*, 230 (1987)
Lead arsenate (*see* Arsenic and arsenic compounds)	
Lead carbonate (*see* Lead and lead compounds)	
Lead chloride (*see* Lead and lead compounds)	
Lead chromate (*see* Chromium and chromium compounds)	
Lead chromate oxide (*see* Chromium and chromium compounds)	
Lead naphthenate (*see* Lead and lead compounds)	
Lead nitrate (*see* Lead and lead compounds)	
Lead oxide (*see* Lead and lead compounds)	
Lead phosphate (*see* Lead and lead compounds)	
Lead subacetate (*see* Lead and lead compounds)	
Lead tetroxide (*see* Lead and lead compounds)	
Leather goods manufacture	25, 279 (1981); *Suppl. 7*, 235 (1987)
Leather industries	25, 199 (1981); *Suppl. 7*, 232 (1987)
Leather tanning and processing	25, 201 (1981); *Suppl. 7*, 236 (1987)
Ledate (*see also* Lead and lead compounds)	*12*, 131 (1976)
Light Green SF	16, 209 (1978); *Suppl. 7*, 65 (1987)
d-Limonene	56, 135 (1993)
Lindane (*see* Hexachlorocyclohexanes)	
The lumber and sawmill industries (including logging)	25, 49 (1981); *Suppl. 7*, 383 (1987)

Luteoskyrin	*10*, 163 (1976); *Suppl. 7*, 65 (1987)
Lynoestrenol (*see also* Progestins; Combined oral contraceptives)	*21*, 407 (1979)

M

Magenta	*4*, 57 (1974) (*corr. 42*, 252); *Suppl. 7*, 238 (1987)
Magenta, manufacture of (*see also* Magenta)	*Suppl. 7*, 238 (1987)
Malathion	*30*, 103 (1983); *Suppl. 7*, 65 (1987)
Maleic hydrazide	*4*, 173 (1974) (*corr. 42*, 253); *Suppl. 7*, 65 (1987)
Malonaldehyde	*36*, 163 (1985); *Suppl. 7*, 65 (1987)
Maneb	*12*, 137 (1976); *Suppl. 7*, 65 (1987)
Man-made mineral fibres	*43*, 39 (1988)
Mannomustine	*9*, 157 (1975); *Suppl. 7*, 65 (1987)
Mate	*51*, 273 (1991)
MCPA (*see also* Chlorophenoxy herbicides; Chlorophenoxy herbicides, occupational exposures to)	*30*, 255 (1983)
MeA-α-C	*40*, 253 (1986); *Suppl. 7*, 65 (1987)
Medphalan	*9*, 168 (1975); *Suppl. 7*, 65 (1987)
Medroxyprogesterone acetate	*6*, 157 (1974); *21*, 417 (1979) (*corr. 42*, 259); *Suppl. 7*, 289 (1987)
Megestrol acetate (*see* also Progestins; Combined oral contraceptives)	
MeIQ	*40*, 275 (1986); *Suppl. 7*, 65 (1987); *56*, 197 (1993)
MeIQx	*40*, 283 (1986); *Suppl. 7*, 65 (1987); *56*, 211 (1993)
Melamine	*39*, 333 (1986); *Suppl. 7*, 65 (1987)
Melphalan	*9*, 167 (1975); *Suppl. 7*, 239 (1987)
6-Mercaptopurine	*26*, 249 (1981); *Suppl. 7*, 240 (1987)
Merphalan	*9*, 169 (1975); *Suppl. 7*, 65 (1987)
Mestranol (*see also* Steroidal oestrogens)	*6*, 87 (1974); *21*, 257 (1979) (*corr. 42*, 259)
Metabisulfites (*see* Sulfur dioxide and some sulfites, bisulfites and metabisulfites)	
Methanearsonic acid, disodium salt (*see* Arsenic and arsenic compounds)	
Methanearsonic acid, monosodium salt (*see* Arsenic and arsenic compounds	
Methotrexate	*26*, 267 (1981); *Suppl. 7*, 241 (1987)
Methoxsalen (*see* 8-Methoxypsoralen)	
Methoxychlor	*5*, 193 (1974); *20*, 259 (1979); *Suppl. 7*, 66 (1987)
Methoxyflurane (*see* Anaesthetics, volatile)	
5-Methoxypsoralen	*40*, 327 (1986); *Suppl. 7*, 242 (1987)
8-Methoxypsoralen (*see also* 8-Methoxypsoralen plus ultraviolet radiation)	*24*, 101 (1980)
8-Methoxypsoralen plus ultraviolet radiation	*Suppl. 7*, 243 (1987)
Methyl acrylate	*19*, 52 (1979); *39*, 99 (1986); *Suppl. 7*, 66 (1987)

5-Methylangelicin plus ultraviolet radiation (*see also* Angelicin and some synthetic derivatives)	*Suppl. 7*, 57 (1987)
2-Methylaziridine	*9*, 61 (1975); *Suppl. 7*, 66 (1987)
Methylazoxymethanol acetate	*1*, 164 (1972); *10*, 131 (1976); *Suppl. 7*, 66 (1987)
Methyl bromide	*41*, 187 (1986) (*corr. 45*, 283); *Suppl. 7*, 245 (1987)
Methyl carbamate	*12*, 151 (1976); *Suppl. 7*, 66 (1987)
Methyl-CCNU [*see* 1-(2-Chloroethyl)-3-(4-methylcyclohexyl)-1-nitrosourea]	
Methyl chloride	*41*, 161 (1986); *Suppl. 7*, 246 (1987)
1-, 2-, 3-, 4-, 5- and 6-Methylchrysenes	*32*, 379 (1983); *Suppl. 7*, 66 (1987)
N-Methyl-*N*,4-dinitrosoaniline	*1*, 141 (1972); *Suppl. 7*, 66 (1987)
4,4′-Methylene bis(2-chloroaniline)	*4*, 65 (1974) (*corr. 42*, 252); *Suppl. 7*, 246 (1987)
4,4′-Methylene bis(*N*,*N*-dimethyl)benzenamine	*27*, 119 (1982); *Suppl. 7*, 66 (1987)
4,4′-Methylene bis(2-methylaniline)	*4*, 73 (1974); *Suppl. 7*, 248 (1987)
4,4′-Methylenedianiline	*4*, 79 (1974) (*corr. 42*, 252); *39*, 347 (1986); *Suppl. 7*, 66 (1987)
4,4′-Methylenediphenyl diisocyanate	*19*, 314 (1979); *Suppl. 7*, 66 (1987)
2-Methylfluoranthene	*32*, 399 (1983); *Suppl. 7*, 66 (1987)
3-Methylfluoranthene	*32*, 399 (1983); *Suppl. 7*, 66 (1987)
Methylglyoxal	*51*, 443 (1991)
Methyl iodide	*15*, 245 (1977); *41*, 213 (1986); *Suppl. 7*, 66 (1987)
Methyl methacrylate	*19*, 187 (1979); *Suppl. 7*, 66 (1987)
Methyl methanesulfonate	*7*, 253 (1974); *Suppl. 7*, 66 (1987)
2-Methyl-1-nitroanthraquinone	*27*, 205 (1982); *Suppl. 7*, 66 (1987)
N-Methyl-*N*′-nitro-*N*-nitrosoguanidine	*4*, 183 (1974); *Suppl. 7*, 248 (1987)
3-Methylnitrosaminopropionaldehyde [*see* 3-(*N*-Nitrosomethylamino)-propionaldehyde]	
3-Methylnitrosaminopropionitrile [*see* 3-(*N*-Nitrosomethylamino)-propionitrile]	
4-(Methylnitrosamino)-4-(3-pyridyl)-1-butanal [*see* 4-(*N*-Nitrosomethylamino)-4-(3-pyridyl)-1-butanal]	
4-(Methylnitrosamino)-1-(3-pyridyl)-1-butanone [*see* 4-(*N*-Nitrosomethylamino)-1-(3-pyridyl)-1-butanone]	
N-Methyl-*N*-nitrosourea	*1*, 125 (1972); *17*, 227 (1978); *Suppl. 7*, 66 (1987)
N-Methyl-*N*-nitrosourethane	*4*, 211 (1974); *Suppl. 7*, 66 (1987)
Methyl parathion	*30*, 131 (1983); *Suppl. 7*, 392 (1987)
1-Methylphenanthrene	*32*, 405 (1983); *Suppl. 7*, 66 (1987)
7-Methylpyrido[3,4-*c*]psoralen	*40*, 349 (1986); *Suppl. 7*, 71 (1987)
Methyl red	*8*, 161 (1975); *Suppl. 7*, 66 (1987)
Methyl selenac (*see also* Selenium and selenium compounds)	*12*, 161 (1976); *Suppl. 7*, 66 (1987)
Methylthiouracil	*7*, 53 (1974); *Suppl. 7*, 66 (1987)
Metronidazole	*13*, 113 (1977); *Suppl. 7*, 250 (1987)
Mineral oils	*3*, 30 (1973); *33*, 87 (1984) (*corr. 42*, 262); *Suppl. 7*, 252 (1987)
Mirex	*5*, 203 (1974); *20*, 283 (1979) (*corr. 42*, 258); *Suppl. 7*, 66 (1987)
Mitomycin C	*10*, 171 (1976); *Suppl. 7*, 67 (1987)

MNNG [see N-Methyl-N'-nitro-N-nitrosoguanidine]
MOCA [see 4,4'-Methylene bis(2-chloroaniline)]

Modacrylic fibres	19, 86 (1979); Suppl. 7, 67 (1987)
Monocrotaline	10, 291 (1976); Suppl. 7, 67 (1987)
Monuron	12, 167 (1976); Suppl. 7, 67 (1987); 53, 467 (1991)
MOPP and other combined chemotherapy including alkylating agents	Suppl. 7, 254 (1987)
Morpholine	47, 199 (1989)
5-(Morpholinomethyl)-3-[(5-nitrofurfurylidene)amino]-2-oxazolidinone	7, 161 (1974); Suppl. 7, 67 (1987)
Mustard gas	9, 181 (1975) (corr. 42, 254); Suppl. 7, 259 (1987)

Myleran (see 1,4-Butanediol dimethanesulfonate)

N

Nafenopin	24, 125 (1980); Suppl. 7, 67 (1987)
1,5-Naphthalenediamine	27, 127 (1982); Suppl. 7, 67 (1987)
1,5-Naphthalene diisocyanate	19, 311 (1979); Suppl. 7, 67 (1987)
1-Naphthylamine	4, 87 (1974) (corr. 42, 253); Suppl. 7, 260 (1987)
2-Naphthylamine	4, 97 (1974); Suppl. 7, 261 (1987)
1-Naphthylthiourea	30, 347 (1983); Suppl. 7, 263 (1987)
Nickel acetate (see Nickel and nickel compounds)	
Nickel ammonium sulfate (see Nickel and nickel compounds)	
Nickel and nickel compounds	2, 126 (1973) (corr. 42, 252); 11, 75 (1976); Suppl. 7, 264 (1987) (corr. 45, 283); 49, 257 (1990)
Nickel carbonate (see Nickel and nickel compounds)	
Nickel carbonyl (see Nickel and nickel compounds)	
Nickel chloride (see Nickel and nickel compounds)	
Nickel–gallium alloy (see Nickel and nickel compounds)	
Nickel hydroxide (see Nickel and nickel compounds)	
Nickelocene (see Nickel and nickel compounds)	
Nickel oxide (see Nickel and nickel compounds)	
Nickel subsulfide (see Nickel and nickel compounds)	
Nickel sulfate (see Nickel and nickel compounds)	
Niridazole	13, 123 (1977); Suppl. 7, 67 (1987)
Nithiazide	31, 179 (1983); Suppl. 7, 67 (1987)
Nitrilotriacetic acid and its salts	48, 181 (1990)
5-Nitroacenaphthene	16, 319 (1978); Suppl. 7, 67 (1987)
5-Nitro-ortho-anisidine	27, 133 (1982); Suppl. 7, 67 (1987)
9-Nitroanthracene	33, 179 (1984); Suppl. 7, 67 (1987)
7-Nitrobenz[a]anthracene	46, 247 (1989)
6-Nitrobenzo[a]pyrene	33, 187 (1984); Suppl. 7, 67 (1987); 46, 255 (1989)
4-Nitrobiphenyl	4, 113 (1974); Suppl. 7, 67 (1987)
6-Nitrochrysene	33, 195 (1984); Suppl. 7, 67 (1987); 46, 267 (1989)
Nitrofen (technical-grade)	30, 271 (1983); Suppl. 7, 67 (1987)
3-Nitrofluoranthene	33, 201 (1984); Suppl. 7, 67 (1987)

2-Nitrofluorene	46, 277 (1989)
Nitrofural	7, 171 (1974); *Suppl. 7*, 67 (1987); 50, 195 (1990)
5-Nitro-2-furaldehyde semicarbazone (*see* Nitrofural)	
Nitrofurantoin	50, 211 (1990)
Nitrofurazone (*see* Nitrofural)	
1-[(5-Nitrofurfurylidene)amino]-2-imidazolidinone	7, 181 (1974); *Suppl. 7*, 67 (1987)
N-[4-(5-Nitro-2-furyl)-2-thiazolyl]acetamide	1, 181 (1972); 7, 185 (1974); *Suppl. 7*, 67 (1987)
Nitrogen mustard	9, 193 (1975); *Suppl. 7*, 269 (1987)
Nitrogen mustard *N*-oxide	9, 209 (1975); *Suppl. 7*, 67 (1987)
1-Nitronaphthalene	46, 291 (1989)
2-Nitronaphthalene	46, 303 (1989)
3-Nitroperylene	46, 313 (1989)
2-Nitropropane	29, 331 (1982); *Suppl. 7*, 67 (1987)
1-Nitropyrene	33, 209 (1984); *Suppl. 7*, 67 (1987); 46, 321 (1989)
2-Nitropyrene	46, 359 (1989)
4-Nitropyrene	46, 367 (1989)
N-Nitrosatable drugs	24, 297 (1980) (*corr.* 42, 260)
N-Nitrosatable pesticides	30, 359 (1983)
N'-Nitrosoanabasine	37, 225 (1985); *Suppl. 7*, 67 (1987)
N'-Nitrosoanatabine	37, 233 (1985); *Suppl. 7*, 67 (1987)
N-Nitrosodi-*n*-butylamine	4, 197 (1974); 17, 51 (1978); *Suppl. 7*, 67 (1987)
N-Nitrosodiethanolamine	17, 77 (1978); *Suppl. 7*, 67 (1987)
N-Nitrosodiethylamine	1, 107 (1972) (*corr.* 42, 251); 17, 83 (1978) (*corr.* 42, 257); *Suppl. 7*, 67 (1987)
N-Nitrosodimethylamine	1, 95 (1972); 17, 125 (1978) (*corr.* 42, 257); *Suppl. 7*, 67 (1987)
N-Nitrosodiphenylamine	27, 213 (1982); *Suppl. 7*, 67 (1987)
para-Nitrosodiphenylamine	27, 227 (1982) (*corr.* 42, 261); *Suppl. 7*, 68 (1987)
N-Nitrosodi-*n*-propylamine	17, 177 (1978); *Suppl. 7*, 68 (1987)
N-Nitroso-*N*-ethylurea (*see* *N*-Ethyl-*N*-nitrosourea)	
N-Nitrosofolic acid	17, 217 (1978); *Suppl. 7*, 68 (1987)
N-Nitrosoguvacine	37, 263 (1985); *Suppl. 7*, 68 (1987)
N-Nitrosoguvacoline	37, 263 (1985); *Suppl. 7*, 68 (1987)
N-Nitrosohydroxyproline	17, 304 (1978); *Suppl. 7*, 68 (1987)
3-(*N*-Nitrosomethylamino)propionaldehyde	37, 263 (1985); *Suppl. 7*, 68 (1987)
3-(*N*-Nitrosomethylamino)propionitrile	37, 263 (1985); *Suppl. 7*, 68 (1987)
4-(*N*-Nitrosomethylamino)-4-(3-pyridyl)-1-butanal	37, 205 (1985); *Suppl. 7*, 68 (1987)
4-(*N*-Nitrosomethylamino)-1-(3-pyridyl)-1-butanone	37, 209 (1985); *Suppl. 7*, 68 (1987)
N-Nitrosomethylethylamine	17, 221 (1978); *Suppl. 7*, 68 (1987)
N-Nitroso-*N*-methylurea (*see* *N*-Methyl-*N*-nitrosourea)	
N-Nitroso-*N*-methylurethane (*see* *N*-Methyl-*N*-methylurethane)	
N-Nitrosomethylvinylamine	17, 257 (1978); *Suppl. 7*, 68 (1987)
N-Nitrosomorpholine	17, 263 (1978); *Suppl. 7*, 68 (1987)
N'-Nitrosonornicotine	17, 281 (1978); 37, 241 (1985); *Suppl. 7*, 68 (1987)
N-Nitrosopiperidine	17, 287 (1978); *Suppl. 7*, 68 (1987)

N-Nitrosoproline	*17*, 303 (1978); *Suppl. 7*, 68 (1987)
N-Nitrosopyrrolidine	*17*, 313 (1978); *Suppl. 7*, 68 (1987)
N-Nitrososarcosine	*17*, 327 (1978); *Suppl. 7*, 68 (1987)
Nitrosoureas, chloroethyl (*see* Chloroethyl nitrosoureas)	
5-Nitro-*ortho*-toluidine	*48*, 169 (1990)
Nitrous oxide (*see* Anaesthetics, volatile)	
Nitrovin	*31*, 185 (1983); *Suppl. 7*, 68 (1987)
Nivalenol (*see* Toxins derived from *Fusarium graminearum*, *F. culmorum* and *F. crookwellense*)	
NNA [*see* 4-(*N*-Nitrosomethylamino)-4-(3-pyridyl)-1-butanal]	
NNK [*see* 4-(*N*-Nitrosomethylamino)-1-(3-pyridyl)-1-butanone]	
Nonsteroidal oestrogens (*see also* Oestrogens, progestins and combinations)	*Suppl. 7*, 272 (1987)
Norethisterone (*see also* Progestins; Combined oral contraceptives)	*6*, 179 (1974); *21*, 461 (1979)
Norethynodrel (*see also* Progestins; Combined oral contraceptives	*6*, 191 (1974); *21*, 461 (1979) (*corr. 42*, 259)
Norgestrel (*see also* Progestins, Combined oral contraceptives)	*6*, 201 (1974); *21*, 479 (1979)
Nylon 6	*19*, 120 (1979); *Suppl. 7*, 68 (1987)

O

Ochratoxin A	*10*, 191 (1976); *31*, 191 (1983) (*corr. 42*, 262); *Suppl. 7*, 271 (1987); *56*, 489 (1993)
Oestradiol-17β (*see also* Steroidal oestrogens)	*6*, 99 (1974); *21*, 279 (1979)
Oestradiol 3-benzoate (*see* Oestradiol-17β)	
Oestradiol dipropionate (*see* Oestradiol-17β)	
Oestradiol mustard	*9*, 217 (1975)
Oestradiol-17β-valerate (*see* Oestradiol-17β)	
Oestriol (*see also* Steroidal oestrogens)	*6*, 117 (1974); *21*, 327 (1979)
Oestrogen–progestin combinations (*see* Oestrogens, progestins and combinations)	
Oestrogen–progestin replacement therapy (*see also* Oestrogens, progestins and combinations)	*Suppl. 7*, 308 (1987)
Oestrogen replacement therapy (*see also* Oestrogens, progestins and combinations)	*Suppl. 7*, 280 (1987)
Oestrogens (*see* Oestrogens, progestins and combinations)	
Oestrogens, conjugated (*see* Conjugated oestrogens)	
Oestrogens, nonsteroidal (*see* Nonsteroidal oestrogens)	
Oestrogens, progestins and combinations	*6* (1974); *21* (1979); *Suppl. 7*, 272 (1987)
Oestrogens, steroidal (*see* Steroidal oestrogens)	
Oestrone (*see also* Steroidal oestrogens)	*6*, 123 (1974); *21*, 343 (1979) (*corr. 42*, 259)
Oestrone benzoate (*see* Oestrone)	
Oil Orange SS	*8*, 165 (1975); *Suppl. 7*, 69 (1987)
Oral contraceptives, combined (*see* Combined oral contraceptives)	
Oral contraceptives, investigational (*see* Combined oral contraceptives)	
Oral contraceptives, sequential (*see* Sequential oral contraceptives)	
Orange I	*8*, 173 (1975); *Suppl. 7*, 69 (1987)

Orange G	8, 181 (1975); *Suppl. 7*, 69 (1987)
Organolead compounds (*see also* Lead and lead compounds)	*Suppl. 7*, 230 (1987)
Oxazepam	13, 58 (1977); *Suppl. 7*, 69 (1987)
Oxymetholone [*see also* Androgenic (anabolic) steroids]	13, 131 (1977)
Oxyphenbutazone	13, 185 (1977); *Suppl. 7*, 69 (1987)

P

Paint manufacture and painting (occupational exposures in)	47, 329 (1989)
Panfuran S (*see also* Dihydroxymethylfuratrizine)	24, 77 (1980); *Suppl. 7*, 69 (1987)
Paper manufacture (*see* Pulp and paper manufacture)	
Paracetamol	50, 307 (1990)
Parasorbic acid	10, 199 (1976) (*corr.* 42, 255); *Suppl. 7*, 69 (1987)
Parathion	30, 153 (1983); *Suppl. 7*, 69 (1987)
Patulin	10, 205 (1976); 40, 83 (1986); *Suppl. 7*, 69 (1987)
Penicillic acid	10, 211 (1976); *Suppl. 7*, 69 (1987)
Pentachloroethane	41, 99 (1986); *Suppl. 7*, 69 (1987)
Pentachloronitrobenzene (*see* Quintozene)	
Pentachlorophenol (*see also* Chlorophenols; Chlorophenols, occupational exposures to)	20, 303 (1979); 53, 371 (1991)
Permethrin	53, 329 (1991)
Perylene	32, 411 (1983); *Suppl. 7*, 69 (1987)
Petasitenine	31, 207 (1983); *Suppl. 7*, 69 (1987)
Petasites japonicus (*see* Pyrrolizidine alkaloids)	
Petroleum refining (occupational exposures in)	45, 39 (1989)
Some petroleum solvents	47, 43 (1989)
Phenacetin	13, 141 (1977); 24, 135 (1980); *Suppl. 7*, 310 (1987)
Phenanthrene	32, 419 (1983); *Suppl. 7*, 69 (1987)
Phenazopyridine hydrochloride	8, 117 (1975); 24, 163 (1980) (*corr.* 42, 260); *Suppl. 7*, 312 (1987)
Phenelzine sulfate	24, 175 (1980); *Suppl. 7*, 312 (1987)
Phenicarbazide	12, 177 (1976); *Suppl. 7*, 70 (1987)
Phenobarbital	13, 157 (1977); *Suppl. 7*, 313 (1987)
Phenol	47, 263 (1989) (*corr.* 50, 385)
Phenoxyacetic acid herbicides (*see* Chlorophenoxy herbicides)	
Phenoxybenzamine hydrochloride	9, 223 (1975); 24, 185 (1980); *Suppl. 7*, 70 (1987)
Phenylbutazone	13, 183 (1977); *Suppl. 7*, 316 (1987)
meta-Phenylenediamine	16, 111 (1978); *Suppl. 7*, 70 (1987)
para-Phenylenediamine	16, 125 (1978); *Suppl. 7*, 70 (1987)
Phenyl glycidyl ether (*see* Glycidyl ethers)	
N-Phenyl-2-naphthylamine	16, 325 (1978) (*corr.* 42, 257); *Suppl. 7*, 318 (1987)
ortho-Phenylphenol	30, 329 (1983); *Suppl. 7*, 70 (1987)
Phenytoin	13, 201 (1977); *Suppl. 7*, 319 (1987)
PhIP	56, 229 (1993)
Pickled vegetables	56, 83 (1993)
Picloram	53, 481 (1991)
Piperazine oestrone sulfate (*see* Conjugated oestrogens)	

Piperonyl butoxide	*30*, 183 (1983); *Suppl. 7*, 70 (1987)
Pitches, coal-tar (*see* Coal-tar pitches)	
Polyacrylic acid	*19*, 62 (1979); *Suppl. 7*, 70 (1987)
Polybrominated biphenyls	*18*, 107 (1978); *41*, 261 (1986); *Suppl. 7*, 321 (1987)
Polychlorinated biphenyls	*7*, 261 (1974); *18*, 43 (1978) (*corr. 42*, 258); *Suppl. 7*, 322 (1987)
Polychlorinated camphenes (*see* Toxaphene)	
Polychloroprene	*19*, 141 (1979); *Suppl. 7*, 70 (1987)
Polyethylene	*19*, 164 (1979); *Suppl. 7*, 70 (1987)
Polymethylene polyphenyl isocyanate	*19*, 314 (1979); *Suppl. 7*, 70 (1987)
Polymethyl methacrylate	*19*, 195 (1979); *Suppl. 7*, 70 (1987)
Polyoestradiol phosphate (*see* Oestradiol-17β)	
Polypropylene	*19*, 218 (1979); *Suppl. 7*, 70 (1987)
Polystyrene	*19*, 245 (1979); *Suppl. 7*, 70 (1987)
Polytetrafluoroethylene	*19*, 288 (1979); *Suppl. 7*, 70 (1987)
Polyurethane foams	*19*, 320 (1979); *Suppl. 7*, 70 (1987)
Polyvinyl acetate	*19*, 346 (1979); *Suppl. 7*, 70 (1987)
Polyvinyl alcohol	*19*, 351 (1979); *Suppl. 7*, 70 (1987)
Polyvinyl chloride	*7*, 306 (1974); *19*, 402 (1979); *Suppl. 7*, 70 (1987)
Polyvinyl pyrrolidone	*19*, 463 (1979); *Suppl. 7*, 70 (1987)
Ponceau MX	*8*, 189 (1975); *Suppl. 7*, 70 (1987)
Ponceau 3R	*8*, 199 (1975); *Suppl. 7*, 70 (1987)
Ponceau SX	*8*, 207 (1975); *Suppl. 7*, 70 (1987)
Potassium arsenate (*see* Arsenic and arsenic compounds)	
Potassium arsenite (*see* Arsenic and arsenic compounds)	
Potassium bis(2-hydroxyethyl)dithiocarbamate	*12*, 183 (1976); *Suppl. 7*, 70 (1987)
Potassium bromate	*40*, 207 (1986); *Suppl. 7*, 70 (1987)
Potassium chromate (*see* Chromium and chromium compounds)	
Potassium dichromate (*see* Chromium and chromium compounds)	
Prednimustine	*50*, 115 (1990)
Prednisone	*26*, 293 (1981); *Suppl. 7*, 326 (1987)
Procarbazine hydrochloride	*26*, 311 (1981); *Suppl. 7*, 327 (1987)
Proflavine salts	*24*, 195 (1980); *Suppl. 7*, 70 (1987)
Progesterone (*see also* Progestins; Combined oral contraceptives)	*6*, 135 (1974); *21*, 491 (1979) (*corr. 42*, 259)
Progestins (*see also* Oestrogens, progestins and combinations)	*Suppl. 7*, 289 (1987)
Pronetalol hydrochloride	*13*, 227 (1977) (*corr. 42*, 256); *Suppl. 7*, 70 (1987)
1,3-Propane sultone	*4*, 253 (1974) (*corr. 42*, 253); *Suppl. 7*, 70 (1987)
Propham	*12*, 189 (1976); *Suppl. 7*, 70 (1987)
β-Propiolactone	*4*, 259 (1974) (*corr. 42*, 253); *Suppl. 7*, 70 (1987)
n-Propyl carbamate	*12*, 201 (1976); *Suppl. 7*, 70 (1987)
Propylene	*19*, 213 (1979); *Suppl. 7*, 71 (1987)
Propylene oxide	*11*, 191 (1976); *36*, 227 (1985) (*corr. 42*, 263); *Suppl. 7*, 328 (1987)
Propylthiouracil	*7*, 67 (1974); *Suppl. 7*, 329 (1987)
Ptaquiloside (*see also* Bracken fern)	*40*, 55 (1986); *Suppl. 7*, 71 (1987)
Pulp and paper manufacture	*25*, 157 (1981); *Suppl. 7*, 385 (1987)

Pyrene	32, 431 (1983); *Suppl. 7*, 71 (1987)
Pyrido[3,4-c]psoralen	40, 349 (1986); *Suppl. 7*, 71 (1987)
Pyrimethamine	13, 233 (1977); *Suppl. 7*, 71 (1987)
Pyrrolizidine alkaloids (*see* Hydroxysenkirkine; Isatidine; Jacobine; Lasiocarpine; Monocrotaline; Retrorsine; Riddelliine; Seneciphylline; Senkirkine)	

Q

Quercetin (*see also* Bracken fern)	31, 213 (1983); *Suppl. 7*, 71 (1987)
para-Quinone	15, 255 (1977); *Suppl. 7*, 71 (1987)
Quintozene	5, 211 (1974); *Suppl. 7*, 71 (1987)

R

Radon	43, 173 (1988) (*corr.* 45, 283)
Reserpine	10, 217 (1976); 24, 211 (1980) (*corr.* 42, 260); *Suppl. 7*, 330 (1987)
Resorcinol	15, 155 (1977); *Suppl. 7*, 71 (1987)
Retrorsine	10, 303 (1976); *Suppl. 7*, 71 (1987)
Rhodamine B	16, 221 (1978); *Suppl. 7*, 71 (1987)
Rhodamine 6G	16, 233 (1978); *Suppl. 7*, 71 (1987)
Riddelliine	10, 313 (1976); *Suppl. 7*, 71 (1987)
Rifampicin	24, 243 (1980); *Suppl. 7*, 71 (1987)
Rockwool (*see* Man-made mineral fibres)	
The rubber industry	28 (1982) (*corr.* 42, 261); *Suppl. 7*, 332 (1987)
Rugulosin	40, 99 (1986); *Suppl. 7*, 71 (1987)

S

Saccharated iron oxide	2, 161 (1973); *Suppl. 7*, 71 (1987)
Saccharin	22, 111 (1980) (*corr.* 42, 259); *Suppl. 7*, 334 (1987)
Safrole	1, 169 (1972); 10, 231 (1976); *Suppl. 7*, 71 (1987)
Salted fish	56, 41 (1993)
The sawmill industry (including logging) [*see* The lumber and sawmill industry (including logging)]	
Scarlet Red	8, 217 (1975); *Suppl. 7*, 71 (1987)
Selenium and selenium compounds	9, 245 (1975) (*corr.* 42, 255); *Suppl. 7*, 71 (1987)
Selenium dioxide (*see* Selenium and selenium compounds)	
Selenium oxide (*see* Selenium and selenium compounds)	
Semicarbazide hydrochloride	12, 209 (1976) (*corr.* 42, 256); *Suppl. 7*, 71 (1987)
Senecio jacobaea L. (*see* Pyrrolizidine alkaloids)	
Senecio longilobus (*see* Pyrrolizidine alkaloids)	
Seneciphylline	10, 319, 335 (1976); *Suppl. 7*, 71 (1987)
Senkirkine	10, 327 (1976); 31, 231 (1983); *Suppl. 7*, 71 (1987)

Sepiolite	42, 175 (1987); *Suppl. 7*, 71 (1987)
Sequential oral contraceptives (*see also* Oestrogens, progestins and combinations)	*Suppl. 7*, 296 (1987)
Shale-oils	35, 161 (1985); *Suppl. 7*, 339 (1987)
Shikimic acid (*see also* Bracken fern)	40, 55 (1986); *Suppl. 7*, 71 (1987)
Shoe manufacture and repair (*see* Boot and shoe manufacture and repair)	
Silica (*see also* Amorphous silica; Crystalline silica)	42, 39 (1987)
Simazine	53, 495 (1991)
Slagwool (*see* Man-made mineral fibres)	
Sodium arsenate (*see* Arsenic and arsenic compounds)	
Sodium arsenite (*see* Arsenic and arsenic compounds)	
Sodium cacodylate (*see* Arsenic and arsenic compounds)	
Sodium chlorite	52, 145 (1991)
Sodium chromate (*see* Chromium and chromium compounds)	
Sodium cyclamate (*see* Cyclamates)	
Sodium dichromate (*see* Chromium and chromium compounds)	
Sodium diethyldithiocarbamate	12, 217 (1976); *Suppl. 7*, 71 (1987)
Sodium equilin sulfate (*see* Conjugated oestrogens)	
Sodium fluoride (*see* Fluorides)	
Sodium monofluorophosphate (*see* Fluorides)	
Sodium oestrone sulfate (*see* Conjugated oestrogens)	
Sodium *ortho*-phenylphenate (*see also ortho*-Phenylphenol)	30, 329 (1983); *Suppl. 7*, 392 (1987)
Sodium saccharin (*see* Saccharin)	
Sodium selenate (*see* Selenium and selenium compounds)	
Sodium selenite (*see* Selenium and selenium compounds)	
Sodium silicofluoride (*see* Fluorides)	
Solar radiation	55 (1992)
Soots	3, 22 (1973); 35, 219 (1985); *Suppl. 7*, 343 (1987)
Spironolactone	24, 259 (1980); *Suppl. 7*, 344 (1987)
Stannous fluoride (*see* Fluorides)	
Steel founding (*see* Iron and steel founding)	
Sterigmatocystin	1, 175 (1972); 10, 245 (1976); *Suppl. 7*, 72 (1987)
Steroidal oestrogens (*see also* Oestrogens, progestins and combinations)	*Suppl. 7*, 280 (1987)
Streptozotocin	4, 221 (1974); 17, 337 (1978); *Suppl. 7*, 72 (1987)
Strobane® (*see* Terpene polychlorinates)	
Strontium chromate (*see* Chromium and chromium compounds)	
Styrene	19, 231 (1979) (*corr.* 42, 258); *Suppl. 7*, 345 (1987)
Styrene-acrylonitrile copolymers	19, 97 (1979); *Suppl. 7*, 72 (1987)
Styrene-butadiene copolymers	19, 252 (1979); *Suppl. 7*, 72 (1987)
Styrene oxide	11, 201 (1976); 19, 275 (1979); 36, 245 (1985); *Suppl. 7*, 72 (1987)
Succinic anhydride	15, 265 (1977); *Suppl. 7*, 72 (1987)
Sudan I	8, 225 (1975); *Suppl. 7*, 72 (1987)
Sudan II	8, 233 (1975); *Suppl. 7*, 72 (1987)
Sudan III	8, 241 (1975); *Suppl. 7*, 72 (1987)
Sudan Brown RR	8, 249 (1975); *Suppl. 7*, 72 (1987)

Sudan Red 7B	8, 253 (1975); *Suppl. 7*, 72 (1987)
Sulfafurazole	24, 275 (1980); *Suppl. 7*, 347 (1987)
Sulfallate	30, 283 (1983); *Suppl. 7*, 72 (1987)
Sulfamethoxazole	24, 285 (1980); *Suppl. 7*, 348 (1987)
Sulfites (*see* Sulfur dioxide and some sulfites, bisulfites and metabisulfites)	
Sulfur dioxide and some sulfites, bisulfites and metabisulfites	54, 131 (1992)
Sulfur mustard (*see* Mustard gas)	
Sulfuric acid and other strong inorganic acids, occupational exposures to mists and vapours from	54, 41 (1992)
Sulfur trioxide	54, 121 (1992)
Sulphisoxazole (*see* Sulfafurazole)	
Sunset Yellow FCF	8, 257 (1975); *Suppl. 7*, 72 (1987)
Symphytine	31, 239 (1983); *Suppl. 7*, 72 (1987)

T

2,4,5-T (*see also* Chlorophenoxy herbicides; Chlorophenoxy herbicides, occupational exposures to)	15, 273 (1977)
Talc	42, 185 (1987); *Suppl. 7*, 349 (1987)
Tannic acid	10, 253 (1976) (*corr. 42*, 255); *Suppl. 7*, 72 (1987)
Tannins (*see also* Tannic acid)	10, 254 (1976); *Suppl. 7*, 72 (1987)
TCDD (*see* 2,3,7,8-Tetrachlorodibenzo-*para*-dioxin)	
TDE (*see* DDT)	
Tea	51, 207 (1991)
Terpene polychlorinates	5, 219 (1974); *Suppl. 7*, 72 (1987)
Testosterone (*see also* Androgenic (anabolic) steroids)	6, 209 (1974); 21, 519 (1979)
Testosterone oenanthate (*see* Testosterone)	
Testosterone propionate (*see* Testosterone)	
2,2',5,5'-Tetrachlorobenzidine	27, 141 (1982); *Suppl. 7*, 72 (1987)
2,3,7,8-Tetrachlorodibenzo-*para*-dioxin	15, 41 (1977); *Suppl. 7*, 350 (1987)
1,1,1,2-Tetrachloroethane	41, 87 (1986); *Suppl. 7*, 72 (1987)
1,1,2,2-Tetrachloroethane	20, 477 (1979); *Suppl. 7*, 354 (1987)
Tetrachloroethylene	20, 491 (1979); *Suppl. 7*, 355 (1987)
2,3,4,6-Tetrachlorophenol (*see* Chlorophenols; Chlorophenols, occupational exposures to)	
Tetrachlorvinphos	30, 197 (1983); *Suppl. 7*, 72 (1987)
Tetraethyllead (*see* Lead and lead compounds)	
Tetrafluoroethylene	19, 285 (1979); *Suppl. 7*, 72 (1987)
Tetrakis(hydroxymethyl) phosphonium salts	48, 95 (1990)
Tetramethyllead (*see* Lead and lead compounds)	
Textile manufacturing industry, exposures in	48, 215 (1990) (*corr. 51*, 483)
Theobromine	51, 421 (1991)
Theophylline	51, 391 (1991)
Thioacetamide	7, 77 (1974); *Suppl. 7*, 72 (1987)
4,4'-Thiodianiline	16, 343 (1978); 27, 147 (1982); *Suppl. 7*, 72 (1987)
Thiotepa	9, 85 (1975); *Suppl. 7*, 368 (1987); 50, 123 (1990)
Thiouracil	7, 85 (1974); *Suppl. 7*, 72 (1987)
Thiourea	7, 95 (1974); *Suppl. 7*, 72 (1987)

Thiram	*12*, 225 (1976); *Suppl. 7*, 72 (1987); *53*, 403 (1991)
Titanium dioxide	*47*, 307 (1989)
Tobacco habits other than smoking (*see* Tobacco products, smokeless)	
Tobacco products, smokeless	*37* (1985) (*corr. 42*, 263; *52*, 513); *Suppl. 7*, 357 (1987)
Tobacco smoke	*38* (1986) (*corr. 42*, 263); *Suppl. 7*, 357 (1987)
Tobacco smoking (*see* Tobacco smoke)	
ortho-Tolidine (*see* 3,3'-Dimethylbenzidine)	
2,4-Toluene diisocyanate (*see also* Toluene diisocyanates)	*19*, 303 (1979); *39*, 287 (1986)
2,6-Toluene diisocyanate (*see also* Toluene diisocyanates)	*19*, 303 (1979); *39*, 289 (1986)
Toluene	*47*, 79 (1989)
Toluene diisocyanates	*39*, 287 (1986) (*corr. 42*, 264); *Suppl. 7*, 72 (1987)
Toluenes, α-chlorinated (*see* α-Chlorinated toluenes)	
ortho-Toluenesulfonamide (*see* Saccharin)	
ortho-Toluidine	*16*, 349 (1978); *27*, 155 (1982); *Suppl. 7*, 362 (1987)
Toxaphene	*20*, 327 (1979); *Suppl. 7*, 72 (1987)
T-2 Toxin (*see* Toxins derived from *Fusarium sporotrichioides*)	
Toxins derived from *Fusarium graminearum*, *F. culmorum* and *F. crookwellense*	*11*, 169 (1976); *31*, 153, 279 (1983); *Suppl. 7*, 64, 74 (1987); *56*, 397 (1993)
Toxins derived from *Fusarium moniliforme*	*56*, 445 (1993)
Toxins derived from *Fusarium sporotrichioides*	*31*, 265 (1983); *Suppl. 7*, 73 (1987); *56*, 467 (1993)
Tremolite (*see* Asbestos)	
Treosulfan	*26*, 341 (1981); *Suppl. 7*, 363 (1987)
Triaziquone [*see* Tris(aziridinyl)-*para*-benzoquinone]	
Trichlorfon	*30*, 207 (1983); *Suppl. 7*, 73 (1987)
Trichlormethine	*9*, 229 (1975); *Suppl. 7*, 73 (1987); *50*, 143 (1990)
Trichloroacetonitrile (*see* Halogenated acetonitriles)	
1,1,1-Trichloroethane	*20*, 515 (1979); *Suppl. 7*, 73 (1987)
1,1,2-Trichloroethane	*20*, 533 (1979); *Suppl. 7*, 73 (1987); *52*, 337 (1991)
Trichloroethylene	*11*, 263 (1976); *20*, 545 (1979); *Suppl. 7*, 364 (1987)
2,4,5-Trichlorophenol (*see also* Chlorophenols; Chlorophenols occupational exposures to)	*20*, 349 (1979)
2,4,6-Trichlorophenol (*see also* Chlorophenols; Chlorophenols, occupational exposures to)	*20*, 349 (1979)
(2,4,5-Trichlorophenoxy)acetic acid (*see* 2,4,5-T)	
Trichlorotriethylamine hydrochloride (*see* Trichlormethine)	
T₂-Trichothecene (*see* Toxins derived from *Fusarium sporotrichioides*)	
Triethylene glycol diglycidyl ether	*11*, 209 (1976); *Suppl. 7*, 73 (1987)
Trifluralin	*53*, 515 (1991)
4,4',6-Trimethylangelicin plus ultraviolet radiation (*see also* Angelicin and some synthetic derivatives)	*Suppl. 7*, 57 (1987)
2,4,5-Trimethylaniline	*27*, 177 (1982); *Suppl. 7*, 73 (1987)
2,4,6-Trimethylaniline	*27*, 178 (1982); *Suppl. 7*, 73 (1987)

4,5',8-Trimethylpsoralen	40, 357 (1986); *Suppl. 7*, 366 (1987)
Trimustine hydrochloride (*see* Trichlormethine)	
Triphenylene	32, 447 (1983); *Suppl. 7*, 73 (1987)
Tris(aziridinyl)-*para*-benzoquinone	9, 67 (1975); *Suppl. 7*, 367 (1987)
Tris(1-aziridinyl)phosphine oxide	9, 75 (1975); *Suppl. 7*, 73 (1987)
Tris(1-aziridinyl)phosphine sulphide (*see* Thiotepa)	
2,4,6-Tris(1-aziridinyl)-*s*-triazine	9, 95 (1975); *Suppl. 7*, 73 (1987)
Tris(2-chloroethyl) phosphate	48, 109 (1990)
1,2,3-Tris(chloromethoxy)propane	15, 301 (1977); *Suppl. 7*, 73 (1987)
Tris(2,3-dibromopropyl)phosphate	20, 575 (1979); *Suppl. 7*, 369 (1987)
Tris(2-methyl-1-aziridinyl)phosphine oxide	9, 107 (1975); *Suppl. 7*, 73 (1987)
Trp-P-1	31, 247 (1983); *Suppl. 7*, 73 (1987)
Trp-P-2	31, 255 (1983); *Suppl. 7*, 73 (1987)
Trypan blue	8, 267 (1975); *Suppl. 7*, 73 (1987)
Tussilago farfara L. (*see* Pyrrolizidine alkaloids)	

U

Ultraviolet radiation	40, 379 (1986); 55 (1992)
Underground haematite mining with exposure to radon	1, 29 (1972); *Suppl. 7*, 216 (1987)
Uracil mustard	9, 235 (1975); *Suppl. 7*, 370 (1987)
Urethane	7, 111 (1974); *Suppl. 7*, 73 (1987)

V

Vat Yellow 4	48, 161 (1990)
Vinblastine sulfate	26, 349 (1981) (*corr.* 42, 261); *Suppl. 7*, 371 (1987)
Vincristine sulfate	26, 365 (1981); *Suppl. 7*, 372 (1987)
Vinyl acetate	19, 341 (1979); 39, 113 (1986); *Suppl. 7*, 73 (1987)
Vinyl bromide	19, 367 (1979); 39, 133 (1986); *Suppl. 7*, 73 (1987)
Vinyl chloride	7, 291 (1974); 19, 377 (1979) (*corr.* 42, 258); *Suppl. 7*, 373 (1987)
Vinyl chloride–vinyl acetate copolymers	7, 311 (1976); 19, 412 (1979) (*corr.* 42, 258); *Suppl. 7*, 73 (1987)
4-Vinylcyclohexene	11, 277 (1976); 39, 181 (1986); *Suppl. 7*, 73 (1987)
Vinyl fluoride	39, 147 (1986); *Suppl. 7*, 73 (1987)
Vinylidene chloride	19, 439 (1979); 39, 195 (1986); *Suppl. 7*, 376 (1987)
Vinylidene chloride–vinyl chloride copolymers	19, 448 (1979) (*corr.* 42, 258); *Suppl. 7*, 73 (1987)
Vinylidene fluoride	39, 227 (1986); *Suppl. 7*, 73 (1987)
N-Vinyl-2-pyrrolidone	19, 461 (1979); *Suppl. 7*, 73 (1987)

W

Welding	49, 447 (1990) (*corr.* 52, 513)
Wollastonite	42, 145 (1987); *Suppl. 7*, 377 (1987)
Wood industries	25 (1981); *Suppl. 7*, 378 (1987)

X

Xylene	*47*, 125 (1989)
2,4-Xylidine	*16*, 367 (1978); *Suppl. 7*, 74 (1987)
2,5-Xylidine	*16*, 377 (1978); *Suppl. 7*, 74 (1987)

Y

Yellow AB	*8*, 279 (1975); *Suppl. 7*, 74 (1987)
Yellow OB	*8*, 287 (1975); *Suppl. 7*, 74 (1987)

Z

Zearalenone (*see* Toxins derived from *Fusarium graminearum*, *F. culmorum* and *F. crookwellense*)	
Zectran	*12*, 237 (1976); *Suppl. 7*, 74 (1987)
Zinc beryllium silicate (*see* Beryllium and beryllium compounds)	
Zinc chromate (*see* Chromium and chromium compounds)	
Zinc chromate hydroxide (*see* Chromium and chromium compounds)	
Zinc potassium chromate (*see* Chromium and chromium compounds)	
Zinc yellow (*see* Chromium and chromium compounds)	
Zineb	*12*, 245 (1976); *Suppl. 7*, 74 (1987)
Ziram	*12*, 259 (1976); *Suppl. 7*, 74 (1987); *53*, 423 (1991)

PUBLICATIONS OF THE INTERNATIONAL AGENCY FOR RESEARCH ON CANCER

Scientific Publications Series

(Available from Oxford University Press through local bookshops)

No. 1 Liver Cancer
1971; 176 pages (*out of print*)

No. 2 Oncogenesis and Herpesviruses
Edited by P.M. Biggs, G. de-Thé and L.N. Payne
1972; 515 pages (*out of print*)

No. 3 N-Nitroso Compounds: Analysis and Formation
Edited by P. Bogovski, R. Preussman and E.A. Walker
1972; 140 pages (*out of print*)

No. 4 Transplacental Carcinogenesis
Edited by L. Tomatis and U. Mohr
1973; 181 pages (*out of print*)

No. 5/6 Pathology of Tumours in Laboratory Animals, Volume 1, Tumours of the Rat
Edited by V.S. Turusov
1973/1976; 533 pages (*out of print*)

No. 7 Host Environment Interactions in the Etiology of Cancer in Man
Edited by R. Doll and I. Vodopija
1973; 464 pages (*out of print*)

No. 8 Biological Effects of Asbestos
Edited by P. Bogovski, J.C. Gilson, V. Timbrell and J.C. Wagner
1973; 346 pages (*out of print*)

No. 9 N-Nitroso Compounds in the Environment
Edited by P. Bogovski and E.A. Walker
1974; 243 pages (*out of print*)

No. 10 Chemical Carcinogenesis Essays
Edited by R. Montesano and L. Tomatis
1974; 230 pages (*out of print*)

No. 11 Oncogenesis and Herpesviruses II
Edited by G. de-Thé, M.A. Epstein and H. zur Hausen
1975; Part I: 511 pages
Part II: 403 pages (*out of print*)

No. 12 Screening Tests in Chemical Carcinogenesis
Edited by R. Montesano, H. Bartsch and L. Tomatis
1976; 666 pages (*out of print*)

No. 13 Environmental Pollution and Carcinogenic Risks
Edited by C. Rosenfeld and W. Davis
1975; 441 pages (*out of print*)

No. 14 Environmental N-Nitroso Compounds. Analysis and Formation
Edited by E.A. Walker, P. Bogovski and L. Griciute
1976; 512 pages (*out of print*)

No. 15 Cancer Incidence in Five Continents, Volume III
Edited by J.A.H. Waterhouse, C. Muir, P. Correa and J. Powell
1976; 584 pages (*out of print*)

No. 16 Air Pollution and Cancer in Man
Edited by U. Mohr, D. Schmähl and L. Tomatis
1977; 328 pages (*out of print*)

No. 17 Directory of On-going Research in Cancer Epidemiology 1977
Edited by C.S. Muir and G. Wagner
1977; 599 pages (*out of print*)

No. 18 Environmental Carcinogens. Selected Methods of Analysis. Volume 1: Analysis of Volatile Nitrosamines in Food
Editor-in-Chief: H. Egan
1978; 212 pages (*out of print*)

No. 19 Environmental Aspects of N-Nitroso Compounds
Edited by E.A. Walker, M. Castegnaro, L. Griciute and R.E. Lyle
1978; 561 pages (*out of print*)

No. 20 Nasopharyngeal Carcinoma: Etiology and Control
Edited by G. de-Thé and Y. Ito
1978; 606 pages (*out of print*)

No. 21 Cancer Registration and its Techniques
Edited by R. MacLennan, C. Muir, R. Steinitz and A. Winkler
1978; 235 pages (*out of print*)

No. 22 Environmental Carcinogens. Selected Methods of Analysis. Volume 2: Methods for the Measurement of Vinyl Chloride in Poly(vinyl chloride), Air, Water and Foodstuffs
Editor-in-Chief: H. Egan
1978; 142 pages (*out of print*)

No. 23 Pathology of Tumours in Laboratory Animals. Volume II: Tumours of the Mouse
Editor-in-Chief: V.S. Turusov
1979; 669 pages (*out of print*)

No. 24 Oncogenesis and Herpesviruses III
Edited by G. de-Thé, W. Henle and F. Rapp
1978; Part I: 580 pages, Part II: 512 pages (*out of print*)

Prices, valid for March 1993, are subject to change without notice

List of IARC Publications

No. 25 Carcinogenic Risk. Strategies for Intervention
Edited by W. Davis and C. Rosenfeld
1979; 280 pages (*out of print*)

No. 26 Directory of On-going Research in Cancer Epidemiology 1978
Edited by C.S. Muir and G. Wagner
1978; 550 pages (*out of print*)

No. 27 Molecular and Cellular Aspects of Carcinogen Screening Tests
Edited by R. Montesano, H. Bartsch and L. Tomatis
1980; 372 pages £29.00

No. 28 Directory of On-going Research in Cancer Epidemiology 1979
Edited by C.S. Muir and G. Wagner
1979; 672 pages (*out of print*)

No. 29 Environmental Carcinogens. Selected Methods of Analysis. Volume 3: Analysis of Polycyclic Aromatic Hydrocarbons in Environmental Samples
Editor-in-Chief: H. Egan
1979; 240 pages (*out of print*)

No. 30 Biological Effects of Mineral Fibres
Editor-in-Chief: J.C. Wagner
1980; Volume 1: 494 pages Volume 2: 513 pages (*out of print*)

No. 31 N-Nitroso Compounds: Analysis, Formation and Occurrence
Edited by E.A. Walker, L. Griciute, M. Castegnaro and M. Börzsönyi
1980; 835 pages (*out of print*)

No. 32 Statistical Methods in Cancer Research. Volume 1. The Analysis of Case-control Studies
By N.E. Breslow and N.E. Day
1980; 338 pages £25.00

No. 33 Handling Chemical Carcinogens in the Laboratory
Edited by R. Montesano et al.
1979; 32 pages (*out of print*)

No. 34 Pathology of Tumours in Laboratory Animals. Volume III. Tumours of the Hamster
Editor-in-Chief: V.S. Turusov
1982; 461 pages (*out of print*)

No. 35 Directory of On-going Research in Cancer Epidemiology 1980
Edited by C.S. Muir and G. Wagner
1980; 660 pages (*out of print*)

No. 36 Cancer Mortality by Occupation and Social Class 1851-1971
Edited by W.P.D. Logan
1982; 253 pages (*out of print*)

No. 37 Laboratory Decontamination and Destruction of Aflatoxins B_1, B_2, G_1, G_2 in Laboratory Wastes
Edited by M. Castegnaro et al.
1980; 56 pages (*out of print*)

No. 38 Directory of On-going Research in Cancer Epidemiology 1981
Edited by C.S. Muir and G. Wagner
1981; 696 pages (*out of print*)

No. 39 Host Factors in Human Carcinogenesis
Edited by H. Bartsch and B. Armstrong
1982; 583 pages (*out of print*)

No. 40 Environmental Carcinogens. Selected Methods of Analysis. Volume 4: Some Aromatic Amines and Azo Dyes in the General and Industrial Environment
Edited by L. Fishbein, M. Castegnaro, I.K. O'Neill and H. Bartsch
1981; 347 pages (*out of print*)

No. 41 N-Nitroso Compounds: Occurrence and Biological Effects
Edited by H. Bartsch, I.K. O'Neill, M. Castegnaro and M. Okada
1982; 755 pages £50.00

No. 42 Cancer Incidence in Five Continents, Volume IV
Edited by J. Waterhouse, C. Muir, K. Shanmugaratnam and J. Powell
1982; 811 pages (*out of print*)

No. 43 Laboratory Decontamination and Destruction of Carcinogens in Laboratory Wastes: Some N-Nitrosamines
Edited by M. Castegnaro et al.
1982; 73 pages £7.50

No. 44 Environmental Carcinogens. Selected Methods of Analysis. Volume 5: Some Mycotoxins
Edited by L. Stoloff, M. Castegnaro, P. Scott, I.K. O'Neill and H. Bartsch
1983; 455 pages £32.50

No. 45 Environmental Carcinogens. Selected Methods of Analysis. Volume 6: N-Nitroso Compounds
Edited by R. Preussmann, I.K. O'Neill, G. Eisenbrand, B. Spiegelhalder and H. Bartsch
1983; 508 pages £32.50

No. 46 Directory of On-going Research in Cancer Epidemiology 1982
Edited by C.S. Muir and G. Wagner
1982; 722 pages (*out of print*)

No. 47 Cancer Incidence in Singapore 1968–1977
Edited by K. Shanmugaratnam, H.P. Lee and N.E. Day
1983; 171 pages (*out of print*)

No. 48 Cancer Incidence in the USSR (2nd Revised Edition)
Edited by N.P. Napalkov, G.F. Tserkovny, V.M. Merabishvili, D.M. Parkin, M. Smans and C.S. Muir
1983; 75 pages (*out of print*)

No. 49 Laboratory Decontamination and Destruction of Carcinogens in Laboratory Wastes: Some Polycyclic Aromatic Hydrocarbons
Edited by M. Castegnaro et al.
1983; 87 pages (*out of print*)

No. 50 Directory of On-going Research in Cancer Epidemiology 1983
Edited by C.S. Muir and G. Wagner
1983; 731 pages (*out of print*)

No. 51 Modulators of Experimental Carcinogenesis
Edited by V. Turusov and R. Montesano
1983; 307 pages (*out of print*)

* Available from booksellers through the network of WHO Sales agents.

† Available directly from IARC

List of IARC Publications

No. 52 Second Cancers in Relation to Radiation Treatment for Cervical Cancer: Results of a Cancer Registry Collaboration
Edited by N.E. Day and J.C. Boice, Jr
1984; 207 pages (*out of print*)

No. 53 Nickel in the Human Environment
Editor-in-Chief: F.W. Sunderman, Jr
1984; 529 pages (*out of print*)

No. 54 Laboratory Decontamination and Destruction of Carcinogens in Laboratory Wastes: Some Hydrazines
Edited by M. Castegnaro et al.
1983; 87 pages (*out of print*)

No. 55 Laboratory Decontamination and Destruction of Carcinogens in Laboratory Wastes: Some N-Nitrosamides
Edited by M. Castegnaro et al.
1984; 66 pages (*out of print*)

No. 56 Models, Mechanisms and Etiology of Tumour Promotion
Edited by M. Börzsönyi, N.E. Day, K. Lapis and H. Yamasaki
1984; 532 pages (*out of print*)

No. 57 N-Nitroso Compounds: Occurrence, Biological Effects and Relevance to Human Cancer
Edited by I.K. O'Neill, R.C. von Borstel, C.T. Miller, J. Long and H. Bartsch
1984; 1013 pages (*out of print*)

No. 58 Age-related Factors in Carcinogenesis
Edited by A. Likhachev, V. Anisimov and R. Montesano
1985; 288 pages (*out of print*)

No. 59 Monitoring Human Exposure to Carcinogenic and Mutagenic Agents
Edited by A. Berlin, M. Draper, K. Hemminki and H. Vainio
1984; 457 pages (*out of print*)

No. 60 Burkitt's Lymphoma: A Human Cancer Model
Edited by G. Lenoir, G. O'Conor and C.L.M. Olweny
1985; 484 pages (*out of print*)

No. 61 Laboratory Decontamination and Destruction of Carcinogens in Laboratory Wastes: Some Haloethers
Edited by M. Castegnaro et al.
1985; 55 pages (*out of print*)

No. 62 Directory of On-going Research in Cancer Epidemiology 1984
Edited by C.S. Muir and G. Wagner
1984; 717 pages (*out of print*)

No. 63 Virus-associated Cancers in Africa
Edited by A.O. Williams, G.T. O'Conor, G.B. de-Thé and C.A. Johnson
1984; 773 pages (*out of print*)

No. 64 Laboratory Decontamination and Destruction of Carcinogens in Laboratory Wastes: Some Aromatic Amines and 4-Nitrobiphenyl
Edited by M. Castegnaro et al.
1985; 84 pages (*out of print*)

No. 65 Interpretation of Negative Epidemiological Evidence for Carcinogenicity
Edited by N.J. Wald and R. Doll
1985; 232 pages (*out of print*)

No. 66 The Role of the Registry in Cancer Control
Edited by D.M. Parkin, G. Wagner and C.S. Muir
1985; 152 pages £10.00

No. 67 Transformation Assay of Established Cell Lines: Mechanisms and Application
Edited by T. Kakunaga and H. Yamasaki
1985; 225 pages (*out of print*)

No. 68 Environmental Carcinogens. Selected Methods of Analysis. Volume 7. Some Volatile Halogenated Hydrocarbons
Edited by L. Fishbein and I.K. O'Neill
1985; 479 pages (*out of print*)

No. 69 Directory of On-going Research in Cancer Epidemiology 1985
Edited by C.S. Muir and G. Wagner
1985; 745 pages (*out of print*)

No. 70 The Role of Cyclic Nucleic Acid Adducts in Carcinogenesis and Mutagenesis
Edited by B. Singer and H. Bartsch
1986; 467 pages (*out of print*)

No. 71 Environmental Carcinogens. Selected Methods of Analysis. Volume 8: Some Metals: As, Be, Cd, Cr, Ni, Pb, Se, Zn
Edited by I.K. O'Neill, P. Schuller and L. Fishbein
1986; 485 pages (*out of print*)

No. 72 Atlas of Cancer in Scotland, 1975–1980. Incidence and Epidemiological Perspective
Edited by I. Kemp, P. Boyle, M. Smans and C.S. Muir
1985; 285 pages (*out of print*)

No. 73 Laboratory Decontamination and Destruction of Carcinogens in Laboratory Wastes: Some Antineoplastic Agents
Edited by M. Castegnaro et al.
1985; 163 pages £12.50

No. 74 Tobacco: A Major International Health Hazard
Edited by D. Zaridze and R. Peto
1986; 324 pages £22.50

No. 75 Cancer Occurrence in Developing Countries
Edited by D.M. Parkin
1986; 339 pages £22.50

No. 76 Screening for Cancer of the Uterine Cervix
Edited by M. Hakama, A.B. Miller and N.E. Day
1986; 315 pages £30.00

List of IARC Publications

No. 77 **Hexachlorobenzene: Proceedings of an International Symposium**
Edited by C.R. Morris and J.R.P. Cabral
1986; 668 pages (*out of print*)

No. 78 **Carcinogenicity of Alkylating Cytostatic Drugs**
Edited by D. Schmähl and J.M. Kaldor
1986; 337 pages (*out of print*)

No. 79 **Statistical Methods in Cancer Research. Volume III: The Design and Analysis of Long-term Animal Experiments**
By J.J. Gart, D. Krewski, P.N. Lee, R.E. Tarone and J. Wahrendorf
1986; 213 pages £22.00

No. 80 **Directory of On-going Research in Cancer Epidemiology 1986**
Edited by C.S. Muir and G. Wagner
1986; 805 pages (*out of print*)

No. 81 **Environmental Carcinogens: Methods of Analysis and Exposure Measurement. Volume 9: Passive Smoking**
Edited by I.K. O'Neill, K.D. Brunnemann, B. Dodet and D. Hoffmann
1987; 383 pages £35.00

No. 82 **Statistical Methods in Cancer Research. Volume II: The Design and Analysis of Cohort Studies**
By N.E. Breslow and N.E. Day
1987; 404 pages £35.00

No. 83 **Long-term and Short-term Assays for Carcinogens: A Critical Appraisal**
Edited by R. Montesano, H. Bartsch, H. Vainio, J. Wilbourn and H. Yamasaki
1986; 575 pages £35.00

No. 84 **The Relevance of N-Nitroso Compounds to Human Cancer: Exposure and Mechanisms**
Edited by H. Bartsch, I.K. O'Neill and R. Schulte-Hermann
1987; 671 pages (*out of print*)

No. 85 **Environmental Carcinogens: Methods of Analysis and Exposure Measurement. Volume 10: Benzene and Alkylated Benzenes**
Edited by L. Fishbein and I.K. O'Neill
1988; 327 pages £40.00

No. 86 **Directory of On-going Research in Cancer Epidemiology 1987**
Edited by D.M. Parkin and J. Wahrendorf
1987; 676 pages (*out of print*)

No. 87 **International Incidence of Childhood Cancer**
Edited by D.M. Parkin, C.A. Stiller, C.A. Bieber, G.J. Draper, B. Terracini and J.L. Young
1988; 401 pages £35.00

No. 88 **Cancer Incidence in Five Continents Volume V**
Edited by C. Muir, J. Waterhouse, T. Mack, J. Powell and S. Whelan
1987; 1004 pages £55.00

No. 89 **Method for Detecting DNA Damaging Agents in Humans: Applications in Cancer Epidemiology and Prevention**
Edited by H. Bartsch, K. Hemminki and I.K. O'Neill
1988; 518 pages £50.00

No. 90 **Non-occupational Exposure to Mineral Fibres**
Edited by J. Bignon, J. Peto and R. Saracci
1989; 500 pages £50.00

No. 91 **Trends in Cancer Incidence in Singapore 1968–1982**
Edited by H.P. Lee, N.E. Day and K. Shanmugaratnam
1988; 160 pages (*out of print*)

No. 92 **Cell Differentiation, Genes and Cancer**
Edited by T. Kakunaga, T. Sugimura, L. Tomatis and H. Yamasaki
1988; 204 pages £27.50

No. 93 **Directory of On-going Research in Cancer Epidemiology 1988**
Edited by M. Coleman and J. Wahrendorf
1988; 662 pages (*out of print*)

No. 94 **Human Papillomavirus and Cervical Cancer**
Edited by N. Muñoz, F.X. Bosch and O.M. Jensen
1989; 154 pages £22.50

No. 95 **Cancer Registration: Principles and Methods**
Edited by O.M. Jensen, D.M. Parkin, R. MacLennan, C.S. Muir and R. Skeet
1991; 288 pages £28.00

No. 96 **Perinatal and Multigeneration Carcinogenesis**
Edited by N.P. Napalkov, J.M. Rice, L. Tomatis and H. Yamasaki
1989; 436 pages £50.00

No. 97 **Occupational Exposure to Silica and Cancer Risk**
Edited by L. Simonato, A.C. Fletcher, R. Saracci and T. Thomas
1990; 124 pages £22.50

No. 98 **Cancer Incidence in Jewish Migrants to Israel, 1961–1981**
Edited by R. Steinitz, D.M. Parkin, J.L. Young, C.A. Bieber and L. Katz
1989; 320 pages £35.00

No. 99 **Pathology of Tumours in Laboratory Animals, Second Edition, Volume 1, Tumours of the Rat**
Edited by V.S. Turusov and U. Mohr
740 pages £85.00

No. 100 **Cancer: Causes, Occurrence and Control**
Editor-in-Chief L. Tomatis
1990; 352 pages £24.00

No. 101 **Directory of On-going Research in Cancer Epidemiology 1989/90**
Edited by M. Coleman and J. Wahrendorf
1989; 818 pages £36.00

---

* Available from booksellers through the network of WHO Sales agents.

† Available directly from IARC

List of IARC Publications

No. 102 Patterns of Cancer in Five Continents
Edited by S.L. Whelan, D.M. Parkin & E. Masuyer
1990; 162 pages £25.00

No. 103 Evaluating Effectiveness of Primary Prevention of Cancer
Edited by M. Hakama, V. Beral, J.W. Cullen and D.M. Parkin
1990; 250 pages £32.00

No. 104 Complex Mixtures and Cancer Risk
Edited by H. Vainio, M. Sorsa and A.J. McMichael
1990; 442 pages £38.00

No. 105 Relevance to Human Cancer of N-Nitroso Compounds, Tobacco Smoke and Mycotoxins
Edited by I.K. O'Neill, J. Chen and H. Bartsch
1991; 614 pages £70.00

No. 106 Atlas of Cancer Incidence in the Former German Democratic Republic
Edited by W.H. Mehnert, M. Smans, C.S. Muir, M. Möhner & D. Schön
1992; 384 pages £55.00

No. 107 Atlas of Cancer Mortality in the European Economic Community
Edited by M. Smans, C.S. Muir and P. Boyle
1992; 280 pages £35.00

No. 108 Environmental Carcinogens: Methods of Analysis and Exposure Measurement. Volume 11: Polychlorinated Dioxins and Dibenzofurans
Edited by C. Rappe, H.R. Buser, B. Dodet and I.K. O'Neill
1991; 426 pages £45.00

No. 109 Environmental Carcinogens: Methods of Analysis and Exposure Measurement. Volume 12: Indoor Air Contaminants
Edited by B. Seifert, H. van de Wiel, B. Dodet and I.K. O'Neill
1993; 384 pages £45.00

No. 110 Directory of On-going Research in Cancer Epidemiology 1991
Edited by M. Coleman and J. Wahrendorf
1991; 753 pages £38.00

No. 111 Pathology of Tumours in Laboratory Animals, Second Edition, Volume 2, Tumours of the Mouse
Edited by V.S. Turusov and U. Mohr
Publ. due 1993; approx. 500 pages

No. 112 Autopsy in Epidemiology and Medical Research
Edited by E. Riboli and M. Delendi
1991; 288 pages £25.00

No. 113 Laboratory Decontamination and Destruction of Carcinogens in Laboratory Wastes: Some Mycotoxins
Edited by M. Castegnaro, J. Barek, J.-M. Frémy, M. Lafontaine, M. Miraglia, E.B. Sansone and G.M. Telling
1991; 64 pages £11.00

No. 114 Laboratory Decontamination and Destruction of Carcinogens in Laboratory Wastes: Some Polycyclic Heterocyclic Hydrocarbons
Edited by M. Castegnaro, J. Barek, J. Jacob, U. Kirso, M. Lafontaine, E.B. Sansone, G.M. Telling and T. Vu Duc
1991; 50 pages £8.00

No. 115 Mycotoxins, Endemic Nephropathy and Urinary Tract Tumours
Edited by M. Castegnaro, R. Plestina, G. Dirheimer, I.N. Chernozemsky and H Bartsch
1991; 340 pages £45.00

No. 116 Mechanisms of Carcinogenesis in Risk Identification
Edited by H. Vainio, P.N. Magee, D.B. McGregor & A.J. McMichael
1992; 616 pages £65.00

No. 117 Directory of On-going Research in Cancer Epidemiology 1992
Edited by M. Coleman, J. Wahrendorf & E. Démaret
1992; 773 pages £42.00

No. 118 Cadmium in the Human Environment: Toxicity and Carcinogenicity
Edited by G.F. Nordberg, R.F.M. Herber & L. Alessio
1992; 470 pages £60.00

No. 119 The Epidemiology of Cervical Cancer and Human Papillomavirus
Edited by N. Muñoz, F.X. Bosch, K.V. Shah & A. Meheus
1992; 288 pages £28.00

No. 120 Cancer Incidence in Five Continents, Volume VI
Edited by D.M. Parkin, C.S. Muir, S.L. Whelan, Y.T. Gao, J. Ferlay & J. Powell
1992; 1080 pages £120.00

No. 121 Trends in Cancer Incidence and Mortality
Edited by M. Coleman, J. Estève and P. Damiecki
1993; approx. 800 pages, approx £95.00

No. 122 International Classification of Rodent Tumours. Part 1. The Rat
Editor-in-Chief: U. Mohr
1992/93; 10 fascicles of 60–100 pages, £120.00

No. 123 Cancer in Italian Migrant Populations
Edited by M. Geddes, D.M. Parkin, M. Khlat, D. Balzi and E. Buiatti
1993; 292 pages, £40.00

No. 124 Postlabelling Methods for Detection of DNA Adducts
Edited by D.H. Phillips, M. Castegnaro and H. Bartsch
1993; approx. 380 pages; approx. £46.00

List of IARC Publications

IARC MONOGRAPHS ON THE EVALUATION OF CARCINOGENIC RISKS TO HUMANS

(Available from booksellers through the network of WHO Sales Agents)

Volume 1 Some Inorganic Substances, Chlorinated Hydrocarbons, Aromatic Amines, *N*-Nitroso Compounds, and Natural Products
1972; 184 pages (*out of print*)

Volume 2 Some Inorganic and Organometallic Compounds
1973; 181 pages (*out of print*)

Volume 3 Certain Polycyclic Aromatic Hydrocarbons and Heterocyclic Compounds
1973; 271 pages (*out of print*)

Volume 4 Some Aromatic Amines, Hydrazine and Related Substances, *N*-Nitroso Compounds and Miscellaneous Alkylating Agents
1974; 286 pages Sw. fr. 18.-

Volume 5 Some Organochlorine Pesticides
1974; 241 pages (*out of print*)

Volume 6 Sex Hormones
1974; 243 pages (*out of print*)

Volume 7 Some Anti-Thyroid and Related Substances, Nitrofurans and Industrial Chemicals
1974; 326 pages (*out of print*)

Volume 8 Some Aromatic Azo Compounds
1975; 357 pages Sw. fr. 36.-

Volume 9 Some Aziridines, *N*-, *S*- and *O*-Mustards and Selenium
1975; 268 pages Sw.fr. 27.-

Volume 10 Some Naturally Occurring Substances
1976; 353 pages (*out of print*)

Volume 11 Cadmium, Nickel, Some Epoxides, Miscellaneous Industrial Chemicals and General Considerations on Volatile Anaesthetics
1976; 306 pages (*out of print*)

Volume 12 Some Carbamates, Thiocarbamates and Carbazides
1976; 282 pages Sw. fr. 34.-

Volume 13 Some Miscellaneous Pharmaceutical Substances
1977; 255 pages Sw. fr. 30.-

Volume 14 Asbestos
1977; 106 pages (*out of print*)

Volume 15 Some Fumigants, The Herbicides 2,4-D and 2,4,5-T, Chlorinated Dibenzodioxins and Miscellaneous Industrial Chemicals
1977; 354 pages Sw. fr. 50.-

Volume 16 Some Aromatic Amines and Related Nitro Compounds - Hair Dyes, Colouring Agents and Miscellaneous Industrial Chemicals
1978; 400 pages Sw. fr. 50.-

Volume 17 Some *N*-Nitroso Compounds
1978; 365 pages Sw. fr. 50.-

Volume 18 Polychlorinated Biphenyls and Polybrominated Biphenyls
1978; 140 pages Sw. fr. 20.-

Volume 19 Some Monomers, Plastics and Synthetic Elastomers, and Acrolein
1979; 513 pages (*out of print*)

Volume 20 Some Halogenated Hydrocarbons
1979; 609 pages (*out of print*)

Volume 21 Sex Hormones (II)
1979; 583 pages Sw. fr. 60.-

Volume 22 Some Non-Nutritive Sweetening Agents
1980; 208 pages Sw. fr. 25.-

Volume 23 Some Metals and Metallic Compounds
1980; 438 pages (*out of print*)

Volume 24 Some Pharmaceutical Drugs
1980; 337 pages Sw. fr. 40.-

Volume 25 Wood, Leather and Some Associated Industries
1981; 412 pages Sw. fr. 60.-

Volume 26 Some Antineoplastic and Immunosuppressive Agents
1981; 411 pages Sw. fr. 62.-

Volume 27 Some Aromatic Amines, Anthraquinones and Nitroso Compounds, and Inorganic Fluorides Used in Drinking Water and Dental Preparations
1982; 341 pages Sw. fr. 40.-

Volume 28 The Rubber Industry
1982; 486 pages Sw. fr. 70.-

Volume 29 Some Industrial Chemicals and Dyestuffs
1982; 416 pages Sw. fr. 60.-

Volume 30 Miscellaneous Pesticides
1983; 424 pages Sw. fr. 60.-

Volume 31 Some Food Additives, Feed Additives and Naturally Occurring Substances
1983; 314 pages Sw. fr. 60.-

Volume 32 Polynuclear Aromatic Compounds, Part 1: Chemical, Environmental and Experimental Data
1983; 477 pages Sw. fr. 60.-

Volume 33 Polynuclear Aromatic Compounds, Part 2: Carbon Blacks, Mineral Oils and Some Nitroarenes
1984; 245 pages Sw. fr. 50.-

Volume 34 Polynuclear Aromatic Compounds, Part 3: Industrial Exposures in Aluminium Production, Coal Gasification, Coke Production, and Iron and Steel Founding
1984; 219 pages Sw. fr. 48.-

Volume 35 Polynuclear Aromatic Compounds, Part 4: Bitumens, Coal-tars and Derived Products, Shale-oils and Soots
1985; 271 pages Sw. fr. 70.-

---

* Available from booksellers through the network of WHO Sales agents.

† Available directly from IARC

List of IARC Publications

Volume 36 Allyl Compounds, Aldehydes, Epoxides and Peroxides
1985; 369 pages Sw. fr. 70.–

Volume 37 Tobacco Habits Other than Smoking: Betel-quid and Areca-nut Chewing; and some Related Nitrosamines
1985; 291 pages Sw. fr. 70.–

Volume 38 Tobacco Smoking
1986; 421 pages Sw. fr. 75.–

Volume 39 Some Chemicals Used in Plastics and Elastomers
1986; 403 pages Sw. fr. 60.–

Volume 40 Some Naturally Occurring and Synthetic Food Components, Furocoumarins and Ultraviolet Radiation
1986; 444 pages Sw. fr. 65.–

Volume 41 Some Halogenated Hydrocarbons and Pesticide Exposures
1986; 434 pages Sw. fr. 65.–

Volume 42 Silica and Some Silicates
1987; 289 pages Sw. fr. 65.

Volume 43 Man-Made Mineral Fibres and Radon
1988; 300 pages Sw. fr. 65.–

Volume 44 Alcohol Drinking
1988; 416 pages Sw. fr. 65.

Volume 45 Occupational Exposures in Petroleum Refining; Crude Oil and Major Petroleum Fuels
1989; 322 pages Sw. fr. 65.–

Volume 46 Diesel and Gasoline Engine Exhausts and Some Nitroarenes
1989; 458 pages Sw. fr. 65.–

Volume 47 Some Organic Solvents, Resin Monomers and Related Compounds, Pigments and Occupational Exposures in Paint Manufacture and Painting
1989; 536 pages Sw. fr. 85.–

Volume 48 Some Flame Retardants and Textile Chemicals, and Exposures in the Textile Manufacturing Industry
1990; 345 pages Sw. fr. 65.–

Volume 49 Chromium, Nickel and Welding
1990; 677 pages Sw. fr. 95.–

Volume 50 Pharmaceutical Drugs
1990; 415 pages Sw. fr. 65.–

Volume 51 Coffee, Tea, Mate, Methylxanthines and Methylglyoxal
1991; 513 pages Sw. fr. 80.–

Volume 52 Chlorinated Drinking-water; Chlorination By-products; Some Other Halogenated Compounds; Cobalt and Cobalt Compounds
1991; 544 pages Sw. fr. 80.–

Volume 53 Occupational Exposures in Insecticide Application and some Pesticides
1991; 612 pages Sw. fr. 95.–

Volume 54 Occupational Exposures to Mists and Vapours from Strong Inorganic Acids; and Other Industrial Chemicals
1992; 336 pages Sw. fr. 65.–

Volume 55 Solar and Ultraviolet Radiation
1992; 316 pages Sw. fr. 65.–

Volume 56 Some Naturally Occurring Substances: Food Items and Constituents, Heterocyclic Aromatic Amines and Mycotoxins
1993; 600 pages Sw. fr. 95.–

Supplement No. 1
Chemicals and Industrial Processes Associated with Cancer in Humans (IARC Monographs, Volumes 1 to 20)
1979; 71 pages (*out of print*)

Supplement No. 2
Long-term and Short-term Screening Assays for Carcinogens: A Critical Appraisal
1980; 426 pages Sw. fr. 40.–

Supplement No. 3
Cross Index of Synonyms and Trade Names in Volumes 1 to 26
1982; 199 pages (*out of print*)

Supplement No. 4
Chemicals, Industrial Processes and Industries Associated with Cancer in Humans (IARC Monographs, Volumes 1 to 29)
1982; 292 pages (*out of print*)

Supplement No. 5
Cross Index of Synonyms and Trade Names in Volumes 1 to 36
1985; 259 pages (*out of print*)

Supplement No. 6
Genetic and Related Effects: An Updating of Selected IARC Monographs from Volumes 1 to 42
1987; 729 pages Sw. fr. 80.–

Supplement No. 7
Overall Evaluations of Carcinogenicity: An Updating of IARC Monographs Volumes 1–42
1987; 440 pages Sw. fr. 65.–

Supplement No. 8
Cross Index of Synonyms and Trade Names in Volumes 1 to 46
1990; 346 pages Sw. fr. 60.–

List of IARC Publications

IARC TECHNICAL REPORTS*

No. 1 Cancer in Costa Rica
Edited by R. Sierra,
R. Barrantes, G. Muñoz Leiva, D.M. Parkin, C.A. Bieber and
N. Muñoz Calero
1988; 124 pages Sw. fr. 30.-

No. 2 SEARCH: A Computer Package to Assist the Statistical Analysis of Case-control Studies
Edited by G.J. Macfarlane,
P. Boyle and P. Maisonneuve
1991; 80 pages (out of print)

No. 3 Cancer Registration in the European Economic Community
Edited by M.P. Coleman and
E. Démaret
1988; 188 pages Sw. fr. 30.-

No. 4 Diet, Hormones and Cancer: Methodological Issues for Prospective Studies
Edited by E. Riboli and
R. Saracci
1988; 156 pages Sw. fr. 30.-

No. 5 Cancer in the Philippines
Edited by A.V. Laudico,
D. Esteban and D.M. Parkin
1989; 186 pages Sw. fr. 30.-

No. 6 La genèse du Centre International de Recherche sur le Cancer
Par R. Sohier et A.G.B. Sutherland
1990; 104 pages Sw. fr. 30.-

No. 7 Epidémiologie du cancer dans les pays de langue latine
1990; 310 pages Sw. fr. 30.-

No. 8 Comparative Study of Anti-smoking Legislation in Countries of the European Economic Community
Edited by A. Sasco, P. Dalla Vorgia and P. Van der Elst
1992; 82 pages Sw. fr. 30.-

No. 9 Epidemiologie du cancer dans les pays de langue latine
1991; 346 pages Sw. fr. 30.-

No. 11 Nitroso Compounds: Biological Mechanisms, Exposures and Cancer Etiology
Edited by I.K. O'Neill & H. Bartsch
1992; 149 pages Sw. fr. 30.-

No. 12 Epidémiologie du cancer dans les pays de langue latine
1992; 375 pages Sw. fr. 30.-

No. 13 Health, Solar UV Radiation and Environmental Change
Edited by A. Kricker, B.K. Armstrong, M.E. Jones and R.C. Burton
1993; 216 pages Sw.fr. 30.-

DIRECTORY OF AGENTS BEING TESTED FOR CARCINOGENICITY (Until Vol. 13 Information Bulletin on the Survey of Chemicals Being Tested for Carcinogenicity)*

No. 8 Edited by M.-J. Ghess,
H. Bartsch and L. Tomatis
1979; 604 pages Sw. fr. 40.-

No. 9 Edited by M.-J. Ghess,
J.D. Wilbourn, H. Bartsch and
L. Tomatis
1981; 294 pages Sw. fr. 41.-

No. 10 Edited by M.-J. Ghess,
J.D. Wilbourn and H. Bartsch
1982; 362 pages Sw. fr. 42.-

No. 11 Edited by M.-J. Ghess,
J.D. Wilbourn, H. Vainio and
H. Bartsch
1984; 362 pages Sw. fr. 50.-

No. 12 Edited by M.-J. Ghess,
J.D. Wilbourn, A. Tossavainen and
H. Vainio
1986; 385 pages Sw. fr. 50.-

No. 13 Edited by M.-J. Ghess,
J.D. Wilbourn and A. Aitio 1988;
404 pages Sw. fr. 43.-

No. 14 Edited by M.-J. Ghess,
J.D. Wilbourn and H. Vainio
1990; 370 pages Sw. fr. 45.-

No. 15 Edited by M.-J. Ghess, J.D. Wilbourn and H. Vainio
1992; 318 pages Sw. fr. 45.-

NON-SERIAL PUBLICATIONS †

Alcool et Cancer
By A. Tuyns (in French only)
1978; 42 pages Fr. fr. 35.-

Cancer Morbidity and Causes of Death Among Danish Brewery Workers
By O.M. Jensen
1980; 143 pages Fr. fr. 75.-

Directory of Computer Systems Used in Cancer Registries
By H.R. Menck and D.M. Parkin
1986; 236 pages Fr. fr. 50.-

---

* Available from booksellers through the network of WHO Sales agents.

† Available directly from IARC

www.ingramcontent.com/pod-product-compliance
Ingram Content Group UK Ltd.
Pitfield, Milton Keynes, MK11 3LW, UK
UKHW051257180426
11947UKWH00020B/1764